Social Credit Rating

Oliver Everling
Hrsg.

Social Credit Rating

Reputation und Vertrauen beurteilen

 Springer Gabler

Hrsg.
Oliver Everling
Frankfurt am Main, Deutschland

ISBN 978-3-658-29652-0   ISBN 978-3-658-29653-7 (eBook)
https://doi.org/10.1007/978-3-658-29653-7

Die Deutsche Nationalbibliothek verzeichnet diese Publikation in der Deutschen Nationalbibliografie; detaillierte bibliografische Daten sind im Internet über http://dnb.d-nb.de abrufbar.

Springer Gabler
© Springer Fachmedien Wiesbaden GmbH, ein Teil von Springer Nature 2020
Das Werk einschließlich aller seiner Teile ist urheberrechtlich geschützt. Jede Verwertung, die nicht ausdrücklich vom Urheberrechtsgesetz zugelassen ist, bedarf der vorherigen Zustimmung des Verlags. Das gilt insbesondere für Vervielfältigungen, Bearbeitungen, Übersetzungen, Mikroverfilmungen und die Einspeicherung und Verarbeitung in elektronischen Systemen.
Die Wiedergabe von allgemein beschreibenden Bezeichnungen, Marken, Unternehmensnamen etc. in diesem Werk bedeutet nicht, dass diese frei durch jedermann benutzt werden dürfen. Die Berechtigung zur Benutzung unterliegt, auch ohne gesonderten Hinweis hierzu, den Regeln des Markenrechts. Die Rechte des jeweiligen Zeicheninhabers sind zu beachten.
Der Verlag, die Autoren und die Herausgeber gehen davon aus, dass die Angaben und Informationen in diesem Werk zum Zeitpunkt der Veröffentlichung vollständig und korrekt sind. Weder der Verlag, noch die Autoren oder die Herausgeber übernehmen, ausdrücklich oder implizit, Gewähr für den Inhalt des Werkes, etwaige Fehler oder Äußerungen. Der Verlag bleibt im Hinblick auf geografische Zuordnungen und Gebietsbezeichnungen in veröffentlichten Karten und Institutionsadressen neutral.

Springer Gabler ist ein Imprint der eingetragenen Gesellschaft Springer Fachmedien Wiesbaden GmbH und ist ein Teil von Springer Nature.
Die Anschrift der Gesellschaft ist: Abraham-Lincoln-Str. 46, 65189 Wiesbaden, Germany

Geleitwort

Die Tücken mit dem Social Credit System für Unternehmen in China
Nicht nur Privatpersonen, sondern auch Unternehmen müssen sich in China bald einem datenbasierten Kontrollsystem unterstellen. Das stellt besonders ausländische Unternehmen vor große Herausforderungen. Westliche Medien haben in den vergangenen Monaten viel über das orwellianische Social Credit System (SCS) in China geschrieben. Immer noch weitgehend unbekannt ist dagegen die Tatsache, dass sich ab 2020 nicht nur Privatpersonen, sondern auch Firmen in der Volksrepublik einem Social Credit System unterstellen müssen. Das ist alarmierend, denn für ausländische Unternehmen drohen umfassende Konsequenzen.

Im Kern ist das geplante Unternehmens-SCS eine neuartige Anwendung von Big-Data-Technologie, die laufend Informationen über das Verhalten von Unternehmen sammelt. Diese werden dann durch undurchsichtige Algorithmen verarbeitet, um festzustellen, wie konform die Firmen mit den chinesischen Vorschriften sind.

Basierend auf der Analyse erhalten Unternehmen in jeder von mehreren Dutzend Kategorien eine Bewertung, die sich dann zu einem Gesamtergebnis zusammenfügen.

Mit dem neuen Social Credit System werden Unternehmen in China auf die Einhaltung von Vorschriften konditioniert. Die Anreize sind deutlich gesetzt: Höhere SCS-Werte können zu niedrigeren Steuersätzen, besseren Kreditbedingungen, einem leichteren Marktzugang und mehr Möglichkeiten für öffentliche Aufträge führen.

Niedrigere SCS-Werte führen zum Gegenteil und können im Extremfall bedeuten, dass ein Unternehmen auf schwarze Listen gesetzt wird. So wird beispielsweise die Häufigkeit von Inspektionen davon bestimmt, wie vertrauenswürdig ein Unternehmen ist – und die Vertrauenswürdigkeit wiederum hängt am SCS-Rating.

Betrachtet man die Zollkontrollen als Beispiel, so meldete die chinesische Zollbehörde eine Inspektionsrate von 98,1 Prozent für Unternehmen, denen der Staat misstraut. Für Unternehmen mit der höchstmöglichen Vertrauensbewertung liegt die durchschnittliche Rate von Zollkontrollen hingegen bei nur 0,5 Prozent.

Direkte Auswirkung auf die Wettbewerbsfähigkeit
Was bedeutet das für ausländische Unternehmen in China?

Es wird wesentlich davon abhängen, wie gut sich ein Unternehmen auf das Corporate-SCS vorbereitet. Teile des Gesamtrahmens sind bereits vorhanden, das gesamte System soll bis 2020 einsatzbereit sein, mit weiteren drei bis fünf Jahren für zusätzliche Anpassungen. Das gibt den ausländischen Firmen Zeit, sich vorzubereiten, aber nicht viel.

Unternehmensführer sollten zunächst verstehen, welche Auswirkungen das Corporate-SCS auf ihre übergeordnete Strategie haben wird. In den meisten Märkten halten sich Unternehmen selbstverständlich an die Vorschriften und bleiben wettbewerbsfähig, indem sie den Marktkräften folgen. Der bevorstehende Paradigmenwechsel in der Marktregulierung in China bedeutet jedoch, dass die Wettbewerbsfähigkeit bald ebenso stark von der Einhaltung des Corporate-SCS bestimmt wird wie von den Marktkräften.

Compliance wird nicht mehr eine eigenständige, binäre Frage der Rechtmäßigkeit sein, sondern wird zu einem Gradienten auf dem Spektrum von „gutem" bis „schlechtem" Verhalten. Das Social-Credit-Rating wird sich künftig direkt auf die Wettbewerbsfähigkeit der Unternehmen auswirken.

Unternehmen müssen noch konkretere Überlegungen anstellen. Die Social Credit Scores von Lieferanten beeinflussen die eigene Bewertung, was eine kontinuierliche Überwachung der Lieferketten entscheidend macht. Die individuellen Social Credit Scores von gesetzlichen Vertretern und hochrangigen Managern wirken sich ebenfalls auf das Firmenrating aus. Global tätige Konzerne werden dadurch gezwungen, darüber nachzudenken, wie sie mit dem persönlichen Verhalten von Mitarbeitern außerhalb des Arbeitsplatzes umgehen sollen – für viele ein Tabu.

Der vernetzte Charakter des Corporate-SCS erfordert auch, dass Firmen ihre interne Kommunikation verbessern, da die Handlungen einer Abteilung das gesamte Unternehmen ernsthaft beeinträchtigen können.

Kein Zurück

Ausländische Unternehmen dürften gegenüber lokalen Konkurrenten grundsätzlich benachteiligt sein, da Letztere Vorteile aus ihren engeren Verbindungen zur Regierung ziehen können. Trotzdem stellt das Corporate-SCS auch für chinesische Unternehmen eine große Compliance-Herausforderung dar. Viele von ihnen haben in der Vergangenheit Vorschriften – beispielsweise rund um das Thema Emissionen – umgangen, weil sie auf den Schutz lokaler Behörden zählen konnten.

In diesem Licht betrachtet können ausländische Unternehmen positive Aspekte aus dem Corporate-SCS ziehen: Wenn die leidenschaftslosen Algorithmen transparent und diskriminierungsfrei funktionieren, dann kann das neue System dazu beitragen, gleiche Wettbewerbsbedingungen zu schaffen.

Internationale Konzerne können sich sogar im Vorteil sehen, da sie in der Einhaltung strenger Vorschriften in den verschiedensten Märkten über einen reicheren Erfahrungsschatz verfügen. Viele von ihnen folgen ohnehin globalen Standards, die die aktuellen Anforderungen in China übertreffen.

Ich kann aus eigener Erfahrung berichten: Nachdem ich einen gründlichen Audit meines eigenen Unternehmens durchgeführt hatte, war ich zunächst schockiert und überwältigt von der Größe des Corporate-SCS und den Veränderungen, die das System notwendig macht. Aber das Gedeihen in China hat von ausländischen Unternehmen schon immer Geschicklichkeit, Flexibilität und einen pragmatischen Ansatz verlangt.

Das Corporate-Social-Credit-System wird nicht verschwinden. Je früher alle Unternehmen auf dem chinesischen Markt dies erkennen und beginnen, ihre internen Prozesse und externen Lieferanten zu überprüfen, desto besser. Sobald das System vollständig implementiert ist, wird es kein Zurück mehr geben: Alle Marktteilnehmer werden entweder nach dem Score leben oder nach dem Score sterben.

Jörg Wuttke
Jörg Wuttke ist Chefrepräsentant der BASF in China. Er ist zudem Präsident der EU-Handelskammer in China – ein Amt, das er bereits von 2007 bis 2010 sowie von 2014 bis 2017 besetzt hatte. Wuttke ist Mitglied des Beratergremiums des Mercator Institute for China Studies (MERICS) in Berlin. Er lebt seit mehr als drei Jahrzehnten in Peking.

Preface

What is China's social credit system?
What are the connotation and content of the social credit system?

In the past ten years, with the rapid development of Chinese society, the phenomenon of dishonesty has from time to time appeared. Incidents such as malicious breach of contract, commercial fraud, counterfeit production and inferior goods were difficult to manage and prohibit, and the public strongly urged to change the social integrity environment.

On June 14, 2014, the State Council of China issued the "Outline of the Plan for the Building of Social Credit (2014–2020)". The "Outline" pointed out that the social credit system is an important part of the socialist market economic system and social governance system. It is based on laws, regulations, standards and contracts, and relies on the credit records system and credit infrastructure network covering members of the society. It is supported by the application of credit information and a credit service system, and the mechanism of reward and punishment for acts of good or bad faith. Its inherent requirements are establishing a culture of integrity and promoting traditional virtue of honesty, and its aims are improving the integrity of the whole society and the level of credit.

In order to implement the "Outline", in recent years departments and local governments in China have issued a lot of guidance, management methods, implementation plans, etc. Social institutions have also taken action to build platforms, collect information, carry out credit evaluations, etc. Comprehensively speaking, the participants in the social credit system include not only the government, but also various service agencies, enterprises, associations, the Chamber of Commerce, and scientific research institutions. They not only set up rules, but also build the information system; not only carry out education and training on honest behavior, but also offer the real preference and convenience according to the credit level; not only punish the dishonest or guide the dishonest to reform, but also provide incentives to keep faith. At present, government departments and market institutions are participating in and exploring various pilot activities of the social credit system. The overall acceptance in the Chinese society is relatively high, and many trustworthy people benefit from it.

From my point of view, the construction of the social credit system has become an important measure to promote the modernization of China's governance capacity. It is an important way for the Chinese government to streamline administration and improve service functions. It is also a big change for the society and the market to establish credit rules and improve the credit environment. In my judgement, the construction of China's social credit system has been gradually incorporated into the top-level design of the country, and has become a new content of structural reform and social governance. The construction of Social Credit System has promoted a qualitative to quantitative shift in the focus of social governance, and at the same time promoted the normalization of social governance. The next development direction will be legalization and functionalization. Under such a policy situation, Chinese and foreign trade associations and chambers of commerce should play a key role in the construction of the social credit system.

How to regard the building of China's social credit system

Some people think that the construction of China's social credit system is a technical means of honesty training; some think it is a digitalization of government management; and some think it is an expanded financial service. In fact, these views all make sense from different perspectives. This is because credit contains many aspects, and the understanding of China's social credit system is related to the perception of credit itself.

In my opinion, credit is the capital to gain trust, which is composed of three dimensions: integrity, compliance and performance. In "Modern Credit Science" published by the Renmin University of China Press in 2009, I put forward and expounded on this "three-dimensional credit" viewpoint. Specifically, the first dimension, Integrity, is the basic capital to gain general trust. It represents the basic integrity quality of the credit subject, and involves the moral cultural ideas, spiritual literacy and code of conduct. It exists in the subconscious mind of the credit subject, and is the soft constraint on morality that society requires from the credit subject. When integrity gradually solidifies and becomes the common value pursuit, spirit and code of conduct of social groups, a social credit culture will be formed and a good credit environment will be constructed. The second dimension, Compliance, is the social capital to gain trust from administrators. It is expressed as the willingness, ability and behavior of credit subjects to comply with social administrative regulations, industry rules, folk practices, and internal management regulations in social activities. It involves the general social activities of the credit subject, reflecting the credit value orientation and credit responsibility, and is perceived as a hard constraint and clear social rule guiding people's behavior. The third dimension, performance, is the economic capital to gain the trust of counterparty. It is represented by the ability of the credit subject to abide by the rules of the transaction in credit trading activities, mainly the ability to complete and exercise the deal, involving the economic activities of the credit subject, and is represented by the credit value orientation and credit responsibility in economic activities. Performance affects the economic relations, trading order and development. The rules of performance are the compulsory legal restrictions, which require the credit subject to respect the spirit of contract, and the transaction behavior is protected by clear legislation.

Looking at the social credit system from the perspective of three-dimensional credit, what should be built? It can be said that credit construction is to improve the level of capital for credit subjects to obtain social basic trust, administrators' trust, and counterparties' trust. Specifically speaking, credit construction work can be summarized into three aspects:

The construction of credit culture of basic quality, the construction of the compliance system for social activities and the construction of contract management of economic trading activities.

Firstly, the integrity part of the social credit system is mainly the construction of standards and a culture of integrity. A series of activities such as joint punishment and "Xinyi +", can affect the psychology and behavior of social subjects, increase their awareness of integrity, and establish integrity constraints. At the same time, the construction of the social credit system also includes widespread integrity education in society, requiring schools to open integrity-related courses in basic education, companies to establish ethical standards, employees to abide by professional ethics, business operations in the market to adhere to the principle of integrity, and civil servants to stick to government departmental integrity standards and make related government affairs public. In these activities typical cases and figures of integrity have been formed, which has played a good guiding role in the society.

Secondly, the compliance part of the social credit system is mainly the construction of compliance systems and credit informatization. The construction of the compliance system is to improve the rules of all aspects of social life on the basis of laws with the government in the leading role. At present, information asymmetry is serious in many fields, the cost of illegal dealings is low, and information on illegal activities is difficult to bring to public attention, which requires government disclosure and involvement in supervision. Therefore, the current work of building a compliance system mainly includes various government departments recording, publicizing and evaluating the social subject's compliance with administrative regulations, industry rules, etc., and carrying out categorized supervision. The establishment of a compliance system can help the government optimize supervision, and the disclosure of compliance information can increase the cost of dishonesty and help the market better identify credit risks. For example, the social credit system requires government regulatory agencies such as industry and commerce, taxation, environmental protection departments, etc., to penalize companies with abnormal operations, dishonest tax payments and environmental pollution and other behaviors. The departments are asked to publicize their information, establish a comprehensive credit record and evaluation system, carry out classified supervision according to past misconduct, and establish a repair mechanism to allow individuals and companies to correct their dishonest behavior. The pressure of public disclosure and punishment can effectively reduce the occurrence of violations and dishonesty. For example, restricting high-consumption, flying, etc., and publicizing them made many "Lao Lai" take the initiative to repay the arrears.

On the other hand, the realization of the compliance part needs considerable informatization to promote the application of compliance rules in society. This includes infrastructure construction, such as building a credit information platform to collect information, establishing a public announcement system, promoting cross-regional information-sharing, designing a compliance evaluation model and compliance information application. Information construction is indispensable, which makes the abstract concept of credit visible in data and facilitates the use of credit information by the government and the market. The government can carry out basic credit evaluation based on compliance data and provide convenience and preferential treatment for subjects with a good compliance record. This is helpful in order to optimize the public service and business environment. The publicity and sharing of compliance information can strengthen the government and society's punishing effect regarding dishonesty. Market institutions can also use the open compliance information to evaluate and determine how much trust will be given to social subjects. Different levels of trust will determine how much convenience they will offer, or how many restrictions they can impose.

Thirdly, the performance part of the social credit system is mainly records and evaluation of economic transactions. At present, the key goal of the performance part is to establish a credit reporting system, comprising a financial and a commercial credit system. The construction of the financial credit system is carried out by The Credit Reference Center, the People's Bank of China, which has been working hard to collect more data and expand its coverage in the area. In the financial system, contract performance has always been the focus of risk prediction and management. It is also an important part of the social credit system in recording and evaluating credit. On the other side, participants in the commercial credit system include various e-commerce platforms, supply chain financing and credit service institutions, etc. They rely on their own resources to establish their databases, launch evaluations and provide credit products and services accordingly for citizens and business entities.

For individuals, enterprises, institutions, and governments, the deepening of the construction of the social credit system will gradually affect their economic and social behavior. A sound social credit system can provide a fairer and more transparent environment. In the case of asymmetric information, it is difficult for enterprises to be known by the society for their faithfulness or dishonesty, so they cannot get corresponding rewards and punishments, which is not conducive to establishing a good business competition environment. The social credit system records, evaluates and rewards credit behavior which enables companies that comply with laws and regulations to accumulate credit capital. This can in turn improve their competitiveness and help them gain more benefits, such as more credit resources, bidding opportunities and simpler approval processes, etc.; companies that do not abide by the law and act honestly will be punished and naturally eliminated. The joint rewards and punishments of society will have strengthened the effect of the survival of the fittest in the market, which in turn can promote enterprises to pay more attention to integrity in internal management, establish internal ethics standards, conduct staff integrity and compliance training, restrict behavior of enterprise personnel, and review the behavior and qualification of partners more strictly, so as to form a healthy and orderly business environment. Of course, these management measures added to adapt to the social credit system may increase the cost of the enterprise in the short term, but in the long run, companies that operate with integrity can obtain greater benefits.

For foreign enterprises or investors in China, the social credit system is a completely new thing. They may not adapt to it, so how do they maintain competitiveness in the environment of the social credit system? Here, I give four suggestions to the European Union

Chamber of Commerce in China and its members. First of all, there is an old saying in China that "When in Rome, do as the Romans do". Foreign institutions themselves should understand the operational rules of the social credit system and know how to avoid punishment or gain advantages. Second, the European Union Chamber of Commerce should actively organize institutions to learn and understand the provisions of the social credit system in China and invite relevant experts and scholars to interpret policies. Third, there are associations in all walks of life in China. The European Union Chamber of Commerce can establish cooperative relationships with corresponding industry organizations in China, so as to exchange policy information and promote a better industry competitive environment. In other words, the Chamber of Commerce can play an organizational role providing some assistance to the members to adapt to China's new social governance mode. Fourth, the Chamber of Commerce, working together with Chinese experts and scholars, can carry out research projects, based on China's actual situation and drawing on advanced international experience. They can put forward suggestions, and actively promote China's social credit system to become better, so that foreign institutions can benefit from it.

Generally speaking, the starting point of the construction of China's social credit system is to solve the problem of lack of integrity in the society. Information recording and evaluation are its means, and management rules are its basis. It is a welcome step to change the current situation of lack of honesty and improve the social credit environment. But this is really different from the past social management, as it uses more information to record and evaluate, makes new management rules, and brings changes to people. People are more worried about their inadaptability to the rules and being recorded or afraid of information leakage or abuse, rather than the credit construction of the system itself. Now China is discussing the legislation of credit information security and privacy protection, constantly improving the laws and regulations to ensure the normal and healthy operation of the social credit system, reducing the unfairness of rule-making and information evaluation, and the excessive use or abuse of credit information, by setting a unified standard. The construction of China's social credit system is essentially a transformation of social governance in the era of big data. At present, this innovative attempt is conducive to the establishment of an open, transparent, and orderly credit environment, which is beneficial for foreign businesses operating in China and it deserves active attention and participation.

Prof. Jingmei Wu 吴晶妹
School of Finance, Renmin University of China. Prof. Wu has engaged in credit research for more than 30 years. She founded modern credit theory and built the theory of credit capital and three-dimensional credit theoretical framework. Her credit theory has a far-reaching influence on the construction of China's social credit system.

Prof. Wu has participated in the formation and argumentation of many credit-related policies of China. As a core member of the drafting team and the leader of the expert argumentation committee, she took part in "Outline of the Plan for the Building of Social Credit (2014–2020)", issued by the State Council of China. She was also the leader of the expert argumentation committee of "Overall Plan for the Building of Unified Social Credit

Code System for Legal Persons and Other Organizations" (approved and transmitted by the State Council of China). In the assessment of the first-batch model cities of social credit system building in China, she was the leader of the expert assessment committee. So far, she has led hundreds of credit research projects of government ministries, local governments and large corporations.

Vorwort des Herausgebers

Social Credit Ratings sind das Ergebnis von Sozialkreditsystemen. Diese umfassen auf verschiedene Datenbanken zugreifende, online betriebene Rating- oder Scoringsysteme, bei denen beispielsweise die Kreditwürdigkeit, das Strafregister und das soziale und gesellschaftliche Verhalten von Personen oder Organisationen wie Unternehmen oder Nichtregierungsorganisationen zur Klassifizierung ihrer Reputation verwendet werden.

Das Fahreignungsbewertungssystem des Kraftfahrt-Bundesamtes, das für jedermann bestimmte Ordnungswidrigkeiten, Fahrverbote oder Straftaten mit Punkten bewertet und speichert, ist in Deutschland ebenso bekannt und anerkannt wie die SCHUFA-Bonitäts Auskunft, der Creditreform-Bonitätsindex, interne Ratings von Banken oder Noten von Ratingagenturen. Ähnliche Systeme wie Verkäuferbewertungen in Online-Shops, Likes, Zertifikate und Zeugniszensuren aller Art sind in Deutschland wie auch in vielen anderen Ländern und weltweit nicht nur für interne Zwecke in Organisationen aller Art, sondern auch öffentlich und in Social Media verbreitet. Solche Beurteilungen verwischen die Grenzen zwischen Urteilen für das Verhalten von Unternehmen und natürlichen Personen.

Sozialkreditsysteme führen solche Klassifizierungen und Punktebewertungen, Ratings und Scorings aufgrund einzigartiger Verknüpfungen in eine neue Dimension, die erst durch die neuen Informations- und Kommunikationstechnologien ermöglicht wurde. Die Ansätze fordern dazu heraus, das Zusammenwirken der Systeme zu überdenken.

Dieses Buch gibt einen Einblick in die verwendeten Daten, Verfahren, Methoden und Modelle sowie diskutiert Bedeutung, Nutzen, Funktionen und Anwendungsbereiche von Social Credit Ratings. Das Buch lässt als Herausgeberwerk die maßgeblichen Akteure, Autoren der Praxis wie auch der Wissenschaft zur Sprache kommen.

Angesichts des entstehenden Sozialkreditsystems in China beeilten sich in den letzten Jahren Journalisten wie auch Consultants, kurze Darstellungen zum neuen „Social Credit Rating" zu liefern oder sich kritisch mit den chinesischen Ansätzen auseinanderzusetzen. Dabei ist das System der Volksrepublik noch weit davon entfernt, eine monolithische Einheit zu bilden. Viele Elemente überdauerten kaum das Experimentierstadium oder sind lediglich Teil von Pilotprojekten.

Das vorliegende Buch greift die Idee eines Social Credit Rating unabhängig von Ort und Zeit auf. Daher reichen die Beiträge von historischen Betrachtungen, über Details des chinesischen Sozialkreditsystems bis hin zu Sentimentanalysen an den Finanzmärkten oder zur neuartigen Kultur aus Bloggern und Influencern, deren Reputation gezielt im Marketing kommerzialisiert wird.

Obwohl führenden, amerikanischen Technologiekonzernen aufgrund ihrer allumfassenden Datensammlung schon ausgeklügelte Systeme des Ratings und Scorings mit (da auch sanktionierend) meinungs- und verhaltenssteuerndem Einfluss auf praktisch alle Bürger erlaubt sind, richten sich die insbesondere in den US-Medien artikulierten Sorgen um ein neues Sozialkreditsystem weniger gegen private, als gegen staatliche Betreiber.

Ähnliche Ideen haben manchmal sehr unterschiedliche Namen: So sind Parallelen zwischen dem chinesischen Anspruch, heimische Unternehmen unter ethischen, ökologischen und sozialen Aspekten zusammengefasst in einem Social Credit Rating zu beurteilen, und den Ansätzen zum Nachhaltigkeitsrating zu erkennen (Environment Social Governance, ESG). Die Ziele für nachhaltige Entwicklung (UN Social Development Goals, SDG) der Vereinten Nationen bedürfen der Operationalisierung. China liefert dazu eine eigene Antwort.

Indem nicht nur Geld die Welt regiert, sondern das Verhalten von Marktteilnehmern (von natürlichen wie auch von juristischen Personen) ausnahmslos einem (öffentlichen) Rating unterzogen wird, gewinnt ein staatlich kontrolliertes Sozialkreditsystem unerhörten Einfluss auf die gesamte Wirtschaft und Gesellschaft. Das Thema „Social Credit Rating", das mit dem vorliegenden Buch erstmals in einem Sammelwerk aufgegriffen wird, muss daher auch mit Konsequenzen auf psychologische Prozesse und rechtliche Beurteilungen untersucht und in einen kulturellen Kontext gestellt werden.

Das vorliegende Buch soll als Diskussionsbeitrag verstanden werden und der Wissenschaft wie auch der Praxis Material an die Hand geben, sich eingehender mit Sozialkreditsystemen zu befassen. Es liegt in der Natur einer ergebnisoffenen Disputation, widersprechende Meinungen zu Wort kommen zu lassen.

Angesichts der Fülle relevanter Fragestellungen wäre der vorliegende Sammelband ohne das Zusammenwirken vieler Personen nicht möglich gewesen. Dieses Werk ist das Ergebnis meiner aktuellen Zusammenarbeit mit mehr als 200 Experten, namhaften Wissenschaftlern und Führungskräften der Wirtschaft. Leider würde es zu weit führen, diese hier namentlich alle zu benennen, wie auch viele Fakten und Ideen bedauerlicherweise noch im Verborgenen bleiben müssen. Allen Autoren sind wir, der Springer-Verlag wie auch ich als Herausgeber, sehr verbunden.

Jörg Wuttke, Präsident der Handelskammer der Europäischen Union in Peking, ist ein Impuls zu verdanken, nicht nur die in China tätigen Firmen, sondern Unternehmen in ganz Europa und darüber hinaus für die praktische Relevanz des neuen Sozialkreditsystems in China sensibilisiert zu haben http://www.europeanchamber.com.cn/en/publications-corporate-social-credit-system. Sein Wirken hatte Einfluss auf die öffentliche Wahrnehmung wie auch auf die Maßnahmen exportorientierter Unternehmen, sich auf ihr Social Credit Rating vorzubereiten und das Sozialkreditsystem auch sonst für ihre Geschäftstätigkeit zu nutzen. Daher freuen wir uns, sein Geleitwort dem Buch voranstellen zu dürfen.

Den wichtigsten Anstoß für dieses Buch gab zweifellos Frau Professor Jingmei Wu von der School of Finance an der Renmin Universität der Volksrepublik China in Peking, die uns seit 2002 nicht nur in vielen persönlichen Begegnungen mit ihr und ihren Mitarbeitern an ihrer Hochschule wie auch in ihrer Firma an ihrer Forschung zum Social Credit System teilhaben ließ, sondern schon vor Jahren auch dazu ermutigte, über eine Publikation nachzudenken, mit der einer deutschsprachigen Fachöffentlichkeit ein tieferes Verständnis des chinesischen Systems ermöglicht würde. Nachdem der Staatsrat der Volksrepublik China 2014 die Weichen hin zu einem Social Credit Rating gestellt hatte, war klar, dass die Arbeiten von Prof. Wu nicht bloß von akademischem Wert sind, sondern ihre Vision von einem Social Credit Rating und die Mission ihres Unternehmens weitreichende Bedeutung über China hinaus haben würden. Daher haben wir ihr Vorwort vorangestellt.

Noch Jahrzehnte nach Beginn der Reform und Öffnungspolitik 1978 galt China als ein Land, das aus dem Westen gleichermaßen Produkte und Ideen kopiert. So schien auch der Aufbau von Kreditbüros und Ratingagenturen in China zunächst nur als Übertragung von Ablauf- und Aufbauorganisationen auf chinesische Verhältnisse. Die Zeit des Kopierens ist aber längst vorbei, von Ideenklau kann keine Rede mehr sein. Ein chinesisches Sozialkreditsystem ist vielmehr ein origineller Eigenbau, der trotz vieler Ähnlichkeiten in Europa wie auch in Amerika seinesgleichen sucht.

Das neue System bringt eine Fülle von neuen Begriffen und Definitionen mit sich. Als wäre das Erlernen der chinesischen Sprache nicht schon schwer genug, sind daher in der fachlichen Auseinandersetzung mit dem Social Credit Rating in China bisher ungekannte Hürden zu überwinden. So wird der Leser Wortschöpfungen begegnen, die zum Zeitpunkt der Drucklegung anderweitig kaum zu finden sind, im Kontext des jeweiligen Buchbeitrags aber selbstverständlich klingen. So konnten wir leider nicht auf verlässliche Standards in der Übersetzung vertrauen. Für die Übersetzungen ins Deutsche ist insbesondere Herrn Kaifei Jin und meiner Frau, Jian Ren, mit je rund 20 Jahren Erfahrung in Deutschland, zu danken.

Darüber hinaus freut es mich, mit meiner Mitherausgeberin unseres Buches „Risk Performance Management", Frau Reavis Hilz-Ward, Geschäftsführerin der Interprojects GmbH, in Bezug auf die englischsprachigen Beiträge erneut zusammengearbeitet zu haben. Gebürtige Amerikanerin, arbeitet sie seit mehr als zwei Jahrzehnten in Europa. Sie hatte verschiedene Senior-Managementpositionen, wie bei einer führenden Geschäftsbank und den Vereinten Nationen, und versteht die Sprachen, die Kulturen, die Geschäfte und hilft, Brücken in internationalen Märkten zu bauen. Darüber hinaus danke ich Herrn Andreas Fornefett dafür, nach unserem Buch „Transparenzrating" auch bei diesem Titel von Anfang an mit Rat und Tat mitgewirkt zu haben.

Dank gilt für die technische Bearbeitung der Texte Frau Nandhini Rajadhanam, stellvertretend für das ganze Team in Indien, die sich mit großer Geduld der deutschen Sprache angenommen hat, um alle Änderungswünsche und Korrekturen der Autoren zu verstehen und in den Druckfahnen umzusetzen.

„Last not least" ist Herrn Guido Notthoff vom Springer-Verlag zu danken, auf dessen professionelle Betreuung wir seit 1999 bei mehr als zwei Dutzend Buchprojekten wieder zählen durften. Kein Projekt ist wie das andere, so hatten wir auch bei diesem Buch zum „Social Credit Rating" mit Besonderheiten zu tun, wie den Übersetzungen aus dem Chinesischen.

Kommentare und Anregungen sind willkommen und erreichen uns bei der RATING EVIDENCE auch unter der Adresse socialcreditrating@gmail.com.

Dr. Oliver Everling

Dr. Oliver Everling ist seit 1998 selbstständig und Geschäftsführer der RATING EVIDENCE GmbH (www.ratingevidence.com). Als Beirat, Berater, Gastprofessor an der Capital University of Economics and Business in Peking, Mitglied von Ratingkommissionen, Chairman des ISO-TC Rating Services, Independent Non-Executive Director nach der EU-Verordnung über Ratingagenturen oder Aufsichtsratsvorsitzender einer Ratingagentur war oder ist er aus unterschiedlichen Perspektiven mit Ratings befasst. Zuvor war er Abteilungsdirektor und Referatsleiter der Dresdner Bank und bis 1993 in der WM Gruppe Geschäftsführer der Projektgesellschaft Rating mbH nach Promotion am Banken- und Börsenseminar der Universität zu Köln.

Frankfurt am Main, Deutschland   Dr. Oliver Everling
August 2020

Inhaltsverzeichnis

Teil I

Chinas Social Credit Rating

Reflections on China's Credit Reporting Practice

1

Jingmei Wu 吴晶妹

Abstract

This article discusses the current situation, logic and development trends of China's credit reporting system. China's credit reporting system is composed of financial, commercial and administrative credit reporting, which were set up according to its resource allocation system divided into financial sector, non-financial sector, and government sector, respectively. The three sectors take credit as one of the referential factors in resource allocation, and establish corresponding credit reporting systems and credit evaluation mechanisms. The high participation of government departments in resource allocation gives the pattern, content, and purpose of the credit reporting systems its Chinese character. China's three major resource allocation systems and three major credit reporting systems complement each other, which is a new type of social governance model to achieve co-governance on credit rules.

1.1 Analysis of China's Current Three Credit Reporting Systems

1.1.1 China Has Formed Three Major Credit Reporting Patterns

At present, the three major credit reporting systems for finance, commerce and administration have been created in China (see Abb. 1.1). In particular, the administrative credit reporting system is quietly emerging and growing rapidly in China, which deserves attention from all around the world.

Jingmei Wu 吴晶妹 (✉)
Beijing, China
E-Mail: wujingmei@ruc.edu.cn

© Springer Fachmedien Wiesbaden GmbH, ein Teil von Springer Nature 2020
O. Everling (Hrsg.), *Social Credit Rating*,
https://doi.org/10.1007/978-3-658-29653-7_1

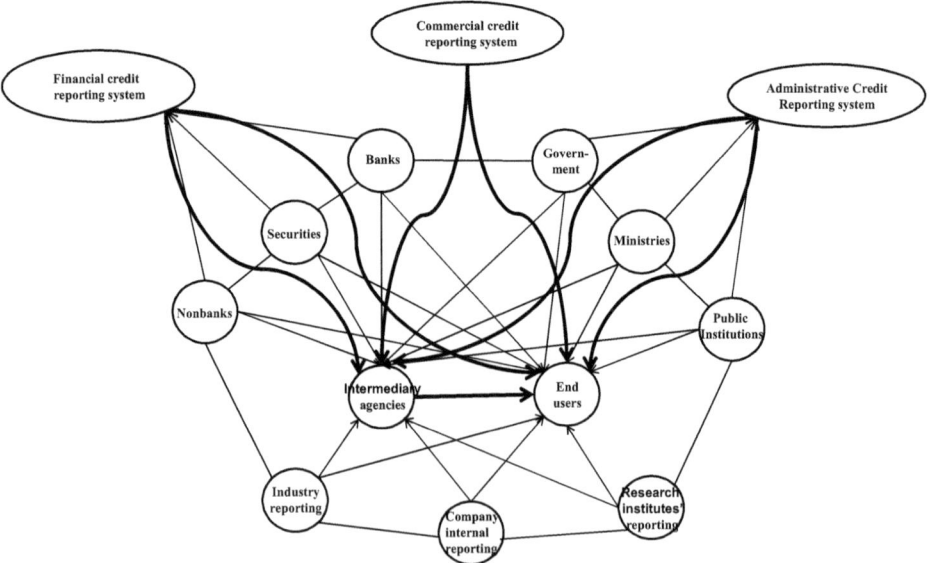

Abb. 1.1 Wu's Credit Theory: Schematic diagram of China's three major credit reporting system (by the author)

Financial Credit Reporting System

The financial credit reporting system serves financial institutions. The Credit Reference Center (the People's Bank of China (CCRC)), which provides services centered on individual and corporate credit reports, is the core of China's financial credit reporting system. For example, a commercial bank can inquire at the CCRC about an individual credit report when conducting a specific business; or an individual can inquire about his own credit report at CCRC or from the Internet.

CCRC has become the world's largest credit reporting system, with the largest number of people, the largest data scale and the widest coverage. As per June 2019, a total of 990 million individuals, 25.91 million enterprises and other organizations were included in CCRC. The average daily query volume of CCRC is 5.5 million and 300,000 for enterprises and individuals, respectively (From: Credit reporting system includes 990 million natural persons [EB/OL]. Beijing Youth Daily June 15, 2019 (A03) http://epaper.ynet.com/html/2019-06/15/content_330351.htm?div=-1).

On May 23, 2018, Baihang Credit Co., Ltd., which is China's only credit reporting agency with an individual credit business license, opened up in Shenzhen. Under the leadership of the People's Bank of China (PBC), Baihang Credit was jointly initiated by the National Internet Finance Association of China (36 percent share) and eight individual credit reporting agencies including Sesame Credit, Tencent Credit, and Qianhai Credit, etc. (8 percent share each). The credit data of the central bank mainly comes from traditional licensed financial institutions such as banks, etc., while the credit data of Baihang Credit also comes from the major Internet platforms, including online loan platforms, Internet companies, small loan companies, etc. As of October 2019, the number of institutions cooperating with Baihang Credit exceeded 1200, among which 750 institutions with signed information-sharing agreements; the number of persons included in its system exceeded 100 million; the number of credit accounts exceeded 120 million; the total number of individual credit report queries exceeded 30 million; and the average number of daily queries was 400,000 (From: Baihang Credit website[EB/OL]. http://www.baihangcredit.com/news/news_31.html). As a supplement to the CCRC, Baihang Credit mainly provides services such as Internet Finance and online lending, rather than traditional finance. At present, China's financial credit system, which is mainly composed of CCRC and Baihang Credit, is widely used in financial institutions' pre-loan approval, risk pricing, and post-loan risk management, which plays an important role in preventing financial risks and promoting the development of the financial industry.

Commercial Credit Reporting System

The commercial credit reporting system provides credit reporting services for various industries in the entire business system. It mainly aims at the interconnection of credit information within the organization and the corresponding market, and joint prevention of credit transactions and management risks. For example, in the case of individuals, commercial credit institutions use credit management means to understand the credit situation of individuals, and provide different services according to different credit levels, so that individual consumers can obtain opportunities and convenience by credit; in the case of enterprises, commercial credit institutions learn about the assets, lending, business operation ability, credit level and other information of enterprises to provide reference for credit management and supply chain financing. For example, Sesame Credit (2017) is a commercial credit company. It uses cloud computing, machine learning and other means to present the commercial credit status of individuals and enterprises. Its products are applied to many business scenarios such as leasing, shopping, transportation and accommodation, so that merchants can provide convenient services for more users. At present, commercial credit is widely used by individuals. Taking the application of the Sesame credit score as an example, "living by credit", which is an application in the accommodation scene, has cumulatively helped 20 million users save 14 million hours of waiting time in line and eliminated 36 billion yuan in accommodation deposits. (From: Feizhu & Sesame Credit, "Travel report of "living by credit""[EB/OL]. http://credit.fzgg.tj.gov.cn/506/22396.html. 2019-03-11.) At present, the national commercial credit reporting center and credit giants have not yet appeared in China. The commercial credit reporting system for corporates is still decentralized. Currently, the credit reporting system for commercial credit transaction recording, such as credit sales in B2B and B2C and credit settlement of suppliers and sales-agents, does not cover regions and is not even close to covering the whole country.

Administrative Credit Reporting System

Administrative credit reporting mainly provides reference for the administrative supervision and public service of government departments, the most representative are the National Credit

Information Sharing Platform and "Credit China" website established under the guidance of the National Development and Reform Commission and the People's Bank of China (PBC). As of the end of July 2019, the total number of credit information on National Credit Information Sharing Platform grew to 37 billion. "Credit China" website published 197 million pieces of credit information, including 156 million pieces of administrative license information and 40.63 million pieces of administrative penalty information (From: The National Development and Reform Commission regularly time and theme press conferences in August. [EB/OL]. https://www.ndrc.gov.cn/xwdt/xwfb/201908/t20190816_954461.html. 2019-08-16). The "Credit China" website also has an individual credit inquiry function, which aggregates four types of personal credit information inquiry channels, including credit information from the PBC and credit scores from credit service agencies, local governments and telecommunication operators.

At present, the administrative credit reporting system, with the National Information Sharing Platform and the "Credit China" website as the core, constitutes the basic project of the construction of China's Social Credit System, and is the main hub for credit information sharing and exchange.

The administrative credit reporting system has formed a cross-regional and cross-sectoral sharing mechanism of public credit information, and has also promoted the disclosure of government information. For example, the National Enterprise Credit Information Disclosure System launched in February 2014 publicizes information on the registration, administrative approval, annual reports, administrative penalties, results of spot checks and abnormal business conditions of market entities, so that society can participate in monitoring market violations. As of the end of August 2019, the total number of visits to the website of the National Enterprise Credit Information Disclosure System reached 1,509.053 billion, the cumulative query volume of market entities reached 22.07 billion and the basic information conveyed by the General Administration for Market Regulation to other government departments reached 60.76 million (From: The National Enterprise Credit Information Disclosure System has exceeded 1.5 trillion visits, and the cumulative query volume of market entities has reached 22.07 billion [EB/OL]. https://www.creditchina.gov.cn/home/zhuantizhuanlan/xinyongdashuju/xinyongdashujuqianyan/201910/t20191008_170420.html. 2019-10-08.). People's attention to corporate compliance and the demand for public credit information are both increasing.

1.1.2 Features of the Three Major Credit Reporting Systems in China

Firstly, China's three major credit reporting systems are not limited to the field of finance, which differentiates them from the definition and understanding of credit reporting in many other countries. For example, Margaret J. Miller (2003) considers credit reporting as an important part of the financial system, which is mainly used to solve the problem of information asymmetry in the lending process. And the core of the financial credit reporting system is to record the payment history of the individual or the enterprise, to enables lenders to more accurately measure credit risk and reduce the time and cost of the lending process.

Secondly, not only financial data, but also a large amount of credit data from administrative management systems are collected in China's three major credit reporting systems, which is different from foreign credit reporting and rating systems based solely on financial indicators. Taking credit ratings and credit reporting in the United States as an example, the individual credit scoring system in the United States is mainly FICO launched by the Fair Isaac Company, which provides different credit models to credit agencies. The FICO scoring model mainly evaluates financial indicators. There are five main types of factors that it focuses on, namely individual credit repayment history, number of credit accounts, the period of use of credit, the type of credit in use, and the newly opened credit account (From: Fico website [EB/OL]. https://www.myfico.com). Therefore, FICO is mainly a part of the financial credit reporting system, which examines financial capabilities.

But in modern times, financial credit reporting is increasingly insufficient to meet people's diverse needs. If market participants want to participate in economic transactions and social resource allocation under the traditional system, they must have financial capabilities, such as physical resources (land, equipment, factory buildings, funds or cash equivalents). Only in this way can production activities be carried out, transactions be concluded, products or services be exchanged, the circulation of commodities be achieved, and profits be realized. On the contrary, it is difficult for market entities lacking physical capital to obtain financial services fairly to meet their funding and resource requirements, which make them unable to carry out subsequent steps in economic activities such as production, exchange, and consumption. These disadvantages give financial credit reporting a strong impetus to collect and use more data, and to break through into various non-financial fields.

Thirdly, the most important characteristic of China's three major credit reporting systems is that they exist not only for the purpose of serving credit risk control but also for the deeper goal of social governance, and support the government's aim to create social equity and civil welfare.

The financial credit reporting must serve the need to allocate resources with physical capital at its core, so a credit reporting framework based on financial information has been formed to predict credit transaction risk and repayment ability. This financial-based credit reporting has high accuracy in preventing and controlling credit risks, predicting defaults, screening customers, and determining credit lines and interest rates. The effectiveness of financial credit reporting has been tested and recognized in credit risk practice for a long time. Therefore, financial credit reporting is an important and irreplaceable part of China's credit reporting system.

However, the over-emphasis on financial data relative to physical capital has effectively screened out a large number of borrowers without or with only small physical capital, which has limited the scope of financial credit resource allocation to a smaller range of people and made a relatively small contribution to social development. Moreover, credit providers such as banks take into account many non-credit financial capacity factors for the credit evaluation of their borrowers, which generally means that the financial needs of low-income people and entrepreneurial enterprises cannot be met. This situation will ultimately promote social injustice.

The development of the administrative credit reporting system has promoted the application and sharing of public credit information in society, which expands the scope of credit information application. In an era of informatization and digitalization, the government's provision of public credit information to the society as a new type of public product constitutes progress for public services, which have diversified credit reporting agencies and their products and services. If there is only a small financial credit reporting system, many people will not be able to obtain resources because of their limited assets and lack of transaction records. However, the data from the administrative credit reporting system can provide support for their credit evaluation, allowing them to obtain opportunities for credit and development, expanding the welfare of the public and promoting social progress.

1.2 Analysis of the Logic for the Formation of the Three Major Credit Reporting Systems in China

1.2.1 The Three Major Credit Reporting Systems Are Rooted in the Three Major Resource Allocation Systems in China

Currently in China, resource-holders have begun to add credit factors to the allocation of resources. Allocation of economic and social resources has increasingly taken credit conditions into account. Credit has become the capital that people rely on to gain trust from distributors, and people participate in the transaction and distribution of social resources with credit capital. To distribute according to credit capital, we need to collect credit information and price credit capital according to this information. In China, the leaders of resource allocation are also the masters of credit information. They can not only provide credit information but also rely on credit capital to allocate resources. The application of this kind of resource allocation in society is all-rounded, involving multiple parties.

In China, there are three major sectors that allocate resources and master credit information. Therefore, three corresponding credit reporting systems have been formed. As is shown in Abb. 1.2.

The first sector is the financial sector. In the past, the allocation of resources was mainly about the allocation of funds. Financial institutions are the ones which have the capital and are qualified to allocate funds. Therefore, financial institutions have always been the core of a market economy. For financial institutions to allocate funds in a safe and profitable manner, it is necessary to conduct a comprehensive and full-process credit risk assessment, to monitor credit applicants and to conduct pre-loan review, loan-in-management, and post-loan tracking. All of these measures require the collection and updating of borrowers' credit information, so a credit reporting system adapted to the allocation of credit funds has been formed, and now the financial credit reporting system has become an important part of the operation of a relatively mature market economy.

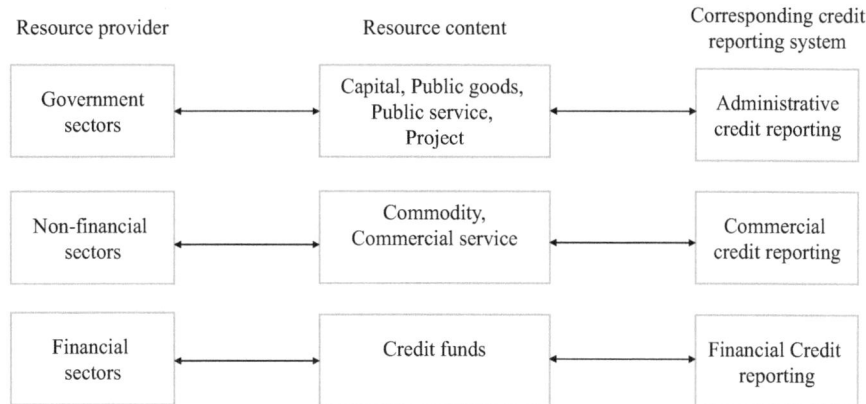

Abb. 1.2 Wu's Credit Theory: Relationship between the three major resource allocation sectors and the three major credit reporting systems (by the author)

The second sector is the non-financial sector, which sector mainly includes enterprises, industry organizations and individuals. The resources held by the non-financial sector are primarily the goods and services they produce, create, accumulate or store.

In the past, companies allocating resources to cooperative firms and consumers (i.e. to provide goods and services), consumers choosing products, and enterprises competing against each other have all relied on factors such as price, quality, and after-sales service, with the credit factor rarely considered.

On the one hand, more and more non-financial institutions, especially enterprises, have begun to consider the credit status of consumers. They are increasingly willing to change their sales methods for goods and services and to provide goods/services based on the credit status of consumers or cooperative enterprises.

On the other hand, consumer demand is becoming more and more diversified, with higher and higher quality expectations. Consumers basically add the pursuit of a better life into the demand for goods and services. Abraham Harold Maslow proposed Maslow's Hierarchy of Needs in "A Theory of Human Motivation" (in 1943), which divides human needs in society into five levels: physiological, safety, belonging, esteem and self-actualization. Nowadays people have reached the needs of being esteemed and self-actualization, which improves people's material and spiritual demand for premium goods and services. The evolution of the citizen's consumption, from mostly food and clothing to mostly spiritual, cultural, educational and entertainment-related, has accelerated the optimization and upgrading of the consumption structure and the development of credit services. The demand for esteem and self-actualization makes people thirst for credit. So, they care not only about price, quality and after-sales service, but also whether they can get preferential treatments such as trust-related benefits and convenience including deposit exemption, installment payments, enjoyment before payment policies, etc.

From the perspective of the enterprises providing resources, if the credit status of consumers upstream and downstream of the supply chain is recognized as good, they will be willing to make some concessions. They are more willing to provide conveniences such as deposit exemption, installment payments, policy of enjoyment before payment, extension of the payment period, etc., to achieve lower management costs. This is actually a credit bonus granted by enterprises to consumers. The premise of these concessions is that the enterprises must fully understand the consumer's credit status and the consumer must have good credit support. As a result, the commercial credit reporting system emerges organically.

The third sector is the government sector. The resources held by the government sector are mainly public resources such as funds, services and projects. The government sector has the right to allocate public resources in accordance with its administrative authority. The purposes of resource allocation by government sector are multiple. The function and role of the government in the allocation of resources are also multi-faceted. Within the allocation process, government departments must take macro-control of the market, the efficiency of resource usage and social equity into consideration. Different perspectives consider different core factors, but the common factor is credit, especially in terms of fairness. The allocation of resources according to the credit status of enterprises or individuals can reduce the cost and improve the efficiency of government supervision and promote social equity and justice. As a result, the administrative credit reporting system emerges.

1.2.2   The Degree of Government Participation in the Allocation of Social Resources is High

The division of the three major credit reporting systems stems from the existence of three major sectors in China's resource allocation. Unlike many countries where resources are allocated primarily by the financial and business systems, in China, government departments have a strong influence in economic and social life, and their participation in the process of resource allocation is very high. Among the three major credit reporting systems in China, the scale and application of the administrative credit reporting system are not inferior to, or stronger than, those of the financial and the commercial credit reporting systems. It can be said that the joining of government departments has changed the pattern of credit reporting in China, expanding its content and deepening its purpose, and realizing the co-governance of society by promoting a wider database and more diverse participation in China's credit reporting systems.

The Chinese government is moving towards a service-oriented government while the administrative credit reporting system has complied with the government's requirements for "fang guan fu". "Fang" means that the old governance style will undergo a fundamental change, and some old management measures will be abolished. For example, in order to break down all kinds of market access barriers and improve the business environment, the Chinese government has lowered many market entry barriers. On December 21, 2018, the National Development and Reform Commission and the Ministry of Commerce released the "Negative List of Market Access (2018 Edition)", which contains two types of lists, including prohibition and permission. The "list" stipulates that: for prohibited items, market entry shall not be allowed; for permitted items, market entry shall be applied by market entities and approved by administrative authorities according to law and regulation, including relevant qualification requirements and procedures, technical standards and licensing requirements; for items of industries, fields, and businesses not in "the List" all types of market entities can equally enter in accordance with the law. On November 22, 2019, the "Negative List of Market Access (2019 Edition)" was issued. A total of 131 items are included in the 2019 version list, 20 items less than the 2018 version list, and the former has the authority of being designated the "unique list in the country".

Reducing restrictions to access means that it is more difficult to manage during and after the entry. In order to be able to manage, it is necessary to build a new credit-based supervision mechanism, so that the liberalization of market access is synchronized with regulatory policies, and the regulatory gap can be eliminated. This could not only streamline processes but also improve regulatory efficiency and optimize the business environment. The new type of credit-based supervision requires corresponding administrative credit reporting system. Therefore, China has established a directory to publicize credit information, and is improving the national credit information sharing and exchange platform across regions and departments. In this way, the credit information between administrative systems across the country can be interconnected, relying on which, categorized supervision and joint reward and punishment can be carried out. This ensures that violators will be punished and the subjects with good credit records enjoy more convenience in public services. In such "fang" and "guan" practices, the government can better serve the public, maintain the social environment, and establish market order by the new credit-based governance methods.

China's administrative credit reporting system, with the goal of social governance incorporated, has collected the public credit information data of various government departments. In addition to providing basic and equal public services, the government system also has a large amount of funds, resources, and opportunities to allocate to society. From the point of view of the government's social management objectives, it is definitely not enough to merely rely on the price mechanism for allocation. Therefore, the governments use public credit information data accumulated in their own system for identification and judgment. They establish a credit-based social evaluation mechanism, conduct reward and punishment practices, and implement differentiated supervision measures. Subjects that violate laws and social order and infringe on the rights of others will be limited in their access to resources and opportunities in the field of administrative services. This is in line with the requirements of social governance and the interests of the vast majority.

The administrative credit reporting formed in China is very different from the traditional credit reporting considered by other countries, but there are similar credit management methods in foreign governance practices. For example, the United States links tax records with residents' credit, investment, and even retirement plans. The Japanese government issues taxpayers blue and white tax application forms according to taxpayers' differentiated tax records. Those with blue tax application forms can receive tax benefits or the preferential treatment of lower frequency of inspections.

On the basis of the administrative credit system, the government has promoted the implementation of regulatory reform and credit facilitation measures for the benefit of the people and enterprises, improving the role of credit elements in capital allocation. In June 2018, the China National Development and Reform Commission announced the launch of a series of "Credit For" projects to expand honesty incentives in the key areas of the people's livelihood, to make the intangible value of honesty into a tangible value, and to make honesty useful and sensible. (From a speech by Lian Weiliang, deputy director of the National Development and Reform Commission, at the opening of China's Credit City Development Forum in 2018.) The "Credit For" series of projects began with five major projects, namely, Credit For Loan, Credit For Lease, Credit For Travel, Credit For Tourism and Credit For Approval.

"Credit For Loan" is aimed at enabling small and micro businesses with good Credit to enjoy more favorable loan interest rates and more convenient loan approval services. The better their performance, the easier it is to get loans. "Credit For Lease" is aimed at making enterprises with good Credit enjoy more preferential rent discounts, longer lease terms, and more convenient lease procedures. "Credit for Travel" is aimed at helping individuals with demonstrated integrity enjoy a lower price for travel services, lower deposit or no deposit privileges, and preferential access to travel activities. The more honest one is, the easier it is to travel. For example, "Credit For Tourism" is aimed at making the tourists with good Credit enjoy more convenient ticket purchase and check-in services, no inspection privileges, line-jumping privileges, etc. "Credit For Approval" is aimed at sending people with good compliance through the green channel, giving them priority access to convenient approval services in government administrative issues. Through these projects, market entities can rely on credit to obtain resources and convenience.

By opening and sharing, the administrative credit reporting system has improved the use of public credit information and public participation in social governance. Government departments record and evaluate the behavior of the management objects in the administrative management process. They make the evaluation standards, evaluation results, and red and blacklists public according to laws and regulations. For example, the General Administration of Customs has publicized the "Customs Certification Enterprise Standards" (From: General Administration of Customs website [EB/OL]. http://www.customs. gov.cn/customs/302249/302266/302267/2127183/index.html), the State Administration of Taxation has publicized the "Tax Credit Evaluation Index and Evaluation Method (Trial)" (From: State Administration of Taxation website [EB/OL] http://www.chinatax. gov.cn/n810341/n810765/n812141/n812242/c1078412/content.html), and the Ministry of Ecology and Environment has publicized the "Enterprise Environmental Credit Evaluation Method (Trial)" (From: Ministry of Ecology and Environment website [EB/OL]. http://www.mee.gov.cn/gkml/hbb/bwj/201401/t20140102_265940.htm) and so on. As a result, third-party credit evaluation agencies can add public credit information data to their evaluation models to achieve multi-dimensional risk assessment, which is a form of participation in the social co-governance of violations.

1.2.3   The Promotion of China's Social Credit System to Establish New Rules

On June 14, 2014, the State Council of China issued the "Outline of the Plan for the Building of Social Credit (2014–2020)". The "Outline" proposes to promote the construction of four key areas: government integrity, business integrity, social integrity and judicial credibility. It defines building tasks that are closely related to the vital interests of the people and the healthy development of the economy and society. In terms of the support system, it is proposed to strengthen the construction of integrity education and integrity culture, accelerate the construction and application of credit information systems, and improve the reward and punishment mechanisms of the social credit system.

The construction goals set out in the "Outline Plan" have mostly been achieved by 2020: "Basic laws and regulations and the standard system on social credit are basically established; the credit information system covering the whole society based on the sharing of credit information resources is basically completed; the credit supervision system is basically sound; the credit service market system is relatively perfect; and the credit reward and punishment mechanisms are fully functioning; Significant progress was made in the building of government integrity, business integrity, social integrity, and judicial credibility, and the satisfaction of market and society have been greatly improved; The social awareness of integrity has generally increased; The credit environment for economic and social development has improved markedly, and the economic and social order has improved significantly."

The building of China's social credit system promotes the establishment of the administrative credit system. Although the administrative credit reporting established by the Social Credit System has been repeatedly criticized for privacy violations abroad, the public is more concerned about the convenience brought and the improvement of social integrity. German scholar Genia Kostka's new research report focuses on the Chinese people's own views on the social credit system. The cross-regional survey found that the groups interviewed showed a surprisingly high degree of recognition of the social credit system, and they were more inclined to use a social benefit framework to interpret the social credit system, that is, they saw that the social credit system can promote honest transactions in society and the economy (from: Kostka, Genia (July 23, 2018), China's Social Credit Systems and Public Opinion: Explaining High Levels of Approval[EB/OL]. https://ssrn.com/abstract=3215138).

The social credit system is a huge and structural system, and many of its key tasks are closely related to the public's social and economic life, such as credit-based benefits to the people and enterprises, credit-based regulation, joint rewards and punishments mechanism, and credit repair, etc. The implementation of these measures requires the establishment of the corresponding credit recording system and rules.

The establishment and improvement of the credit reward and punishment mechanism is the key step of the social credit system. In May 2016, the State Council issued "Guidelines on Establishing and Perfecting the Joint Credit Reward and Punishment System to Accelerate the Construction of Social Integrity", which put forward systematic guidance on the joint credit reward and punishment mechanism. This is the first institutional arrangement to comprehensively guide and regulate the joint rewards and punishments at the national level, and is an important top-level design for the construction of China's social credit system. It points out that it is necessary to build a cross-regional, cross-departmental and cross-domain reward and punishment mechanism, which is jointly participated in by the government and the society, through the incentive and restraint means in accordance with laws and regulations. This will then promote the operation of the market subject to law and credit, maintain the normal order of the market, and build a social environment of integrity.

How to implement the joint credit reward and punishment mechanism within various social programs? On January 19, 2018, the National Development and Reform Commission and the People's Bank of China jointly issued "the Guiding Opinions on Strengthening and Regulating the Management of the List of Subjects for Credit Reward and Punishment". The "Guiding Opinions" proposed specific suggestions on the identification standards, identification procedures, and sharing and release of the redlists and blacklists, as well as on how to implement joint rewards and punishments by government departments and social forces, etc. It also put forward specific requirements concerning the protection of the rights and interests of credit subjects and the withdrawal mechanism of the joint rewards and punishments list. The "Guiding Opinions" require member units of the joint meeting of the social credit system and other relevant departments and units to jointly sign a memorandum of cooperation on credit reward and punishment, and clarify the specific measures for red and blacklists in various departments (units) and fields.

From guidance to memorandum and to the specific implementation, relevant departments need to formulate corresponding measures. For example, the "Memorandum of Cooperation on Implementing Joint Incentive Measures for A-Class Taxpayers" should be based on the assessments of enterprises with levels of A, B, M, C, and D, according to "Tax Credit Evaluation Indicators and Evaluation Methods (Trial)", "Tax Credit Management Measures (Trial)" and "Announcement on Tax Credit Evaluation and Related Matters", etc., to implement incentive measures.

The "Memorandum of Cooperation on the Implementation of Joint punishment for Customs Dishonest Enterprises" and the "Memorandum of Cooperation on the Implementation of Joint Incentives for Advanced Customs Certification Enterprises" should be based on the assessment of enterprises and their categorization into certified enterprises, general credit enterprises and untrustworthy enterprises, according to the "Customs Certification Enterprise Standards" and "Customs Enterprise Credit Management Measures" published by the General Administration of Customs to implement corresponding punishment or reward measures.

Penalties and incentives, which run through the daily administrative supervision of government departments, are manifested as the differentiation measures of credit management, that is, the idea of categorized supervision. For example, the "Customs Enterprise Credit Management Regulation of the People's Republic of China" announced by the General Administration of Customs on January 29, 2018, lists the conditions that advanced certification enterprises, general certification enterprises, general credit enterprises and dishonest enterprises need to meet, and offers the corresponding regulations as follows:

The average inspection and quarantine sampling rate for general certified enterprises is less than 50 percent of the average sampling rate for general credit enterprises;

In addition to the general certification enterprise management measures, advanced certification enterprises also benefit from an average inspection rate of import and export goods below 20 percent of the average inspection rate of general credit enterprises; it can apply to the customs for exemption of guarantees; it can reduce the frequency to be inspected, etc.;

For dishonest companies, the average inspection rate of imported and exported goods is more than 80 percent; lifting, relocation, and warehousing costs cannot be exempted from inspection; the clearance measures of keeping samples and images cannot be applied; the full amount of guarantees for processing trade operation is required; the frequency of audits and inspections is increased.

In short, these classification measures are to give certified companies, especially advanced certified companies, preferential facilities and implement strict supervision and punishment for untrustworthy companies.

However, for regulators, punishment is a means, not the purpose. It is more important to guide society to improve integrity. The punishment of dishonesty makes more and more credit subjects hope to repair their credit as soon as possible and reduce credit losses through active error correction. In July 2019, the General Office of the State Council issued the "Guiding Opinions on Accelerating the Construction of the Social Credit System and Building a New Credit-based Supervision Mechanism" and proposed that government departments should explore the establishment of a credit repair mechanism, which means if the untrustworthy market entities have corrected the behaviors within the prescribed period and the adverse effects are eliminated, credit repair can be carried out by making credit commitments and completing credit rectification.

Successively, various departments will introduce relevant standards on credit repair. For example, the State Administration of Taxation issued the "Announcement on Matters Relating to Tax Credit Repair" on November 7, 2019, which specifically explained the conditions and standards, time limits and procedures for tax credit repair.

In accordance with the principle of limited repair, the "Announcement" clarified 19 types of tax credit dishonesty with minor or no serious social impact and corresponding restoration conditions. The Announcement provides for eligible taxpayers to apply to the tax authorities for tax credit repair. Credit repair is premised on correcting dishonest behaviors. For example, according to the "Announcement", if taxpayers, who fail to handle tax declarations, payment of taxes and data filing within the legal time limit, have corrected this failure, the credit points will be recalculated based on the time interval between the time of making up and the dishonest behavior being recorded by the tax authorities. If the interval is within 30 days, within this year, and within the following year, the taxpayer will be able to recover 80 percent, 40 percent, and 20 percent of the penalty points respectively. If an enterprise does not report or pay the declared tax within the prescribed time limit, if the tax amount does not exceed 1,000 yuan and the taxpayer can make up in time within 30 days, 100 percent of the points will be recovered.

The tax credit repair system is an important innovation in the institutional rules, which shows that the purpose of disciplinary measures is to correct rather than punish. Therefore, for tax credit problems caused by non-subjective malice in practice, insufficient tax law knowledge, or poor awareness of tax compliance, even if the punishment is given, taxpayers are allowed to recover tax credits in compliance with the law.

Foreign investors in China are not observers of the Chinese social credit system, but participants in it. Those who do not understand or follow the rules of the social credit system may be recorded and may face some problems. On March 15, 2019, the Second Session of the Thirteenth National People's Congress voted to adopt "the Foreign Investment Law of the People's Republic of China" which came into force from January 1, 2020. The "Foreign Investment Law" stipulates that the state establishes a foreign investment information reporting system, and foreign investors or foreign-invested enterprises shall submit investment information to the competent commercial department. On November 8, 2019, the Ministry of Commerce issued the "Foreign Investment Information Reporting Measures (Draft for comments)". The "Draft" clearly states that the competent commercial department shall establish a foreign investment integrity filing system, and the information that reflects the integrity status of foreign investors or foreign-invested enterprises in the supervision and inspection processes shall be recorded in the foreign investment integrity filing system in a timely manner, and the list of dishonest behaviors shall be publicized through the foreign investment information publicity platform, and market supervision, foreign exchange, customs, tax and other relevant departments can be notified. The Ministry of Commerce has stated that violations of these measures by foreign investors and foreign-invested enterprises would be dealt with by the integrity information system according to law. Local governments and relevant departments may use information in the system as reference factors for allocating preferential policies in finance, taxation, land, finance, and recruitment, etc.

Of course, foreign investors can also use the social credit system to safeguard their rights and enjoy the benefits brought by a better social business environment. For example, in response to the problem of intellectual property rights infringement in China, on November 24, 2019, the General Office of the Central Committee and the General Office of the State Council jointly issued the "Opinions on Strengthening the Protection of Intellectual Property Rights". The "Opinions" propose that by 2022, the phenomenon of infringement will be more effectively curbed, and the situation that rights holders have difficulty in defending their rights will be significantly improved. The Opinions propose four specific directions for the protection, one of which is to establish strict protection policy guidance for intellectual property, strengthen case enforcement measures, establish and improve the "blacklist" system, conduct social publicity for the list of enterprises which repeatedly and intentionally infringe, and improve the joint punishment mechanism for dishonesty. In other words, information disclosure, classified supervision, and joint punishment in the social credit system will all play a role in protecting intellectual property rights and provide support and guarantee for all enterprises, including foreign companies, to carry out innovation and research and development.

Generally speaking, the social credit system has formed the administrative credit system with Chinese characteristics, and established the corresponding credit rules. It is necessary to understand these policies of China's social credit system before we can understand the administrative credit system.

1.3   Analysis of the Development Trends of China's Three Major Credit Reporting Systems

1.3.1   Promoting the Allocation of Resources According to Credit Capital

In a broad sense, credit is a relationship of trust. The establishment of trust comes from economic and social life. Wu (2002, 2009) believes that the basic connotation of "credit" in the broadest sense is trust. Various performance and application of "credit" are based on trust.

From the study of trust, we can find out the interlinked social characteristics between credit and trust. Arrow (1974) argues that trust is an indispensable part of the market economy and a special commodity. Barber (1983), in his book "The Logic and Limitations of Trust", divides trust into three categories from the perspective of individual expectations: the first is a general trust generated by the recognition of the morality, literacy and code of conduct of the trustee; the second is a skill-based trust generated by the trustee's professional ability in the act of trust; the third is an obligation-based trust that other people can be responsible without involving their own interests. Putnam (1993) believed that honesty is a component of social capital. The main form of social capital is credit level. A higher individual credit level can effectively improve social operation efficiency, reduce information adoption costs, ensure the safety of the interpersonal environment and promote social trust. Luhmann (1995) divided trust into the trust of traditional society and the trust of modern society. He believes that the trust of traditional society is familiar, but the trust of modern society is complex and abstract. Weber (1997) considers that trust has two types: one is special trust and the other is general trust. Special trust is mainly based on blood, and general trust is reflected through the interests and value community.

Wu (2002, 2009) believes credit is the capital to gain trust. Credit capital is a kind of wealth that can be traded, measured, and managed. People can use credit capital to participate in social transactions and resources allocation.

Other scholars also pay attention to the capital attributes of credit. Sun (2002) argues that if "credit" can affect economic and production efficiency and investment in "credit" can bring back market benefits, then this "credit" has the nature of capital. Yan (2005) believes that "credit", "credit reporting" and morality all interact with each other. Morality is the norm that the whole society must abide by. "Credit" is the support to maintain the normal operation of society. "Credit reporting" can relieve information asymmetry. The three lay a solid foundation for a good social "credit" environment together. Shi et al. (2002) argued that the participation of "credit" in the market economy is indispensable, and it can be transformed into "credit" information and generates "credit" capital.

There are many ways to allocate resources: some are allocated equally according to the principle of fairness, some are allocated according to the size of physical capital, and some are allocated according to efficiency, each with different rules. When a party chooses to allocate resources according to credit capital, the rule is the condition that the counterparty can gain his trust. Such rules are commonly agreed, which is that the resource holder allocates the resources and takes the laws and regulations as rules (or formulates rules according to his own needs) and the resource seeker agrees to abide by the rules; or the resource seeker makes a commitment and the resource holder tracks the implementation of the commitment. After such an agreement or commitment process, the resource holder can allocate resources to a party he trusts via credit management.

In the process of allocating resources according to credit capital, the rules for distribution are basically legal, open and generally recognized. Evaluation of the trustworthiness of the credit subject is not made by the resource allocator subjectively and operated in a dark box. What actions have caused one not to be trusted by the resource allocator and thus unable to obtain resources and opportunities? What rules are used to measure credit capital? This is what the credit subject should and can know. Only in this way can the purpose of resource owners be achieved through the reward and punishment mechanism, such as reducing risks or achieving a more orderly social environment.

Though many rules in commercial credit reporting system are not written regulations, they are also generally accepted industrial and social conventions. For example, in the process of housing leasing, in addition to paying rent on time, tenants should also pay attention to hygiene, care of furniture, and regular time schedules. As another example, when using a shared bicycle, in addition to paying the ride fee on time, the user cannot destroy the bicycle, install a private lock, modify the bicycle or keep the bicycle for exclusive usage. These regulations are important for businesses and enterprises. They will record related behaviors and eventually decide whether this customer is trustworthy or not, and whether or not to provide him with beneficial products and preferential services.

On the whole, it is a Pareto Improvement if the society and the market can allocate resources according to both traditional physical capital and credit capital. As long as people can adapt to credit rules, they can get resources and opportunities, and they can rely on credit capital to achieve their goals. In such an environment, the old sayings of "the poor are getting poorer and the rich richer" will be broken. The poor who lack physical capital will still have the opportunity to accumulate credit capital and obtain social resources. This reflects social equity and solves the Matthew Effect, and enriches the means and field of social resource allocation. In fact, the general public obtaining more resources and opportunities constitutes a substantial improvement of the welfare of society.

1.3.2   Developing in Mutual Integration

Although all are allocated by credit capital, the rules of the three major resource allocation sectors each have their own focus, and the information collected by the corresponding credit reporting system is also different.

The financial sector attaches importance to risks and returns when allocating resources, so it mainly focuses on the financial strength and repayment history of the credit subject in credit reporting and evaluation. Under the condition that there is no infringement on citizens' equal rights, the government sector tends to allocate public resources and opportunities to subjects who comply with social laws and regulations and not to those who violate the normal order of society. Therefore, the administrative credit information system and corresponding evaluations established by government departments will mainly focus on the compliance of credit subjects with administrative regulations and social rules in social life. In addition, the non-financial sector has a large number of commercial resources and opportunities, such as exemption from deposits, discounts, and enjoyment before payment policy. It also allocates these resources considering its own interests, and it will tend to allocate resources to those who can bring them low risks, high returns and low costs. Therefore, in commercial credit reporting, a large amount of behavioral information will be recorded to predict the trends of future payment and the consumption behavior of the subjects.

But the three major credit reporting systems are not isolated from each other. The information they collect can complement each other, and the evaluations based on this can benefit from each other. For example, the financial sector has mastered information in terms of credit performance. At the same time, it also needs information on administrative management and commercial transactions as auxiliary judgments on credit decisions. The CCRC of People's Bank of China explained the concept of credit reporting (From: China's Credit Reference Center of People's Bank of China website [EB/OL]., http://www.pbccrc. org.cn/zxzx/zxzs/201401/87814073facf4b9795480d40fd626467.shtml): the most important role of credit reporting is to prevent losses in credit exchanges. Obviously, historical performance records can best reflect a person's ability and willingness to perform. When there is no historical record of performance, we can only rely on non-economic information to judge. This is also why some of the world's major transnational credit agencies are more and more comprehensive in their information collection. The collection of information from multiple angles is mainly to confirm each other, to comprehensively and accurately evaluate the credit status of the credit subject, such as collecting various types of registration information, administrative penalties, information, etc. At the same time, it is also helpful to motivate the credit subject to comply with their commitments in the above-mentioned aspects.

Therefore, the credit data collected and used by the credit reporting system is continuously expanding from the financial field to the public and commercial fields and bringing together more diversified credit data.

1.3.3   Playing a Role in Solving Social Trust Issues

China's three major credit reporting systems were formed to adapt to China's economic and social conditions, and are reasonable and necessary. They solve the problem of trust in social development. In China, with the continuous advancement of industrialization and urbanization, more and more people have left the acquaintance society and entered a stranger society. The trust relationship based on blood, kinship and geography has encountered difficulties in supporting a wider range of interpersonal communication, and meeting the needs of the rapid development of the market economy. Different countries have different methods to improve the environment of social trust, some rely on religion, some rely on centralized or imperial power, and some rely on law, but China faces an altogether different situation.

On the one hand, China's legal system is still not perfect. In the past two or three decades, China has rapidly moved from a planned economy and a small farmer economy to a multi-regional and multi-national market economy system. However, the legal system for protecting the spirit of a contract has yet to be improved. Some laws are still at the principal level, and it is difficult to refine the implementation.

On the other hand, the traditional concept of honesty and credit is eroded by dishonest events that have happened from time to time, and the binding force of morality is declining, which makes social governance more difficult. The lack of credit and integrity has led to various social problems, such as malicious tax evasion, factory accidents, food and drug safety panics, academic fraud and product counterfeiting, and so on. These untrustworthy and dishonest phenomena have seriously affected the social environment and social order.

No matter whether it is from a legal or moral perspective, China is facing the reality that the social governance and economic development trend do not match. The establishment of legislation and moral constraints is a slow process. In fact, the three credit reporting systems make up for the gap between law and morality. The participants in these systems achieve a common governance of society with their own commitment and constraint rules (Wu 2013).

The "two-line theory" in Wu's credit theory (see Abb. 1.3) explains that the law is the bottom line and most people will not violate or touch it; morality is the high line and not everyone can fully achieve it. What should people do, when caught between law and morality, that is, between breaking the law and not meeting moral requirements? How

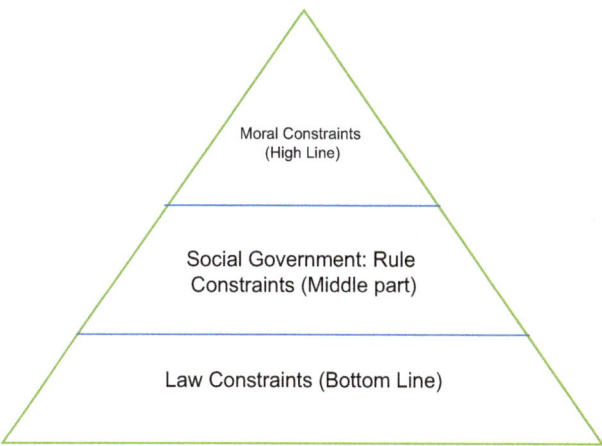

Abb. 1.3 Wu's Credit Theory: the role of Social Credit System—constraints between two lines (by the author)

should social governance be managed? For these questions, the Social Credit System, as a new institutional arrangement, has a lot of room to function. It creates a series of rule constraints between legal constraints and moral constraints, making social governance more refined, executable, and specific. For example, with regard to the enforcement against dishonest persons, the previous situation was that some of the persons refused to execute the enforcement, and the courts had no way to change the situation. However, relying on the credit system to implement joint rewards and punishments, as well as supervision and guidance, relatively obvious results are achieved.

It was mentioned at a press conference held by the National Development and Reform Commission on July 16, 2019 that, since the implementation of the information disclosure system for the list of dishonest Executees in October 2013 as of the end of June 2019, the national courts have cumulatively released 14.43 million names of persons who performed dishonesty, banned 26.82 million persons from the purchase of air tickets and limited 5.96 million persons from buying high-speed rail tickets; and as a result 4.37 million people who failed to perform honesty were encouraged by credit discipline to fulfill their legal obligations (From: The national courts have issued a total of 14.43 million person-times for breach of trust [EB/OL]. http://news.china.com.cn/2019-07/16/content_74997606. htm). Punishing the dishonest executees not only protects the legitimate rights and interests of the applicant, but also improves the social trust relationship and optimizes the business environment.

This is how credit builds rules between the bottom line of law and the high line of morality, and plays a role in establishing and maintaining social order. In the end, credit rules may be moralized or legalized, but this process is very long, so the relevant credit reporting systems and credit institutions are necessary in China's current economic development and social governance process.

In general, China's current trust issue is multifaceted and complex. The resource allocators on each side cannot solve this big social problem all by themselves. However, the combination of the three major resource allocation systems and the three major credit reporting systems can solve the trust problem facing China. It is a new type of social governance, which has achieved social co-governance on credit rules, established and maintained order and a positive environment in the economic and social fields. China's ongoing attempts and experience gained from this innovation are worthy of attention, research, and reference.

References

Arrow, K J. (1974): The limits of organization. New York: Norton.

Barber, B. (1983): The Logic and Limits of Trust. New Jersey: Rutgers University Press.

Kostka, Genia (July 23, 2018): China's Social Credit Systems and Public Opinion: Explaining High Levels of Approval. Available at SSRN: https://ssrn.com/abstract=3215138.

Luhmann, N. (1995): Social Systems. California: Stanford University Press.

Max Weber (1997): Economy and Society. Translated by Li Rongyuan. Commercial Press.

Margaret J. Miller (2003): Credit reporting systems and the international economy. London: Massachusetts Institute of Technology,

Putnam, R D. (1993): Making Democracy Work: Civic Traditions in Modern Italy. New Jersey: Princeton University Press.

Sun Zhiying (2002): Economic Analysis on Credit Problems. China City Press.

Shi Qin, Dai Xian, Yang Tao (2002): The Establishment of Transmission Mechanism of Credit Information – from the Angle of Information Economics. Finance and Accounting Monthly, (12)13–14.

Sesame Credit (2017): Report on China's deposit-exempting Xinyong service. [EB/OL].http://finance.ifeng.com/a/20170323/15254033_0.shtml.

Wu Jingmei (2013): The future of China's Xinyong reporting: Three database system. Credit Reference, (1). 4–12.

Wu Jingmei (2009): Modern Xinyong Science (2nd ed.). Peking: Renmin university press.

Wu Jingmei (2002): Modern Xinyong Science (1nd ed.). Peking: China financial publishing house.

Yan Yanyang, Liu Yan (2005): Morality, Trust, Credit and Credit Reporting. The Theory and Practice of Finance and Economics, (4): 43–48.

Jingmei Wu 吴晶妹, School of Finance, Renmin University of China, PhD. tutor, leader of credit management discipline

Zur historischen Entwicklung des Kreditbegriffs

2

Xinzhong Shi 石新中

Zusammenfassung

Mit dem Wandel der menschlichen Gesellschaft hat der Begriff des Kredits in verschiedenen historischen Perioden unterschiedliche Konnotationen. Die ethische Bedeutung von Kredit erschien zuerst, anschließend die wirtschaftliche Bedeutung und schließlich die rechtliche Bedeutung. Das Konzept des Kredits in der modernen Gesellschaft ist eine Vereinigung verschiedener Bedeutungen in der Geschichte. Die Bedeutung von „Kredit" unterscheidet sich von „Reputation" und „Vertrauen", hängt jedoch von ihnen auch zusammen.

Wie bei vielen anderen Konzepten wurden dem Kredit auf verschiedenen Stufen der Menschheitsgeschichte unterschiedliche Bedeutungen beigemessen. Das chinesische Wort für Kredit, Xin信yong用, enthält das Zeichen Xin信, was sich bei alten chinesischen Klassikern hauptsächlich auf das Einhalten von Versprechen in sozialen Interaktionen bezieht. Aus „LunYu" (The Analects of Confucius) stammt das Sprichwort: „Wenn jemand sein Versprechen nicht einhält, was darf man sonst von ihm erwarten?" Im modernen Kontext hat das Wort „Kredit" mehr Bedeutungen als nur die ethische. Zum Beispiel die Interpretation von „Kredit" in Cihai (Miniaturversion 1999) lautet: 1. „Verwendung von Vertrauen", 2. „Versprechen werden eingehalten und Vereinbarungen werden ausgeführt, um das Vertrauen der Gegenseite zu gewinnen", 3. „Die auf der Rückzahlung beruhenden Sonderformen der Wertentwicklung, die hauptsächlich aus Währungsleihen, Kreditfinanzierung und Vorauszahlung von Warengeschäften beruhen".

Xinzhong Shi 石新中 (✉)
Editorial Department of Journal of Capital Normal University, Beijing, China

Das Cihai (chinesisch 辭海/辞海 – „Meer der Wörter") ist ein modernes umfassendes Lexikon und Wörterbuch für einzelne Schriftzeichen und mehrsilbige Begriffe der chinesischen Sprache. (wikipedia)

Die Hauptformen sind Staatskredit, Bankkredit, Geschäftskredit und Verbraucherkredit. „Es ist ersichtlich, dass „Kredit" im modernen Kontext mindestens zwei Sinngehalte hat: Der eine ist moralisch, der andere ist wirtschaftlich. Bisher wurde die Kreditforschung in akademischen Kreisen hauptsächlich unter ethischen und wirtschaftlichen Gesichtspunkten durchgeführt, die gleichzeitig den logischen Ausgangspunkt für die Kreditforschung darstellen. In den letzten Jahren sind Kreditfragen mit der Entwicklung der Gesellschaft schrittweise in das Blickfeld des Rechtskreises getreten. Viele Juristen haben „Kredit" auch aus rechtlicher Sicht definiert. So sind einige Wissenschaftler der Ansicht, dass sich „Kredit" auf die soziale Bewertung und das Vertrauen bezieht, das natürliche Personen, juristische Personen und andere Organisationen aufgrund ihrer wirtschaftlichen Fähigkeiten und ihrer Leistungsbereitschaft erhalten" (vgl. 杨立新 2005, S. 139). Es ist ersichtlich, dass der Begriff „Kredit" in der modernen Gesellschaft aus rechtlicher Sicht neu interpretiert wird. Dieser Artikel verwendet den dialektischen und historischen Materialismus von Karl Marx, um den Prozess zu erläutern, wie sich das Konzept von Kredit mit den Veränderungen der menschlichen Gesellschaft entwickelt und bereichert hat.

2.1 Arbeitsteilung und Zusammenarbeit der Urmenschen: die Entstehungsphase des Kreditbegriffs

In der Altsteinzeit war die Überlebensfähigkeit der Menschen sehr gering. Einzelpersonen konnten nur durch Gruppenarbeit Subsistenzmittel beschaffen, um das Leben zu halten und die grundlegendste Sicherheitsgarantie zu erhalten. Wie Marx sagte: „Je tiefer wir in der Geschichte zurückgehen, je mehr erscheint das Individuum, daher auch das produzierende Individuum, als unselbstständig, einem größeren Ganzen angehörig" (vgl. 马克思 恩格斯全集(第46卷上册) 1972, S. 21). Um zusammen zu überleben, brauchen die Menschen im primitiven Stamm eine einfache Arbeitsteilung. Im Prozess der Arbeitsteilung und Zusammenarbeit muss jeder eine vernünftige Erwartung an die Handlungen anderer haben. Nur wenn die Mitglieder der Gruppe sich gegenseitig vertrauen, können sie gemeinsam etwas erreichen. Beispielsweise sind bei Jagdaktivitäten ein oder mehrere Mitglieder einer Gruppe so angeordnet, dass sie die Beute zu einem bestimmten Ort locken oder treiben, während andere Mitglieder darauf warten, sie an vorbestimmten Orten zu töten. Diese Zusammenarbeit ist die Grundlage für die Existenz von Urmenschen. Es ist zu erkennen, dass in der frühen menschlichen Gesellschaft das gegenseitige Vertrauen der Mitglieder in einer Gruppe die grundlegendste Voraussetzung war. Ohne gegenseitiges Vertrauen kann die gesamte Gruppe nicht überleben. Zu dieser Zeit war „Versprechen muss eingehalten werden" fast ein bewusster menschlicher Instinkt.

In der Tat haben viele vergleichende Anthropologen und Biologen auch bei der Untersuchung primitiver Stämme und fortgeschrittener Tiergruppen in Afrika, Amerika und

Australien entdeckt: Je geringer die unabhängige Überlebensfähigkeit einer Gruppe ist, desto eher tendiert das Wesen des Gruppenverhaltens zu Altruismus als Egoismus. Unter dem Druck der Bedürfnisse zum Überleben nehmen Individuen instinktiv altruistisches Verhalten und selten opportunistisches Verhalten an. Wenn sich die unabhängige Überlebensfähigkeit des Einzelnen verbessert, steigt die Häufigkeit verschiedener Streite und Interessenkonflikte (vgl. F.普洛格.D.G.贝茨 1988, S. 348). Es ist ersichtlich, dass Egoismus und Opportunismus die Produkte von erhöhter Produktivität, Entstehung von Überschussprodukten und erhöhter Wahrscheinlichkeit individuellen unabhängigen Überlebens sind. Es ist kein unveränderliches Merkmal des menschlichen Verhaltens.

Aus Obigem ist ersichtlich, dass das Prinzip „Versprechen muss eingehalten werden" mit der Entstehung der menschlichen Gesellschaft einhergeht. In den frühen Stadien der menschlichen Gesellschaft bedeutete Kredit nur, „halte Dein Versprechen, führe Deine Vereinbarung aus". Kredit hatte nur ethische Bedeutung und regulierte die Beziehungen zwischen den Stämmen der primitiven Gesellschaft und den Menschen innerhalb des Stamms.

2.2 Entwicklung des Kreditkonzepts in der naturökonomischen Zeit

In der mittleren und späten Phase der primitiven Gesellschaft wird mit der Verbesserung der Produktionsmittel die Überlebensfähigkeit der Menschheit immer stärker. Zusätzlich zur Aufrechterhaltung ihres eigenen Überlebens können die Menschen eine kleine Menge an überschüssigen Produkten haben. So begannen innerhalb oder zwischen Stämmen, zufällige, sporadische Aktivitäten zum Austausch überschüssiger Produkte aufzutreten. Dieser Austausch ist nur ein sehr einfacher Tauschhandel, aber wenn dieser Austausch stattfindet, haben beide Parteien implizit Vertrauen in die Produkte des anderen. Bei der oben genannten Situation des Jagens vertrauen die Menschen dem Verhalten des anderen, während zu diesem Zeitpunkt das Objekt des Vertrauens der Menschen die Produkte des anderen sind. Natürlich ist die Qualität dieser Art von Waren mit geringem technischen Gehalt leicht zu beurteilen. Dieses Vertrauen in die Produkte des anderen hat jedoch eine große Bedeutung für die historische Entwicklung der Menschheit: Es zeigt, dass Menschen als ein soziales Tier, in der Phase der Zusammenarbeit ein neues Niveau erreicht hat: vom Vertrauen in das Verhalten des anderen zum Vertrauen zugleich in das Verhalten und die Produkte des anderen. Vertrauen ist zu diesem Zeitpunkt selbstverständlich relativ einfach und unkompliziert.

Der direkte Tauschhandel ist zwar die direkteste und transparenteste Form des Austauschs, hat jedoch auch seine Grenzen. Viel Zeit wird verbracht, um ein passendes Objekt für die Transaktion zu finden. Das schränkt den Umfang und die Anzahl der Transaktionen erheblich ein. Langsam wird eine Art von Ware mit der Überzeugung entdeckt, dass diese Art von Ware allgemein akzeptiert werden kann, wird sie als Zwischenware relativ unveränderlich angesehen. Diese Art von Ware ist zu einem allgemeinen Äquivalent geworden, das mit vielen Waren ausgetauscht werden kann. Später, weil Metalle wie Gold und Silber leicht zu transportieren, leicht zu trennen, schwer zu zersetzen sind und gleichzeitig hohen Wert haben, verwendeten zivilisierte Länder auf der Welt Metalle als solche allgemeinen Äquivalente. Dies ist die Geburt des Geldes. Von direktem Austausch von Dingen zu indirektem Austausch mit Geld zeigt es, dass sich die menschliche Zusammenarbeit zu einem neuen Stadium entwickelt hat: Vertrauen von Menschen in bestimmte Personen und Dingen hat sich zu Vertrauen in eine relativ feste Sache, nämlich Währung, gewandelt.

Austausch basierend auf Währung ist einfacher zu implementieren als direkter Tauschhandel. Im realistischen Transaktionsprozess kann Kauf gegen Geld jedoch aufgrund von Produktknappheit oder unzureichender Geldmittel häufig nicht sofort durchgeführt werden. „Das Problem und die Mittel zur Lösung des Problems treten gleichzeitig auf." (vgl. 马克思恩格斯全集(第23卷) 1972, S. 106.) Um das Geschäft trotzdem zu ermöglichen, können die Parteien der Transaktion in dem Fall vereinbaren, die Waren erst dem Käufer zu übergeben, während die Zahlung zu einem späteren vereinbarten Zeitpunkt erfolgt. Dies bedeutet, dass Transaktionen zwischen Personen realisiert werden können, indem der Gegenpartei das Vertrauen hat, dass die Rückzahlung zu einem bestimmten Zeitpunkt in der Zukunft wie versprochen erfolgen wird. Die Partei, die erwartet, dass die andere Partei zu einem bestimmten Zeitpunkt in der Zukunft zurückzahlt, wird als Gläubiger, die andere Partei als Schuldner und die Beziehung zwischen ihnen als Schuldverhältnis bezeichnet. Das Entstehen dieses Schuldverhältnisses ist für die Entwicklung der menschlichen Gesellschaft von außerordentlicher Bedeutung. Dazu hatte der japanische Gelehrte わがつま さかえ (我妻荣 Wagatsuma Sakae) eine klassische Darstellung: Die Forderung beruht auf gegenseitigem Vertrauen zwischen Menschen. In der Geschichte der menschlichen Kultur entwickelte sich die Forderung nach dem dinglichen Recht. Dank der Anerkennung der Forderung ist das Wirtschaftsleben der Menschheit reichhaltiger. Menschen konnten nur in der Vergangenheit und Gegenwart leben, als Eigentumsverhältnis nur auf dinglichem Recht beruhte und nur das als Gegenstände eines Eigentumsobjekts angesehen wurde. Durch Anerkennung der Forderung bekommt die vereinbarte zukünftige Zahlung einen äquivalenten Gegenwert. Im Wirtschaftsleben der Menschheit können neben früheren und gegenwärtigen Vermögen somit auch zukünftige Vermögen existieren. Mit den Worten von Kola durch Entstehung von Kredit (das heißt das Auftreten von Forderung), „die Vergangenheit kann der Zukunft dienen. Die Zukunft kann der Vergangenheit dienen. Die Zeitbarriere wird durchbrochen und die Menschen können Zeit und Raum frei erobern." (vgl. 我妻荣 1999, S. 5–6). Aufgrund der Entstehung des Schuldverhältnisses kann eine Transaktion trotzdem abgewickelt werden, obwohl eine Partei vorübergehend nicht in der Lage ist, die Zahlung zu leisten. Dadurch können soziale Ressourcen mehr vernünftig verteilt werden. Als Währung zu einem universellen Tauschmittel geworden ist, ist es auch ein Zahlungsmittel geworden. Nachdem sich die Währung selbst dem Transaktionsprozess angeschlossen hat. sind Darlehensaktivitäten aufgetreten. Der Grund, warum das Darlehen entstehen kann, ist, dass der Gläubiger dem Versprechen des Schuldners auf künftige Rückzahlung und Zinszahlung vertraut hat. Aus demselben Grund übergibt eine Seite der Transaktion der anderen Seite Waren ohne sofortige Zahlung, da sie glaubt, dass die Zahlung in Zukunft erfolgen wird. Unabhängig von Sach- oder Geldkredite, handelt es sich darum, dass wirtschaftlicher Gegenstandswert vorzeitig von einer Partei an eine andere Partei übertragen wird, weil Vertrauen existiert. Anders als Tauschhandel oder Bezahlung bei Lieferung ist es das sogenannte Kreditgeschäft, das heißt eine Art von Geschäft, bei dem Waren, Geld oder Dienstleistungen nicht unmittelbar vor Ort getauscht werden.

In dieser Zeit der menschlichen Gesellschaft bezieht sich das Wort „Kredit" neben der ethischen Bedeutung von „Versprechen halten und Vereinbarung ausführen" auch auf die besondere Form von Wertbewegungen mit der Voraussetzung von Kreditrückzahlungen und Zinszahlungen, was eine bedingte Übergabe der Zahlungsmittel oder Ware darstellt. Diese Bedeutung ist auch die übliche Bedeutung von Kredit, die Ökonomen in der heutigen Zeit gebrauchen. Anhand der Kreditmedientheorie kann man sehen, dass David Ricardo, Adam Smith und John Muller den Begriff Kredit in diesem Sinne verwendeten (vgl. 曾康霖,王长庚 1993, S. 127–137). Marx erklärte die Bedeutung des Kredits unter seinem Verständnis mit einem Zitat von Tooke: „Der Kredit, in seinem einfachsten Ausdruck, ist das wohl oder übel begründete Vertrauen, das jemanden veranlaßt, einem anderen einen gewissen Kapitalbelauf anzuvertrauen, in Geld oder in, auf einen bestimmten Geldwert abgeschätzten, Waren, welcher Betrag stets nach Ablauf einer bestimmten Frist zahlbar ist." (vgl. 马克思 1975, S. 452)

2.3   Kreditbegriff während der Entwicklung der Marktwirtschaft

Wie oben beschrieben hat die Einführung von Kredit Transaktion die Realisierung von Markttransaktionen erheblich gefördert. Die sozialen Ressourcen wurden dadurch effizienter genutzt. Die Entwicklung von Markttransaktionen hat die Arbeitsteilung in der Gesellschaft weiter gefördert, die wiederum den Austausch zwischen Menschen vertieft und befruchtet hat. Gerade aufgrund der schrittweisen Weiterentwicklung der sozialen Arbeitsteilung und der zunehmenden Entwicklung des Austauschs tritt die Menschheit endlich in die marktwirtschaftliche Gesellschaft ein. In der Entwicklungsphase der Marktwirtschaft stießen Kreditgeschäfte jedoch auf neue Herausforderungen:

Erstens hat der Austausch mit seinem reichen Inhalt, der Ausweitung des Spielraums und der Diversifizierung der Formen eine bedeutende Stellung im gesamten wirtschaftlichen Betrieb genommen. Der Austausch ist für die Wirtschaftssubjekte zu einem unverzichtbaren Bestandteil des Wirtschaftslebens geworden. Ohne Tausch kann fast kein Wirtschaftssubjekt sein Wohlstand effektiv ausbauen. Selbst das Überleben kann zu einem Problem werden. In diesem Zusammenhang ist es notwendig fürs Überleben, durch eigene Arbeit Produkte anderer auszutauschen. Um Produkte anderer sorglos zu benutzen, muss man diese gut kennen. So ist zu fragen, ob die Produkte qualifiziert sind, wie man sie verwendet, wie gut sie funktionieren und so weiter. Doch so wie alles zwei Seiten hat, hat die feinkörnige Aufteilung der Arbeit auch negative Auswirkungen: Jeder Marktteilnehmer versteht immer weniger oder sogar nichts mehr Dinge außerhalb des eigenen Berufsfeldes.

Das heißt, die ständige Weiterentwicklung der Arbeitsteilung und die daraus resultierende Wissensverteilung führen zur zunehmenden Informationsasymmetrie über Produkte zwischen den beiden Parteien der Transaktion. Zur Ermöglichung einer fairen und reibungslosen Transaktion muss mehr Zeit investiert werden, um die Produkte der anderer zu verstehen. Das heißt, die Asymmetrie der Marktinformationen zwischen den beiden Parteien führt letztendlich zu immer höheren Kosten für das gegenseitige Verständnis der Produktinformationen.

Zweitens stützten sich in der naturökonomischen Periode die Subjekte von Kreditgeschäften durch Bekanntheit gegeneinander oder Garantien durch zwischengeschaltete Makler oder andere kontrollierbare Mittel als Kaution. „Der Mönch kann weglaufen, der Tempel aber nicht mit ihm" ist eine Metapher für Kredittransaktion, die in vollem Umfang garantiert wird. In der Entwicklungsphase der Marktwirtschaft mit der Ausweitung des Marktes und des Transaktionsumfangs gehen die Transaktionsgegenstände jedoch weit über den für die Marktteilnehmer bekannten Umfang hinaus. Die Transaktionsgegenstände könnten aus fernen Regionen oder sogar aus dem Ausland stammen. In der Realität kommt es mit der Bildung des Weltmarktes immer häufiger vor, dass die Objekte von Transaktionen vom fernen Ausland kommen. Zu diesem Zeitpunkt ist es zu einem neuen Problem für Marktteilnehmer geworden, die Krediteigenschaften des Handelspartners genau zu kennen und ausreichende Mittel zur Genüge zu haben, um den Handelspartner zu zwingen, Versprechen zu halten.

In diesem neuen historischen Zustand stellt die Asymmetrie der Kreditinformationen (einschließlich Informationen zur Produkte und Krediteigenschaften) zwischen den beiden Transaktionsparteien den Hauptwiderspruch dar. Wie die Kommunikation von Kreditinformationen gefördert werden kann, ist zur Hauptüberlegung des Marktteilnehmers geworden. Um Hindernisse im Informationsaustauschprozess zu beseitigen, werden die Investitionen zur Beschaffung der Informationen weiter erhöht, sodass die Transaktionskosten weiter steigen. Wenn die Marktsubjekte hierfür Mittel bis auf bestimmten Höhe aufwenden müssen, wird der Abschluss von Markttransaktionen erheblich beeinträchtigt. Die Gesellschaft braucht dringend einen neuen Mechanismus, um dieses Problem der asymmetrischen Information zwischen den beiden Parteien zu lösen. Wie Zhang Weiying sagte: „Informationsasymmetrie schafft eine Nachfrage nach Reputation. Je asymmetrischer die Informationsstruktur in einer Branche ist, desto wichtiger ist die Reputation in dieser Branche (vgl. 张维迎 2001, S. 144–145)". Dies erfordert, die Kreditinformationen von Marktteilnehmern der Gesellschaft offenzulegen. In der Tat sind in Ländern mit entwickelter Marktwirtschaft seit den 1970er-Jahren Sozialkreditsysteme zur Lösung dieses Problems schrittweise eingeführt und verbessert worden: Die Gesetzgebung schreibt vor, dass Marktkreditinformationen für die Außenwelt offen sein müssen. Kreditvermittlerorganisationen wie Auskunfteien zur speziellen Sammlung und Verarbeitung der Kreditinformationen von Marktteilnehmern und Ratingagenturen zur Bewertung der Kreditwürdigkeit von Marktteilnehmern wurden ins Leben gerufen. Sie haben einen ausgereiften Betriebsmechanismus entwickelt. Die Verwaltung und Überwachung der Kreditbranche wurde durch die Regierung verbessert. Die Berufsausbildung entwickelt sich im Kreditmanagement schrittweise weiter. Kreditvermittlerorganisationen sind bereits notwendiger Bestandteil für die Aufrechterhaltung des normalen Betriebs des sozioökonomischen Mechanismus geworden. Kreditinformationen des Transaktionsobjekts können leicht abgerufen werden, sodass eine Geschäftsentscheidung schnell getroffen werden kann. Insbesondere mit der Entwicklung des Internets können die Kreditinformationen von Marktteilnehmern schneller und bequemer in der gesamten Gesellschaft verbreitet werden und sind in der Tat zu einer Informationsquelle geworden, die von der gesamten Gesellschaft gemeinsam genutzt wird. Dank dieses modernen Kreditsystems schätzen die Marktteilnehmer sehr ihren Ruf und stellen sicher, dass ihre Produkte oder Dienstleistungen den Bedürfnissen der Gesellschaft erfüllen.

Unter diesen Umständen profitieren die Subjekte moderner Marktwirtschaft wie Personen, Unternehmen und Länder, die entwickelte Kreditbeziehungen und gute Bonitätsnachweise genießen. Es kommt denen zugute, mehr effizientere Kombinationsmöglichkeiten für dezentrale soziale Ressourcen zu finden und eine höhere soziale Produktivität zu erreichen. Einzelperson kann gute Kreditressourcen nutzen, um die Interaktions- und Kooperationschance zu erhöhen und somit eigenes Einkommen zu steigern. Unternehmen können Kreditressourcen für Finanzierung, Finanzmanagement einsetzen, mit maximaler Nutzung externer Ressourcen mehr Handelsmöglichkeiten zu gewinnen und somit mehr Profit zu realisieren. Mit Kreditressourcen können Länder die Attraktivität des Außenhandels erhöhen und mehr Nutzen aus der internationalen Arbeitsteilung ziehen. Im Gegenteil, wenn der Kredit eines Marktteilnehmers stark beeinträchtigt ist, stoßen seine wirtschaftlichen Aktivitäten überall auf Hindernisse.

Angesichts dieses neuen gesellschaftlichen Wandels glauben sensible Ökonomen, dass das traditionelle Konzept des Kredits die Konnotation des Vertrauens nicht mehr vollständig reflektieren kann (vgl. 骆玉鼎 2000, S. 19). Zusammenfassend aus den Ansichten dieser Ökonomen hat das Wort „Kredit" in der modernen Gesellschaft folgende Bedeutungen: (1) eine psychologische Aktivität, das heißt das gegenseitige Vertrauen. Es erfolgt hauptsächlich aus ethischer und moralischer Sicht; (2) eine Kreditaktivität, eine besondere Form der Wertbewegung. Es erfolgt hauptsächlich aus der Perspektive der traditionellen Ökonomie; (3) eine Fähigkeit, einen bestimmten Wert ohne sofortige Zahlung zu erwerben; (4) ein System, eine Reihe von Normen und Grundsätzen für Kreditgeschäfte und Kreditaktivitäten in der modernen Gesellschaft. Die beiden letzteren Bedeutungen sind neue Ideen, die von zeitgenössischen Ökonomen zusammengefasst werden. Kredit als eine Art von Fähigkeit zu betrachten, stammt daraus, dass in modernem Wirtschaftsleben Kredit zu einem wichtigen Aspekt seiner wirtschaftlichen Wettbewerbsfähigkeit für Marktteilnehmer geworden ist. Heutzutage gewinnt die Institutionenökonomik an Popularität, und es ist auch logisch, Kredit als institutionelles Arrangement zu betrachten.

Juristen im heutigen China haben auch die Bedeutung von Krediten aus rechtlicher Sicht zusammengefasst (vgl. 杨立新 2005, S. 538). Zusammenfassend aus den Verständnissen dieser Juristen von Krediten aus rechtlicher Perspektive, ist der Autor der Ansicht, dass die Bedeutung von Krediten wie folgt lauten sollte: Bewertung der Leistungsfähigkeit und Leistungsbereitschaft von Zivilpersonen durch die Gesellschaft. Seine Hauptmerkmale sind: (1) Das Kreditsubjekt ist eine Zivilperson, inklusive natürlicher Personen, juristischer Personen und anderer Organisationen. In einigen Fällen kann auch die Regierung (einschließlich der Zentral- und Lokalregierungen) Gegenstand von Krediten werden. (2) Es ist eine soziale Bewertung, keine Selbstbewertung. (3) Der Inhalt der Bewertung umfasst die Leistungsfähigkeit und Leistungsbereitschaft des Zivilsubjekts und beinhaltet keine politischen Einstellungen und allgemeinen moralischen Eigenschaften. Der Inhalt der Leistungsfähigkeit ist für verschiedene Subjekte sehr unterschiedlich: Für eine natürliche Person bezieht es sich hauptsächlich auf ihr Eigentum, ihr Einkommen usw. Für ein Unternehmen umfasst es Produktqualität, Kundenservice, Finanzkapazität, Zahlungsfähigkeit und Rentabilität sowie Aktiv-Passiv-Verhältnis in der Bilanz usw. Die Leistungsbereitschaft bezieht sich auf die Haltung zur Leistungserbringung und Vertrauenswürdigkeit des Zivilsubjekts, die anhand einschlägiger öffentlicher und kaufmännischer Kreditunterlagen überprüft werden kann, beispielsweise durch die Kreditunterlagen der Behörden für Industrie- und Handelsverwaltung, Steuern, Zoll, Transport, Warenkontrolle und Quarantäne usw. oder durch Dokumentationen über Wirtschaftsstreitigkeiten und Kreditaufzeichnungen in gewerblichen Bereichen (wie bei Banken und in der Telekommunikation). Es ist ersichtlich, dass für Unternehmen die „Bindung" in der „Leistung" hier nicht nur die greifbare Form ist, die von dem Unternehmen unterzeichnet wurde, mit der es eine Schuldverhältnis unterhält, sondern sich auch gegenüber den anderen sozialen Einheiten bindet, wie zum Beispiel gegenüber dem Verbraucher durch immaterielle „Bündnisse" in verschiedenen Verantwortlichkeiten. (4) Es ist eine Kombination der geistigen Interessen und der Eigentumsinteressen des Zivilsubjekts.

Es zeigt sich, dass die Definitionen von Krediten in Wirtschafts- und Rechtswissenschaft nicht vollständig übereinstimmen, was naturgemäß mit den jeweiligen Merkmalen der beiden Disziplinen zusammenhängt. „Die Ökonomie untersucht, wie eine Gesellschaft knappe Ressourcen nutzt, um wertvolle Güter und Dienstleistungen zu produzieren und unter verschiedenen Menschen zu verteilen." (vgl. 保罗·萨缪尔森,威廉·诺德豪斯 2004, S. 2). Die Rechtswissenschaft untersucht, Rechte und Pflichten in den gegenseitigen Beziehungen der Menschen zu bestimmen, um soziale Gerechtigkeit und Gleichheit zu erreichen. Die Auslegung des Kredits in diesen beiden Disziplinen ist jedoch nicht unabhängig. Die neue Interpretation zu dem Worte „Kredit" durch zeitgenössische chinesische Juristen passt sich der neuen Bedeutung des Kredits in der realen Gesellschaft an. Wie bereits erwähnt, ist der Kredit in der modernen Gesellschaft zu einem wichtigen Aspekt der Wettbewerbsfähigkeit der Marktteilnehmer geworden. Der Kredit selbst ist zum Inhalt mit wirtschaftlichem Wert und einem messbaren Objekt geworden. Daher ist für Marktteilnehmer die Bewertung ihrer Leistungsfähigkeit und Leistungsbereitschaft von außerordentlicher Bedeutung. Diese logische Beziehung ist deutlich in der Definition des Wortes „Kredit" im „Oxford Dictionary of Law" zu sehen: „ein Ansatz, womit Waren oder Dienstleistungen erhalten oder geliefert werden, wobei die Zahlung nicht sofort sondern in Zukunft versprochen wird. Ob eine Partei Transaktionen mit der anderen Partei durch Kredit tätigt, hängt von seiner Einschätzung der Eigenschaften, der Fähigkeit zur Rückzahlung und der geleisteten Garantie des Schuldners ab." Diese „Einschätzung" ist die

Bedeutung von Kredit im rechtlichen Sinne. Diese Definition verbindet den von der traditionellen Ökonomie definierten Kredit mit dem Kredit der modernen Rechtsprechung und deckt die interne Beziehung zwischen beiden auf. Es ist eine vorbildliche Interpretation vom modernen Kreditkonzept.

2.4 Unterscheidung zweier Konzepte im Zusammenhang mit Krediten

In der Alltagssprache und in der akademischen Forschung werden die Wörter „Ruf" und „Vertrauen" oft im Zusammenhang mit dem Wort „Kredit" erwähnt. Durch Erläuterung der Konzepte von diesen beiden Wörtern sollten Unterschied und Verbindung zwischen ihnen und Kredit offenbart werden.

2.4.1 Ruf

„Modern Chinese Dictionary" interpretiert Ruf als „Kredit und Reputation". Der Autor hält diese Erklärung nicht für geeignet. Laut „Shuowen Jiezi Zhu" wird „Ruf" auch „Ansehen" genannt und hat einen Bezug zu „Schönheit".

Shuowen Jiezi Zhu 说文解字注 ist ein Wörterbuch zur Interpretation der chinesischen Zeichen, veröffentlicht im Jahr 1815.

Es ist ersichtlich, dass Ruf eine Bedeutung von gutem Kredit hat. Es ist gleichbedeutend mit „Prestige" oder „Reputation" in Englisch. Daher stimmt der Autor der Ansicht zu, dass Ruf sich auf die Schönheit des Kredits richtet, das heißt ein guter Kredit. „Seine Beziehung zu „Kredit" besteht darin, dass das Wort „Ruf" eine positive Bedeutung hat, während Kredit ein neutrales Wort ist." Zu Kredit gehört Ruf, aber auch das allgemeine Kredit. Ob es sich um eine gute Kredit oder das allgemeine Kredit handelt, beides wird als objektive soziale Bewertung unter dem Begriff von Kredit zusammengefasst (vgl. 杨立新 2006, S. 617).

In den letzten Jahren haben einige Ökonomen den Begriff „Kredit" selbst definiert. Li Shimei meint zum Beispiel: „In der theoretischen Kategorie der Marktwirtschaft sollte sich der Ruf auf die Bewertung und Reputation beziehen, die ein Subjekt oder ein Akteur (eine Einzelperson oder ein Kollektiv oder eine Organisation) in einem bestimmten sozialen Austauschverhältnis auf der Grundlage einer langfristigen Berechnung des Eigeninteresses und seiner Fähigkeit zur Vertragserfüllung etabliert hat." (vgl. 李士梅 2005, S. 44). Diese Definition für Ruf entspricht offensichtlich auch der Bedeutung von „Kredit" im rechtlichen Sinne. Es ist nicht überraschend, denn in der Vergangenheit gab es einen Konsens über den Begriff „Kredit" in der Wirtschaftsgemeinschaft, nämlich „eine Art von Kreditverhalten". Wirtschaftswissenschaftler sind der Ansicht, dass es nach der neuen Konnotation von Krediten in der modernen Gesellschaft unangebracht ist, das ursprüngliche Wort „Kredit" zu verwenden. Daher wird der Begriff Ruf (信誉 Xìnyu) benutzt, um diese neue Bedeutung zusammenzufassen. Das ist auch der Grund, warum das Wort „Ruf" (信誉) statt „Kredit" in den gegenwärtigen wirtschaftlichen Schriften häufig verwendet wird. Nach Ansicht des Autors ist es in der Tat unpraktisch, dasselbe Vokabular für Kredite im juristischen Sinne und Kredite im Sinne der traditionellen Ökonomie zu verwenden. Es gibt im Chinesischen noch kein geeignetes neues Wort, um den Begriff des Kredits im rechtlichen Sinne zu ersetzen.

Zur Zeit wird der Begriff „Kredit" 信用 sowohl in der Wirtschaft als auch im Recht häufig verwendet, wenn es sich um „Aufbaus eines Sozialkreditsystems" angeht, zum Beispiel kam der Begriff „Sozialkreditsystem" im Buch von Cheng Minxuan 2006 schon vor (vgl. 程民选 2006, S. 256). Ökonomen verwenden im Allgemeinen nicht den Begriff „Aufbau eines Sozialrufsystems" (vgl. 李士梅 2005, S. 179). „Kredit" als Bestandteil des Wortes „Sozial-Kredit-System" ist eindeutig im rechtlichen Sinne. Es ist ersichtlich, dass die rechtliche Bedeutung des Begriffs „Kredit" tatsächlich von allen Gesellschaftsbereichen akzeptiert wurde. Aus linguistischer Sicht ist es auch für andere Sprachen nicht ungewöhnlich, dasselbe Vokabular zu verwenden, um mehrere Bedeutungen auszudrücken. Aus diesem Grund ist es nach Ansicht des Autors angemessener, „Kredit" anstelle von „Ruf" zu verwenden, wenn zu diesem Zeitpunkt über Kredit im rechtlichen Sinne gesprochen wird.

2.4.2   Vertrauen 信任

„Vertrauen" ist ein weiterer Begriff, der eng mit „Kredit" zusammenhängt. In einigen Fällen können die beiden austauschbar verwendet werden. Einige Wissenschaftler haben „Kredit" als „Vertrauen" definiert. Aus soziologischer Sicht kann Kredit als ein psychologisches Phänomen definiert werden, wobei man anderer Subjekte vertraut, ihre zukünftigen Handlungen seiner Erwartung entsprechen werden. Ohne das Vertrauen der Menschen ist keine Kreditaktivität möglich. Das Vertrauensgefühl zwischen Menschen ist die Grundlage jeder Kreditaktivität. Kredit ist eine der Manifestationen des Vertrauens zwischen Menschen.

Vertrauen stellt ein wichtiges Thema in der Ethik- und Soziologieforschung dar. Seit den 1950er- und 1960er-Jahren, insbesondere in den letzten zehn Jahren, ist Vertrauen für viele Soziologen zu einem wichtigen Thema geworden. „Das Studium des Vertrauens wurde plötzlich Mainstream. Unzählbare Monografien und Thesen sind entstanden (vgl. 郑也夫 2001, S. 18)". Das Forschungsspektrum über Vertrauen in Soziologie und Ethik ist jedoch breiter als das in Wirtschaft und Recht. In Wirtschafts- und Rechtswissenschaften werden hauptsächlich aus wirtschaftlicher Perspektive die Aktivitäten von Marktsubjekten untersucht, während Soziologie und Ethik neben der Untersuchung der Kreditaktivitäten zwischen Marktsubjekten auch das Vertrauen in die allgemeine zwischenmenschliche Kommunikation, Vertrauen zwischen Familienmitgliedern, Vertrauen unter den Mitgliedern einer Organisation, Vertrauen zwischen der Regierung und den Bürgern, Vertrauen zwischen relevanten Gesellschaftssystemen und so weiter erforschen. Da Soziologie und

Ethik das Thema Vertrauen unter Menschen aus einer breiteren Perspektive untersuchen, haben ihre Forschungsergebnisse wichtige inspirierende Bedeutungen auf Wirtschaft und Recht. Zum Beispiel schlug der deutsche Soziologe Niklas Luhmann vor, dass Vertrauen einer der Mechanismen zur Vereinfachung der Komplexität sei (vgl. 尼克拉斯·卢曼, S. 30–40). „Vereinfachung ist eine Strategie, die sich aus Lebewesen, einschließlich Menschen, entwickelt hat. Der Grund ist, dass die Welt oder die Umwelt zu komplex ist und Ungewissen und Änderung enthält, während die menschliche Vernunft sowie die Fähigkeiten anderer Arten begrenzt sind. Ob es sich um vollständige Informationen (exklusiv Ungewissen) oder unvollständige Informationen (einschließlich Ungewissen und Änderungen) in der Umwelt handelt, es zwingt Menschen und andere Arten dazu, eine spezielle Strategie zu entwickeln, um damit umzugehen." (vgl. 郑也夫 2001, S. 99). Vertrauen ist eine Brücke, die Vergangenheit und Zukunft verbindet und die Menschen dazu bringt, sich auf die Vergangenheit zu verlassen, um auf die Zukunft zu schließen. Die vereinfachte Funktion des Vertrauens ähnelt dem Prinzip, dass der von Wirtschaft und Recht befürwortete Kreditmechanismus die Informationsasymmetrie zwischen Marktteilnehmern wirksam lösen kann.

Ein weiteres Beispiel für die Erkenntnisse der Soziologie, die Auswirkungen auf Wirtschaft und Recht haben, ist, dass Soziologen soziales Vertrauen in spezielles Vertrauen und allgemeines Vertrauen oder Persönlichkeitsvertrauen und Systemvertrauen aufteilen. Spezielles Vertrauen bezieht sich auf Vertrauen, das auf familiären Beziehungen, Blutlinie, Freunden und Regionen basiert. Es wird durch nicht-institutionalisierte Gegenstände wie Moral und Ideologie garantiert. Gegenstand des Vertrauens können Personen, Familien, Verwandte oder ein Ort sein. Es zeichnet sich durch großes Verständnis unter den Subjekten aus. Emotionen sind die Grundlage dieses Kredits. Allgemeines Vertrauen ist das Vertrauen, das auf Vertrag basiert und durch Rechtssystem gewährleistet wird. Es kann gegenüber anderen Personen angewendet werden, die keine Blutsverwandtschaft oder persönliche Beziehung zu sich haben, das heißt, es gibt keine spezielle Beziehung zwischen der vertrauenswürdigen Person und der Person, die das Vertrauen schenkt. Die Beziehung und das Vertrauensverhältnis beruhen auf einem Vertrag oder einem System, um dies zu gewährleisten. Die traditionelle Gesellschaft wird von speziellem Vertrauen dominiert, während die moderne Gesellschaft mehr auf allgemeines Vertrauen angewiesen ist. Persönlichkeitsvertrauen ist das Vertrauen in eine bestimmte Person, Verwandte, Territorien. Ähnlich ist auch das Vertrauen bei Heimatgemeinde und Gilden. Systemvertrauen bezieht sich auf das Vertrauen eines institutionellen Systems, das sich aus Anonymen zusammensetzt. Soziologen sind der Ansicht, dass aufgrund der Ausweitung von Kommunikation und Mobilität die moderne Gesellschaft viele Probleme, die früher dem Persönlichkeitsvertrauen galten, auf das Systemvertrauen übertragen hat. Darunter sind Geld und Experten zwei größte Systeme. Diese Klassifizierung des Vertrauens durch Soziologen fällt tatsächlich mit der Entwicklung des Kreditkonzepts zusammen. Vertrauen in der naturökonomischen Periode ist meistens das spezielle Vertrauen oder Persönlichkeitsvertrauen. Vertrauen in der marktwirtschaftlichen Entwicklungsperiode ist meistens das allgemeine Vertrauen oder Systemvertrauen. Soziologen glauben, dass die Beziehung zwischen traditionellem Persönlichkeitsvertrauen und modernem Systemvertrauen Vererbung, Korrespondenz und Koexistenz ist und in der erste Linie Vererbung ist. Sie glauben, dass eine Nation, die ihre traditionellen spirituellen Ressourcen vernichtet hat, keine glänzende Zukunft haben kann. Wenn die traditionellen Organisationsressourcen nicht mehr existieren, kann sie nur auf die lange Ruhe, Erholung und Genesung in Kultur und Zivilisation warten. Diese Idee hat sehr starke Auswirkungen auf den Aufbau des gegenwärtigen chinesischen Kreditsystems. Es zeigt sich, dass bestimmte Gepflogenheiten oder Praktiken in Chinas privater Kreditvergabe von formellen Finanzinstituten übernommen werden können, nachdem sie revidiert wurden. Dadurch wird das informelle System zu einem Teil des formellen Systems. Zum Beispiel kann man von traditioneller Vertragsgarantie lernen, Identität und Vertrag zu kombinieren (vgl. 龚汝富 2003, S. 12–16(9)). Nur auf diese Art und Weise können Sozialkredite mit chinesischen Merkmalen wiederhergestellt werden. Da der Wert der ethischen und soziologischen Forschung zum Vertrauen für die Forschung in Wirtschaft und Recht nicht Gegenstand dieses Artikels ist, werden hier nur diese Beispiele aufgeführt.

2.5   Schlussfolgerung

Das Konzept des Kredits wurde durch die Entwicklung der menschlichen Sozialpraxis bereichert. In der primitiven Gesellschaft, in der keine wirtschaftliche Interaktion stattfand, bezog sich die Bedeutung des Kredits auf das Einhalten von Versprechen in der Kommunikation und Kooperation der Menschen. Es ist nur eine Bewertung der Persönlichkeit und hat nur ethische Bedeutung. Als nicht sofortige Transaktionen im menschlichen Wirtschaftsleben stattfanden, gewann der Begriff des Kredits neben den ethischen Bedeutungen auch wirtschaftliche Konnotationen und bezog sich auf eine besondere Form der Wertentwicklung, die von der Rückzahlung von Darlehen und Zinszahlungen abhängig war. Sie ist eine bedingte Übertragung von Zahlungsmitteln oder Waren. Diese Bedeutung des Kredits wird als die wirtschaftliche Bedeutung des Kredits genannt. Während der Entwicklung der Marktwirtschaft, als die Arbeitsteilung in der Gesellschaft ein gewisses Maß an Präzision erreichte, als Austausch zum Hauptinhalt der menschlichen Wirtschaft und Gesellschaft wurde und als die Entwicklung des Marktes den lokalen Rahmen weit überstieg, kam das moderne Kreditsystem ins Leben, um die Asymmetrie der Kreditinformationen zu beheben. Zu der Zeit war für die Marktteilnehmer die Bewertung der Leistungsfähigkeit und Leistungsbereitschaft durch die Gesellschaft von außerordentlicher Bedeutung. Sie wurde sogar zur Grundlage für Überleben in der Gesellschaft. Diese Bedeutung manifestiert sich hauptsächlich in drei Aspekten: Erstens ist es die Grundlage für einen Marktteilnehmer, ob bzw. wie viel soziale Ressourcen er im Voraus beschaffen kann. Es ist ein wichtiges Merkmal für seine Wettbewerbsfähigkeit auf dem Markt. Zweitens ist es die Grundlage für den Zugang zu Handelsmöglichkeiten auf dem Markt. Drittens ist es ein wichtiger Aspekt der Persönlichkeitsbewertung. Diese Art der sozialen Bewertung zur Leistungsfähigkeit und Leistungsbereitschaft von Marktteilnehmern wird als rechtliche Bedeutung von Krediten bezeichnet. Nachdem Kredit eine neue Bedeutung gewonnen hat, bleiben die ursprüngliche ethische und wirtschaftliche Bedeutung bestehen. Die Bedeutung des modernen Sozialkredits ist die Sammlung der verschiedenen Bedeutungen von Kredit in der Geschichte. Es ist hier anzumerken, dass hier die historischen Veränderungen des Kreditbegriffs analysiert werden. Man muss allerdings nicht zu mechanisch sein, um diesen Prozess zu verstehen. Weil „die Geschichte die Grundlage der Logik ist, muss der Verlauf der Logik dem Verlauf der Geschichte im Allgemeinen entsprechen und diesen widerspiegeln" (vgl. 李秀林,王于,李淮春 1990, S. 274). „Der logische Weg ist jedoch nicht das einfache Wiedererscheinen der historischen Route, aber eine abstrakte Form, die zufällige Störung der Geschichte losgeworden ist." (vgl. 李秀林,王于, 李淮春 1990, S. 275) Gleiches gilt für den Wandel des Kreditbegriffs.

Literatur

杨立新 (2005). 中国人格权法立法报告[M].北京:知识产权出版社.
中共中央马克思、恩格斯、列宁、斯大林著作编译局 (1972) 马克思恩格斯全集(第46卷上册)[M]. 北京:人民出版社.
F.普洛格.D.G.贝茨 (1988). 文化演进与人类行为[M].吴爱明,邓勇译.沈阳:辽宁人民出版社.
中共中央马克思、恩格斯、列宁、斯大林著作编译局 (1972) 马克思恩格斯全集(第23卷)[M]. 北京:人民出版社.
我妻荣(1999).债权在近代法中的优越地位[M].王书江、张雷译,谢怀校.北京:中国大百科全书出版社.
曾康霖,王长庚 (1993). 信用论[M].北京:中国金融出版社.
马克思 (1975). 资本论(第3卷)[M].北京:人民出版社.
张维迎(2001). 产权、政府与信誉[M].北京:生活·读书·新知三联书店.
保罗·萨缪尔森,威廉·诺德豪斯 (2004). 经济学(第17版)[M].萧琛主译.北京:人民邮电出版社.
杨立新 (2006). 人身权法论[M].北京:人民法院出版社.
李士梅 (2005). 信誉的经济学分析[M].北京:经济科学出版社.
程民选等 (2006). 信誉与产权制度[M].成都:西南财经大学出版社.
郑也夫 (2001). 信任论[M].北京:中国广播电视出版社.
尼克拉斯·卢曼 (2005). 信任:一个社会复杂性的简化机制[M].瞿铁鹏,李强译.上海:世纪出版集团、上海人民出版社.
龚汝富 (2003- 12- 16(9)). 古代怎样治理诚信问题[N].人民日报.
李秀林,王于,李淮春主编 (1990). 辩证唯物主义和历史唯物主义原理(第3版)[M].北京:中国人民大学出版社.
骆玉鼎 (2000). 《信用经济中的金融控制》上海财经大学出版社.

Xinzhong Shi 石新中，Master of Philosophy, Doktor der Rechtswissenschaften, Postdoktorand in Wirtschaftswissenschaften. Er ist derzeit Direktor des Forschungszentrums für Kreditgesetzgebung und Kreditbewertung der Capital Normal University, Distinguished Professor der School of Risk Management der Xiangtan University, Vorsitzender des Zhongguancun Kredit Förderung- und Standardisierungsvereinigung für Unternehmen. Forschungsschwerpunkte: Kreditgesetzgebung und Kreditsystem. Er hat zahlreiche wissenschaftliche Artikel zur Kreditgesetzgebung in Fachzeitschriften wie dem Journal of Peking University (Social Science Edition) und der People's Daily Theoretical Edition veröffentlicht und war Co-Vorsitzender des vom National People's Congress Finance Committee vorgeschlagenen und von der National Development and Reform Commission in Auftrag gegebenen Projekts „Credit Legislation Research". Derzeit leitet er das Schlüsselprojekt des Justizministeriums „Forschung zum Rechtssystem der Kreditrepair". Er ist Teilprojektleiter von zwei Großprojekten des Nationalen Sozialwissenschaftlichen Fonds „Forschung zum Aufbau des sozialen Systems für Vertrauen in China" und „Forschung zum chinesischen Vertrauenskultur und Aufbau des chinesischen Sozialkreditsystems". Er hat zahlreiche Forschungsprojekte zur Kreditgesetzgebung durchgeführt, die vom Ministerium für Wissenschaft und Technologie, dem Handelsministerium und anderen Abteilungen in Auftrag gegeben wurden.

Normalisierung koordinierter Sanktion von Unehrlichkeit

Jiaping Han韩家平, Didi Xu 许荻迪 und Yuanyuan Guan 关媛媛

Zusammenfassung

Basierend auf der Definition des Systems der gemeinsamen Sanktion von Unehrlichkeit werden in diesem Artikel systematisch der Evolutionsprozess, die wichtigsten Methoden, Auswirkungen, Probleme und die internationalen Erfahrungen des Systems aufgeführt. Mit der Förderung der koordinierten Sanktion von Unehrlichkeit in größerem Umfang und Ausmaß haben sich an einigen Stellen unfaire Bestrafungen, verallgemeinerte Bestrafungsmaßnahmen und unsachgemäße Verwaltung von schwarzen Listen herausgebildet, was eine hohe Aufmerksamkeit benötigt. Angesichts der Tatsache, dass einige der von China durchgeführten administrativen Maßnahmen zur koordinierten Sanktion von Unehrlichkeit bereits im Rahmen der administrativen Bestrafung liegen, sollten sie gesetzlich verankert werden. Gleichzeitig wird vorgeschlagen, den Mechanismus für die koordinierte Sanktion von Unehrlichkeit unter den folgenden drei Gesichtspunkten zu verbessern: Erstens soll die Gesetzgebung über die koordinierte Sanktion von öffentlichen Krediten auf drei Wegen beschleunigt werden, nämlich Überarbeitung des Verwaltungsstrafgesetzes, Beschleunigung des Gesetzgebungsprozesses für das Sozialkreditgesetz sowie Überarbeitung bzw. Verbesserung der Vorschriften für die damit verbundenen Felder, und die Klauseln über die koordinierte Sanktion von Krediten darin einzubetten usw. Die zweite besteht darin, ein Listenverwaltungssystem für die koordinierte Sanktion einzuführen, um die gemeinsamen Be-

Jiaping Han 韩家平 (✉) · Didi Xu 许荻迪 · Yuanyuan Guan 关媛媛
Beijing, China
E-Mail: han@creditcn.com

© Springer Fachmedien Wiesbaden GmbH, ein Teil von Springer Nature 2020
O. Everling (Hrsg.), *Social Credit Rating*,
https://doi.org/10.1007/978-3-658-29653-7_3

strafungsmaßnahmen explizit, offen und vorhersehbar zu machen. Drittens ist die Einrichtung eines koordinierten Sanktionsmechanismus für Marktkredite zu beschleunigen, um mit dem gemeinsamen Sanktionsmechanismus für Öffentlichen Vertrauensmissbrauch interoperabel zu sein und koordiniert entwickeln zu können.

In den letzten Jahren hat Chinas Aufbau eines Sozialkreditsystems in einigen grundlegenden Bereichen und Knotenpunkten wichtige Fortschritte und (Zwischen-) Ergebnisse erzielt, darunter gilt die überregionale, branchenübergreifende und disziplinübergreifende koordinierte Sanktion von Unehrlichkeiten als wichtigste institutionelle Innovation bisher. Da eine Reihe von besonders schwierigen Problemen im Bereich der Wirtschafts- und Sozialverwaltung (zum Beispiel die schwierige Vollstreckung gerichtlicher Entscheidungen) gelöst wurden, wurde die koordinierte Sanktion auch von den Sozialkreditaufbaubehörden auf allen Ebenen als die nützlichste und wirksamste Methode zum Aufbau vom Vertrauen angesehen, und diese wird in immer größeren Umfang und Ausmaß gefördert werden. Gleichzeitig tauchte aber an einigen Stellen das Phänomen der ungerechten Bestrafung, der Verallgemeinerung von Disziplinarmaßnahmen und der unsachgemäßen Verwaltung von schwarzen Listen in der Praxis gemeinsamer Disziplinarmaßnahmen gegen Unehrlichkeit auf, die zu Sorge Anlass geben müssen. Daher ist es notwendig, die Einführung und Verbesserung der einschlägigen Gesetze zu beschleunigen und die gemeinsame disziplinarische Arbeit gegenüber Unehrlichkeit in die legalisierte und standardisierte Spur zu bringen.

3.1   Definition und Klassifizierung der koordinierten Sanktion von Unehrlichkeit

3.1.1   Definition

Gegenwärtig hat die chinesische Behörde für den Aufbau des Sozialkreditsystems oder auch die chinesische Wissenschaft noch keine einheitliche Definition für „gemeinsame Bestrafung gegen Unehrlichkeit" festgelegt. Sowohl die herausgegebenen normativen Dokumente als auch die zwischen den Sektoren unterzeichneten Memoranden zur koordinierten Sanktion nennen nur die praktischen Elemente, wie die Teilnehmer, Objekte bzw. Disziplinarmaßnahmen, die für eine koordinierte Sanktion gegen Unehrlichkeit bestimmt sind, während die Definition, die Essenz und die Eigenschaften davon unberührt bleiben.

Die Leitmeinung des Generalsekretariats des Staatsrates zur Beschleunigung des Aufbaus des Sozialkreditsystems und zum Aufbau eines neuen kreditbasierten Aufsichtsmechanismus (国办发 [2019] Nr. 35) (im Folgenden: 国办 Nr. 35) wurde am 9. Juli 2019 herausgegeben. Darin hat der Artikel Nr. 10 spezifische Anforderungen für koordinierte Sanktion gegen Unehrlichkeit aufgestellt. In dem Dokument wurde Folgendes betont: Beschleunigung der Einrichtung eines regionen-, branchen-, und sektorenübergreifenden, gemeinsamen Disziplinarstrafmechanismus, um das Problem der wiederkehrenden und ex-situ-Vertrauensbrüche vom Grunde her zu lösen. In Übereinstimmung mit den Gesetzen und Vorschriften soll eine Liste gemeinsamer Disziplinarmaßnahmen erstellt werden. Diese soll ständig aktualisiert und der Gesellschaft offengelegt werden, sodass administrative, marktbezogene und industrielle Disziplinarmaßnahmen gegen Unehrlichkeit mit der umfassenden Beteiligung sozialer Kräfte zugleich eingesetzt werden können.

Aus der Sicht der gegenwärtigen Richtlinien und Praktiken bezieht sich die koordinierte Sanktion von Unehrlichkeit auf die Verwaltungsmaßnahmen, mit denen mehrere Parteien, die innerhalb einer bestimmten Frist, direkt oder indirekt vielfältige Kreditbeschränkungen für die Subjekte durchführen, die schwere Vertrauensbrüche begangen haben, wodurch die Kosten für Unehrlichkeit steigen und die Wirtschafts- und Sozialordnungen reguliert werden. Es ist anzumerken, dass sich die Unehrlichkeit hier auf den Zustand der Nichterfüllung der gesetzlichen bzw. der vereinbarten Verpflichtungen der Subjekte bezieht, während die Kreditbeschränkungen hier nicht nur die Offenlegung und Weitergabe von Kreditinformationen des Vertrauen brechenden Subjekts zwischen den Verwaltungssektoren, sondern auch administrative, marktbezogene bzw. industriebezogene Disziplinarmaßnahmen umfassen. Beispielsweise umfassen die vom 国办 Nr. 35 aufgestellten gemeinsamen Bestrafungsmaßnahmen administrative Verwaltungssanktionen wie Beschränkungen der Emission von Aktien, öffentlicher Ausschreibung, Beantragung von Finanzmitteln und Inanspruchnahme von Steuervorteilen, marktbezogene Verwaltungsaktionen wie Beschränkungen des Zugangs zu Krediten, Flugreisen, Schnellzügen und Sitzplätzen, und industriebezogene Verwaltungssanktionen wie Zirkulation von Kritik und öffentliches Anprangern. Gleichzeitig wurde in dem 国办 Nr. 35 klargestellt, dass sich strenge Aufsicht und gestärkte Bestrafung besonders auf die Bereiche zu konzentrieren haben, die in direktem Zusammenhang mit der Sicherheit von Leben und Eigentum der Bevölkerung stehen, wie Lebensmittel und Medizin, ökologische Umwelt, Konstruktionsqualität, Produktionssicherheit, Senioren- und Kinderbetreuung und Sicherheit im städtischen Betrieb. Gegen die Marktteilnehmer und ihre Verantwortlichen, die sich weigern, gerichtliche Entscheidungen oder verwaltungsrechtliche Sanktionen zu befolgen, wiederholt Verstöße begehen und erhebliche Verluste verursachen, müssen, gemäß den Gesetzen und Vorschriften, die Markt- bzw. Branchenverbotsmaßnahmen innerhalb eines bestimmten Zeitraums konsequent umgesetzt werden bis hin zum schlimmsten Fall einer endgültigen Verdrängung aus dem Markt. Der gesetzliche Vertreter oder Hauptverantwortliche des Subjektes des Vertrauensbruchs ist nach den gesetzlichen Bestimmungen zu ermitteln und zur Verantwortung zu ziehen.

3.1.2 Koordinierte Sanktion von Unehrlichkeit ist der Kernmechanismus und das wesentliche Merkmale des Sozialkreditsystems

3.1.2.1 Kernmechanismus des Sozialkreditsystems

Regulierung gegen Vertrauensbruch ist der Kernmechanismus des Sozialkreditsystems. Der Mechanismus der Regulierung gegen Vertrauensbruch besteht in der Umsetzung von (koordinierten), dem Grad des Vertrauensbruchs entsprechenden Disziplinarmaßnahmen in genau bestimmten Umfang und genau bestimmter Form (mit der wirtschaftlichen und sozialen Entwicklung ändern sich der Umfang und die Form ständig). Der Zweck der Regulierung gegen Vertrauensbrüche besteht darin, das Vertrauen der Mitglieder der Gesellschaft wiederherzustellen, die gewünschte normale wirtschaftliche und soziale Ordnung aufrechtzuerhalten und die langfristigen Interessen der Mitglieder der Gesellschaft insgesamt zu optimal wahrzunehmen.

Sogar in den Anfängen der menschlichen Gesellschaft werden sich aufgrund menschlicher sozialer Eigenschaften einzelne Kreditinformationen (insbesondere schlechte Informationen) unter den Mitgliedern der Gesellschaft weiter verbreiten.

Basierend auf vorherige Vereinbarung oder gemeinsame Erkenntnisse der Mitglieder werden Disziplinarmaßnahmen eingesetzt, indem Transaktionen und/oder Kommunikation abgelehnt werden oder die Bedingungen für Transaktionen bzw. für Kommunikation verschärft werden, sodass ein Regulierungsmechanismus gegen Vertrauensbruch gebildet wird. In diesem Sinne ist der Kreditbeschränkungsmechanismus selbst eine subjektübergreifende, koordinierte Sanktion von Unehrlichkeit und dieser wird zu einem der Kernmechanismen des Sozialkreditsystems.

3.1.2.2 Ein wesentliches Merkmal des Sozialkreditsystems

Die Spieltheorie besagt, dass nur im Fall von langfristig wiederholten Spielen die Menschen sich dafür entscheiden würden, bei Markttransaktionen vertrauenswürdig zu sein, um langfristig zukünftige Vorteile zu erzielen. Daraus wird ein Reputationsmechanismus generiert. In Wirklichkeit sind jedoch Umfang und Anzahl der Transaktionen aller Individuen begrenzt, weshalb das Kreditsystem der modernen Marktwirtschaft einen langfristigen Mechanismus für Anreize zu Vertrauenswürdigkeit bzw. Disziplinarstrafen zu Vertrauensbruch schaffen soll, sodass der Vertrauensbrecher bereits nach dem Bruch der Vertrauenswürdigkeit in einem Bereich überall und für einen bestimmten Zeitraum durch eine akteurenübergreifende, koordinierte Sanktion beeinschränkt wird (一处失信, 处处受限, 一次失信一定时间内受限, 实行跨主体的联合惩戒). Durch die koordinierte Sanktion werden die Kosten für Unehrlichkeit viel höher sein als die Gewinne aus Unehrlichkeit, was die Wiederholung von Vertrauensbruch vom Grunde her stoppt und eine Rolle zur Kreditregulierung spielt. Daher kann gesagt werden, dass die koordinierte Sanktion von Unehrlichkeit ein wesentliches Merkmal des Kreditsystems ist.

3.1.3 Arten der gemeinsamen Disziplinarstrafe

Je nach Teilnehmer und Anwendungsbereich kann die koordinierte Sanktion für Unehrlichkeit in eine für Marktkredite sowie eine für öffentliche Kredite unterteilt werden. Der Mechanismus für die koordinierte Sanktion von Marktkreditstörungen kann wiederum in marktbezogene, branchenbezogene und sozial koordinierte Sanktion unterteilt werden. Der Mechanismus für die koordinierte Sanktion von öffentlichen Kreditstörungen kann in administrative und juristische Mechanismen für die koordinierte Sanktion differenziert werden. Die Merkmale dieser fünf Mechanismen zur Bestrafung von Vertrauensbruch werden nachstehend analysiert.

3.1.3.1 Marktbezogener Mechanismus der gemeinsamen Bestrafung

Bei Markttransaktionen treffen Handelspartner Entscheidungen hauptsächlich aufgrund des Kreditstatus der Gegenpartei, wodurch die Transaktionen positiv oder negativ entwickelt werden. Wie bereits erwähnt, durch die Ablehnung von Transaktionen und die Verschärfung der Transaktionsbedingungen kann eine effektive Kreditstrafe für unehrliche Subjekte eingesetzt werden. Beispielsweise tauschen Geschäftsbanken Kreditinformationen von Kreditnehmern aus, und Versicherungsunternehmen tauschen Kreditinformationen von Versicherungsnehmern aus, sodass sie relativ konsistente, harmonisierte Handelsentscheidungen treffen, um eine marktkreditbezogene, koordinierte Sanktion von Unehrlichkeit durchzuführen.

3.1.3.2 Branchenbezogener Mechanismus der gemeinsamen Bestrafung

Einerseits können Branchenverbände, Handelskammern usw. eine entsprechende Kreditstrafe für die Unehrlichkeit ihrer Mitglieder auf der Grundlage von Selbstdisziplinierungsregeln der Branche verhängen und die Integrität der Branche und standardisierte Betriebe fördern, andererseits kann durch den Austausch und die Weitergabe von Kreditinformationen von Kunden und Lieferanten durch die Branchenverbänden das allgemeine Kreditrisiko der Branche wirksam reduziert oder vermieden und die Mitgliedsunternehmen nicht wiederholt getäuscht werden.

3.1.3.3 Sozialbezogener Mechanismus der gemeinsamen Bestrafung

Der sozialbezogene Mechanismus umfasst hauptsächlich Kreditstrafen und Aufsichtsmechanismen, an denen Kreditinstitute, soziale Organisationen und die Öffentlichkeit beteiligt sind. Durch professionelle Auskunftsteien, Rating- und Kreditmanagementdienste bieten Kreditservices von Drittanbietern ihren Kunden den Kreditstatus der untersuchten (bewerteten) Unternehmen an, wodurch Kreditentscheidungen der Kunden beeinflusst werden. Durch Anzeige über Unehrlichkeit, Medienaufsicht und Rechtsstreitigkeiten von öffentlichem Interesse können sich soziale Organisationen und die Öffentlichkeit wirksam daran beteiligen und mit kontrollieren, sodass Unehrlichkeit bestraft werden kann.

Die oben genannten drei Arten gemeinsamer Bestrafung werden alle unter Beteiligung gleichberechtigter zivil- und handelsrechtlicher Körperschaften eingerichtet und beruhen hauptsächlich auf dem Markt. Daher werden sie alle als marktkreditbezogene Mechanismen für koordinierte Sanktion bezeichnet. Diese Mechanismen wurden im Allgemeinen in entwickelten westlichen Ländern schon etabliert und spielen die zentrale Rolle in den Kreditmechanismen.

3.1.3.4 Administrativer gemeinsamer Strafmechanismus

Die administrative koordinierte Sanktion ist eine Art öffentlichkeitsbezogene, gemeinsame Bestrafung und eine der Hauptsonderheit im Aufbau des chinesischen Sozialkreditsystems. China hat einen überregionalen, branchenübergreifenden und sektorenübergreifenden gemeinsamen administrativen Disziplinarmechanismus eingerichtet und in der Marktaufsicht und im öffentlichen Dienst umgesetzt. Zum Beispiel werden laut der Kooperationsvereinbarung über die kooperative Aufsicht und die koordinierte Sanktion gegenüber unehrlichen Unternehmen, unterzeichnet von 38 staatlichen Einheiten wie der ehemaligen staatlichen Verwaltung für Industrie und Handel, der Nationalen Entwicklungs- und Reformkommission und dem Obersten Volksgerichtshof, der Vertrauensbrecher bezüglich des Konsums mancher Luxusprodukte und -dienstleistungen und der Qualifikation von Führungspositionen in entsprechenden Branchen in Schranken verwiesen.

3.1.3.5 Juristischer Mechanismus für koordinierte Sanktion

In China umfasst der öffentlichkeitsbezogene gemeinsame Strafmechanismus auch die koordinierte Sanktion durch die Justiz. Einerseits kann die Justiz durch professionelle Justiztätigkeiten gegen Unehrlichkeit vorgehen, andererseits tauscht die Justiz durch die Unterzeichnung gemeinsamer Disziplinarvermerke mit den zuständigen Regierungsstellen und öffentlichen Stellen die Informationen über Vertrauensbrecher mit anderen zuständigen Abteilungen aus, die die gerichtlichen Entscheidungen der Vollstreckungsbehörden nicht umsetzen, um eine koordinierte Sanktion nach dem Gesetz durchzuführen.

Aufgrund der enormen Unterschiede zwischen der Konnotation und Ausweitung des Sozialkreditsystems zwischen China und dem Westen (vgl. 韩家平 2018, S. 5) und der Merkmale im gegenwärtigen Stadium des Aufbaus des chinesischen Sozialkreditsystems ist der marktbezogene Mechanismus der gemeinsamen Bestrafung in westlichen Ländern häufiger, während der öffentlichkeitsbezogene Mechanismus der gemeinsamen Bestrafung derzeit in China mehr gebraucht wird. Gegenwärtig ist die Anwendung des marktkreditbezogenen Mechanismus der gemeinsamen Bestrafung auf dem chinesischen Markt noch nicht weit verbreitet und hat im Grunde genommen keine sozialen Zweifel hervorgerufen. Die juristisch koordinierte Sanktion ergibt sich aus dem übergeordneten Zivilprozessrecht und hat die damit zusammenhängenden Auslegungen der Justiz als Rechtsgrundlage, und es gibt weniger soziale Fragen zu solchen Bestrafungsmaßnahmen. Der soziale Verdacht fokussiert hauptsächlich auf administrative, koordinierte Sanktion, da einige der administrativen gemeinsamen Bestrafungsmaßnahmen bereits in den Bereich der verwaltungsrechtlichen Sanktionen fallen, aber es keine klare rechtliche Grundlage dafür gibt.

3.2 Lernen von den wichtigsten Methoden und Erfahrungen der gemeinsamen Bestrafung von Unehrlichkeit im In- und Ausland

3.2.1 Hauptpraktiken und Ergebnisse der gemeinsamen Bestrafung von Unehrlichkeit in China

Mit der Evolution des Aufbaus des Sozialkreditsystems entwickeln und ändern sich Umfang und Form gemeinsamer Bestrafung von Unehrlichkeit ständig.

3.2.1.1 Verkündung und Entwicklung der gemeinsamen Bestrafung von Unehrlichkeit

Seit 2002 China offiziell vorgeschlagen hat, ein Sozialkreditsystem einzurichten, hat sich während der Vertiefung der praktischen und theoretischen Forschungen das Verständnis für den Mechanismus der Kreditbeschränkung schrittweise von „Disziplinarstrafe von Unehrlichkeit" zu „koordinierte Sanktion von Unehrlichkeit" vertieft.

Erste Stufe: Vorschlag zur Disziplinarstrafe von Unehrlichkeit
Im Jahr 2006 hieß es im „11. Fünfjahresplan": „Der Aufbau eines Sozialkreditsystems soll beschleunigt werden und das Bestrafungssystem für Unehrlichkeit soll vervollständigt werden." Die Kreditstrafe hat sich zum ersten Mal zu einer wichtigen institutionellen Regelung für den Aufbau eines Sozialkreditsystems in amtlichen Dokumenten entwickelt, wobei die zu diesem Zeitpunkt erwähnte Kreditstrafe nur die Bestrafung unehrlicher Subjekte betont, aber nicht weiter ausgeführt wurde.

Zweite Stufe: von der Disziplinarstrafe zur gemeinsamen Bestrafung
Während verschiedene Regionen und Sektoren die Praxis des Kreditaufbaus weiter explorieren, sind die Ideen, Methoden und Wege für die Disziplinarstrafe nach und nach klar geworden. Um die Intensität der Bestrafung zu erhöhen, wurde begonnen, auf gemeinsames Handeln umzusteigen. In den „Stellungnahmen des Generalsekretärs des Staatsrates zum Aufbau des Sozialkreditsystems" (国发办 [2007] Nr. 17) heißt es: „Einen gemeinsamen Bestrafungsmechanismus für Unehrlichkeit zu bilden, sodass Vertrauensbrecher überall und in jeglichen Perspektiven beeinschränkt werden".

Im Jahr 2014 wurde im Entwurf des Konstruktionsplans für das Sozialkreditsystem (2014–2020) Folgendes vorgeschlagen: „Vervollständigung des Sozialkreditemechanismus zur verknüpfenden Belohnung und Bestrafung", „Stärkung der gemeinsamen Bestrafung", „Intensivierung der gemeinsamen Bestrafung", „Einrichtung eines sek-

torens- und regionenübergreifenden koordinierten Kreditbelohnungs- und Bestrafungsmechanismus", „Erzielen von sektorenübergreifenden, überregionalen Kreditbe-
lohnungs- und Strafverknüpfungen, damit vertrauenswürdige Personen überall davon
profitieren können, während Vertrauensbrecher überall mit Schwierigkeiten zu konfrontieren sind." Darüber hinaus werden spezifische Maßnahmen vorgeschlagen, wie
zum Beispiel „Verschärfung der administrativen Aufsichtsregulierungen und -stra-
fen", „Verbesserung des disziplinarischen Strafsystems auf der Grundlage bestehender Verwaltungsstrafen und Einrichtung eines Systems schwarzer Listen und eines
Marktaustrittsmechanismus für verschiedene Branchen". All diese bilden eine Grundlage für das Follow-up der Praxis koordinierter Bestrafung bzw. Belohnung. Zu dieser
Zeit haben sich zwar die Begriffe wie „koordinierte Sanktion von Unehrlichkeit" formal nicht gebildet, aber die Entwicklung in Richtung koordinierten Handelns war eingeleitet.

Dritte Stufe: Koordinierte Sanktion gegen Unehrlichkeit formell verkündet
Auf der Exekutivsitzung des Staatsrates am 22. Juli 2015 wurde die Einrichtung von Inte-
gritätsakten, koordinierte Sanktionsmaßnahmen und „schwarzen Listen" im System vorgeschlagen. Die Formulierung einer „koordinierten Sanktion" wurde formell festgelegt. In
der Leitmeinung des Staatsrates zur „Verbesserung einer koordinierten Förderung von
Vertrauenswürdigkeit und Sanktion von Unehrlichkeit und zur Beschleunigung des Aufbaus Sozialkredit" (国发 [2016] Nr. 33) wurde im Jahr 2016 der Mechanismus der koordinierten Belohnung und Bestrafung präzisiert und für die Publizität der „schwarzen Listen" und entsprechende Disziplinarmaßnahmen geregelt.

Im Jahr 2016 gaben das Generalbüro des Zentralkomitees der Kommunistischen Partei
Chinas und das Generalbüro des Staatsrates die „Stellungnahmen zur Beschleunigung des
Aufbaus von Kreditüberwachungs-, Warn- und Disziplinarmechanismen für verurteilte
Vertrauensbrecher" (中办发 [2016] Nr. 64) heraus, um den Aufbau sektorübergreifender
koordinierter Aufsicht und Bestrafung zu beschleunigen und einen Mechanismus für die
Kreditüberwachung, Warnung und Bestrafung zu schaffen, sodass bei „gebrochenem Vertrauen überall eingeschränkt wird". Diese innovative Maßnahme hat später dazu geführt,
dass „Lao Lai" (Bezeichnung für diejenigen, die zur Zahlungspflicht vom Obersten Volksgericht verurteilt wurden, aber dieser Pflicht nicht nachgekommen sind; Anm. d. Hrsg.) in
Bezug auf Reisen, Ausschreibung und Angeboten, Investition und Finanzierung, Aus-
zeichnung sowie Beschäftigung „überall eingeschränkt" war. Dies veranlasste einen erheblichen Teil der verurteilten Vertrauensbrecher, die Initiative zu ergreifen, um ihren
rechtlichen Verpflichtungen nachzukommen, was ein typischer Fall für koordinierte Sanktion darstellt.

3.2.1.2 Wichtigste Methoden der koordinierten Sanktionen
Voraussetzung für koordinierte Sanktionen von Unehrlichkeit ist die von den zuständigen
Sektoren als „schwarze Liste" bezeichneten Namenlisten von schwerwiegenden Vertrau-
ensbrüchen. Nach der Veröffentlichung auf der Website von Credit China (https://www.

creditchina.gov.cn/) werden koordinierte Sanktionen gemäß den geltenden Vorschriften und dem zwischen den Sektoren unterzeichneten gemeinsamen Disziplinarmemorandum durchgeführt.

Identifizierung der „schwarzen Liste"

Außer den von Gericht verurteilten Vertrauensbrechern, bezieht sich die „schwarze Liste" im Bereich des öffentlichen Kredits auf die Dokumente von schwerwiegend vertrauensbrechenden Subjekten, die von Regierungsbehörden gemäß Gesetzen, insbesondere Verwaltungsgesetzen, behördlichen Vorschriften oder behördlichen Dokumenten über die Ebene der Provinzen zum Zweck der Einschränkung der Kredite und sozialen Bestrafung erstellt werden, und in der Gesellschaft veröffentlicht werden. Deren Träger verfügen in der Regel über Verzeichnisse und Datenbanken. Die „Leitmeinung der Nationalen Entwicklungs- und Reformkommission und Zentralbank zur Stärkung und Regulierung der Verwaltung der Liste für koordinierte Belohnung und Bestrafung" (发改财经规 [2017] Nr. 1798) legt die Grundlage für die Ermittlung von „schwarzen Listen" klar fest. Die Identifizierung auf der „Schwarzen Liste" in allen Sektoren setzt im Prinzip national einheitliche Standards um. Die Standards werden von Mitgliedern der Allianz des interministeriellen Aufbaukomitee des Sozialkreditsystems oder anderer nationaler Industriebehörden im Einklang mit der Marktaufsicht, der sozialen Steuerung und den Verantwortlichkeiten des öffentlichen Dienstes recherchiert und entwickelt. Relevante Sektoren auf Provinzebene können nach Bedarf lokale Standards formulieren und diese nach Prüfung und Genehmigung durch die höheren Behörden und die Regierungen der Provinzen umsetzen.

Zu den derzeit weit verbreiteten „schwarzen Listen" zählen: schwarze Listen von Vertrauensbrechern, die von Gericht verurteilt wurden (Laolai-Liste), Unternehmen, die schwerwiegende Vertrauensbrüche begangen haben, schwarze Listen bezüglich Vertrauensbrüche bei Steuerangelegenheiten, schwarze Listen wegen Vertrauensbruch in der Produktionssicherheit und schwarze Listen wegen Vertrauensbruch in Umweltschutz, E-Commerce-Unehrlichkeit usw.

Veröffentlichung einer „schwarzen Liste" der Unehrlichkeit

Nachdem die schwarze Liste der Unehrlichkeit erstellt wurde, muss sie veröffentlicht werden, damit soziale Aufsicht eingesetzt werden kann und die Fälle der Vertrauensbrüche offengelegt und verbreitet berücksichtigt werden können. Nachdem die Informationen der schwarzen Liste von dem entsprechenden Sektor (Einheit) festgestellt wurden, müssen sie dem übergeordneten Sektor und der zuständigen Einheit des Aufbaus des Sozialkreditsystems auf derselben Ebene übermittelt werden. Innerhalb von zehn Arbeitstagen ab dem Datum der Festlegung müssen sie der nationalen Plattform für den Austausch von Kreditinformationen weitergeleitet werden. Über die Kredit-Websites der lokalen Regierung und über die Websites von „Credit China" (https://www.creditchina.gov.cn/) werden die schwarzen Listen an die Öffentlichkeit gegeben.

Durchführung koordinierter Sanktion von Unehrlichkeit

In der „Leitmeinung des Staatsrates zur Schaffung und Verbesserung des Systems zur koordinierten Belohnung von Vertrauenswürdigkeit und koordinierten Bestrafung von Unehrlichkeit" wird darauf hingewiesen, dass koordinierte Belohnungs- und Strafmaßnahmen, die auf schwarzen Listen basieren, im Allgemeinen in Verwaltungs-, Markt-,
Sozial- und Industriekategorien unterteilt sind. Verwaltungsmaßnahmen beziehen sich
hauptsächlich darauf, dass die schwerwiegend das Vertrauen Brechenden von entsprechenden Regionen und relevanten Sektoren als Hauptaufsichtsziele identifiziert werden,
und gemäß Gesetzen und Verordnungen verwaltungsbezogene Beschränkungen und Disziplinarmaßnahmen diesen gegenüber durchführen. Marktmaßnahmen beziehen sich hauptsächlich darauf, dass relevante Sektoren und Einheiten nach einheitlichen Sozialkreditcode indiziert werden, um relevante Informationen rechtzeitig offenzulegen, damit
Vertrauensbrüche im Markt erkannt und Kreditrisiken verhindert werden können. Soziale
Maßnahmen sollten die sozialen Kräfte zur Teilnahme an koordinierten Sanktionen von
Unehrlichkeiten anleiten. Industriemaßnahmen beziehen sich auf die Festlegung und Verbesserung von Selbstdisziplinierungskonventionen und Berufsethikstandards in der Industrie, um die Vertrauensbildung in der Industrie zu fördern. Darüber hinaus können auch
gerichtliche Disziplinarmaßnahmen in Fällen durchgeführt werden, bei denen der Vertrauensbruch eine schwerwiegende Straftat ist.

3.2.1.3 Ergebnisse koordinierter Disziplinarmaßnahmen gegen Unehrlichkeit

Die Daten in diesem Abschnitt stammen hauptsächlich aus dem „Jährlichen Analysebericht der Schwarzen Liste der Vertrauensbrüche von 2018" (2018年失信"黑名单"年度分析报告) des National Public
Credit Information Centre und dem „Monatlichen Analysebericht über den Vertrauensbruch im Juni 2019"
(2019年6月失信治理月度分析报告).

Bisher wurde Chinas koordinierte Sanktion von öffentlichen Krediten hauptsächlich in
Schlüsselbereichen wie Strafverfolgung, Marktaufsicht, Steuererhebung und -verwaltung,
Import und Export (Zoll), Finanzen, Transportwesen und medizinische Ordnung (Social
Governance „Psoriasis") angewandt. In den letzten Jahren haben die zuständigen Sektoren
eine Reihe koordinierter Bestrafungsmaßnahmen gegen diejenigen ergriffen, die gemäß
dem Memorandum der koordinierten Bestrafung gesetzlich in die schwarze Liste aufgenommen wurden. Die Disziplinarmaßnahmen wurden kontinuierlich verstärkt und die sozialen Reaktionen waren beachtlich.

Weitere Ausweitung des Geltungsbereichs der koordinierten Bestrafung

Bis Ende 2018 wurden insgesamt 51 gemeinsame Memoranden zur koordinierten Belohnung und Bestrafung unterzeichnet. Die Anzahl der Sektoren, die mit unterzeichnet haben,
belief sich auf 51 und verschiedene Arten von koordinierten Belohnungs- und Bestrafungsmaßnahmen wurden auf über 100 erweitert. Im Jahr 2018 wurden in 18 Bereichen
gemeinsame Memoranden über Belohnung und Bestrafung unterzeichnet, darunter Ein-

reise- und Ausreisekontrolle und Quarantäne, Spenden für wohltätige Zwecke und Registrierung von Ehen. Zum ersten Mal wurden schwarze Listen in vier Bereichen veröffentlicht, darunter Spenden für wohltätige Zwecke und wissenschaftliche Forschung. Auf der nationalen Plattform für den Austausch von Kreditinformationen wurden etwa 14,21 Millionen Informationen von Vertrauensbrüchen gesammelt, von denen im Jahr 2018 4,096 Millionen neue Informationen hinzugefügt wurden und die von 3,594 Millionen vertrauensbrechenden Subjekten handeln.

Die Stärke der koordinierten Bestrafungsmaßnahmen nimmt weiter zu
Bis Ende 2018 haben die Gerichte im ganzen Land insgesamt 12,77 Millionen verurteilte vertrauensbrechende Personen angezeigt, den Kauf von Flugtickets von 17,46 Millionen Personen beschränkt, den Kauf von Hochgeschwindigkeits-Bahntickets bei 5,47 Millionen beschränkt und 290.000 mal verurteilte Vertrauensbrecher daran gehindert, als gesetzliche Vertreter und leitende Angestellte aufzutreten.

Das Finanzamt hat 16.642 Fälle von Steuerverstößen veröffentlicht, 12.920 Finanzierungen und Darlehen verhindert, und 128 mal wurden Steuerzahler, die ihre Zahlungsverpflichtungen nicht erfüllt haben, daran gehindert, das Land zu verlassen. Der Zoll führte koordinierte Sanktionsmaßnahmen gegen 19.180 vertrauensbrechende Unternehmen in verschiedenen Bereichen durch und senkte die Ratingstufen von 188 Unternehmen. 19.180 Unternehmen wurden daran gehindert, zollzertifizierte Unternehmen zu werden. Bei diesen Unternehmen werden Risikokontrollen durchgeführt und Inspektionsbemühungen gestärkt.

Erhebliche Wirkung der koordinierten Bestrafung
Bis Juni 2019 gab es 1045 Fälle von vertrauensbrechenden Subjekten, die aktiv auf den Kreditportalen von verschiedenen Ebenen oder auf „Credit China" Versprechen zur Ehrlichkeit abgegeben haben, und 1240 Fälle, die aktiv an Kreditreparaturschulungen teilgenommen haben. Bis Ende Mai 2019 haben 4,22 Millionen Vertrauensbrecher, abgeschreckt durch Kreditstrafe, aktiv ihre gesetzlichen Verpflichtungen erfüllt.

Seit der Einführung des koordinierten Bestrafungsverfahrens hat die Gesamtzahl der von Gerichten im ganzen Land verhandelten Fälle 4,4 Billionen Yuan erreicht. Insgesamt sind 3,51 Millionen Vertrauensbrecher ihren gesetzlichen Verpflichtungen aktiv nachgekommen. 1417 Fälle aus der schwarzen Steuerliste ergriffen die Initiative, um Steuern, verspätete Gebühren und Geldbußen zu zahlen. Nur ein halbes Jahr nach dem Inkrafttreten der koordinierten Disziplinarmaßnahmen wegen besonders schwerwiegender Unehrlichkeit auf dem Wertpapier- und Terminmarkt ergriffen 37 verurteilte Parteien die Initiative, Bußgelder zu zahlen, was insgesamt in Höhe von fast 150 Millionen Yuan erreichte. Unter den Personen, die in der schwarzen Liste beschränkter Nutzung von Zügen und Zivilluftfahrzeugen geführt werden, haben 92 die Initiative ergriffen, die entsprechenden Ticketkosten zu erstatten und ihre gesetzlichen Verpflichtungen zu erfüllen. Die Zahl der Fälle, die aus der schwarzen Liste gelöscht werden konnten, nahm weiter zu: 2018 zogen sich insgesamt 2,175 Millionen Fälle aus der schwarzen Liste der Unehrlichkeit zurück.

3.2.2 Hauptprobleme der koordinierten Bestrafung und die Ursachen dafür

3.2.2.1 Hauptprobleme bei der koordinierten Bestrafung von Unehrlichkeit

Während koordinierte Bestrafung von Unehrlichkeit weiterhin wichtige Fortschritte und bedeutende Ergebnisse erzielt hat, gab es in einzelnen Orten jedoch Phänomene wie unfaire Bestrafung, Verallgemeinerung koordinierter Disziplinarmaßnahmen und unsachgemäße Verwaltung von schwarzen Listen. Der Missbrauch (insbesondere die administrativen „Blacklist"-Maßnahmen) hat große Bedenken in Wissenschaft und Öffentlichkeit ausgelöst, wie auch Anschuldigungen der internationalen Gesellschaften hinsichtlich der Legitimität des Aufbaus von Sozialkrediten in China. Manche Rechtswissenschaftler sind der Ansicht, dass China illegales Verhalten nicht in einem Unehrlichkeitsprotokoll aufführen sollte, geschweige denn den Aufbau von Rechtsstaatlichkeit durch den Aufbau des Sozialkreditsystems zu ersetzen. Es wurde sogar vorgeschlagen, die koordinierten Bestrafungsmaßnahmen wie die administrativen schwarzen Listen, die derzeit keine klare Rechtsgrundlage haben, unverzüglich einzustellen und erst nach Entschluss über einschlägige Vorschriften weiter voranzutreiben. Es sollte gesagt werden, dass diese Meinungen repräsentativ sind und zu Sorge Anlass geben müssen.

Übermäßige Bestrafung

Bei manchen Fällen wurden zu strenge Strafen für eine Unehrlichkeit verhängt, was weitaus höher zu bewerten ist als die Ernsthaftigkeit der Unehrlichkeit selbst. Nicht schwerwiegende Unehrlichkeiten wurden trotzdem durch koordinierte Disziplinarverfahren bestraft. In der Leitmeinung des Staatsrates zur „Verbesserung einer koordinierten Förderung von Vertrauenswürdigkeit und Sanktion von Unehrlichkeit und zur Beschleunigung des Sozialkreditaufbaus" wurde festgestellt, dass nur schwerwiegende Vertrauenbrecher als Fälle in die koordinierte Bestrafung einbezogen werden dürfen.

Im Mittelpunkt stehen „Handlungen, die die Gesundheit und das Leben des Volks ernsthaft gefährden", „die die Ordnung des fairen Wettbewerbs im Markt und die sozialpolitische Ordnung ernsthaft untergraben", „die Weigerung, rechtliche Verpflichtungen zu erfüllen und die Glaubwürdigkeit von Justiz- und Verwaltungsorganen ernsthaft zu beeinträchtigen, sowie die Weigerung, nationale Verteidigungsverpflichtungen zu erfüllen". In der Praxis des Kreditaufbaus an einzelnen Orten (da keine passenden disziplinarischen Maßnahmen ergriffen werden, die die Unehrlichkeit dem Verhalten selbst entsprechen) wurden nicht nur Subjekte der Unehrlichkeit geschädigt, sondern auch der koordinierte Bestrafungsmechanismus und seine Auswirkungen verzerrt.

Verallgemeinerung koordinierter Bestrafungsmaßnahmen

Die Verallgemeinerung koordinierter Bestrafungsmaßnahmen bezieht sich hauptsächlich auf Disziplinarmaßnahmen, die nicht mit Unehrlichkeit verbunden sind. Zum Beispiel verlangt das Aufnahmeverfahren an einer weiterführenden Schule, dass „Eltern von bewerbenden Schülern keine Unehrlichkeitsaufzeichnungen haben dürfen", oder der Sohn von einem „Lao Lai" (siehe oben) einen Antrag für Luftfahrthochschulen stellte und die Flugprüfung nicht bestand, weil ihre Eltern in die Liste der unehrlichen Aufzeichnung aufgenommen wurden. Das Oberste Gesetz hat in „Bestimmungen zur Beschränkung des relevanten hohen Konsums von verurteilten Vertrauensbrechern" lediglich vorgesehen, dass „Lao Lais Kinder keine Privatschulen mit hohen Schulgebühren besuchen dürfen", das Recht der Kinder aber auf Bildung an Schulen mit normalen Schulkosten weiterhin gesetzlich geschützt bleibt. Wenn das Recht der Kinder von Lao Lai auf Bildung generell eingeschränkt wäre, hieße das nichts anderes, als dass die anderen Subjekte, die keinen Vertrauensbruch begangen haben, mit bestraft würden, und die Disziplinarmaßnahmen stünden in keinem offensichtlichen Zusammenhang mit dem Verhalten von „Lao Lai". Die koordinierten Bestrafungsmaßnahmen gegen bestimmte Arten von Unehrlichkeit sollten eindeutig mit dem unehrlichen Verhalten in Zusammenhang stehen, da sonst die Bestrafung verallgemeinert würde und es schwierig wäre, als „rechtlich fundiert und vernünftig" zu gelten.

Das Verwaltungssystem der „schwarzen Listen" muss verbessert werden
Obwohl die zuständigen Ministerien und Kommissionen des Staatsrates und einiger Provinzialregierungen (Districtions, Städte) eine Reihe von Maßnahmen zur Verwaltung der schwarzen Listen erlassen haben, hat das Fehlen einheitlicher Standards und Normen dazu geführt, dass die derzeitigen schwarzen Listen zu kompliziert waren, und die Kriterien zur Aufnahme und die folglichen Bestrafungsmaßnahmen wie auch die Anlässe zur Entfernung aus den Listen nicht standardisiert sind. Außerdem können die Blacklist-Informationen verschiedener Sektoren nicht untereinander geteilt werden. Diese Probleme haben dazu geführt, dass verschiedene Regierungsstellen nicht in der Lage waren, vertrauensbrechende Unternehmen effektiv koordiniert zu bestrafen, und ihnen sogar erlaubt haben, die Lücken des Systems zu nutzen und die vorgesehenen Schranken zu überwinden. Dies hat zu einer Abnahme der Glaubwürdigkeit des Blacklist-Systems geführt und die Kosten der staatlichen Aufsicht erhöht.

3.2.2.2 Problemursachenanalyse
Gegenwärtig besteht der Hauptzweck der Umsetzung der koordinierten Sanktionen bezüglich der öffentlichen Kredite in China darin, Vertrauensbrechern die Kosten für Gesetzesverstöße und Vertrauensbrüche zu erhöhen. Derzeit sind die Kosten für Illegalität und Unehrlichkeit in China im Allgemeinen relativ niedrig, was zu wiederholter Illegalität und eben Vertrauensbrüchen führt. Ausgehend von einer umfassenden Analyse sind die Hauptgründe für die geringen Kosten von Vertrauensbrüchen folgende: Erstens sind die bestehenden Vorschriften schlecht zu bedienen, die Bestrafung ist zu gering und die Vollstreckung bleibt dauerhaft schwierig. Dies unterscheidet sich erheblich von der Situation in den westlichen Industrieländern. Zum Beispiel werden Gerichtsurteile in den Vereinigten Staaten in hohem Maße respektiert, und es gibt im Grunde kein solches Problem wie die „schwierige Vollstreckung" von Gerichtsurteilen. Nach den vom Obersten Volksgericht

veröffentlichten Daten haben die chinesischen Volksgerichte auf allen Ebenen im Zeitraum von 2016 bis 2018 insgesamt 20,435 Millionen Vollstreckungsverfahren angenommen. Bis Ende 2018 sind insgesamt 19,361 Millionen Fälle vollsteckt worden, nachdem
die Maßnahmen wie koordinierte Bestrafung von Vertrauensbruch ergriffen wurden. Die
Vollstreckungsrate erreichte 94,7 Prozent. Aber es besteht immer noch eine Lücke zu den
Industrieländern. Zweitens, aufgrund der Intransparenz, unzureichenden Austauschs, Tei-
lens und Veröffentlichung der Kreditinformationen Chinas kann der marktbasierte Reputationsmechanismus nicht vollständig funktionieren. Drittens haben Marktteilnehmer ein
geringes Compliance-Bewusstsein. Die Kreditkultur ist unterentwickelt.

Die Praxis hat gezeigt, dass die Einrichtung eines koordinierten Bestrafungsmechanismus gegen Unehrlichkeit in einigen Bereichen die oben genannten Probleme wirksam
gelöst hat. Dieser Ansatz steht im Wesentlichen im Einklang mit Chinas derzeitiger
Situation, wie mangelnde Integrität und notwendige Kontrolle von Unehrlichkeit. Aufgrund der Verzögerung bei der Ausarbeitung von Gesetzen und Normen haben die koordinierten Bestrafungsmaßnahmen gegen Unehrlichkeit jedoch unweigerlich zu unzurei-
chendem Verständnis, Mangel an Beweisen und unangemessenen Maßnahmen während
des Umsetzungsprozesses geführt. Diese Phänomene sollten rechtzeitig unter Rückgriff
auf inländische und ausländische Erfahrungen korrigiert werden, und die koordinierten
Bestrafungsmaßnahmen, die im administrativen Kontext angeraten sind, sollten so bald
wie möglich auf den Weg der Rechtsstaatlichkeit gebracht werden.

3.2.3 Lehren aus internationalen Erfahrungen zur koordinierten Bestrafung von Unehrlichkeit

3.2.3.1 Wichtigste Maßnahmen der Bestrafung bezüglich Kredite im Ausland

Listenentwicklung

Gegenwärtig haben viele Länder und Regionen das Blacklist-System als wichtige Verwal-
tungs- und Disziplinarmaßnahme in Bereichen mit großen sozialen und öffentlichen Interessen eingeführt (vgl. 王伟 2018). Die Vereinigten Staaten haben ihr Blacklist-System in
verschiedenen Bereichen eingerichtet, einschließlich der schwarzen Liste gegen Einwan-
derung, der schwarzen Liste mit Namen von Lieferanten des öffentlichen Beschaffungswesens, der schwarzen Liste des Office of Foreign Assets Management (OFAC) und der
schwarzen Liste der Food and Drug Administration (FDA), schwarze Liste der Anwälte,
„berüchtigter Markt,,-Liste des US-Handelsbeauftragten (USTR) usw. Darüber hinaus gehören zu den typischen schwarzen Listen im Ausland: die schwarze Liste des British Taxation Office (HMRC), das deutsche Auskunftssystem, das Blacklist-System des südko-
reanischen Börse Kosdaq und so weiter. Zusätzlich zu den von den einzelnen Ländern
festgelegten Blacklist-Systemen haben einige internationale Organisationen nacheinander
ähnliche Systeme etabliert. Das seit vielen Jahren implementierte „Debarment Processes"
-System der Weltbank ähnelt auch dem Blacklist-System.

Das Blacklist-System stellt eine sehr seriöse Bestrafungsmaßnahme dar. Es ist erforderlich, die Gründe und Kriterien für die Aufnahme in die schwarze Liste klar festzulegen, was häufig gesetzlich präzisiert werden muss. Am Beispiel des Blacklist-Systems von US-amerikanischen Lieferanten für das öffentliche Beschaffungswesen gibt es nach dem US-amerikanischen Gesetz für das öffentliche Beschaffungswesen zwei Hauptgründe, die dazu führen können, dass Lieferanten von der Teilnahme am öffentlichen Beschaffungswesen ausgeschlossen werden. Die erste Kategorie sind Beschaffungsgründe, zu denen gehören: durch Straf- oder Zivilgerichte verurteilte Straftaten im Zusammenhang mit der Beschaffung, die darauf hinweisen, dass dem Lieferanten die wirtschaftliche Integrität fehlt, der Lieferant die Vertragsbedingungen ernsthaft verletzt, vorsätzlich gegen den Vertrag verstoßen oder den Vertrag nicht ausgeführt hat. Zum anderen handelt es sich um gesetzliche Gründe, zu denen hauptsächlich Verstöße gegen Gleichstellungsklauseln, Arbeitsgesetze und Umweltschutzgesetze und -vorschriften gehören. Entscheidungen bezüglich „Beschaffungsgründen" werden vom Leiter der Beschaffungssektoren oder einem vom Leiter bestimmten Debarring Officer getroffen. Die Entscheidungen bezüglich „rechtlicher Gründe" werden von den Leitern der Strafverfolgungsbehörden der jeweiligen Gesetze getroffen.

Offenlegung von Informationen
Wie die verschiedenen Länder die schwarzen Listen offenlegen, lässt sich in zwei Arten unterteilen: zum einen wird die Liste durch die Abteilung angekündigt, die die Liste erstellt hat. Zum Beispiel wird die Blacklist des US-amerikanischen öffentlichen Beschaffungswesens von der Federal Services Administration (GSA) veröffentlicht. Die Veröffentlichung einer schwarzen Liste erfolgt in der Regel über eine feste Zeitspanne. Zum Beispiel macht die Blacklist der HMRC vierteljährliche Ankündigungen öffentlich zugänglich. Eine andere Sorte ist, dass Anfragen zugelassen werden, die Gesamtliste jedoch nicht veröffentlicht wird. Beispielsweise kann die deutsche Schuldnerliste von jedermann konsultiert werden, sie kann jedoch nicht als Ganzes heruntergeladen werden. Der Zweck der Untersuchung und der verfügbare Inhalt sind begrenzt, das heißt, die Untersuchung muss für die in Artikel 915 Abs. 3 der deutschen Zivilprozessordnung (ZPO) genannten Zwecke verwendet werden, und die durch die Untersuchung offengelegten Informationen sind begrenzt. Man kann nur nachschlagen, wer wegen welcher Angelegenheit auf die Schuldnerliste aufgenommen wurde.

Bestrafungsmaßnahmen
Koordinierte Bestrafungsmaßnahmen, die ausländischen schwarzen Listen entsprechen, werden im Allgemeinen in drei Kategorien unterteilt. Das erste sind Maßnahmen zur Wahrung des öffentlichen Interesses von dem entsprechenden Sektor, der eine schwarze Liste führt. Das Blacklist-System für das Beschaffungswesen der US-Regierung diente zum Beispiel dem Schutz der vertraglichen Interessen der Regierung. Die Lieferanten auf der schwarzen Liste sind bei der Teilnahme am Beschaffungswesen der Regierung eingeschränkt. Die von der US-amerikanischen Food and Drug Administration herausgegebene

Blacklist verbietet Unternehmen, die schwerwiegende Verstöße gegen Arzneimittelvor-
schriften begangen haben, die Teilnahme an Aktivitäten im Zusammenhang mit vermark-
teten Arzneimitteln in der Pharmaindustrie. Das zweite betrifft kreditwirtschaftliche Maßnahmen. Beispielsweise haben Personen, die in die deutsche Schuldnerliste aufgenommen
werden, innerhalb von drei Jahren keinen Anspruch auf Kreditaufnahmen wie Bankdarlehen, Ratenzahlungen und Versandhandelswaren. Drittens hat sich aufgrund der Offenlegung von Blacklist-Informationen eine starke Abschreckung gegen nicht vertrauenswürdige Subjekte durch rechtlichen, wirtschaftlichen, moralischen, sozialen und öffentlichen
Druck ergeben. Außerdem bildet sich Marktdruck dadurch, dass andere Marktteilnehmer
aufgrund der schwarzen Liste autonom entsprechende Handelsentscheidungen treffen.

Bei diesen Bestrafungen beschränken sich die Disziplinarmaßnahmen grundsätzlich
auf den Verantwortungsbereich der entsprechenden öffentlichen Sektoren und stellen im
Wesentlichen keine koordinierten Disziplinarmaßnahmen zwischen öffentlichen Sektoren
dar. Ob Marktteilnehmer auch Disziplinarmaßnahmen ergreifen, liegt im Ermessen der
Marktteilnehmer selbst.

3.2.3.2 Vergleich chinesischer und ausländischer koordinierter Bestrafungsmaßnahmen und Erfahrungen daraus

Unterschiede in den Bereichen und Methoden der chinesischen und ausländischen koordinierten Bestrafung

Durch Vergleich kann man feststellen, dass es keinen wesentlichen Unterschied zwischen
dem inländischen Ansatz und Ansätzen im Ausland zur Entwicklung von schwarzen Listen und zur Offenlegung von Informationen gibt. Was jedoch die Bereiche und Methoden
der koordinierten Bestrafung anbelangt, so hat das Ausland mehr im Marktgebiet koordinierte Disziplinarmaßnahmen ergriffen, während in China mehr im Bereich der Einhaltung von Gesetzen und Vorschriften durch öffentliche Sektoren direkt die administrative
koordinierte Sanktion herbeigeführt wird. Mit anderen Worten, das Hauptmerkmal von
Chinas koordinierter Bestrafung ist die von mehreren öffentlichen Sektoren gemeinsam
durchgeführte kreditbezogene Sanktion, und der verwendete Bereich hat sich auch von
wirtschaftlicher Governance zu sozialer Governance ausgeweitet.

Rolle des Marktkreditmechanismus voll ausschöpfen

Die Hauptmethode zur kreditbezogenen Bestrafung in den westlichen Industrieländern
besteht darin, die Rolle der Marktmechanismen voll auszuschöpfen. Erstens wirkt die von
Regierungsstellen veröffentlichte schwarze Liste in der Regel nur auf die Regierung selbst.
Ob Marktteilnehmer (auf welche Art und Weise auch immer) Sanktionsmaßnahmen
ergreifen, liegt in der Hand der Marktteilnehmer. Beispielsweise beschränkt das Blacklist-
System für das öffentliche Beschaffungswesen in den USA nur das öffentliche Beschaffungswesen. Geschäfte bezüglich der auf der Liste aufgeführten Waren und Dienstleistungen mit anderen Marktteilnehmern werden nicht eingeschränkt, die Entscheidung
machen die Marktteilnehmer selbst. Zweitens werden Aufzeichnungen zur Marktkredit-
historie in Kreditreferenzdatenbanken und Kreditberichten aufgenommen und wirken in

den Markt. In den Industrieländern des Westens werden die meisten Kreditinformationen von marktorientierten Drittorganisationen bereitgestellt, und die Marktteilnehmer können nach eigenem Bedarf selber Informationen beschaffen, verwenden und basierend auf eigene Interesse Entscheidungen treffen. Die Märkte für Kreditauskünfte in diesen Ländern sind sehr entwickelt, und die Marktkreditmechanismen spielen eine entscheidende Rolle. Drittens arbeitet die Regierung auf einer festen Grundlage, um den Marktkreditmechanismus gut umzusetzen. Viele Länder verfügen über ein hohes Maß an sozialer Integrität, ein gutes Fundament für Rechtsstaatlichkeit und einen hohen Digitalisierungsgrad. Auf dieser Grundlage unternimmt die Regierung weiterhin Anstrengungen zur weiteren Digitalisierung, zur Steuerung der Integritätsatmosphäre und zur Schaffung von Rechtsstaatlichkeit, um die Rolle von Marktkreditmechanismen weiter zu fördern.

Festlegung und Vervollständigung einschlägiger Gesetze und Vorschriften
Eine weitere Erfahrung, die für die koordinierte Sanktion von Unehrlichkeit in westlich entwickelten Ländern herangezogen werden sollte, ist die Verbesserung der Rechts- und Verwaltungsvorschriften. Die schwarze Liste und die koordinierte Sanktion müssen eine klare rechtliche Grundlage für die Praxis haben. So gibt es in Deutschland zwar kein eigenes Gesetz zur Kreditverwaltung (Kreditverwaltung im chinesischen Sinne), die gesetzlichen Regelungen dazu sind jedoch in Gesetzen wie Handelsrecht, Zivilrecht, Kreditrecht und Datenschutzrecht verstreut. Das in § 915 bzw. neu § 882b der deutschen Zivilprozessordnung (ZPO) festgelegte System der „Schuldnerliste" hat klare Richtlinie für die Aufstellung, Veröffentlichung und Entfernung aus der Schuldnerliste. Das deutsche Schuldnerlistensystem verfügt über eine solide rechtliche Grundlage. Die entsprechende Gesetzgebung befindet sich auf einem höheren Niveau.

3.3   Schlussfolgerungen und Vorschläge

3.3.1   artielle koordinierte Bestrafung ist bereits eine Verwaltungsstrafe

Gemäß den einschlägigen chinesischen Gesetzen bezieht sich die Verwaltungsstrafe auf die spezifische verwaltungsrechtliche Handlung, die von Verwaltungsorganen oder anderen Verwaltungsbehörden gegen eine Gegenpartei verhängt werden, die gegen Verwaltungsvorschriften verstößt, jedoch keine Straftat im Sinne der gesetzlichen Funktionen und Verfahren darstellen. Verwaltungsstrafen stellen eine Form von Verwaltungssanktion dar. Der Zweck von Sanktionen besteht in der Aufrechterhaltung der sozialen Ordnung sowie im Schutz der nationalen Sicherheit und der Bürgerrechte.

Basierend auf der obigen Analyse ergibt sich das Konzept der koordinierten Bestrafung von Unehrlichkeit aus der Politik und Praxis des Aufbaus des Sozialkreditsystems, während die Verwaltungsstrafe ein rechtliches Konzept darstellt. Obwohl der Inhalt und die Erweiterung der beiden unterschiedlich sind, gibt es auch einen Schnittpunkt zwischen den beiden. Die Überschneidung der beiden liegt in der Tatsache begründet, dass manche koordinierte administrative Disziplinarmaßnahmen von Unehrlichkeit tatsächlich zu Entzug der Rechte und Erhöhung der Pflichten des Subjekts führen, die eigentlich zu Verwaltungsstrafen zählen. Während die koordinierten Markt-, Industrie- und sozialen Bestrafungsmaßnahmen wegen Unehrlichkeit zu Marktkreditbeschränkungen bei gleichberechtigten zivilen und kommerziellen Einheiten gehören, hängen sie nicht mit Verwaltungssanktionen zusammen. Nach dem Rechtsgrundsatz der Verwaltungsstrafe müssen die einschlägigen Rechtsvorschriften so bald wie möglich verbessert werden, sodass die auf Verwaltungssanktionen bezogenen, koordinierten Bestrafungsmaßnahmen durch Rechtsstaatlichkeit legitimiert werden.

3.3.2 Beschleunigung der Gesetzgebung über die koordinierte Bestrafung bezüglich öffentlicher Kredite

Das 国办 Nr. 35 forderte eindeutig: „Die Formulierung von Gesetzen in Bezug auf den Aufbau des Sozialkreditsystems ist zu fördern und das Erforschen und die Festlegung von Vorschriften über Verwaltung der öffentlichen Kreditinformationen und einheitlichen Sozialkreditcodes sind zu beschleunigen." Am 30. August 2019 veranstaltete Lian Weiliang, Vizedirektor der Nationalen Entwicklungs- und Reformkommission, ein Gesetzgebungssymposium für Sozialkredite. Er wies darauf hin, dass mehr als zwei Drittel der Provinzen, Regionen und Städte die Einführung lokaler Kreditvorschriften vorangebracht haben oder diese derzeit untersuchen. Es gibt schon 26 Gesetze und 28 Verwaltungsvorschriften, die Kreditklauseln enthalten, die eine solide Grundlage für die Rechtsstaatlichkeit des Kreditrechts bilden.

Es gibt drei konkrete Gesetzgebungspfade: Erstens, in Verbindung mit der Überarbeitung des Verwaltungsstrafrecht, sind die einschlägigen Bestimmungen über die koordinierte Bestrafung bezüglich der Verwaltungssanktion zu ergänzen; zweitens, das Gesetzgebungsverfahren für das „Kreditgesetz" ist zu beschleunigen und der maßgebliche Inhalt der koordinierten Bestrafung in den Artikeln ist festzulegen; drittens sollen Regelungen in verwandten Geschäftsfeldern geändert und Klauseln zur koordinierten Bestrafung von Unehrlichkeit hinzugefügt werden. Unabhängig von der Methode und dem Pfad sollten bezüglich der schwarzen Liste die folgenden Inhalte wie Identifizierungskriterien, -verfahren, -weg und der Prozess der koordinierten Bestrafung und das Entlastungsverfahren festgelegt werden. Außerdem sind in der Situation, in der die gesetzlichen Bestimmungen in der Rechtspraxis nicht gut bedienbar sind und die Bestrafung zu leicht ist, die einschlägigen gesetzlichen Bestimmungen zu überarbeiten und zu verbessern, zum Beispiel die Einrichtung eines Entschädigungssystems, um die Effizienz und das Niveau der Rechtsstaatlichkeit des Landes weiter zu verbessern.

Einige soziale Probleme, die durch die koordinierte Sanktion bezüglich öffentlicher Kredite entstehen können, wie die Arbeitslosigkeit von vertrauensbrechenden Personen, der Konkurs von vertrauensbrechenden Unternehmen usw., sollten durch den staatlich geführten Mechanismus der sozialen Sicherheit gelöst oder behoben werden. Hier können wir von den Gesetzgebungskonzepten der Vereinigten Staaten und anderen westlichen Ländern über „Rechtshoheit und soziale Sicherheit" lernen. Einerseits werden strenge rechtliche Sanktionen und koordinierte Strafen gegen unehrliche Subjekte verhängt, was auch den Druck auf den Vertrauensbrecher erhöht, die Verpflichtungen zu erfüllen. Andererseits werden eine entsprechende soziale Sicherheit hergestellt, die grundlegenden Menschenrechte gewährleistet und soziale Stabilität gewahrt (vgl. 余成群 2012, http://bjgy. chinacourt.gov.cn/article/detail/2012/06/id/886394.shtml).

3.3.3 Vorschlag der Einführung eines Verzeichnisses für koordinierte Bestrafung

Im Allgemeinen sind eine Reihe von Bestimmungen im Zusammenhang mit dem Unehrlichkeitsbestrafungsmechanismus in China in verschiedenen Gesetzen und Verordnungen oder Grundsatzdokumenten verstreut, und es gibt noch kein ausgebautes logisches System bezüglich Bestrafung von Unehrlichkeit. Aufgrund der Erfahrungen, die an verschiedenen Orten gesammelt wurden, wird empfohlen, bei der Umsetzung koordinierter Bestrafungsmaßnahmen gegen Unehrlichkeit die Arten von Unehrlichkeitstaten, die entsprechenden Disziplinarmaßnahmen, Fristen, Methoden und Ansätze für Credit Repair sowie die wichtigsten Rechtsgrundlagen klar aufzulisten, sodass die koordinierten Bestrafungsmaßnahmen explizit, offen, vorhersehbar sein werden.

3.3.4 Beschleunigung der Einrichtung eines koordinierten Bestrafungsmechanismus auf dem Markt

In der Praxis konzentriert sich Chinas koordinierter Bestrafungsmechanismus zu sehr auf die administrative Bestrafung, die gerichtliche Bestrafung und andere Bestrafungen, die von der öffentlichen Gewalt ausgehen. Die Funktionen der Marktbestrafung, der industriellen Bestrafung und der sozialen Bestrafung sind weit davon entfernt, allumfänglich angewendet zu werden. Als nächstes sollte China Wert auf den Aufbau von Marktkreditmechanismen legen, schrittweise zum marktorientierten Charakter von Kreditmechanismen zurückkehren und ein neues Muster der koordinierten Bestrafung etablieren, bei dem Marktkreditmechanismen und öffentliche Kreditmechanismen interoperabel und koordiniert sind.

Zusammenfassend hat sich herausgestellt, dass die koordinierten Bestrafungsmaßnahmen gegen Unehrlichkeit sich als wichtige institutionelle Innovation erwiesen haben, die für die aktuelle Situation in China geeignet ist. Das einschlägige Gesetzgebungsverfahren sollte beschleunigt werden, und die koordinierten Bestrafungsmaßnahmen, die auf verwaltungsrechtliche Sanktionen bezogen sind, sollten in der Rechtsstaatlichkeit enthalten sein, sodass das Aufbau des chinesischen Sozialkreditsystems nachhaltig gefördert wird und es sich gesund entwickelt.

Literatur

韩家平：《中国社会信用体系建设的特点与趋势分析》，《征信》2018年第5期.

王伟：《论失信""黑名单""制度的法治化》，《人民法治》2018年21期.

余成群：《美国强制执行法律制度》，2012-06-05，北京法院网http://bjgy.chinacourt.gov.cn/article/detail/2012/06/id/886394.shtml

国务院办公厅《关于社会信用体系建设的若干意见》（国办发（2007）17号），http://www.gov.cn/zhuanti/2015-06/13/content_2879028.htm

国务院《关于建立完善守信联合激励和失信联合惩戒制度加快推进社会诚信建设的指导意见》（国发（2016）33号），http://www.gov.cn/zhengce/content/2016-06/12/content_5081222.htm

中共中央办公厅 国务院办公厅印发 《关于加快推进失信被执行人信用监督、警示和惩戒机制建设的意见》 （中办发 （2016）64号），https://m.sohu.com/a/343305963_120207161

国家发展改革委人民银行《关于加强和规范守信联合激励和失信联合惩戒对象名单管理工作的指导意见》（发改财金规（2017）1798号），https://m.sohu.com/a/202096411_269004

国务院办公厅《关于加快推进社会信用体系建设构建以信用为基础的新型监管机制的指导意见》（国办发（2019）35号） ，http://www.gov.cn/zhengce/content/2019-07/16/content_5410120.htm

Jiaping Han 韩家平, Direktor des Kreditforschungsinstituts für internationalen Handel und wirtschaftliche Zusammenarbeit im Handelsministerium, Direktor des Komitees für kommerzielles Factoring der China Service Trade Association, Vizedirektor des Kreditarbeitsausschusses der China Marketing Association, Direktor des Technischen Komitees für soziale Kreditstandardisierung der Stadt Peking. Gastprofessor an der School of Finance der Renmin Universität der Volksrepublik China und an der Fakultät für Wirtschaft und Handel an der Universität für Außenwirtschaft und Handel. Er befasst sich hauptsächlich mit der Erforschung des Sozialkreditsystems, der Kreditwirtschaft, der New Economy, des Kreditmanagements und der Lieferkettenfinanzierung und ist einer der ersten, die sich mit Kreditforschung in China befassten.

Didi Xu 许荻迪 ist Assoziierter Forscher des Kreditforschungsinstituts am Institut für internationalen Handel und wirtschaftliche Zusammenarbeit, Handelsministerium.

Yuanyuan Guan 关媛媛 ist Assistant Forscher des Kreditforschungsinstituts am Institut für internationalen Handel und wirtschaftliche Zusammenarbeit, Handelsministerium.

Bewertung der Integrität und Zustand des chinesischen städtischen Kreditsystems mit CEI-Index

Junyue Lin 林钧跃

4.1 Einleitung

Die Theorie des chinesischen Sozialkreditsystems wurde 1999 begründet. Im selben Jahr begann die chinesische Regierung mit dem Aufbau des städtischen Kreditsystems. Im Herbst dieses Jahres wurde Shanghai in einem durch den damaligen Premierminister des Staatsrates, Zhu Rongji erteilten Untersuchungsbericht der Regierung als Pilot für den Bau des städtischen Kreditsystems der Zentralregierung bestimmt.

Ein staatlicher Untersuchungsbericht über die Probleme der Geschäftsethik in Bezug auf Diskreditierung und Unordnung wurde von einem Teamleiter erstellt von der Volksbank von China.

Wir können daher annehmen, dass der Aufbau des sozialen Kreditsystems vom städtischen Kreditsystem ausgeht. Natürlich basiert diese Ansicht auf der technischen Perspektive der „System" -Konstruktion.

Der Aufbau des chinesischen Sozialkreditsystems erfolgt „von Punkt zu Fläche", das heißt, es gibt zuerst Pilotädte. Ihre Erfahrungen sind weitergetragen worden, nachdem das Sozialkreditsystem bei ihnen gelungen ist. Diese Vorgehensweise entspricht der Routine der chinesischen Regierung und hat sich als vernünftig und praktikabel erwiesen.

Das städtische Kreditsystem stellt den Inbegriff des Sozialkreditsystems dar. Um das städtische Kreditsystem zu verstehen, müssen die Grundfunktionen und Funktionsweisen des sozialen Kreditsystems vom Konzept her verstanden werden.

In methodischer Hinsicht ist die wichtigste theoretische Grundlage für die Gestaltung eines Sozialkreditsystems die Theorie des Unternehmens- und Verbraucherkreditmanagements, einschließlich einiger Grundsätze der Kreditökonomie. Dabei sind die Verwendung von Unternehmenskrediten, Privatkrediten und Kreditbewertungstechniken die wichtigsten

Junyue Lin 林钧跃 (✉)
Beijing, China

© Springer Fachmedien Wiesbaden GmbH, ein Teil von Springer Nature 2020
O. Everling (Hrsg.), *Social Credit Rating*,
https://doi.org/10.1007/978-3-658-29653-7_4

technischen Instrumente. Das soziale Kreditsystem ist daher im Wesentlichen eine „offene" Verwaltungsmethode, die systembasiert funktioniert und deren Datenbasis von einer riesigen Informationsinfrastruktur bzw. einer Big-Data-Umgebung unterstützt werden.

Aus makroökonomischer Sicht hat die Theorie der „Kreditökonomie" auch die Gestaltung des sozialen Kreditsystems beeinflusst. Die sogenannte „Kreditökonomie" bezieht sich speziell auf zwei Theorien, die mit dem Nobelpreis für Wirtschaftswissenschaften ausgezeichnet wurden: Zum einen auf die von Oliver Williamson in den 1980er-Jahren vorgeschlagene Theorie des „opportunistischen Verhaltens von Menschen". Diese Theorie erklärt die Ursachen wirtschaftlicher Unehrlichkeiten. Die zweite ist die von Joseph E. Stiglitz und George A. Akerlof vorgeschlagene Theorie der „Informationsasymmetrie" (vgl. Junyue Lin 2015, S. 1–39).

In Chinas Markt- und Sozialpraxis sind Dutzende von Methoden zur Kreditrisikokontrolle für Unternehmen und Einzelpersonen in das Sozialkreditsystem integriert. Es gibt zwei Hauptgründe, warum der Kreditrisikokontrollmechanismus als „System" ausgelegt ist: Der erste basiert auf dem Konzeptentwurf „durch Struktur bestimmte Funktion", wobei die Praktikabilität der Funktion des Systems in vollem Umfang berücksichtigt wird. Der zweite besteht darin, die erfolgreiche Erfahrung von PDCA (Plan-Do-Check-Act-Zyklus, Planen-Umsetzen-Prüfen-Handeln) im Qualitätsmanagementsystemen zu nutzen (Deming Cycle of TQM, der die Plan-Do-Check/Study-Act-Prozesse umfasst). Aus diesem Grund weist das soziale Kreditsystem alle Systemmerkmale auf, die im ISO9000-Standard definiert sind, einschließlich der Missionsziele, der Kreditinformationsinfrastruktur (inklusive aller Arten von Produkten und Dienstleistungen zur Meldung von Kreditinformationen), des Funktionsrahmens, der Regeln (zum Beispiel Gesetze, Standards und Regierungsrichtlinien) Subsystem- oder Funktionsblöcke, Talent- und Community-Unterstützungssysteme.

In Bezug auf die Ingenieuraufgaben und -ziele zeigt die schematische Darstellung des Social-Credit-System-Modells (Abb. 4.1) die Aufgaben und Ziele der Bauarbeiten auf (vgl. Abb. 4.1).

In Bezug auf die Grundfunktionen, die auf dem ursprünglichen Entwurf von 1999 basieren, sollte das Sozialkreditsystem die folgenden vier Grundfunktionen haben, wie in Abb. 4.2 dargestellt.

In Bezug auf den Aufbau der Informationsinfrastruktur hat China zwei öffentliche Kreditregister eingerichtet, eines ist die „Basisdatenbank für Finanzkreditinformationen". Es wird durch das Kreditauskunftszentrum der Zentralbank von China betrieben, genau wie die öffentlichen Kreditauskunftssysteme anderer Zentralbanken in der Welt. Der andere ist die „National Credit Information Sharing Platform". Diese wird seit 2017 gebaut und betrieben durch ein nationales öffentliches Kreditinformationszentrum unter der Nationalen Entwicklungs- und Reformkommission (NDRC). Natürlich haben die Regierungsbehörden in Städten im ganzen Land auch Kreditinformationsinfrastruktur aufgebaut, um ihre eigenen Kreditsysteme zu unterstützen. Andererseits ist die Förderung der Entwicklung der Kreditdienstleistungsbranche eine der Aufgaben des Sozialkreditsystems, und diese Branche ist auf Informationen essenziell angewiesen. Sie verfügt über eine große Anzahl kommerzieller Datenquellen und bietet Kreditinformationsprodukte und -dienstleistungen an. Diese Branche besteht aus mehr

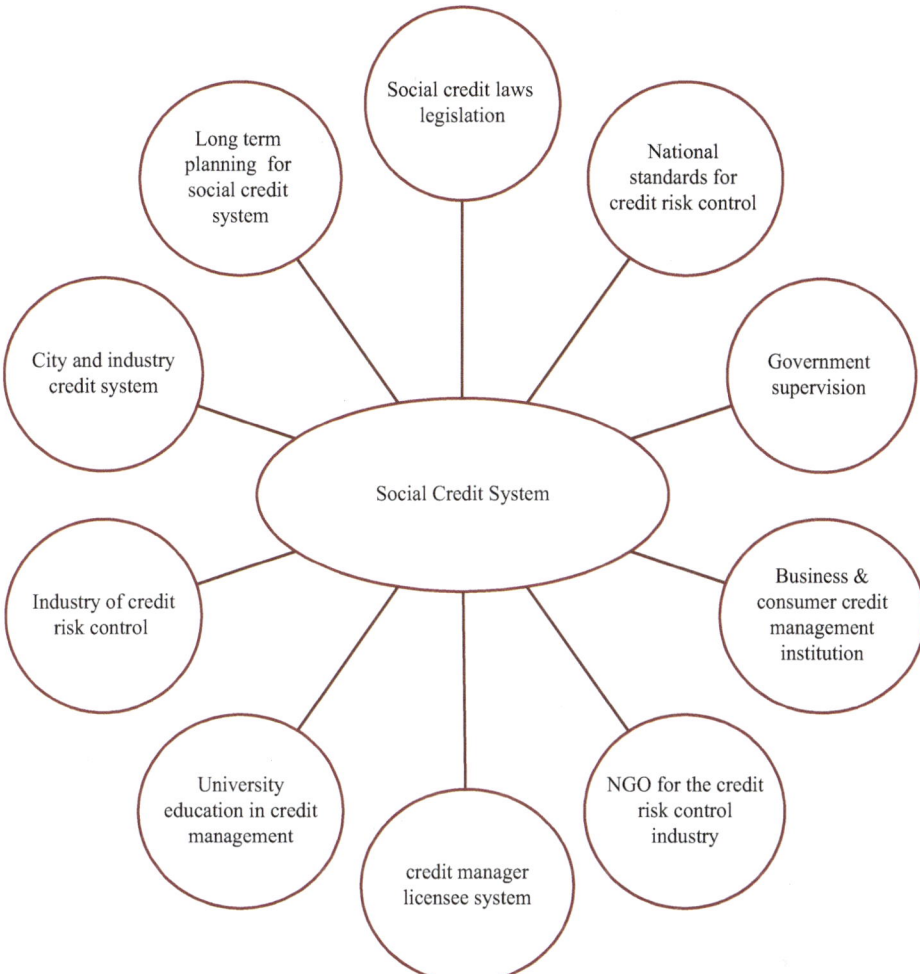

Abb. 4.1 Prägnantes Muster eines sozialen Kreditsystems. (Quelle: Junyue Lin 2012, S. 12)

als zehn Zweigen, die eine vollständige Palette von Kreditinformationsberichten, mathematischen Modellen und Softwareprodukten umfasst (vgl. Abb. 4.3).

Als Risikomanagementsystem kann das Sozialkreditsystem zur Steuerung von Markt- und Finanzkreditrisiken sowie als Instrument der sozialen Governance zur Steuerung verschiedener moralischer Risiken in der Gesellschaft eingesetzt werden. Von 1999 bis 2011 spielte das Sozialkreditsystem hauptsächlich im Bereich der Marktwirtschaft eine Rolle, wurde jedoch auch auf soziale Bereiche wie Krankenhäuser und Schulen ausgeweitet. Im Oktober 2011 wurde die Sechste Plenartagung des Siebzehnten Zentralkomitees der Kommunistischen Partei Chinas einberufen. Auf dieser Tagung wurde ein Beschluss mit dem Titel „Decision of the CPC Central Committee on several major issues concerning deepening the reform of the cultural system and promoting the great development and prosperity

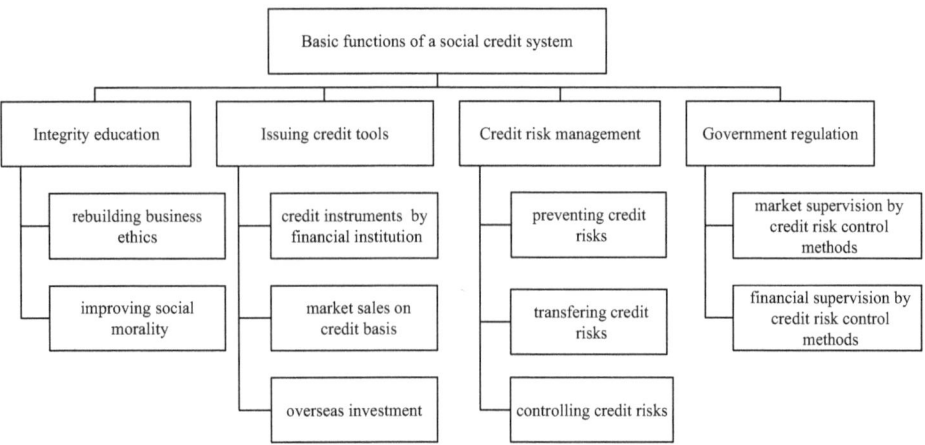

Abb. 4.2 Grundlegende Funktionen, die ein Sozialkreditsystem wahrnehmen sollte. (Quelle: Junyue Lin 2002, S. 44)

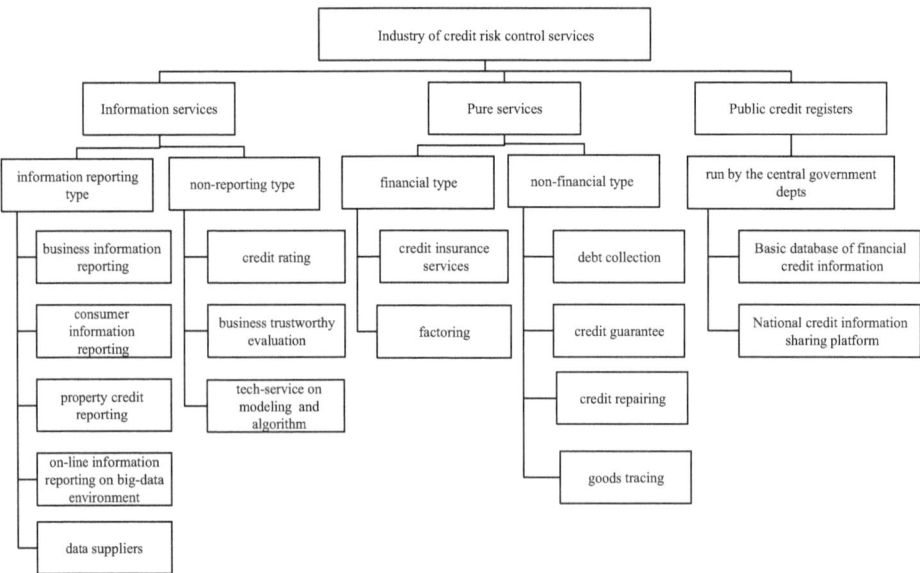

Abb. 4.3 Branche der Kreditrisikokontrolle und ihre Zweige in China. (Quelle: Junyue Lin 2007, S. 17)

of socialist culture" angenommen. Der Geltungsbereich des Systems der sozialen Kredite wird auf vier Bereiche ausgeweitet: Integrität der Governance, Integrität der Geschäfte, soziale Integrität und juristische Vertrauenswürdigkeit. Seitdem hat die Nationale Entwicklungs- und Reformkommission NDRC einen „Social credit system construction planning outline (2014–2020)" erstellt und Missionsziele in den oben genannten vier Bereichen festgelegt.

Die „Planning Outline" wurde im Sommer 2014 umgesetzt. Der Geltungsbereich des Sozialkreditsystems hat sich erheblich erweitert. Der Aufbau des von der NDRC geführten Sozialkreditsystems hat sich im sozialen Bereich rasant vollzogen und wurde auf den Bereich der Netzwerkökonomie ausgedehnt, was durch den elektronischen Handel repräsentiert wird. Aus technischer Sicht befinden sich daher seit 2014 im Wesentlichen drei verschiedene Systeme unter dem Namen „Sozialkreditsystem" in Bau und Betrieb. Die drei „Subsysteme" und ihre Eigenschaften unter dem „Big System" sind in Abb. 4.4 dargestellt.

Die chinesische Regierung hat die Aufgabe zum Wiederaufbau der sozialen Moral dem Sozialkreditsystem zugeschrieben. Da die technischen Methoden und die Informationsinfrastruktur des Systems für Zwecke der sozialen Governance eingesetzt werden können, hat die Regierung nebenbei einen „Freeride" genommen. Obwohl der Einsatz des Sozialkreditsystems zur Verwaltung der sozialen Integrität und der öffentlichen Ethik effektiv ist, führt er unweigerlich zu Problemen der Verwechslung von Kreditkonzepten und der Generalisierung des Kredits, was sich auf die internationale Reputation des „Credit-Sub-System" auswirkt. Daher steht die Regierung vor der schwierigen Entscheidung, den WIederaufbau der sozialen Moral weiterhin im Rahmen des Sozialkreditsystems umzusetzen oder als separate Aufgabe anzugehen.

Aus Abb. 4.4 ist ersichtlich, dass es unabhängig vom Subsystem äußerst wichtig ist, einen Disziplinarstrafmechanismus einzurichten. Daher wird in der Theorie des Sozialkreditsystems der Mechanismus der Sanktion gegen Unehrlichkeit als die wichtigste „Komponente" des Systems angesehen (vgl. Junyue Lin 2002, S. 1–6). Der Mechanismus der Unehrlichkeitsbestrafung besteht aus zwei Teilen: „Bestrafung gegen Unehrlichkeit"

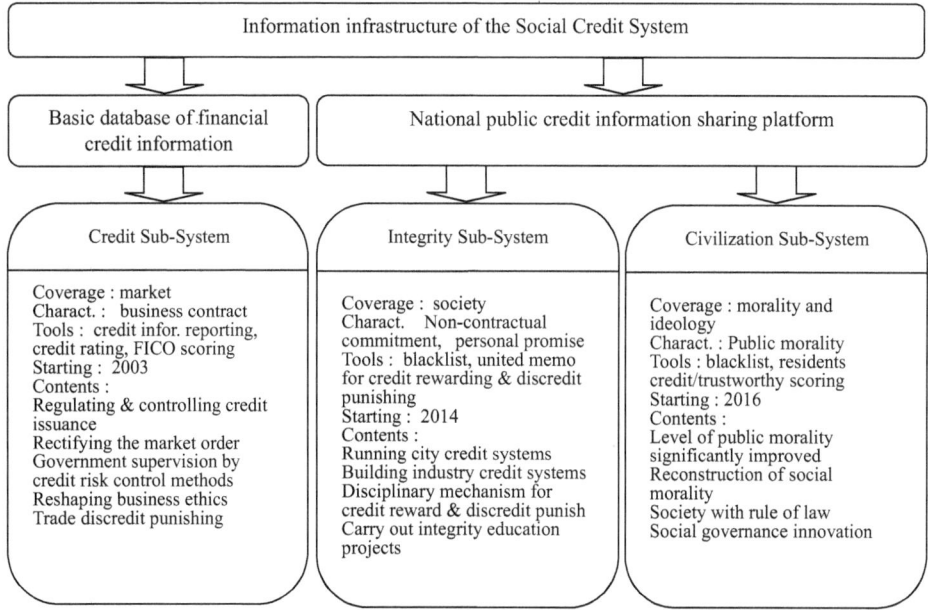

Abb. 4.4 Technische Analyse für das Sozialkreditsystem Chinas. (Quelle: Junyue Lin 2019)

und „Anreize für Treue". Der Konstruktionspfad besteht aus der Sammlung negativer Informationen, der Anwendung des Blacklist-Systems, der Umsetzung der kollektiven Unehrlichkeitsbestrafung und Anreize für Treue durch die Regierung und der Einrichtung von Markt-Joint-Defense-Mechanismen.

Außer materiellen Bestrafungen gegen Unehrlichkeit ist ein weiterer wichtiger „Bestandteil" des Sozialkreditsystems das Integritätserziehungsprojekt. Ein effektives Integritätserziehungsprojekt unterscheidet sich vom traditionellen Lehransatz, der pfadabhängig ist. Das Integritätserziehungsprojekt beruht darauf, durch den Unehrlichkeitsbestrafungsmechanismus Fälle mit Abschreckungseffekt zu entwickeln, sodass die Ergebnisse der Bestrafung gegen Unehrlichkeit und der Belohnung von Vertrauenswürdigkeit in materieller Ebene auf das spirituelle Niveau der Bürger gehoben wird. Schließlich sollte das Verhalten der Ehrlichkeit zur „Gewohnheit" aller Bürger gemacht werden. Aus makroökonomischer Sicht sollte durch die Umsetzung des Integritätserziehungsprojekts der soziale Effekt erzielt werden, die soziale Moral und Integrität als Ausgangspunkt wiederherzustellen.

Die Einrichtung eines Bestrafungsmechanismus gegen Unehrlichkeit erfordert eine Reihe technischer Instrumente: Blacklist-Systeme, Kreditratings und Kreditbewertungen gehören zu den am häufigsten verwendeten Instrumenten. Die Theorie des Blacklist-Systems wurde im Jahr 2002 aufgestellt. Das Blacklist-System setzt sich aus fünf Arten von Listen zusammen. Die Listen und ihre Funktionen sind in Abb. 4.5 dargestellt.

Das Kreditrating ist hier eine technische Methode zur Identifizierung und Klassifizierung des Kreditrisikos eines Subjektes. Der international gängige Einsatz der Ratingmethode dient hauptsächlich dem Kapitalmarkt. Chinas Ratingagenturen können auch Emit-

Abb. 4.5 Vollständiges Blacklist-System mit seinen fünf Funktionen. (Quelle: eigene Darstellung)

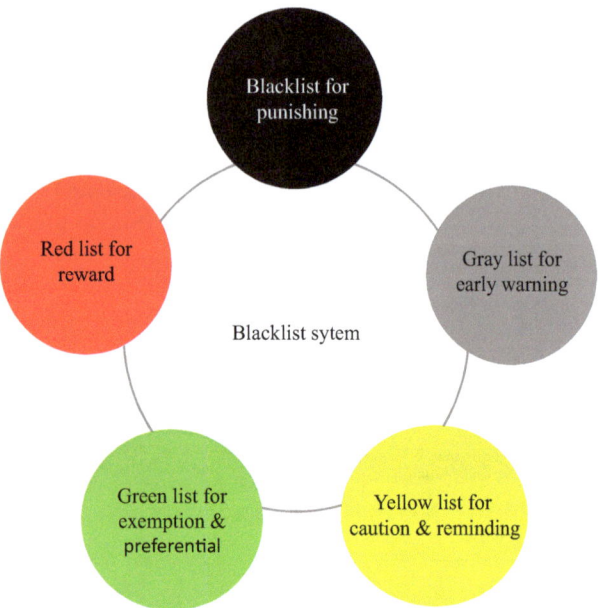

tentenratings, Anleiheratings und Länderratings bereitstellen. Auf dem chinesischen Markt hat sich seit 2005 eine „Branche zur Bewertung der Vertrauenswürdigkeit von Unternehmen" („Business Trustworthiness Evaluation Industry") herausgebildet. Die verwendeten trustworthiness assessment (vgl. Credit–General vocabulary 2018, S. 16) hat sich aus der Bonitätsbewertungsmethode entwickelt und dient der Beurteilung, ob nichtfinanzielle Unternehmen oder Organisationen ehrlich und vertrauenswürdig im Geschäftsprozess sind und auf die Werte der Environmental and Social Responsibility (ESR) geachtet haben. Ihre theoretischen Ansichten sind jedoch bisher nicht vereinheitlicht worden. Die Integritätsbewertungsmethode dient hauptsächlich dem Aufbau des Kreditsystems in verschiedenen Industrien. Die ursprüngliche Absicht der Konzeption dieser Methode besteht darin, Branchenverbände zu ermutigen, die Kreditbewertungsmethode zu verwenden, um die Unternehmen in der Branche zu bewerten.

Scoring für Kredite werden mit den auf logistischen Regressionsmethoden basierenden mathematisch-statistischen Modellen berechnet, zum Beispiel FICO-Punkte. Alle Geschäftsbanken in China nutzen dieses Kredit-Scoring-Tool, das vom Credit Reference Center der Zentralbank, der People's Bank of China (PBC), bereitgestellt wird. Laut dem „China Inclusive Financial Development Report 2019", der am 29. September 2019 gemeinsam von der China Banking and Insurance Regulatory Commission und der Zentralbank von China (PBC) veröffentlicht wurde, sind bereits Informationen von 999 Millionen natürlichen Personen im Land in das öffentliche Auskunfteisystem eingetragen worden. Wir können daraus schließen, dass das Kreditauskunftszentrum der Zentralbank von China (PBC) über Daten verfügt, mit denen die Kreditscores dieser 999 Millionen Einwohner berechnet werden kann.

In der Praxis des Sozialkreditsystems wurden jedoch aus der Kreditbewertungsmethode zwei andere Bewertungsmethoden verfremdet. Diese beiden Bewertungsmethoden können sowohl der privaten Finanzierung dienen, als auch verwendet werden, um die vom städtischen Kreditsystem geforderte „Anwohner-Kreditscores" zu erstellen. In Bezug auf die Methodik basiert einer dieser Art von Kredit-Scores auf der Index-Methode, und der Sesame Credit Score ist ein typischer Vertreter. Sesame Credit Score beruht auf einem Indexsystem, das auf fünf Dimensionen basiert: Benutzer-Bonitätshistorie, Verhaltenspräferenzen, Leistungsfähigkeiten, Identitätsmerkmale und Netzwerkbeziehungen.

Der Sesame Credit Score diente ursprünglich nur als Online-Zahlungssystem der Alibaba Group, wurde jedoch von einigen Städten als Grundlage für die Erstellung des städtischen Kreditscoressystem verwendet. Eine andere Methode zur Erstellung der Kreditscores für Anwohner besteht darin, eine „Positiv- und Negativliste" der Verhaltensweisen der Anwohner zu erstellen. Basierend auf einem einfachen Algorithmus werden die Kreditscores für Anwohner berechnet, nämlich extra Punkte für vertrauenswürdiges Verhalten und Punktabzug für Vertrauensmissbrauch. Die Stadt Rongcheng in der Provinz Shandong, eine der Vorzeigestädte, ist ein typisches Beispiel für Anwendung dieser Methode. Gegenwärtig gibt es in China nur ein Dutzend Städte, die Kreditscoresysteme eingesetzt haben. In Anbetracht der Tatsache, dass die Kreditscores der Anwohner verwendet werden können, um eine moralische Beurteilung über jeden Bürgern zu bilden und entsprechende

Belohnungen oder Strafen zu verhängen, hat dieses Verfahren international große Aufmerksamkeit auf sich gezogen.

Der Aufbau des Sozialkreditsystems hat sich sukzessive in zwei Richtungen entwickelt: „Stadt" und „Industrie". „Industrie" bezieht sich auf das Kreditsystem der Branchen und in den Schlüsselbereichen. Die Regierung hat Anfang 2018 mit dem Aufbau begonnen. „Stadt" bezieht sich auf das städtische Kreditsystem. Technisch ist das „städtische Kreditsystem" zum Synonym für eine Ingenieurmethode geworden. Die sogenannten „Städte" lassen sich in große Regionen (Beijing-Tianjin-Hebei, Jangtse-Delta und Pearl-River-Delta), Provinzen, Städte, Landkreise, Gemeinden sowie Wissenschafts- und Technologieparks unterteilen. Beim Aufbau von Kreditsystemen in den oben genannten Regionen mit unterschiedlichen Größen werden als technische Methoden Entwurf und Bauweisen von städtischen Kreditsystemen verwendet.

Um die in der „Planning Outline" festgelegten Aufgaben und Ziele umzusetzen, wurde im Jahr 2015 der Bau eines groß angelegten städtischen Kreditsystems vollständig in Angriff genommen, das zum zweiten Mal einen Höhepunkt beim Aufbaus städtischer Kreditsysteme darstellt. Bisher haben fast alle Städte mit dem Aufbau städtischer Kreditsysteme begonnen. „Zuerst pilotieren, Beispiele auswählen und dann popularisieren" ist die grundlegende Methode zur Förderung des Aufbaus städtischer Kreditsysteme. Daher haben die Nationale Entwicklungs- und Reformkommission (NDRC) und die Zentralbank von China (PBC) 28 Städte in zwei Chargen nacheinander als „Modellstädte für den Aufbau eines Sozialkreditsystems" im Januar 2018 ausgerufen: Hangzhou, Nanjing, Xiamen, Chengdu, Suzhou, Suqian, Huizhou, Wenzhou, Weihai, Weifang, Yiwu und Rongcheng, im August 2019 Qingdao, Wuhan, Anshan, Pudong New District, Shanghai, Jiading District, Shanghai, Wuxi, Hefei, Huaibei, Wuhu, Anqing, Fuzhou, Putian, Zhengzhou, Yichang, Xianning und Luzhou.

Kurz gesagt, das Sozialkreditsystem ist ein funktional integriertes Managementsystem, das zur Steuerung von Markt- und Sozialkreditrisiken ebenso verwendet wird wie auch für die soziale Verwaltung. Das Sozialkreditsystem trägt ein technisches Modell, was in jedem Land oder jeder Region der Welt durch individuelle Gestaltung und Anpassung verwendet werden könnte.

Städte stellen die Grundeinheit des Aufbaus von Sozialkreditsystemen dar. Die Förderung des Aufbaus städtischer Kreditsysteme ist eine der beiden Hauptrichtungen des Aufbaus von Sozialkreditsystemen. Da fast alle Städte städtische Kreditsysteme aufbauen, müssen alle Gesellschaftsbereiche den Fortschritt des Systemaufbaus und die Funktionsweise des Systems in jeder Stadt verstehen. Insbesondere müssen die Stadtverwaltungen und die für den Systembetrieb zuständigen Sektoren die Vollständigkeit und die Wirksamkeit des Systems kennen und aus den Erfahrungen anderer Städte lernen, um „die Schwächen auszugleichen". Aus diesem Grund hat 2010 das Credit Academic Committee der China Marketing Association, mit der Finanzierung von Beijing Wharf Think Tank, ein Forschungs- und Entwicklungsteam gebildet, um den „China City Commercial Credit Environment Index (CEI)" zu entwickeln. CEI füllt die technische Lücke bei der Bewertung der Wirksamkeit des städtischen Kreditsystems und erfüllt auch die Bedürfnisse der gesamten Gesellschaft, über den Fortschritt des Aufbaus des städtischen Kreditsystems informiert zu werden.

4.2   Definition

CEI ist ein kreditwirtschaftlicher Index, der zur Messung des kreditwirtschaftlichen Entwicklungsstands und des kommerziellen Kreditumfelds großer und mittlerer Städte in China herangezogen wird. Er kann die Vollständigkeit der Funktion und die operativen Auswirkungen des städtischen Kreditsystems in China widerspiegeln. Daher kann CEI als Indexinstrument angesehen werden, das technische Leitlinien für den Aufbau eines städtischen Kreditsystems liefern kann.

Der Name CEI stammt aus der Präfix Abkürzung von der englischen Übersetzung, „China Commercial Credit Environment Index".

Das Umfeld für städtische Wirtschaftskredite ist sehr kompliziert. Es bestehen immer viele wirtschaftliche und soziale Faktoren, einschließlich der Auswirkungen staatlicher Richtlinien und des Einflusses der regulatorischen Maßnahmen. Viele Einflussfaktoren sind umfassend und mehrfach überlagert. Um den Nebel komplexer Phänomene zu lichten und seinem Wesen als Kreditwürdigkeitsmaßstab zu entsprechen, muss ein mathematisches Werkzeug gewählt werden, das mit mehrdimensionalen Informationen umgehen kann. Die Indexmethode ist ein solches Werkzeug, das die oben genannten komplexen Phänomene beschreiben und analysieren kann. Ähnlich wie bei anderen Wirtschaftsindizes handelt es sich beim CEI um einen speziellen Satz aus Verhältniszahlen. Bei der Festlegung von Indikatoren misst CEI das gewerbliche Kreditumfeld der Stadt anhand von sieben Dimensionen: Einführung von Kreditmarktinstrumenten, Funktionen für das Kreditmanagement von Unternehmen, Aufbau von Auskunfteisystemen, staatliche Kreditaufsicht, Verbesserung der Integrität in Schlüsselbereichen, Integritätserziehung und Einschätzungen von Unternehmen über das Kreditumfeld des lokalen Marktes. Auf dieser Grundlage beurteilt CEI das Wirtschaftskreditumfeld der Städte und erstellt ein differenziertes Ranking.

Im Oktober 2011 wurde in dem Beschluss der 6. Plenartagung des 17. Zentralkomitees der Kommunistischen Partei Chinas der Geltungsbereich des Sozialkreditsystems auf vier Bereiche erweitert: Integrität in Regierungsangelegenheiten, Geschäftsintegrität, soziale Integrität und Justizintegrität. Die Bauarbeiten der „Planning Outline" wurden ebenfalls auf vier Bereiche ausgedehnt. Ausgehend von dieser Aufteilung kann CEI als Kreditindex angesehen werden, der sich auf den Bereich „Geschäftsintegrität" konzentriert.

Seit 2014 hat die interministerielle gemeinsame Sitzung des Staatsrates für den Aufbau des Sozialkreditsystems den Aufbau des städtischen Kreditsystems nachdrücklich vorangetrieben. Die Führungseinheiten davon, nämlich die Nationale Entwicklungs- und Reformkommission (NDRC) und die Zentralbank von China (PBC), haben Piloten und Vorführungen des Aufbaus städtischer Kreditsysteme in vier Stufen nominiert. Gleichzeitig wurden große Fortschritte beim Aufbau und bei der Anwendung der öffentlichen Kreditinformationsinfrastruktur in verschiedenen Städten erzielt. Der Grad der Plattformisierung und Standardisierung hat sich ebenfalls stark verbessert. Das Umfeld für Kreditinformationen auf dem Markt und in der Gesellschaft hat sich stark verändert. Es wurden kontinuierlich neue Quellen für Kreditinformationen geschaffen. Es wurde eine kommerzielle Big-Data-Umgebung generiert. Gleichzeitig sind Datenverarbeitungstechnologien und -algorithmen ausgereifter. Die Entwicklung des CEI profitiert von der Verbesserung der oben genannten Situation. Den Erfordernissen der Indexerstellung im Hinblick auf die Informationsbeschaffung kann immer besser entsprochen werden. Seit der neunjährigen Entwicklung des CEI haben sich mit der Zunahme der Datenquellen und der Verbesserung der Datenqualität auch die Objektivität und Wissenschaftlichkeit des CEI verbessert und die Kompilierungstechnologie und die -methoden wurden immer perfekter.

Es ist erwähnenswert, dass sich die CEI-Zusammenstellung zwar hauptsächlich auf Wirtschaftsindikatoren konzentriert, Daten aus den Bereichen Governance, Soziales und Justiz jedoch nicht vollständig ausgeschlossen wurden. Die Forschungsgruppe hat selektiv relevante Daten aus den drei Bereichen einbezogen und sich bemüht, die Struktur und Gewichtung von Indikatoren wissenschaftlicher und plausibler zu gestalten und die Genauigkeit von CEI-Messungen zu verbessern. Deshalb wurden im Vergleich mit dem anfänglichen Indikatorensystem nach 2017 mehrere neue Indikatoren zu den Sekundärindikatoren des CEI hinzugefügt, wodurch sich die Gesamtzahl der Sekundärindikatoren auf 20 erhöhte. Zum Beispiel wurden die Verschuldung der Städte (einschließlich Quasi-Kommunalanleihen) und Informationen zur Bestrafung von Unehrlichkeiten auf der Website „Credit China" (https://www.creditchina.gov.cn/) auch veröffentlicht.

4.3   Funktionen

Das städtische Kreditsystem kann die Entwicklung der lokalen Kreditwirtschaft fördern, die Vitalität der Stadtentwicklung stimulieren und die Wettbewerbsfähigkeit der Städte erhöhen. In Bezug auf die Umgestaltung der Geschäftsethik kann das städtische Kreditsystem verwendet werden, um die Ordnung in der lokalen Marktwirtschaft zu berichtigen, das Geschäftsumfeld der Stadt zu verbessern und ein ehrliches und vertrauenswürdiges soziales Umfeld zu schaffen, das die Grundlage für die Etablierung städtischer Marken bildet.

Genau gesagt ist CEI ein „Barometer" zur Messung der Qualität der Geschäftskredittransaktionen einer Stadt. Es kann Anlegern, Finanzinstituten oder staatlichen Aufsichtsbehörden helfen, die Qualität des Marktkreditumfelds einer Stadt sowie deren Schwankungen und Trends zu beurteilen. Mit anderen Worten, CEI kann Anlegern eine Referenz für Investitionssicherheit bieten, wie auch staatlichen Regulierungsbehörden eine Entscheidungsgrundlage liefern, um die Allokation von Regulierungsschwerpunkten zu optimieren.

In Zusammenhang mit Instituten, die die Aufgabe haben, das städtische Kreditsystem aufzubauen und zu betreiben, ermöglicht CEI der Stadtverwaltung, die kreditwirtschaftlichen Probleme in ihrem Gerichtsstand zu verstehen, indirekt auf die potenziellen Risiken von Unehrlichkeitsereignissen hinzuweisen und die Mängel der Systemfunktionen zu überprüfen. Kurz gesagt, als Wirtschaftsindex kann CEI aus Sicht der Analyse des Geschäftskreditumfelds verwendet werden, um Folgendes zu bewerten:

a. die Lebendigkeit der städtischen Kreditwirtschaft;
b. den Betrieb und die Fluktuation der städtischen Kreditwirtschaft;

c. das Gleichgewicht der Verbreitung von Finanzkreditinstrumenten vor Ort und deren Fairness und Gerechtigkeit;

d. die Größe des lokalen Kreditgeschäftsmarktes und sein Entwicklungspotenzial;

e. die Höhe des Kredittransaktionsrisikos auf dem lokalen Markt;

f. den Entwicklungsgrad der lokalen Kreditdienstleistungsbranche;

g. das Niveau von Aufbau des lokalen Unternehmenskreditsystems und den Mangel an Arbeitskräfte für das Kreditmanagement;

h. die Auswirkungen des Kreditumfelds im lokalen Markt auf das Überleben und die Entwicklung von Unternehmen;

i. den Betriebsstatus jedes Teilsystems des städtischen Kreditsystems;

j. die Unterschiede und die Lücken in der kreditwirtschaftlichen Entwicklung zwischen Städten;

k. die Frühwarnung vor Märkten mit chaotischer urbaner Wirtschaftsordnung.

Aus Sicht der staatlichen Kreditaufsicht kann der CEI-Index verwendet werden oder hilfreich sein, um Folgendes zu untersuchen:

a. die Förderung des Aufbaus städtischer Kreditsysteme durch die lokale Regierung;

b. die Funktion des lokalen Marktkreditaufsichtssystems;

c. die Umsetzung kollektiver staatlicher Sanktionsmaßnahmen;

d. den Schuldenstand der lokalen Regierung und die Möglichkeit einer technischen Insolvenz der Staatsfinanzen;

e. die Verbesserung des Geschäftsumfelds auf dem lokalen Markt;

f. die Entwicklung und die Umsetzung lokaler Kreditvorschriften, lokaler Kreditstandards und Branchen-/Gruppenkreditstandards;

g. den Stand der Einrichtung und der Umsetzung des Integritätsmechanismus der Regierung;

h. den Umfang der Aktivitäten zur Aufklärung über Integrität durch Partei- und Regierungsabteilungen.

Der CEI-Index entwickelt die Kreditökonomie und die Kreditmanagementtheorie mit Methoden- und Werkzeuginnovationen, die sich hauptsächlich in folgenden Aspekten widerspiegeln:

a) Untersuchung von Grenzfragen der Marktwirtschaft von städtischen Perspektiven;

b) eine besondere Perspektive zur Beobachtung der Funktionsweise der chinesischen Marktwirtschaft;

c) Datenunterstützung für neue wirtschaftspolitische Forschung;

d) Bereitstellung von Testmethoden für die Wissenschaftlichkeit der Theorie des Sozialkreditsystems;

e) Bereitstellung praktischer Experimente und Fallstudien zur Innovation im Bereich Social Governance.

4.4   Objektstädte

Die Bewertungsobjekte von CEI sind Chinas Städte auf Präfekturebene, Provinzhaupt-
städte, Kommunen mit eigenständigem Planungsstatus und Städte direkt unter der Zentralregierung (außer Hongkong, Macao und Taiwan).

Im Jahr 2010 wurden bei der Vorbereitung des ersten CEI aufgrund des Einflusses der
damaligen Richtlinien und Datenerfassungsbedingungen die Städte in den Autonomen Regionen Tibet und Xinjiang nicht in die Bewertung einbezogen. Die Gesamtzahl der in die
CEI-Bewertung 2010 einbezogenen Städte betrug 284.

2012 bezog das Forschungsteam Urumqi, die Hauptstadt der autonomen Region Xinjiang, und Karamay, eine bezirksfreie Stadt, in die CEI-Bewertung ein. Lhasa, die Hauptstadt der Autonomen Region Tibet, konnte jedoch nicht in die Bewertung einbezogen werden, da keine vollständigen und zuverlässigen Daten erhoben werden konnten. Daher
wurden 2012 insgesamt 286 Städte in die CEI-Bewertung einbezogen.

Im Jahr 2013, als die Stadt Chaohu mit der Provinzhauptstadt Hefei in Anhui ver-
schmolz, wurde die eigene Evaluierung der Stadt Chaohu eingestellt. Infolgedessen betrug
2013 eine Anzahl von 285 Städten in dem CEI-Bewertungssystem.

Im Jahr 2015 betrug die Gesamtzahl der bezirksfreien Städte in China nach Änderungen in den nationalen administiven Einteilungen 290. Aufgrund fehlender Daten und anderer Gründe wurden die Städte Bijie und Tongren in der Provinz Guizhou, Haidong in der
Provinz Qinghai und die neu gegründete Stadt Sansha in der Provinz Hainan nicht in die
Bewertung einbezogen. Daher ist die Anzahl der Städte, die in die CEI-Bewertung einbezogen wurden, im Jahr 2015 bei 285 geblieben.

2017 wurde Lhasa, die Hauptstadt der autonomen Region Tibet, in die CEI-Bewertung
einbezogen, somit alle großen Städte des Landes in die CEI-Bewertung enthalten sein
würden. Mit Ausnahme der Stadt Sansha, Provinz Hainan, wurden alle anderen bezirksfreien Städte auch in die Bewertung integriert. Die Städtenamen wurden ebenfalls entsprechend den Änderungen in den nationalen administrativen Einteilungen geändert.

Im Jahr 2019 wurde die Stadt Laiwu der Provinz Shandong aufgrund von Änderungen
in den nationalen administrativen Einteilungen in die Stadt Jinan eingemeindet. Daher
wurde Laiwu aus der CEI-Bewertung gelöscht und ihre Informationen mit denen in Jinan
zusammengefasst. Den jüngsten Verwaltungsabteilungen zufolge gibt es landesweit 293
bezirksfreie Städte und provinzunmittelbare Verwaltungszonen. Aufgrund der kurzen
Gründungszeit und anderer Gründe bestehen jedoch relativ große Datenlücken in neun
Städten, Sansha und Danzhou in Hainan, Turpan und Hami in Xinjiang, Shigatse, Changdu,
Nyingchi, Shannan und Naqu in Tibet. Diese Städte wurden vorerst nicht in die CEI-Bewertung einbezogen. Im Jahr 2019 erreichte die Anzahl der einbezogene Städt 288.

Um die Auswirkungen von Stadtgröße und Entwicklungspolitik zu eliminieren, hat das
Forschungsteam 288 Städte in zwei Kategorien unterteilt: Die erste Kategorie umfasst
regierungsunmittelbare Städte, provinzunmittelbare Verwaltungszonen und Provinzhaupt-
städte. In dieser Kategorie gibt es 36 Städte. Die zweite Kategorie sind bezirksfreie Städte
mit insgesamt 252 Städten. Alle Städte, die 2019 in die CEI-Bewertung einbezogen wurden, sind in Tab. 4.1 aufgelistet.

Tab. 4.1 Chinese cities included in 2019 CEI evaluation

Nr.	Provinz/ regierungsunmittelbare Stadt	Stadt	Anzahl
1	Beijing	Beijing	1
2	Tianjin	Tianjin	1
3	Shanghai	Shanghai	1
4	Chongqing	Chongqing	1
5	Provinz Hebei	Shijiazhuang, Zhangjiakou, Chengde, Qinhuangdao, Tangshan, Langfang, Baoding, Hengshui, Cangzhou, xingtai, Handan	11
6	Provinz Shanxi	Taiyuan, Shuozhou, Datong, Yangquan, Changzhi, Jincheng, Xinzhou, Jinzhong, Linfen, Lüliang, Yuncheng	11
7	Innere Mongolei	Hohhot, Baotou, Wuhai, Chifeng, Tongliao, Hulunbuir, Ordos, Ulanqab, Bayannaoer	9
8	Provinz Liaoning	Shenyang, Chaoyang, Fuxin, Tieling, Fushun, Liaoyang, Anshan, Dandong, Dalian, Yingkou, Panjin, Jinzhou, Huludao	14
9	Provinz Jilin	Changchun, Baicheng, Songyuan, Jilin, Nippen, Liaoyuan, Tonghua, Baishan	8
10	Provinz Heilongjiang	Harbin, Qitaihe, Qiqihar, Heihe, Daqing, Hegang, Yichun, Jiamusi, Shuangyashan, Jixi, Mudanjiang, Suihua	12
11	Provinz Jiangsu	Nanjing, Xuzhou, Lianyungang, Suqian, Huaian, Yancheng, Yangzhou, Taizhou, Nantong, Zhenjiang, Changzhou, Wuxi, Suzhou	13
12	Provinz Zhejiang	Hangzhou, Huzhou, Jiaxing, Zhoushan, Ningbo, Shaoxing, Shengzhou, Jinhua, Taizhou, Wenzhou, Lishui	11
13	Provinz Anhui	Hefei, Suzhou, Huaibei, Luzhou, Fuyang, Bengbu, Huainan, Luzhou, Maanshan, Wuhu, Tongling, Anqing, Huangshan, Lu'an, Chizhou, Xuancheng	16
14	Provinz Fujian	Fuzhou, Nanping, Putian, Sanming, Quanzhou, Xiamen, Zhangzhou, Longyan, Ningde	9
15	Provinz Jiangxi	Nanchang, Jiujiang, Jingdezhen, Yingtan, Xinyu, Pingxiang, Ganzhou, Shangrao, Fuzhou, Yichun, Ji'an	11
16	Provinz Shandong	Jinan, Liaocheng, Dezhou, Dongying, Zibo, Weifang, Yantai, Weihai, Qingdao, Rizhao, Linyi, Zaozhuang, Jining, Tai'an, Binzhou, Heze	16
17	Provinz Henan	Zhengzhou, Sanmenxia, Luoyang, Jiaozuo, Xinxiang, Hebi, Anyang, Liyang, Kaifeng, Shangqiu, Xuchang, Luohe, Pingdingshan, Nanyang, Xinyang, Zhoukou, Zhumadian	17
18	Provinz Hubei	Wuhan, Shiyan, Xiangfan, Jingmen, Xiaogan, Huanggang, Ezhou, Huangshi, Xianning, Jingzhou, Yichang, Suizhou	12

(Fortsetzung)

Tab. 4.1 (Fortsetzung)

Nr.	Provinz/ regierungsunmittelbare Stadt	Stadt	Anzahl
19	Provinz Hunan	Changsha, Zhangjiajie, Changde, Yiyang, Yueyang, Zhuzhou, Xiangtan, Hengyang, Shengzhou, Yongzhou, Shaoyang, Huaihua, Loudi,	13
20	Provinz Guangdong	Guangzhou, Qingyuan, Shaoguan, Heyuan, Meizhou, Chaozhou, Shantou, Jieyang, Shanwei, Huizhou, Dongguan, Shenzhen, Zhuhai, Zhongshan, Jiangmen, Foshan, Zhaoqing, Yunfu, Yangjiang, Maoming, Zhanjiang	21
21	Provinz Guangxi	Nanning, Guilin, Liuzhou, Wuzhou, Guigang, Yulin, Qinzhou, Beihai, Fangchenggang, Chongzuo, Baise, Hechi, Laibin, Hezhou	14
22	Provinz Hainan	Haikou, Sanya	2
23	Provinz Sichuan	Chengdu, Guangyuan, Mianyang, Deyang, Nanchong, Guang'an, Suining, Neijiang, Leshan, Zigong, Luzhou, Yibin, Panzhihua, Bazhong, Dazhou, Ziyang, Meishan, Ya'an	18
24	Provinz Guizhou	Guiyang, Liupanshui, Zunyi, Anshun, Bijie, Tongren	4
25	Provinz Yunnan	Kunming, Qujing, Yuxi, Baoshan, Zhaotong, Lijiang, Simao, Lincang	8
26	Provinz Shanxi	Xi'an, Yan'an, Tongchuan, Weinan, Xianyang, Baoji, Hanzhong, Yulin, Ankang, Shangluo	10
27	Provinz Gansu	Lanzhou, Jiayuguan, Jinchang, Baiyin, Tianshui, Wuwei, Jiuquan, Zhangye, Qingyang, Pingliang, Dingxi, Longnan	12
28	Provinz Qinghai	Xining, Haidong	1
29	Autonomes Gebiet Ningxia	Yinchuan, Shizuishan, Wuzhong, Guyuan, Zhongwei	5
30	Autonomes Gebiet Xinjiang	Urumqi, Karamay	2
31	Autonomes Gebiet Tibet	Lhasa	
Total	**288 Städte**		

4.5   Das Indikatorsystem

Aufgrund der Grundfunktionen des Sozialkreditsystems und unter Berücksichtigung der Verfügbarkeit von Informationen umfasst das von CEI festgelegte Bewertungssystem sieben Teile, nämlich: Kreditverwendung, Corporate Credit Management-Funktionen, Aufbau des Kreditsystems, staatliche Kreditaufsicht, Unehrlichkeit und Verstöße, Integritätserziehung und Unternehmenserfahrung. Sie bilden Indikatoren der ersten Ebene im Indikatorsystem.

Da die Indikatoren der ersten Ebene relativ abstrakt sind, werden mehrere Indikatoren der zweiten Ebene darunter, das heißt der „Kriteriumsebene" des Indikators, eingerichtet. Mit anderen Worten, die Primärindikatoren und Sekundärindikatoren bilden den Rahmen des gesamten Indikatorsystems.

Schließlich werden unter der zweiten Ebene noch mehrere Indikatoren der dritten Ebene gesetzt, um die entsprechenden Indikatoren der zweiten Ebene zu beschreiben und auf konkrete Datenquellen zu verweisen. Das umfassende Bewertungsindexsystem von CEI ist in Tab. 4.2 dargestellt.

Tab. 4.2 Three-level indicators of CEI

Primärindikatoren	Sekundärindikatoren	Tertiärindikatoren
Kreditplatzierung	Platzierung von finanziellen Kreditinstrumenten	Verbraucherkreditsaldo der Geschäftsbanken im Verhältnis zum BIP Anteil der KMU-Kreditlinien an den Bankkreditlinien
	Kommerzielle Kreditverkäufe	Anteil der Kreditverkaufsverträge an den Verkaufsverträgen Betrag der Kreditverträge im Verhältnis zum BIP Erfüllungsbetrag von Kreditverkaufsverträge im Verhältnis zu Kreditvertragssumme
	Schuldenlast des Staates	Verhältnis von Staatsverschuldung zu Steuereinnahmen
Unternehmenskreditmanagement	Einrichtung der Unternehmenskreditmanagement-Abteilung	Anteil der Unternehmen mit Kreditmanagementabteilungen Anzahl der Kunden großer Rating und Auskunfttei-agenturen
	Human resources für Unternehmenskreditmanagement	Anzahl der Beteiligte an der nationale Berufsqualifikationsprüfung für KreditmanagerKreditmanager im Verhältnis zu den Praktizierenden Anzahl der Absolventen aus dem Fach Kreditmanagement in den lokalen Universitäten
	Finanzielles Unternehmensrisiko	**Themis** Risikowarnindex für börsennotierte Unternehmen
Auskunfteisystem	Entwicklung der Auskunftei-Branche	Vollständigkeit des Typs von Auskunfteien Anzahl der Auskunfteien
	Produkte und Dienstleistungen von Auskunfteien	Vollständigkeit der Arten von Kreditwirtschaft erbrachten Dienstleistungen
	Nutzung des öffentlichen Auskunfteisystems	Das Verhältnis der jährlichen Anzahl der Anfragen bei den Auskunfteisystem zur Anzahl der Unternehmen Das Verhältnis der jährlichen Anzahl der Anfragen bei den Auskunfteisystem zur Anzahl der Bevölkerung

(Fortsetzung)

Tab. 4.2 (Fortsetzung)

Primärindikatoren	Sekundärindikatoren	Tertiärindikatoren
Staatliche Kreditaufsicht	Aufsichtsbehörden	Anzahl der an kollektiven Sanktionen beteiligten Regierungssektoren
	Aufsichtseinrichtungen	Upgrade der staatlichen Kreditaufsichtseinrichtungen und funktionale Optimierung Verwendungshäufigkeit der verschiedenen Listen im Blacklist-System und Aktualisierung des Publizitätssystems
	Formulierung und Durchsetzung von Kredittransaktionsregeln	Anzahl lokaler Kreditvorschriften Anzahl lokal festgelegter Kreditstandards Pilotfälle der Umsetzung nationaler Kreditstandards
Integrität in Schlüsselbereichen	Integrität der Governance	Existenz eines Mechanismus zum Austausch von Kreditinformationen und die Anzahl der unterzeichneten Verträge Anzahl schwerwiegender Fälle von Vertrauensbrüche der Regierung
	Geschäftsintegrität	Anzahl der Fälle von Wirtschaftsbetrug Produktqualität und Anzahl der Fälle von gefälschten und minderwertigen Waren Anzahl der Unternehmen, gegen die eine kollektive Sanktion eingeleitet wurde Anzahl der natürlichen Personen, die einer kollektiven Sanktion unterliegen Anzahl der problematischen Online-Kreditplattformen
	Soziale Integrität	Anzahl der Fälle von schwerwiegenden Vertrauensbrüchen in den Bereichen Bildung, Wissenschaft, Kultur und Gesundheit Anzahl der pseudosozialen Gruppen und bestraften sozialen Gruppen
	Justizielles Vertrauen	Anzahl der Fälle von schwerwiegenden Vertrauensverletzungen durch Gesetzesvertreter
Integritätserziehung	Aufbau und Verwaltungsorganisationen des Kreditsystems	Ob die Partei- und Regierungsorgane führende Institutionen für Integritätserziehungsprojekte aufbauen
	Entwicklung der Integritätserziehung	Anzahl der jährlichen Propaganda-Aktivitäten zur Integritätserziehung, die von Einheiten auf Kreisebene und aufwärts durchgeführt werden Erstellung von Websites und Videos zur Integritätserziehung von Partei- und Regierungsbehörden

4.6   Angewandte mathematische Methode

CEI muss einerseits alle Aspekte des Wirtschaftskreditumfelds einer Stadt vollständig abbilden, andererseits die Sammlung von maßgeblichen Informationen sichern. Dafür müssen Dutzende von Einflussfaktoren berücksichtigt werden. Daher muss die Berechnung
des CEI mit einer sogenannten „umfassenden Bewertungsmethode" durchgeführt werden.

Die sogenannte umfassende Bewertungsmethode bezieht sich auf eine Methode, die
mehrere Indikatoren verwendet, um mehrere Einheiten zu bewerten, und wird als „Multi-
Index Comprehensive Evaluation Method" bezeichnet. Die Grundidee dieser Bewertungsmethode besteht darin, mehrere Indikatoren in einen Indikator umzuwandeln, der die Ge-
samtsituation für die Bewertung widerspiegeln kann.

In dem umfassenden Bewertungsprozess sind zwei Schritte erforderlich: Zum einen
müssen die Indikatorwerte verschiedener Maßeinheiten mit den gleichen Basisgrößen ent-
dimensionalisiert werden, zum anderen müssen die verarbeiteten Indikatorwerte zusammengefasst werden, um den Gesamtbewertungsindex oder die Gesamtbewertungsnote zu
berechnen. Für eine umfassende Auswertung müssen daher zunächst die Methode zur
Entdimensionalisierung („non-dimensionalized method") der einzelnen Indikatoren sowie
zur Zusammenfassung der einzelnen Indikatorwerte festgestellt werden. Auf dieser Grundlage muss bei Verwendung einer synthetischen Methode (wie der linearen Gewichtung)
das Gewicht jedes Indikators im Indikatorsystem bestimmt werden, um die Wissenschaftlichkeit der Bewertung sicherzustellen. Um den CEI-Wert jeder Stadt zu berechnen, muss
das Forschungsteam daher drei Hauptprobleme lösen: die Wahl der geeigneten „non-dimensionalized"-Methode, die Wahl der Synthesemethode mit Multi-Indikatoren und die
Bestimmung der Indikatorgewichte.

Durch Anwendung der Efficacy Coefficient Method (ECM) kann man die Gesamtbewertungsnote erhalten. Zuerst werden Ober- und Untergrenze jedes einzelnen Indikators
bestimmt. Mit Hilfe des Efficacy Koeffizienten wird jeder einzelne Indikator in einen
messbaren Bewertungswert, die sogenannten einzelnen Bewertungswerte, umgewandelt.
Der tatsächliche Wert des i-ten Index sei X_i. X^i_{min} und X^i_{max} sind die Maximal- und Mini-
malwerte des Indikator. Z_i ist der entdimensionalisierte Bewertungswert. Für positive Indikatoren (das heißt, je größer der Indikatorwert, desto besser der Indikator) lautet die
allgemeine Formel für die ECM:

$$Z_i = \frac{x_i - x^i_{min}}{x^i_{max} - x^i_{min}} \times 100$$

Die verbesserte Formel für die ECM lautet:

$$Z_i = \frac{x_i - x^i_{min}}{x^i_{max} - x^i_{min}} \times 40 + 60 \tag{1}$$

Für die verbesserte ECM liegen die einzelnen Bewertungswerte zwischen 60 und 100.

Für den negativen Indikator (je kleiner der Indikatorwert, desto besser der Indikator) lautet die verbesserte Formel für die ECM:

$$Z_i = \frac{x^i_{max} - x_i}{x^i_{max} - x^i_{min}} \times 40 + 60 \tag{2}$$

Der Bewertungswert mit der ECM kann die Größe des Bewertungsindex widerspiegeln und Abstand zwischen den Bewertungsobjekten zeigen. Darüber hinaus kann es im Vergleich zur Indexmethode den Bereich einzelner Auswertungswerte einschränken und die Unterschiede zwischen einzelnen Auswertungswerten kontrollieren.

Aufgrund den Sonderheiten und Erfordernissen der CEI-Forschung enthält das CEI-Bewertungsindexsystem sowohl quantitative als auch eine große Anzahl qualitativer Indikatoren. Um anschauliche Ergebnisse zu erhalten, die direkt verglichen werden können und selbsterklärend und bedienbar sind, plant das Forschungsteam die Verwendung der linearen Gewichtungsmethode zur Bewertung des Geschäftskreditumfelds verschiedener Städte. Die spezifische Methode ist wie folgt:

Das Gewicht jedes Indikators sei W_i. Der CEI-Index ergibt sich aus der Berechnung des entdimensionalisierten Bewertungswerts Z_i und seiner Gewichtung von 34 Indikatoren im Bewertungsindexsystem für das städtische Geschäftskreditumfeld nach Formel (3):

$$CEI = \frac{\sum_{i=1}^{34} Z_i W_i}{\sum_{i=1}^{34} W_i} \tag{3}$$

Bei der Ermittlung der Gewichte verwendete das Forscherteam den analytischen Hierarchieprozess (AHP) in der subjektiven Gewichtungsmethode. Um den Einfluss menschlicher Faktoren einzelner Experten auf die subjektive Gewichtungsmethode zu überwinden, wird in die Datenverarbeitung eine weiter fortgeschrittene Gruppen-AHP-Methode eingeführt.

4.7   Datenquellen

CEI ist eine „Post-hoc-Bewertung" des Zustands des geschäftlichen Kreditumfelds der Stadt, das heißt, die Daten, die für die Erstellung der in diesem Jahr veröffentlichten Liste verwendet wurden, sind Daten für das vorangegangene Geschäftsjahr und spiegeln den Zustand des Kreditumfelds der Stadt im vorangegangenen Jahr wider. Es ist eine Konvention für Bearbeitung von Wirtschaftsindizes. Das im Juli 2012 veröffentlichte Indexranking verwendet beispielsweise Daten, die vom 1. April 2011 bis zum 1. April 2012 generiert wurden. Mit zunehmender Anzahl von Jahren und Datenmengen können jedoch Prämissen bezüglich der Daten zur Verwendung von Regressionsanalysen erfüllt werden, um künftige Trends vorherzusagen.

Angesichts der hohen Relevanz von CEI ist die Zuverlässigkeit und Autorität der Datenquelle von großer Bedeutung. Das Forschungsteam hat strenge Standards für die Auswahl der Datenquellen und die Datennutzung festgelegt. Da die Forschungs- und Entwicklungsorganisation von CEI zur Kreditrisikokontrollbranche gehört, bietet die Informationssammlung einige einzigartige Vorteile. Im Laufe der Jahre wurden nicht nur die von den zuständigen zentralen Regierungsstellen und den lokalen Regierungen veröffentlichten Daten verwendet, sondern es gibt auch viele inländische etablierte Kreditauskunfteien, die zuverlässige und informative Informationen für die Erstellung von CEI zur Verfügung gestellt haben. Darüber hinaus bemüht sich die Forschungs- und Entwicklungseinrichtung des CEI, in jeder Stadt mit mehreren Forschungsinstitutionen Verträge zu unterzeichnen, um ein Netz von Datenquellen aus erster Hand einzurichten.

Die Suche und Erfassung von Informationen muss nach einheitlichen Standards und streng nach technischen Definitionen erfolgen. Das Forschungsteam hat die Definitionen und Standards der verschiedenen Basisindikatoren strikt geregelt, um die Konsistenz und Vergleichbarkeit der Daten mit dem Bewertungsindex der einzelnen Städte sicherzustellen. Zu den Informationsquellen, die seit 2012 für die Erstellung von CEI verwendet werden, gehören:

a. Mainstream-Kreditauskunfteien: Indikatordaten vom Niveau des Unternehmenskreditmanagements und der Anzahl der Fachleute, Verwendungsverhältnis der Berichte von Auskunfteien in verschiedenen Regionen und Aufbauzustand von Auskunftssystemen. Diese Informationen stammen hauptsächlich aus diesem Kanal.
b. Regierungsstatistik: Die statistischen Indikatoren für die sozioökonomische Entwicklung jeder Stadt werden aus diesem Kanal abgeleitet.
c. Staatliche Kreditaufsichtsbehörden: Indikatordaten von staatlichen Kreditaufsichtseinrichtungen und -maßnahmen, von staatlichen Sanktionen und Gerichtsentscheiden in Bezug auf Unehrlichkeit und Verstöße des Unternehmens sowie von Aktivitäten der Integritätserziehung stammen im Wesentlichen aus diesem Kanal. Die gesammelten Informationen stammen hauptsächlich aus der Suche auf kreditbezogenen Websites lokaler Parteigliederungen und Regierungsstellen.
d. Indirekte maßgebliche Daten: Zum Beispiel die Daten von Messung der Unternehmensfinanzrisiken; das Research-Team verwendete die Daten und Bewertungsergebnisse der Finanzrisikomessung von börsennotierten chinesischen Unternehmen aus dem Themis-System.
e. Daten aus der Fragebogen der Feldstudie durch die Forschungsinstituten.

Das finanzielle Risikofrüherkennungsmodell von ThemisThemis ist eine Reihe empirischer Analysemodelle zur Beurteilung des finanziellen Risikos von Unternehmen und zur Vorhersage des Konkurses von Unternehmen, die auf der theoretischen Analyse der anormalen Finanzindikatoren des Konkursunternehmens auf der Grundlage der betrieblichen Merkmale des Unternehmens vor dem Konkurs basieren. Das Modell zur Warnung vor Finanzrisiken wurde 1987 von Herrn Shoji Ito, einem maßgeblichen Experten für finanzielle Risikofrüherkennung in Japan, erstellt.

Natürlich gibt es während der Entwicklung von CEI immer noch einige Schwierigkeiten und Probleme bei der Erhebung von Kreditinformationen. Betrachtet man beispielsweise die Datenquellen, die den Indikatoren der dritten Ebene entsprechen, so gibt es fast ein Viertel der Arten von Informationen, die schwer zu erfassen sind, nicht erfasst werden können, oder die Datenquelle ist nicht maßgeblich oder nicht zuverlässig.

4.8   Bewertungsergebnisse für 2019

Das CEI-Forschungsteam veröffentlichte jedes Mal vier umfassende Ranaglisten mit den Bezeichnungen „CEI-Ranglisten für Großstädte 2019" (Tab. 4.3), „CEI-Ranglisten für bezirksfreie Städte 2019" (Tab. 4.4), „CEI-Ranking von provinziellen Durchschnittswerten 2019" und „CEI-Ranking von Durchschnittswerten aus sieben großen Regionen 2019".

Tab. 4.3 2019 the big city ranking list

City	2019 score	2019 ranking
Beijing	87.085	1
Shanghai	85.488	2
Guangzhou	80.181	3
Hangzhou	79.727	4
Chongqing	78.636	5
Shenzhen	78.446	6
Tianjin	78.294	7
Nanjing	77.971	8
Xiamen	77.524	9
Wuhan	77.352	10
Jinan	76.962	11
Chendu	76.838	12
Hefei	76.734	13
Fuzhou	76.602	14
Dalian	76.529	15
Haikou	76.528	16
Zhengzhou	76.502	17
Qingdao	76.499	18
Xining	77.295	19
Ningbo	76.083	20
Shenyang	75.820	21
Nanning	75.136	22
Xi'an	74.985	23
Lanzhou	74.639	24
Hohhot	74.311	25
Shijiazhuang	73.663	26

City	2019 score	2019 ranking
Changsha	73.553	27
Guiyang	73.143	28
Urumqi	72.901	29
Nanchang	72.539	30
Taiyuan	72.185	31
Harbin	72.166	32
Changchun	71.560	33
Lhasa	71.529	34
Kunming	70.897	35
Yinchuan	70.523	36

Tab. 4.4 2019 the prefecture-level city ranking list

City	CEI score	2019 ranking
Zhuhai	78.950	1
Suzhou	77.573	2
Yantai	77.301	3
Foshan	76.089	4
Jinhua	76.058	5
Weifang	75.964	6
Jiaxing	75.711	7
Dongguan	75.597	8
Weihai	75.495	9
Wuhai	75.450	10
Taizhou	74.964	11
Changzhou	74.898	12
Sanya	74.762	13
Suqian	74.721	14
Chaozhou	74.665	15
Shaoxing	74.660	16
Anshan	74.545	17
Linyi	74.481	18
Huizhou	74.459	19
Zibo	74.162	20
Jiangmen	74.147	21
Wenzhou	74.144	22
Wuxi	74.109	23
Quanzhou	74.059	24
Wuhu	74.045	25
Xuancheng	74.004	26
Putian	73.997	27
Nantong	73.942	28
Taizhou	73.805	29

City	CEI score	2019 ranking
Jinmen	73.352	30
Shantou	73.345	31
Luzhou	73.248	32
Yangzhou	73.190	33
Huainan	73.157	34
Anqing	73.076	35
Fuyang	72.978	36
Liaoyang	72.962	37
Huaibei	72.920	38
Yichang	72.872	39
Huangshi	72.725	40
Quzhou	72.716	41
Taian	72.701	42
Xiangtan	72.628	43
Suzhou	72.621	44
Lishui	72.531	45
Langfang	72.502	46
Zaozhuang	72.397	47
Maanshan	72.347	48
Zhenjiang	72.343	49
Tongling	72.270	50
Wuwei	63.560	251
Suihua	62.848	252

Im CEI-Indikatorensystem gibt es insgesamt sechs First-Level-Indikatoren und eine Tabelle mit Korrekturwerten. Es wurden Ranglisten für die Städte mit den fünf oben genannten Indikatoren veröffentlicht, nämlich „Ranking des Niveaus der Kreditverwendung 2019", „Ranking des durchschnittlichen Niveaus der Qualität des Unternehmenskreditmanagements 2019". „Ranking der Perfektion des Auskunfteimarktes 2019", „Ranking des Niveaus der Kreditaufsicht der städtischen Verwaltungen 2019" und „Ranking der Entwicklung der Integritätserziehung 2019".

Das „Blue Book of CEI China City Commercial Credit Environmental Index (2019)" wurde Ende des Jahres veröffentlicht, und die vollständigen Listen wurden gleichzeitig auf der offiziellen CEI-Website veröffentlicht.

Das Forschungsteam hat die sieben Regionen, Provinzen und bezirksfreien Städte nach ihren CEI-Werten sortiert und drückt sie in Histogrammen und Tabellen aus. Die Forschungsgruppe gab den durchschnittlichen CEI-Wert der First-Level-Indikatoren im Land an. Mit Radardiagramm werden die Differenzen zwischen den Werten der sechs First-Level-Indikatoren in den Provinzen, regierungsunmittelbaren Städten und autonomen Regionen und dem nationalen Durchschnitt widergespiegelt. Die CEI-Werte und -Rankings der sieben großen Regionen des Landes im Jahr 2019 sind in Abb. 4.6 dargestellt.

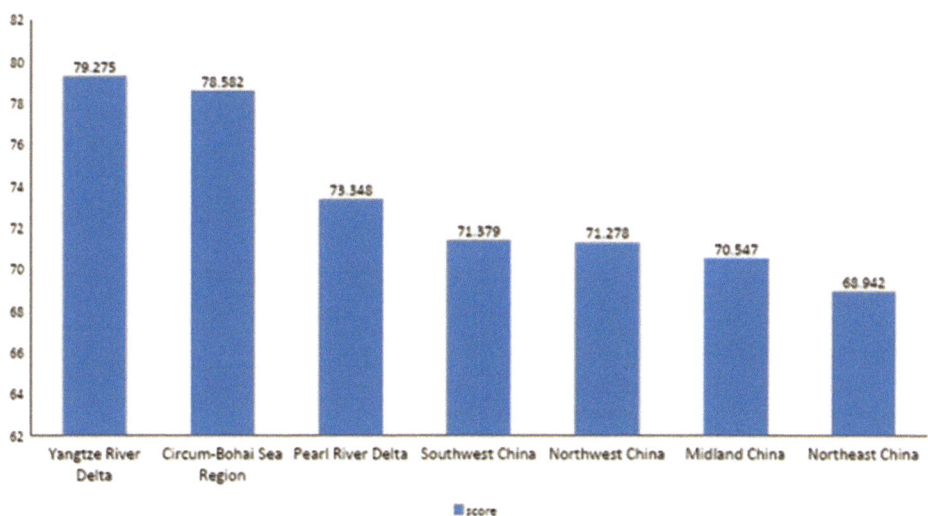

Abb. 4.6 CEI-Werte für die sieben großen Gebiete 2019. (Quelle: eigene Darstellung)

4.9   Leistung der Pilotstädte

Seit Januar 2018 haben die leitenden Behörden der interministeriellen gemeinsamen Konferenz zum Aufbau des Sozialkreditsystems des Staatsrates, das heißt die Nationale Entwicklungs- und Reformkommission (NDRC) und die Zentralbank von China (PBC), 28 Städte als „Modellstädte für den Aufbau eines landesweiten Sozialkreditsystems" anerkannt. Besondere Aufmerksamkeit widmete das Forscherteam dem Aufbau und dem Betrieb des städtischen Kreditsystems in Städten oberhalb der Bezirksebene unter diesen Vorzeigestädten.

Im CEI-Ranking der 36 Großstädte, ohne die vier regierungsunmittelbaren Städte, fallen die Pilotstädte, die auch Großstädte sind, unter die besten 50 Prozent auf dieser CEI Rankingliste, während die Modellstädte, die bezirksfreie Städte sind, alle im besten Drittel auf der entsprechenden CEI Rankingliste zu finden sind. Unter ihnen steht Hangzhou an erster Stelle unter den großen Städten und sein CEI an zweiter Stelle. Nachdem es sich den regierungsunmittelbaren Städten angeschlossen hatte, belegt es den vierten Platz. Der letzte Platz in den großen Städten ist Qingdao, das in CEI den 14. Platz belegt. Nachdem regierungsunmittelbare Städte mitgezählt werden, belegt es den 18. Platz, genau in der Mitte der Liste.

Gemessen am CEI-Ranking von 252 Städten auf Präfekturebene belegen die Pilotstädte Suzhou und Weifang die vordersten Plätze, die CEI-Rankings sind 2. und 6. Die letztplatzierte Pilotstadt ist Xianning mit einem CEI-Ranking von 72 (Tab. 4.5).

Warum sind einige Modellstädte niedriger eingestuft? Aus Sicht des CEI liegt dies daran, dass die Mängel beim Aufbau des Kreditsystems in diesen Städten dieselben sind:

Tab. 4.5 2019 the prefecture-level city ranking list

City	2019 CEI ranking
9 Modell cities in CEI big city ranking	
Hangzhou	4
Nanjing	8
Xiamen	9
Wuhan	10
Chengdu	12
Hefei	13
Fuzhou	14
Zhengzhou	17
Qingdao	18
13 Modell cities in CEI prefecture-level city ranking	
Suzhou	2
Weifang	6
Weihai	9
Suqian	14
Anshan	17
Huizhou	19
Wenzhou	22
Wuxi	23
Wuhu	25
Putian	27
Anqing	35
Yichang	39
Xianning	72

Erstens wird die „grundlegende Politik" zum Aufbau des Unternehmenskreditsystems nicht ausreichend durchgeführt, und zweitens ist die mangelnde lokale Entwicklung der Kreditdienstleistungsbranche zu beklagen, beispielsweise die mangelnde Bereitstellung von Produkten und Dienstleistungen zur Kreditrisikokontrolle aller Art. Da die Kreditdienstleistungen die notwendige externe technische Unterstützung für den Aufbau des Kreditsystems eines Unternehmens darstellt, hängen diese beiden Indikatoren eng zusammen.

Insgesamt ist der durchschnittliche Status des Aufbaus und Betriebs des städtischen Kreditsystems der Demonstrationsstädte nach dem CEI-Stadtranking im Jahr 2019 gut und kann als Vorbild und Modell dienen.

4.10   Analyse kreditwirtschaftlicher Phänomene

Die neun CEI-Listen des China City Commercial Credit Environment Index werden offiziell als „Blue Book" veröffentlicht. Das Blue Book bietet nicht nur Listen, sondern auch einen Überblick über die kreditwirtschaftliche Entwicklung in jeder Region und Provinz sowie eine sequenzielle Analyse der CEI-Werte, damit die Leser den Fortschritt oder den

Rückschritt jeder Stadt verstehen und Kommentare abgeben können. Daher liefert das Blue Book den Bau- und Betriebseinheiten für die Kreditsysteme in den Städten eine wichtige technische Referenz, um aktiv von fortschrittlicheren Städten als Benchmark zu lernen, und hilft ihnen dabei, die Verbesserungen beim Bau des Kreditsystems in der Stadt zu bestimmen.

Darüber hinaus enthält jedes Blue Book Artikel zur Analyse der Kreditwirtschaft, um den Lesern das Verständnis des CEI und der Entwicklung der inländischen Kreditwirtschaft zu erleichtern. In dem 2013 veröffentlichten „Bluebook of China Commercial Credit Environment Index 2013" stellten die Experten im Forschungsteam beispielsweise zwei Artikel zur Kreditwirtschaftsanalyse zur Verfügung. Ein Artikel ist „Analysis of the relationship between size of financial credit supply in a city and its economic growth" von Dr. Qingyan Shi und der andere Artikel ist „CEI: a new indicator measuring city commercial credit risk initiated in China" (in Englisch) von Dr. Ruonan Lin und Yi Gu. 2015 lieferten die Experten des Forschungsteams im veröffentlichten „Bluebook of China Commercial Credit Environment Index 2015" zwei Artikel zur Kreditökonomie. Ein Artikel ist „The significance of Hu Huanyong line to the development of credit economy of China" von Professor Junyue Lin. Ein weiterer Artikel ist „On the relationship between size of sales on credit basis and total patent owners in a city", der von Beijing Chenxin Credit Information Co., Ltd. zur Verfügung gestellt wird.

So hat das Forschungsteam im Jahr 2012 es unter Beweis gestellt, dass das Geschäftskreditumfeld einer Region und das Niveau der wirtschaftlichen Entwicklung eine stark positive Korrelation aufweisen. Mit dem Pro-Kopf-BIP jeder Stadt als Abszisse und dem CEI als Ordinate wird ein Streudiagramm zwischen der wirtschaftlichen Entwicklung der chinesischen Städte und dem Geschäftskreditumfeld erstellt (vgl. Abb. 4.7).

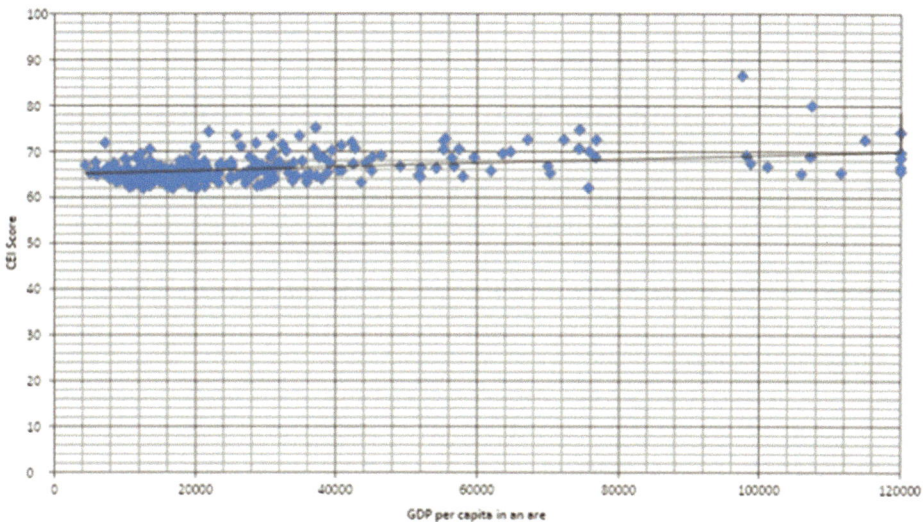

Abb. 4.7 Zusammenhang zwischen der wirtschaftlichen Entwicklung der Stadt und dem Wirtschaftskreditumfeld

Aus Abb. 4.7 geht hervor, dass im Allgemeinen eine positive Korrelation zwischen dem Grad der städtischen Wirtschaftsentwicklung in China und dem Geschäftskreditumfeld besteht (diese verstreuten Punkte können nicht mit dem Städtenamen gekennzeichnet werden). Um diesen Zusammenhang quantitativ zu beschreiben, berechnete das Forschungsteam den Korrelationskoeffizienten zwischen dem Pro-Kopf-BIP und dem Commercial Credit Environment Index (CEI) mit R = 0,4274. Dies deutet in gewissem Maße darauf hin, dass ein moderater positiver Zusammenhang zwischen dem wirtschaftlichen Entwicklungsstand und dem Geschäftskreditumfeld besteht.

Aus dem gesamten CEI-Ranking der Städte geht jedoch hervor, dass das Wirtschaftskreditumfeld einiger wirtschaftlich gut entwickelter Regionen nicht gut ist, während einige Regionen mit einem niedrigeren regionalen Pro-Kopf-BIP ein besseres Wirtschaftskreditumfeld aufweisen. Einige Städte mit einem hohen Pro-Kopf-BIP weisen aber ein niedriges CEI-Ranking auf. Zum Beispiel rangierten einige Städte 2010 in Bezug auf das Pro-Kopf-BIP unter den Top Ten Chinas, aber ihr Ranking in der CEI-Rangliste ist sehr niedrig oder sogar eines der Vorletzten. Einige Städte mit einem niedrigen Pro-Kopf-BIP weisen ein hohes CEI-Ranking auf. Bei Betrachtung der Einflussfaktoren liegt der Hauptgrund für dieses Phänomen darin, dass Kultur und Tradition in einigen relativ rückständigen Regionen funktionieren und diese Städte relativ „ethisch und ehrlich" sind.

Um das oben genannte Phänomen weiter zu analysieren, legte das Forscherteam Pro-Kopf-BIP (29210 Yuan/Person) und den durchschnittlichen CEI-Wert (66.037) aller Städte (mit Ausnahme derjenigen mit unvollständigen Daten) im Jahr 2010 als Trennlinie fest. Alle Städte können in die folgenden vier Kategorien eingeteilt werden: Die erste Art von Städten sind Städte mit einem überdurchschnittlich hohen Pro-Kopf-BIP und überdurchschnittlich hohen CEI-Wert, davon gibt es 60 Städte. Die zweite Art von Städten mit einem überdurchschnittlich hohen Pro-Kopf-BIP, aber niedrigeren CEI-Werten als Durchschnitt, davon gibt es 33 Städte. Die dritte Kategorie sind Städte mit einem niedrigeren Pro-Kopf-BIP als Durchschnitt und überdurchschnittlich hohen CEI-Wert, davon gibt es 54 Städte. Die vierte Kategorie von Städten sind Städte mit unterdurchschnittlichen Pro-Kopf-BIP und CEI-Werten, davon gibt es insgesamt 133 Städte. Durch Datenanalyse stellte das Forschungsteam fest, dass zwischen den verschiedenen Städten in derselben Region erhebliche Unterschiede in Bezug auf das wirtschaftliche Entwicklungsniveau und das geschäftliche Kreditumfeld bestehen. Gleichzeitig ist auch festzustellen, dass bei einer Ausweitung des Erhebungsbereichs auf ein größeres Gebiet als die Städte auch große Unterschiede im Geschäftskreditumfeld zwischen den verschiedenen Regionen festzustellen sind.

Das Forscherteam wählte zur Untersuchung vier typische Wirtschaftsregionen aus: Nordwestchina, Nordostchina, Yangtze River Delta und Pearl River Delta. Um die Unterschiede im geschäftlichen Kreditumfeld zwischen den Städten in jeder Region zu untersuchen, berechnete das Forschungsteam den Durchschnitt der CEI-Werte und den Variationskoeffizienten jeder Stadt in diesen vier Regionen, wie in Tab. 4.6 gezeigt.

Tab. 4.6 CEI mean and coefficient of variation of the four regions

Region	CEI mean	CEI coefficient of variation
Northeast	65.712	0.03114
Northwest	67.210	0.04543
Yangtze River Delta	68.730	0.1712
Pearl river delta	67.303	0.05431

Das Verhältnis der Standardabweichung zum Mittelwert wird als Variationskoeffizient bezeichnet und als C · V aufgezeichnet. Der Variationskoeffizient ist eine Statistik, die den Variationsgrad aller beobachteten Daten misst. Es kann die Auswirkung verschiedener Einheiten und/oder Durchschnitt auf den Vergleich des Variationsgrades von zwei oder mehr Daten eliminieren.

Aus Tab. 4.6 geht hervor, dass der durchschnittliche CEI-Wert der Region Jangtse-Delta mit 68.730 am höchsten ist, der Variationskoeffizient jedoch auch am höchsten ist. Dies zeigt, dass das Geschäftskreditumfeld der Region am besten ist, es jedoch große Unterschiede zwischen den verschiedenen Städten gibt. Die Region Nordosten weist den niedrigsten Durchschnitts-CEI (65.712) auf, der niedriger ist als der Durchschnittswert aller Städte (66.037), und ihr Variationskoeffizient ist gering, was darauf hinweist, dass die CEI-Werte der Städte in der Region relativ konzentriert sind und die meisten von ihnen unter dem Durchschnittswert aller Städte liegen. Die CEI-Werte der Pearl River Delta-Region und der nordwestlichen Region liegen über dem Durchschnittsniveau aller Städte. Da ihre Variationskoeffizienten gering sind, sind die CEI-Werte der Städte in diesen beiden Regionen ausgeglichener.

Im Jahr 2012 untersuchte das Forschungsteam die institutionelle Tragkraft (Institutional Thickness) des regionalen Wirtschaftskreditumfelds. Es wurde festgestellt, dass es zwischen verschiedenen Städten innerhalb einer Region erhebliche Unterschiede im Geschäftskreditumfeld gibt. Wenn der Umfang der Untersuchung auf ein größeres Gebiet wie eine große administrative Region ausgedehnt wird, kann festgestellt werden, dass es auch große Unterschiede im Geschäftskreditumfeld zwischen verschiedenen Regionen gibt, und dieser Unterschied weist eine gewisse Stabilität auf. Mit anderen Worten, wie viele wirtschaftliche und soziale Phänomene, hat auch das Geschäftskreditumfeld einen gewissen Grad an räumlicher Agglomeration.

Viele sozioökonomische Phänomene weisen räumliche Korrelationen auf. Aufgrund der Auswirkungen von räumlichen Wechselwirkungen und räumlicher Diffusion sind diese Phänomene möglicherweise nicht mehr unabhängig voneinander, sondern hängen miteinander zusammen und bilden die sogenannte „räumliche Agglomeration". Begrifflich bezieht sich „räumliche Autokorrelation" auf die mögliche gegenseitige Abhängigkeit einiger Variablen in den Beobachtungsdaten desselben Gebiets. Sie wird verwendet, um eine grundlegende Eigenschaft von geografischen Daten zu messen, nämlich der Grad der gegenseitigen Abhängigkeit zwischen Daten an einem Ort und Daten an anderen Orten, diese Art von Abhängigkeit wird in der Regel als „Raumabhängigkeit" bezeichnet. Das Forscherteam verwendete den Moran-Index, um die räumliche Autokorrelation der sieben Regionen Chinas zu charakterisieren. Mithilfe der mathematischen Statistik kann analysiert werden, ob eine Korrelation zwischen Änderungen in zwei Phänomenen (Statistik) vorliegt. Um die Reichweite des Versammlungsortes abzuschätzen, wurde die LISA-Methode von Anselin aus der University of Illinois in den USA verwendet.

L. Anselin von der Universität von Illinois schlug 1995 die Theorie der lokalen Indikatoren der räumlichen Assoziation (LISA) vor.

4.11 Fazit

Aus sozialer Verantwortung hat sich das Forscherteam beharrlich bemüht, CEI zu erstellen. Es soll aus Sicht privater Think Tanks ein Beitrag zum Aufbau des städtischen Kreditsystems geleistet werden. Das Ziel des Teams liegt darin, ein einfaches und quantitatives Analysewerkzeug für Provinz- und Stadtverwaltungen, den Aufbau von Sozialkreditsystemen, verschiedene Arten von Investoren und sozioökonomische Forscher bereitzustellen, um dynamisches Tracking der Auswirkungen des Betriebs städtischer Kreditsysteme und der Entwicklung der Kreditwirtschaft zu ermöglichen. Es bietet einige technische Unterstützungen für die Verbesserung und Optimierung des städtischen Kreditsystems. CEI kann manche Phänomene und Probleme erklären, die bei der Entwicklung lokaler Kreditwirtschaften auftreten, und in gewissem Maße das moralische Niveau von Ehrlichkeit und Integrität der lokalen Gesellschaft sowie die Investitionskosten widerspiegeln.

Obwohl das städtische Kreditsystem nur eine Einheit des sozialen Kreditsystems ist, handelt es sich auch um ein riesiges soziales Systemprojekt mit „all things in one", das den Aufbau und Betrieb einer umfangreichen Informationsinfrastruktur sowie eine Vielzahl technischer Tools und vollständiger technischer Richtlinien umfasst. Es werden Methoden und Instrumente benötigt, um den Fortschritt und die Qualität des Aufbaus der städtischen Kreditsysteme zu testen. CEI ist ein nützliches Instrument zur Bewertung der Entwicklung und des Funktionszustands des Sozialkreditsystems.

Die Verzögerung der globalen kreditwirtschaftlichen Forschung stellt ein weiteres Problem dar. Die Entwicklung des CEI wird eine technische Methode für die theoretische Erforschung der Kreditwirtschaft und des Sozialkreditsystems anbieten. Darüber hinaus können die Ergebnisse der umfassenden Bewertung von Städten durch CEI einige wirtschaftliche und soziale Phänomene genauer erklären. Daher darf man zuversichtlich sein, dass mit kontinuierlichen und eingehenden Untersuchungen CEI auch Analysen und interessante Fälle aus speziellen Perspektiven für die theoretische Forschung der Kreditökonomie liefern kann.

Literatur

Lin, Junyue, "The theory of credit economy (Chapter 1)". Credit Economics, Higher Education Press, Aug., 2015, S. 1–39
Lin, Junyue, "The social credit system model of China and its analyzing". An Exploration in the Mode of Social Credit System of China, China Fangzheng Press, July, 2012, S. 12

Lin, Junyue, "The structure of social credit system". Principle of Social Credit System, China Fang-
   zheng Press, Jan., 2002, S. 44
Lin, Junyue, "The generalized credit risk control industry". Fundamental of Credit Information Re-
   porting Technology, China Renmin University Press, Aug., 2007, S. 17
Lin, Junyue, "On the position and function of the government to promote the construction of social
   credit system". Collected Papers: Credit World of Art of Swordsmanship in Xitang, by Shanghai
   Credit Services Trade Association, July, 2019
Lin, Junyue, "Designing and maintenance of a disciplinary mechanism for discredit punishing".
   Economic and social system comparison, No. 3, May 2002, S. 1–6
Trustworthiness assessment. Credit–General vocabulary, National Standard of 22117-2018, S. 16

Junyue Lin 林钧跃，Präsident des China Market Credit Academic Council und außerordentlicher
Professor mehrerer chinesischer Universitäten. Seine akademischen Veröffentlichungen zu „National Credit Management System (2000)" und „Principle of Social Credit System (2003)", die auf
seinen frühen Forschungen aus dem Jahr 1999 beruhen, gelten als Grundlage der Theorie des sozialen Kreditsystems. Er ist auch der früheste Experte für die Einführung von Business Credit Manage-
ment-Methoden in China.

Teil II

Social Credit Rating Governance

A Study on the Typological Regulation of the Dishonesty Punishment

5

Wei Wang 王伟

Abstract

The punishment of dishonesty is an important content of the credit rule of law. The punishment of dishonesty is not a precise legal concept, so we must use legal technology to analyze and define it. Among the market punishment, industrial punishment, social punishment, administrative punishment, judicial punishment and other punishment mechanisms, the punishment by the public power is the focus of the credit rule of law. Administrative disciplinary measures are not all right-limited measures. In essence, the administrative blacklist measure of right restriction belongs to administrative punishment, but it does not violate the principle of "one punishment for one violation". In the future, when legislating the social credit law, we should typologically regulate dishonesty punishment and focus on the administrative punishment regulating the constitutive elements, punishment measures and procedures, credit restoration mechanism and other issues.

With the rapid development of social credit construction in China, the legislation of social credit law is in progress. At the level of national legislation, the Social Credit Law has been listed as Type III legislation by the National People's Congress, which is a legislative project that needs to be further studied and demonstrated. At the level of local legislation, Shanghai, Hubei, Hebei, Zhejiang, Shaanxi and Xiamen have formulated regulations on credit information management or comprehensive social credit law while Guangdong, Hainan, Guizhou and Jiangsu have been promoting comprehensive credit legislation con-

Wei Wang 王伟 (✉)
Beijing, China

tinuously. Since 2017, the author has led the drafting of the "Social Credit Law of the People's Republic of China (expert proposal text)". It is worth noting that, whether it is in the social credit legislation text or in the academic community's suggestions, the dishonesty punishment mechanism is an indispensable important content, and also the key to the entire social credit legislation. Strictly speaking, dishonesty punishment is only an abstract practical term, not a strictly legal one. Therefore, we must use legal technology to deliberately define "dishonesty punishment", and make a typological analysis in order to clarify its connotation and extension. Only in this way can we regulate the dishonesty punishment by typological classification. This paper analyzes the legal logic and basis of the dishonesty punishment mechanism, and expounds on the design ideas of the legal rules of the dishonesty punishment mechanism combined with the formulation of the social credit law.

5.1   The Reason and Premise of the Typological Regulation of Dishonesty Punishment

The dishonesty punishment mechanism is the foundation and focus of the social credit system. However, recently in the field of credit punishment, the phenomenon of disproportionate violation & punishment and joint punishment has caused great concern in the public about the abuse of credit punishment measures (especially the administrative "blacklist" measures), and even caused the international community's unreasonable criticism of the legitimacy of China's social credit construction. In view of this, it is important for China's social credit construction to bring the dishonesty punishment mechanism into the orbit of the rule of law and to construct a laws-based dishonesty punishment mechanism as soon as possible.

From the perspective of local legislative practice of social credit, some local legislatures have tried to bring the dishonesty punishment mechanism into the orbit of the rule of law, and made a lot of useful attempts to design the rules. However, due to the limitation of their legislative power, the relevant attempts have remained at the level of "local experiments", and failed to form a unified law system of dishonesty punishment. The key to building a systematic dishonesty punishment mechanism is to have a full understanding of the legal connotation and extent of the dishonesty punishment. Strictly speaking, the concepts of honesty rewards and dishonesty punishments are both derived from relevant credit policies and used in practical operations, which belong to relatively abstract practical terms rather than rigorous legal concepts.

It is not realistic to try to build a legalized dishonesty punishment mechanism only based on abstract concepts or language. In the process of constructing the rules for the dishonesty punishment system, we should always emphasize the use of typological thinking, typological research and legal regulation. Typological thinking is the bridge between abstract concepts and concrete facts. The essence of typology is a way of thinking to bridge the gap between highly abstract concepts and specific trivial facts. The relationship between concept and typology is important, just as Kaufman said: "concept without typology is empty, typology without concept is blind" (Artur Kaufman 2004, p. 192). Therefore, it is an important legal step to study the dishonesty punishment mechanism by typology and objectification. The following two basic premises should be observed in the typological classification of the dishonesty punishment mechanism.

The construction of the dishonesty punishment mechanism should not be limited to traditional economic credit, but also be based on the reality of China's social credit construction. In the market economy countries, "credit" as a concept has a specific context, which mainly refers to the credit relationship governing creditor's rights and debts, while the term "dishonest behavior" mainly refers to the behavior of not respecting agreed obligations. In this context, the punishment of dishonesty mainly refers to the distrust of or refusal to cooperate with dishonest subjects by the counterparty or other members of society. Thus, traditional dishonesty punishment is mainly a state of economic governance. In the context of China's credit construction nowadays, the connotation of "credit" has gone far beyond the scope of economic credit, including not only the state of performing contractual obligations in the traditional sense, but also the requirement to comply with legal obligations. For example, Article 2 of the "Regulation on Social Credit of Shanghai" stipulates that social credit refers to the state in which natural persons, juridical persons, unincorporated organizations and other subjects with full civil capacity abide by legal obligations or perform agreed obligations in social and economic activities. In this context, "social credit", which has the two aspects of economic governance and social governance, includes the abiding by the law and the performance of contractual obligations, and belongs to a broad concept of credit (roughly corresponding to the meaning of integrity, reputation, honor and other words in English) rather than only economic credit or obligations. Thus, the dishonesty punishment mechanism should be constructed in both the economic and the social fields.

In order to construct the dishonesty punishment mechanism, we need to classify dishonesty punishment from the perspective of public powers and private rights. The dishonesty punishment mechanism includes not only a variety of punishment subjects (high integration of public power subjects and private rights subjects), but also a variety of punishment methods (prohibitive, restrictive, cautionary punishment and even general administrative activities), which can be classified into moral punishment, social punishment, market punishment, administrative punishment and other types. Some scholars also summarize the dishonesty punishment methods as personal freedom punishment, property punishment, qualification punishment and moral punishment. In my opinion, the dishonesty punishment as currently practiced in China can be categorized as the broadest level of dishonesty punishment. This level of dishonesty punishment can be defined as the process and behavior in which the subject (government departments, social organizations, contract counterparties, etc.) has the right to punish dishonest behaviors through legal punishment, economic sanctions, moral condemnation and other means. Some people claim from a comparably broad sense that the function of the dishonest punishment mechanism is to punish dishonesty in market activities by legal, economic and moral means, and to exclude enterprises and individuals with serious dishonest behaviors from the mainstream market. This view is more in line with the actual situation of China's social credit system construction. To sum up, it is necessary to decompose the dishonesty punishment mechanism according to its legal nature. Only when the connotation and extension of credit punishment are discussed in a specific context can legal methodology really be applied to the construction of system and rules. According to the typological legal technology and thinking mode, the conceptualized method plays a leading role in the core determination level, while the typological method plays a major role in the boundary delimitation level. They complement each other and build an internal organic typological system.

In this paper, the dishonesty punishment is classified into two categories: one enforced by the civil and commercial subjects with equal status, and the other enforced by public power. Among them, the dishonesty punishment enforced by public authorities may involve impairing others' rights and increasing others' obligations, which is the focus of legal regulation and needs to be brought into the strict orbit of the rule of law. The dishonesty punishment enforced by civil and commercial subjects with equal status can be normalized by civil and commercial law, social organization law, credit reporting law, etc.

5.2 The Legal Logic of the Typological Regulation of Dishonesty Punishment

5.2.1 On the Dishonesty Punishment System in China

Historically, the development and improvement of the credit system has laid a solid foundation for the economic development of western countries, and a dishonesty punishment mechanism is the cornerstone of the credit system, allowing it to play an effective role. It is also an important content of credit construction and credit rule of law to build an effective dishonesty punishment mechanism in China. Through years of social credit construction practice, China has gradually formed a systematic dishonesty punishment mechanism. At the Third Plenary Session of the 18th CPC Central Committee, the Central Committee put forward the important policy of "dishonesty punishment". The 13th Five-Year-Plan also clearly puts forward the "perfect honesty rewards and dishonesty punishments mechanism". The report of the 19th National Congress of the Communist Party of China, which also emphasizes the construction of the dishonesty punishment system, puts forward the requirements of "perfecting the credit evaluation of environmental protection, information compulsory disclosure, severe punishment and other systems". In China, there are a lot of policy documents about credit construction and joint punishment memorandum signed by relevant departments, all of which refine the rules of dishonesty punishment, and these rules have achieved remarkable results in the implementation. In recent years, the Anti Unfair Competition Law, E-commerce Law and other laws of China have made provisions on dishonesty punishment. At present, this is based on the credit reporting system of the People's Bank of China, the national credit information sharing platform and the national enterprise credit information publicity platform, China has built the information platform for the dishonesty punishment, which provides a strong technical support for implementation. From the perspective of the Party's and the Government's credit policies and relevant practices, China's dishonesty punishment measures generally involve the following five aspects:

The first is market punishment. In market transactions, the counterparties can punish dishonest subjects by refusing the transaction and increasingly demanding the transaction terms. Examples of this are the dishonesty punishment imposed by commercial banks on borrowers and insurance companies on policy holders.

The second is industrial punishment. Trade associations, chambers of commerce and other social organizations can implement appropriate dishonesty punishments on their members based on the self-discipline rules of the industry. For example, according to Article 36 of the Assets Appraisal Law of China, a trade association shall establish a member's credit file and record, and publicize the member's compliance with laws and administrative regulations as well as appraisal standards. Accordingly, the trade association may impose dishonesty punishment on its members.

The third is social punishment. Social organizations and the public can punish dishonesty by effective social participation and supervision through mechanisms such as dishonesty reporting, media supervision, public interest litigation. For example, the information disclosure and public participation system in the Environmental Protection Law and the public interest litigation system in the Civil Procedure Law provide an important legal basis for the implementation of social punishment.

The fourth is administrative punishment. China has implemented dishonesty punishment in public service and market supervision by establishing an administrative punishment mechanism between central and local governments and one between horizontal departments. For example, the "Memorandum of Cooperation on Collaborative Supervision and Joint Punishment of Dishonest Enterprises" signed by the former State Administration for Industry and Commerce, the National Development and Reform Commission, the Supreme People's Court and 38 other departments, implements joint punishment and restricts the dishonest subjects' enjoyment of some high consumption services, and their ability to qualify for senior positions in enterprises in related industries. Local governments are also exploring innovations in the dishonesty punishment system, such as the "Measures for Dishonesty Punishment of Natural Persons (Provisional)" issued by Jiangsu Province, which clarifies dishonest behavior and punishment measures; and the "Measures for Determination and Treatment of Bad Behaviors in Bidding and Tendering" issued by Nanjing, which requires the publication of dishonest behavior in bidding and tendering.

The fifth is judicial punishment. In China, the dishonesty punishment mechanism actually includes punishment by judicial organs. Judicial organs should carry out professional judicial activities to crack down on dishonesty. The current laws and policies on judicial punishment are mainly: the Civil Procedure Law, the "Opinions on Accelerating the Construction of Credit Supervision, Warning and Punishment Mechanism for Dishonest Executees" and the "Memorandum of Cooperation on Joint Punishment for Dishonest Execu-tees" issued by the general office of the CPC Central Committee and the general Office of the State Council, the "Provisions on the Publication of Information on the List of Dishonest Executees" issued by the Supreme People's court. The Supreme People's court has also established a judicial punishment mechanism for dishonesty through measures such as the disclosure of judgment documents and the disclosure of information on disho-nest executees.

Generally speaking, China's dishonesty punishment mechanism involves a series of institutional arrangements scattered in various laws and regulations or policy documents, and has not formed an internal logical system of dishonesty punishment. At the same time, the implementation of the dishonesty punishment mechanism in practice places emphasis on administrative and judicial punishments which stem from the public power. The market punishment, industrial punishment and social punishment mechanisms are far from play-ing their roles, and it is difficult for social forces to effectively supervise and punish dishonesty.

5.2.2 The Legal Logic of the Typological Regulation of Dishonesty Punishment

The enforcers of dishonesty punishment can be divided into public power enforcers and private right enforcers, each of which has different power sources. The legal basis for the implementation of dishonesty punishment can be public law norms, private law norms or social norms (articles of association, self-discipline provisions, etc.) which are not co-vered by formal legislation. Therefore, the legal sources of dishonesty punishment are different. As pointed out by some scholars, the dishonesty punishment measures can be divided into private law measures and public law measures. The former consist of civil liability methods. The latter refer to government intervention, sanctions or punishments, such as disclosing the integrity records of the dishonest according to law or imposing administrative punishments such as warning, confiscation of property, deprivation of qua-lification and restriction of rights of the dishonest (Hu Chaoyang 2012). The following summarizes the legal logic of different classifications of the dishonesty punishment mechanism.

5.2.2.1 The Analysis of Source and Essence of Dishonesty Punishment Power

The credit subject is in a specific social network, which needs to trade or communicate with public authorities, market subjects, the public and even relevant industry organiza-tions. The important purpose of dishonesty punishment is to ensure the trust and safety of social members, maintain social solidarity and promote social cooperation. Some scholars believe that when the credit commitment relationship between members of society is destroyed, the law must reaffirm the relationship where punishment is an effective means (Wang Lifeng 2013, p. 124). From the perspective of retributive punishment, dishonesty punishment can make the dishonest pay the corresponding price for their behavior and purify the credit environment; from the perspective of utilitarian punishment, dishonesty punishment can create a certain deterrent for both dishonest people and other members of society, realize the purpose of "deterrence" through "punishment" and maintain social credit. It is worth noting that dishonesty punishment has its limitations. Dishonesty punishment functions by impugning the reputation of the dishonest, based on the premise that reputation has great value to the actor. However, for those who don't care about reputation at all or those who will not be substantially impacted when losing reputation, dishonesty punishment has no use (Li Fuying et al. 2012).

In my opinion, credit restriction and joint punishment mechanism have at least two levels of meaning. The first level meaning is that the punishment mechanism dominated by law (i.e. clearly stipulated by law) can realize dishonesty punishment through civil, administrative and criminal sanctions. The second level meaning is that the punishment mechanism enforced by market factors (business reputation, etc.) can increase the cost of dishonesty and realize dishonesty punishment through market means, with the credit information database as the link between the two. Market subjects restrict the dishonest through financing, contracts and other aspects according to the credit data (Wang Wei et al. 2016). Some argue that it is up to the market, society and industry to make use of the records of the dishonest subjects and to treat the subjects who have been dishonest (Chen zhe 2017). From this point of view, the legal source of dishonesty punishment can be not only the legal provisions, but also the agreement of the related parties, and even the rules and business practices in market transactions. Specific to different classifications of dishonesty punishment measures, the source of the rights (power) of the punisher can be analyzed as follows.

Firstly, the right source of market punishment and social punishment. According to the traditional legal system of credit reporting, counterparties to a transaction can ensure the transaction safety by understanding the credit status of the credit subject according to the relevant credit information records, taking corresponding defensive measures (such as increasing the transaction terms, raising loan interest and insurance rates, etc.), or refusing to trade. This kind of punishment measure is based on the market reputation mechanism, which is a rational choice made by the equal trading subject. In essence, it is a form of market self-organization, which in principle should be adjusted by Civil and Commercial Law, Credit Reporting Law and other laws. The author believes that: for the regulation of such punishment measures, the main function of the government is to provide the corresponding laws, market-oriented and socialized credit mechanisms so as to form the social level credit legal mechanism (Wang Wei 2016, p. 140). In practice, if a party to a transaction (such as a financial institution) has a more advantageous market trading position or even a dominant market position, or belongs to a monopoly public institution, its behavior must conform to the relevant provisions of the Anti-Monopoly Law, the Anti Unfair Competition Law and the Public Utilities Law. Currently, there are many laws and regulations in this respect. For example, Article 13 of China's "Anti-Monopoly Law" stipulates that operators with competitive relationships are prohibited from carrying out boycott transactions through monopoly agreements; Article 17 stipulates that operators with a dominant position in the market are prohibited from abusing their position. As another example, Paragraph 2, Article 26 of the "Regulations on the Administration of Social Credit Information of Hubei Province" stipulates that information on arrears of public utilities and property management, that has not been confirmed according to law, shall not be used as the basis for the implementation of joint punishment. Social punishment has the same legal basis as market punishment, which is implemented between subjects with equal status.

Secondly, the power source of industrial punishment. In the operation of the modern market economy, social organizations (industry associations, chambers of commerce, etc.) have corresponding industry self-discipline and supervision responsibilities for their members, based on the self-discipline rules signed between members. As to the nature of the punishment power of trade associations, some scholars claim that it is a kind of social right, in other words, that social subjects make other social subjects obey by controlling social resources (Tan Jiusheng 2010, p. 30). This kind of punishment power is obviously different from state power. The nature of the measures (publicizing information on dishonest behavior, publicly criticizing and liquidated damages) taken by trade organizations against their members who violate the law and break good faith agreements, is based on the punishment provided for in their own autonomous rules. For example: according to Article 6 of the "Credit Self Discipline Convention in Electric Power (Provisional)" issued by the China Power Enterprise Federation, if the member units of the Convention violate the self-discipline clauses such as the responsibility to act in good faith, the list of names shall be publicized to all member units, reflected in the credit evaluation and published to the society if necessary; for the enterprises listed in the "blacklist", measures such as canceling the qualification for industrial evaluation can also be taken (Zhang Jianhua and Wang Wei 2018, p. 107–108).

Thirdly, the power source of administrative punishment and judicial punishment. Administrative punishment is a kind of dishonesty punishment implemented by public authorities based on administrative relations. Based on public power, the public authority can inquire or disclose the relevant information about the credit subject, understand the current or past compliance and performance of the credit subject, and then undertake corresponding dishonesty punishment measures. As opposed to the punishment between equal subjects, administrative punishment refers to the administrative management, administrative penalties and administrative enforcement behaviors implemented by means of public power. The power source of judicial punishment is judicial power. Because the exercise of judicial power has a higher level of "Civil Procedure Law" as the legal basis, so there are few social doubts about such punishment measures.

5.2.2.2 The Analysis of Ways and Legal Consequences of Dishonesty Punishment

Market punishment, industrial punishment and social punishment are three kinds of dishonesty punishment between equal subjects. Their legal basis is mainly the contract between the parties, the autonomous rules agreed between members, and the Credit Reporting Law. The corresponding legal consequences are refusing to conclude the contract, demanding the trading terms, reducing the rights of the dishonest as a member or making them bear the corresponding losses respectively. The result of this kind of dishonesty punishment is more uneconomical and unprofitable conditions for people acting in bad faith in civil affairs.

The dishonesty punishment implemented by public power subjects mainly aims at the dishonest behavior that does not comply with legal obligations, resulting in adverse consequences in public law. It covers a wide range of activities and the restrictions on the dishonest are more extensive, such as forbidding or restricting market access, strengthening management, directly limiting the rights or increasing the obligations, and listing in the "blacklist". Some of these dishonesty punishment measures will not directly limit the activities of credit subjects, while others will have legal consequences similar to administrative penalties and administrative enforcement. For the latter, the authorities or organizations that implement dishonesty punishment may bear the corresponding administrative legal responsibility. Therefore, the requirement for legalization of such punishment measures is higher, and needs to be regulated within the scope of public law.

5.3   The Key Point of Typological Regulation of Dishonesty Punishment – Administrative Punishment

Among the current dishonesty punishment measures implemented in China, the most noteworthy one is the punishment measures taken by the public authority, especially the legality of such punishment measures. Concerning the dishonesty punishment implemented by administrative organs, we need to explore its legal basis and understand the corresponding legal regulations.

5.3.1   Whether All Administrative Punishments Constitute Administrative Penalties

In practice, there are many types of administrative punishments, but not all of them are high-intensity administrative penalties and enforcements. The management intensity of administrative punishment measures is different, including management strengthening measures, non-benefit granting measures, "blacklisting" measures, restriction or impairment of qualification rights, etc. Whether administrative punishment measures constitute administrative penalties or not should be determined according to the nature of the measures. The following is an analysis of several important administrative punishment measures.

5.3.1.1 Management Strengthening Measures and Non-benefit Granting Measures

Management strengthening measures, such as increasing the frequency of supervision, canceling the corresponding convenience (such as green channel service), etc., do not levy substantial administrative constraints on the activities of the credit subject, nor constitute the restriction or impairment of the rights of the credit subject. Non-benefit granting measures including non-granting honorary titles, which are restrictions on non-core rights and interests of enterprises and do not belong to the scope of administrative penalties or enforcements. In their nature, they can be understood as administrative treatment acts that bring adverse burdens to credit subjects. Administrative processing can be understood as a kind of specific administrative act in theory, which refers to the unilateral administrative act that an organization with administrative power uses to set, change or eliminate the rights and obligations of a specific subjects (Jiang Mingan 2011, p. 191). Although such acts do not constitute administrative penalties, they are actionable because they are specific administrative acts.

5.3.1.2 Administrative "Blacklist" Measures

The administrative "blacklist" measure is an important form of dishonesty punishment, which aims at serious and/or illegal dishonest behavior. Some scholars believe that the "blacklist" system as an innovation of the administrative supervision mode, is conducive to the realization of the dynamic of administrative supervision, which enables the government to shift from direct supervision to indirect supervision of market subjects, and also urges the government to shift from pre-supervision to post-supervision while streamlining administrative and delegation power (Zhao Xudong et al. 2018, p. 32–33). In addition, administrative "blacklist" measures are also conducive to promoting the rule of law in a market economy, strengthening the credit constraints on market subjects, so as to ensure the safety of transactions. From the perspective of legal attributes, the administrative "blacklist" measure is to a large extent a kind of administrative penalty measure. In theory, administrative penalties can be divided into personal penalties, property penalties, behavior penalties and reputation penalties. It will be found that some administrative "blacklist" measures are similar to reputation penalties in administrative law, the purpose of which is to let social members know as much as possible about the illegal and dishonest behaviors of credit subjects, so as to take preventive measures in investment, transactions or civil activities. Some administrative "blacklist" measures are similar to a qualification penalty. Through information sharing, more administrative or judicial organs implement credit restrictions and dishonesty punishments on those who are listed in the "blacklist". The main reasons that administrative "blacklist" measures are essentially equivalent to administrative penalties are as follows.

1. The administrative "blacklist" measures constitute strong negative judgments against the dishonest. The administrative "blacklist" is the result of the administrative organ's judgement based on the enterprise's credit information, which has a strong sense of value judgment, and has the function of prompting and warning the administrative counterpart and the public. The value judgment attribute of administrative "blacklist" measures makes it completely different from the public reflection of basic information on enterprises and other credit subjects. For example, according to the "Interim Regulations on the Publicity of Enterprise Information" promulgated by the State Council, the publicity of enterprise basic credit information does not involve value judgments by administrative organs.

2. Administrative "blacklist" measures will reduce market influence and social reputation of the dishonest, leading to the loss of rights. The "blacklist" itself does not directly have the restriction and punishment effect of administrative law. However, because of the superior position and authority of the administrative organ, a "blacklist", once published, will have a realistic derogatory effect on the reputation and public opinion image of the dishonest, and have a great influence and substantial restriction on its business and life.

3. Administrative "blacklist" measures are often accompanied by the following measures, such as the derogation of the rights and the expansion of the obligations by the public authority. After the administrative organ publishes the "blacklist", it will usually attach more severe supervision or disciplinary measures to those on the "blacklist". It can be said that the "blacklist" is not only the result of the integration of the relevant credit information by the administrative organ, but also the basis for other punishment measures for the dishonest. From the perspective of practical effect, being blacklisted is even more serious than being warned or criticized.

It is worth noting that there is also a kind of non-right-limited so-called "blacklist" measures in the construction of social credit system in China. The list system of abnormal business operating stipulated in the "Interim Regulations on Enterprise Information Publicity" is a typical non-power-limited measure. Because the fundamental goal of the system is not to punish, but to publicize to the society the minor violations such as the failure of annual returns so as to urge the enterprises to publicize the relevant information as soon as possible. From the original intention of the rule design, the system is a "fast in and fast out" system. As long as the enterprise corrects its violation, it should be removed from the list without bearing the consequences of punishment (Article 17 of "Interim Regulations on Enterprise Information Publicity"). Therefore, the system is essentially a non-power-limited measure.

5.3.1.3 Measures to Restrict Qualification and Rights

These measures constitute substantive restrictions on the legal rights of dishonest subjects, and also restrictions on their core rights and interests. Although this kind of punishment does not belong to the measures listed in the Administrative Punishment Law, it is essentially an administrative punishment due to its obvious punitive effect on the qualification and rights of relevant subjects. For example, restricting the qualification of enterprises to participate in certain projects, or reducing their qualification rating, will have a great or even serious impact on the production and operational activities of enterprises. The implementation of such measures should be in line with the provisions of the Legislative Law, and be included in the specific administrative acts that are actionable, which shall be subject to the corresponding judicial review.

To sum up, administrative punishment consists of administrative "blacklist" measures and restrictions on qualification and rights, while management strengthening measures and non-benefit granting measures which are not aimed at reducing or impairing rights do not constitute administrative punishment.

5.3.2 Whether the Dishonesty Punishment Measures That Constitute Administrative Punishment Violate the Principle of "One Punishment for One Violation"

The next important question to be discussed is: whether the dishonesty punishment measures that constitute administrative punishment violate the principle of "one punishment for one violation". As far as the legal value of the principle of "one punishment for one violation" is concerned, its purpose is to avoid two or more administrative punishments for the same violation. At present, there is still a great controversy in theoretical circles about what is "one violation", what is "one punishment" and even concerning the legal connotation of "one punishment for one violation", which leads to a different understanding of administrative punishment. One view is that administrative punishment is under the principle of "one punishment for one violation". Some scholars believe that the principle of "one punishment for one violation" aims to prevent repeated punishments, reflects the legal value of "fair penalty" to protect the legitimate rights and interests of the subjects: when the violation and dishonest behavior of the credit subject has been punished by an administrative punishment, then the implementation of an additional dishonesty punishment constitutes double punishment, which violates the principle of "one punishment for one violation". Another view is that administrative punishment does not fall under the principle of "one punishment for one violation". This view tends to understand the principle of "one punishment for one violation" in a narrow sense, and holds that "one punishment for one violation" stipulated in Article 24 of the Administrative Punishment Law of China only refers to "no two fines can be imposed for the same fact", and there is no restriction or prohibition on other administrative punishments for the same violation; if the dishonesty punishment is not a fine, it is not included in the conditions for "one punishment for one violation". I believe that the administrative dishonesty punishment measures which restrict qualification and rights do not violate the principle of "one punishment for one violation", and the main reasons are as follows.

5.3.2.1 From the Perspective of Credit Law, Dishonesty Punishment is Essentially a Kind of Social Evaluation Ex-post

Dishonesty punishment mainly refers to the social evaluation of the specific subject after the credit granter makes a comprehensive judgment on the credit information of the credit subject in terms of performance and compliance with the law. The purpose is to prevent or deter the subject from damaging private or public interests, and to advocate for integrity and abidance with law and contracts. After making such a comprehensive credit evaluation and judgment, the punishment measures taken are not aimed at a specific violation or breach of contract, whose nature has been transformed into the social evaluation of credit. For the public authorities, the dishonesty punishment is based on the judgment about non-performance of a legal obligation, and is necessary to safeguard the social public interest. According to the "Interim Measures for the Administration of the List of Serious Illegal and Dishonest Enterprises" issued by the former State Administration for Industry and Commerce, enterprises that have received more than three administrative punishments in two years will be included in the list of illegal and dishonest enterprises, which is the judgment of non-performance of legal obligations. As one scholar holds, the dishonesty punishment measures are aimed at the credit status of the subjects (i.e. the status of breaching law and contracts) rather than the specific behaviors, and the credit status is the integration of different pieces of information on honest and dishonest behaviors, which is the comprehensive cumulative result of multiple behaviors and events (Li Zhenning 2018). Therefore, dishonesty punishment is not aimed at the same behavior, nor based on the same fact or evidence, and does not violate the principle of "one punishment for one violation".

5.3.2.2 From the Perspective of Administrative Law, the Principle of "One Punishment for One Violation" Should Be Interpreted in a Restrictive Way

The principle of "one punishment for one violation" originates in criminal procedure law. With the development of society, more and more countries have extended the application of the principles from early criminal procedure law to a broader legal field (including administrative law), and even regarded them as important in order to protect human rights. One disadvantage here is that Article 24 of the Administrative Punishment Law of China limits "penalty" to "fine". However, modern needs for economic and social development cannot be met by expanding infinitely the application scope of the principle of "one punishment for one violation". From the perspective of extraterritorial legislation, the application of the principle of "one punishment for one violation" in the field of administrative management has heretofore usually been through a relatively cautious and realistic approach, striving to leave the necessary elastic space for government management. For example, German legislation limits the scope of application of the principle of "one punishment for one violation". According to the Order Violation Law of Germany, the principle of "one punishment for one violation" in administrative punishment holds that if the same act violates several administrative laws and regulations, a fine shall be imposed according to the law with the highest amount, and additional measures stipulated by other laws and regulations may be imposed at the same time. In another example, Austrian legislation does not apply the principle of "one punishment for one violation". According to the Administrative Punishment Law of Austria, when the administrative defendant violates different administrative obligations with various independent acts or incurs several charges in the same act, he or she shall be punished separately. Referring to the practices of foreign legislation, combined with the implementation of China's dishonesty punishment, I believe that the principle of "one punishment for one violation" should be interpreted in a restrictive way. The dishonesty punishment by administrative organs is "multiple punishments", not "re-punishment", which does not violate the principle of "one punishment for one violation".

5.3.2.3 There Are a Lot of Ex Post Evaluation or Re-evaluation Norms in Other Laws, Which Do Not Violate the Principle of "One Punishment for One Violation"

Article 146 of the Company Law of China stipulates that those who are sentenced to criminal punishment for embezzlement, bribery, misappropriation of property, or sabotage of the order of the socialist market economy, shall not serve as directors, supervisors or senior managers of a company within five years of committing the crime. Article 15 of the Securities Investment Fund Law stipulates that lawyers, certified public accountants, practitioners of asset appraisal and verification institutions, as well as investment consulting practitioners whose licenses have been revoked or who have been disqualified due to illegal acts shall not serve as directors, supervisors, senior managers and practitioners of fund managers who raise funds publicly. Article 22 of the Government Procurement Law lists the specific situations and stipulates that suppliers should have good credit and moral character to participate in government procurement activities. According to these Provisions, if the administrative organ puts forward higher requirements for the credit and moral status of the relevant subjects in the fields of financial enterprise operation, government procurement or major trading activities involving public interests, this is also within the scope of the dishonesty punishment, but is not a violation of the principle "one punishment for one violation".

5.4 The Legislative Reform of the Typological Regulation of Dishonesty Punishment – On the Rule Design of the Social Credit Law

The making of the Social Credit Law has been included in the legislative plan of the National People's Congress, and the dishonesty punishment mechanism is an important part of this credit legislation. In the legislative process for creating the Social Credit Law, we should clarify the legal logic of different classifications of dishonest punishment measures, and establish different regulatory principles and rules on the basis of their typological classifications, so that all kinds of dishonest punishment mechanisms can play their respective roles. This is an important mission and task for a Social Credit Law with Chinese characteristics.

5.4.1   On the Typological Adjustment of the Dishonesty Punishment Mechanism

There must be a legal basis for dishonesty punishment. China's Legislative Law, Contract Law, Regulations on the Administration of Credit Reporting and other relevant legislation provide an important legal basis for dishonesty punishment. I suggest that the Social Credit Law should systematically and comprehensively divide the dishonesty punishment mechanisms, including market punishment, industrial punishment, social punishment, administrative punishment, judicial punishment and other punishment mechanisms into the category of typological adjustment, and clarify the rights (powers) of the decider of the relevant punishment. Among them, the provisions on market punishment should be connected with the relevant laws such as Civil and Commercial Law, Credit Reporting Law, Anti-Monopoly Law, etc.; the provisions on social punishment should be connected with the provisions on information disclosure, social participation, social supervision in the current laws, and the relevant systems in the Civil Procedure Law, etc.; The provisions on industrial punishment should make clear that the articles of association or other self-discipline rules can incorporate necessary dishonesty punishment items or rules.

The administrative punishment should follow the rule of law principles that "Government departments must not do anything unless it is mandated by the law" and "Government departments must fulfill their legal duties". The implementation of administrative punishment should meet legal requirements which include punishment standards, punishment objectives, punishment subjects, punishment measures, punishment procedures, punishment periods and relief mechanisms. The legal requirements for administrative punishment should not only show the values of honesty and credit, fairness and justice, but also ensure the stability of the law to meet the reasonable expectations of the public. In case the law does not clarify sufficiently the relevant issues of dishonesty punishment, the public will not have reasonable expectations concerning the legal consequences of dishonesty or even enough information to accept the validity and legitimacy of dishonesty punishment.

To solve this problem, we must realize the legalization, standardization and due process of the dishonesty punishment mechanism. After the legalization of the dishonesty punishment mechanism, members of society will have reasonable legal expectations on the consequences of dishonesty due to the publication of legal standards in advance and the general binding force of legal rules, which will strengthen the stability of law.

5.4.2   On the Legal Definition of Dishonesty in Administrative Punishment

For dishonesty which is included in the scope of administrative punishment, we should review the seriousness and subjective intention of the ill behavior of the credit subject according to the requirements of legal punishment. Firstly, to punish dishonest subjects, we should aim at the dishonest behavior with a certain amount of seriousness. Dishonesty punishment shall not be directly applied to minor or corrected behaviors (Article 8 of the "Social Credit Regulations of Xiamen Special Economic Zone" publicized in 2019). This is because the intensity of dishonesty punishment is very strong, so we must consider its serious adverse consequences to ensure fair penalties. Secondly, the dishonest behavior to be punished should have objective intent. Dishonesty punishment should not be applied to negligent acts, especially to generally negligent acts. Other ways can be adopted to make the dishonest bear corresponding civil, administrative and criminal legal liabilities. This is mainly to consider that, from the perspective of the requirements of credit rule of law, being honest mainly means that people should be willing to abide by the law or contract objectively, and should not intentionally violate the law or breach the contract. In the case of negligence, there is no close relationship between dishonest behavior and the willingness to be honest. Although the subject should also bear the corresponding legal liability for his negligent behavior, he should not be punished by joint punishment which is too severe.

The author believes that the Social Credit Law should clearly stipulate that the authorities and organizations that have the right to implement dishonesty punishment should determine whether the credit subject has carried out dishonest behaviors within the scope of legal authority according to the severity, objective breach and other factors concerning the credit subject's behavior. The identified dishonest actions shall be collected, shared and disclosed in accordance with the law, and shall serve as an important basis for the implementation of dishonesty punishment.

5.4.3   On the Systematization of the Administrative Punishment Mechanism

The dishonesty punishment implemented by the administrative organs will have a great influence on the rights, obligations and responsibilities of the credit subject. Therefore, it should follow the corresponding principles of the rule of law, and realize the powers, measures and procedures of punishment according to the provisions of the law. For this reason, the following three aspects should be covered by the legislation.

5.4.3.1 Establish Administrative Punishment Mechanisms by Typological Classification

We should regulate administrative punishment measures according to their legal nature. The Social Credit Law should emphasize administrative punishments which will restrict or impair the rights of the dishonest, clarify the legal attributes, and stipulate the corresponding enforcement and relief procedures. For the non-power-limited "blacklist" measures (such as the list of abnormal business operations stipulated in the Interim Regulations on Enterprise Information Publicity), it should be clearly stipulated that they should not be used as the basis for the implementation of administrative punishments.

In the process of allocating administrative punishments, the credit subject is in a relatively weak and passive position. In order to effectively restrain the public power and prevent it from causing unnecessary damage to the credit subject, it is necessary to regulate administrative punishments. According to Article 82 of the Legislative Law of China, administrative punishments should not infringe on the law to derogate rights or increase obligations, and should be within the local government's administrative scope.

5.4.3.2 Implement Key Regulation Measures of the Administrative "Blacklist"

The administrative "blacklist" measures of power restriction, which impose many restrictions on the rights of credit subjects and have a significant impact on their interests, are the top priority of the Social Credit Law. Compared with the implementation of administrative penalties, the establishment of the administrative "blacklist" should be subject to more strict restrictions and "legal reservations". It is necessary to legalize the blacklist measures strictly. The application of the blacklist should only be possible in conformity with specific relevant laws. It is also necessary to ensure the operation of the measures within the rule of law, through information disclosure, democratic decision-making, effective relief and other mechanisms. (Wang Wei 2018).

5.4.3.3 The Joint Disciplinary Mechanism Shall Not Establish Right-Limited Administrative Punishment Measures Without Legal Authorization

It is common practice to sign a memorandum of joint punishment through the ministries and local relevant departments, to clear guidelines for joint dishonesty punishment. The cooperation of public authorities to accuse the related credit subjects of bad faith dealings can increase the constraints and enhance the effectiveness of joint credit punishment. However, as far as the legal nature of the joint punishment cooperation memorandum is concerned, it can belong to the category of normative documents which is legal documents at a lower level. The joint punishment should not only comply with the requirements of the

the Law on Legislation (Article 80, 82 of the Law on Legislation of China), but also with the requirement that administrative punishment, as well as the legal punishment methods, punishment procedures and relief procedures, should be determined by law, etc. Some local governments list the types, the measures, and the main legal basis of dishonesty by publishing a joint punishment catalogue, which makes the joint punishment measures explicit, open and predictable, so as to clarify the boundary between public power and private rights, and achieve better legal effects.

Based on the experience of local credit legislation, the author suggests that the Social Credit Law should make the following provisions: The state builds a joint credit rewards and punishments mechanism across regions, departments and domains; The people's governments at or above the county level establish a list system of joint credit reward and punishment measures, and clarify the compulsory and recommended punishment measures in accordance with law; Unless otherwise provided by laws and regulations, administrative punishment shall not be implemented in addition to the measures specified in the list.

5.4.4 On the Legitimacy of Administrative Punishment Measures

If administrative punishment measures are to be legitimate, they should not blindly pursue retribution against the dishonest, nor can they neglect personal interests solely for the purpose of maintaining social credit order. Administrative punishment measures should balance the relationship between punishment and guarantee to ensure the legitimacy of implementation.

5.4.4.1 There Should Be a Reasonable Connection Between Administrative Punishment and Illegal Acts

The implementation of dishonesty punishment by administrative organs must meet the legal requirements of reasonable administration. According to the principle of irrational-relationship-forbidden in modern administrative law, when public power restricts private rights, administrative organs can't act arbitrarily and must prove that the benchmark of restriction has reasonable necessity, substantiality, legitimacy and relevance (Ou Aimin 2008). Specifically, in the field of social credit rules of law, when the public authority punishes the illegal and dishonest behavior of the subject, it must be determined that there is a reasonable connection according to law between the bad faith behavior and the disciplinary measures. According to the requirements of the rule of law of administrative punishment, if the credit subject is accused of illegal and dishonest behavior in Place A, it is necessary to establish the relationship between A and B to impose dishonesty punishment on him or her in Place B. For example, if a company's motor vehicles have multiple violations, the traffic management department may impose necessary restrictions when handling traffic-related matters, and the commercial insurance company may increase the insurance rate. However, if the company applies for an IPO, the relevant departments shall not refuse its application on the ground that its vehicles have violated traffic regulations.

Social Credit Law should establish the Relevant Standard of Credit Information, and enhance transparency and predictability of application of this standard. In order to avoid the influence of improper factors on administrative management, the Social Credit Law should require administrative organs to determine the scope of credit information related to their own administrative matters in accordance with the principle of reasonable administration, and to use and publicize the scope as the basis for carrying out classified management (Luo Peixin 2016). The author believes that the Social Credit Law should establish the principle of relevance on dishonesty punishment, especially the list of punishment measures, and the credit management department should take the lead to set clear punishment measures and publicize them to society. Such a clear list can, on the one hand, provide clear legal guidance for the credit subject to understand the legal consequences of the behavior; on the other hand, it can enhance the transparency of dishonesty punishment and restrict the punishment discretion of administrative organs.

5.4.4.2 The Implementation of Administrative Punishment Measures Should Conform to the Principle of Proportionality

The implementation of administrative dishonesty punishment should conform to the principle of proportionality in Administrative Law, reflecting fair penalties. Generally, strengthening measures of managerial administrative punishment will not damage the core rights and interests of the credit subject, so it does not have the actionability in Administrative Law. However, measures considered to be administrative penalties, such as restrictions on market access, qualifications and public resources access, may seriously impact the rights and interests of credit subjects. This kind of administrative punishment should be strictly regulated by law, and corresponding administrative reconsideration, litigation and other means of relief shall be provided to the affected credit subjects.

For this reason, I suggest that Social Credit Law should make the following provisions: The dishonesty punishment measures implemented by the administrative organ should conform to the principle of proportionality; Administrative punishment should adapt to the nature, circumstances and social impact of the dishonest behavior; Administrative punishment should not go beyond the conditions, types and range of legal penalties.

5.4.5 On the Credit Repair System of the Dishonest Actor

The punishment and correction mechanism to address dishonesty are two sides of a coin. In social credit construction, the dishonesty punishment is the core system, but its purpose is not to permanently nail the dishonest subject to the stigma pillar, but to let him or her have the opportunity to correct mistakes and rehabilitate him/herself. In the long run, it is beneficial for the reconstruction of social credit. There are two kinds of credit repair in the practice of social credit construction: one is the rehabilitation after contract breach, and the other is rehabilitation after breaking the law. The author suggests that the Social Credit Law should not only construct the typological classification system of dishonesty punishment, but also construct the credit repair system in an all-round way and form the special legal rules for credit repair.

The relevant provisions include but are not limited to: Firstly, for the breach of contract obligations, the credit subject can correct his or her dishonest behavior by means of repaying debt, providing a guarantee, reaching a settlement with the creditor, etc. Secondly, for the violation of legal obligations, if this is repairable, the credit subject should apply for credit restoration, make a commitment to abide by the law and implement rectification and other restoration actions. After examination, the relevant departments shall make a decision as to whether the conditions for credit restoration have been met. The decision on credit restoration shall be publicized and subject to social supervision. Thirdly, when the credit subject has corrected the dishonest behavior, he or she may request the provider of information on his/her dishonesty to issue a credit repair certificate or the creditor to issue a statement of understanding. Relevant authorities and organizations may, in accordance with the law, delete the information on the dishonest behavior, annotate the credit information, etc., so as to restore the credit record for the dishonest subject. (The above suggestions refer to the relevant proposed article on "Credit Repair in Social Credit Law of the People's Republic of China (expert proposal text)", which has been drafted by the author.)

References

Artur Kaufman (2004): Philosophy of law, translated by Liu Xingyi, etc., Law Press, p. 192.

Chen zhe (2017): Dishonesty Punishment and Credit Restoration, Zhejiang Economy, issue 8.

Hu Chaoyang (2012): Legal Regulation of Social Dishonesty, Law and Business Research, issue 6.

Jiang Mingan (2011): Administrative Law and Administrative Procedure Law, Peking University Press, p. 191.

Luo Peixin (2016): Good Governance Needs Good Law: On Social Credit Legislation, Law, Issue 12.

Li Zhenning (2018): Characteristics of Dishonesty Punishment and Its Enlightenment on Local Legislation, Journal of Nanjing Municipal Party School of the Communist Party of China, Issue 2.

Li Fuying, Wang Zhongju, Guo Wenshu, et al (2012): The Application of Credit Punishment in the Design of Local Legislative System, Beijing People's Congress, issue 6.

Ou Aimin Xie Xiongjun (2008): the Principle of Irrational-Relationship-Forbidden and its Application, Journal of Social Sciences of Hunan Normal University, issue 5.

Tan Jiusheng (2010): On the Public Law Regulation of Punishment Power of Professional Associations, Doctoral Dissertation of Wuhan University, p. 30.

Wang Wei (2018): On the Legalization of the Blacklist System of Dishonesty, the People's Rule of Law, Issue 11.

Wang Wei (2016): Legal Logic and Institutional Mechanism of Market Supervision – an analysis based on the reform of commercial system, Law Press, p. 140.

Wang Wei, Hou Jiangshan, Guo Fuwei (2016): The Top Priority of Enterprise Credit Construction: Credit Restriction and Joint Punishment, the People's Rule of Law, Issue 9.

Wang Lifeng (2013): Philosophy of punishment (2nd Edition), Tsinghua University Press, p. 124.

Zhao Xudong, et al (2018): Blacklist System, China Legal Publishing House, p. 32–33.

Zhang Jianhua and Wang Wei (2018): China Enterprise Credit Construction Report (2017–2018), China Legal Publishing House, p. 107–108.

Wei Wang 王伟, male, PH.D, professor, director of Civil, Commercial & Economic Law Division attached to Politics and Law Department of Central Party School of CPC (Chinese Academy of Governance) (Beijing 100091). He graduated from Law School of Renmin University of China, mainly engaged in the research of social credit law. company law, market regulation law. In recent years, he focuses on the research of social credit law and involved in the projects from National Development and Reform Commission, State Administration for Market Regulation, and so on. He has participated in drafting a number of expert proposals on social credit legislation. This paper was originally published in Zhongzhou academic journal, issue 5, 2019, and reprinted in issue 8 of constitutional law and administrative law in Periodical Literatures by Renmin University of China.

The Social Credit System and China's Rule of Law

Marianne von Blomberg

Abstract

In 2014, the People's Republic of China central government formally declared the construction of a Social Credit System (SCS) a national task. Meanwhile, government-designated localities and companies are experimenting with scoring systems for businesses, citizens and the administration. The government's initiative introduces mechanisms for a massive aggregation and exchange of data about 'credit subjects', pushes for the application of such credit information in the decision-making processes in both the public and the private sector, and elevates the punishment of naming and shaming to new prominence. Its conceptual heritage is social management, a governance strategy born in the political apparatus of the PRC that does not operate with the traditional notion of law. The SCS' potentially heavy impact, as well as its conceptual heritage in social management, begs the question of what difference it makes to the rule of law in the PRC. A legal framework for the SCS does not (yet) exist. It has been held that the SCS is a powerful tool to strengthen the rule of law. However, this thesis aims to bring to light challenges that arise from the SCS for the rule of law. It does so by considering the SCS' conceptual cradle, and further mapping what has surfaced of the SCS to date in policy and legislative documents, the commercial credit market, and local pilot projects. Drawing on this comprehensive picture of the SCS, elements which appear at odds with rule of law will be pointed out. They include a lack of legal definitions for SCS key terms such as 'trustworthiness', opaque procedures and possible penalties that bypass the law. Ways to integrate them into the Rechtsstaat will be considered, all of which necessitate a re-definition of what law is. Finally, the angle of social management offers a meta-social credit system as a solution to conciliate the SCS and rule of

M. von Blomberg (✉)
Rinteln, Deutschland

© Springer Fachmedien Wiesbaden GmbH, ein Teil von Springer Nature 2020   111
O. Everling (Hrsg.), *Social Credit Rating*,
https://doi.org/10.1007/978-3-658-29653-7_6

law. The question remains whether the SCS can truly solve all problems that it brings about merely by means of its own conceptual heritage, social management theory – or whether an independent organ outside of itself is indispensable.

"Sincerity, the Way of Heaven. Thinking about sincerity is the way of man."—孟子—

6.1   Introduction

As the PRC remains shaken by fraud, bribery, and weak implementation of laws, as well as court judgments, Xi Jinping's administration has taken the fight against what it summarizes as 'trust-breaking behaviour' to new heights. Making use of the new possibilities that digitalization has opened up, it pushes forward the concept of 'social management' (社会管理) to build a moderately wealthy, harmonious socialist society. Because sharing information about a person or an organization's trustworthiness is now possible effortlessly and in real time, the consequences of (un)trustworthy behaviour can follow legal and natural persons further into other areas of their lives than ever before. Based on this, the State Council published its Planning Outline for the Construction of a Social Credit System in 2014. It shares a vision of how social management can help the PRC's government to achieve its goals. 'Social credit' (社会信用) is a record (not necessarily a score!) of trustworthiness of an individual, a private organization, or a public organ. These so-called credit information subjects (信用信息主体) shall enjoy benefits in all areas of life if behaving in a 'trustworthy' manner and face restraints and punishments when engaging in 'untrustworthy' conduct. 'The trustworthy shall roam everywhere under heaven, while those who breach trust shall not be able to move a single step,' as Xi Jinping puts it (NDRC, PBOC 2017).

An ambitious governance project, the Social Credit System (社会信用体系, hereafter SCS) is developed in several realms simultaneously. This paper will focus on the SCS development in the legal realm.

Despite its major impact on every actor in society, no national legislation for the SCS exists to date. Merely the people's congresses of some pilot provinces have passed relevant local regulations. The Party and the government have declared the establishment of such the first of four basic principles for the acceleration of the SCS (State Council and CPC Central Committee 2016). However, a legislative framework would not necessarily change the fact that the very structure and function of the SCS as it is planned call into question the currently existing legal system as such. This paper suggests that by its conceptual nature, the SCS fundamentally challenges the rule of law.

The SCS does so firstly by introducing new forms of punishment parallel to what is laid down by law. In setting up institutional cooperation among government departments, Party organs, and companies, those who break trust in one department's field of responsibility are denied services in another organ's or company's sphere of influence. It is the result of such partnerships, for example, that in 2018, citizens have been blocked from buying plane tickets 17.5 million times (McDonald 2019). Further, naming and shaming as a mechanism of punishment is explicitly called for and implemented through various channels.

Secondly, as the groundwork for the former, government bodies and agencies as well as companies are held to extensively collect information relevant to a (legal) person's credibility. This encompasses the set-up of credit evaluation mechanisms to aggregate such information, which also entails the creation of rules and standards defining trustworthiness. As one's credit record is, according to the plan, to become a central criterion for decisions on recruitment, grants of financial support for a project, and others, such rules will inevitably gain substantial importance.

This underlines the descriptions of the SCS as "evolving practice of control" (Creemers 2018, p. 1) which "ought to be understood as a system of governance with its own organizing principles and characteristics" (Backer 2019, p. 209). Therefore, we asked: What difference does the SCS make to the rule of law – and what difference does the rule of law make to the SCS?

To explore this question, the SCS will be evaluated against the backdrop of the rule of law as it exists in the PRC. Randall Peerenboom's application of a thin rule of law framework to the PRC will be used as the theoretical basis, in which the idea of rule of law as 'legal systems in which the law imposes meaningful limits on state actors' (Peerenboom 2003: fn. 81) serves as the starting point.

As to a conceptualization of the SCS, this paper will turn to its conceptual cradle, namely social management theory. Through this lens, the SCS appears as one flexible element in the whole of the government apparatus and its legislative framework may appear as a dynamic operational plan, rather than a static law providing safeguards for subjects.

6.2   Theory

6.2.1   Thin Rule of Law

Several attempts have been made to describe the rule of law as it exists in the PRC. Donald Clarke reflects on the frequently applied, but only seldomly explicitly stated framework of PRC law analysis, which he terms the 'Ideal Western Legal Order' (Clarke 2003, p. 95). This model describes the end state of a legal system, in particular 'what Western legal orders would look like if their perceived imperfections were eliminated' (Clarke 2003, p. 96). Rule of law therein is closely related to democratic institutions, such as a parliament as the legislative body. When applied to the PRC, this model cannot but detect several flaws, which are largely due to the lack of such institutions in the PRC. It is blind to historical, cultural and political particularities of the PRC system and thus fails to accurately describe its crucial elements.

Clarke further points to a possible and somewhat diametrical opposite alternative, drawing on Thomas Stephens' (1992) work. He suggests that in the pursuit of not imposing a Western ideal on the PRC, we must radically put aside any assumption of the existence of legal institutions in the PRC. If we do not assume, for example, that a judge ought to decide independently, we can observe the nature of influences on courts in more detail. This 'disciplinary model' suggests replacing the notion of order as a liberal democratic one by the kind of order that exists, for example, in a family, a nursery or the army. Such a framework, however, ignores that the majority of Chinese voices do regard their institutions as equivalents to those in the West, just as the PRC Constitution and other legal documents do (Chen 2016, p. 7).

A way out of this trap of theories is introduced by Peerenboom: Thin theories of rule of law avoid negating country-specific particularities and can at the same time accept that rule of law in the PRC is possible. In Peerenboom's words, thin rule of law is the necessary requirement and foundation for thick theories; it is a concept of which thick theories are different conceptions of (Peerenboom 2003, p. 5). Thin rule of law stresses those features 'that any legal system allegedly must possess to function effectively as a system of laws' (Peerenboom 2002, p. 3). He adds that any 'legal systems in which the law imposes meaningful limits on state actors merit the label rule of law' (Peerenboom 2003: fn. 81). Thin rule of law comprises only positive law (law posited by humans), not natural law (Peerenboom 2002, p. 127). Such law must be general, public, prospective, clear, consistent, capable of being followed, stable, and enforced (following Ion Fuller 1964). A thin rule of law can exist regardless of the ideological, political and economic systems of a state, which frees the term of many values such as those a liberal democratic order is associated with. This notion of rule of law is useful in the analysis of the SCS as it allows for more variations and possible outcomes.

Peerenboom argues that even though many refuse to accept that the government ever intended to introduce rule of law, the discourse among PRC officials in preparation of its introduction clearly shows that what is strived for cannot be mistaken for rule by law. This can be held even though from its very beginnings with the judicial reforms in 1978, it was clear that the relationship between the judiciary, the public security organs and the procuratorates shall be one of 'dialectical unity', functional through mutual restraint and cooperation at the same time (People's Daily 1978). In its fundamental idea, this has not changed – courts safeguard not only the law, but also government policies (Chen 2011), hence law in the PRC should not be regarded 'as an autonomous sphere, but one intimately connected with politics and governance' (Creemers 2018, p. 5). However, a body of law is continuously expanded, amended, and more vigorously enforced than ever before (Chen 2016).

As administrative law describes legal relations between the state and the societal actors in it, it appears as a key area when observing the rule of law as it stands. After the development of administrative law began to set off in the 1980s and since 1990, when the first Administrative Procedure Law went into effect, citizens could sue public administrative organs (not, however, party organs). Since then, significant steps forward have been taken with the introduction of the principles of due process (Law of Administrative Licensing 2003, Law of Administrative Coercion 2011) and the enactment and 2019 revision of the Regulations on Openness of Government Information (2007) which gives rise to administrative litigation.

Legal principles central to the rule of law further include nullum crimen, nulla poena sine lege (as first so phrased by Feuerbach, 1801): only if it is stipulated by law may behaviour be regarded as criminal and the state may only punish in strict accordance with a law stipulating the respective punishment and linking it to a certain crime. The PRC's Criminal Law stipulates it most clearly in Article 3. The less severe forms of punishment by the state, administrative penalties, are, since 1996, listed and laid down in the respective law as well. PRC law further adds another, positive dimension to this basic principle which it calls the 'relative nulla poena'. It stipulates that 'where acts are explicitly defined as criminal acts in law, the offenders shall be convicted and punished in accordance with law' (Li 2010, p. 657).

Lastly, a significant principle which serves as a fruitful dimension of rule of law to observe the SCS against is ne bis in idem, the ban of double jeopardy. Even though not explicitly laid down in Chinese Law, it has been found to play a vital role in legal practice and research in China (Zhang et al. 2002).

Carl Minzner (2011, 2018) points out that in recent years, the development of rule of law has fallen prey to a revival of authoritarianism. He argues that the government calls to avoid litigation and turn to mediation and adopts an overall rhetoric emphasizing that adjudication ought to represent the 'mass line'. These two changes however are not novel to the PRC's legal system, but rather date back to the Mao era and might as well 'be interpreted as a moderate policy adjustment in response to changing social circumstances' (Chen 2011, p. 15). The legal system as it stands remains.

6.2.2   Social Management: Systems Engineering for Governance

The SCS may be placed in its broader conceptual cradle of social management, a concept born in the PRC's political apparatus. Since the 1950s, scientific concepts from systems engineering are being brought to the sphere of governance, with pioneering work done by Yu Guangyuan, Qian Xuesen and Song Jian (Hoffman 2017a; Creemers 2018). The notion of 'social management' derived from neoliberal ideas of public management (as the better and newer form of public administration) and in the party-state context has been aerated with the PRC's focus on social stability and security (Pieke 2012, p. 155). Governance was 'increasingly seen as management rather than politics involving both state and non-state institutions' (Pieke 2012, p. 155). The result was the understanding of social management as a comprehensive net of mechanisms outside of the public administration which bridge the gap between the people and the CPC and thus act to help the government resolve upcoming stability problems (Hu 2005). In this understanding, non-state forces are no longer a potential threat, but an asset to party rule.

Social management took a substantial step forward when the theory of cybernetics and complex systems management emerged. Complex systems are, amongst others, defined by the fact that they cannot be described by observing their component parts. Instead, the multiple variables that determine the overall dynamic of the system by interacting with each other need to be observed. It is the investigation of these variables' interaction that explains the behaviour of a complex system. Despite a strong sensitivity to starting conditions, complex systems also display a long-term stability as they steadily reflect their environment (Svyantek and Brown 2000; Gallagher and Appenzeller 1999).

These concepts elevated social management to another level, both in terms of its importance to the PRC leadership – The 12th Five-Year Plan (2011) dedicates a whole chapter to the discussion of the necessity to innovate social management as a major target of the government – and in terms of what it can achieve. Social management in the age of digitalization in the PRC is described as 'complex systems management process through which the Party leadership interacts with both the entire Party and society' (Hoffman 2017a, p. 47).

Hoffman's work demonstrates that a useful framework to describe social management is complex systems management theory. It should be noted at this point that the author does not agree with and does not rely on what she wrote about the SCS in particular (Hoffman 2018).

An autonomic system is defined by its capability to operate in a constantly changing environment and perform operations necessary to preserve itself, thereby substituting any need for a system administrator manually performing these tasks. It combines several elements, all of which 'must be self-managing and must function interdependently for the system to achieve equilibrium' (Hoffman 2017a, p. 59). Autonomic computing is thus the application of 'technology to manage technology' (IBM 2005, p. 4), installing control loops which collect and forward information to the next cycle. More precisely, autonomic computing's characteristics can be summarized as self-configuring, self-healing, self-optimizing, and self-protecting. Self-configuring is the process of readjusting to a changing environment. Self-healing describes the capability of the system to detect and repair faults that appear as well as predict threats. Self-optimization refers to a constant search for possibilities to enhance the system's capabilities and make it more efficient. Finally, self-protection is the capability to make out potential dangers in the environment of the system. All four processes support each other and are strongly interdependent (Hoffman 2017a; IBM 2005). They each operate by engaging in the following four operations: monitoring their own operations, analysing the environment, planning and deciding, and finally, executing the changes in accordance with the knowledge won in the previous steps (as laid down by IBM 2005). The system is thus designed to constantly learn and adapt to changing circumstances, thereby keeping itself fit.

Hoffman finds that these four characteristics directly correlate with the CPC's social management objectives as stipulated in the 12th Five Year Plan: a social management system, combining source governance, dynamic management, and emergency response (2017a) and therefore speaks of the PRC's 'autonomic nervous system' (p. 12). She situates the SCS in the system's objective of self-optimization. It is part of the responsibility mechanism built to pre-emptively shape and direct behaviour (Hoffman 2017a, p. 106).

6.3   Methodology

Other than the theoretical framework, to date, there exists no unified national SCS. The SCS today is a combination of firstly, the central plan designed at the national level, secondly, a range of public and commercial pilots (leading Kostka (2018) to speak of SCSs and Dai (2018) of the SCS Project instead of SCS) and finally, the specific local regulations of some pilot provinces. The three elements do not necessarily relate to each other. For example, the Shanghai Sincerity App as a public pilot project seems alien to the municipality's social credit regulation.

Due to its fragmented and dynamic nature, attempts to create comprehensive accounts of the SCS(s) are sparse. This paper hence first sets out to draw together the different strands about the SCS. Even though not every piece of information gathered will reappear in the analysis, because of the lack of comprehensive accounts of the SCS and the many faulty descriptions of it in the media, it is crucial to lay down what this paper refers to when it discusses the SCS in its conclusion.

To provide an idea of where the term and the concept came from, the fourth chapter will contextualize the SCS in its conceptual and historical heritage and identify and explain three main roots: the Dang'an system, the financial credit mechanisms imported to the PRC from the USA, and finally, the discourse around public security informatization.

In the fifth chapter, the PRC's plan for the SCS is investigated. The aims of the system will be summarized and the novel mechanisms it introduces to reach these aims, including mechanisms of data collection, sharing and usage, mechanisms of joint punishment and reward, as well as the mechanisms of creating trustworthiness standards, are laid down. For this purpose, available policy documents related to the SCS and the respective pilot regulations of three SCS pilot provinces are analysed.

Besides the government's planning at the central level, numerous localities and companies have developed pilot systems which are operating in their respective realms. This paper will investigate one example of each, a public pilot system (Suining County's "the Masses Credit") and a commercial pilot system (Alibaba's "Sesame Credit") regarding experiences they offer for a deeper understanding of the SCS's practical dimension.

The final chapter will draw together these findings in the context of the SCS as a measure of social management and evaluate its impact on thin rule of law. A significant short-coming of this research design is that it lacks precision due to the fact that there are numerous open questions about the very concept of the SCS, which is reflected in the gaps between what can be observed on the ground and what is stipulated by the (local) government(s). This paper is thus a rather flawed pioneer, limited to pointing out what appears most striking in the field of tension between the SCS and the rule of law.

6.4   Conceptual and Historical Roots of the SCS

The SCS came into being as a measure of social management as explained above. More precisely, it serves the end of "dynamic management" (in the language of automatic computing: self-optimization) by constantly optimizing the overall system's operations. It helps guarantee continued Party control by on the one hand strengthening individual responsibility for state security, and on the other hand enforcing the former through a system of joint punishments and rewards that reaches far beyond the public administration's boundaries (Hoffman 2017a).

An early instance reflecting CPC-designed social management is the Dang'an (档案, files/records) system, which began operating in 1953. A file about every worker was created, only higher cadres and farmers were excluded. The work unit (单位) administered the file and a copy was kept at the local police station. From an elementary school teacher's assessment, children, hospital visits, and reports by employers, several pieces of personal and professional information were stored in the file. It could not be accessed by the individual it concerned (Yang 2011). In the 1970s, this first instance of social management had to make way for the market economy.

The beginnings of credit in China merely focused on the concept of financial credit known from Fico and Schufa. Lin Jinyue, referred to as a pioneer of the "Theory of the Social Credit System," (Lin 2012; Ye 2015; Meissner 2017, p. 6) introduced concepts of credit practices in the USA to China. In the late 1990s, at the Institute of World Economics and Politics of the Chinese Academy of Sciences, his group developed the basic concept of evaluating credit at a central level. Pilot projects began from 2000 and it was through their interpretation of credit evaluations that the notion was further expanded to include its social dimension.

In 2007, the Interministerial Conference on Social Credit System Construction (社会信用体系建设部际联席会议) was set up by the government, comprised of 46 members from the CPC, the Ministry of Finance, the State Administration for Industry and Commerce, and the Ministry of Public Security. The members that where added later, such as the Central Discipline Committee, the Central Leading Group for Spiritual Civilization Construction, the Central Propaganda Department, the Supreme People's Court and the Supreme People's Procuratorate, display how the SCS's focus shifted from financial credit to a more encompassing idea of social credit (Creemers 2018, p. 11). Another driving factor was the overall strategic shift of CPC governance towards morals to counter the moral vacuum which the CPC had diagnosed. In the course of this shift, the concept of credit as a measure of social trustworthiness appeared just in time (Pieke 2012, p. 155; Creemers 2018, p. 12). The main coordinator of the Conference today is the Central Leading Small Group for Comprehensively Deepening Reforms, headed by Xi Jinping himself (Yuandiancredit 2016).

Another ideological forerunner of social management is the discourse about public security informatization that started in 2008, pushed particularly by Zhou Yongkang, an official who ironically was later sentenced to life in prison because of corruption (Hoffman and Peter 2016). He coined the term 'social management system', which would measure not only indicators of criminal activity, but also happiness levels and encourage compliance, and presented it as a solution to any kind of social stability problem (Zhou 2006, 2011). He suggested innovative methods of surveillance by using new technology. Zhou's ideas experienced a backlash when he was dismissed from the Party leadership. However, while the discussion of his concepts was criticized (Hoffman and Peter 2016), under Xi Jinping, initiatives to combat attacks on public safety stepped up. Social management and the idea of the SCS with it soon moved closer to the state security system than it was the case under Hu Jintao (Hoffman 2017b, p. 6). The 2017 Cybersecurity Law, the 2015 Internet Plus initiative, and the 2006 Golden Shield project (Schwarck 2018) are all closely intertwined with the SCS.

6.5 The Central Plan and the State of the Art[2]

Four target fields for the SCS are stipulated. The first, judicial credibility (司法公信), entails strengthening the (to date rather poor) enforcement of court decisions and 'safeguard[ing] judicial authority' (State Council & CPC Central Committee 2016). The second, commercial sincerity (商务诚信), is mainly directed at fighting fraud, which is the second most common type of crime in the PRC overall (National Bureau of Statistics 2017). Further, this aim includes avoiding financial risk, which is prevalent in the absence of a free flow of information (Sapio 2017c); and enforcing regulations directed at companies (Jingji Ribao 2017; Meissner 2017, p. 3).

Societal sincerity (社会诚信) is another aim and needs to be considered against the background of a uniquely grave trust problem in PRC civil society. The frequently shared cases of traffic accidents, where for minutes, every passer-by refuses to help, afraid of being sued, illustrate the issue (Ahmed 2017b). Deception and scams, such as fake QR-codes on bikes, flourish (Hawkins 2017). The SCS is designed to strengthen civil society and services in areas such as labour, welfare, public health and education. The overarching goal is creating 'an upward, charitable, sincere and mutually helpful social atmosphere' (State Council and CPC Central Committee 2016).

Finally, sincerity in government affairs (政务诚信) is to be strengthened. State agencies and administrative entities, including SCS administrators, shall themselves become credit subjects. To this end, the SCS may be a tool to combat corruption and reinstall confidence in public institutions, while at the same time observing changes in the public opinion (Shteyngart 2016).

All four aims are regarded as keys to further economic growth in the PRC (Wang 2017; Meissner 2017; Zhang Zheng in Sapio 2017d), which complements Hoffman's perspective of the SCS as crucial to ensuring state security.

6.5.1 Groundwork

6.5.1.1 Data Aggregation

The analysed local pilot regulations stipulate that public credit information be collected at newly set up public credit information management bodies (Hubei Provincial People's Congress, 2017 (hereafter HU): Art. 9 ff.; Hebei Provincial People's Congress, 2017 (hereafter HE): Art. 10 ff.; Shanghai Municipal People's Congress, 2017 (hereafter SH): Art. 9 ff). All information subjects are encouraged to actively provide their social credit information and are held responsible for the correctness and completeness of their information. Legal persons are further called to share credit information they have accumulated about other (legal) persons during their operations.

The type of information collected is restricted to categories stipulated in the public credit information catalogues. They shall include 'information that reflects the credit subject's basic circumstances in public administration and services' (SH: Art. 11 I; HE: Art. 11; HU: Art. 9 II), such as administrative punishments, payments received, volunteer work, and refusals to perform on legal documents.

Some limits on information collection are listed. A legal person may only collect credit information about private individuals if the latter consents. "Personal data" (for example, Art. 6 VI SPC on JDL) may not be collected; however, this very notion is nowhere defined, neither in the context of the SCS, where it is so frequently used, nor elsewhere in national laws (Chen and Cheung 2017, p. 2). In Shanghai, collection is also limited by 'the principles of truthfulness, objectivity, and necessity' (Art. 6). It is prohibited to collect information on religious faith, genetics, fingerprints, blood type, illnesses, and medical history.

Information collection under the SCS however is most likely not limited to what the pilot regulations stipulate. Plans for a wide usage of big data in connection with the SCS have been announced in the context of the 13th Five Year Plan in 2016. According to the definition provided in the policy paper, this would include administrative data, such as medical information and school records, transactional data, sensor data, tracking data (such as GPS information from mobile phones), and data on online behaviour such as searches and social media comments (Cheng 2014, p. 3).

6.5.1.2 Data Sharing

In pilot regulations and policy documents, credit information data sharing among public and commercial entities is a key point to building the SCS. Most credit information is not aggregated at the central level but scattered across local administrations and other businesses that directly interact with citizens. However, the whole project is heavily relying on a functioning flow of information. Information inside the public administration shall be shared horizontally and vertically 'to make it so that social credit information is interconnected and intercommunicating, and used across departments, fields, and regions' (HU: Art. 6 I). Further, state organs at all levels are called to cooperate with enterprises for credit information exchange.

A first step was the assignment of a unified social credit code to every credit subject, the ID number in case of citizens and a new number for legal persons (Ministry of Civil Affairs 2017; State Council 2015). All credit information about one legal entity is exchanged under this number across administrative levels and regions.

A second significant step toward the end of sharing credit information are the governmental Public Credit Information Sharing Platforms, the 'data backbone of the social credit system' (Meissner 2017, p. 6). The National Credit Information Sharing Platform (全国信用信息共享平台), for example, gathers credit information from more than 30 central ministries and government agencies, overall around 400 datasets. The central policy documents explicitly call upon all government institutions to forward information to the national platform (for example in State Council and CPC Central Committee 2016, 3 III).

6.5.2   Penalizing

6.5.2.1   Joint Punishment

As to the application of credit information, policy documents and pilot regulations provide for the establishment of 'joint social credit rewards and punishments that cross departments, fields, and regions, and is jointly participated in by administrative organs, judicial organs, and market entities (SH: Art. 21). The facilitating mechanism behind this is a 'name list system for persons subject to enforcement for trust-breaking' (State Council and CPC Central Committee 2016, part 3). Such blacklists for trust breakers and redlists for the particularly trustworthy have been set up in different contexts and on several levels. Where foreign media have falsely explained that a 'credit score' is calculated, at least in the current layout of the SCS, there is merely the comparably simple concept of blacklists and redlists. The central level blacklist that has brought about most penalties so far is the Supreme People's Court's List of Judgement Defaulters (hereafter LJD), introduced in 2013. It includes individuals who have been ordered by a court to a certain conduct (payment of compensation, for example), are able to comply with it but do not do so and waived or exhausted all chances for legal recourse (SPC Provisions on the LJD 2016). To be removed from the LJD, one must either fulfil the court's order or file a valid complaint. In any case, the name will be removed from the LJD after two years.

The blacklists provide the ground for a new type of state punishment enabled through partnerships between the private sector and state organs. 'Joint punishments' are the result of such partnerships. To date, 44 departments and 60 companies participate in the trial for this mechanism. 30 'joint incentives and joint disciplinary mechanisms' have been set up (Zhang 2018). For those enlisted in the LJD for example, this means that they cannot purchase train and plane tickets (ID is required in the PRC for such purchases), book a room in hotels above the standard class, or send their children to private school. Due to a memorandum signed in 2016 between Alibaba and the NDRC, the blacklisted cannot purchase luxury goods on Tmall or Taobao (Jing 2016).

In February 2017, the SPC announced that since the introduction of the blacklist system, 6.72 million people had faced LJD-related penalties, an increase of six million in two years (Jing 2016; Jing and Caoyin 2015). The NDRC is working on a wider range of punishments (Yang 2017).

More than individuals, companies have faced blacklist-based punishments. The requirements for companies to be entered on the most relevant blacklists are lower. The decision of an administrative department on whether a company violated a rule suffices. Prior notification upon which the company could possibly defend itself is not part of the process. Punishments include higher taxes, denial of licenses, lower chances to gain public contracts, participation in publicly-funded projects, mandatory government approval for investments in sectors where market access is not usually regulated, and several joint disciplinary measures. Meissner (2017) found that these punishments already lead companies to regulate themselves. She and Sapio (2017) point out that old rules now become self-enforcing and government interference invisible. 'If implemented as planned, the system has the potential to become the most globally sophisticated and fine-tuned model for IT-backed and big data-enabled market regulation' (Meissner 2017, p. 11).

6.5.2.2 Naming and Shaming

Public shaming has long been a measure to exert additional pressure on suspects and, more importantly, to make an example out of a person as to warn the public from violating rules (Whitman 1998; Global Times 2018; Lim 2018). Nearly every official document on the SCS stipulates such measures to publicly shame the untrustworthy. It is rarely discussed as a formal form of punishment. However, in connection with the blacklist-based punishment mechanism, naming and shaming gains striking importance.

The Credit Information Sharing Platforms display 75 percent of the credit information they share, with only the remaining 25 percent designated for limited sharing (Meissner 2017, p. 7). Most importantly, the blacklists are made publicly accessible online as well.

Courts are explicitly encouraged to use the media and widely forward information on who got blacklisted (SPC Provisions on the LJD, Art. 7) and a website listing both blacklisted individuals and companies was set up. Information to be entered besides the name, ID number, and in some cases, photos and home addresses, include the delinquency, the responsible people's court, the court's order that is to be fulfilled, and 'the situation of the person' (Art. 6 III). Additionally, the court may enter every information that it deems useful, only excluding information relating to 'national secrets, commercial secrets or personal privacy', the latter of which remains undefined (Art. 6 VI). Furthermore, in the case of public servants and politicians, the names of the blacklisted shall be forwarded directly to their respective work units (Art. 8), leading to a direct impact on careers and work life even if their superiors do not proactively search them on the list.

The official call for measures of naming and shaming has borne fruit in several local projects as well. Sanmen County in Zhejiang and Dengfeng County in Henan, for example, started experimenting with a partnership between courts and telecommunication companies in June 2017. The latter installed a dial tone on the phones of the blacklisted informing the caller that the person they are calling has not complied with court orders, and further encouraging the caller to convince the defaulter to cooperate (Ohlberg et al. 2017, p. 12). As a result, not only business deals but also marriages have been cancelled because people have found their partners to be blacklisted.

Reactions by the public display a strong reluctance to accept reputational damaging as a legitimate punishment. When in Huilai County, Guangdong, police sprayed "drug crime related household" on homes with a family member who was found guilty of drug crimes, a public outcry for privacy ensued. Only a few days later, the police returned to remove the writing on the wall (Global Times 2018).

In the economic sphere, naming and shaming is more prevalent and less criticized. Companies face a new, non-market yardstick they will be measured by, as any failure to comply with environmental protection policies, for example, could be widely published immediately. Wang (2017) speaks of a 'transformation from the mechanism of enterprise competition to the mechanism of competition on credit'.

6.5.3   Rule Making

6.5.3.1   Using Credit Information as Bases When Evaluating Credit Subjects

Policy documents and pilot legislation has also called for companies, the public administration, as well as industry associations, to widely make use of the credit information available. They are to integrate the credit information they have access to in their daily operations and decision making. Administrative organs at all levels are called upon to make inquiries into credit information in the course of personnel decisions, granting permits, the transfer of land use rights, administrative punishments, the acceptance of residency cards, financial project support, and so on, shall be influenced by public credit information (SH: Art. 28; HU: Art. 24; HE: Art. 35). Market entities are also called upon to make use of available social credit information when negotiating terms of trade, in production, the provision of services and others, and market entities, favouring high scorers over others (SH: Art. 26; HE: Art. 38).

6.5.3.2 Formulating Trustworthiness Standards

Media reports have wrongly described the SCS as already calculating trustworthiness scores for citizens according to their behaviour. Such a central calculation does not happen, however, public organs, market entities, and industry associations are called to develop credit evaluation mechanisms and to experiment with grading systems in their respective realms. For the target area 'government affairs' for example, the Planning Outline encourages the creation of 'credit dossiers' for civil servants which, among others, shall list 'violations of discipline' and 'sincerity and cleanliness in government affairs' (part 2, section 1). Targeting the civil society, evaluation systems that are to be set up mainly concern organizational structures in health care, labour and others. Industry associations shall work out a plan for carrying out credit grading and classification for their respective areas and develop possible rewards and punishments (HU: Art. 14; SH: Art. 27; HE: Art. 33).

Any development of such grading systems will entail the formulation of standards according to which trustworthiness will be evaluated. This adds a crucial dimension to the SCS, expanding its role from a mere law enforcement tool to a source of norms and rules itself.

6.6   The Pilots

The term pilot suggests centrally designated pilot SCSs in different localities. Although such pilots exist, with Rongcheng being the most prominent example, a number of schemes of various structures have been tested even before the central government released the 2014 Planning Outline. Further, several grassroot initiatives such as the Banks of Virtues (道德银行) were, even though the initiators know nothing about the SCS, designated pilots in upper-level policy documents. Also, apart from local public systems, corporations experiment with commercial credit rating systems, such as Wechat Pay Points and the below described Sesame Credit.

6.6.1   A Public Pilot: Suining's "The Masses Credit"

One of the very first experiments with social credit was conducted in Suining (睢宁) county in Jiangsu. Starting in 2010, a score was calculated at the local administration according to trial rules laid down by the county government. Credit subjects started with the highest possible score (1000 points, category A), and could reach the lowest category (D, less than 599 points) when not living up to the laid down rules. For example, a maximum of 150 points was given on the grounds of whether one was indebted or not (with banks or other private persons). The range of relevant actions was broad, including for instance not living up to family virtues and jaywalking. The system subjected local party cadres as well as companies and included provisions that made a promotion of politicians with a score lower than A or B impossible. Rewards ranged from receiving loans from banks with favourable conditions, preferable treatment in public services, to receiving scholarships. Penalties included restrictions on lending money, receiving social welfare, and the renewal of business qualifications and licenses (Suining County Government 2010; Xindong Bao 2014).

Major point deductions were disclosed to the public, as were the actions taken against those with low scores (Yan 2010). Others could retrieve their own score at the local public administration. With every change in one's credit score, a text message would be sent out to the respective credit subject explaining the reason. If credit subjects found a mistake, they could seek correction with the local administration or the court, though the provisions provide little detail on this process. By 2014, 85 percent of Suining's residents had a social credit score.

The system led to a wave of criticism from state media and individuals. They found that the collected data was misleading, that the rating was more directed towards citizens and not seriously implemented towards cadres and government entities, and that the local government had no qualifications to rank according to their self-made categories hidden behind catch-all phrases such as 'public order'. For example, citizens who filed complaints were given lower scores (Boxun 2014; Economist 2016). Also, it was criticized that those promoting investment effectively received points. Furthermore, and despite being guaranteed by law, citizens had difficulties accessing their own scores and seeking corrections (Boxun 2014; Economist 2016).

Even though the criticism resulted in an overall turning-away of the idea of coercively ranking citizens for the time being, the very concept of the system itself was not challenged. Scrutiny was rather directed at its poor implementation. Later experiments covering all of a locality's citizens and commercial entities did not engage in score calculation but merely recorded incidents of law-breaking or non-compliance with court orders. In the absence of such incidents, later pilot systems assume trustworthiness.

6.6.2   A Commercial Pilot: Sesame Credit

Developing a credit service industry is a crucial part of SCS construction and a central theme in related documents. Commercial actors are not only regarded as credit subjects themselves, they are also critical in developing technology, aggregating credit information and act as agents for enforcement through partnerships with courts as discussed above. Local regulations call upon companies to innovate by participating in international cooperation projects, making use of technological innovations and establishing experimental zones. Creemers (2018) points to the long-standing tradition of such blurring of public and non-public actors in China. He holds that 'the "social" dimension of the SCS also entails that members of society create the incentives for each other to act in the desired manner, without direct intervention of State actors' (p. 8).

On January 5th 2015, the PBOC granted official licenses to eight private companies to conduct social credit pilot tests. Among them was Ant Financial (subsidiary of Alibaba), who rolled out its pilot called Sesame Credit (芝麻信用) only weeks later. The smartphone application is integrated in Alipay and calculates a score from 300 to 950 for every user. Categories of information include credit history, fulfilment capacity, and personal characteristics, which are applied by traditional credit rating services in other countries as well. A novel category Sesame Credit applies is interpersonal relationships. Being friends with a low scoring peer on Sesame Credit will lower one's own score. A fifth category is called 'behaviour and preferences'. It is, for example, comprised of a person's shopping habits (Hatton 2015). Taking cheap offers on Alibaba's shopping platforms will lower a person's score, while buying sport and kitchen equipment or handicrafts will raise it. People buying diapers are more trustworthy than those playing computer games, Li Yingyun explained (in Lee 2017).

With the category on behaviour and preferences, Sesame Credit is conducting a rating of trustworthiness that clearly exceeds financial credibility. The exact procedures, algorithms and categories used by the rating remain undisclosed, termed as 'trade secrets' (Ahmed 2017a; Yang 2015). 30–40 percent of the data used to calculate the score comes from Alibaba-owned companies. The remaining amount is contributed by any other service connected to the Alipay app and from government bureaus (Ahmed 2017a). As Alibaba has partnered up with the dating platform Baihe and Didi Chuxing, the country's most-used ride-hailing service, the amount of data at its disposal is vast.

Sesame Credit offers several benefits to high scorers, including booking a rental car or a hotel room without paying a deposit, receiving loans at a lower interest rate, having online purchases delivered for trial without having to pay, and skipping the queue when applying for visas to Luxembourg, Japan and Singapore (Ahmed 2017a). The credit service industry is to date nearly entirely in the hands of Alibaba's Sesame Credit. In July 2017, the PBOC decided not to grant licenses to any of the companies for a continuation of the experiment (Hornby 2017), saying commercial actors 'cannot be trusted' (Ohlberg et al. 2017, p. 12). The failure was due to insufficient privacy protection, but also to the fact that all eight competed with each other, presenting 'a potential impediment to the centralized vision of social credit that would require these firms to share their proprietary data with one another' (Ohlberg et al. 2017, p. 12).

In 2017, a silent conflict between commercial credit entities and the state over their definitions of trustworthiness ensued. Reports pointed out that, very different from the government, commercial credit rating businesses interpreted creditworthiness as loyalty to their services (Ohlberg et al. 2017). Still however, the state relies on the companies' credit information data and scoring technology.

For now, this problem seems to be solved. On January 4th 2018, the National Internet Finance Association (中国互联网金融协会), a PBOC initiative, 'in cooperation with' the former eight credit information service pilots founded Baihang Credit Scoring (百行征信) (Guo 2018). The pilots each hold eight percent of the shares. The National Internet Finance Association is the largest shareholder with 36 percent. With a seed capital of one billion Yuan, Baihang will draw on credit information it receives from the eight former pilot companies, all of which are big-data-driven, to create a profile for every credit subject. It is yet unclear whether there will be a unified score (China Daily 2018; Xinhua 2018). The close connection of this association to the government and the fact that eight big data driven, credit information service companies are its shareholders both indicate that Baihang Credit Scoring is to be the company under which a unified credit information service under the SCS will evolve (Caijing Toutiao 2018). In PRC media, questions about how coercive this step is are vigorously cast aside. Alibaba and others may have to share their data, but in return 'receive a good credit environment and better conditions for operating' (Caijing Toutiao 2018). In late 2019 it was reported that Tencent and Alibaba refuse to share their customers' loan data with Baihang (Yang and Liu 2019), casting doubts on the future success of the effort.

6.7   Analysis

6.7.1   The SCS in the Law

It might be possible to integrate the SCS into the legal system. In the following, the elements of the SCS that appear at odds with the rule of law will be discussed.

6.7.1.1 Social Credit Norms and the Law

When considering an integration of the SCS as described above into the existing legal framework, a first impediment for a national legislation appears to be the lack of a clear definition of the key concept of the SCS: trust. Not keeping trust (失信), keeping trust (守信), trustworthiness and credit (信用), social credit (社会信用), and sincerity (诚信) all are key terms in every legislative and policy document, depicting the overall goal that all SCSs aim to reach. None of them are unambiguously defined.

All pilot regulations formally describe trustworthiness in the SCS sense as law-abiding. Provisions on the blacklists also define trust-breaking entities as those who broke the law, or indirectly did so by refusing to comply with court orders. This definition is what several spectators argue for. Luo (2016), observing the Suining experiment, brings up the example of a credit punishment for citizens who do not visit their parents regularly. This may only be implemented because a respective clause exists in the PRC's Law of the Protection of Rights and Interests of the Elderly. In this scenario, the SCS does not necessarily touch upon rule of law.

It remains nebulous however why the new term "trust" is even needed if it can be translated to "law" without losing any of its meaning. If trustworthiness was to be defined as law obedience, the SCS was merely an innovative law enforcement system.

It was shown that in the name of the SCS, rules are to be created to evaluate trustworthiness in different areas. While some of these rules impacting social credit have the quality of law – such as the requirements to be entered on the pilot region's public credit blacklists – some do not. In several instances, rules relevant to social credit evolve which are not law in the formal sense as stipulated in the Legislation Law. For example, local public credit bodies are to draft specifications according to which public credit information is to be evaluated. Such 'specifications' will possibly be the de facto-rules according to which punishments and rewards will be granted, although they don't classify as law. That the SCS shall reach beyond what is stipulated by law is also indicated by the educational measures on trustworthiness introduced in the relevant documents: Shanghai and Hebei's education plans, for example, will cover 'social morals, professional ethics, family virtues and individual morals' (SH: Art. 47; HE: Art. 9). Further, in the name of trustworthiness, companies are steered according to their compliance with the policy directives shown above. All these findings suggest that the SCS makes and enforces norms that are not law. This reflects the finding that in the PRC, several policies have a de-facto power of law (Chen 2011).

Chen (2011) argues that a clear distinction between what is law and what not is under way. Following his path, a possible way out of this collusion of trustworthiness definition is to clearly define areas where trustworthiness rules emerge (for example inside of the social credit evaluation systems that are to be developed in the public health sector). Then, procedures could be stipulated which will elevate the new rules or criteria to the status of formal law. The definition of trustworthiness as law-abiding, which is so frequently proclaimed, could be kept – and the SCS could, at least in this regard, smoothly be integrated in the legal system.

However, the question regarding the utility of the "trust" terminology would remain. The Fourth Plenum, and particularly Xi Jinping himself, has repeatedly insisted that the rule of law and the rule of virtue need to go hand in hand, suggesting that law shall not stand alone but be supplemented by extra-legal and non-defined guidelines. Viewed through the lens of social management, this lack of clarity might be intentional. It offers space for the leadership to fill with norms that situationally appear necessary, keeping the system from growing stiff. For the social management system to operate automatically, it needs to engage in self-learning. The lengthier the process to formulate trustworthiness rules, the smaller the probability that such rules will effectively serve the ends of ensuring the survival of the system. Static trustworthiness rules halt the control loop mechanism which constantly tests and reacts to new rules. Hence, when viewing the lack of definitions from the perspective of social management, it appears incompatible with thin rule of law, if keeping the above mentioned definition of law as prospective, clear, and stable.

6.7.1.2 Social Credit Penalties and the Law

The potential power of the SCS to punish untrustworthy behaviour together with the vagueness about what can be regarded as untrustworthy behaviour cast doubts on whether the SCS can be realized in adherence with ne bis in idem (double jeopardy) and nullum crimen, nulla poena sine lege (no crime without a law stipulating it as such, no punishment without a law laying it down).

It remains questionable whether ne bis in idem, no second punishment for a conduct an entity already has been convicted for, can be upheld as long as blacklisting occurs on the grounds of criminal conduct that is simultaneously already dealt with through the regular legal criminal procedure.

Punishments that are part of the SCS are mostly laid down in Memoranda of Understanding[1] and largely do not exceed what has been stipulated as administrative penalty before social credit came to life. However, measures against blacklisted entities are not laid down the way nullum crimen requires. While stipulated in the respective memoranda of understanding, they are not necessarily connected to a specific conduct but rather apply to the entire group of enlisted entities regardless of the behaviour which had them be entered on the list in the first place. Nullum crimen sine lege however requires any punishment to be directly bound to a certain legal norm describing the conduct.

More importantly, the penalty of naming and shaming is systematically employed, even though it causes significant harm to not only the perpetrator, but surrounding friends, colleagues and family members as well. It cannot be brought into compliance with nullum crimen sine lege because the deterrence effect it unfolds through punishing for instance the perpetrator's children lies at the core of naming and shaming. As reputational damaging cannot be contained, it is in any case at odds with thin rule of law.

---

[1] Examples for such include the State Administration for Industry and Commerce (September 15th 2015): Joint Memorandum of Cooperation on coordinated oversight and joint disciplinary action against unreliable Enterprises; China Securities Regulatory Commission (December 24th 2015): Memorandum of Understanding (MoU) on the Implementation of Joint Disciplinary Actions for Entities Responsible for Unlawful and Untrustworthy Listed Companies; similarly, policies on joint incentives have been passed, such as the Customs General Administration (October 19th 2016): Memorandum on cooperation for the implementation of Joint Incentives for Enterprises with High-grade Customs Certification.

As a measure of social management, punishment is not an act conducted in a separate space, concerning the state and the subject of punishment only. Being one element of the automatic system of the state, punishment must be influencing and itself be influenced by the other elements. In this light, the deterrent power of a punishment heavily defines its usefulness for the system. In other words, the very purpose of punishment is less compensation of guilt or the re-installation of justice (as absolute theories of punishment hold), but the prevention of future crimes (relative theories of punishment). Against this background, a SCS legislative framework from a social management perspective would not control punishment through linking it to a defined set of offences. Rather, punishment is checked through integrating the system threatening external influences, such as an instance of public anger against a particular act of punishment. Not justice, but the equilibrium (critical to autonomic systems) would determine punishment. The legislative framework in this context would consist of operational clauses stipulating the punishment system's setup, unfolding applicability only to this very system rather than to societal actors and their lawyers.

Again, the social management perspective reveals that thin rule of law with a Fullerian conceptualization of law as described above is incompatible with the concept of the SCS.

6.7.1.3 Transparency

If the trustworthiness definition will in fact reach beyond norms laid down by law, a transparency problem arises. To meaningfully restrict state actors by law, it must be available to the subjects wishing to invoke it (in Fuller's words: public and clear). The Dang'an 'haunting' (Yang 2011) its respective subjects, blurry score calculation methods throughout the Suining pilot project, and the poor procedures for complaints and corrections have demonstrated that the SCS has the potential to impede the rule of law through obscuring norms and procedures.

All three pilot regulations stipulate a right to know one's own credit information. It can be enquired through a platform managed by the public credit information service centres. The procedure does not mention courts, which would probably only come into play if an administrative lawsuit was filed against the centres. Also, no accepted cases brought under the Open Government Information Regulations have involved disclosure of social credit-related information yet.

6.7.2 The Law in the SCS

Several of the observed strands of the SCS point to a future where trust-related acts will not directly be linked to a penalty. They might rather be drawn together to create a more or less complex trustworthiness record, according to which credit subjects will be awarded or punished. This is the concept that was tested in Suining and is now, on a voluntary basis and with rewards instead of punishments, tested with Sesame Credit and the Sincere Shanghai App. It is the alternative to the existing, binary blacklist model which only knows the categories trustworthy and untrustworthy. It resonates with both, foreign media reports painting the Orwellian dystopia of an all-powerful social credit score, but also social management. Precisely because it appears to be in line with the latter, even though technologically and institutionally far from reality, the following paragraphs will consider implications for rule of law for this controversial scenario.

Legislation for a score-based SCS would entail quantifying a catalogue of behaviour, which means attaching certain amounts of points to certain behaviours. This extra step between behaviour and punishment would effectively decouple the act from the punishment that the criminal code or administrative penalty law provides. This carries a problem with the blacklist mechanism one step further: the blacklist mechanism already enacts the same group of punishments on everyone on the list regardless of their respective trust-breaking action. Under umbrella terms such as "judgment defaulters", someone blacklisted for not visiting his or her parents regularly faces the same penalties as the owner of a factory who did not pay his employees because both of them lost the respective lawsuit. This categorization by lists would become superfluous if a quantified score was introduced. The score could take the place of lists and further generalize them. It would further disconnect the action of a credit subject and the punishment that awaits it.

This would not change even if the calculation methods including how which act translates into which amount of points were clearly laid down in law. The laid down calculation methods are likely to become overly complex and barely comprehensible for non-specialists as the system grows more sophisticated. Transparency might be seriously impeded, posing a major obstacle to an effective rule of law. As a solution within the universe of social management, a meta-social credit system which allows for a transparent rating of those who rate can be considered (Backer 2017a, in line with Zhu's thoughts in Schmitz 2017). Anyone and any mechanism involved in evaluating social credit, constructing measurement criteria or others would be subjected to such a 'social credit system of social credit systems' (Backer 2017a, p. 15). In social management terminology, it closes the control loop and thereby guarantees continuous self-monitoring and the other three procedures connected to it. Thereby, potential threats, such as a large number of demands for an explanation of score calculation, would be forwarded to the meta-system which, in its constant effort to reply to environmental challenges, would perhaps pressure the lower level to solve the problem by for example drafting explanations or even adjusting the score calculation itself. Similarly, threats in the shape of administrative abuse would be found and reported to the next cycle, which would activate a self-configuring process to adapt and keep harm as low as possible. A thin understanding of rule of law would accept such a mechanism as an effective way to exert control over the rulers, even though it is subject to debate whether its means may be called law.

This would respond to the fourth aim put forward in the Planning Outline 2014, which is to strengthen sincerity in government affairs (政务诚信) by subjecting state agencies including SCS administrators to the SCS in order to combat corruption. The pilot regulations already stipulate a 'administrative veil piercing' civil servants may be held personally responsible if they disobey rules and the principle of reasonable administration when handling social credit information.

It is questionable whether the detected problems the SCS poses to thin rule of law can entirely be solved from within the thinking of social management itself, fully substituting traditional legal procedures. Several observers doubt this (Backer 2017a, Chen and Cheung 2017).

Liberal democracies define the quality of their legal systems through the independence of the organs. The notion of the courts as guardians of the law outside of the system is held so dear that legal systems elsewhere are regularly judged based on this criterium. Only through the independence of the judiciary can the abuse of power be prevented. From this perspective, the faultiness of an SCS is assumed. Therefore, solutions from within the project itself are delusive. The premise for an effective safeguard for credit subjects must be the likeliness of errors in the system, the procedures behind it and by system designers themselves, regardless of system immanent solutions.

6.7.3   Reconsidering the Concept of Law

The SCS is compatible with thin rule of law when ignoring the law's traditional character. As the previous chapters have shown, the greatest difference the SCS makes to the thin rule of law is its challenging of the influential notion of law Fuller (1964) put forward (particularly it being public, prospective, clear, and stable). Integrating the SCS in such a legal system 'means to unify and integrate systems of monitoring, of transparency and of compliance within the traditional law-administrative regulation construct of state systems, [and] appears to be one of the most innovative and interesting efforts of this decade' (Backer 2017a, p. 2). It suggests that if the notion of law was opened up, the SCS in its most radical scenario could become a part of the Rechtsstaat as found in the PRC. Any case filing, court procedure and decision would have to directly inform the rule makers, and so on.

6.8   Conclusion and Outlook

This work set out to explore a possible integration of the SCS into the rule of law and found that this is only possible after a thorough reconsideration of the very concept of law. Taking the thin rule of law as a starting point, we found it to not only be in imbroglio with the mother of the SCS, namely social management, but also with the government's design of the SCS as well as with the first instances of the SCS on the ground in China. Problems arise especially from the lack of a clear concept of trust. Understanding trust breaking as law breaking, as some government sources suggest, begs the question why the term is necessary in the first place. There are indicators for the creation of a set of rules beyond law, punishments might be meted out based on mere norms. Additionally, a change in the nature of punishments is beginning to show particularly through the application of naming and shaming in the context of the SCS.

In an experiment of thought, we considered a meta social credit system as a possible solution to the problems the SCS was found to be posing to the rule of law. Such a meta system to supervise the supervisors would itself be rooted in social management and presses an observer viewing it from the angle of liberal democracies to comment: A mechanism of punishment (and reward) such as the SCS can only be effectively controlled by an independent outside organ with a different underlying logic.

This paper has shown that the SCS questions the concept of law, while at the same time stating to enforce it. Further research might consider taking one step back from the SCS and investigating the compatibility of social management itself with the rule of law.

Such work might start with the finding that social management is a holistic approach (Hoffman 2017b, p. 4). it aims to solve problems by taking into consideration the overall situation only, instead of looking at the details of one problem. A rule of law system on the other hand traditionally solves problems by granting rights to subjects and stipulating legal procedures, on the grounds of which issues can be solved via court judgements, case by case. It appears to be a reductionist approach and does not resonate with the nature of complex systems (Hoffman 2017a). Complex systems are not manually managed but are instead self-learning and thus able to handle changing dynamics and threats, such as identifying unsatisfied credit subjects internally. However, neither the problems brought before a court nor the results are systematically re-integrated in the SCS. They are mainly designed to unfold effects outside of the SCS, in the realm of the individual or entity which filed the case. At most, it signals a warning to the public, and sets a possible precedent for future court decisions (referring to precedents is, although widely informally practised, not mandatory for judges in the PRC's legal system). Since neither the problem nor the solution is directly reported back to the system, the self-learning process is impeded.

References

Ahmed, Shazeda (2017a) Cashless Society, Cached Data Security Considerations for a Chinese Social Credit System. Citizenlab. Available from: https://citizenlab.ca/2017/01/cashless-society-cached-data-security-considerations-chinese-social-creditsystem/

Ahmed, Shazeda interviewed by Wessling, Claudia (2017b) Shazeda Ahmed about Chinas 'Societal System of Rewards' (Podcast in English). Mercator Institute for China Studies. Available from: https://www.merics.org/index.php/de/podcast/shazedaahmed-ueber-chinas-gesellschaftliches-bonitaetssystem

Backer, Larry Catá (2017a) Measurement, Assessment and Reward: The Challenges of Building Institutionalized Social Credit and Rating Systems in China and in the West, Draft for the Conference 'The Chinese Social Credit System 2017' at Shanghai Jiao Tong University, Working Papers – Coalition for Peace & Ethics. 9 (2). Available from: https://ssrn.com/abstract=3040624

Backer, Larry Catá (2017b) Closing Remarks: International Symposium on Rule of Law & Credit System. In: Foundation for Law and International Affairs: International Symposium on Rule of Law & Credit System, 23 September 2017, Shanghai Jiao Tong University. Available from: https://flia.org/closing-remarks-internationalconference-international-symposium-rule-law-soci-al-credit-systems/

Backer, Larry Catá (2019) China's Social Credit System: Data-driven governance for a new Era. Current History. September 2019, 209–214.

Boxun (2014) 江苏睢宁给市民定信用等级 [Jiangsu's Suining Town gives citizens credit grades]. Available from: https://www.peacehall.com/news/gb/china/2014/06/201406230107.shtml

Caijing Toutiao (2018) 央行下了铁命令：马云正式被'收编' [The Central Bank gives a strong command: Jack Ma officially gets incorporated]. Available from: http://cj.sina.com.cn/articles/view/5617133798/14ecea8e6001003fgx?Cre=sinapc&mod=g&loc=2&r=0&doct=0&rfunc=84&tj=none

Chen, Albert Hung-Yee (2011) An introduction to the legal system of the People's Republic of China. 4th edition, Butterworths Asia.

Chen, Albert Hung-Yee (2016) China's Long March towards Rule of Law or China's Turn against Law? The Chinese Journal of Comparative Law. 4, 1–35.

Chen, Yongxi & Cheung, Anne Sy (2017) The Transparent Self under Big Data Profiling: Privacy and Chinese Legislation on the Social Credit System. University of Hong Kong Faculty of Law Research Paper. 2017/011. Available from: https://ssrn.com/abstract=2992537

Cheng, Jackie Hoi Wai (2014) Big Data for Development in China. UNDP China Working Paper. Available from: http://www.cn.undp.org/content/dam/china/docs/Publications/UNDP%20Working%Paper_Big%20Data%20for%20Development%20in%20China_Nov%202014.pdf

China Daily (2018) China to Launch First Unified Personal Credit Platform for Online Lending. China Daily. Available from: http://www.chinadaily.com.cn/a/201801/05/WS5a4ef1aba31008cf16da54b4.html

Clarke, Donald (2003) Puzzling observations in Chinese law: When is a riddle just a mistake? In: Hsu, C. Stephen (2003, ed.): Understanding China's legal system. Essays in Honor of Jerome A. Cohen. New York University Press, 93–121.

Creemers, Rogier (2018) China's Social Credit System: An evolving practice of control. SSRN. Available from: https://papers.ssrn.com/sol3/papers.cfm?abstract_id=3175792

Dai, Xin (2018) Toward a Reputation State: The Social Credit System Project of China. SSRN. Available from https://papers.ssrn.com/sol3/papers.cfm?abstract_id=3193577.

The Economist (2016) China Invents the Digital Totalitarian State. Available from: https://www.economist.com/news/briefing/21711902-worrying-implications-itssocial-credit-project-china-invents-digital-totalitarian

Von Feuerbach, Paul Johann Anselm (1801) Lehrbuch des gemeinen in Deutschland geltenden Peinlichen Rechts. Gießen. Available from: http://www.deutschestextarchiv.de/feuerbach_recht_1801/48

Fuller, Lon L (1964) The Morality of Law. Revised edition 1969, Yale University Press.

Gallagher, Richard & Appenzeller, Tim (1999) Beyond Reductionism. Science. 02: (284)/ (5411), 69–79.

Global Times (2018) Police spray paint 'drug-dealing family' on homes in shaming campaign, then netizens shame police. Available from: http://www.globaltimes.cn/content/1101692.shtml

Guo Yuhao (2018) 央行批准在深圳设百行征信'为首家个人征信牌照' [The Central Bank approved the business license of 'Baihang Zhengxin' in Shenzhen as the first personal credit information license. Techweb. Available from: http://www.techweb.com.cn/it/2018-02-23/2639679.shtml

Hatton, Cecilia (2015) China 'social credit': Beijing sets up huge system. BBC News. Available from: http://www.bbc.com/news/world-asia-china-34592186

Hawkins, Amy (2017) Chinese Citizens Want their Government to Rank Them. Foreign Policy. Available from: http://foreignpolicy.com/2017/05/24/chinese-citizenswant-the-government-to-rank-them/

Hebei Provincial People's Congress. (2017) Hebei Social Credit Information Regulations, 河北省社会信用信息条例. Adopted 28 September 2017 and effective 1st January 2018. Available from: http://hbrb.hebnews.cn/pc/paper/c/201710/01/c25267.html

Hoffman, Samantha & Mattis Peter (2016) What Could China's 'Social Credit System' Mean for its Citizens? Foreign Policy. Available from: http://foreignpolicy.com/2016/08/15/what-could-chinas-social-credit-system-mean-for-its-citizens/

Hoffman, Samantha R. (2017a) Programming China: the Communist Party's autonomic approach to managing state security. PhD thesis, University of Nottingham. Available from: http://eprints.nottingham.ac.uk/48547/1/Hoffman%2C%20Samantha%20Student%20ID%204208393%20PHD%20THESIS %20Post%20Viva%20copy.pdf

Hoffman, Samantha R. (2017b) Programming China: The Communist Party's autonomic approach to managing state security. Mercator Institute for China Studies. Available from: https://www.merics.org/sites/default/files/2017-12/171212_China_Monitor_44_Programming_China_EN__0.pdf

Hoffman, Samantha R. (2018) Social Credit. Australian Strategic Policy Institute. Available from: https://www.aspi.org.au/report/social-credit

Hornby, Lucy (2017) China changes tack on „social credit" scheme plan. Financial Times. Available from: https://www.ft.com/content/f772a9ce-60c4-11e7-91a7502f7ee26895

Hu, Jintao (2005) 提高构建社会主义和谐社会的能力 [Raise the capacity of building a socialist harmonious society]. [Speech] Beijing,19th February. Available from: http://politics.people.com.cn/GB/1024/3497566.html

Hubei Provincial People's Congress. (2017) 湖北省社会信用管理条例 [Hubei Social Credit Management Regulations]. Adopted 30 March 2017 and effective on 1st July 2017. Available from: http://www.hubei.gov.cn/zwgk/fgwj/201705/t20170502_988638.shtml

IBM (2005) An architectural blueprint for autonomic computing. White Paper Autonomic Computing. Available from: http://www-03.ibm.com/autonomic/pdfs/AC%20Blueprint%20White%20Paper%20V7.pdf

Jing, Meng (2016) NDRC to Collaborate with Alibaba to Build Credit System. China Daily. Available from: http://www.chinadaily.com.cn/business//tech/201612/03/content_27558688.htm

Jing, Meng & Caoyin (2015) Justice comes to the virtual world. China Daily. Available from: https://www.chinadailyasia.com/business/2015-07/02/content_15284668.html

Jingji Ribao (2017) 社会信用体系的内涵和外延 [The social credit system's implications and denotations]. Available from: http://www.gov.cn/xinwen/2014-07/15/content_2717489.htm

Kostka, Genia (2018) China's Social Credit Systems and Public Opinion: Explaining High Levels of Approval. Forthcoming. Available from https://www.researchgate.net/publication/326625329_China%27s_Social_Credit_Systems_and_Public_Opinion_Explaining_High_Levels_of_Approval

Lee, Felix (2017) Die AAA-Bürger. Die Zeit. Available from: http://www.zeit.de/digital/datenschutz/2017-11/china-social-credit-systembuergerbewertung/seite-2

Li, Li (2010) Nulla Poena Sine Lege in China: Rigidity or Flexibility? Suffolk University Law Review. 43, 654–667.

Lim, Nicole (2018) Jaywalkers punished with social media shame. SixthTone. Available from: https://www.sixthtone.com/news/1002025/jaywalkers-punished-withsocial-media-shame

Lin, Junyue (2012) 社会信用体系理论的传承脉络与创新 [Development and Innovation of the Social Credit System Theory].

Luo, Peixin (2016) 社会信用体系不是道德档案 [The SCS is no virtue Dang'an]. 文汇学人. Available from: http://yuandiancredit.com/h-nd-475.html

McDonald, Joe (2019) China bans millions from travel for 'social credit' offenses. AP News. Available from: https://apnews.com/9d43f4b74260411797043ddd391c13d8.

Meissner, Mirjam (2017) China's Social Credit System. A big-data enabled approach to market regulation with broad implications for doing business in China. Mercator Institute for China Studies, available at: https://www.merics.org/sites/default/files/201709/China%20Monitor_39_SOCS_EN.pdf

Ministry   of   Civil   Affairs   (2017)
关于全面推进社会组织统一社会信用代码制度建设有关事项的通知 [Notice on matters relating to the overall advancement of the standardization of social organizations under the social credit number system]. Available from: http://www.shui5.cn/article/53/114401.html

Minzner, Carl (2011) China's Turn against Law. The American Journal of Comparative Law. 25, 935–984.

Minzner, Carl (2018) End of an Era. How China's authoritarian revival is undermining its rise. Oxford University Press.

National Bureau of Statistics of China. (2017) Number of crimes in China in 2017, by type. Statista. Available from https://www.statista.com/statistics/224776/number-ofcrimes-in-china-by-type/

NDRC & PBOC (2017) 关于印发首批社会信用体系建设示范城市名单的通知 [Notice on publishing the first list of Social Credit System construction model cities.]. Available from: http://www.gov.cn/xinwen/2018-01/09/content_5254715.htm

Ohlberg, Mareike; Lang, Bertram & Ahmed, Shazeda (2017) Central Planning, Local Experiments: The Complex Implementation of China's Social Credit System. Mercator Institute for China Studies. Available from: https://www.merics.org/sites/default/files/2017-12/171212_China_Monitor_43_Social_Credit_System_Implementation.pdf

Peerenboom, Randall (2002) China's Long March toward Rule of Law. Cambridge University Press.

Peerenboom, Randall (2003) The X-Files: Past and Present Portrayals of China's Alien 'Legal System'. Global Studies Law Review 2003, retrieved as part of the UCLA Research Paper Series, 3 (2), available from: http://ssrn.com/abstract=374040

People's Daily (1978) Special Commentator. Democracy and the legal system, p. 1–2.

Pieke, Frank N. (2012) The Communist Party and social management in China. China Information. 26 (2), 149–16.

Sapio, Flora (2017) The Meanings and Extensions of 'Social Credit System'. China Social Credit System. Available from: https://china-social-credit.com/2017/07/meaningextension-social-credit-system/.

Schmitz, Rob (2017) What's Your Public Credit Score? The Shanghai Government Can Tell You. NPR.   Available   from: http://www.npr.org/sections/parallels/2017/01/03/507983933/whats-your-public-credit-score-the-shanghaigovernment-can-tell-you

Schwarck, Edward (2018) Intelligence and Informatization: The Rise of the Ministry of Public Security in Intelligence Work in China. The China Journal. 80 (1), 1–23.

Shanghai Municipal People's Congress (2017) 上海市社会信用条例 [Shanghai Social Credit Regulations]. Adopted 23 June 2017 and effective 1st October 2017. Available from: http://www.saic.gov.cn:9080/fgs/zcfg/201707/t20170706_267361.html

Shteyngart, Gary (2016) China Invents Digital Totalitarian State. The Economist. Available from: https://www.economist.com/news/briefing/21711902-worryingimplications-its-social-credit-project-china-invents-digital-totalitarian

SPC (2016) 关于对失信被执行人实施联合惩戒的合作备忘录 [Memorandum of understanding on joint incentives and discipline for judgment defaulters]. Available from: http://ezszy.hbfy.gov.cn/DocManage/ViewDoc?docId=6b06ac35-ef07-4c5d-b8f45edd15ac05c0

State Administration for Industry and Commerce (2015) 失信企业协同监管和联合惩戒合作备忘录 [Joint Memorandum of Cooperation on coordinated oversight and joint disciplinary action against unreliable enterprises]. Available from: http://www.hbcredit.gov.cn/xyjs/wjzl/20152016/201801/t20180104_25502.shtml

State Council (2014) 社会信用体系建设规划纲要 (2014—2020年) [The planning outline for the establishment of a Social Credit System (2014–2020)]. Available from: http://www.gov.cn/zhengce/content/2014-06/27/content_8913.htm

State   Council   (2015)
关于批转发展改革委等部□法人和其他组织统一社会信用代码制度建设总体方案的通知 [Note

on endorsing the Development and Reform Commission and other Departments' overall plan for the establishment of a uniform system of credit numbers for legal persons and other organizations]. Available at: http://www.gov.cn/zhengce/content/201506/17/content_9858.htm

State Council & CPC Central Committee (2016) 关于加快推进失信被执行人信用监督警示和惩戒机制建设的意见 [Opinions on accelerating and advancing the construction of credit supervision, warning and punishment mechanisms for judgement defaulters]. Available from: http://www.gov.cn/gongbao/content/2016/content_5120693.htm

Stephens, Thomas B. (1992) Order and Discipline in China: The Shanghai Mixed Court 19111927. University of Washington Press.

Suining County Government (2010) 睢宁县大众信用管理试行办法 [Suining Countytrial management rules formass credit]. Available from: http://jssnfy.chinacourt.org/public/detail.php?id=388

Svyantek, Daniel J. and Brown, Linda L. (2000) A Complex-Systems Approach to Organization, in: Current Directions in Psychological Science, 9 (2), 69–74.

Wang, Xiaohui (2017) 诚信是最强的软实力, 专家呼吁完善诚信长效机 [Sincerity is the strongest soft power, experts call for the perfection of an effective long-term system], 华复时报口, available from: http://www.chinatimes.cc/article/68866.html

Whitman, James Q. (1998) What is wrong with inflicting shame sanctions? Yale Law Journal,107 (5), 1052–1092.

Xindong Bao (2014) 江苏睢宁信用评级运行4年民众不知闹访要扣分 [Jiangsu's Suining credit rating has been running for 4 years, and people are unaware that their complaining about it lead to point deductions]. 人民网. Available from: http://society.people.com.cn/n/2014/0702/c1008-25226558.html

Xinhua (2018) 信联'来了央行受理'百行征信'申请 [Xinlian arrived, the Central Bank handles the application of 'Baihang Zhengxin']. Available from: http://www.xinhuanet.com/fortune/201801/05/c_1122212406.htm

Yan, Weiliu (2010) 睢宁县大众信用管理试行办法'主要内容 [Suining County's 'Mass Credit Management Trial Rules' main substance]. 滕讯评论. Available from: http://view.news.qq.com/a/20100404/000008.htm

Yang, Jie (2011) The Politics of the Dang'an: Spectralization, Spatialization, and Neoliberal Governmentality in China. Anthropological Quaterly. 84 (2), 507–533.

Yang, Xiaoya (2015) 马云开始给个人信用打分, 快来看看你在朋友圈中排第几 [Jack Ma begins to give credit marks to individuals, come check how you rank among your contacts]. 澎湃新闻. Available from: http://www.thepaper.cn/newsDetail_forward_1298828.

Yang, Yuan (2017) China penalises 6.7m debtors with travel ban. Financial Times. Available from: https://www.ft.com/content/ceb2a7f0-f350-11e6-8758-6876151821a6.

Yang, Yuan; Liu, Nian (2019) Alibaba and Tencent refuse to hand loans data to Beijing. Financial Times. Available from: https://www.ft.com/content/93451b98-da12-11e9-8f9b-77216ebe1f17.

Ye, Xiangrong (2015) 中国模式社会信用体系建设的创新与挑战 [Innovation and Challenges: Building up a Chinese Social Credit System]. Innovation. 4 (9), 25–32.

Yuandiancredit (2016) 中国社会信用体系建设全景报告 [Overview Report on the Construction of China's Social Credit System]. Available from: http://yuandiancredit.com/hnd-1312-2_347.html

Zhang, Jun; Shan, Changzong; Miao, Youshui (2002) China's theory and practice on ne bis in idem. Revue Internationale de Droit Pénal. 3 (73), 865–872.

Zhang, Yong (2018) '信用中国' 兀站震槭了法院执行难等失信行为 ['Credit China' Website frightens off difficult implementation of court judgements and other trust breaking acts].中国青年网. Available from: http://news.china.com/domesticgd/10000159/20180306/32161054.html

Zhou, Yongkang (2006) 加强和改进社会管理促进社会稳定和谐 [Strengthen and improve social management, promote social stability and harmony]. 中国城市经济. 12, 4–9. Available from: http://www.china.com.cn/law/txt/2006-10/25/content_7274178.htm

Zhou, Yongkang (2011) 社会管理创新要破本制障碍 [Social management shall defeat systemic obstacles]. 中国改革论坛. Available at: http://www.chinareform.org.cn/society/manage/Speech/201112/t20111204_129129.htm

Marianne von Blomberg is doing a PhD in Chinese Law at Zhejiang University in Hangzhou, from which she also holds an LL.M Degree, and is a Research Associate at the University of Cologne's Institute for East Asian Studies. During her undergraduate studies at Zeppelin University, she discovered her passion for ancient Chinese rhetorics and wrote about the concept of originality in China in the face of architectural mimicry and counterfeit products. After having interned at the Mercator Institute for China Studies, the Future Research Department of Volkswagen AG, and the German Embassy in Ottawa, she moved to Hangzhou in 2016, where she continues to take a sociological perspective in her observations of the PRC legal system and social credit in particular.

Part of this research was funded by the Fritz Thyssen Foundation under grant 10.19.2.003RE.

Toward A Reputation State: A Comprehensive View of China's Social Credit System Project

7

Xin Dai 戴昕

Abstract

China's Social Credit System Project (the "SCSP") is one of the most misunderstood recent developments in China's law and policy. This chapter offers a comprehensive conceptual thesis that explains the SCSP as the Chinese government's multi-faceted strategy to use reputation in law and governance. The SCSP envisions that reputation mechanisms such as blacklisting, rating, and scoring be used to tackle a range of the country's intractable governance problems in its social and economic realms. While knowing no apparent equivalent elsewhere in the world, the SCSP portends the rise of the "reputation state" on a wider scale, as government authorities outside of China will also increasingly seek to use reputation mechanisms and technologies in the spheres of law and governance. And as it both raises high hopes and stokes grave fears, the SCSP has so far been shaped and limited by the institutional and market forces that animate it in the first place.

7.1 Introduction

One of the most misunderstood developments about China's law and policy in recent years has been the Social Credit System Project (the "SCSP"). Often reflecting its own anxiety over the increasingly data-driven society (Zuboff 2019), the West's perception about the SCSP has for years been dominated by sensationalized media accounts that compare it to a techno-dystopia themed episode from the British sci-fi show *Black Mirror* (Bruney

Xin Dai 戴昕 (✉)
Beijin, China
E-Mail: xindai@pku.edu.cn

© Springer Fachmedien Wiesbaden GmbH, ein Teil von Springer Nature 2020
O. Everling (Hrsg.), *Social Credit Rating*,
https://doi.org/10.1007/978-3-658-29653-7_7

2018). The influence of such mis-informed portrayal first culminated in a hostile reference by U.S. Vice President Mike Pence in his widely noted hawkish speech on China (Pence 2018). And despite recent reflections over the possibility of misunderstanding (Mastsakis 2019), the Western political narrative has so far clung firmly to perceiving the SCSP as the Chinese government's deliberate, systematic attack on human rights (evidenced most recently in the SCSP's notable reference in the U.S. "Hong Kong Human Rights and Democracy Act" of 2019).

Aside from misinformation and bias, one must also be right to attribute the lack of proper understanding of the SCSP to its externally presented complexity, consisting of almost countless moving parts and local variations. In recent years, law and policy scholars have made initial efforts to produce analytical and empirical research that aims to shed a more systemic light on the SCSP. Early contributions, informative and insightful as they are, tend to focus on one aspect of the behemoth, such as the SCSP's plan for business regulation reform (Meissner 2017), the role played by technology and tech firms (Creemers 2018), or challenges to data privacy protection (Chen and Cheung 2017). Few attempt to capture the SCSP in a comprehensive manner (Chen et al. 2018). But the SCSP is neither about the market nor politics alone; its policy approaches involve more than "big data;" and its implications for law and governance also go beyond concerns about data privacy. The critical research challenge, therefore, lies precisely in canvassing the project's vast contours.

This chapter offers a comprehensive conceptual framework for understanding the SCSP, which is given a shorthand as the "*reputation state.*" It demonstrates that in designing and implementing the SCSP, the Chinese government intends to strategically use reputation mechanisms, such as blacklisting, rating, and scoring, to tackle many of the country's intractable governance problems in the social and economic realms, ranging from fraudulent behaviors in the marketplace, to difficulties in enforcing court judgments, to corruption in the government, and to professional malpractices and even plagiarism in scholarship.

Reputation as a force of behavioral regulation has in recent decades been extensively discussed in the socio-economic literature. Much ink has already been spilled on the rise of the "reputation society," referring to the phenomenon that tech-empowered, systemized reputation mechanisms, such as credit scores and customer reviews, have become widely adopted in many market sectors and significantly affected the economic and social welfare of individuals and firms (Masum and Tovey 2011). Nonetheless, the emergence and the rise of the "reputation state" remain inadequately explored. This chapter clarifies that using reputation for law and governance is not a uniquely Chinese practice. Before all lights shone down on China, the strategic use of reputation mechanisms by government had already been explored in the West. What the SCSP offers, meanwhile, is a highly illustrative case for studying the past, present and future of this ongoing trend.

This chapter in Section 7.2 draws on literature on reputation, law, and governance to explain the conceptualization of the reputation state. Section 7.3 unpacks the SCSP's major policy initiatives and examines the same as government strategies to use reputation in law and governance. Section 7.4 explains the critical institutional and market forces that have motivated and shaped the SCSP. Section 7.5 identifies the SCSP's three structural implications that deserve future deliberation. Section 7.6 briefly concludes.

7.2   Reputation for Regulation

7.2.1   Conceptualizing Reputation

Reputation is a perennial human phenomenon. As commonly understood, reputation entails a set of descriptive and/or evaluative information about an actor, including but not limited to his or her past conduct, which in one way or another tells us about such actor's characteristics and helps us predict its future act or performance (Dellarocas 2011, p. 4; Goldman 2011, pp. 51–52).

In the pre-digital age, reputation-related information was collected and disseminated through mechanisms of offline interpersonal networks, such as word-of-mouth, gossip, shared memories and community norms (Macaulay 1963; Ellickson 1991). Today such information can be generated with data that are collected with sophisticated surveillance and behavioral tracking technologies and processed through computer algorithms. Sometimes reputation information takes the form of direct feedback, such as through grapevine gossip and eBay consumer reviews (Cascio 2011, pp. 185–186). Other times it is indirectly generated through a mediating process that further aggregates and analyzes the relevant data, as illustrated by FICO scores for financial credibility and Slashdot's karma points for content quality.

Yet reputation information about one actor matters primarily because it is used by other actors to make decisions. For example, patrons check Yelp! and similar local business review sites so as to decide whether to dine at a restaurant or shop in a store; gig workers read Turkopticon reviews on job requesters to decide whether a job request posted on Amazon Mechanical Turks is worthwhile (Silberman and Irani 2016). These private decisions made upon reputation information affect the welfare of both the decision-makers themselves and the reputation subjects. Reputation subjects face consequences that take such varied forms as ostracism or esteem from communities, and increased or reduced economic and social transactional opportunities with potential counterparties. The prospect of being subjected to reputation-based consequences creates behavioral incentives. For example, when Uber adopts the policy to suspend a driver for falling below a certain rating, that offers a good reason for other Uber drivers to provide satisfactory services (Mohammed 2013).

Existing popular and academic literature on reputation tend to focus more on the information element of reputation. A useful conceptualization of reputation, however, must highlight factors and dynamics related to both aspects of reputation-related information and reputation-based decision-making. Many reputation systems today are the product of a deliberate system design that seeks to establish a specific connection between particular types of reputation information and decision-making activities. For one example, Alibaba's online shopping platforms, Taobao and Tmall, operate rating systems that specifically tie eligibility for platform-sponsored promotion opportunities and transaction restrictions for vendors to their consumer satisfaction scores (Liu and Weingast 2018).

7.2.2 The Conventional Reputation State: Regulating Reputation

What is the state's role where the reputation society is on the rise? In law and policy scholarship, one common perspective is to view reputation systems as primarily a spontaneous force of self-regulation. For example, Eric Goldman (2011, p. 53) theorizes that, by reducing information asymmetry in the marketplace, reputation systems improve the efficient functioning of the Smithian "invisible hand". Lior Strahilevitz (2006) further posits that, through mobilizing crowdsourced information and informal social sanctions, reputation mechanisms induce socially efficient behaviors in less costly ways than the state's formal regulatory tools.

The intuitive implication of these optimistic accounts, accordingly, is to allow reputation to flourish in the market; government and law, as an exogenous imposition, are only relevant when the reputation market fails. Specifically, legal scholars have considered potential failures of the reputation market in the following senses. First, private reputation systems could cause information harm to individuals, including data privacy violations and discriminatory treatments (Citron and Pasquale 2014, p. 14). Second, the market for reputation may fail as it may become plagued by such strategic behavior problems as fake reviews, astroturfing, and malicious inputs. Third, as today's digital reputation systems are sometimes monopolized by companies who master black box algorithms, these monopolists could manipulate opaque reputation systems to exploit citizens for self-profit (Pasquale 2015). Fourth, given that socially desirable reputation systems can sometimes function as a public good, the market alone may fail to produce them in optimal numbers (Strahilevitz 2006, pp. 1717–1719).

The state's role is thus envisioned primarily as an external regulator that intervenes to address market failures. What's expected for the government in a reputation society is to set up regulatory regimes to regulate the potential externalities arising from the generation, dissemination, and use of consumer credit history and scores. It may also, through legislation, private litigation and regulatory enforcement actions, tackle problems such as fraudulent reviews and contractual suppression of reviews.

7.2.3 The Strategic Reputation State: Reputation as Regulation

Less intuitive is that government actors may also strategically interject their own information and power resources to more assertively use reputation in regulatory contexts. At least three types of government strategies are plausible.

7.2.3.1 The "Searchlight" Strategy

The term "searchlight" strategy was coined by Strahilevitz (Strahilevitz 2011b, p. 158). Government actors may channel public sector information in different formats into the reputation market for private actors' use in their decision-making activities. Such information allows the reputation market to more effectively sort reputation subjects and thus create stronger incentives. The government thus expects some of its regulatory goals to be achieved without necessarily invoking otherwise costlier interventions.

One well-known U.S. example for such a searchlight strategy concerns regimes for the public disclosure of criminal history. Under many states' Megan's Laws, for example, authorities impose mandatory registration requirements for sex offenders and publicize their personal information on dedicated websites, with which communities, employers, and potential romantic suits could make more informed decisions in managing their safety risks. The rise of the Open Government Data ("OGD") regimes suggests that the searchlight strategy may be pursued in an even more wholesale format. For example, U.S. government authorities under OGD regimes typically set up websites, APIs, and data repository platforms to allow access to a large quantity of government data in both raw and processed forms (Altman et al. 2015, p. 2059).

7.2.3.2 The "Incorporation" Strategy

Government actors may also actively acquire and incorporate reputation information generated in the market to inform public decision-making, in particular with respect to the allocation of its abundant possession of "carrots" and "sticks." Such a government strategy is referred to here as "incorporation."

In the U.S. context, it may be difficult to imagine that authorities check on an individual's FICO scores or social media popularity ratings in order to decide how large a fine to impose on such individual for a violation of, say, a parking ordinance. Nonetheless, after 9/11, U.S. national security authorities have leveraged a range of informal collaborative arrangements with private data controllers and data brokers, through which they procure private-sector data in both raw and processed forms and aggregate the same in "fusion centers" established across regions (Michaels 2008, pp. 908–919). In the civilian context, the U.S. Securities Exchange Commission has rules that require financial firms to use credit ratings issued by a "nationally recognized statistical rating organization" to fulfill a range of regulatory requirements (US SEC 1975). Besides, government actors have also sought outside feedback and reviews from citizens in the process of overseeing public officials. Police authorities, for example, have long been known to use a system of civilian complaint reviews to collect citizen reports of police abuse.

7.2.3.3 The "Institutionalization" Strategy

In pursuing either the searchlight or the incorporation strategy, the state essentially transforms reputation into a public-private joint venture. But government actors may also set up reputation profiles and/or generate reputation scores on individual and corporate citizens, which are then used by the same and other government actors in making decisions on these subjects in future administrative and regulatory encounters with the same. In so doing the state effectively introduces the operational logic of reputation systems into the formal institutional process of law and regulation. Such a strategy may thus be called "institutionalization."

There are quite a number of scenarios where authorities in the U.S. have adopted reputation tracking, interagency data sharing and/or data mining practices. For example, in government procurement contexts, some agencies track and score vendors on their past performance for use in future bidding and contracting processes (Picci 2010). The surge of big data technologies has further expanded data analytics efforts within the government systems. Tax, health, security, and education authorities have frequently sought to mine digitized personal data held in networked public databases (Zarsky 2013, p. 1511). Actual or conceptual blacklists have resulted from these data mining efforts, which the authorities use to make decisions about whether to subject relevant individuals to working, voting, or flying restrictions, immigration detention and deportations, and counter-terrorism surveillance (Hu 2015, pp. 1742–1744). Moreover, in connection with the recent development in "smart cities," U.S. local governments have increasingly deployed predictive algorithms in performing many types of public functions, including but not limited to decisions on the pretrial release of a defendant, child custody, and police patrol allocations (Brauneis and Goodman 2018, pp. 114–115). While government use of predictive algorithms is not identical to its use of institutionalized reputation, they are closely related.

7.2.4   A Conceptual Framework for the Reputation State

Table 7.1 presents the ideal types that capture the previously described government strategies.

The reputation state consists of government practices that in various ways use these reputation-based strategies in law and governance contexts. Existing legal and public policy scholarship on reputation has thus far focused primarily on regulation and searchlight strategies. In practice, meanwhile, government actors have already made attempts at taking advantage of the more assertive incorporation and institutionalization strategies. Each policy initiative or program in the real world does not have to fall exclusively into one but not any other quadrant in the above matrix. Instead, a reputation-inspired government initiative could well combine the use of more than one type of strategy. The prospect of combination foreshadows even more novel and potentially controversial government practices.

Table 7.1 Four government strategies using reputation

	Private sector information	Public sector information
Private actors' decision-making	Regulation	Searchlight
Public authorities' decision-making	Incorporation	Institutionalization

7.3 The SCSP Unpacked

What does the Chinese government exactly envision for the SCSP? For those who start with the project's caption for a clue, the English term of "social credit" is far from perfect a translation for the Chinese phrase "*shehui xinyong*." While the word "credit" in English does not have such a broad connotation, its Chinese counterpart is associated with a host of lofty moral virtues such as trustworthiness, promise-keeping, norm abiding, integrity and general courtesy. Meanwhile, the modifier "social" underscores, within the Chinese language, that "credit" here concerns not only financial and commercial transactions but all other contexts of societal interactions as well.

Such a broad and vague moral framing appears, however, the only visible overarching theme that unifies the various cohorts of policy initiatives included in the SCSP. Otherwise, to many observers the SCSP inevitably appears to be a hodgepodge of reform initiatives. In 2014, China's highest executive authority, the State Council, issued a top-level policy document, the State Council Blueprint for Building the Social Credit System (国务院《社会信用体系建设规划纲要》(2014–2020年)) (hereinafter the "2014 Blueprint") that offered a lengthy description of a blueprint for the SCSP. According to the 2014 Blueprint, the project packages, under the caption of "government honesty," almost every type of policy initiative plausibly pertinent to improving public administration. It proposes to control, as issues of "business honesty," a wide array of questionable or illegal market behaviors, ranging from selling unsafe consumer products to antitrust violations. Initiatives for improving the overall public image and credibility of the judicial system are included in the SCSP as its "judicial credibility" component. Whatever else is not grouped into the foregoing categories, such as regulating medical professionals with ethical rules and restraining academics from plagiarism, among others, are lumped together under the caption of "societal honesty." Besides, in characteristically Chinese fashion, the SCSP also calls for dedicating resources to education and propaganda programs to promote a general culture of trust and honesty.

This chapter proposes a unifying theme that underlies all these seemingly unrelated policy efforts: The SCSP, as unpacked below, consists of the Chinese government's multifaceted efforts at utilizing the effects, resources, and mechanisms of reputation to upgrade its law and governance apparatus.

7.3.1 Building Reputation Infrastructure

The Chinese government often emphasizes that the country's reputation society is generally underdeveloped. The SCSP thus includes first a set of initiatives for upgrading China's relevant market, technical and institutional infrastructures.

7.3.1.1 Making a Reputation Market

The origin of the SCSP lies in a much more narrowly conceived government project to create modern credit reporting and rating practices for bank lending. The momentum for expanding the credit information service market started in the late 2000s thanks to the boom in China's big data and Internet-based finance. After promulgating in 2013 a long overdue nationwide regulation, the *Regulation on the Administration of Credit Investigation Industry* ("**RACII**"), the People's Bank of China, the country's central bank, greenlighted in 2015 eight firms, including affiliates of such dominant Internet giants as Alibaba and Tencent, to "prepare for operating" consumer credit systems on a pilot basis. The PBOC initially planned to issue formal licenses to these pilot firms after a six-month period. In an unexpected turn, however, no license ended up being issued to any of the pilot firms, as the PBOC found that none had met its expectations on fairness, institutional independence and data privacy (Wan 2017). Such development reflected a high-level shift of the Chinese government's policy stance in 2016 and 2017 towards more cautiously managing risks from Internet-based credit practices. By early 2018, the authorities had determined to run the consumer credit reporting and rating market in a more tightly controlled model: The first formal license for consumer credit business, issued by the PBOC in February 2018, was to a joint venture named "Baihang Credit Scoring," newly formed with the government-backed National Internet Finance Association of China as the largest shareholder with 36 percent interest in equity, and the eight pilot firms each as a shareholder with 8 percent (Xinhua News 2018).

Cautious with licensing consumer credit reporting agencies, the Chinese government has meanwhile sought to cultivate private firms that specialize in servicing the SCSP's demands for reputation-related technologies and mechanism designs. In October 2018, the National Development and Reform Commission (the "**NDRC**"), the leading central agency for the SCSP alongside PBOC, designated 27 pilot firms to run "comprehensive credit services" (Credit China 2018a). These firms have in recent years been very active in undertaking hardware and software procurement deals with local governments in connection with SCSP programs.

7.3.1.2 Information Infrastructure

The government acknowledges that the underdevelopment of China's reputation market is attributable to a serious shortage in credit-related information. This perceived shortage stems from issues on both sides of demand and supply.

On the demand side, banks and other financial institutions are looking to expand their reach to borrowers who until recently have little to none credit history, including rural residents, young urban residents and small and micro business firms (Shen 2013). On the supply side, government efforts have continued to be viewed as lacking in making public sector data accessible. Starting from 2012, more than a dozen localities, including Shanghai, Beijing, Foshan, and Wuhan, have set up OGD portals, and the plan for a nationwide

OGD platform has also been on the central government's policy agenda since 2015. The more sophisticated among them, constructed with cloud computing technologies, offer tools and interfaces for analytics. Close observers have pointed out, however, that existing programs are still limited in their scopes and features (Zheng and Xiong 2017, p. 57). In particular, data collected in these programs often have not been processed into machine-readable formats, are not updated in a timely manner and frequently, or are subject to restrictive terms on secondary use.

Through the SCSP, the Chinese government aims to further upgrade its efforts to exploit the value in public sector information. The novel construction of "public credit information," which includes broadly any reputation-relevant data collected or generated by public agencies during the course of government business, has paved the way for a new wave of government ICT investments in dedicated national, local, and sectoral "credit information" systems that all consist of standardized data portals and underlying databases. As of January 2020, the nationwide system, operated by the National Public Credit Information Center, has integrated 36 sectoral systems, 32 provincial systems, and 39 municipal systems (Credit China 2020). Such an infrastructure certainly supports the government's use of the searchlight strategy. It also enables data sharing among government actors for internal coordination in institutionalization programs.

7.3.1.3 Backbone Data Regulation

Establishing an effective regulatory framework to cover the imminent and potential data risks is also a major agenda the government has inserted into the SCSP. The previously mentioned RACII sets up basic data protection norms for credit agencies servicing the financial market. An increasing number of provincial and lower level authorities have also enacted regulatory regimes for their local SCSP programs that, among others, set up backbone data privacy norms that have long been absent in China. For example, Shanghai's Social Credit Regulation, the first comprehensive provincial regulation, incorporates such progressive data privacy norms as, among others, a "right to be forgotten" for reputation subjects from the government social credit data system, a right to dispute an erroneous record and to restore credit status, and a reasonableness limit on administrative agencies' queries over citizens' social credit information.

Indeed, these rules alone are far from enough to tackle the looming data privacy and security challenges from the SCSP (Chen and Cheung 2017). News reports have also repeatedly revealed that many government data platforms and systems are highly vulnerable to hacker attacks and security breaches (e.g. Netease News 2015). That said, many of China's government data risks predate, as opposed to newly arise from, the SCSP; what the SCSP has brought about is thus not only incremental risk but also opportunities toward more stringent regulation. Besides, along with the SCSP's implementation, cultural norms and social attitudes in China have also evolved towards favoring more robust privacy regulations, triggering visible reactions from market players toward improving their data practices.

7.3.2   Equipping the State with Reputation Tools

While infrastructure building remains a work in progress, the Chinese government has proceeded with a series of regulatory initiatives that are premised on the strategic use of reputation mechanisms.

7.3.2.1 Reputation for Everything

The SCSP unequivocally supports establishing reputation systems in virtually all sectors of economic transactions and social interactions. The 2014 Blueprint overtly instructs, for example, all industry organizations and professional service associations to establish credit profiles and rating systems on member firms and individuals as they exercise their industry self-regulatory functions.

Encouraged by the government's supportive stance, tech-based credit systems had been eager to extend their applications to scenarios beyond finance. Before failing to acquire a formal license in 2017, Sesame Credit, for example, had once embraced an extended concept of "social credit," expanding the use of its credit scores to scenarios ranging from travel privileges to online dating. Qianhai Credit, another firm then in the PBOC's pilot program, incorporated into its consumer credit model one's record on the malicious use of bike sharing services, which was lauded by the Shanghai government as an exemplary social credit initiative (Eastern Daily 2016).

Overall, however, the government did little to specifically nudge, support or even compel autonomous reputation systems to actually get set up and run their course in the market. Instead, the SCSP embellishes the importance of propaganda and social meaning creation, and it calls for government bodies to carry out promotional programs to create the salience of "honesty and trustworthiness" as a universal norm. Such government propaganda has apparently succeeded in creating a focal point that market players visibly utilized in promoting their own products to consumers. For example, before it took note of the associated negative international publicity (Ming 2017), Sesame Credit had deliberately invested in a public image that its credit information products serve the SCSP's public interest goals of restoring societal trust and creditworthiness (Netease News 2016).

7.3.2.2 Searchlight with Blacklists

The SCSP has directed government actors to produce specific types of reputation information. Such government-generated reputation, in the spirit of the searchlight strategy, is both generally publicized and specifically disseminated to financial institutions, credit agencies, and other intended users. The most common formats of government generated reputation information are blacklists for subjects of negative reputation and "redlists" for subjects of good reputation. Pioneering such strategies is the Supreme People's Court, which operates a nationwide blacklist program for "trust-breaking enforcement subjects," referring to individuals and firms that refuse to observe effective court decisions. The blacklisted defaulters, having their information publicized through such program, are reported to have faced restrictions widely from commercial banks, business partners, and consumer credit agencies. They also suffer public shaming as localities display such blacklists on large LED screens in public places or on screens at theatres before movies start (Credit China 2018b). Some local courts even partnered with telecom carriers to unilaterally change the blacklisters' cell phone ringtones so that whenever such individuals receive calls their phones ring to announce their blacklisted status (Tencent News 2017).

Inspired by the courts' blacklist program, administrative agencies have also compiled and publicized blacklists on matters ranging from "irregular" corporate operating status, serious tax violations, material breach of government procurement contracts and rules, to general records of administrative sanctions. Building on these offense-based blacklists and other government records, some local authorities have sought to generate subject-based credit reports for private sector users. For example, Wuxi City has since 2011 produced "baseline credit evaluation reports" on local business firms, which are provided to many commercial lenders.

One issue often raised is whether the naming and shaming in such format as the blacklists constitutes duplicative and excessive punishment. Some legal scholars argue that, as a matter of legal nomenclature, "publication" is distinguishable from "punishment" (Luo 2016, p. 111). Such formalist interpretation hardly resolves the underlying fairness conundrum. On the one hand, certain punitive effects from publicizing records of past offenses seems inevitable, and in fact should even be reasonably expected by anyone who ever contemplates breaking the law. On the other hand, publicizing government blacklists surely creates the risk that the information will be taken out of its original context and abused. The fairness question thus commands contextual consideration. Many probably have good reason to protest, for example, if China's State Tourism Bureau issues a rule that directs the publication of a blacklist of tourists engaging in "uncivilized conduct," such as littering and graffiti at sites of attractions. By contrast, given China's notorious problem with enforcing court decisions, more could agree that some overt public shaming is not a bad idea if it takes that to enhance the judiciary's stature.

7.3.2.3 Incorporating Outside Reputation

The SCSP emphasizes that government actors consult reputation information, such as blacklists, redlists, and credit scores, which is created by community organizations, financial institutions, credit bureaus, rating agencies, industrial organizations and chambers of commerce. One early example was reported in Hangzhou where, in January 2017, the municipal business regulatory and supervisory authorities entered into an agreement with Sesame Credit, through which the authorities, in carrying out compliance inspections, are given access to Sesame's information on "small and micro-enterprises" (People.cn 2017). In a more blanket fashion, Shanghai encourages public agencies to "query credit information and use credit products," including those produced by market entities, when making decisions relating to regulatory approval, inspection, sanctions, government procurement, bidding, project funding, land transfer, residential management, civil servant hiring and promotion, recognition with awards and other administrative matters.

In pursuing the incorporation strategy, government actors face an increased risk from the uncertain quality and reliability of such information and data, which potentially harms, instead of improves, government decisions. Some national and local regulations in China have set out basic procedural mechanisms for addressing issues relating to data errors. Provincial regulations often require that private sources of reputation information give broad warranties on such information's legality and integrity.

7.3.2.4 Government Reinforcing Norms

An exceptionally conceptually novel use of the incorporation strategy can be observed in certain eye-catching local SCSP programs, which involves the government's attempt at reinforcing, with government incentives, reputation mechanisms already embedded in local norm communities. For example, through its "Honest Farmer" Project, the Qingzhen City government in Guizhou Province, one of the least developed provinces located in Southwestern China, has since 2010 aimed to incentivize rural residents to comply with "the good moral value of promise keeping." Local authorities have sought to condition government incentives on evaluations of individual rural residents and households generated through community monitoring and peer review mechanisms. In its more recent effort to upgrade such program, the Qingzhen government also sought to further systemize the operational procedure in such a way that each village runs peer review sessions so as to score individual residents' behaviors according to a set of publicized "village norms." The scores so generated were to go into the local government's basis for distributing agricultural loans, subsidies and multiple types of local and residential benefits (Qingzhen Government Site 2017).

"Village norms," as in this context, are adopted through self-governance procedures that have existed for decades in China's rural communities. Such norms regulate matters both within and outside of the ordinary reach of formal laws. Some village norms reiterate legal requirements for observing contractual obligations, family planning regulations, construction and transportation safety regulations, and the like. Other norms, by contrast, such as those on "harmony within the family," "good neighborliness," "diligent work ethics," or against hosting lavish banquets and littering, do not reflect typical governmental concerns. Since in recent decades the constraining force of such village norms' has been much weakened, Qingzhen authorities hope to reinvigorate these norms by backing them with high-powered incentives.

Western audiences may perceive such programs as merely authoritarianism with Chinese characters. But the essential logic here is about addressing what used to be characterized in the law and social norms literature as "normative failure," meaning that substantive norms whose content reflects community consensus are underenforced, with legal coercion (Cooter 1997). Nonetheless, using government-imposed reputation sanctions and rewards to reinforce moral norms could be practically tricky. Authorities first must be able to correctly identify and target the type of "failing" norms that are actually worth reinforcing. Second, government incentives used to induce optimal behavior must be neither inadequate nor excessive. Third, practical and concrete logistics are needed for collecting the behavioral data and running the relevant community review and feedback mechanisms. None of the local incorporation/reinforcement programs, however, is known to have developed coherent solutions to those issues.

A related development concerns the Chinese government's attempt at legislating moral norms. Since 2013, many provincial and municipal governments have adopted local regulations on "promoting civilized behaviors" (Cai 2018). Such regulations formally prescribe norms applicable to a variety of rather petty daily life matters, such as elevator etiquette and doing what Romans do when being a tourist abroad. One obvious concern for such legislation is that it may at times reduce considerably citizens' practical obscurity with respect to their norm transgressions.

7.3.2.5 Institutionalizing Reputation: "Breaking Trust in One Place, Facing Restrictions in all Places"

The SCSP's most ambitious agenda is to transform China's regulatory state with institutionalized reputation. That envisions that information which each government body normally collects on regulatory subjects becomes redesignated, processed and aggregated as reputation information, shared across multiple government agencies, and a particular subject, based on its reputation profile, will face in the future additional sanctions or rewards from government agencies with which it previously may not have dealt.

Representing such comprehensive institutionalization are a range of "joint sanctions and rewards" schemes (hereinafter "**JSRs**") that originated again from the judgment defaulter blacklist program run by the court system. Under the nationwide scheme, the SPC? SCSP? shares its data on blacklisted individuals and entities with over 40 other central government agencies that all commit to imposing sanctions within their respective authority against those blacklisted individuals and entities. Among others, firms defaulting on court judgments face restrictions in financing, doing business in the financial sectors and receiving government subsidies or other handouts; individuals are banned from government employment, prohibited from travelling by air or on high-speed trains, and their children may not enroll in expensive private schools.

Besides the judgment defaulter blacklists, at the national level JSRs have now been set up against "trust breaking behaviors" in almost all regulatory spheres. At the local level, provincial and municipal governments across the country have also set up JSRs and similar institutionalization schemes. In some cases, multiple local agencies join forces to crack down on a particular type of regulatory violation. For example, in Huaibei City of Anhui Province, nine government agencies jointly set up a JSR in 2017 that targets employers who default in bad faith on workers' salaries. In other cases, multiple departments set up a JSR scheme that covers more than one single regulatory issue. For example, in Shanghai, 37 agencies participated in the municipal comprehensive JSR regime in 2018, and each participating agency in carrying out its own regulatory mandates may request JSR collaboration from other agencies on a particular subject.

In yet a smaller number of cases, local governments seek to set up comprehensive rating and scoring systems on regulatory subjects as the foundation for their JSR schemes.

In Suining County of Jiangsu province, the local social credit program, which had commenced in 2010 and subsequently received international media attention, introduced a system with both points and ordinal grades for assessing individual residents. Despite a minority approach nationwide, more than twenty localities have rolled out their social credit scores for individual citizens (see Lin's contribution in this volume). Many of these local scoring systems are known use a linear model, whereby a score is generated through addition and subtraction of points based on a set of rules. In a few of the well-publicized cases, such as Suzhou's "Fragrans Score," municipal scoring schemes apparently employ multi-dimensional, black box scoring models resembling those used for financial credit evaluation. By far, the application of city credit scores in the context of JSRs has been confined to limited and often low-stakes public service benefits, such as borrowing books from public libraries without paying a deposit fee.

Overall, the massive roll-out of JSRs could increase risks of data security breaches. Yet the greater concern often cited is about potential abuse and detriment to fairness. Under the JSR schemes, a regulatory subject now may face sanctions or rewards for who he or she is, not just what he or she does, and a common worry is that sanctions could as a result risk being blown out of proportion. That said, despite the self-aggrandizing propaganda from national and local government authorities, there has not been any validation study to systemically support the JSR regimes' actual behavioral effect. Conceivably, designing rules and formulae for effectively generating and using reputation in the institutionalization contexts is no easy task. By far, coherent logic has hardly been unveiled or explained with respect to how points are assigned under these scoring systems in order to effectively differentiate behaviors.

7.3.2.6 Government Internal Control with Reputation

The Chinese authorities view reputation as not only a tool they can apply to regulate the market and the society. They also expect reputation to play an important role in policing the behavior of government agencies and officials. For starters, the SCSP searchlight is intended to shine not only on the private sector but also on government actors. Beside reiterating textbook government transparency requirements, some novel initiatives call for publicizing information about the credit status of government agencies and officials, especially in relation to how well they observe contractual obligations and judicial decisions. According to a senior NDRC official, more than 5,000 government entities were blacklisted for defaulting on court decisions, and 99.7 percent of such delinquency cases had subsequently been "corrected" (Lian 2018).

Other SCSP policy proposals apply the incorporation strategy to government actors. These proposals encourage the generation of reputation information on government and official behaviors by outside sources. Besides calling for strengthening conventional mechanisms such as petitioning and whistleblowing, the SCSP directs governments at all levels to support outside credit agencies and research institutions in producing and publicizing reputation scores and ratings on government bodies. One of the few notable third-party ratings on government performance has been a check on the "searchlights": The NDRC in recent years has commissioned private agencies to evaluate how well provincial governments followed through on requirements to timely publicize administrative approval and sanction decisions.

The SCSP also aspires to institutionalize reputation into the government's internal oversight and discipline systems. Most importantly, credit profiles established through the SCSP on government bodies and individual officials are expected to be used by higher authorities in reviewing the subsidiaries' performance and making funding, promotion and other similar decisions. While similar oversight and human resource management practices have long existed (Schlaeger 2013, pp. 67–75), what the SCSP now aims to bring to these pre-existing internal control practices are new types of information that trigger a wider array of disciplinary incentives. The actual effect of these reputation-based internal control initiatives is yet to be seen. After all, as the Chinese government has carried out its aggressive anti-graft campaign during the past years, it did not really rely much on the SCSP-related novel monitoring and control arrangements.

7.4 Animating the Reputation State

The foregoing descriptive account demonstrates the reputation state as the unifying institutional logic for the SCSP. In putting together its reputation state, China has resorted to both conventional and innovative strategies.

As noted along the way, the SCSP faces multiple practical challenges to its effective implementation, the resolution of which require the government's continued investmenton a not insignificant scale. Why, then, has China turned to the reputation state as its mode of governance? Inevitably, a project as ambitious and multifaceted as this is the product of diverse interests and institutional forces.

7.4.1 Developmental Interests

"Developmental interests" here broadly include government interests in taking policy measures with a view to improving economic performance and public welfare. For starters, underlying the SCSP is a clear state interest in market reform. As the country's pace of economic growth has notably slowed since 2012 due to languishing external demand, sustainable future growth, as policymakers see it, has to rely on structural market reforms to exploit domestic potentials in investment and consumption. Expanding access to financing for SMEs and consumers thus appears a natural policy option, and credit expansion, in turn, is predicated on credit systems that serve traditionally under-financed parties.

Yet the Chinese state's need for novel approaches to government reform is another important developmental interest. The internal perspectives on the shortcomings of China's government system have often focused on both the failure to effectively wield "sticks" against opportunistic violations and to allocate "carrots" in a "targeted and precise" manner. Such shortcomings are attributed to three general pathologies. First, China's government information and data practices have long been plagued by the so called "information silos" problem, which causes significant internal transaction costs in delivering public services and administering policies. Second, many regulatory agencies responsible for enforcing laws and regulations on the frontlines have complained about being understaffed, underfunded, or not being given the necessary authority to use coercive power. Third, the usual agency control problems, such as corruption, arbitrariness, shirking, or simply official incompetence, are still considered prevalent.

These pathologies are familiar to students of law and government in low-income and middle-income countries (Peerenboom and Ginsburg 2014), and the SCSP represents the Chinese government's direct responses to the same: It addresses the deficiencies in government information technology and data practices by investing in better ICT infrastructure, straightening out government data collection practices, and promoting inter-agency data sharing; it tackles regulatory under-enforcement by both mobilizing private sanctions and pooling resources of multiple public agencies; and it approaches the government integrity and competence problems by incorporating certain data-driven and automation decision techniques, and bolstering external and internal oversights on government behavior.

In this vein, the SCSP reflects China's continued interest in exploring plausible technocratic solutions to modernizing its "governing capabilities." Notably, unlike what many have suggested, the SCSP's technocratic approach consists of not merely deploying ICT but also rearranging its institutional resources.

7.4.2 Authoritarian Interests

Western observers more commonly subscribe to the theory that the SCSP is premised upon the Chinese Party-state's interest in solidifying its political control over its citizens and containing opponents and challengers. Without question, the state will have an interest in applying, to solidify its political control, the augmented capabilities in surveillance, data mining and inter-agency coordination that it expects to acquire through the SCSP. For example, the national Internet content regulator issued rules in 2017 and 2019 that seek to use blacklisting and JSRs to tighten control over operators and users of Internet content services. Also, for those initiatives that insert government incentives to back moral norms, requirements for ideological commitments to the party line could become part of the "moral" norms and forcefully imposed upon individuals. From the state's perspective, the reputation techniques may serve as a useful tool for augmenting behavioral control as it sees fit.

More subtly, the SCSP serves the Chinese state's authoritarian interests through not only deterrence but also expressive mechanisms. The word "credit" (xin) in the Chinese language carries a thick moral valence, with which the state may deliberately seek to soften the image of political control by appealing to the generally favorable cultural attitudes towards this moral idea. In addition, the appearance of the state deploying sophisticated social control tools, such as big data and predictive analytics, is also likely expected to help create the effect of a "security theater" and achieve a certain level of symbolic deterrence (Zarsky 2013, p. 1568).

Nonetheless, the Party-state's authoritarian interests are not separate or all divergent from its developmental interests. If the SCSP generates welfare gains through bringing about positive reforms in market and government, that will validate the regime's claim of adaptability to the country's evolving developmental challenges through technocratic strategies as opposed to political liberalization (Zhao 2009).

7.4.3   Bureaucratic Interests

The institutional structure for the SCSP's implementation is highly decentralized, with nearly all government bodies along different lines of vertical and horizontal authorities being more or less involved. Local and sectoral government actors are interested in carrying out the SCSP policy initiatives because, first, the relevant programs generate particular gains in capabilities and power for these actors. Such gains may include a single agency's access to new sources of data and information by virtue of enhanced surveillance or compelled inter-agency sharing. It may also include such agency's augmented coercive power by virtue of its ability to mobilize private reputation sanctions or to tap into enforcement power from other agencies.

Second, local officials may find it easier to claim credits for themselves by working on SCSP-related programs, in particular as compared with driving hard economic growth. Through "constructing social trust," in particular with cutting-edge technological solutions and novel ideas of institutional arrangements, local officials can showcase to superiors their obedience to the top-down directive as well as their prowess in "governance innovation."

Third, and more easily understandable, the SCSP policy initiatives imply requirements for new investment in upgrading government ICT infrastructure. That provides a venue of public spending for local officials.

These bureaucratic interests help explain why a considerable amount of government resources have been channeled to the SCSP policy initiatives. However, being parochial in nature, these bureaucratic interests also contribute to collective action problems that undermine interagency sharing and collaboration. Moreover, as officials are tempted to self-promote with presentable, short-term "progress" in implementing the SCSP, they may as a result devote less than necessary efforts to figuring out the harder design questions, such as those in relation to the optimal, or at least reasonably useful scoring formula. Further, while the SCSP's implementation requires proper and efficient technological solutions, the rent-seeking dynamics may also cause less than optimal solutions to become adopted. The presence of the bureaucratic interests, therefore, most likely contribute to many of the practical challenges the SCSP currently faces, although government reform itself is, ironically, the project's key target.

7.4.4 Private Business Interests

China's increasingly influential private ICT companies have been rightfully noted as a critical player in the SCSP. Home to some of the world's leading ICT giants, China in recent years has seen a salient increase of such companies' political clout. Success stories and rosy visions from the private tech sector have had much credence with China's government authorities (Creemers 2016). Such input has shaped the government authorities' generally positive outlook for using reputation tracking and analytics in government. From the authorities' perspective, since behemoth commercial platforms such as Taobao, TMall, and JD.com can achieve efficiency and order through managing a vast amount of transactions with digitized reputation mechanisms, there is no principled reason why government agencies, police forces and courts cannot do the same.

Besides the tech giants, as earlier noted, dozens of small to mid-size tech firms, as earlier mentioned, have become drawn to SCSP-related opportunities and devoted their primary business to undertaking related government contracts. An emerging business interest group, therefore, has already become created in support of the SCSP's implementation.

The influence of the business interests presents itself in high expectations for and reliance by government actors on the firms' capabilities and expertise in overcoming the SCSP's implementation challenges. On the one hand, authorities from time to time are emboldened by the lobbying firms to delve into implementing SCSP projects that obviously require greater technological infrastructure and sophistication than they currently possess, counting largely on the firms to bridge the gap with novel solutions. On the other hand, as firms lobby government actors to adopt their tech solutions, or to subsidize their reputation systems with public sector information and incentives, they also tend to oversell the utility of their technologies, fanning further interest from officials. Moreover, since systematic validation tests have not been required for government officials to claim success, spending on conceptually novel technologies, regardless of their practical utility or suitability, coincides with the government officials' interest in presenting themselves as progressive and sophisticated.

The influence, meanwhile, does not go in just one direction, as private firms also actively attempt to formulate and adjust their own strategies in accordance with their perception of government interest. For example, since late 2016, the tech sector had started to pitch blockchain technologies to China's national and local government authorities. While blockchain technologies are generally understood to be associated with a distributed data structure, fintech entrepreneurs, knowing the Chinese authorities' obvious preference for centralization, strategically highlighted to government authorities blockchain's virtue primarily from the angle of data accuracy, reliability, and security, significantly downplaying or even omitting the implications relating to public blockchain's inherent logic of decentralization.

7.5   Structural Implications

This section does not dwell on the possible human rights implications of the SCSP that have already been visited frequently (e.g. Chen et al. 2018). Instead it points observers to the SCSP's three structural implications, relating in broad terms to its efficient mode of governance, optimal approach to enforcement, and intragovernmental agency control. The primary goal here is to identify these perspectives for advancing related normative deliberations.

7.5.1   More Reputation, More Regulation?

Legal theories have predominantly adopted a dichotomous model of the sources of order, contrasting the state's centralized legal commands and the spontaneous, decentralized private ordering, including those facilitated by social norms and reputation (Richman 2004). The paradigmatic neoliberal thinking has emphasized the efficiency advantages of private ordering (Cooter 1996, pp. 1644–1646). As facilitated by new technologies, reputation has become increasingly a prevalent force undergirding self-governance. As previously noted, this led some to argue that the normatively desirable governance pattern may shift to one of "less regulation, more reputation," as the state could reduce direct intervention in socioeconomic activities and instead support reputation-based private ordering (Strahilevitz 2011a, p. 72).

One proclaimed objective of the SCSP, according to the 2014 Blueprint, is indeed to "reduce government interventions in economic affairs" through promoting credit agencies and reputation systems. But what the Chinese government overall expects from reputation is a vehicle for expanding the reach of government and law. "More reputation," as the Chinese state envisions, is to eventually serve its purpose to have "more regulation," or greater capabilities to arrange social order from the power center.

Is this plausible, even just in theory? As private reputation continues to proliferate and systematize, the increasing amount of reputation by itself could indeed make the state ever more informed. And if the state is technically capable of not only monitoring private reputation but also applying predictive analytics to understand, in a timely manner, how the underlying behavioral patterns evolve, then over the course of a dynamic process, "more reputation" does seem to plausibly allow "more regulation."

In more abstract terms, the equilibrium of how efficient ordering consists of public versus private governance often shifts, and reputation technologies may add to the reason for portending a shift in a direction different from what the neoliberal theories predict and recommend. Such a possibility has been foreshadowed by scholars who in recent years attempted to conceive some radical prospects for legal developments resulting from the use of big data analytics (Casey and Niblett 2017). Provided that laws are to be designed and implemented by legal authorities that have low-cost access to information, computing power, and communication infrastructure, as these scholars explain, the legal system will become more capacious as it rids itself of much of the currently intractable tradeoffs, including but not limited to those associated with arbitrary line drawings, *post hoc* discretions, and sacrificing contextual flexibility for formal certainties.

Those theoretical discussions seem to suggest that, with powerful technological tools, the *modus operandi* of law could be transformed into something very much resembling reputation: The centralized legal system regulates in a manner that is more informed, sensitive to context, and dynamic, than how it has been typically perceived. The reputation state, its key logic embodied in the SCSP, may thus become the plausible path law and governance will take towards their much speculated future of personalization.

7.5.2 Enforcement: Perfection or Selection?

The SCSP aims to use reputation to tighten the Chinese government's loose ends in enforcing its laws and regulations. How effectively that may be achieved must be investigated in concrete, empirical contexts. But a more forward-looking question concerns what government actors may actually do with their potentially augmented enforcement capabilities: Even under many of the smaller JSRs, regulatory subjects expect to face more powerful incentives now than before.

There are certainly benefits from having law and regulation in some areas bite harder; similarly to many other low and middle-income countries, China still wrestles with the problem of deficient legal protection of persons, property, and the environment (van Rooij and McAllister 2014, p. 288). But all cases of under-enforcement are not equal. Near perfect enforcement of laws, even were it attainable, would seldom be optimal. Not only the direct cost of resources spent on enforcement actions could be high (Polinsky and Shavell 2000), but the welfare effects of legal rules also cannot be presumed to be static; some breathing room therefore is often needed to accommodate subversive but creative responses (Cohen 2013, p. 1918). Both individuals and business firms in China, as is well known, have long managed to take the authorities' laxity in enforcing formal laws and regulations as a breathing room to exercise individual liberties, make practical business arrangements, and explore market and technological innovations. For one example, the rise of China's Internet giants in the late 1990s and early 2000s itself, as some have argued, is attributable to the lack of stringent enforcement of copyright laws and regulations (Hu 2016, pp. 124–125).

Such laxity in enforcement has stemmed from the authorities' lack of both capability and will. The new enforcement paradigm, based on reputation tracking and higher-powered incentives, could eventually render the first hindrance obsolete. As a result, the authorities' strategic selection about where to deploy the reputation-enabled regulatory capacities will be more consequential than ever.

In the near term, the SCSP's implementation has already led to the proliferation of JSRs. Since the expansion of JSRs has outpaced the growth of the actual enforcement capabilities, many JSRs established are likely superfluous and wasteful. In the long run, near perfect enforcement through JSRs may become more practically attainable for the government in an increasing number of regulatory areas. In both scenarios, however, the overall welfare result from the state's regulatory enforcement will depend upon how high-level policy makers work out the state's regulatory priorities. Specifically, what remains critically lacking is coherent thinking over the following question: What conduct norms, either legal or extra-legal, will produce the most socially beneficial results once supported by the high-powered enforcement tools that will be made available through the SCSP?

7.5.3   Intragovernmental Agency Control

One objective of China's national authorities is to strengthen their control over lower level government bodies and officials. Some types of SCSP-directed disclosures, such as those about government actors' particular "trust-breaking conduct," reportedly have created notable pressure on local officials. But short of pushing for even more generalized open government data than that under the current freedom of information regime, the SCSP's incremental contribution to improving government transparency and strengthening external constraints on government conduct may not be significant for its own sake.

Yet reputation could serve as a tool for agency control also because it creates a plausible venue for higher authorities to check on lower-level agents. Provided it is generated upon aggregating and analyzing a large amount of behavioral and circumstantial information, reputation is, in most cases, less arbitrary a basis for decision-making than, say, an official's snap judgment, selfish interest, or favoritism based on personal and familial ties. The SCSP's requirement for local government actors to consult reputation thus functions as a useful constraint on their discretion over, for example, who among the local residents will have access to government incentives, contracting opportunities, or expedited processing for administrative approvals.

In this vein, the SCSP may add a further layer of discipline over government officials' exercise of power: If reputation becomes increasingly factored into local government actors' decision-making process, the room for arbitrary and abusive official behaviors, which the Chinese public has long complained, could be considerably reduced. By way of extension, contrasting the Western aversion to automated government decision-making (Eubanks 2018), there is even good reason to think that in the Chinese context the governmental processes could change for the better with more digitization (Schlaeger 2013) and even automation.

That said, without a structural legislative framework, there is currently much uncertainty with respect to how powers of defining reputation, allocating reputation incentives, and controlling data storage and use are divided between central and local authorities. In practice, if the SCSP implementation continues to be fragmented along geographical and sectoral lines, that could render oversight by central government even more difficult. This suggests that the SCSP may potentially serve a new battleground in the perennial struggle for power, both along horizontal (between ministerial agencies) and vertical (between the center and peripheral) lines of authorities, within China's political system.

7.6 Conclusion

China's SCSP offers the world a unique opportunity to observe the promises and challenges of the rising reputation state. This chapter aims to provide observers in general with a comprehensive and also empirically informed perspective so that they avoid the risk of being misled by partial or inaccurate impressions while missing the real action.

In the 2014 Blueprint, the Chinese government set the end of 2020 as the deadline for a range of SCSP milestones. These targeted implementation achievements consist of formal establishment of credit data systems and credit-inspired regulatory mechanisms across industries, commercial sectors, and provincial and municipal localities, along with the beneficial regulatory effects these systems are expected to produce. Although many of the foregoing objectives have been fulfilled, what has already become clear is that by the end of 2020 the SCSP will not achieve all milestones envisioned. In its current form, the SCSP has a reasonable chance to facilitate the state's developmental objectives, such as fostering a more functional reputation market and modernizing government data practices. The pursuit of transformative changes to the economy and government, meanwhile, is facing challenges that call for difficult mechanism designs and institutional rearrangements.

As the State Council reaffirmed in the middle of 2019, the SCSP expects to remain central in the coming years to China's endeavor to level up its governance capabilities. For further research on the SCSP's process and impact, this chapter recommends to consider the following lines of efforts: First, systematic empirical investigations on the actual behavioral and market effect of the various SCSP initiatives may become feasible in the future; second, normative theories about the reputation state may be further developed as the SCSP's actual welfare and distributive impact become manifested; Third, comparative studies can be performed on similar developments involving the use of reputation-based techniques in more countries from both the West and the developing world.

References

Altman, M., Alexandra Wood, David R. O'Brien, Salil Vadhan, and Urs Gasser. 2015. "Towards A Modern Approach to Privacy-Aware Government Data Releases". *Berkeley Technology Law Journal*, 30(2):1967–2072.

Brauneis, R. and Ellen P. Goodman. 2018. "Algorithmic Transparency for the Smart City". *Yale Journal of Law and Technology*, 20:103–176.

Bruney, G. 2018. "A 'Black Mirror' Episode Is Coming to Life in China", *Esquire*, 17 March 2018, https://www.esquire.com/news-politics/a19467976/black-mirror-social-credit-china/.

Cai, J. 2018. "Possibility and Limitation of Legislation of Civilized Behavior (文明行为立法的可能与限度)". *Northern Legal Science* (北方法学), (3): 105–117.

Cascio, J. 2011. "The Future of Reputation Networks", in *The Reputation Society*.

Casey, A. J. & Anthony Niblett. 2017. "The Death of Rules and Standards". *Indiana Law Journal*, 92(4):1401–1447

Chen, Y. and Anne S. Y. Cheung. 2017. "The Transparent Self under Big Data Profiling: Privacy and Chinese Legislation on the Social Credit System". *Journal of Comparative Law*, 12(2): 356–378.

Chen, Y., Ching-Fu Lin & Han-Wei Liu. 2018. "'Rule of Trust': The Power and Perils of China's Social Credit Megaproject". *Columbia Journal of Asian Law*, 32(1):1–36.

Citron, D. K. and Frank Pasquale. 2014. "The Scored Society: Due Process for Automated Predictions". *Washington Law Review*, 89(1):1–34

Cohen, J. E. 2013. "What Privacy Is for". *Harvard Law Review*, 126(7): 1904–1933.

Cooter, R. D. 1997. "A Normative Failure Theory of Law". *Cornell Law Review*, 82(5): 947–979.

Cooter, R. D. 1996. "Decentralized Law for a Complex Economy: The Structural Approach to Adjudicating the New Law Merchant". *University of Pennsylvania Law Review*, 144(5): 1643–1696.

Credit China website. 2018a."Publicized List of Comprehensive Credit Service Agency Pilot Firms (关于综合信用服务机构试点名单的公示)". Oct. 18, 2018, https://www.creditchina.gov.cn/xinyongdongtai/buwei/201810/t20181017_128369.html.

———. 2018b. "Theatres in Hengshui City Put Information of Blacklisted Defaulters on Screens before Movies Start (衡水:影院电影放映前播放110名"老赖"信息)". July 9, 2018, https://www.creditchina.gov.cn/home/dianxinganli1/201807/t20180706_119891.html.

———. 2020. https://www.creditchina.gov.cn/xinyongfuwu/quanguogeshengxinyongdaohang/.

Creemers, R. 2016. "Disrupting the Chinese State: New Actors and New Factors." SSRN Scholarly Paper ID 2978880. https://ssrn.com/abstract=2978880.

Creemers, R. 2018. "China's Social Credit System: An Evolving Practice of Control." SSRN Scholarly Paper ID 3175792. Rochester, NY: Social Science Research Network. https://papers.ssrn.com/abstract=3175792.

Dellarocas, C. 2011. "Designing Reputation Systems for the Social Web", in *The Reputation Society*.

Eastern Daily. 2016. "Abusing Rideshare Services Will Leave One Negative Credit Record and Subject Him to Restrictions in Procuring Financial Services (滥用单车者征信将留污 获得金融服务等将受限)". November 17, 2016, http://wap.eastday.com/node2/node3/n5/u1ai702476_t72.html.

Ellickson, R. C. 1991. *Order Without Law: How Neighbors Settle Disputes*. Harvard University Press.

Eubanks, V. (2018). *Automating Inequality: How High-Tech Tools Profile, Police, and Punish the Poor*. New York: St. Martin's Press.

Goldman, E. 2011. "Regulating Reputation", in *The Reputation Society*.

Hu, L. (胡凌). 2016. "Rising Illegally: One Way to Understand the Evolution of Internet in China (非法兴起:理解中国互联网演进的一个视角)". *Beijing Cultural Review* (文化纵横), (5): 120–125.

Hu, M. 2015. "Big Data Blacklisting". *Florida Law Review*, 67(5):1735–1807.

Lian, W. (连维良). 2018. "Carrying Out the Social Credit System Project Creatively Under the New Circumstances (创造性地做好新形势下的社会信用建设工作)". *China Credit* (中国信用), (10), http://www.creditsx.gov.cn/gndt/6132506.jhtml.

Liu, L. and Barry Weingast. 2018. "Taobao, Federalism, and the Emergence of 'Law, Chinese Style'". *Minnesota Law Review*, 102(4):1563–1590.

Luo, P. (罗培新). 2016. "Good Governance Requires Good Law: General View on Social Credit Legislation (善治须用良法:社会信用立法论略)" *Legal Science (法学)*, (12):104–112.

Macaulay, S. 1963. "Non-Contractual Relations in Business: A Preliminary Study". *American Sociological Review*, 28(1):55–67.

Mastsakis, L. 2019. "How the West Got China's Social Credit System Wrong". *Wired*, July 29, 2019, https://www.wired.com/story/china-social-credit-score-system/.

Masum, H. and Mark Tovey (eds.). 2011. *The Reputation Society: How Online Opinions Are Reshaping the Offline World*. The MIT Press.

Meissner, M. 2017. "China's Social Credit System: A Big-Data Enabled Approach to Market Regulation with Broad Implications for Doing Business in China", MERICS, https://www.merics.org/en/microsite/china-monitor/chinas-social-credit-system.

Michaels, J. D. 2008. "All the President's Spies: Private-Public Intelligence Partnerships in the War on Terror". *California Law Review*, 96(4):901–966.

Ming, C. 2017. "FICO with Chinese Characteristics: Nice rewards, But Punishing Penalties". *CNBC*, March 16, 2017, https://www.cnbc.com/2017/03/16/china-social-credit-system-ant-financials-sesame-credit-and-others-give-scores-that-go-beyond-fico.html.

Mohammed, R. 2013. "Uber's 'Price Gouging' Is the Future of Business", *Harvard Business Review*, December 16, 2013, www.hbr.org/2013/12/ubers-price-gouging-is-the-future-of-business?utm_source=feedburner.

Netease News website. 2015. "Social Security Information Portal Exposed to Severe Security Risks (社保信息门户大开:政府网站安全堪忧)". May 4, 2015, http://tech.163.com/15/0504/09/AOOTLV4Q000915BF.html.

———. 2016. "State Council's Annual Work Report Mentions Again SCSP Again and Sesame Credit Pulls Forward in the Related Market (政府工作报告再提"信用建设"芝麻信用市场前行)". March 10, 2016, http://news.163.com/16/0310/14/BHQ8JBRP00014AED.html.

Pasquale, F. 2015. *The Black Box Society: The Secret Algorithms That Control Money and Information*. Harvard University Press.

Peerenboom, R. and Tom Ginsburg (eds.). 2014. *Law and Development of Middle Income Countries: Avoiding the Middle-Income Trap*. Cambridge University Press.

Pence, M. 2018. "Vice President Mike Pence's Remarks on the Administration's Policy Towards China". *Hudson Institute*, October 4, 2018, https://www.hudson.org/events/1610-vice-president-mike-pence-s-remarks-on-the-administration-s-policy-towards-china102018.

People.cn. 2017. "Hangzhou Municipal Regulatory Bureau Joined Force with Sesame Credit to Build Small and Micro Enterprise Social Credit System (杭州市市场监管局联手芝麻信用共筑小微企业诚信体)", January 11, 2017, http://zj.people.com.cn/n2/2017/0111/c186327-29583894.html.

Picci, L. 2010. *Reputation-Based Governance*. Stanford University Press.

Polinsky, A. M. and Steven Shavell. 2000. "The Economic Theory of Public Enforcement of Law". *Journal of Economic Literature*, 38(1):45–76.

Qingzhen Government Site. 2017. "'Gov Tech' Deployed to Produce Result in Qingzhen's SCSP as the Provincial Pilot Program ("治理科技"引领试点强推进 "六项行动"保障试点出成效——清镇市作为省级社会信用体系建设示范点初现成效)", August 2, 2017, http://www.gzqz.gov.cn/ztzl/cxjs/201708/t20170802_2595283.html.

Richman, B. D. 2004. "Firms, Courts, and Reputation Mechanisms: Towards a Positive Theory of Private Ordering", *Columbia Law Review*, 104(8):2328–2368.

Silberman, M. S. and Lilly Irani. 2016. "Operating an Employer Reputation System: Lessons from Turkopticon, 2008–2015", *Comparative Labor Law & Policy Journal*, 37:505–542.

Schlaeger, J. 2013. *E-Government in China: Technology, Power and Local Government Reform*. Routledge.

Shen, W. 2013. "Shadow Banking System in China – Origin, Uniqueness and Governmental Responses". *Journal of International Banking Law and Regulation*, 28(1):20–26.

Strahilevitz, L.J. 2006 "'How's My Driving?' For Everyone (and Everything?)". *New York University Law Review*, 81(5):1699–1765.

———. 2008. "Reputation Nation: Law in an Era of Ubiquitous Personal Information". *Northwestern University Law Review*, 102(4):1667–1738.

———. 2011a. "Less Regulation, More Reputation", in *Reputation Society*.

———. 2011b. *Information and Exclusion*. Yale University Press.

Tencent News. 2017. "Court Makes Customized Cellphone Ringtone for Defaulters So That Everyone Hears That He Broke Trust (法院给老赖手机"定制"彩铃 谁打都能听到他失信)", June 14, 2017, http://tech.qq.com/a/20170614/046644.htm.

U.S. Securities Exchange Commission. 1975. "Adoption of Amendments to Rule 15c3-1 and Adoption of Alternative Net Capital Requirement for Certain Brokers and Dealers". *Exchange Act Release 34-11497*, June 26, 1975, https://www.sec.gov/ocr/ocr-learn-nrsros.html.

van Rooij, B. and Lesley K. McAllister. 2014. "Environmental Challenges in Middle-Income Countries: A Comparison of Enforcement in Brazil, China, Indonesia, and Mexico", in *Law and Development of Middle Income Countries*.

Wan, C. 2017. "Why the Eight Agencies Have Not Been Licensed As Being More Than Two Years in the Pilot Program (万存知:为什么八家机构两年多还没拿到个人征信牌照?)". *Peking University Institute of Digital Finance*, May 31, 2017, http://idf.pku.edu.cn/index/point/2017/0531/29490.html.

Xinhua News. 2018."First Consumer Credit License Formally Issued to Baihang Credit Reporting" (首张个人征信牌照正式下发 百行征信获牌), Feb. 24, 2018, http://www.xinhuanet.com/money/2018-02/24/c_129815794.htm.

Zarsky, T. Z. 2013. "Transparent Predictions". *University of Illinois Law Review*, 2013:1503–1569.

Zhao, D. 2009. "The Mandate of Heaven and Performance Legitimation in Historical and Contemporary China", *American Behavioral Scientist*, 53(3):416–433.

Zheng, L. & Xiong Jiuyang (郑磊 熊久阳). 2017. "Study on Open Local Government Data in China: Technical and Legal Characteristics (中国地方政府开放数据研究:技术与法律特性)". *Public Administration Review (公共行政评论)*, (1):53–206.

Zuboff, S. 2019. *The Age of Surveillance Capitalism: The Fight for a Human Future at the New Frontier of Power*. Public Affairs.

Xin Dai 戴昕 is associate professor (tenured) at Peking University Law School. His primary areas of research interests include legal theories, law and economics, information privacy, internet law, and digital governance. He received his LL.B from Peking University, J.D from Duke University, and J.S.D from the University of Chicago. Xin practiced corporate and securities laws with Shearman & Sterling's New York and Hong Kong offices, and taught previously at Ocean University of China Law School where he served as an associate dean.

Forschung zum städtischen Personal Credit Scoring im Kontext der Rechtsstaatlichkeit

8

Minghua Lin 林明华

Zusammenfassung

Im Zuge der Beschleunigung des Aufbaus und der Vervollständigung des Sozialkreditsystems in China führen immer mehr Regionen innovative Mechanismen mit lokalen Besonderheiten zum Personal Credit Scoring ein. Mit Belohnungs- oder Bestrafungsmaßnahmen gemäß Credit Scores ist es ein Mittel der Sozialgovernance geworden. Als ein neues Governanceinstrument spielt Personal Credit Scoring eine wichtige Rolle in der Verbesserung von Zielgerichtetheit und Effektivität der Sozialgovernance. In der Praxis entstehen allerdings Probleme wie eine zu breit angelegte Datenerfassung, eingeschränkte Transparenz der Bewertungsregeln sowie unklare Festlegung der Anwendungsbereiche. Vollständige Umsetzung der Verfahrensanforderungen während Etablierung und Durchführung der Regeln zur Ermittlung der Credit Scores, Optimierung der Bewertungsdimensionen sowie Einschränkung des Verwendungsumfangs etc. sind einerseits unumgänglich für die Gewährleistung der Rechtmäßigkeit und Wirksamkeit des Personal Credit Scorings als neues Governanceinstrument und sind andererseits erforderlich zur Förderung der Rechtsstaatlichkeit sowie dem Aufbau einer Kreditgesellschaft.

In den letzten Jahren hat China im Aufbau eines Sozialkreditsystems dank Einsatzes moderner Informationstechnologien wie Big Data in sämtlichen grundlegenden Bereichen und Schlüsselsegmenten wesentliche Fortschritte erzielt. Damit beschreitet China den Weg des Aufbaus eines Sozialkreditsystems mit nationalen und regionalen Charakteristiken.

Minghua Lin 林明华 (✉)
Xiamen, China

© Springer Fachmedien Wiesbaden GmbH, ein Teil von Springer Nature 2020   165
O. Everling (Hrsg.), *Social Credit Rating*,
https://doi.org/10.1007/978-3-658-29653-7_8

Diesbezüglich ist eine Reihe politischer Initiativen ergriffen worden. Auf der National-ebene wurde das Sozialkreditsystem nach Beschleunigung der Konzeption auf der höchs-ten Ebene und Einführung umfangreicher Regelwerke ein System mit einzigartigen Di-mensionen des Gesetzesvollzugs, das in gewissem Maße traditionelle Mechanismen von Justiz- und Verwaltungsaufsicht, regulatorischer Revision und Verfahrenstransparenz in-tegriert. Auf der Regionalebene wird durch Pilotprojekte bei der Regierung und im kommerziellen Bereich ein auf moderner Technologie und quantifizierten wissenschaftli-chen Modellen basierendes, umfassendes Kreditbewertungssystem etabliert, um ein „städ-tisches Personal Credit Scoring" mit regionalen Charakteristiken einzuführen, das als in-novatives Mittel zur Sozialgovernance genutzt wird. Während das Sozialkreditsystem bemerkenswerte Erfolge erzielte, stießen in der Praxis das von der Regierung geführte städtische Personal Credit Scoring auch auf Probleme wie zu breit angelegter Datenerfas-sung, eingeschränkter Transparenz der Bewertungsregeln sowie unklare Festlegung der Verwendungsgrenze. Dieser Artikel untersucht das städtische Personal Credit Scoring und analysiert dessen Grundfunktionen, potenzielle Risiken und gesetzliche Bestimmungen.

8.1 Funktionsentwicklung des städtischen Personal Credit Scorings

8.1.1 Vom eindimensionalen Risikomanagement zum multidimensionalen Governanceinstrument

Traditionelles Personal Credit Scoring basiert auf einem streng wissenschaftlichen, ma-thematischen Modell, um persönliche Kreditinformationen umfassend auszuwerten und allgemein anwendbare, quantitative Ergebnisse zu erzielen. Infolgedessen kann die Boni-tät einer Person zur Erfüllung ihrer geschäftlichen Verpflichtungen umfassend beurteilt werden. In den Industrieländern erfasst das Personal Credit Scoring alle Arten von Kon-sum. In den Vereinigten Staaten beispielsweise werden FICO-Scores generell anerkannt und von der Gesellschaft akzeptiert. In den Bereichen Vergabe von Bankdarlehen, Versi-cherung, Rekrutierung und sogar Partnersuche kann auf FICO-Scores zurückgegriffen werden. Höhere Scores bedeuten höhere Kreditwürdigkeit und geringere Risiken.

In Bezug auf institutionelle Auskunfteien über Individuen, beginnend im Januar 2015, bereiteten acht Pilot-Ratingagenturen bzw. Auskunfteien (Sesam Credit, Tencent Credit, Shenzhen Qian Hai Credit, Peng Yuan credit, Zhong Cheng Xin Credit, Zhong Zhi Cheng Credit, Kao La Credit und Beijing Hua Dao Credit), gemäß der von der chinesischen Zen-tralbank (People's Bank of China, PBC) herausgegebenen „Mitteilung über die Vorberei-tung auf die Einführung von Auskunfteien über Individuen", ihren Einsatz des Personal Credit Scorings vor. „Sesame Credit Score", „Tencent Credit Score", „Wan Xiang Credit Score", „Koala Credit Score" usw. wurden eingeführt und in vielen alltäglichen Szenarien integriert. Jedoch erhielt keine der acht Pilotinstitute die von der Zentralbank (PBC) aus-zustellende institutionelle Lizenz für Auskunfteien über Individuen. Im Jahr 2018 erhielt das Joint Venture aus den vorgenannten acht Diensten unter anderen, „Bai Hang Kredit" (百行征信), die erste institutionelle Lizenz Chinas für Personal Credit Scoring, wobei es bis dahin noch keine Personal Credit Scores veröffentlicht hatte.

In den letzten Jahren führten auf der Regionalebene viele Städte in China durch Weiterentwicklung des Sozialkreditsystems Personal Credit Scorings mit regionalen Charakteristiken ein, zum Beispiel „Qian Jiang Credit Score" in Hangzhou, „Bai Lu Credit Score" in Xiamen, „Gui Hua Credit Score" in Suzhou, „Xi Chu Credit Score" in Suqian, „Hai Bei Credit Score" in Weihai, „Mo Li Credit Score" in Fuzhou und so weiter (vgl. Tab. 8.1). Der Ausbau des Personal Credit Scorings in verschiedenen Städten führt dazu, dass aus dem Credit Scoring als einem traditionellen Kreditrisikomanagementinstrument nun ein umfassendes öffentlich-rechtliches Governanceinstrument wird.

Von März bis Oktober 2019 führten der Autor und das Institut für Kreditinnovationsforschung von Xiamen Guoxin eine landesweite Untersuchung zu Personal Credit Scorings durch. Die Grafiken sind Statistiken und Vergleiche auf der Grundlage der Untersuchung (Statistiken bis zum 31. Oktober 2019).

Tab. 8.1 Einführungszeit von Personal Credit Scorings in verschiedenen Städten

Nr.	Stadt	Personal Credit Score	Einführungsdatum
1	Rongcheng	Rong Cheng Credit Score	2014.05.05
2	Suzhou	Guihua Credit Score	2016.11.24
3	Guiyang Baiyun	Yun Xin Credit Score	2017.01.01
4	Dongying Kenli	Cheng Ken Credit Score	2017.11.23
5	Wuhu	Le Hui Credit Score	2017,11
6	Suqian	Xi Chu Credit Score	2018.03.23
7	Fuzhou	Mo Li Credit Score	2018.06.03
8	Xiamen	Bai Lu Credit Score	2018.07.05
9	Wuxi	Chengxin Afu Credit Score	2018.07.16
10	Fuzhou	Yu Ming Credit Score	2018.08.20
11	Zhaoqing	Qi Xing Credit Score	2018.09.29
12	Weihai	Hai Bei Credit Score	2018.11.13
13	Ordos	Tian Jiao Credit Score	2018.11.14
14	Hangzhou	Qian Jian Credit Score	2018.11.16
15	Liyang	Longdu Credit Score	2018.12.28
16	Yiwu	N.N	2019.01.01
17	Hohot	Ding Xiang Credit Score	2019.08.01
18	Yulin	Tao Hua Credit Score	2019.08.01
19	Zheng Zhou	Shang Ding Credit Score	2019.08.06
20	Quzhou	Xin An Credit Score	2019.08.07
21	Wuhan	Huang He Credit Score	2019年左右
22	Nanjing	N.N	steht noch aus
23	Shenyang	Sheng Jing Credit Score	steht noch aus
24	Zhoushan	Zi Zai Credit Score	steht noch aus

Das klassische Credit Scoring dient hauptsächlich als technisches Mittel oder Instrument im Risikomanagement von Kreditinstituten während des Kreditvergabeprozesses. Das städtische Personal Credit Scoring basiert auf persönlichen Kreditinformationen. Durch Optimierung und Integration dieser Informationen wird der Gesamtkreditstatus einer Einzelperson quantitativ bewertet und entsprechend Belohnungs- oder Bestrafungsmaßnahmen eingeleitet. Dadurch lässt sich ein Sozialgovernance-Mechanismus „Belohnung für Kredittreue und Bestrafung von Kreditbruch" aufbauen. Mit der kontinuierlichen Anreicherung der Daten, Verbreiterung des Implementierungsumfangs sowie Verbesserung des Anwendungsmechanismus wurde eine Reihe standardisierter Normen, Regelwerke und Systeme etabliert, die sich auf alle Segmente der Sozialgovernance auswirken. Das Scoring wandelt sich allmählich von einem Instrument der Sozialgovernance auf einer taktischen und infrastrukturellen Ebene in einen solideren Governance-Mechanismus auf höher Regelebene um (vgl. 胡凌 2019, S. 31).

8.1.2   „Unsichtbarer Aufseher" in der Stadtverwaltung und Erhöhung der Effektivität der Sozialgovernance

Das klassische Credit Scoring dient hauptsächlich dem Risikomanagementzweck, insbesondere im Rahmen von Geschäftskredit. Mögliche Risiken wie Vertragsbruch und Forderungsausfall, die in Zukunft auftreten können, können im Voraus durch objektive Bewertungsmaßstäbe gesteuert werden, wie zum Beispiel Leistungs- und Rückzahlungsnachweise des Kreditnehmers in der Vergangenheit. Nachdem sich der Umfang des Personal Credit Scorings jedoch von einzelnen Geschäftsbereichen zu einem breiteren und umfassenderen sozialen Kredit ausgeweitet hat, sind die Funktion und die Zielsetzung des Credit Scorings nicht nur auf die Ermittlung der Ausfallwahrscheinlichkeit beschränkt (vgl. 姜明辉 2017, S. 70). Für eine lange Zeit konnten Gesetze und soziale Normen auf traditionelle Art und Weise soziale Probleme wie Lao Lai, Gewalttätigkeiten gegen Ärzte, Fußgänger, die rote Ampel nicht beachten etc. nicht effektiv lösen, was Schwierigkeit und Schmerzpunkt bei der Sozialgovernance darstellt.

Lao Lei (老赖): Verurteilte Personen, die zwar Fähigkeit zur Leistung haben, jedoch diejenigen Verpflichtungen nicht erfüllen, die durch wirksame Rechtsinstrumente festgelegt wurden.

Der Erfassungsumfang des städtischen Personal Credit Scorings nimmt sukzessive zu, der sich von der sesshaften und haushaltsregistrierten Bevölkerung zu den Migranten und ausländischen Bevölkerung erweitert (vgl. Tab. 8.2). Das Personal Credit Scoring entwickelt sich zum „unsichtbaren Aufseher" in der Stadtverwaltung. Es stellt verschiedenen Anwendern einfache und wirksame Steuerungsinstrumente bereit und verbessert die Wirksamkeit der Sozialgovernance. Angeleitet durch das Personal Credit Scoring werden dessen Subjekte als Beaufsichtigte im Rahmen der administrativen Verwaltung gesetzestreuer sein in Bezug auf Steuerzahlung, Transport, Umweltschutz, Brandschutz und Si-

Tab. 8.2 Übersicht der Großstadteinwohner, für die Personal Credit Scores vergeben werden

Eigenschaften der Objektkategorie	Stadt	Personal Credit Score	Objektkategorie	Anzahl der Einwohner, für die Scores vergebenen wurden
Nur haushaltsregistrierte Bevölkerung	Fuzhou	Mo Li Credit Score	Einwohner innerhalb des Verwaltungsgebiets von Fuzhou, die in Fuzhou sesshaft und haushaltsregistriert sind, das 18. Lebensjahr vollendet haben und geschäftsfähig sind	Keine Daten
	Suzhou	Gui Hua Credit Score	Einwohner innerhalb des Verwaltungsgebiets von Suzhou, die in Suzhou sesshaft und haushaltsregistriert sind, das 18. Lebensjahr vollendet haben und geschäftsfähig sind	1,63 Millionen
	Hohhot	Ding Xiang Credit Score	Einwohner innerhalb des Verwaltungsgebiets von Hohhot, die in Hohhot sesshaft und haushaltsregistriert sind, das 18. Lebensjahr vollendet haben und geschäftsfähig sind	1,811 Millionen
Nur sesshafte Bevölkerung	Weihai	Hai Bei Credit Score	Einwohner innerhalb des Verwaltungsgebiets von Weihai, die in Weihai sesshaft sind, das 18. Lebensjahr vollendet haben und geschäftsfähig sind	2,32 Millionen
	Zhengzhou	Shang Ding Credit Score	Einwohner innerhalb des Verwaltungsgebiets von Zhengzhou, die in Zhengzhou sesshaft sind, das 18. Lebensjahr vollendet haben und geschäftsfähig sind	12.000

(Fortsetzung)

Tab. 8.2 (Fortsetzung)

Eigenschaften der Objektkategorie	Stadt	Personal Credit Score	Objektkategorie	Anzahl der Einwohner, für die Scores vergebenen wurden
Einschließlich schwebender Bevölkerung	Hangzhou	Qian Jiang Credit Score	Sesshafte und schwebende Einwohner innerhalb des Verwaltungsgebiets von Hangzhou, die das 18. Lebensjahr vollendet haben und geschäftsfähig sind	11 Millionen
	Rongcheng	Rong Cheng Credit Score	Sesshafte und schwebende Einwohner innerhalb des Verwaltungsgebiets von Rongchen, die das 18. Lebensjahr vollendet haben und geschäftsfähig sind	811.000
	Xiamen	Bai lu Credit Score	Sesshafte einschließlich haushaltsregistrierter und schwebender Einwohner innerhalb des Verwaltungs-gebiets von Xiamen, die das 18. Lebensjahr vollendet haben	207.000
Einschließlich Ausländer	Yiwu	Nicht Benannt	In- und ausländische natürliche Personen im Verwaltungsgebiet von Yiwu, die das 18. Lebensjahr vollendet haben und geschäftsfähig sind	2,25 Millionen

cherheit usw. Dies führt zur Reduzierung illegaler Handlungen und somit Reduzierung der relevanten Verwaltungsbelastung und -kosten.

8.1.3 Positiver Anreiz zu Kredittreue zum Aufbau einer Kreditgesellschaft

Ob eine Person sich entscheidet, kredittreu zu sein oder nicht, hängt vom externen Kontrollmechanismus und internen Überzeugungen ab (vgl. 龙西安 2004, S. 203). Zum einen muss man aufgrund der Kontrolle des externen Systems kredittreu sein, um Bestrafung zu vermeiden, zum anderen kann man aufgrund Konsens und Verantwortungsgefühl bewusst die Entscheidung treffen, kredittreu zu sein. Aufgrund der förderlichen Wirkung motiviert der Anreiz zu Kredittreue die Menschen mehr dazu, sich Kredittreue als ihre eigene Wertorientierung zueigen zu machen, sodass sich Kredittreue von externer Kontrolle in interne bewusste Entscheidungen umwandelt. Personal Credit Scores werden in verschiedenen konkreten Szenarien angewendet, wie öffentlichem Verkehr, medizinischer Versorgung, einigen öffentlichen Dienstleistungen (zum Beispiel ohne Bürgschaft) und Strom-, Wasser- sowie Gasversorgung usw. Die Bürger profitieren immer mehr von Nutzung und Belohnung durch Kredit, was eine solide praktische Grundlage und Unterstützung innerhalb der Bevölkerung geschafft hat (vgl. 王伟 2019, S. 111). Ein konkretes Beispiel ist die Stadt Xiamen. Seit der Einführung des „Bai Lu Credit Score" am 5. Juli 2018 hat das System 4,11 Millionen sesshafte Einwohner erfasst. Mehr als 230.000 Benutzer haben sich registriert und mehr als 2,2 Millionen Abfragen wurden bearbeitet. 23 innovative Anwendungsdienste zugunsten der Bürger in Bereichen wie medizinischer Vorsorge, Zahlungsverkehr, Finanzdienst, Justiz, öffentlichem Verkehr, Bildung, Wohnen sowie weiteren öffentlich-rechtlichen Diensten wurden eingeführt. Für Bürger mit guten Personal Credit Scores wurde bereits kumulativ eine Vergünstigung von mehr als 30 Millionen Yuan und zusätzlich 200.000 Parkminuten kostenlos gewährt. Dadurch wurde eine Atmosphäre bei der Bevölkerung initial geschaffen, kredittreu zu sein und davon zu profitieren.

Auf der Exekutivsitzung unter dem Vorsitz von Premierminister Li Keqiang am 05. Mai 2019 entschied der Staatsrat über die sukzessive Reduzierung der Anzahl zu prüfender Bescheinigungen und Standardisierung des Prüfprozesses durch systematische Innovationen, um die öffentlich-rechtlichen Dienstleistungen und wirtschaftliche Rahmenbedingungen zu verbessern. Zum ersten Mal erschienen Begrifflichkeiten wie „Aufbau einer Kreditgesellschaft" und „vom Aufbau sozialer Kredite zum Aufbau der Kreditgesellschaft" in einem wichtigen Partei- und Staatsdokument, welches die präzedenzlose Betonung des Aufbaus von Sozialkrediten durch die Zentralregierung widerspiegelt (vgl. 王淼 2019). Jede Einzelperson stellt das Subjekt des Aufbaus einer „Kreditgesellschaft" dar. Die Modernisierung der Sozialgovernance durch Kreditkultur und der Aufbau einer Kreditgesellschaft lassen sich durch die Verwendung des Personal Credit Scores als Governanceinstrument besser fördern.

8.2 Potenzielle Risiken und Hindernisse des städtischen Personal Credit Scorings

Im Kontext der Rechtsstaatlichkeit in der Staatsverwaltung stehen Objekte, Faktoren, Umfang der Kreditinformationen, Anwendungsumfang und rechtliche Konsequenzen des Scorings sowie das Recht der Öffentlichkeit auf Informationen über das Scoring im rechtsstaatlichen Rahmen. Mit Blick auf die Credit Scorings verschiedener Regionen als Governanceinstrument werden sie meistens von den Regionalregierungen oder den von diesen eingerichteten Credit-Scoring-Zentren ermittelt. Das von der Regierung durchgeführte Scoring kann das Kreditbewusstsein der Einzelpersonen in kurzer Zeit wirksam motivieren, sodass sich das Vertrauensniveau der gesamten Gesellschaft erhöht. Ein Individuum darf allerdings nicht selbst auswählen und entscheiden, ob und wie es bewertet werden soll. In Bezug auf die Rechtsnatur stellt das allgemeine Scoring durch Regionalregierungen bereits einen konkreten Verwaltungsakt dar. Mangelnde Autorisierung durch Gesetze und Richtlinien kann Verwaltungsstreitigkeiten auslösen. In der gegenwärtigen Praxis des Personal Credit Scorings gibt es jedoch Probleme wie unvollständig geregelte Abläufe und unangemessene Erweiterung des Umfangs der eingesammelten Kreditinformationen sowie Anwendungsszenarien und Gültigkeitsgrenzen des Scorings usw.

8.2.1 Beeinträchtigung des Rechts der Öffentlichkeit auf Information über Scoring aufgrund mangelnden Regelwerks

Personal Credit Scores haben rechtliche Auswirkungen, die die Rechte und Interessen undefinierter Gruppen erheblich beeinträchtigen können. Die meisten Städte, mit Ausnahme weniger Städte wie Rongcheng und Fuzhou, haben aufgrund kontinuierlicher Optimierung und Verbesserung noch keine Scoringregeln veröffentlicht, sodass das Recht der Öffentlichkeit auf Information über Scoring nicht gewährleistet werden kann.

Um die Wissenschaftlichkeit, Relevanz und Bedienbarkeit des Aufbaus des Sozialkreditsystems weiter zu verbessern, hat die Stadt Rongcheng folgende sechs städtische Richtlinien und behördliche Dokumente novelliert und erneut veröffentlicht: „Governance-Richtlinie des Credit Scoring natürlicher Personen", „Governance-Richtlinie zur Auskunfteien juristischer Personen", „Vorschrift zum Credit Scoring natürlicher und juristischer Personen", „Richtlinie der Belohnungs- und Bestrafungsmaßnahmen anhand Personal Credit Scorings", „Implementierungsrichtlinie zum Kreditmanagement in Partei- und Verwaltungsgovernance sowie anderen Bereichen" und „Entwurf der Vorschrift zur doppelten Veröffentlichung der administrativen Lizenzierung und Bestrafung".

Bis Oktober 2019 führten insgesamt 24 Städte städtisches Personal Credit Scoring ein, wobei lediglich zehn Städte klare Scoringregeln und Grundlagen zum Nachschlagen veröffentlichten (vgl. Tab. 8.3).

8.2.2 Auswirkung auf Gültigkeit und Richtigkeit des Personal Credit Scores aufgrund uneinheitlicher Scoringmodelle

Das Personal Credit Scoring wird anhand eines quantitativen Modells, das auf einem bestimmten Algorithmus und Regelwerk basiert, kalkuliert und durch Indexberechnung skaliert. Da das Ergebnis des Personal Credit Scorings in verschiedenen Bereichen der administrativen Verwaltung verbreitet angewandt wird, wirkt es sich direkt auf die Interessen der Bewerteten aus. In der Praxis kann sich der Personal Credit Score einer Person durchaus ändern, sodass sie gewisse Chancen oder Rechte auf bestimmte Ressource gewinnt oder verliert, nur weil sich der Kalkulationsalgorithmus oder die Skalierungsmethode ändert, wobei ihr Kreditstatus gleich bleibt. In diesem Sinne sind die Genauigkeit, Stabilität und Interpretierbarkeit der Modelle für Personal Credit Scoring besonders wichtig (vgl. 张涛 2019, S. 122).

Tab. 8.3 Rechtliche Grundlage der städtischen Personal Credit Scores

Nr.	Stadt	Personal Credit Score	rechtliche Grundlage	Inkrafttreten am
1	Rongcheng	Rong Cheng Credit Score	Maßnahmen zur Verwaltung des Credit Scores und Credit Scorings von Einwohnern der Stadt Rongcheng	2014.05.05
2	Suzhou	Gui Hua Credit Score	vorübergehend keine	2016.11.24
3	Guiyang Baiyun	Yun Xin Credit Score	vorübergehend keine	2017.01.01
4	Dongying Kenli	Cheng Ken Credit Score	vorübergehend keine	2017.11.23
5	Wuhu	Le Hui Credit Score	vorübergehend keine	2017.11
6	Suqian	Xi Chu Credit Score	Vorläufige Vorschriften zum Aufbau des Personal Credit Scoring Systems und zur Beurteilung der Scores natürlicher Personen in Suqian	2018.03.23
7	Fuzhou (Provinz Fujian)	Mo Li Credit Score	Vorläufige Maßnahmen zur Verwaltung von Personal Credit Scores in Fuzhou	2018.06.03
8	Xiamen	Bai Lu Credit Score	vorübergehend keine	2018.07.05
9	Wuxi	Cheng Xin Afu Credit Score	vorübergehend keine	2018.07.16
10	Fuzhou (Provinz Jiangxi)	Yu Ming Credit Score	vorübergehend keine	2018.08.20
11	Zhaoqing	Qi Xing Credit Score	vorübergehend keine	2018.09.29
12	Weihai	Hei Bei Credit Score	Vorläufige Maßnahmen zur Verwaltung von Personal Credit Scores in Weihai	2018.11.13
13	Ordos	Tian Jiao Credit Score	Vorläufige Maßnahmen zur Verwaltung von Personal Credit Scorings in Ordos	2018.11.14
14	Hangzhou	Qian Jiang Credit Score	vorübergehend keine	2018.11.16
15	Liyang	Long Du Credit Score	Vorläufige Maßnahmen zur Verwaltung der Kreditinformationen natürlicher Personen in Liyang Vorläufige Maßnahmen zur Verwaltung der Personal Credit Scores und Beurteilung der Scores natürlicher Personen in Liyang	2018.12.28
16	Yiwu	Noch nicht benannt	Vorläufige Maßnahmen zur Verwaltung von Personal Credit Scorings in Yiwu	2019.01.01

(Fortsetzung)

Tab. 8.3 (Fortsetzung)

Nr.	Stadt	Personal Credit Score	rechtliche Grundlage	Inkrafttreten am
17	Hohhot	Ding Xiang Credit Score	Vorläufige Maßnahmen zur Verwaltung von Personal Credit Scores in Hohhot	2019.08.01
18	Yulin	Tao Hua Credit Score	Vorläufige Maßnahmen zur Verwaltung von Personal Credit Scorings in Yulin	2019.08.01
19	Zhengzhou	Shang Ding Credit Score	Maßnahmen zur Verwaltung von Personal Credit Scores in Zhengzhou	2019.08.06
20	Luzhou	Xin'an Credit Score	vorübergehend keine	2019.08.07
21	Wuhan	Huang He Credit Score	vorübergehend keine	ca. 2019
22	Nanjing	Noch nicht benannt	vorübergehend keine	N/A
23	Shenyang	Sheng Jing Credit Score	vorübergehend keine	N/A
24	Zhoushan	Zi Zai Credit Score	vorübergehend keine	N/A

8.2.2.1 Uneinheitliche Scoringstandards und -algorithmen

Verschiedene Städte haben unterschiedliche Bewertungsstandards und -algorithmen (vgl. Tab. 8.4). Als Indikatordimensionen wählte die Stadt Xiamen für ihr Bai Lu Credit Scoring grundlegende Informationen, kredittreue Verhalten, Kreditbruch, Kreditreparatur und Kreditverwendung, wobei der Schwerpunkt bei der Kreditverwendung und Kreditreparatur bei der Anwendung liegt. Die Indikatoren für Fuzhou Mo Li Credit Score bestehen aus sechs Dimensionen: Grundlegende persönliche Informationen, persönlicher Sozialkredit, persönlicher beruflicher, persönlicher wirtschaftlicher, persönlicher administrativer und persönlicher gerichtlicher Kredit. Weihai Hai Bei Credit Score besitzt drei Dimensionen: grundlegende persönliche Informationen, positive pers önliche Kreditinformationen und negative persönliche Kreditinformationen. Ordos Tian Jiao Credit Score verwendet Kombinationen verschiedener Arten von Kreditinformationen parallel: Betrachtung unterschiedlicher Kategorien der Kreditinformationen, mit hoher Gewichtung die Kreditverwendung und Unterscheidung der Kreditinformationstypen. In Bezug auf Bewertungsalgorithmen verwendet Suqian Xi Chu Credit Score Additions- sowie Subtraktionsmethoden und Weihai Haibei Credit Score differenzierte Klassifizierungs- und Zuordnungsmethoden. Ordos Tianjiao Credit Score verwendet Indexgewichtungsadditions- und – subtraktionsmethode sowie direkte Anpassungsmethode. Dies spiegelt zwar besondere Umstände verschiedener Regionen und ihrer Zweckmäßigkeit wider, führt allerdings zu mangelnder gegenseitiger Anerkennung und Verwendungsschwierigkeiten der Scores sowie Verschwendung von personellen und materiellen Ressourcen (vgl. 张涛 2019, S. 127).

Tab. 8.4 Vergleich der Modelle und Algorithmen zur Berechnung der Personal Credit Scores in Schlüsselstädten

Stadt	Personal Credit Score	Verwendete Dimensionen	Einstufung der Scores		Anmerkung
			Score-Intervall	Interpretation der Intervalle	
Fuzhou	Mo Li Credit Score	6 Dimenstionen: – Persönliche Kompetenz – Beruf – öffentlicher Kredit – finanzieller Kredit – administrativer Kredit und – gerichtlicher Kredit	1000-850 Punkte	hervorragend	1000-Punkte-System inkl. 6 Stufen; Bewertungsmodell: Initialguthaben + Zusatzscore + Jahresscore
			849-750 Punkte	sehr gut	
			749-650 Punkte	gut	
			649-550 Punkte	durchschnittlich	
			549-350 Punkte	schlecht	
			349-0 Punkte	sehr schlecht	
Hangzhou	Qian Jiang Credit Score	5 Dimensionen: – grundlegende Informationen – Einhaltung von Gesetzen und Vorschriften – Kreditverwendung in der Gesellschaft – Wirtschaftliche Kreditverwendung – Pro-soziales Verhalten	750 Punkte oder mehr	hervorragend	1000-Punkte-System inkl. 5 Stufen
			700-750 Punkte	sehr gut	
			600-700 Punkte	gut	
			550-600 Punkte	durchschnittlich	
			550 Punkte oder weniger	zu verbessern	
Hohhot	Ding Xiang Credit Score	6 Dimensionen: – persönliche Kreditfähigkeit – persönlicher öffentlicher Kredit – persönlicher beruflicher Kredit – persönlicher finanzieller Kredit – persönlicher administrativer Kredit – persönlicher gerichtlicher Kredit	850 Punkte oder mehr	hervorragend	1000-Punkte-System inkl. 6 Stufen
			849-750 Punkte	Sehr gut	
			749-650 Punkte	gut	
			649-550 Punkte	durchschnittlich	
			549-350 Punkte	schlecht	
			349-0 Punkte	Sehr schlecht	

(Fortsetzung)

Tab. 8.4 (Fortsetzung)

Stadt	Personal Credit Score	Verwendete Dimensionen	Einstufung der Scores		Anmerkung
			Score-Intervall	Interpretation der Intervalle	
Rong-cheng	Rong Cheng Credit Score	5 Dimensionen, 61 Kategorien und 530 Indikatoren: – Businessver-halten – Sozialgover-nance – administrative Governance – Justiz – Gesellschaft-licher Beitrag	1050 Punkte oder mehr	AAA Level (vorbildlich hervorragend)	1000-Punkte-System inkl. 4 Hauptstufen
			1049-1030 Punkte	AA Level(her-vorragend)	
			1029-1001 Punkte	A+Level(gut)	
			1000 Punkte	A Level(gut)	
			999-960 Punkte	A-Level(gut)	
			959-850 Punkte	B Level(durch-schnittlich)	
			849-600 Punkte	C Level(Alarm-stufe)	
			599 Punkte oder weniger	D Le-vel(schlecht)	
Suzhou	Gui Hua Credit Score	5 Hauptdimensio-nen basierend auf grundlegenden, moralischen, Stabilitäts-, Vermögens- und sonstigen Informationen Darunter 243 Indexpunkte in 22 Kategorien je nach Haushaltsregistrie-rung, Alter, Familienstand, Bildung und Zustand der Sozialversicherung	200-150 Punkte	hervorragend	200-Punkte-System inkl. 6 Stufen
			150-100 Punkte	gut	
			100-50 Punkte	durchschnittlich	
			50-0 Punkte	zu verbessern	
Weihai	Hai Bei Credit Score	5 Dimensionen umfassen 56 Kategorien: – öffentliche Dienste – Moral und Gemeinwohl – Compliance – soziale Verantwortung und – Belobigungs-preise	1150 Punkte oder mehr	AAA Le-vel(vorbildlich hervorragend)	1000-Punkte-System mit Initialguthaben von 1000 Punkten, inkl. 6 Stufen (AAA, AA, A, B, C, D)
			1149-1050 Punkte	AA Level(her-vorragend)	
			1049-1000 Punkte	A Level(gut)	
			999-950 Punkte	B Level(gut)	
			949-801 Punkte	C Level(Alarm-stufe)	
			800 Punkte oder weniger	D Le-vel(schlecht)	

Tab. 8.4 (Fortsetzung)

Stadt	Personal Credit Score	Verwendete Dimensionen	Einstufung der Scores		Anmerkung
			Score-Intervall	Interpretation der Intervalle	
Xiamen	Bai Lu Credit Score	5 Dimensionen, einschließlich 57 Hauptkategorien von Indikatoren und mehr als 750 Datenelementen: – Grundlegende Informationen – gutes Kredit-verhalten – kreditbrechen-des Verhalten – Kreditreparatur und – Kreditverwen-dung	1000-900 Punkte	hervorragend	Punktsystem mit einem Intervall von 150–1000, inkl. 5 Stufen
			900-700 Punkte	Sehr gut	
			700-600 Punkte	gut	
			600-550 Punkte	durchschnittlich	
			550-150 Punkte	Zu verbessern	
Yiwu	Noch nicht benannt	5 Dimensionen einschließlich 106 Hauptkategorien und 175 Datenele-mente: – Identität – Einhaltung von Gesetzen und Vorschriften – Vertragstreue – soziale Aktivitäten und – moralisches Verhalten	150 Punkte oder mehr	hervorragend	1000-Punkte-Sys-tem inkl. 4 Hauptstufen
			150-120 Punkte	sehr gut	
			120-100 Punkte	gut	
			100-80 Punkte	leichter Vertrauensbruch	
			80-50 Punkte	schlecht	
			50 Punkte oder weniger	sehr schlecht	
Zheng-zhou	Shang Ding Credit Score	5 Dimensionen einschließlich 135 Indikatoren: – grundlegende persönliche Informationen – Auszeichnung – Berufsethik – Familientugend – Gesetze und Vorschriften	150 Punkte oder mehr	hervorragend	200-Punkte-System inkl. 5 Stufen
			150-120 Punkte	Sehr gut	
			120-100 Punkte	gut	
			100-80 Punkte	leichter Vertrauensbruch	
			80 Punkte oder weniger	schlecht	

8.2.2.2 Fehlendes wissenschaftliches Assessment der tatsächlichen Wirkung der Scoring-Indikatoren

Am Beispiel von Stadt X umfasst das Modell fünf Dimensionen: grundlegende Informationen, Zivilisation des öffentlichen Wohls, Einhaltung von Gesetzen und Vorschriften, Kreditreparaturen und Kreditverwendung. Indikatoren der Dimensionen sind:

- Grundlegende Informationen: kumulierte und ununterbrochene Arbeitszeiten in der Stadt, sesshafter Einwohner, Sozialversicherungsstatus usw.
- Zivilisation des öffentlichen Wohls und soziales Engagement: verschiedene öffentliche Auszeichnungen, freiwillige Blutspenden, Freiwilligendienste, Spenden und Rettungsaktionen usw.
- Einhaltung von Gesetzen und Vorschriften: kreditbrechendes Verhalten, Verstöße gegen Gesetze und Vorschriften, Zahlungsrückstände, verwaltungsrechtliche Sanktionen usw.
- Kreditreparatur: Verbesserung des schlechten Kredits aus unzivilisiertem und illegalem Verhalten
- Kreditverwendung: Verhalten in Kreditverwendungsszenarien im Sozialleben.

Die Gewichtung der Dimensionen beträgt: Einhaltung von Gesetzen und Vorschriften (35,35 Prozent), soziales Engagement (25 Prozent), Zivilisation des öffentlichen Wohls (23 Prozent), Kreditverwendung (zwölf Prozent), Kreditreparatur (fünf Prozent). Da die Dimensionen Kreditreparatur und Kreditverwendung in der Praxis nicht tatsächlich in Betracht gezogen werden, führt es zu einer Gewichtung von 45,33 Prozent für sozialen Beziehungen, 31,73 Prozent für Einhaltung von Gesetzen und Vorschriften und 22,94 Prozent für Zivilisation des öffentlichen Wohls, welches der ursprünglichen Konzeption nicht entspricht. Die tatsächliche Berechnung bzgl. sozialer Beziehung hängt hauptsächlich von Daten der Sozialversicherung und der öffentlichen Sicherheit ab. Dabei werden die Sozialversicherungsdaten verwendet, um Scores drei Dimensionen, nämlich Dauer der Ansässigkeit in der Stadt, Beschäftigungsstatus und Sozialversicherungsstatus, zu berechnen. Der gültige Zeitraum der Sozialversicherungsdaten umfasst grundsätzlich die letzten sechs Jahre bis heute und bei Beschäftigungsstatus wurde die letzten drei Jahre berücksichtigt. Im Rahmen der Skalierung der Scores wird der Beschäftigungsstatus allerdings ignoriert. Die durch Sozialversicherungsdaten ermittelten Scores von regulären Arbeitnehmern, Rentnern oder Arbeitnehmern kurz vor Rente wirken sich aufgrund unterschiedlichen Beschäftigungsstatus stark auf ihre städtischen Personal Credit Scores aus. Infolge des Mangels an wissenschaftlicher Nachverfolgung und Auswertung weicht der ermittelte Score von dem echten ab, was dessen Zuverlässigkeit und Autorität beeinträchtigt.

8.2.3 Unzulängliche und wissenschaftlich verbesserungswürdige Zusammensetzung der Kreditinformationen für Scoring

Kreditinformationen müssen drei grundlegende Merkmale aufweisen: Relevanz, Objektivität und Rückverfolgbarkeit. Der Schwerpunkt liegt dabei auf der Relevanz, das heißt starke Korrelation mit Kredit. Das ist das essenzielle Merkmal von Kreditinformationen. Sie bietet einen wesentlichen Maßstab für zukünftige Entwicklung von Scoring-Standards an und balanciert zwischen zu breiter und zu enger Erfassung.

8.2.3.1 Externalisierung des internen Verwaltungsverhaltens

Der Anhang 1 „Standard zur Evaluation der persönlichen Kreditinformation" von „Maßnahmen zur Verwaltung des Credit Scores und Credit Scorings von Einwohnern der Stadt Rongcheng" bezieht die Personalverwaltung von Mitarbeitern in Unternehmen in Kreditinformationen ein. Beispielsweise werden zwei Punkte für normale Verstöße gegen die Vorschriften des Werks abgezogen, 20 Punkte für stärkere Verstöße und 50 Punkte für schwerwiegende Verstöße.

8.2.3.2 Administration des zivilen Verhaltens

Der Community-Management-Teil von Rongchengs „Standard zur Evaluation der persönlichen Kreditinformation" bezieht eine große Anzahl von Beziehungen zwischen natürlichen Personen als zivilrechtlichen Subjekten in die Kreditinformationen für die administrative Bewertung ein, zum Beispiel „Ablehnung von Mediation zwischen streitenden Nachbarn" und so weiter.

8.2.3.3 Schrumpfen mehrdimensionaler Werte auf einen einzigen Score

Die „Vorläufigen Maßnahmen zur Verwaltung von Personal Credit Scores (Bai Lu Credit Score) in Fuzhou" umfassen sieben Kategorien von persönlichen öffentlichen Kreditinformationen: grundlegende persönliche Informationen, persönlicher Sozialkredit, persönlicher beruflicher, persönlicher wirtschaftlicher, persönlicher administrativer, persönlicher gerichtlicher Status sowie weitere relevante persönliche Kreditinformationen bgzl. Gesetze, Verordnungen und Richtlinien. Die Kategorie „persönlicher beruflicher Kredit" beinhaltet positive Anrechnungspunkte für gutes berufliches Verhalten und negative Anrechnungspunkte für schlechtes berufliches Verhalten. Darunter sind Kriterien wie „Innovation in der Arbeit oder bedeutende Erfolge usw." Offensichtlich gute berufliche Verhaltensweisen, die zu positiver Bewertung führen, haben keine starke direkte Korrelation zur Kreditbeurteilung. Zum persönlichen administrativen Kredit zählen Informationen über verwaltungsrechtliche Sanktionen der zuständigen Behörden in Bereichen wie Straßenverkehr, Transport, Umweltschutz, Kultur und Sport, Arbeitssicherheit, Sozialsicherheit, Städtebau, Stadtverwaltung, soziale Ordnung, behördliche Genehmigung und Steuererhebung. Zum Beispiel sieht Artikel 7 der „vorläufigen Maßnahmen zur Verwaltung von Personal Credit Scores (Bai Lu Credit Score) in Fuzhou" Folgendes vor: Persönlicher beruflicher Kredit umfasst Auszeichnungen und Bestrafungen, die Einzelpersonen in ihrem Beruf erhalten. Zu guten persönlichen beruflichen Kreditinformationen gehören hauptsächlich Hingabe an die Arbeit, Tätigkeiten zum Dienst am Volke, Sozialengagement, vorbildliches Verhalten als Sozialmodell sowie Innovation und bedeutende Erfolge in der Arbeit usw. Schlechte persönliche berufliche Kreditinformationen umfassen hauptsächlich Bestrafung aufgrund von Verstößen gegen Parteidisziplin oder politische Disziplin, Bestechung oder

Annahme von Bestechung sowie Erlangen beruflicher Lizenzen oder Qualifikationen durch unzulässige Mittel. Zum persönlichen administrativen Kredit zählen Informationen über verwaltungsrechtliche Sanktionen der zuständigen Behörden in Bereichen wie Straßenverkehr, Transport, Umweltschutz, Kultur und Sport, Arbeitssicherheit, Sozialsicherheit, Städtebau, Stadtverwaltung, soziale Ordnung, behördliche Genehmigung und Steuererhebung. Am Beispiel des Straßenverkehrs werden in der Dokumentation „Vorfälle und Standardskala (negative Vorfälle und Minuspunkte) zur Berechnung der Personal Credit Scores in Fuzhou" verschiedene Verstöße und entsprechende Minuspunkte aufgeführt. Bei sieben Arten von Verkehrsverstößen wie Rotlichtverstoß von nicht motorisierten Fahrzeugen und Mopeds, unerlaubtes Mitnehmen von Personen, Fahrten gegen Fahrtrichtung, Fahrten auf falscher Fahrspur, Verfälschung von Autokennzeichen, Fahrten in gesperrten Bereichen, unerlaubte Änderung von Kraftfahrzeugen und unerlaubtes Parken von Kraftfahrzeugen werden fünf Punkten abgezogen. Bei Fahrern und Bevollmächtigten, die während der Anmeldung eines Kraftfahrzeuges oder nicht motorisierten Fahrzeugs relevante Informationen nicht wahrheitsgemäß zur Verfügung stellen oder verfälschte Anmeldungsunterlagen (einschließlich verfälschter Informationen und verfälschter Bescheinigungen usw.) abgeben, sowie bei denjenigen, die Unfallstellen oder Unfallhergang im Straßenverkehr gefälscht haben, werden 150 Punkte abgezogen. Das Einbeziehen von Verhalten wie Unterlagenfälschung als kreditbrechendes Verhalten entspricht dem Grundsatz der Ordnungsmäßigkeit im Verwaltungsrecht. Aber die Klassifizierung nicht schwerwiegender Verstöße gegen Straßenverkehrsvorschriften, wie zum Beispiel Fahrten auf falscher Fahrspur, die durch Unvertrautheit mit Straßen oder unklare Verkehrszeichen verursacht sein können, als kreditbrechendes Verhalten entspricht weder dem Kreditgedanken noch der Ordnungsmäßigkeit im Verwaltungsrecht. Gleichzeitig können illegale Handlungen nicht einfach mit kreditbrechendem Verhalten verwechselt werden. Nicht jede illegale Handlung ist im Allgemeinen ein Verstoß gegen den „Gesellschaftsvertrag", sodass man zu der Schlussfolgerung gelangt, dass illegale Handlungen Kreditbruch sind. Dies wird zur Verwirrung in der Rechtslogik führen. Das Rechtssystem ist von mehrdimensionalen Werten geprägt. In den Augen des Gesetzgebers werden solche mehrdimensionalen Werte jedoch auf einen einzigen Wert reduziert, nämlich auf Kredit, der dazu führt, dass „Sozialkredit" allgegenwärtig und allmächtig wird. Aus Kreditakten werden schließlich „Mülleimer der Gesetzwidrigkeiten" (vgl. 沈毅龙 2019, S. 112).

8.2.4 Risiko der Generalisierung von Kredit aufgrund ungenauer Definition der Anwendungsszenarien und Grenzen von Anwendbarkeit des Scorings

Nach den traditionellen Regeln für die Kreditwürdigkeitsprüfung sind die Folgen von Kreditausfällen messbar und vorhersehbar. Die geltenden Bereiche sind ebenfalls eingeschränkt. Die Anwendungsszenarien für städtische Personal Credit Scores sind nicht streng begrenzt und die geltenden Bereichen werden ständig ausgedehnt und erweitert, sodass die rechtlichen Konsequenzen unvorhersehbar sind. Dies steht im Widerspruch zur Stabilität und Vorhersehbarkeit des Gesetzes und vermischt in gewissem Maße die Grenze zwischen Kredit- und Strafverfolgungsmechanismus. Der Strafverfolgungsmechanismus entsteht durch gesetzliche Autorität, folgt rechtlichen Verfahren und besitzt eine verbindliche Zwangsgewalt. Für die betroffenen Personen ist dies vorhersehbar und erlösend. Das städtische Personal Credit Scoring dient dazu, Verhalten der Stadtbewohner in der Vergangenheit zu klassifizieren und diese nach Maßstäben zu quantifizieren. Für die betroffenen Personen ist es nicht bekannt, welches Verhalten seinerzeit Kreditbruch wäre, welche Bestrafung das kreditbrechende Verhalten verursachen würde und ob die Bestrafung durchgeführt würde. Das ist hinderlich für den Schutz von Rechten, verstößt gegen das Strafgesetz und bringt Marktsubjekten Unsicherheit (vgl. 周汉华 2016). Es ist nicht zu leugnen, dass die kontinuierliche Erweiterung der Anwendungsszenarien und Bereichen des Personal Credit Scorings einen guten Demonstrationseffekt auf die Schaffung einer sozialen Atmosphäre von „Belohnung von Kredittreue und Bestrafung von Kreditbruch" hat. Allerdings handelt es sich bei den für Personal Credit Scoring verwendeten Kreditinformationen um historische Informationen, die eine rückwirkende Wirkung und eine unerwartet strengere Restriktion für das beurteilte Subjekt haben. Jeder Versuch, die ursprünglich für Kreditrisikomanagement entwickelten Basisinstrumente als Bewertungsinstrumente zu Integrität und Ethik zu verwenden, insbesondere wenn solche Bewertungen von der Regierung bzw. öffentlichen Behörden durchgeführt werden und mit öffentlichen Diensten verbunden sind, ist eine Entfremdung und Generalisierung von Credit Scoring (vgl. 汪路 2018, S. 114).

Darüber hinaus beschließt die Regierung Maßnahmen zur Belohnung von kredittreuen Subjekten. Zusätzlich zur Bereitstellung verschiedener Arten von Unterstützung und Bequemlichkeit in Bereichen wie öffentlich-rechtlicher Verwaltung, öffentlichen Diensten und finanziellen Subventionen gewähren einige Regionalregierungen sogar direkte materielle Belohnungen. Zum Beispiel schreiben die „Maßnahmen zur Verwaltung von Personal Credit Scores in Rongcheng – Belohnungs- und Bestrafungsmaßnahmen" in Artikel 14 Nr. 4 vor, dass „Personen, die mindestens ‚A' geratet werden, bekommen, … in der Aufnahme als Parteimitglieder, Nominierung als Arbeitnehmer mit hervorragender Leistung, Qualifikationsbewertung, Berufung für bestimmte Positionen, Verleihung von beruflichen Titeln und sonstigen Aktivitäten, vorrangige Empfehlung." Gemäß Artikel 12 „… genießen sesshafte Einwohner mit einem Rating von ‚AA' oder ‚AAA', die keine Beamten sind … folgende Belohnungen: (1) 200 Yuan Zuschuss bei Zahlung der Grundrentenversicherung; (2) 100 Yuan Zuschuss bei Zahlung der Grundkrankenversicherung; (3) fünf Prozent Zuschuss auf den geltenden Erstattungsgrundsätzen an Versicherten durch die Krankenversicherung; (4) 100 Yuan Zuschuss für sesshafte Einwohner unter dem vollendeten 64 Lebensjahr für öffentliche Verkehrsmittel …". Obgleich es sich bei den derzeitigen, auf Personal Credit Scores basierenden Governance-Maßnahmen um Bevorzugung und Belohnung handelt, haben Städte wie Suqian und Weihai festgelegt, dass sie das Prinzip „grundsätzlich Belohnung beim guten Kredit und ausnahmsweise Bestrafung bei schlechtem Kredit" verfolgen. Nicht auszuschließen ist, dass es während der Anwendung von Bestrafungsmaßnahmen weiterhin Verdacht auf Verletzung des Grundsatzes von Verhältnismäßigkeit im Verwaltungsrecht besteht (Tab. 8.5).

Tab. 8.5 Vergleich von Anwendungsszenarien für Personal Credit Scores in den Schlüsselstädten

Stadt	Persönliche Kredit-Score	Maßnahmen	Anwendungsszenario
Fuzhou	Mo Li Credit Score	Belohnung	*verwendete Bereiche*: 1. Tourismus, 2. medizinische Versorgung, 3. Bildung, 4. Finanzen, 5. Kultur und Unterhaltung, 6. öffentlicher Verkehr, 7. administrative Dienstleistungen.
Hangzhou	Qian Jiang Credit Score	Belohnung	*verwendete Bereiche*: 1. Befreiung von Überprüfung für Campus-Fitness, 2. Nachträgliche Zahlung für Busse und U-Bahnen, 3. komfortable medizinische Behandlung mit nachträglicher Zahlung, 4. 50 % Rabatt auf Kaution für öffentliche Mietwohnungen; 5. Befreiung von Kaution für Bibliotheksausweis; 6. Befreiung von Kaution für Leasing von Büromaterialien. *Zukünftige Bereiche*: 1. Automietung, 2. Altersversorgung, 3. Credit Score Etikette auf Partnerbörsen.
Hohhot	Ding Xiang Credit Score	Belohnung	*verwendete Bereiche*: 1. Nominierung zum Arbeitnehmer mit hervorragender Leistung, 2. Verwaltungslizenz, 3. Öffentliche Dienste, 4. Allokation von staatlichen Finanzmitteln.
Rongcheng	Rong Cheng Credit Score	Belohnung	*verwendete Bereiche*: 1. Nominierung zum Arbeitnehmer mit hervorragender Leistung; 2. Aufnahme als Parteimitglieder; 3. Ernennung zum Kader; 4. Verleihung von beruflichen Titeln; 5. Empfehlung zu Volksdeputierten, Parteivertreter oder Mitglieder der politischen Konsultativkonferenz des chinesischen Volkes; 6. Wahl des Kaders in ländlichen Regionen; 7. Rekrutierung; 8. Projektantrag; 9. Auszahlung der staatlichen Finanzmitteln; 10. Politische Unterstützung; 11. Transaktionen der öffentlichen Mitteln; 12. Gewerbliche Kredite; 13. Administrative Überprüfung und Genehmigung (Xin Yi Pi); 14. Öffentlicher Verkehr und Automietung (Xin Yi Xing); 15. Tourismus (Xin Yi You); 16. Medizinische Behandlung (Xin Yi Yi).
Suzhou	Gui Hua Credit Score	Belohnung	*verwendete Bereiche*: 1. Verlängerung der Mietdauer für öffentliche Fahrräder, 2. Erhöhung der Anzahl der auszuleihenden Bücher, 3. Ermäßigung der Bürger- und Freizeitkarte. *Zukünftige Bereiche*: 1. Öffentlicher Verkehr, 2. Stadterscheinung und Stadtverwaltung, 3. Bildung, 4. Stadtwerke, 5. Gesundheitswesen, 6. Talenteinführung, 7. Unternehmenskooperation.

(Fortsetzung)

Tab. 8.5 (Fortsetzung)

Stadt	Persönliche Kredit-Score	Maßnahmen	Anwendungsszenario
Weihai	Hei Bei Credit Score	Belohnung und Bestrafung	verwendete Bereiche: 1. Tourismus (Xin Yi You); 2. Kreditgenehmigung (Xin Yi Dai); 3. Auto- und Fahrradmietung (Xin Yi Xing); 4. Rabatt beim Einkaufen (Xin Yi Gou); 5. Mitarbeiterrekrutierung, unternehmensinterner Stellenwettbewerb, Wohlfahrtsverteilung; 6. Kommerzielle Darlehen; 7. Öffentlicher Verkehr; 8. Gesundheitswesen.
Xiamen	Bai Lu Credit Score	Belohnung	*Verwendete Bereiche*:1. Fahrradmietung ohne Kaution, 2. Nachträgliche Zahlung beim Einkauf in unbemannten Supermärkten, 3. Befreiung von Kaution für Bibliotheksausweis, 4. späte Zahlung der Mieten, 5. Gulangyu-Fähre ohne Anstehen, 6. Berufsbezeichnungen und Antrag auf berufliche Beförderung, 7. behördliche Überprüfung und Genehmigung (auch bei mangelnden Unterlagen).
Yiwu	Noch nicht benannt	Belohnung ergänzt durch Bestrafung	*verwendete Bereiche*:1. Behördliche Genehmigung; 2. Kreditfinanzierung; 3. Staatsfinanzen; 4. Öffentliche Ausschreibungen; 5. Transport und Tourismus; 6. Kautionsfreies Leasing; 7. Gesundheitlich Behandlung; 8. Wohnungssicherheit; 9. Nominierung zum Arbeitnehmer mit hervorragender Leistung
Zhengzhou	Shang Ding Credit Score	Belohnung ergänzt durch Bestrafung	*Verwendete Bereiche*:1. Wohnungsmieten (Xin Yi Zu), 2. Gesundheitswesen (Xin Yi Yi), 3. Bibliotheken (Xin Yi Yue), 4. Tourismus (Xin Yi You), 5. Altersversorgung (Xin Yi Yanglao), 6. Verkehr (Xin Yi Xing), 7. 信易跑 (Xin Yi Pao), 8. 信易京东(Xin Yi Jingdong); 9. Kreditgenehmigung (Xin Yi Dai)
Suqian	Xi Chu Credit Score	Belohnung und Bestrafung	*Verwendete Bereiche*:1. Finanzielle Aktivitäten; 2. Öffentlicher Verkehr; 3. Öffentliche Kultur- und Sportdienstleistungen; 4. Tourismus; 5. Gesundheitswesen; 6. Verwaltungs- und behördliche Genehmigung; 7. Beschäftigung (Selbstständigkeit oder Existenzgründung); 8. Altersversorgung; 9. Öffentliche Projektausschreibung; 10. Öffentliches Beschaffungswesen

8.3 Gesetzliche Regelung des städtischen Personal Credit Scorings im Kontext der Rechtsstaatlichkeit

In der modernen Gesellschaft hat das von der Regierung durchgeführte Personal Credit Scoring einen direkten Einfluss auf die sozialen öffentlichen Interessen der Gesellschaft sowie die Rechte und Pflichten bestimmter verwalteter Personen. Während des beschleunigten Aufbaus des Sozialkreditsystems auf der nationalen und regionalen Ebene mit kontinuierlicher Verbesserung des Personal Credit Scorings sowie Erweiterung der Anwendungsszenarien auf alle Aspekte der Sozialgovernance sollen Probleme und Risiken des Personal Credit Scorings zur gleichen Zeit klar identifiziert werden. Aufgrund dessen ist die Einbeziehung der gesetzlichen Vorschriften eine wesentliche Voraussetzung der Konzeption des Sozialkreditsystems auf höchster Ebene.

8.3.1 Verbesserung des Mechanismus vom Personal Credit Scoring

Als neue Methode der Sozialgovernance wirkt das Personal Credit Scoring aus formaler Sicht durch Verwendung von Reputations- und anderen Umsetzungsmechanismen auf das Verhalten der Beurteilten und deren Konsequenz, und es hat daher ähnliche soziale Funktionen wie Gesetze (vgl. 胡凌 2019, S. 26). Um die Wissenschaftlichkeit und Rechtmäßigkeit zu gewährleisten, sollten die Transparenz- und Verfahrensanforderungen bei der Erstellung und Umsetzung von Regeln des Personal Credit Scorings vollständig umgesetzt werden.

8.3.1.1 Befolgung der prozeduralen Anforderungen an die Erstellung der Scoring-Regeln, um möglichst großen Konsens zu erreichen

Da das von der Regierung geführte Personal Credit Scoring eine vergleichbare Funktion wie Gesetze hat und daher sich auf die Rechte und Pflichten unspezifischer Gruppen auswirkt, müssen bei der Erstellung von Scoring-Regeln entsprechende Verfahren eingehalten werden. Erstens soll die führende Behörde während der Konsultationsphase Einwände weiterer Verwaltungsbehörden hinreichend berücksichtigen. Zweitens ist die Professionalität der Scoring-Regeln durch Teilnahme von Experten am gesamten Erstellungsprozess zu gewährleisten. Drittens sind die Scoring-Regeln der Öffentlichkeit zur Stellungnahme zu veröffentlichen, damit ein möglichst großer Konsens darüber gebildet werden kann.

8.3.1.2 Verbesserung des Offenlegungsverfahrens von Personal Credit Scoring-Regeln und Schutz des Auskunftsrechts des betroffenen Kreditsubjektes

Das Recht auf Information ist eine wichtige Grundlage und Voraussetzung für den Schutz des Kreditsubjekts, seine Rechte auszuüben. Erst durch Aufbau von entsprechenden Systemen und Regeln können die Kreditsubjekte vollständig Verständnis über die Art und Weise gewinnen, wie deren Kreditinformationen eingesammelt, zusammengefasst, verteilt und verwendet werden und wie der Score ermittelt wird (vgl. 王伟 2019, S. 111). Einerseits soll die Verwaltungsbehörde einen Datenkatalog zum Personal Credit Scoring und entsprechende Implementierungsregeln umgehend offenlegen. Alle relevanten Eigenschaften, Abfragemethoden und Fristen sowie die Aktualisierungshäufigkeit der offengelegten Informationen sind ausführlich zu erläutern. Artikel 38 der „Suqian Sozialkreditrichtlinie" sieht vor, dass Sozialkreditsubjekte das Recht haben, die Erhebungs- und Verwendungsverfahren ihrer Sozialkreditinformationen, ihre Kreditauskunft oder ihren

Kreditstatus sowie die Quelle und die Gründe der Scoreänderung zu erfahren. Andererseits ist durch den Offenlegungsprozess von Kreditinformationen die Regelkonformität des Scorings zu verbessern und das Recht der Kreditsubjekte auf Informationen sicherzustellen, sodass die Prognose für und (Verhaltens-) Auswahlfähigkeiten der Kreditsubjekte gewährleistet werden können.

8.3.1.3 Aufbau eines Mechanismus zum Einspruch, Reparatur und Löschung im Personal Credit Scoring

Der Einspruchsmechanismus konzentriert sich auf die Gewährleistung des Rechts von Kreditsubjekten auf Auskunft und Berichtigung. Es soll ermöglicht werden, dass einzelne Personen das Recht haben, ihre Kreditdaten zu erkunden. Haben sie Fragen und Einwände über aufgenommene Daten, soll deren Recht gewährleistet sein, Einspruch bei der zuständigen Behörde für Credit Scoring zu erheben. Nach der Überprüfung dieser Behörde sind Maßnahmen situationsabhängig einzuleiten, wie Korrektur, Löschen oder Ergänzung. Der Schwerpunkt des Reparaturmechanismus besteht darin, das Recht auf Vergessenwerden durch vordefinierte Gültigkeitsdauer und automatische Löschvorgänge von persönlichen Kreditdaten zu realisieren. Da Personal Credit Scoring für die Verwaltungsbehörden ein Sozialgovernanceinstrument und ein Verwaltungsakt ist, liegt der Fokus des Löschungsmechanismus darauf, die Rechte der Kreditsubjekte zur Nachprüfung von Rechtsstreitigkeiten und Klagebefugnis zu gewährleisten, um gesetzliche Rechte der Kreditsubjekte ausreichend zu schützen.

8.3.2   Optimierung der Dimensionen des Personal Credit Scorings

Wenn die Dimensionen des Credit Scorings nicht wissenschaftlich konzipiert sind, wirkten sich diese unmittelbar auf den Rang der Kreditsubjekte im Anreizprozess aus. Aus diesem Grund müssen die Legitimität und Rationalität der Bewertungsdimensionen der Personal Credit Scorings kontrolliert werden.

8.3.2.1 Legitimitätskontrolle

Indikatoren und Dimensionen des Scorings dürfen nur auf objektive Daten und Unterlagen basieren, die zur Identifizierung und Analyse des Status der Erfüllung von Verpflichtungen einer Einzelperson verwendet werden können. Informationen, die nicht damit zusammenhängen, insbesondere Verbotsdimensionen gemäß Gesetz wie Religion, Gene, Fingerabdrücke, Blutgruppen und Krankengeschichte sind auszuschließen.

8.3.2.2 Rationalitätskontrolle

Das Personal Credit Scoring als Kreditinstrument muss dem Verbot einer unzulässigen Verbindung unterliegen. Das Verhalten von einem Kreditsubjekt soll nicht einfach als „kreditwürdig" oder „kreditbrechend" eingestuft werden. Ebenfalls sollen umfangreiche Maßnahmen zur Belohnung und Bestrafung nicht verallgemeinert auf die beiden Klassen

verteilt werden. Ein angemessener Zusammenhang zwischen dem Verhalten des Kreditsubjekts und den von der Verwaltungsbehörde vorgeschriebenen Belohnungs- und Bestrafungsmaßnahmen ist sorgfältig zu prüfen (vgl. 王瑞雪 2017, S. 171). Die Dimensionen
des Personal Credit Scorings unterliegen ebenfalls dem Grundsatz der Verhältnismäßigkeit. Das Gesetz soll die Verwaltungsbehörden dazu verpflichten, eine Liste von Bewer-
tungsindikatoren auf der Grundlage der Erfordernisse und der Aktualität der Governance
zu erstellen.

8.3.3 Beschränkung der Anwendungsbereiche des Personal Credit Scorings gemäß Gesetz

Im Jahr 2009 wurde gegen die US-amerikanische Ratingagentur, TransUnion, aufgrund
Verwendung ihrer Credit Scores im Recruitingprozess in Connecticut eine Klage erhoben.
Viele US-Bundesstaaten haben durch Gesetze Kredituntersuchungen (einschließlich Informationen über Kreditwürdigkeit) während des Recruitingprozess verboten. Die Verwendung von Credit Scores seitens Versicherungen oder Vermieter verursachte auch behördliche Überprüfungen und gesellschaftliche Kontroversen. Viele meinen, dass der
Kreditstatus eines Verbrauchers wenig damit zu tun, eine stabile Arbeit zu erwerben, eine
komfortable Wohnung zu mieten oder einen Versicherungsvertrag abzuschließen, da sie
fundamentale Bestandteile einer modernen Gesellschaft sind. Dafür sollte der Credit
Score nicht als Hindernis gelten (vgl. 刘新海 2016, S. 161).

8.3.3.1 Anwendungsbereiche definieren und Belohnung von Kredittreue als Hauptprinzip

Gegenwärtig wird das Personal Credit Scoring hauptsächlich auf der Belohnungsebene
eingesetzt. Mit der Vertiefung der Kredit-Governance werden Verwendungsumfang und
Grenzen von Personal Credit Scorings immer erweitert. Es ist ebenfalls zunehmend in
verschiedene Ebenen der Sozialgovernance eingebettet. Nach dem Grundsatz der gesetz-
mäßigen Verwaltung darf das Gesetz nicht ohne Ermächtigung durchgeführt werden.
Wenn Personal Credit Scorings im Bereich der Bestrafung verwendet werden, muss dies
ausdrücklich vom Gesetz ermächtigt werden und gleichzeitig gesetzlich gebunden sein,
um sich nicht zu einem übermächtigen und leicht missbräuchlichen Instrument zu entwickeln.

8.3.3.2 Geltende Anwendungsbereichen definieren, Ermessensspielraum der Verwaltungsorgane beschränken nach dem Prinzip der starken Korrelation zu Kredit

Unterschiedliche Scoringsanwender haben unterschiedlichen Bedarf. Wer Credit Scoring
verwendet, soll ihn auch ermitteln. Die Kreditklassifizierung wird von Scoringsanwendern
aufgrund Verwendungsszenarien selbstständig festgelegt, was wissenschaftlichem Prinzip
entspricht. Zum Beispiel achtet das Market Supervision Bureau genau auf die illegalen

Vorfälle der überwachten Objekte, um die Genauigkeit der Strafverfolgung zu verbessern. Abteilungen wie die Nationale Entwicklungs- und Reformkommission und die Wissenschafts- und Technologiekommission, die mehr staatliche Finanzmittel einsetzen, legen mehr Wert auf die Historie der betrügerischen Beschaffung von Finanzmitteln durch Verwaltungsobjekte (vgl. 罗培新 2017, S. 10). Das Gesetz soll vorschreiben, dass nach dem Grundsatz der Verhältnismäßigkeit sowie der starken Korrelation zu Kredit die Verwaltungsbehörden in Bezug auf ihren eigenen Verwaltungsbereich, basiert auf ihren eigenen Verwaltungspraktiken, den Umfang der anwendbaren Bereiche und eine Liste dazu feststellen.

Mit der weit verbreiteten Verwendung der Personal Credit Scorings infiltrieren und beeinflussen die Mechanismen ihrer Einstellung und Funktionsweise immerfort die traditionellen Governance-Prozesse und spielen eine wichtige Rolle bei der Verbesserung der Relevanz und Wirksamkeit von Sozialgovernance. Als ein neues Governanceinstrument besteht auch Probleme wie zu breit erfassende Kreditinformationen, nicht ausreichende Transparenz der Bewertungsregeln und unklar definierte Verwendungsbereichen. Aus diesem Grund sind vollständige Umsetzung der Verfahrensanforderungen bei der Erstellung und Implementierung der Regeln von Scoring, Optimierung der Bewertungsdimension und Einschränkung der Verwendungsbereichen des Personal Credit Scorings unvermeidlich, um die Rechtmäßigkeit und Wirksamkeit dieses neuen Governanceinstruments sicherzustellen. Diese sind ebenfalls Anforderungen für Förderung der Rechtsstaatlichkeit bei der nationalen Governance und Aufbau einer Kreditgesellschaft.

Literatur

胡凌 (2019):《数字社会权力的来源:评分、算法与规范的再生产》,《交大法学》2019年第1期.
姜明辉 (2017):《优化个人信用评价 促进社会治理创新》,《社会治理》2017年第9期.
罗培新 (2017):《信用治理别用„大炮打蚊子"》,《解放日报》, 2017年07月02日.
龙西安 (2004):《个人信用、征信与法》中国金融出版社.
刘新海 (2016):《征信与大数据 移动互联网时代如何重塑„信用体系"》,中信出版社.
沈毅龙 (2019):《公共信用立法的合宪性考察与调整》,《行政法学研究》, 2019年第1期.
汪路(2018):《征信若干基本问题及其顶层设计》,中国金融出版社.
王瑞雪 (2017):《政府规制中的信用工具研究》,《中国法学》2017年第4期.
王伟 (2019):《个人信用分的法治维度》,《中国信用》2019年第3期.
王淼 (2019):《从„社会信用建设"到"信用社会建设"》, http://www.creditsd.gov.cn/28/83138.html. 2019-05-07.
周汉华 (2016):《社会信用体系建设的法律问题》, http://www.rmlt.com.cn/2016/1027/443557.shtml. 2016-10-27.
张涛 (2019):《个人信用评分的地方实践与法律控制 – 以福州等7个城市为分析样本》,《行政法学研究》2019年第2期.

Minghua Lin 林明华, Associate Professor der Lehr- und Forschungsabteilung des Xiamen Administration Instituts, Forschungsschwerpunkt: Verwaltung gemäß Gesetz und Kreditgovernance.

Risk Culture as a Means of Mitigating Conduct Risk

9

Similarities with and Differences to the China Social Credit System

Thomas Kaiser and Tatjana Schulz

Abstract

Banks around the world have been exposed to numerous cases of misconduct at individual as well as on a systemic level, inflicting harm on single customers and the society as a whole. This includes inappropriate product design (e.g. securitizations which led to the financial crisis), large-scale market manipulations (LIBOR and other reference rates), as well as other fraudulent activities (e.g. creation of accounts without the knowledge of the affected clients). Regulatory bodies have reacted with a broad range of requirements and recommendations. A key tool in fighting misconduct is the strengthening of risk culture as an "institution's norms, attitudes and behaviours related to risk awareness, risk-taking and risk management, and the controls that shape decisions on risks"(EBA 2017). Implementing risk culture frameworks is a means of influencing behaviour of employees to mitigate those risks in individual banks and thus ultimately to improve the reputation of the banking sector as a whole. While banks have made progress in designing those frameworks, the maturity of this particular discipline is still at a moderate level and full-scale implementation is not yet common. The China Social Credit System also aims at improving behaviour of individuals and corporations by setting clear expectations and measuring compliance with those. Chinese authorities have gathered substantial experience with this methodology during pilot implementations and are refining the approach further during the rollout throughout the country. A

T. Kaiser
Goethe-Universität Frankfurt, Frankfurt am Main, Deutschland
E-Mail: kaiser@finance.uni-frankfurt.de

T. Schulz (✉)
KPMG AG, München, Deutschland
E-Mail: tatjanaschulz@kpmg.com

© Springer Fachmedien Wiesbaden GmbH, ein Teil von Springer Nature 2020
O. Everling (Hrsg.), *Social Credit Rating*,
https://doi.org/10.1007/978-3-658-29653-7_9

comparison of those two approaches leads to suggestions on what the two approaches could learn from each other.

9.1   The Need for a More Sustainable Risk Culture in Banks

Questions of behaviour and misconduct have never attracted as much attention in the financial services sector as they do today. Misconduct in banking creates a wide variety of potential risks. These range from adverse customer outcomes to weakening the resilience of individual institutions. More broadly, misconduct damages public trust in the banking sector and can even contribute to systemic instability (FSB 2018). The experiences from the financial market crisis in 2008 and further trust-shattering developments show that misconduct not only can lead to the failure of individual institutions, but to chain reactions throughout the whole world. It is therefore not surprising that research identified corporate governance deficiencies, poor cultural foundations, and significant ethical failure as the main causes of the financial market crisis in 2008 and the scandals of subsequent years (Cohn et al. 2014). Reputational damage and the loss of public confidence in the banking sector as a whole resulted in further significant financial losses for banks in terms of fines, litigation and regulatory action, as well as for society as a whole. An entire industry that was once kept in high regard and trustworthiness made negative headline news by being involved in mailbox companies ("Panama Papers"), "dividend stripping", "cum-ex-deals" or other morally questionable behaviours (KPMG 2019).

This was caused by major failure of the prevailing corporate and risk culture, i.e. the norms and informal rules in the banking industry. Those should have served as a prerequisite for value-based business practices and risk-adequate behaviour by their employees, but instead have sometimes been found to tolerate or even encourage dishonest behaviours rather than preventing them (Cohn et al. 2014). In the course of analysing and remediating the financial market crisis and financial scandals, commentators and scientific investigations have attributed these scandals to the financial sector's business culture and have thereby launched a discussion on culture that was previously unknown in the banking sector.

International standard setters as well as numerous regulators in countries all over the world (e.g. FSB 2012, 2014) have started to pay close attention to this rather vague and intangible topic over the years. A number of European initiatives focused on banking conduct and behaviour, complemented by the Financial Stability Board's toolkit (FSB 2014) aimed at reducing misconduct risks and urged banks to take an active, self-critical approach to misconduct. Ultimately, banks are expected to actively engage with their own risk culture, understand it as a critical tool for proper risk management, and use it accordingly.

The concept of a "sound and consistent risk culture" (EBA 2017) – adopted into German legislation with the 5th amendment to the Minimum Requirements on Risk Management (MaRisk) – does not aim for a new risk management approach (BaFin 2017a, b).

Rather, banking regulators expect institutions to deal with the issue more closely and define for themselves which businesses, behaviours and practices are ultimately considered desirable and which are not (Hannemann et al. 2019).

The term "risk culture" is defined as an "institution's norms, attitudes and behaviours related to risk awareness, risk-taking and risk management, and the controls that shape decisions on risks" (EBA 2017). Four indicators, which however are neither meant to be exhaustive nor as a checklist for supervision, look into the appropriateness of an institution's risk culture:

- Tone from the Top
- Effective Communication and Challenge
- Clear Accountability
- Appropriate Incentive Structures (EBA 2017).

In this respect, risk culture aims to consciously deal with risks in daily business. Its firm anchoring in the institutes' corporate culture should ensure that risk awareness is created which shapes the daily thinking and actions of management and employees at all levels of the institute. Risk culture thus is the crucial link between existing operational and methodical risk management procedures and their actual effectiveness and sustainability (KPMG 2019). This finding is also confirmed by the latest review by the Group of Thirty (G30) (G30 2018), which makes it clear that focusing on behaviour and culture neither is an add-on nor a reluctant response to regulatory requirements nor can it be. Despite the challenges that arise for institutions from establishing an appropriate risk culture, it can ultimately be critical to a bank's economic success and the sustainability of the entire sector.

Studies show that many institutions have so far focused mainly on the qualitative description of their risk culture. Often it is not clear from the relevant documents whether the description is the target state or the currently prevailing risk culture. Instruments for the systematic monitoring and evaluation of defined cultural indicators have so far been used only sporadically (KPMG 2019). For a holistic implementation of a risk culture framework however, it is not enough to just create documents referring to existing methods, processes, and other policies and procedures. In order to achieve true cultural change, behaviour and attitude of people at all levels must be influenced. Ultimately, it requires a system that can collect and evaluate information about behaviour.

9.2   Key Elements of a Risk Culture Framework

Improving risk culture has to be based on the four key building blocks. It is important to link those building blocks closely with each other as well as with the traditional processes and governance structures in banks.

Tone from the Top

The leadership culture of an institution, i.e. the credible commitment of its management to risk-adequate behaviour plays an important role in the establishment of a viable risk culture. The role-model behaviour of all executives is an essential building block – if even board and senior management do not advocate their own corporate values and behave contrary to the agreed-upon code of conduct, employees also will find little reason to behave in a risk-adequate and ethical manner. Ensuring the understanding and awareness of risks should not only be part of the initial professional and personal aptitude test for key individuals ("fit and proper") but should be continuously scrutinized and promoted (G30 2015).

The behaviour of the members of the executive board should therefore reflect the value system defined by them, which in turn forms the basis for the behaviour of all employees and the desired risk culture. For this, the code of conduct imposed by the management determines which behaviour is acceptable and which is not. In particular, the code of conduct should clarify that illegal activities are explicitly disapproved of and that ethical behaviour is expected. The latter should not solely be based on legal requirements, but should also be shaped to a considerable extent by social expectation (Lo 2015).

However, to ensure lasting and continuous change to a more sustainable risk culture, it is not just the top management of an institute that has a duty. Rather, culture and responsibility has to be created at all levels of the organization. Particular emphasis should be placed on middle management and the so-called first line of defence, i.e. operational areas such as loan processing. A recent study on "Banking Conduct and Culture" conducted by the G30 (2018) therefore recommends to make cultural responsibility a permanent and integral part of business models and to include it into day-to-day processes in order to avoid regression to old behavioural patterns or even cultural fatigue.

"Tone from the top" also includes the management's task of communicating the value system within the institution, ensuring its consideration when taking risks as well as its tight link with risk management and internal controls. Again, the behaviour of middle management plays a crucial role as they, representing the link between top management and employees in the various business units, have to transport and communicate the value system, the risk culture, as well as its principles and objectives.

Effective Communication and Challenge

A sound risk culture promotes an environment that fosters divergent views throughout decision-making processes and enables the questioning of current practices. The positive critical attitude of employees finds support in an environment of open and constructive engagement ("speak-up culture" and "listening culture"). To achieve this, transparency and an open dialogue are necessary – at all levels and at all times. Between management and the supervisory body as well as between management, middle management, and employees, open communication of alternative views, constructive suggestions, and criticism should be encouraged (G30 2015, 2018).

Closely linked to this element of a sustainable risk culture are other cultural elements such as communication, leadership and error culture. Individuals must be able to express their concerns about practices that they consider illegal, unethical or at least questionable in confidence and without concern for reprisals (Fed New York 2017). Furthermore, mistakes should not be concealed or even covered up for fear of consequences, but openly communicated and discussed. The benefits of an open error culture can be made visible to all employees by identifying which mitigation actions could have been derived from open communication about errors. In this respect, the establishment of an appropriate risk culture poses great challenges to the managers, who are ideally guided by an open and cooperative management style (KPMG 2019).

Such a culture also challenges traditional hierarchies and role models. After all, its implementation means that control functions are enabled to work independently and have the same significance as business units. This is a new approach for many institutions, as control functions traditionally play a minor role. Regular channels of communication between units of the first and second lines of defence, in which all participants meet at par and openly discuss with each other, have to be set up.

At a formal level, this element also means establishing risk culture reporting that informs senior management about the current risk culture, its deviation from the target risk culture, and planned measures to promote it at a regular basis.

Employees, in turn, need to be clearly informed about the target risk culture and the requirements they have to fulfil in this context. This is where the aforementioned cascaded risk appetite comes into play again. It is not enough to communicate risk appetite at an aggregated level via the risk strategy. Instead, the risk appetite should be specified transparently and comprehensibly and cascaded to all levels – also with regard to risk culture. Institutions need to be aware that tangible risk appetite statements need to be defined for non-financial risks in particular, as a zero-tolerance approach is not realistic. Only then can the desired risk appetite influence "decisions of management and employees during the day-to-day activities" and have "an impact on the risks they assume" (EBA 2017).

In order to foster a culture of open communication, institutions should set up regular channels of communication between business units. There should also be incentives for openness about errors (error culture) and clear responsibilities. The middle management plays a major role as on the one hand it serves as role model for the actual style of communication in the institution. On the other hand, managers are accountable for identifying, assessing, and managing risks within their areas of responsibility. Finally, it has to network across all neighbouring business units in order to break up "silo mentalities" or not to let them arise in the first place (KPMG 2019). In doing so, the risk limits and the value system of the institute must be observed and clear tasks and responsibilities assigned.

Clear Accountability

Effective risk management is not just the responsibility of top and middle management. Rather, successful risk management requires that individuals at all levels understand the core values of the institution along with its approach to risk, that they are capable of fulfilling the roles assigned to them, and that they are aware of their responsibility for their risk behaviour and that they will be held accountable (G30 2015, 2018).

A positive risk culture is supported by clearly defined roles and responsibilities. Thus, the alignment of an individual's actions with the value system, the defined risk appetite, and the existing risk limits lie within the responsibility of each employee.

To achieve this responsibility, employees need to know the risk-related goals and associated values and be aware of the imminent consequences following the failure to meet the expected behaviour (KPMG 2019).

This shared responsibility is an essential prerequisite for promoting an appropriate risk culture whose principles and objectives should be communicated consistently by all members of the management to all employees. Only in this way both management and employees will be able to align their activities with the value system, the defined risk appetite, and the existing risk limits. Likewise, everyone must be aware of the imminent consequences. To support this, there needs to be an incentive system that rewards desired behaviours and penalizes unwanted ones.

Appropriate Incentive Structures

A positive risk culture is supported by clearly defined roles and responsibilities. It is strengthened by appropriate compensation and performance structures. In such cultures, executives and employees have a clear understanding of their roles, performance expectations, and goals. Likewise, the consequences that follow unwanted behaviour are transparent. Achievement and reward can therefore support the desired risk culture, if incentives and consequences are triggered by clearly defined behavioural expectations (G30 2015, 2018).

For example, taking on excessive or undesirable risks as well as developing unacceptable business activities and practices, may lead to disciplinary action such as bonus cuts, warnings or, in extreme cases, terminations. On the other hand, desired behaviour must also be rewarded. This is the key to effective behavioural change: Identifying the key positive behaviours and strengthening them through incentives is instrumental in promoting tone, direction, and momentum for change, ultimately resulting in a sustainable risk culture (Khan 2018).

Results of the aforementioned study on risk culture in German banks (KPMG 2019) show that especially in terms of positive incentives, there seems to be a lot of catching up to do in the German financial sector. Positive incentives rewarding desirable behaviour are rarely found in the industry. If present, these are mainly monetary incentives such as profit sharing. Non-monetary positive incentives, such as better career opportunities for individuals who align their actions with the value system, the set risk appetite, and the existing risk limits are very rarely found in the industry.

The reason for this certainly is not only the lack of awareness that incentives and performance structures are at the very centre of many questions concerning risk culture and risk behaviour. Rather, the identification of key positive behaviours poses particular challenges to institutions, because behavioural monitoring is of crucial importance. Governing bodies and persons with governance roles need to have access to information that integrates data from the areas of finance, risk management, sales, and human resources (Kellermann et al. 2013). Few banks invest in systems that provide this information in real time to quickly identify and manage areas of risk and to develop methods for more effective coaching of their employees. In addition, the integration of non-monetary incentives would have an impact on other HR processes. Personnel managers would have to examine the risk behaviour of potential employees during the recruitment process (as part of a fit and proper routine). They would have to communicate the defined value system as a basis for the desired behaviour and risk culture not only during the onboarding process, but continuously and establish regular monitoring processes for risk behaviour (DNB 2013).

9.3   Challenges in Establishing a Sustainable Risk Culture

Changing risk culture is a major endeavour. Challenges both on an organizational as well as on a psychological level have to be overcome in order to succeed.

Organizational Hurdles
Many executives have a good intuitive understanding of their organizations' risk culture. A definition of the target risk culture that is concrete and tangible enough to be put down in writing is nevertheless a great challenge (Khan 2018). But the very lack of a clear and holistic understanding of the organisation's risk culture often leads to risks being approached by an inappropriately narrow approach. For instance, Chief Risk Officers or the entire risk functions are equipped with more and more competencies or incentive instruments such as the deduction or postponement of bonus payments (Kellermann et al. 2013). Although these methods can be helpful, they only partially address the risk culture of an institution. Because culture works on different levels, changing it generally takes place slowly and can be very challenging (Kaiser and Saleheyan 2019). To capture the whole concept of risk culture, a holistic definition of the target risk culture of an institution is needed (DNB 2013). This target risk culture will most certainly be influenced by factors such as business areas and target customer orientation, thus creating a high variability in the different types of risk cultures across institutions.

Another challenge on the way to establishing and maintaining a sustainable risk culture is its operationalization. This should include a process of defining and tracking suitable indicators which make it clearly distinguishable, measurable, and understandable by empirical observation. In this way the current risk culture of the organization can be made transparent, a comparison can be made to the defined target risk culture, and finally an implementation programme for achieving the target culture can be designed including concrete measures and methods. The continuous tracking of the achievement of objectives is thus made possible.

Although the topic of risk culture has gradually arrived in the banking landscape and efforts are already underway to improve culture and behaviour, the establishment of a sustainable risk culture is still considered a major challenge. This is not only due to the described methodological difficulties. Rather, results of a recent survey in Germany (KPMG 2019) show that banks identify numerous other obstacles.

While there is a general agreement with the need for action and regulatory influence, the ambiguity with which regulators approach the issue has been criticised. Banks ask for more concrete guidance, e.g. in the form of case studies concerning the day-to-day work of employees, the provision of toolkits as well as an open dialogue between institutions and regulators. Some institutions feel that there is not sufficient initiative for change.

However, surveyed institutes recognize that the major obstacles to establishing a sustainable risk culture are to be found internally. Difficulties have built up and manifested in the respective institutes for years. Traditional role models and hierarchical structures, for instance, not only prevent open communication between employees and managers, but also hinder accountability. Equally obstructive is the so-called silo mentality, which even fuels the frequently observed tensions between various departments.

Many banks acknowledge that the establishment and maintenance of a sustainable risk culture is an ongoing task, not a project completed after a defined time period. This is reflected by the G30's view, which, for good reason, subtitled their most recent publication "A Permanent Mindset Change" (G30 2018). Thus it appears that in practice human and financial resources must be provided permanently for this topic, which again is seen as one of the obstacles.

Overall, ... banks identify the biggest obstacles and challenges amongst their own workforce. A close connection with the readiness and speed of change of employees is established. Above all, some institutions complain about the "persistence" of many seasoned employees who at times oppose cultural change. They thus aim at hiring new employees, which in turn is perceived difficult due to the tense situation in the labour market in general and in particular in some locations.

In connection with the problems of attracting applicants and the retention of qualified personnel, some banks also mention the alleged "mini-me-hiring phenomenon". Many managers are subject to the similarity bias, which means that candidates are considered to be all the more sympathetic and qualified the more similar they are to their own person. In consequence, people are recruited for visual, cultural, and demographic factors that are not related to their job performance. For corporate and risk culture, this means that new employees do not automatically initiate cultural change when selected by people whose attitudes to culture are very similar to their own.

In addition, many institutions would like to see greater diversity across employees both in the whole workforce as well as in specific functions, committees, and project teams. Interestingly enough, respondents in the above-mentioned survey (KPMG 2019) do not only speak about the very obvious characteristics of diversity such as age or gender of the workforce – although there is an explicit desire to position more female employees both in management and in the first line of defence. Rather, many banks spoke in favour of putting together teams in which long-time employees and newcomers as well as people with typical and atypical careers would work together, so that many different aspects can be incorporated into discussions and taken into account in decision-making.

Given that institutions feel that major obstacles to establishing a sustainable risk culture are to be found internally, i.e. difficult personnel selection and the mini-me-hiring phenomenon, disruption of open communication, and lack of willingness to change or slow rate of change within the workforce, banks need to look at ways to influence the behaviour of their employees. Some of them have been consulting specialists in governance, risk management, change management, and organizational psychology and are trying to apply concepts such as "nudging" to influence the behaviour and decision making of the whole workforce or select individuals by positive reinforcement and indirect suggestions. A so-called nudge is "any aspect of the choice architecture that alters people's behaviour in a predictable way without forbidding any options or significantly changing their economic incentives. To count as a mere nudge, the intervention must be easy and cheap to avoid. Nudges are not mandates. Putting the fruit at eye level counts as a nudge. Banning junk food does not" (Thaler und Sunstein 2008). Some banks use posters, placemats or similar means to transport the intended values and elements of the risk culture framework.

In a more pragmatic approach, boards and senior management need to provide clear guidance, e.g. by cascading the overall risk appetite statement to all business levels or by communicating clear "Dos and Don'ts" via a code of conduct which provides examples of desired everyday behaviour. This for instance comprises qualitative risk appetite statements based on error rates, customer complaints, or other metrics.

Banks therefore need specifications that clearly determine which behaviours are desired and which are not.

Psychological Limitations to Changing Behaviour
The effectiveness of influencing behaviour of individuals (and employees in particular) depends on the degree with which the desired change is embraced. One specific component is called reactance. Reactance is a psychological phenomenon which describes an unpleasant motivational reaction to offers, persons, rules, or regulations that threaten or eliminate behavioural freedom (Steindl et al. 2015). That means, reactance occurs when humans feel that they are being deprived of their choices or limited in the range of alternatives (Brehm und Brehm 1981). It is most likely to arise when massive pressure is applied to people to accept a certain view or attitude. This can cause them "to adopt or strengthen a view or attitude that is contrary to what was intended, and also increases resistance to persuasion" (Roeckelein 2006).

Although there are still too few reliable intercultural comparative studies on reactance (Ng et al. 2019), it seems reasonable to assume that concepts such as self-determination and freedom of choice are more salient in individualistic cultures than in collectivist cultures. From this it can be deduced that persons from individualistic cultures feel restricted in their self-determination and freedom of choice earlier on, and thus reactance develops earlier and more pronouncedly, especially when the range of actions is significantly reduced.

Institutions might want to avoid reactance amongst their employees and thus need to be particularly careful when introducing new sets of behavioural rules. In consequence, while illegal and immoral behaviour should be clearly prohibited and lead to strict consequences, employees should still not feel limited in their choices. Applying a risk appetite that has been cascaded to all levels can be a good solution, as it sets clear boundaries for employees, but provides them some leeway within the appetite, leaving them control over their actions. The sense of being able to determine one's own actions and having available a range of solutions will avoid reactance (Steindl et al. 2015).

The example of reactance gives evidence that a too rigorous system cannot be applied to establish a sustainable risk culture in financial institutions. While behaviour should be influenced by clear guidelines, it is necessary to leave sufficient room for manoeuvre and to give employees the opportunity to voluntarily adhere to the given rules.

9.4   What Could Risk Culture Frameworks and the China Social Credit System Learn from Each Other?

China as a country has identified related issues of misconduct by citizens and companies and is in the process of rolling out a process for mitigating those phenomena. The declared goals of this China Social Credit Systems (CSCS; Cheng 2019) are an improvement of the state leadership and the prevention of fraud. Consequently, in an attempt to make individuals as well as companies behave in a certain way, a reward and penalty system is meant to promote trustworthiness, integrity, and mutual trust. The design of the CSCS is discussed in detail in other chapters of this book and hence not outlined in this chapter.

Hence, ... there are two frameworks for shaping behaviour of individuals, one designed by individual banks (guided by common regulatory requirement), the other designed by the government of a country. A comparison of the features of those two approaches appears to be interesting. Thus, conclusions... as to what the two might learn from each other can be drawn.

Table 9.1 aims to analyse key features of the two systems. It has to be considered though that there is no unique Risk Culture framework and CSCS has been adjusted frequently, hence only directional statements could be given.

As it becomes evident, while the goals of the two systems are similar, the means of achieving them are fairly different. It appears that the two systems could learn from each other to some degree, bearing in mind the overall differences in cultural embedding.

As mentioned above, current risk culture frameworks are somewhat hindered by their rather subjective nature and the relatively infrequent assessment. Using readily available, higher frequency data sources/indicators as in the CSCS might improve the system. In contrast to CSCS, which imposes rules on people and reinforces these rules by reward and

Table 9.1 Comparison between risk culture frameworks and the China Social Credit System

Feature	Risk culture	CSCS
Goals	Avoiding conduct risk in a bank	Improving state leadership and preventing fraud
Scope	Narrow: all employees of a bank	Very broad: all citizens and corporations in China
Owner of system	Management board	State
Principles vs. rules	Principle-based detailed by select rules	Rule-based with large number of individual rules
Degree of freedom	Defined ranges (e.g. risk appetite)	Very limited/digital
Risk of reactance	Limited, as substantial degrees of freedom exist	Relatively high, as degrees of freedom are very limited
Speed of change	At most annual adjustments to system	Very flexible with score components added on demand
Frequency of observations	Typically annual assessment and monthly tracking of indicators	Near-time, depending on data source
Degree of objectivity	Limited, strong influence of judgement	High, based on observable data/indicators
Tone from the top	Role modelling and cascading down	System applies to everybody; synchronized implementation
Effective communication and challenge: Challenging of rules	Encouraged within certain limits	Not provided
Accountability	Key design element	No key characteristic ("check the box")
Incentives: Incentives and sanctions	Confined to workplace environment	Broad range across personal and professional life

punishment, a sound risk culture in financial institutions can only be implemented and prevail in other ways. Ultimately, for an appropriate risk culture it is essential to motivate employees to behave in accordance with the value system and the code of conduct and to act within the defined risk tolerances. Material and immaterial incentives can be useful additions here. However, the decisive factor is the attitude of persons (Lo 2015), which leads to desired behaviour resulting from genuine conviction and is not motivated solely by career opportunities or fear of negative consequences. While a change in structures, methods, or processes is certainly necessary to ensure the support of an adequate risk culture, a genuine cultural change can only take place, if employees are convinced that it is necessary and right.

On the other hand, risk culture frameworks seem to have advantages when looking at accountability and reactance. CSCS might benefit from shifting its focus towards principles instead of rules, which might increase the "buy-in" and thus would make individuals feel more accountable for their actions.

9.5 Risk Culture Frameworks and CSCS as Examples for a Broader Range of Social Credit Systems

In addition to the systems discussed so far, incentive systems in a broader sense (including social credit systems) have been developed and are expanding more and more in Western societies. This includes tracking of driving habits to obtain lower insurance premiums for cars, fitness trackers for health and life insurance at reduced rates and so forth. Those systems promise benefits for following defined rules. A question of morality arises when insurance companies and other providers of those services stop offering products to individuals which do not meet defined criteria. This on the one hand would make a voluntary effort a mandatory requirement and on the other hand would exclude certain persons from services, ultimately leading to the question whether the underlying promise of an insurance product (protecting members of a defined cohort from financial perils) still holds true.

Furthermore, companies make use of publicly available information more and more. This includes deriving credit worthiness scores derived from information such as provider of mail address, operating system, and type of device used etc. Hence, the way an individual behaves in the real or in cyber worlds affects his or her prospects to obtain loans and other products and services, not necessarily being aware of or even having consented with those types of analyses.

Incentive systems have been around for a very long time. With modern technologies, readily available data, and instant communication channels, it is possible to refine those systems even more. While there clearly are benefits in making people follow rules and behave in ways which at a minimum do not harm the lives of others, there are ultimately ethical, philosophical, and political questions to answer on how widely those tools should finally be used.

References

BaFin (2017a). *Anschreiben zur Veröffentlichung der MaRisk 6.0*, GZ: BA 54-FR 2210-2017/0002, 27.10.2017.

BaFin (2017b). *Rundschreiben 09/2017 (BA) vom 27.10.2017*, Anlage 1: Erläuterungen zu den Ma-Risk in der Fassung vom 27.10.2017, AT 3, Tz. 1 und AT 5, Tz. 2.

Brehm, S. S., Brehm, J. W. (1981). Psychological reactance: A theory of freedom and control.

Cheng, E. (2019). China is building a 'comprehensive system' for tracking companies' activities, report says. Retrieved from: https://www.cnbc.com/2019/09/04/china-plans-for-corporate-social-credit- system-eu-sinolytics-report.html

Cohn, A., Fehr, E. & Maréchal, M. A. (2014). Business culture and dishonesty in the banking industry. *Nature.* 19, 2014.

De Nederlandsche Bank (2013). *Methods for supervising behaviour and culture*, Amsterdam.

European Banking Authority (2017). *Guidelines on internal governance*, London.

Federal Reserve Bank of New York (2017). *Misconduct risk, culture and supervision*, New York.

FSB (2012). Increasing the Intensity and Effectiveness of SIFI Supervision – Progress Report to the G20 Ministers and Governors, Basel.

FSB (2014). *Guidance on Supervisory Interaction with Financial Institutions on Risk Culture – A Framework for Assessing Risk Culture*, Basel.

FSB (2018). *Strengthening Governance Frameworks to Mitigate Misconduct Risk: A Toolkit for Firms and Supervisors*, Basel.

Hannemann, R., Steinbrecher, I., Weigl, T. (2019). *Mindestanforderungen an das Risikomanagement (MaRisk) – Kommentar*. 5. überarbeitete und erweiterte Auflage 2019, Stuttgart.

Kaiser, T., Saleheyan, A. (2019), Risikokultur als Change-Aufgabe, CGO – das Govenance-Magazin 06/2019, KPMG AG, Frankfurt a. M.

Kellermann, A. J., de Haan, J., de Vries, F. (2013). *Financial supervision in the 21st century*, Berlin.

Khan, A. (2018). *A Behavioral Approach to Financial Supervision, Regulation, and Central Banking*. International Monetary Fund, Washington, DC.

KPMG (2019). *Es ist Bewegung unter der Oberfläche. Studie zur Unternehmens- und Risikokultur in deutschen Banken*. KPMG AG, Frankfurt a. M.

Lo, A. W. (2015). *The Gordon Gekko effect: The role of culture in the financial industry*, National Bureau of Economic Research, Cambridge, MA.

Ng, A. H., Kermani, M. S., Lalonde, R. N. (2019). Cultural differences in psychological reactance: Responding to social media censorship. Current Psychology, 1–10.

Roeckelein, J. E. (Ed.). (2006). Elsevier's dictionary of psychological theories. Elsevier.

Steindl, C., Jonas, E., Sittenthaler, S., Traut-Mattausch, E., Greenberg, J. (2015). Understanding psychological reactance. Zeitschrift für Psychologie.

Thaler, R. H., Sunstein, C. R. Nudge: Improving decisions about health, wealth, and happiness.

The Group of Thirty (G30) (2015). *Banking Conduct and Culture – A Call for Sustained and Comprehensive Reform*, Washington, DC.

The Group of Thirty (G30) (2018). *Banking Conduct and Culture – A Permanent Mindset Change*, Washington, DC.

Prof. Dr. Thomas Kaiser has been working in risk management for more than 20 years. He is Director in the Financial Services division of KPMG AG Wirtschaftsprüfungsgesellschaft in Frankfurt/M. and honorary professor for risk management at the Goethe University Frankfurt. After studying business administration in Saarbrücken and completing a doctorate in the field of financial econometrics in Tübingen, Prof. Kaiser held a managerial role in risk controlling at four major German banks. He is co-editor of the Journal of Operational Risk and the author of numerous essays and books on risk management topics.

Tatjana Schulz works for KPMG AG Wirtschaftsprüfungsgesellschaft in Munich. As a psychologist with a focus on risk research and a banker with many years of experience in the financial services sector, she deals with the qualitative elements of risk management at KPMG. Among other things, she supports the audit teams in the areas of operational risk and risk culture and has valuable insights into the current state of implementation of the regulatory requirements in the German banking landscape.

Überwachungsstaat China

10

Theo Sommer

Zusammenfassung

Die Kommunistische Partei Chinas (KPCh) ist darauf aus, das Verhalten und die Gesinnung der Bürger nach den Vorgaben der Führung gleichzuschalten. Wichtigster Hebel dabei ist das geplante Sozialkreditsystem – eine von Algorithmen gesteuerte Maschinerie, die Bürger, Unternehmen, Institutionen und Behörden überwacht, bewertet und, je nachdem, belohnt oder bestraft. Anders als in der deutschen Schufa geht es dabei nicht nur um Bonitätsauskünfte über Kreditverträge und Zahlungsverhalten, sondern um sämtliche Daten. Das System ist als „Totalitarismus im digitalen Gewand" beschrieben worden. Bundeskanzlerin Angela Merkel soll dazu gesagt haben, George Orwell's „1984" sei im Vergleich dazu nur ein „laues Lüftchen". Das SKS stellt auch die ausländischen Unternehmen in China vor große Herausforderungen.

In der Volksrepublik China entsteht derzeit der technologisch fortgeschrittenste Überwachungsstaat der Weltgeschichte. Die Kommunistische Partei ist darauf aus, nicht nur das Verhalten, sondern die Gesinnung der Bürger nach den Vorgaben der Führung gleichzuschalten. Ein Hebel ist dabei das „Sozialkreditsystem", auch „Bonitätssystem" genannt, ein beispielloses Experiment des ehrgeizigen *social engineering*. Konkret geht es um eine von Algorithmen gesteuerte Maschinerie, die das Verhalten aller Bürger, Unternehmen, Institutionen und Behörden überwacht, bewertet und, je nachdem, belohnt oder bestraft. In diesem „System der gesellschaftlichen Vertrauenswürdigkeit" erfasst der staatliche Datenkrake

T. Sommer (✉)
DIE ZEIT, Hamburg, Deutschland
E-Mail: sommer@zeit.de

© Springer Fachmedien Wiesbaden GmbH, ein Teil von Springer Nature 2020   203
O. Everling (Hrsg.), *Social Credit Rating*,
https://doi.org/10.1007/978-3-658-29653-7_10

sämtliche Lebensbereiche. Als sich Bundeskanzlerin Merkel im Frühsommer 2018 über Chinas Digitalisierungsstrategie informierte, entfuhr ihr beiläufig der Kommentar, George Orwells *1984*-Fantasien seien gegen die chinesische Realität bloß „ein laues Lüftchen".

Zensur und Überwachung sind seit jeher Pfeiler der chinesischen Parteiherrschaft. Seit Xi Staatspräsident ist, das heißt seit 2013, sind sie massiv verstärkt worden. Das Bildungsministerium sieht junge Lehrer und Studierende als Ziele ausländischer Infiltration und verbannt daher westliche Konzepte wie Menschenrechte, Rechtsstaat und Zivilgesellschaft aus den Lehrbüchern. Allein 2017 wurden 128.000 Webseiten abgeschaltet, berichtet Kai Strittmatter. Eine Unzahl ausländischer Medien ist blockiert, darunter Google, Twitter, Facebook und die *New York Times*. Vom Bildschirm verbannt sind Künstler, die als „unanständig, vulgär oder obszön" gelten, Rapper und sogar Sportler mit Tattoos – überhaupt alle, die den „Kernwerten der Partei" entgegenstehen. Unliebsame Plattformen werden geblockt. Wo die „Scheren im Kopf" nicht funktionieren, wird ohne Zögern robust nachgeholfen.

Die Suchmaschinen Baidu, der Onlinehändler Alibaba und der Chat-Dienst Tencent beschäftigen längst viele Tausend von Schnüfflern – Sieber und Löscher von Text- und Bildmaterial. Angeblich durchstöbern zwei Millionen Zensoren permanent das Internet. Die achtzehn größten sozialen Medien geben dafür jährlich schätzungsweise 2,5 Milliarden Dollar aus. Herausgefiltert und geahndet wird dabei Anstiftung zur Verletzung von Verfassung und Gesetzen, zum Sturz der Regierung oder des sozialistischen Systems, zur Spaltung des Landes, zum Schüren von Hass und zum Terrorismus; ferner die Verbreitung von Fake News, Gerüchten, feudalistischem Aberglauben, sexuell anzüglichem Material; schließlich die Verführung zu Spielsucht, Gewalt und Kriminalität. Ohnehin ist, worauf Elizabeth Economy in *Foreign Affairs* hingewiesen hat, die Polizei ermächtigt, Daten aus privaten Mobiltelefonen abzugreifen. Neuerdings gibt es auch Berichte, dass die Handys von einreisenden Ausländern an Grenzstationen ausgelesen werden.

Als biete all dies nicht schon ausreichenden Schutz vor elektronischer Unterwanderung und genug Instrumente der Überwachung, soll 2020 das „Sozialkreditsystem" oder „Bonitätssystem" zur Bewertung der „gesellschaftlichen Vertrauenswürdigkeit" eingeführt werden. Ursprünglich war es – wie die deutsche Schufa – einfach ein Informationssystem über Kreditnehmer und Schuldner. Gegründet 2006 von der Volksbank Chinas mit anfänglich 500 Millionen Scores, hatte es bis Mitte 2019 Daten von 990 Millionen Bürgern und fast 26 Millionen Unternehmen und anderen Einrichtungen gesammelt.

Doch nun geht es – anders als bei der Schufa – nicht nur um Bonitätsauskünfte über Kreditverträge und Zahlungsverhalten, sondern Gerichts- und Gesundheitsakten, biometrische Angaben, Reisepläne, Typ, Wagenfarbe und Nummernschilder der Autos, die Nutzung der sozialen Medien, Einkäufe per Kreditkarte oder Bezahl-App. Dazu kommt die Bild-Erfassung durch Gesichtserkennungskameras, von denen es 2016 schon 176 Millionen gab; inzwischen sollen 600 Millionen installiert sein. Die 15,35 Millionen Einwohner von Chongqing werden laut *South China Morning Post* von 2,58 Millionen Kameras mit Gesichtserkennungsfähigkeit überwacht, das sind 168 Kameras auf tausend Einwohner. Die Ausbeute des Systems Xue Liang – „Adlerauge" – wird von Supercomputern gesiebt, die bis zu 100.000 Kameras automatisch durchsuchen können; sie finden aus 50.000 Besuchern eines vollen Sportstadions in Sekundenschnelle eine gesuchte Person heraus. Außerdem hat das „Amt für Ehrlichkeit" natürlich Zugriff auf alle Daten der chinesischen IT-Giganten. Die Chinesen leben ständig unter der Lupe des Staates.

Ähnlich wie bei den Schufa-„Scores" werden Punkte und Noten vergeben – von AAA („vorbildlich") bis D („unehrlich"). Gesetzestreue, moralisches Wohlverhalten, soziales Engagement bringen Punkte, säumiges Zahlen von Rechnungen und Verkehrsstrafzetteln, Betrug und politische Abweichlerei haben Punkte-Abzug zur Folge. In einigen Pilotprogrammen gelten 1000 Punkte als optimal; wer unter 600 Punkte fällt, wird nach der Devise behandelt „Einmal unehrlich, überall eingeschränkt". Privatpersonen, die gegen irgendwelche Regeln verstoßen haben, kommen auf eine schwarze Liste, auf der schon zehn Millionen Namen stehen sollen; für die in der Wirtschaft tätigen Verantwortungsträger gibt es eine rote Liste.

Dem Regime geht es dabei um die Niederhaltung und Gängelung jeglicher demokratischen Anwandlungen und aller Anfälle „westlicher Dekadenz". Xis formierte Gesellschaft stellt den totalitären Überwachungsstaat, den George Orwell in seinem Roman *1984* beschrieben hat, weit in den Schatten. Kai Strittmatter, der dem System in seinem Buch „Die Neuerfindung der Diktatur" ein ganzes Kapitel widmet, nennt es „Totalitarismus im digitalen Gewand". Es gibt nichts an Tugenden und Untugenden, was der Staat nicht wissen, bewerten, bestrafen oder belohnen will. So wird zur Rechenschaft gezogen, wer bei Rot über eine Ampel fährt, zu oft hupt, nicht den Zebrastreifen benutzt, nicht angeschnallt fährt oder in Parkverbotszonen parkt. In Schanghai sollen 18.000 Roboter als „E-Polizisten" die Verkehrsverstöße festhalten und zur Ahndung gleich an die staatlichen Datenbanken weiterreichen. Desgleichen wird belangt, wer sich regierungskritisch äußert (Orwell nennt derlei Äußerungen „Gedankendelikte") oder an einer Protestdemonstration teilnimmt; wer Hundehaufen nicht beseitigt oder sich weigert, „freiwillig" beim Pflanzen von Bäumen mitzuhelfen; wer Pornos schaut oder zu viel Zeit mitComputerspielenverdaddelt; wer seine Eltern nicht regelmäßig besucht; mancherorts sogar, wer allein in einer großen Wohnung lebt oder ein großes Auto fährt. („Wer in einem riesigen Mercedes zur Arbeit kommt, erhält weniger Punkte als derjenige, der ein Leihfahrrad nimmt", wird ein Beamter aus Xiongang zitiert.)

Drastische Sanktionen erwarten die Übeltäter. Sie werden auf vielfältige Weise benachteiligt. Sippenhaft ist an der Tagesordnung: Ihren Kindern wird der Besuch eines Kindergartens oder der nahen Grundschule verwehrt. Sie dürfen keine Hochgeschwindigkeitszüge und Flugzeuge mehr benutzen; bereits 2017 wurde sechs Millionen Menschen der Kauf von Flugtickets verwehrt und weiteren 17 Millionen die Ausgabe von Eisenbahnfahrkarten. Die Sozialsünder dürfen auch nicht mehr in Hotels der gehobenen Klasse absteigen. Zudem erhalten sie weder Kredite noch Kreditkarten; im schlimmsten Fall können sie ihren Job verlieren. Und sie werden – *naming and shaming* – in Wandzeitungen, im Internet und per telefonischer Information des Bekanntenkreises öffentlich an den Pranger gestellt. Den Punktsiegern hingegen – etwa Knochenmarkspendern und Blutspendern – winken öffentliche Belobigung, Beförderung, vergünstigte Kredite und bessere

Krankenversicherungen. Doch der Staat verlässt sich nicht nur auf elektronische Überwachung: Wie zu Maos Zeiten, werden ältere Leute dafür bezahlt, dass sie den Nachbarn nachspionieren und Berichte über sie liefern.

Erstaunlicherweise befürworten nach einer repräsentativen Online-Umfrage, die Genia Kostka vom Institut für Chinastudien der Freien Universität Berlin durchgeführt hat, 80 Prozent der chinesischen Internetnutzer das repressive Punktesystem, zumal die Gebildeteren, Wohlhabenderen und die ältere Generation. Sie versprechen sich davon eine transparente Messung ihrer Kreditwürdigkeit und damit eine Steigerung ihrer Lebenschancen. Auch kennen sie derlei Systeme schon von großen Online-Händlern wie Alibaba. Den Orwell-Effekt ignorieren sie. Ausserdem geht ihnen Sicherheit über Privatheit. So zitierte die *South China Morning Post* den Taxifahrer Liu Gangqiang; „Die Überwachungskameras geben einem ein Gefühl der Sicherheit. Es gibt weniger Verbrechen." Sein Kollege Wu stimmte ihm zu: „Du bist ja nicht betroffen, wenn Du nicht stiehlst oder die Gesetze brichst, Solange die Kameras nicht in meinem Schlafzimmer oder Bad installiert sind, macht es doch nichts aus. Was soll denn persönliche Privatheit in öffentlichen Räumen?"

Im Niemandsland südwestlich von Peking baut sich Xi Jinping ein 300 Milliarden Dollar teures städtebauliches Denkmal – die Großstadt des 21. Jahrhunderts: durchdigitalisiert, mit sauberen Industrien, Supermärkten, die per Gesichtserkennung Zugang gewähren, Parkplätzen für selbstfahrende Autos und mit öffentlicher Kontrolle durch zigtausend Kameras – ein Prototyp überwachter Urbanität.

Der Aufbau des chinesischen Techno-Polizeistaates mit Hilfe digitaler Bilderfassung, verbesserter Datenanalyse und Künstlicher Intelligenz ist übrigens nicht nur ein innerchinesischer Vorgang, er eröffnet auch eine neue Front geopolitischer Rivalität. Die Autokraten der Welt werden Chinas Orwell-Technologie begierig übernehmen, um ihre Bürger schärfer an die Kandare zu nehmen. Laut Freedom House haben mindestens 18 autokratische Regime bereits chinesische Überwachungstechnologie importiert (*Financial Times*, 21.Januar 2019).

Seit 2014 wird das Sozialkreditsystem in 43 Distrikten oder Kommunen erprobt. „Rasen Sie nicht und fahren Sie nicht unter Alkoholeinfluss. Gefährden Sie ihren Sozialkredit nicht" – Schilder mit solchen und ähnlichen Aufschriften stehen heute schon an vielen Straßen. Aber damit hat es nicht sein Bewenden. Das Kontrollmonstrum, das der Staatsrat 2020 landesweit einführen will, geht weit über derlei vernünftige Mahnungen hinaus. Dies stellt nicht nur die chinesische Privatwirtschaft vor Probleme, sondern auch die in der Volksrepublik tätigen ausländischen Unternehmen. Ihnen drohen umfassende Konsequenzen, sagt Jörg Wuttke, der Präsident der EU-Handelskammer in China. Auch sie sollen nach „Vertrauenswürdigkeit" und „Aufrichtigkeit" bewertet werden. Ein Kriterium: die Zahl der von ihnen beschäftigten Parteimitglieder.

Jörg Wuttke hat keine Illusionen. Er sagt: „Höhere SKS-Werte können zu niedrigeren Steuersätzen, besseren Kreditbedingungen, einem leichteren Marktzugang und mehr Möglichkeiten für öffentliche Aufträge führen. Niedrige SKS-Werte führen zum Gegenteil und können im Extremfall bedeuten, dass ein Unternehmen auf schwarze Listen gesetzt wird. So wird beispielsweise die Häufigkeit von Inspektionen davon bestimmt, wie vertrauenswürdig ein Unternehmen ist – und die Vertrauenswürdigkeit wiederum hängt am SKS-Rating."

Es stehe ein Paradigmenwechsel bevor, argumentiert Wuttke weiter, da die Wettbewerbsfähigkeit bald ebenso stark von der Einhaltung des Corporate-SCS bestimmt würden wie von den Marktkräften. Zudem würden in Zukunft auch die individuellen Bewertungen der Manager, also ihr persönliches Verhalten außerhalb des Arbeitsplatzes, sich auf das Firmenrating auswirken. Sein abschließender Befund: „Ausländische Unternehmen dürften gegenüber lokalen Konkurrenten grundsätzlich benachteiligt sein, da Letztere Vorteile aus ihren engeren Verbindungen zur Regierung ziehen können."

Das Sozialkreditsystem ist denn in der Tat eine große Compliance-Herausforderung für sämtliche Unternehmen in China, auch für die ausländischen. „Alle Marktteilnehmer werden entweder nach dem Score leben oder nach dem Score sterben.", spitzt Jörg Wuttke seinen Befund zu.

In unserem Verhältnis zu dem dynamischen Aufsteiger China könnte sich nach der Einführung des Sozialkreditsystems eine neue Front auftun. Unsere Unternehmen in China plagen sich schon heute mit lästigen und oft recht vagen Regeln und Beschränkungen, die sich dauernd ohne Vorwarnung ändern. Und noch sind die alten Streitfälle nicht erledigt: die Forderung von Chinas Handels-und Wirtschaftspartnern nach einem „ebenen Spielfeld", nach Gegenseitigkeit in den Beziehungen: leichterem Marktzugang, mehr Teilhabe an öffentlichen Ausschreibungen, Aufhebung des zwangsweisen Technologietransfers in *joint ventures*, Respektierung der intellektuellen Eigentumsrechte und beschleunigter Aushandlung eines Investitionsabkommens zwischen China und der EU.

Noch freilich ist das Sozialkreditsystem nicht Realität. Seine endgültige Form und Gestalt steht bisher nicht fest. Die Technologie des Systems befindet sich erst in der Entwicklung. Noch lange wird die Master-Datenbasis nicht von Künstlicher Intelligenz gesteuert werden, sondern von alten Konsolidierungstechniken. Auf jeden Fall sind weitere drei bis fünf Jahre für zusätzliche Anpassungen vorgesehen, doch bis zur vollen Ausbildung des Systems mag es nach anderen Einschätzungen fünfzehn bis zwanzig Jahre dauern. So bleibt durchaus Raum für Verhandlungen darüber, wie weit Peking sein Orwell-Projekt den ausländischen Partnern aufzwingen kann.

Theo Sommer schreibt seit 62 Jahren für Die ZEIT. Er war deren Chefredakteur von 1973 bis 1993 und danach bis 2000 neben Marion Gräfin Dönhoff und Helmut Schmidt ihr Herausgeber. Asien ist eines seiner Lebensthemen. Seine Doktorarbeit schrieb er über „Deutschland und Japan zwischen den Mächten, 1935–1940". China bereiste er zum ersten Mal 1975 und seitdem immer wieder. 1979 veröffentlichte ein Buch über die sich reformierende Volksrepublik, „Die chinesische Karte"; 2019 erschien sein Werk „China First".

Social Credit, Sicherheit und Freiheit

Katika Kühnreich

Zusammenfassung

Die Nachrichten über die sogenannten Social Credit Systeme verbreiteten sich im Westen verhältnismäßig spät und entwickelten sich seither in der Berichterstattung zu einer Art negativem Faszinosum und Clickbait-Thema ohne Tiefgang. In ihrem Beitrag betrachtet die Politikwissenschaftlerin und Sinologin Katika Kühnreich die westliche Kritik der Systeme, arbeitet oft vernachlässigte ideologische Wurzeln der Systeme heraus und legt Verwandtschaften zu Tendenzen im westlichen Machtgefüge frei. Neben dem Eingehen von Aspekten der Gamifizierung innerhalb dieser Systeme zeichnet sie die Wege, auf denen Menschen in solche Systeme geführt werden nach und zeigt internationale Parallelen in den Veränderungen von Herrschaft. Zum Ende wirft sie die Frage nach Konsequenzen, aber auch dem Energie- und Ressourcenverbrauch solch technisierter Systeme der Kontrolle auf.

11.1 Social Credit Systeme in China und westliche Reaktionen auf die Einrichtung

Nachrichten über die Einrichtung von sogenannten „Social Credit Systemen" 社会信用体系 (SCS) in der erstarkten Volksrepublik China machten in den vergangenen Jahren international Schlagzeilen. Während die chinesische Regierung ihre auf die Einbeziehung von analogen wie digitalen Daten und auf Automatisierung ausgelegten Systeme als holistisch und als Maßnahme der Vertrauensbildung bezeichnet, werden sie in der westlichen Berichterstattung meist mit George Orwells „1984" (Orwell 1949) verglichen und als Daten-

K. Kühnreich (✉)
Freiberufliche Autorin und Referentin, Berlin, Deutschland

© Springer Fachmedien Wiesbaden GmbH, ein Teil von Springer Nature 2020
O. Everling (Hrsg.), *Social Credit Rating*,
https://doi.org/10.1007/978-3-658-29653-7_11

diktatur bezeichnet. Im westlichen Diskurs wird die Implementation der Systeme meist abgelehnt. Deren Annahme im Land sowie eine als fehlend wahrgenommene Gegenwehr aus der chinesischen Bevölkerung wird kritisiert. Weitere Kritik fokussiert sich auf die Regierung, die durch die Nutzung digitaler Methoden im Allgemeinen und die SCS-Daten im Speziellen enorme Macht über ihre Bevölkerung gewänne. Gründe für die Einführung der sozialen Kontrollsysteme werden oft im Konfuzianismus verortet. Neben den hier aufgeführten Ängsten existiert aber auch ein häufig geäußertes Misstrauen, welches die Fähigkeit der chinesische Regierung, solch umfangreiche Systeme in ihrem reich bevölkerten Herrschaftsgebiet tatsächlich einrichten zu können, in Frage stellt. Begründet ist dies zum einen in der Unkenntnis der langen Vorbereitungszeit, die die chinesische Regierung in die SCS und ihrer Umsetzung investierte, zum anderen aber einer fehlenden Beachtung der propagandistischen Vorbereitung und Einbettung der SCS.

Existenzielle Fragen, wie die nach den Hintergründen der Einführung solch umfassender Systeme, nach ihren ideologischen Wurzeln und Einflüssen sowie der propagandistischen Vermittlung werden hingegen selten aufgeworfen.

Diesen Fragestellungen soll in diesem Beitrag nachgegangen und der Vergleich mit Orwells Dystopie im Hinblick auf die Systeme untersucht werden. Zugleich wird die Zuhilfenahme neuer, digitalisierter Machtquellen bezüglich des Gesamtkomplexes und nicht nur im Hinblick auf China betrachtet, sondern auch auf internationaler Ebene analysiert.

Dieser Text nähert sich den chinesischen SCS, indem zu Beginn Beweggründe für die Implementation der SCS herausgearbeitet werden. Darauf folgend wird untersucht, welche ideologischen Einflüsse hierzu führten und weshalb eine technologische Konfrontation sozialer Probleme gewählt wurde. Daran anschließend werden Wege der Vermittlung gegenüber der Bevölkerung analysiert. Nachgezeichnet wird dabei auch der Einfluss neuer Technologien wie Internet, mobile Endgeräte und maschinelles Lernen. Abschließend werden Potenziale, Grenzen und Gefahren solcher Systeme ausgelotet und angerissen, wie der neuen Macht durch Daten international begegnet werden kann. Die westliche Kritik am Vorgehen der chinesischen Regierung wird in diesem Beitrag zum Anlass genommen, das Interesse westlicher Regierungen an Daten und ihren Einsatz von Technologien zur gesellschaftlichen Lenkung zu eruieren.

11.2   Die Situation in China – Ein kurzer Überblick über den Stand der Systeme

Nach jahrelangen Vorbereitungen wurde 2014 die erste Phase der Einführung der SCS in China von 2014 bis 2020 verkündet. China demonstriert die technologische Überwachung offen und unter großer staatlicher Aufsicht. Privatwirtschaftliche Firmen sind aber beteiligt. Anders als in westlichen Berichterstattungen meist dargestellt, existieren in China zur Zeit verschiedene SCS. Ende 2019 ist die Entwicklung der Systeme für den kommerziellen Bereich fast abgeschlossen und Individuen durch die staatlichen Testsysteme immer enger erfasst. Die zudem existierenden privatwirtschaftlichen SCS, wie das „Sesame Credit System" Alibabas, wurden hingegen 2018 stärkerer staatlicher Aufsicht untergeordnet.

Viel weiter als die im Westen meist debattierten SCS für Individuen, deren Testsysteme zahlreich und zum Teil sehr unterschiedlich sind, ist also die Entwicklung der Systeme für Firmen. An deren Beispiel lässt sich das Gegenteil der These, der chinesische Staat sei nicht fähig, seine Pläne umzusetzen, substanziieren. Dieses legen u. a. Mirjam Meissner und Björn Conrad von Sinolytics in ihrem Bericht „*The Digital Hand – How China's Corporate Social Credit System Conditions Market Actors*" (vgl. Meissner und Conrad 2019) für die Europäische Handelskammer in China detailliert dar. Auch wenn beide Systeme nicht unabhängig von einander betrachtet werden können, da sie auf bestimmten Ebenen miteinander verwoben sind, sind die Coporate Social Credit Systeme in ihrer Umsetzung viel weiter und ausgereifter.

11.3   Gründe für die Einrichtung der Systeme

Gründe für die Errichtung der SCS können etwa in der seit Jahren schwelenden Unruhe innerhalb der Bevölkerung gesehen werden, in Unzufriedenheit und Misstrauen gegenüber den Behörden, etwa wegen dem oft als lax wahrgenommenen Umgang mit der Lebensmittelsicherheit oder im Umgang mit Epidemien. Sie wurzelt in den Folgen der wirtschaftlichen Umformung des Landes, in der verbreiteten Korruption oder ethnischen Spannungen. Diese Unzufriedenheit spiegelt sich nicht selten im Ignorieren von Regeln, in Dissens bis hin zu Aufständen wider, die von der Regierung seit einiger Zeit als „Massenzwischenfälle" 群体性事件 bezeichnet werden. Diese „Massenzwischenfälle" können zehntausende von Teilnehmerinnen und Teilnehmer haben und mehrere Tage bis Wochen andauern. Allein 2014 kam es nach offiziellen Angaben zu 90.000 solcher „Zwischenfälle", was einen Durchschnitt von ca. 247 pro Tag ergibt (vgl. Reuters 2014). Die Kommunistische Partei (KP) begegnete solchen Widerständen aus der Bevölkerung auf verschiedenen Ebenen, auch, indem sie aus dem Militärischen stammende Aufstandsbekämpfungstaktiken einsetzt (vgl. Kühnreich 2014, 2018). In der Betrachtung der gesellschaftlichen Unruhe der letzten Jahre und dem Umgang mit ihr kann die These aufgestellt werden, dass die chinesische Regierung die SCS als technologisches Mittel zur Lösung sozialer Probleme implementiert.

11.4   Legitimation und ideologischer Wandel

Entscheidend bei einer Untersuchung der Implementierung der SCS ist eine Betrachtung der Systeme im Hinblick auf die Legitimation der Herrschaft der KP. Denn gleich jeder anderen muss sich auch die chinesische Regierung vor ihrer Bevölkerung legitimieren. Die Begründung der Herrschaft der KP hat seit der Gründung der Volksrepublik 1949 unterschiedliche Wandlungen durchlaufen, von klassischen sozialistischen Begründungen für die Macht der Partei bis hin zu einer immer größeren Öffnung gegenüber dem klassischen Sozialismus fremden Legitimationen. Seit den 1980er-Jahren gehören dazu beispielsweise Nationalismus und Konfuzianismus, was sich anhand der Aufnahme des Forschungsbereichs Konfuzianismus in den VII. Fünfjahresplan von 1986 ablesen lässt.

Die heutige Einbeziehung der digitalen Sphäre in die eigene Macht sollte im Bezug auf die KP nicht überraschen. Die chinesische Regierung erkannte das Internet und die durch die Nutzung entstehenden Daten schon früh als Form der Kommunikation, die somit aufgezeichnet, ausgewertet und gesteuert werden kann. Zu der Einsicht, dass der gesamte vom Menschen erzeugte digitale Datenstrom als Machtmittel genutzt werden kann, war es nur noch ein kleiner Schritt.

Obwohl es sich bei den SCS um technologische Lenkungssysteme handelt, scheint der Anteil kybernetischen Denkens in der Auseinandersetzung mit den Systemen überraschender Weise meist übersehen zu werden. Denn die Systeme verfügen durch ihre vier Grundaspekte (Selbstkonfiguration, Selbstheilung, Selbstoptimierung und Selbstschutz) über implementierte Anpassungsfähigkeit an neue Gegebenheiten. Durch ihr flexibles Design und ihrer ebensolchen Ausrichtung auf die Anpassung an neue Gegebenheiten, wurden die chinesischen SCS von Samantha Hoffman als kybernetische Systeme analysiert und als „Autonome Nervensysteme" bezeichnet (vgl. Hoffman 2017). Die seltene Nennung kybernetischer Anteile an den chinesischen SCS mag aber auch in einer fehlenden gesellschaftlichen Debatte um Kybernetik und instrumentelle Vernunft begründet liegen, die es im deutschsprachigen Raum letztmalig in den 1970er-Jahren gab.

11.5   Gedanken hinter Systemen – die tiefen Wurzeln der Kybernetik

Zu Beginn der Schaffung eines Systems steht der Gedanke. Dieser Gedanke muss Notwendigkeiten und Vorteile seiner Umsetzung in einem neuen Ordnungsprinzip darlegen und den enormen Energie- und Kostenaufwand, den die Einrichtung solch umfassender, holistischer Strukturen bedeutet, rechtfertigen. Zudem muss die Einführung neuer Systeme propagandistisch vorbereitet und begleitet werden. Ein solcher Energie- und Kostenaufwand auf so vielen Ebenen muss in jedem System gerechtfertigt werden, was bedeutet, dass der dahinter stehende Gedanke diesen Aufwand rechtfertigen muss. Im Bezug auf politische Ordnungssysteme sind die Sichtweise der Natur des Menschen für den Aufbau der angestrebten Gesellschaft und den Umgang mit Minderheiten von hoher Relevanz.

Noch lange bevor von Social Credit Systemen gesprochen wurde, prägte ein anderer Begriff die Sichtweise chinesischer Politikerinnen und Politiker auf Politik und Menschen: der des 社会管理 „Social Management" oder gesellschaftlicher Lenkung. Anhand dieses Begriffs kann die Herkunft der dahinterliegenden Ideologie sehr viel besser betrachtet werden als an dem des Social Credit, denn die Lenkung weist auf den dem Altgriechischen entlehnten Begriff der Kybernetik hin.

Bei der Kybernetik handelt es sich um eine Denkweise oder Hypothese, deren Ursprung im Zweiten Weltkrieg liegt. Im von deutschen Luftschlägen angegriffenen Großbritannien suchte der Mathematiker Norbert Wiener nach einem technologischen Weg, den fliegenden Kriegsmaschinen der Nazis, unter anderem den V2-Raketen, zu begegnen. Das Ziel war eine Maschine, der es gelänge, auf alle Bewegung der Angriffsmaschine zu reagieren um sie unschädlich zu machen. Auch wenn es Wiener nicht gelang, diese Verteidigungsmaschine zu realisieren, zeigt sich in diesem Ursprung ein originäres Kennzeichen der Kybernetik: Die sich selbst regulierende Anpassung ihrer lenkenden Einflussnahme. Unter dem Titel „Cybernetics, or the Control and Communication in the Animal and the Machine" (Wiener 2013) veröffentlichte Wiener 1948 seine Gedanken.

Schon in ihrem Namen ist dieses Programm verankert: „Kybernetik" ist dem Altgriechischen entlehnt, was „ein Schiff zu steuern" bedeutet. Dieses perfekte Bild enthält das Programm der klassischen Kybernetik: Ein Ziel anzusteuern, dabei aber um die Existenz unterschiedlichster Einflüsse zu wissen und zu akzeptieren, dass diese Auswirkung auf den Kurs des Schiffs haben werden. Sie könnten ein erfolgreiches Ansteuern auf das angestrebte Ziel verhindern, plante man sie nicht mit ein. Im Falle eines Schiffes könnten dies Strömungen, Tiden, Untiefen und Winde sein. Durch die Einberechnung dieser Kräfte und die direkte Reaktion in Form eines Ausgleichens wird das Ziel trotz all dieser Einflüsse erreicht.

Relevant im Hinblick auf die politische Philosophie sowie deren Umsetzung in reale Politik ist wie bei jeder Denkweise die Sichtweise der Natur des Menschen und folglich der Aufbau der Gesellschaft. Stark vereinfacht gesagt betrachtet die klassische Kybernetik den Menschen als Transmitter. Sie sieht ihn als durch äußere Gegebenheiten beeinfluss- und lenkbar. Werden die korrekten Reize gesetzt, zeigt der Mensch das zum Ziel gesetzte Verhalten. Dieses stellt eine Sichtweise des Menschen dar, die sich auch im Behaviorismus wiederfindet.

In den dem Zweiten Weltkrieg folgenden Jahrzehnten machte die Kybernetik unterschiedliche Entwicklungen durch und wurde sowohl von militärischen als auch zivilen Einflussnehmern geschätzt und verbreitet. Entwicklungen, die Thomas Rid in seinem lesenswerten Werk „Rise of the Machines" (Rid 2016) nachvollzieht. Verschiedene Zweige der kybernetischen Denkweise bildeten sich aus. Auf Zusammenkünften wie den Macy-Konferenzen wurde etwa entschieden, die Kybernetik im angloamerikanischen Raum mit einflussreichen Personen aus der Kultur zusammenzubringen, um die Kybernetik so breiter zu implementieren. Durch das Zusammentreffen von militärischem Denken, der Hippie-Bewegung und LSD in den 1970er-Jahren kam es zu ungeahnten Wendungen und Zweigen der Kybernetik und letztendlich zur Geburt des Cyberspaces. Das Wort „Cyber" ist dem englischen „Cybernetics" entlehnt. Viele technologische Entwicklungen, die heute unsere Leben beeinflussen und zu unserer Normalität geworden sind, wurden von Anhängerinnen und Anhängern der Kybernetik und nicht zuletzt von denen der sogenannten Palo-Alto-Kybernetik aus dem Silicon Valley erdacht.

11.6 Cyber China – Das Aufeinandertreffen der Kybernetik mit dem chinesischen System

Wie kam es aber dazu, dass die Kybernetik zu einem wichtigen Einfluss im heutigen Denken der chinesischen KP wurde? Das Interesse einer asiatischen sozialistischen Partei an der westlichen Ideologie der Kybernetik wirkt auf Menschen im Westen oft irritierend,

wobei verschiedene Aspekte der politischen Theorie und Philosophie in Vergessenheit geraten zu sein scheinen: Auch der Sozialismus ist ursprünglich eine westliche Ideologie.
Am Beispiel Großbritanniens im 19. Jahrhunderts entwickelt, wurde dieses politisch-
philosophisch Denkkonzept bei seiner Reise gen Osten immer wieder übersetzt und verändert. Unbestritten hat der Sozialismus Chinas in seiner Geschichte große Veränderungen
durchgemacht und war von Beginn auch auf Änderungen und nicht den politischen Stillstand, ausgerichtet. Wie fast jedem Sozialismus wohnt auch dem chinesischen ein positiver Zukunfts- und Technikglaube inne. Die Hoffnung des 19. Jahrhunderts, dass Maschinen und Technologie den Menschen vom Joch der Arbeit befreien, spiegelt sich sowohl
im kapitalistischen Glauben als auch dem sozialistischen wider.

In seiner Anfangszeit prägten den chinesischen Sozialismus eine überhöhte Verehrung
westlicher Denkweisen sowie ein kritischer Umgang mit den chinesischen Traditionen,
bis hin zu Abgrenzungen und Abneigungen, wie sie sich auch in der Neue-Kultur-
Bewegung 新文化运动 in den 1910er- und 1920er-Jahren und der 4.-Mai-Bewegung
五四运动 1919 zeigt. Geprägt und vorangebracht wurde der frühe chinesischen Sozialismus aber meist von Denkerinnen und Denkern, die eine klassisch konfuzianische Bildung
durchlaufen hatten, bevor sie sich sozialrevolutionären Philosophien anschlossen.

Das Verhältnis der KP gegenüber den chinesischen Traditionen veränderte sich, was
sich auch in den großen Kampagnen und deren Namen zeigt. Während die 100-Blumen-Be-
wegung 百花运动 noch mit einem poetischen Namen bedacht wurde, wandte sich die Große
Proletarische Kulturrevolution 文化大革命 mit einem klaren Namen in aller Härte gegen
Traditionen, Religionen und gegen die neugebildeten Eliten.

Über den Zeitverlauf hinweg lässt sich aber eine vermehrte Sinisierung der marxistisch-
leninistischen Philosophie beobachten. Hierbei folgt der chinesische Sozialismus interessanterweise einer chinesischen Tradition: Bei einem Aufeinandertreffen mit einflussreichen fremden Denkeinflüssen werden Elemente dieser Philosophien in die eigene
Denkweise integriert, statt sich, eventuell erfolglos, komplett gegen die konkurrierende
Denkschulen zu stellen. So im Konfuzianismus Einflüsse des Daoismus, des Buddhismus
und später auch der Aufklärung aufgenommen worden waren, änderte sich auch die in
China dominierende Auslegung des Sozialismus durch das Zusammentreffen und die Auseinandersetzung mit chinesischen Denktraditionen.

Dass die Kybernetik und der chinesische Sozialismus auf eine besondere Weise miteinander verschmolzen, steht im Zusammenhang mit den anti-linken Verfolgungen in den
USA der McCarthy-Ära und dem sogenannten „Second Red Scare" in den 1940ern und
1950ern. In einer politischen Stimmung, die Arthur Miller zu seinem Stück „Hexenjagd"
inspirierte, wurden zahllose Personen aufgrund von anti-linken, anti-liberalen und anti-
kommunistischen Vorurteilen verfolgt. Verschwörungstheorien reichten aus, um selbst
hochstehende Persönlichkeiten aus ihren beruflichen und sozialen Stellungen zu entfer-
nen. So erging es auch dem chinesischstämmigen Physiker, Mathematiker und Rakenten-
ingenieur Qian Xuesen 钱学森 und seiner Frau, der Opernsängerin und Musikdozentin
Jiang Ying 蒋英, einer Absolventin der Berliner Universität der Künste sowie der Musik-

hochschule Luzern. Der angesehene Wissenschaftler Qian, der während des Zweiten Welt-kriegs im Rahmen des Manhattan Projekts an der Entwicklung der ersten Atombomben mitgearbeitet hatte, wurden 1950 Sympathien für den Kommunismus vorgeworfen, er wurde erniedrigt, interniert und nach Jahren des Hausarrests 1955 zusammen mit Jiang Ying nach China verschifft.

In der jungen Volksrepublik engagierte sich Qian in seinen Forschungsgebieten und brachte neben seinem wissenschaftlichen Fachwissen und seiner Praxis in Bereichen wie der Raketentechnik und dem Bau atomarer Bomben auch die Ideologie der Kybernetik mit.

Qian war seit Jahren Kybernetiker. Wie später in Allendes Chile wurde die kyberneti-sche Hypothese auch in China mit Interesse aufgenommen. Das leninistisch geprägte Sys-tem des chinesischen Sozialismus, welches wie jeder Sozialismus der Theorie nach auf Partizipation und mehr oder minder geführte (Selbst-) Organisation der Bevölkerung setzt, auf deren Rückmeldungen angewiesen ist, bekam ein Hilfsmittel, das unter der Zuhilfe-nahme von Technologien ungeahnte Möglichkeiten versprach. Mit dem Verschmelzen der beiden Theorien entstand die Keimzelle, aus der sich die heutigen Social Credit Systeme entwickelten.

So brachten die aus den USA Deportierten die Kybernetik als neuen Denkansatz mit. Mit den Jahren entwickelte sich dieser im Zusammenspiel mit den vorhandenen Ideolo-gien zum Social Management Gedanken, der unter Xi Jinping als „Social Governance" 社会治理 bezeichnet wird. Der oft als Einfluss genannte Konfuzianismus hat hingegen bei weitem keinen überwältigenden Einfluss auf den Aufbau der Systeme.

Seit der Rückkehr von Qian Xuesen sind Jahrzehnte vergangen, in denen sich die Welt stark veränderte. Im Zuge der ideologischen Umwälzungen nach Maos Tod, die von der KP als „Politik der Reform und Öffnung" 改革开放 bezeichnet werden, wurde China gegenüber der Welt und somit auch deren kapitalistischen Ländern geöffnet. In dieser Zeit begann auch eine vermehrte Umsetzung von Theorien des „Social Management" oder ge-sellschaftlicher Lenkung inklusive ihrer kybernetischen Einflüsse (vgl. Hoffman 2017, S. 55). Damit einher ging die Anpassung der Lehre. In den 1980ern erlebte China, wie so viele andere sozialistische Länder in diesem Jahrzehnt, immer wieder gesellschaftliche Bewegungen für mehr Mitbestimmung. 1989 startete in China eine Bewegung, die die Machtzusammenhänge vor den Augen der Weltöffentlichkeit während des Gorbatschow-Besuchs in Beijing konfrontierte. Nach dieser Auseinandersetzung zwischen Parteispitze und Teilen der Bevölkerung begann eine erneute Umformung im Apparat selbst, aber ins-besondere in der Kommunikation zwischen dem Herrschaftsapparat und der Bevölkerung.

Zudem wurde der Konsum als Mittel entdeckt und mit 让一部分人先富起来 „Lasst einen Teil zuerst reich werden" eine chinesische Version der Trickle-Down-Economics propagiert (vgl. Leonard 2009, S. 45). Die starken Veränderungen der ideologischen Aus-richtung wurden unter dem Begriff „Sozialismus chinesischer Prägung" 中国特色社会主义 zusammengefasst und mit einer Anpassung an speziell chinesische Verhältnisse be-gründet.

11.7   Propaganda: Kommunikation und Macht

In verschiedenen westlichen Studien, Interviews und Berichten wurde bisher eine (zumindest geäußerte) Zustimmung aus der chinesischen Bevölkerung gegenüber der Implementierung der SCS festgestellt, welche im westlichen Diskurs kritisiert wird. Für eine Analyse der Gründe für diese geäußerte Zustimmung bietet sich die Betrachtung der Kommunikation zwischen Regierung und Bevölkerung an.

In jedem Machtsystem, sei es politischer, religiöser oder anderer Ausrichtung, besteht die Notwendigkeit, die Beteiligten auf die eine oder andere Weise von der Teilnahme zu überzeugen. Während im frühen Sozialismus der Sowjetunion „Agitprop", Agitation und Propaganda, das probate Mittel der Unterrichtung der Massen über die eigenen Ziele war, um sie auf diese Weise auf die eigene Seite zu bewegen, entwickelte die chinesische KP nach und nach ihre eigenen Wege der Propaganda, später auch durch die Untersuchung der Machtstrategien politischer Gegner. So ließ die Regierung nach 1989 etwa die Wege westlicher Staaten untersuchen, an der Macht zu bleiben (Ai 2008). Die dabei entdeckte Public Relations wurde als relevantes Machtmittel herausgearbeitet und in die chinesische Politik integriert. Im Westen oft wenig beachtet richtete die chinesische Regierung das von sich selbst und dem Land vermittelte Bild nach Innen und Außen neu auf. Anne-Mary Brady ist eine der wachsamsten Beobachterinnen der Veränderungen der chinesischen Propaganda und Public Relations, die sie in ihren Veröffentlichungen nachzeichnet (Brady 2014). Die Einbeziehung neuer Technologien wie des Internets wird von der Regierung in ihren propagandistischen Methoden immer weiter verfeinert, was auch Roger Creemers, einer der aufmerksamsten westlichen Beobachter der Entwicklung der SCS, beschreibt (Creemers 2015).

11.8   Unbewusste Partizipation – die „neue Freiwilligkeit"

In westlichen Berichten über die SCS der Volksrepublik werden Vergleiche zu George Orwells autoritärer Überwachungsdystopie „1984" nahezu inflationär verwendet. Dies kann zum einen in den anti-kommunistischen Öffentlichkeitskampagnen innerhalb der westlichen Länder in vergangenen Jahrzehnten begründet sein, zum anderen aber auch in der Unkenntnis des Romans selbst. In einem Vergleich des in dem Roman selbst geschilderten Systems mit den SCS fallen mehr Unterschiede als Ähnlichkeiten auf. Zwar beschreibt Orwell die Überwachung als technologisch gestützt, mit interaktiven Überwachungskameras sowie anderen Geräten und Massenkampagnen sowie die Fähigkeit der Regierung, die Vergangenheit durch eine Änderung der Aufzeichnungen zu ändern, was auch in heutigen Systemen zu finden ist, vor allem aber viel Zwang. Die Tendenz heutiger Methoden sozialer Kontrolle, und auch der chinesischen, geht dagegen in eine andere Richtung, nämlich, die Bevölkerung nicht durch den bei Orwell genutzten Zwang, sondern durch später noch angesprochene Lockungen in die Systeme zu integrieren, was als eine „neue Freiwilligkeit" bezeichnet werden könnte. Diese „neue Freiwilligkeit" basiert aber nicht auf einem freien Willen, sondern ist in ihrem Wesen eine unbewusste, von außen gelenkte Partizipation am System.

Große Teile der heutigen Überwachung basieren auf dieser unbewussten Partizipation der Überwachten, meist fälschlicherweise als „Freiwilligkeit" bezeichnet wird. Menschen partizipieren nicht freiwillig an den Systemen, ergo durch bewusstes Abwägen zwischen Vor- und Nachteilen sowie deren Konsequenzen. Auf diese Weise ermöglichen Menschen unbewusst neue Überwachungsarchitekturen und deren Verbreitung, etwa, indem sie Technologien nutzen, die neue Formen der Überwachung erst ermöglichen. Meist geschieht dies, ohne dass die Überwachten die Überwachungspotenziale der Technologien überdenken. Ein einfaches Beispiel sind sogenannte „Smartphones". Durch unsere Verwendung der Geräte akzeptieren wir eine immer umfassender werdende Überwachung des öffentlichen und privaten Raums. Die neuen Machtkomplexe sind partizipativ-spielerische Mitmachsysteme der Überwachung, keine in erster Linie harten und abschreckenden Mangelsysteme, vor denen Orwell warnt. Möchte man im Vergleich bei klassischen Utopien bleiben, bietet Huxleys „Brave New World" (Huxley 1932) bessere Parallelen.

Zudem beschreibt Orwell einen rein staatlichen Machtkomplex. Heutzutage finden wir aber Mischsysteme aus staatlicher und privater Überwachung vor. Weder der US-amerikanische noch der chinesische Staat können eine derart komplexe Überwachung aus der eigenen Infrastruktur heraus stemmen: Staaten und Privatwirtschaft gehen in unserer Zeit Bünde ein und nutzen die gesammelten Daten gleichermaßen. Ein Umstand, der sich am DE-CIX problematisieren lässt, dem weltgrößten Internetknotenpunkt in Frankfurt am Main. Um die über diesen physischen Knotenpunkt laufenden Informationen abhören zu können, erklärte der deutsche Auslandsgeheimdienst BND, der ausschließlich im Ausland tätig sein darf, den Knotenpunkt kurzerhand zum „virtuellen Ausland". Eine Praxis, die vom Bundesverwaltungsgericht als legal betrachtet wird.

11.9   Gamifizierung

Der spielerische und partizipative Charakter neuer Systeme sozialer Kontrolle zeigt sich auch in der Integration von Techniken wie der Gamifizierung. Die kommerziellen Social Credit Systeme, wie etwa Alibabas „Sesame Credit", sind gamifizierte Systeme. Ursprünglich aus der Computerspielentwicklung stammend, nutzt diese Technik im Menschen angelegtes Verhalten dazu aus, die Aufmerksamkeit der Zielpersonen zu binden, positive Emotionen gegenüber dem Gegenstand zu erzeugen und so Verhaltensänderungen zu erreichen. Verkürzt gesagt nutzt die Gamifizierung den im Gehirn mit Erfolg und Belohnung verknüpften Dopaminausstoß, um das menschliche Verhalten zu beeinflussen – ohne, dass dies der betroffenen Person bewusst ist. Gamifizierte Systeme werden nicht kenntlich gemacht, ihre Art der Wirkung nicht vermittelt. In gamifizierten Systemen sind die Teilnehmerinnen und Teilnehmer eine durch das System zu trainierende Masse, deren Teile individuell anvisiert werden können. Der Umstand, dass sie konditioniert werden, bleibt den Betroffenen verborgen, ebenso die Regeln, nach denen die Konditionierung verläuft.

Gamifizierte Systeme werden im Westen wie im Osten eingesetzt, vom Militär bis in die Werbung, im kommerziellen Bereich besonders gerne, um an persönliche Daten von Individuen zu gelangen. Bei dem in Deutschland weit verbreiteten „Payback"-System handelt es sich um das erste kommerzielle gamifizierte System. Allein das Versprechen von Firmen, Konsumentinnen und Konsumenten durch die Teilnahme an ihren Systemen, denen Gamifizierung zu Grunde liegt, Vorteile und Geschenke zu verschaffen, beschert ihnen ein Millionenpublikum. Die Frage, weshalb eine Firma im Kapitalismus Menschen einfach so Geschenke machen sollte und wie Firma und System finanziert werden, scheint sich nur eine Minderheit der Bevölkerung zu stellen.

11.10 Daten, Ablenkung und die Disparität von Macht

An der autoritären Macht durch Daten und dem Einsatz von Techniken wie der Gamifizierung zeigt sich ein Wandel in der Ausübung von Macht, der sich in den letzten Jahrzehnten vollzogen hat: Menschen werden in der neuen digitalisierten Machtwelt heute weniger gewaltsam durch die Androhung oder den tatsächlichen Einsatz von Gewalt zu einem gewissen Verhalten genötigt, wie Orwell es in seiner Dystopie beschreibt, sondern eher in einen freudigen Taumel versetzt, in dem sie motiviert oder unbewusst „genudged" werden und das Dopamin als Huxleys Soma eingesetzt wird. „Nudgen", zu Deutsch „stupsen", beschreibt einen Prozess, in dem die zu beeinflussende Person mit leichtem, evtl. wiederholtem Stupsen zu der gewünschten Verhaltensweise animiert wird.

Die Frage, woher die Legitimation stammt, Menschen auf diese Weise zu beeinflussen, wird kaum aufgeworfen.

Technologie beinhaltet in der heutigen Zeit aber noch eine weitere Dimension: Wer die Hardware oder auch Software herstellt, also Zugriff auf Teile oder die Gesamtheit einer Datenverarbeitungsanlage hat, kann Hintertüren einbauen, die jederzeit Einblick in Daten und Prozesse erlauben. Es kann aber auch zu einer Nutzung der Schnittstelle für Updates kommen, um an diese Informationen zu gelangen. Das Misstrauen gegenüber der in China hergestellten Hardware wird in diesem Kontext in vielen Ländern gerade in Bezug auf den Netzausbau geäußert.

Auf der anderen Seite steht einem wachsenden Einfluss Chinas keinerlei geschlossener Umgang westlicher Staaten damit entgegen. Immer mehr Länder setzten chinesische Überwachungstechnologie ein: Nicht nur in Südafrika, auch in Belgrad stehen „intelligente" Überwachungskameras des chinesischen Herstellers Huawei. In Duisburg vereinbarte der Konzern im Rahmen eines „Safe City" Konzeptes durch ein sogenanntes „Memorandum of Understanding" eine Zusammenarbeit vertraglich, auch Gelsenkirchen arbeitet mit Huawei zusammen. Der Öffentlichkeit gegenüber werden die komplett überwachten „Safe Cities" meist als „Smart Cities" bezeichnet. Auch hier fehlt es an Transparenz. Welche Daten weitergegeben werden, bleibt geheim. Versuche von Datenschützern, die Vereinbarung einzusehen, wurden bisher rundweg abgelehnt.

Das faszinierendste an der westlichen Kritik ist aber die Blindheit gegenüber Fakten und Tendenzen in den eigenen, westlichen Staaten. Denn die Tendenz, Menschen zu immer größeren Datenproduktionen zu animieren, alle nur anfallenden digitalen Daten zu sammeln und die Gesellschaft im Gesamten zu überwachen, ist keine chinesische Erfindung, sondern alarmierenderweise die Routine fast aller Staaten, auch der westlichen, aus denen der Einsatz in China kritisiert wird. Spätestens seit den Enthüllungen Edward Snowdens 2013 ist dies bekannt. Auch im Einsatz von Technologien der sozialen Kontrolle in westlichen Ländern ist der politische Leitgedanke, für soziale Probleme technologische Lösungen zu suchen, zu finden. Statt aufgrund der Einrichtung der chinesischen SCS den eigenen Einsatz von Technologien zur gesellschaftlichen Lenkung zu hinterfragen und die internationale Entwicklung, gesellschaftliche Selbstorganisation nicht mehr zu fördern sowie die Unschuldsvermutung aufzulösen, wird bloß deren Einsatz in China, oft in Unkenntnis der Kultur und Gegebenheiten des Landes, verurteilt.

Was in China kritisiert wird, wird im eigenen Zusammenhang oft akzeptiert.

11.11 Kategorien & Unsichtbarkeit

Dass Daten eine solche Macht beinhalten können, wird immer noch von vielen in Frage gestellt. Dahinter lässt sich die Problematik des menschlichen Umgangs mit haptisch für ihn nicht wahrnehmbaren Bedrohungen vermuten, wie sich schon vor Jahrzehnten im Umgang mit der Radioaktivität zeigte. Genau wie atomare Strahlung sind digitale Daten für uns weder ertast- noch auf eine andere Art körperlich wahrnehmbar. Eine Person, die uns physisch wahrnehmbar überwacht, löst in uns mindestens Unruhe aus. Gegen die nicht wahrnehmbare digitale Beobachtung warnt uns hingegen keine Regung unseres Körpers.

Dagegen kann die Ankündigung von Überwachung sehr wohl als Disziplinierungsmaßnahme eingesetzt werden. Gerade hier wird die Unmerkbarkeit von Überwachung bewusst eingesetzt, da sich die zu disziplinierenden Personen wie im Bentham'schen Gefängnis nie sicher sein können, ob sie gerade unter Beobachtung stehen oder nicht.

Mit der Digitalisierung kann die Überwachung zusätzlich noch automatisiert werden. Automatisierung von Überwachung ist auch in den chinesischen Systemen angelegt und soll mit dem Stand der Technik wachsen. Diese Automatisierung von Überwachung, wie sie auch in westlichen Staaten mehr und mehr praktiziert wird, birgt jedoch hohe ethische und moralische Problematiken, wie sie etwa von Zygmunt Bauman und David Lyon (vgl. Bauman und Lyon 2018), aber auch Joseph Weizenbaum (Weizenbaum 1994) und Shoshana Zuboff (vgl. Zuboff 2018) seit Jahren beschreiben.

Oft wird die Frage gestellt, wie die als „Datendiktatur" angesehene chinesische Regierung, aber auch andere Akteure an die gewünschten Daten gelangen oder auch die Ansicht geäußert, dass niemand aus der Masse der Daten relevante Daten heraussuchen könne. Bewegten wir uns noch im menschengestützten Bereichen von Kontrolle und Beeinflussung, könnte dies zutreffen, bei digitalisierten Systemen ist es nicht mehr so, wie im Folgenden dargestellt werden soll. Um mit den Datenbergen umzugehen, die täglich erzeugt werden, sind Programme notwendig, die diese verarbeiten. Einer der Verarbeitungsschritte ist die Bildung von Kategorien. Unter der Zuhilfenahme von Algorithmen und maschinellen Lernens werden die ergatterten Daten sortiert und aufbereitet, maschinell Kategorien gebildet. Diese Programme unterliegen meistens dem Firmengeheimnis, sodass es unmöglich gemacht wird, Vorgänge des Sortierens und Bewertens oder auch die gewählten Kategorien nachzuvollziehen. Zudem ist zu beachten, dass sowohl Algorithmen als auch maschinelles Lernen vom Menschen erdacht und entwickelt sind. Eine relevante Frage ist die nach dem Umgang mit einprogrammierten Fehlern in den zum Teil sehr mächtig gewordenen maschinellen Entscheidern.

Informationen über Individuen zu sammeln, auch die intimster Art, bedeutet im Zeitalter der Digitalisierung weder Anstrengung noch Herausforderung. Alles, was nicht durch Verschlüsselungs- und Anonymisierungstechniken geschützt im Internet eingegeben, jede Suchanfrage, jede Formulareingabe, jede Email, jeder online gepflegte Kontakt, wird gespeichert. Nicht geschieht nicht nur in China, sondern überall. Sobald wir uns im Netz bewegen, heften sich im kommerziellen Internet sogenannte „Tracker", Verfolger, an uns, die unsere Aktivitäten verfolgen und aufzeichnen. Hinter diesen Trackern sitzen kommerzielle Betreiber, die Informationen schürfen und verkaufen. Auch Geheimdienste und Polizeien nutzen die Informationsspuren, die wir unwissentlich hinterlassen. Die schiere Menge stellt für die eingesetzten Maschinen keine Hinderung dar, im Gegenteil benötigen Prozesse wie das maschinelle Lernen möglichst viele Daten, um zu Ergebnissen zu kommen.

Um die ungeheure Menge an digitalen Daten, die durch die Überwachung unserer Internetnutzung erzeugt werden, verarbeiten zu können, werden aus ihnen Kategorien gebildet. Diese Kategorien werden beispielsweise genutzt, um die Kaufkraft, die Stimmung oder politisch Einstellung einer Person zu ermitteln, aber auch ihre sexuelle Orientierung, ihre Haltung zur Religion oder ihr Gesundheitszustand. Aus unserem digitalen Verhalten kann überdies ermittelt werden, ob jemand als terroristische Gefahr gilt, die zu eliminieren sei.

Gerade in diesem Hinblick wird die Bedeutung der Korrektheit der über das Individuum gesammelten Informationen deutlich. Ob diese korrekt oder inkorrekt sind, kann von den Betroffenen nicht kontrolliert werden, da es keinerlei gesetzliche Grundlage für derartige Überwachungen gibt und sie, wie gamifizierte Techniken, nicht beworben wird.

Zudem können Kategorien im Lauf der Zeit verändert werden. War ein Interesse oder eine Einstellung vor einigen Jahren noch als positiv gewertet worden, kann sich dies in der Zukunft ändern. Wer über Kategorien und ihre Bewertung herrscht, hat in einem digitalisierten Herrschaftssystem die Macht, auch noch im Nachhinein über Gut und Böse zu entscheiden.

Zudem hat sich durch die allgegenwärtige Überwachung die skurrile Situation entwickelt, in der niemand mehr weiß, welche Daten über ihn oder sie von wem gesammelt und wie diese Daten eventuell gegen ihn oder sie einsetzt werden könnten.

11.12 Probleme des Scoring – die Macht durch Daten

Gerade durch die Unsichtbar- und Unmerkbarkeit von Daten ist ihre Sammlung und Auswertung in unserer Zeit zur wenig beachteten Selbstverständlichkeit geworden. Die wirtschaftliche Entwicklung hat die technologische Infrastruktur immer weiter verbilligt, von „Smartphones" als Endgeräten bis hin zu Serverfabriken, in denen die angehäuften Daten dezentral gespeichert werden. Eine Verbilligung, die auch durch ungleiche Verträge und auf Kosten der Umwelt zustande kommt. Unsichtbar und von den meisten nicht beachtet, produzieren Menschen, spätestens wenn sie ein „Smartphone" besitzen, fast ununterbrochen Daten. Daten, die hinsichtlich ihrer wirtschaftlichen Bedeutung seit Jahren als „neues Öl" oder „neues Gold" bezeichnet werden. Diese Daten sind global zu einem Wirtschafts- und Machtfaktor geworden. Eine genauere Kenntnis des Gegenübers ermöglicht eine bessere Grundlage für dessen Einschätzung, aber auch für eine gezielte Manipulation.

Eine Scoring- oder Bewertungsindustrie ist nicht neu, in Deutschland existiert seit 1927 die Schufa. Hier handelt es sich um ein rein privatwirtschaftlich betriebenes Unternehmen, das Wirtschaftsdaten von Privatpersonen und Firmen beurteilt. Auch der von der Schufa genutzte Algorithmus wird den von den Bewertungen Betroffenen nicht mitgeteilt. Zudem sind sowohl Algorithmen als auch maschinelles Lernen vom Menschen erdacht und entwickelt. Eine relevante Frage ist danach mit dem Umgang mit einprogrammierten Fehlern in den zum Teil sehr mächtig gewordenen maschinellen Entscheidern.

Neu sind aber der Umfang der Daten und die Ausbreitung von Verhaltensbewertung im ehemals privaten Bereich. Denn auch der Musik- und Buchgeschmack, die konsumierten Videos und favorisierten Spiele werden bewertet. Die Uhrzeiten der Onlineaktivitäten werden zu einer Berechnung des Lebensrhythmus genutzt. Die Stimme über mit dem Internet verbundene Mikrofone mit gefälligen Frauennamen dazu genutzt, die momentane persönliche Stimmung der Zielperson zu bewerten.

Ziel von Scoringverfahren ist aber nicht nur eine Nachvollziehbarkeit der Vergangenheit der beobachten Zielperson oder Firma, sondern anhand der Daten eine Vorhersagbarkeit des Verhaltens in der Zukunft zu ermöglichen. Diese im Englischen mit „predictive algorithms" bezeichneten Prozesse finden abseits eines gesellschaftlichen Bewusst-·seins statt.

In China bewirbt die Regierung ihr System mit staatlicher Kontrolle der Daten. Aber auch einer umfassenden Kontrolle von Industrie und Gesellschaft, die Xi Jinping als holistisch beschreibt. Dabei greifen chinesische Systeme aber auch auf die in der oben beschriebenen Weise aggregierten Informationen der Privatwirtschaft zu.

Westliche Reaktionen auf die SCS beinhalten neben der Kritik aber auch ein Hinterfragen der Möglichkeiten solch monumentaler Systeme. Einwände, dass die chinesischen SCS lange nicht so mächtig seien, zeigen bei näherer Betrachtung meist wenig Konsistenz. Der Einwand etwa, dass nicht alle Überwachungskameras in China mit Gesichtserkennung ausgestattet bzw. mit einem SCS verbunden seien, ist korrekt, ignoriert aber vollkommen, dass die Regierung zum einen schon vor langem eine Erweiterung und Mo-

dernisierung der eingesetzten Kameras verkündete und zum anderen, dass die eigentliche
Hürde nicht die Fähigkeit der eingesetzten Kamera ist, sondern die Akzeptanz von Kameras als Machtmittel zur ständigen Überwachung des öffentlichen und inzwischen auch
zunehmend des privaten Raums durch Regierung und Bevölkerung. Ist diese erst genommen und sind Kameras als Schutz und nicht Gefahr der eigenen Freiheit akzeptiert, kann
die Kameraflotte jederzeit aktualisiert werden.

11.13 Beeinflussung

Drei Faktoren machen das Internet oder den Cyberspace zu einem extrem mächtigen Instrument der Herrschaft, welches eine Beeinflussung des Menschen in einem nie da gewe-
senen Maß ermöglicht: Zum einen die schiere Menge der Informationen, zum zweiten die
Geschwindigkeit der Verbreitung und als drittes die Individualisierbarkeit. Gesellschaftlich trifft dieser Umstand auf einen Verlust der traditionellen Identitäten oder Zugehörig-
keiten in den meisten heutigen Gesellschaften, etwa aufgrund einer Klasse, einer Zunft
oder eines Glaubens.

Erschwerend kommt hinzu, dass es sich beim Internet selbst und seinen oben beschriebenen Möglichkeiten um ein gesellschaftlich gesehen sehr neues Instrument handelt.
Seine Grundlagen und die durch es ermöglichten Steuerungsmöglichkeiten sind den meisten Menschen unbekannt, sie sind bloße Nutzerinnen und Nutzer. Durch die Fülle der
Informationen im Cyberspace, deren Quellen die meisten nicht überprüfen können, entstehen enorme Probleme der Beeinflussung. Im Westen können die Schlachten von Lob-
bygruppen und zivilgesellschaftlichen Akteuren um Wikipedia-Artikel als öffentlich nach-
vollziehbares Beispiel gelten. Die Geschwindigkeit der Informationsverbreitung ist ein
weiterer Faktor: Durch das Internet können Informationen jeder Art in Millisekunden um
den Globus und in fast jeden Winkel der Welt verbreitet werden. Während es sich bei den
ersten zwei Faktoren, der Menge an Informationen und der Geschwindigkeit ihrer Verbreitung, um eine Steigerung von lange, wenn nicht immer, vorhandenen Faktoren handelt, hat die Individualisierbarkeit, der persönliche Zuschnitt der Information, die das Individuum erreichen, in der heutigen Zeit eine nie da gewesene Qualität erreicht.

Zwar gab es auch in früheren Zeiten Informationen, die auf soziale Gruppen und politische Meinungen zugeschnitten worden waren, was sich in westlichen Ländern an den auf
unterschiedliche Zielgruppen zugeschnittenen Tageszeitungen von ca. Mitte des 19. Jahrhunderts bis Mitte des 20. Jahrhunderts ablesen lässt. Mit der Verbreitung von sozialrevolutionären Philosophien hatten sich auch die Druckerzeugnisse diversifiziert, die sich an
neue Zielgruppen jenseits der wirtschaftlich privilegierten Schichten richteten. Diese Art
der Beeinflussung richtete sich aber an Gruppen, nicht an Individuen, wie die als Klasse
angesehenen Arbeiterinnen und Arbeiter, an Soldaten, an Landarbeiterinnen und Landar-
beiter oder die Massen, die Heimarbeitverrichtenden.

Möglich wird extreme Individualisierung im heutigen Ausmaß durch extrem verklei-
nerte Maschinen, die wir unter dem Namen „Smartphone" verkauft bekamen und seither

in fast jeder Lebenslage mit uns tragen, sowie den Aufbau und die Nutzung des Internets. Während Menschen schon in der frühen Kybernetik als Transmitter gesehen wurden, entwickeln wir uns durch die (durch kybernetisches Denken geprägte) technologischen Entwicklungen tatsächlich immer mehr in diese Richtung. Ausgestattet mit einer kleinen Maschine als Endgerät, die uns weniger in unserem Alltag einschränken soll als der Forschungssensor, der einem wild lebenden Pinguin umgeschnallt wird, um dessen Bewegungen für Studien nachzuvollziehen, produzieren wir fast ständig maschinell auslesbare Daten. Durch „Smartphone"-Nutzung zeigen wir ein elektronisch errechenbares und speicherbares Bild unseres Schlafrhythmus', unseres Bewegungsbildes, unserer elektronisch nachvollziehbaren sozialen Kontakte, unserer Wissens- und Unterhaltungsinteressen, unserer physischen Bewegungen, unserer politischen Einstellungen und unserer gesundheitlichen Lage sowie unserer psychischen Verfassung. Algorithmen errechnen aus unserem Surfverhalten die Veränderungen unserer Gemütslage, des Fruchtbarkeitszykluses und unsere Ängste und Wünsche (vgl. Zuboff 2018).

Das heutige Internet ist „umsonst", weil wir durch dessen Nutzung Daten liefern, die uns beherrschbar machen.

Natürlich gab es auch schon zu früheren Zeiten individuell Beeinflussung von Personen. Gegnerinnen und Gegner oder auch als verbündet wahrgenommene Individuen wurden auch in der Vergangenheit gezielt mit ausgewählten Informationen bzw. Fehlinformationen versorgt, um sie zu einem gewünschten Verhalten zu animieren. Geheimdienste wurden gegründet und ihre Netze ausgebaut, um Informationen im Verborgenen zu sammeln, auszuwerten und Menschen und Gruppen mit oder ohne deren Wissen zu beeinflussen. Im Gegensatz zu heute bedeutete dies aber eine enorme Anstrengung. Hohe Investitionen mussten getätigt werden, um den damals existierenden privaten Raum, die Privatsphäre der Zielpersonen zu penetrieren. Zudem bedurfte es für das Ausführen dieser Aufgaben menschlicher Spezialistinnen und Spezialisten, die oft lange Ausbildungen hinter sich hatten. Aufwand und Kosten sind heute auch daher niedrig, weil die notwendigen Daten meist von den Betroffenen selbst zur Verfügung gestellt werden und die Beeinflussung durch den Einsatz von Maschinen geringe Personalkosten bedeutet.

11.14 Internationale Tendenz: Das Bestreben, soziale Probleme technologisch lösen zu wollen

Der Volksrepublik China wird im Westen oft mit Vorurteilen begegnet, wie sich auch an vielen westlichen Reaktionen auf die Einrichtung der SCS ausdrückt. Denn die hier herausgearbeitete Präferenz der SCS, sozialen Problemen technologisch zu begegnen, zeigt sich ebenso in westlichen Staaten sowie denen des globalen Südens. Auch außerhalb Chinas werden Methoden digitaler Verwaltung, Überwachung und Manipulation eingesetzt. Auf beiden Seiten des Globus entsteht eine immer größer werdende Macht- und Informationsasymmetrie zwischen einer überwachten Gesellschaft auf der einen und überwachenden und manipulierenden staatlichen und wirtschaftlichen Akteuren auf der anderen Seite.

Aspekte und Möglichkeiten der Diskriminierung von nicht-akzeptierten Bevölkerungsgruppen wird in der Entwicklung der Systeme nicht unterbunden.

Die heutige digitalisierte Welt, in nicht wenigen Aspekten von Kybernetikerinnen und
Kybernetikern verschiedenster Couleur geprägt, birgt ein Potenzial an Chancen und Risiken bezüglich gesellschaftlicher Organisation. Einerseits ließen sich Entscheidungsprozesse durch die Formulierung von Bedürfnissen und Meinungen in einem gemeinsam
genutzten Raum, dem Cyberspace, formulieren und sammeln. Bestünde Vertrauen in gesellschaftliche Fähigkeiten und Verantwortung, böte dieser Raum vollkommen neue
Chancen für politische Systeme der gesellschaftlichen Mitbestimmung. Durch die internationale Ausrichtung nach dem Primat der Sicherheit wird das Netz und die in ihm ge-
fischten Informationen hingegen genutzt, um das als Gefahrenquelle gesehene Individuum
zum Objekt von Überwachung und Manipulation zu machen.

Ausschlaggebend für die Ausrichtung eines politischen Systems ist die Sichtweise seiner Entwicklerinnen und Entwickler auf die Natur des Menschen, da diese die Ausformung des Systems prägt. Wird sie vom Primat der Sicherheit beherrscht, werden die Menschen als zu kontrollierende Manipulationsmasse gesehen. International lässt sich in
diesem Zusammenhang beobachten, wie sich autoritäre Systeme immer weicheren Machtmitteln bedienen, um dieses zu erreichen, während sich in als nicht-autoritär gesehenen
Systemen immer totalitärere Menschenbilder und Maßnahmen durchsetzen. Die Auflösung von Unschuldsvermutung und Privatsphäre sind hier alarmierende Entwicklungen.
Die Machtabgabe von politischen Entscheidungsprozessen an Maschinen sollte kritisch
gesehen werden.

Fragen, die im Hinblick auf stark technologische Systeme beachtet werden müssen,
sind nicht, ob diese Systeme chinesisch sind, sondern, welche Macht an Maschinen übertragen wird und wie transparent der Umgang mit neuen Machtmitteln wie Algorithmen ist.
Die heutigen Möglichkeiten von Überwachung und (Massen-)Manipulation auf der Individualebene wurde von den Vordenkerinnen und Vordenkern unserer derzeitigen politischen Systeme nicht erahnt. Es mag an der Zeit sein, die aktuellen Entwicklungen zum
Anlass zu nehmen, die Frage nach Legitimation dieser Machtmittel und nach Verantwortung zu stellen. Denn welche Gefahren Systeme bergen, in denen Verantwortung für Abläufe im Gesamtprozess einfach an andere weitergegeben werden kann und in denen das
Individuum oder ganze Bevölkerungsgruppen hinter Nummern verschwinden, hat die
deutsche Vergangenheit in aller Härte demonstriert. Es waren Maschinen, die die Verwaltung der ersten industriellen Vernichtung von Menschen durch die Lochkartentechnik
übersichtlich hielten und dadurch mit ermöglichten. Und es waren die kleinen Rädchen,
deren kleine Bewegungen angetrieben von einer menschenverachtenden Ideologie einen
Massenvernichtungsapparat zum Laufen brachten.

Anschließend daran kann aufgrund der technologischen und gesellschaftlichen Entwicklungen der letzten Jahrzehnte jedes System hinsichtlich der folgenden Fragen geprüft
werden: Welchen Stellenwert hat die Freiheit des Individuums in diesem System inne und
wie wird diese durch die Gesellschaft bzw. den Staat begrenzt? Wie ist der Umgang mit
„nichtleistungsfähigen" Menschen, die etwa eine körperliche oder geistige Behinderung

haben? Wie heterogen darf die Gesellschaft sein? Wie können zudem Fehler in dem System korrigiert werden? Welchen Einfluss haben gesellschaftliche Akteure darauf und wie können sich Erneuerungen in diesem System oder gegen dasselbe durchsetzen? Und, gerade in Bezug darauf, dass Menschen Fehler machen – wie ist das System wieder rückholbar? Verfügt es sozusagen über einen Not-Ausschalter, wenn sich gravierende Fehler zeigen? Und, da wir auf einem Planeten mit endlichen Ressourcen leben, wie regenerativ ist das System? Welche Ressourcen benötigt es zum Überleben?

Wir leben in einer Welt, in der das Wissen um ihre begrenzten und schwindenden Rohstoffe heute fast jedem und jeder zugänglich ist. Bestimmt wird diese Welt aber von einem Wirtschaftssystem, welches auf ständiges Wachstum und Konsum, statt auf Nachhaltigkeit aufbaut. Gesellschaften sind heute nicht durch Autonomie und Verantwortung geprägt, sondern von den Entscheidungen höherer Gremien abhängig, sind überwacht und kontrolliert, neben dem Konsum wird Ablenkung als Machtmittel eingesetzt.

International ist zu beobachten, wie der Begriff der Sicherheit den der Freiheit seit den 1970ern aus der Debatte verdrängte. Es könnte die These aufgestellt werden, dass erst dieser Wandel im Denken die technologischen Systeme der Kontrolle ermöglichte. Ob der Einsatz von Maschinen zur Bevölkerungskontrolle langfristig taugt, wird sich herausstellen. Wie diese Systeme langfristig mit Energie- und Rohstoffen versorgt werden sollen, ist ungeklärt.

Statt China aufgrund seines Einsatzes von Technologie zur Bevölkerungskontrolle zu kritisieren, sollte die gegen China geäußerte Kritik auch im Hinblick auf die eigenen Gesellschaften ernst genommen und die Einrichtung der SCS als Anlass genommen werden, um den Einsatz von automatisierter Überwachungstechnologie und manipulierenden Techniken wie des Nudgens und der Gamification, im Besonderen aber auch Vorhersagealgorithmen offen gesellschaftlich zu diskutieren. Und in diesem Prozess zudem nicht aus den Augen zu verlieren, dass technologische Überwachung mehr Energie und Rohstoffe kostet, als sie einspart und damit sehr viel mehr, als der Planet unter unseren Füßen zur Verfügung stellt.

Literatur

Ai, J. (2008) The Refunctioning of Confucianism: The Mainland Chinese Intellectual Response to Confucianism since the 1980s. *Issues and Studies* 44, 2, 29–78.

Bauman, Z., Lyon, D. (2018) *Daten, Drohnen, Disziplin ein Gespräch über flüchtige Überwachung.*

Brady, A.-M. (2014) *China's thought management.* London; New York: Routledge.

Creemers, R. (2015) *Cyber China: Updating Propaganda, Public Opinion Work and Social Management for the 21st Century.* Rochester, NY: Social Science Research Network. https://papers. ssrn.com/abstract=2698062 [Stand 2020-01-24].

Hoffman, S. R. (2017) *Programming China: The Communist Party's autonomic approach to managing state security.* University of Nottingham.

Huxley, A. (1932) *Brave new world, and Brave new world revisited.* New York: Harper & Brothers.

Kühnreich, K. (2014) *Die äußere Harmonisierung des inneren Aufstands – „Harmonische Gesellschaft" und „Massenzwischenfälle."* text.thesis.master. Universität zu Köln. http://www.uni-koeln.de/ [Stand 2020-01-24].

Kühnreich, K. (2018) „Kick it like China" – Riots, Action, and Repression. In A. Starodub, A. Robinson, hg. *Riots and militant occupations: smashing a system, building a world – a critical introduction*. London, New York: Rowman & Littlefield International.

Leonard, M. (2009) *Was denkt China?*. München: Dt. Taschenbuch-Verl.

Meissner, M., Conrad, B. (2019) *The Digital Hand – How China's Corporate Social Credit System Conditions Market Actors*. https://sinolytics.de/wp-content/uploads/2019/08/Sinolytics_The-Digital-Hand-How-Chinas-Corporate-Social-Credit-System-Conditons-Market-Actors.pdf [Stand 2020-01-17].

Orwell, G. (1949) *Nineteen eighty-four*. London: Secker & Warburg.

Reuters (2014) Chinese police hunt protesters after waste plant clash – Reuters. *Reuters*. https://in.reuters.com/article/us-china-protests/chinese-police-hunt-protesters-after-waste-plant-clash-idINKBN0DS05020140512 [Stand 2020-01-29].

Rid, T. (2016) *Rise of the machines: a cybernetic history*. First edition. New York: W. W. Norton & Company.

Weizenbaum, J. (1994) *Computer power and human reason: from judgment to calculation*. San Francisco, Ca.: S.H. Freeman.

Wiener, N. (2013). *Cybernetics or control and communication in the animal and the machine*. 2. ed. Cambridge, Mass: MIT Press.

Zuboff, S. (2018) *Das Zeitalter des Überwachungskapitalismus/Shoshana Zuboff; aus dem Englischen von Bernhard Schmid*. [1. Auflage]. Frankfurt: Campus.

Katika Kühnreich studierte Politikwissenschaften und Moderne und Ältere Sinologie in Köln und Kunming/Yunnan. Ihre Forschungsschwerpunkte sind neben den gesellschaftlichen Auswirkungen der Digitalisierung der Umgang mit Dissens. Sie arbeitet seit Jahren zu Formen der digitalisierten sozialen Kontrolle und erreichte mit ihren Vorträgen und Veröffentlichungen zu den chinesischen Social Credit Systemen internationale Bekanntheit.

Europäisches Datenschutzrecht und Bonitätssysteme in China

Barbara Kirchberg-Lennartz

Zusammenfassung

Bonitätssysteme in China sind zum einen auf in China tätige Unternehmen gerichtet und sammeln Informationen über das Geschäftsgebaren, die Einhaltung von gesetzlichen Regelungen insbesondere zu Compliance oder auch zu Umweltschutz. Die daraus abgeleiteten Bewertungen qualifizieren Unternehmen als gute oder schlechte Geschäftspartner oder Wirtschaftsakteure. Damit sind entsprechende Vor- oder Nachteile für die Unternehmen verbunden.

Zum anderen gibt es zahlreiche Ansätze, das Verhalten von natürlichen Personen auf der Grundlage umfassender Informationssammlung aus allen Lebensbereichen zu bewerten und ihnen entsprechend der Ergebnisse den Zugang zu sozialen Leistungen zu erschweren oder zu erleichtern und im Allgemeinen ihre Stellung in der Gesellschaft durch den generierten „Score" zu beeinflussen.

Die europäischen Datenschutzgesetze stellen an alle Unternehmen, Behörden und öffentlichen Einrichtungen umfangreiche Anforderungen, wenn personenbezogene Daten verarbeitet werden. Ziel ist es, den natürlichen Personen ein Selbstbestimmungsrecht über sie betreffende Informationen zu gewähren. Voraussetzung dafür ist eine hohe Transparenz über die Verarbeitung personenbezogener Daten und umfangreiche Rechte der Betroffenen auf Auskunft, Widerspruch und Löschung ihrer Daten.

Die europäischen Datenschutzgesetze binden in der EU ansässige Unternehmen vor allem auch hinsichtlich der Übermittlung von personenbezogenen Daten in Länder außerhalb der EU. Die Anforderungen des europäischen Datenschutzes gelten deshalb für eine Übermittlung von personenbezogenen Daten an Bewertungssysteme in China in

B. Kirchberg-Lennartz (✉)
Frankfurt, Deutschland
E-Mail: info@dataprotect21.de

© Springer Fachmedien Wiesbaden GmbH, ein Teil von Springer Nature 2020   229
O. Everling (Hrsg.), *Social Credit Rating*,
https://doi.org/10.1007/978-3-658-29653-7_12

vollem Umfang. Das hat zur Folge, dass ein in der EU ansässiges Unternehmen für ein Bewertungssystem relevante Daten über seine Geschäftätigkeit in China nur dann zur Verfügung stellen kann, wenn diese Informationen keinerlei Bezug zu natürlichen Personen aufweisen. Ein Rückschluss auf in den Unternehmen agierende Personen muss also völlig ausgeschlossen sein.

Chinesische Unternehmen, die in der EU Produkte und Dienstleistungen natürlichen Personen anbieten oder deren Tätigkeit auf die Beobachtung natürlicher Personen ausgerichtet ist, sind im Hinblick auf die damit im Zusammenhang stehende Verarbeitung von Kundendaten ebenfalls an die europäischen Gesetze gebunden. Dies schließt die Weitergabe personenbezogener Daten an Bewertungssysteme in China völlig aus.

Die Lösung dieser Problematik könnte in der Schaffung einer in der EU wirksamen gesetzlichen Grundlage für die Informationsbereitstellung an Bonitätssysteme in China sein. Voraussetzung dafür wäre eine Etablierung von Datenschutzregeln in der Struktur der chinesischen Bewertungssysteme, die ein EU-adäquates Datenschutzniveau garantieren.

12.1 Zielsetzung

Dieser Beitrag zielt darauf ab, die Bonitätssysteme in China den Grundsätzen des europäischen Datenschutzes gegenüberzustellen. Das europäische Datenschutzrecht beruht auf der europäischen Charta der Grundrechte zum Schutz der Rechte und Freiheiten der europäischen Bürger. Es gewährt den Menschen ein Recht auf Selbstbestimmung hinsichtlich der Verwendung von Informationen zu ihrer Person. Die Anlässe für die Verarbeitung personenbezogener Daten sind eng limitiert. Es schützt natürliche Personen vor dem Missbrauch ihrer Daten, in dem es für die Offenlegung von Verarbeitungen personenbezogener Daten sorgt. Es räumt den Menschen Rechte ein, über die Verarbeitung ihrer Daten zu entscheiden und diese zu kontrollieren. Sämtliche öffentlichen Bereiche und staatliche Institutionen sowie Unternehmen, die in Europa niedergelassen oder tätig sind, müssen europäischen Datenschutz erfüllen. Jeglicher Transfer von Daten europäischer Bürger in Länder außerhalb des Geltungsbereichs europäischer Gesetze ist davon erfasst.

Aus der Anwendung der grundlegenden europäischen Datenschutzregeln auf die Bewertungssysteme für Unternehmen und Personen lassen sich Anforderungen an europäische Unternehmen ableiten, die den chinesischen Institutionen Daten zur Verfügung stellen.

Außerdem wird dargestellt, welche (Mindest-)Anforderungen durch die Bonitätssysteme erfüllt werden müssten, um sie in der EU zu legitimieren. Dadurch wird zudem deutlich, welche Grenzen in Europa bestehen, um Unternehmen und natürliche Personen auf ähnliche Weise zu bewerten.

12.2 Die Grundsätze des europäischen Datenschutzrechts

Es werden die Kernbegriffe des Datenschutzes, die Anwendbarkeit und die Grundsätze des europäischen Datenschutzes kurz erläutert, um den Rahmen für die Einordnung der Bonitätssysteme aufzuspannen.

Verordnung (EU) 2016/679 des Europäischen Parlaments und des Rates vom 27. April zum Schutz natürlicher Personen bei der Verarbeitung personenbezogener Daten, zum freien Datenverkehr zur Aufhebung der Richtlinie 95/46/EG (Datenschutz-Grundverordnung). Im Folgenden DSGVO genannt.

Der Begriff der personenbezogenen Daten
Nach Artikel 4 Ziffer 1 Datenschutzgrundverordnung (im folgenden DSGVO), sind personenbezogene Daten alle Informationen, die sich auf eine identifizierte oder identifizierbare natürliche Person (betroffene Person oder Betroffener) beziehen. Eine Person ist demnach direkt oder indirekt identifizierbar durch Zuordnung zu einer Kennung, wie einen Namen, zu einer Kennnummer, zu Standortdaten, zu einer Online-Kennung oder zu einem oder mehreren besonderen Merkmalen, die Ausdruck der physischen, physiologischen, genetischen, psychischen, physischen, wirtschaftlichen, kulturellen oder sozialen Identität dieser natürlichen Person sind. Die Reichweite dieses Begriffs wird durch die weiteren Erläuterungen in der DSGVO in den Erwägungsgründen des Gesetzes deutlich. Denn um festzustellen, ob eine natürliche Person identifizierbar ist, sind alle Mittel zu berücksichtigen, die von dem Verantwortlichen oder einer anderen Person nach allgemeinem Ermessen wahrscheinlich genutzt werden, um die natürliche Person direkt oder indirekt zu identifizieren, wie zum Beispiel das Aussondern. Bei der Wahl der Mittel für eine Identifizierung sind alle objektiven Faktoren, wie die Kosten und der erforderliche Zeitaufwand unter Berücksichtigung der zum Zeitpunkt der Verarbeitung verfügbaren Technologien und technologischen Entwicklungen zu berücksichtigen.

Das bedeutet, alle Informationen, die allein oder in Kombination dazu geeignet sind, eine natürliche Person aus einer Menge von Personen auszusondern und namhaft zu machen, sind personenbezogene Daten. Dies muss nicht bereits zum Zeitpunkt der Erhebung gegeben sein, sondern kann auch zu einem späteren Zeitpunkt und durch andere als den ursprünglich Verantwortlichen erst möglich werden und erfolgen. In Bezug auf die Bonitätssysteme sind zwei Kategorien von Informationen relevant: Erstens solche, die natürliche Personen als „Autoren" oder Verantwortliche für Unternehmenssachverhalte ausweisen und zweitens, die natürliche Personen beschreiben oder identifizierbar machen.

Verantwortlicher
Jede natürliche oder juristische Person, Behörde, Einrichtung oder andere Stelle, die allein oder gemeinsam mit anderen über die Zwecke und Mittel der Verarbeitung personenbezogener Daten entscheidet ist nach Artikel 4 Ziffer 7 Verantwortlicher, der zur Einhaltung der DSGVO verpflichtet ist. Davon sind Behörden, die Bewertungssysteme betreiben, ebenso umfasst, wie privatwirtschaftliche Plattformbetreiber.

Verarbeitung personenbezogener Daten

Unter Verarbeitung fasst Artikel 4 Ziffer 2 DSGVO jeglichen Vorgang des automatisierten oder manuellen Umgangs mit personenbezogenen Daten zusammen. Dazu gehören das Erheben und das Erfassen, die Organisation und das Ordnen, die Speicherung, die Anpassung und Veränderung, das Auslesen, das Abfragen, die Verwendung, die Offenlegung durch Übermittlung, Verbreitung oder andere Form der Bereitstellung, den Abgleich oder die Verknüpfung, die Einschränkung, das Löschen oder die Vernichtung. Damit ist für die Verarbeitung personenbezogener Bonitätsdaten der Anwendungsbereich eröffnet. Für jede dieser Formen der Nutzung personenbezogener Daten ist deren Rechtmäßigkeit der Nutzung nach den nachfolgend beschriebenen Grundsätzen durch den Verantwortlichen nachzuweisen.

Grundsätze des Artikels 5 DSGVO

In Artikel 5 werden Grundsätze für die Verarbeitung personenbezogener Daten benannt, die nachweislich durch den Verantwortlichen einzuhalten sind. Verarbeitungen sind nur zulässig, wenn sie auf einer der in Artikel 6 DSGVO genannten Rechtsgrundlagen beruht. Solche Verarbeitungen müssen die Grundsätze des Artikels 5 einhalten, die bestimmen wie eine zulässige Verarbeitung personenbezogener Daten auszugestalten ist, um den Datenschutz einzuhalten.

Zuerst müssen personenbezogene Daten auf rechtmäßige Weise nach Treu und Glauben und für die betroffene Person auf nachvollziehbare Weise verarbeitet werden. Das bedeutet, die Verarbeitung muss auf einer gültigen Rechtsgrundlage beruhen, sie darf keine aus Sicht des Betroffenen überraschenden (Weiter-)Verarbeitungen beinhalten, mit denen er nach dem angegebenen Zweck der Verarbeitung nicht rechnen muss. Voraussetzung für eine solche Beurteilung ist eine klare und verständliche Information des Betroffenen über die Zwecke, den Umfang und alle Umstände der beabsichtigten Verarbeitung (**Transparenz**).

Die Verarbeitung personenbezogener Daten ist eng an einen festgelegten, eindeutigen und legitimen Zweck gebunden. Sie darf nicht um mit dem ursprünglichen Zweck nicht zu vereinbarende Zwecke erweitert werden (**Zweckbindung**). Ein nicht zu vereinbarender Zweck wäre zum Beispiel die Weiterleitung von Informationen über Online-Bestellungen über Produkte, die als gesundheitsschädlich eingestuft werden, an ein Institut für Gesundheits- oder Ernährungsberatung. Über eine solche Übermittlung müsste der Kunde im Rahmen seiner Geschäftsbeziehung zu dem Online-Händler ausdrücklich informiert worden sein und er müsste ihr explizit zugestimmt haben.

Die Verarbeitung personenbezogener Daten muss im Hinblick auf den Zweck erforderlich und angemessen sein sowie auf das notwendige Maß beschränkt sein (**Datenminimierung**). Gegen diesen Grundsatz spräche zum Beispiel, wenn für die online-Bestellung von Schuhen neben den erforderlichen Schuhgrößen, Adress- und Rechnungsangaben zum Beispiel Größe und Gewicht oder Haarfarbe des Kunden abgefragt würden.

Die **Richtigkeit** und die **Aktualität** der verarbeiteten personenbezogenen Daten muss jederzeit gegeben sein. Unrichtige Daten sind entweder zu berichtigen oder zu löschen.

Dieser einfach klingende Grundsatz ist nicht ohne konkrete organisatorische und technische Vorkehrungen umsetzbar und erzeugt einen entsprechenden Aufwand für den Verantwortlichen.

Der Grundsatz der **Speicherbegrenzung** legt fest, dass personenbezogene Daten dann zu anonymisieren oder zu löschen sind, wenn der ursprüngliche Zweck ihrer Verarbeitung erloschen ist. Das bedeutet, dass für personenbezogene Daten Löschkonzepte erstellt und umgesetzt werden müssen, die eine Löschung nach Zweckentfall bzw. nach dem Ende gesetzlich definierter Aufbewahrungsfristen, die der Verantwortliche möglicherweise erfüllen muss, sicherstellen. Ausnahmen bestehen für Archivzwecke im öffentlichen Interesse oder für wissenschaftliche, historische sowie statistische Zwecke. In diesen Fällen sind Maßnahmen zur Sicherung der Daten vor Zweckentfremdung, unbefugten Zugriffen, Verfälschungen und Zerstörung zu treffen.

Abschließend beinhaltet der Grundsatz der **Integrität** und **Vertraulichkeit**, dass personenbezogene Daten sicher verarbeitet werden müssen, insbesondere vor unbefugter und unrechtmäßiger Verarbeitung, vor unbeabsichtigtem Verlust, Zerstörung und Beschädigung organisatorisch und technisch zu schützen sind.

Die DSGVO enthält weitere Vorschriften, die die Anforderungen hinsichtlich technischer und organisatorischer Gewährleistung der Legitimität der Datenverarbeitung (Artikel 24), Datenschutz durch Technikgestaltung und durch datenschutzfreundliche Voreinstellungen (Artikel 25) sowie explizit zur Sicherheit der Verarbeitung (Artikel 32) ergänzen. Die Maßnahmen zur Erfüllung dieser Anforderungen sollen an dem Risiko einer Einschränkung der Rechte und Freiheiten des Betroffenen durch die Verarbeitung seiner Daten ausgerichtet werden. Je höher dieses Risiko einzuschätzen ist, zum Beispiel, wenn sensible Daten über die Gesundheit, die ethnische Zugehörigkeit, die sexuelle Orientierung, die politische Meinung oder die Religionszugehörigkeit verarbeitet werden, desto umfangreichere Schutzmaßnahmen sind zu treffen. Risiken, die dadurch weitestgehend auszuschließen sind, bestehen zum Beispiel im Identitätsdiebstahl, in der Verunglimpfung und damit einhergehendem Reputationsverlust und in der möglichen Diskriminierung eines Betroffenen.

Rechtmäßigkeit der Verarbeitung (Verbot mit Erlaubnisvorbehalt)
Die DSGVO verwirklicht das sogenannte Verbot mit Erlaubnisvorbehalt. Das heißt, die Verarbeitung von personenbezogenen Daten ist grundsätzlich verboten, bzw. bis auf die wenigen in der DSGVO in Artikel 6 abschließend genannten Erlaubnisse beschränkt. Genau betrachtet führen die einzelnen Erlaubnisnormen zu umfangreichen Möglichkeiten der Verarbeitung personenbezogener Daten:

- Der Betroffene willigt in die Verarbeitung der Daten ein. Die rechtskonforme Einwilligung bedarf einer spezifischen Ausgestaltung nach Artikel 7 DSGVO bzw. nach Artikel 9 DSGVO für die Verarbeitung sensibler Daten. Insbesondere muss die Einwilligung freiwillig, auf der Grundlage einer genauen Information über die Zwecke der Datenverarbeitung erteilt und durch den Verantwortlichen nachgewiesen werden. Freiwillig ist eine Einwilligung nur, wenn der Betroffene sie aus einer autonomen Position erteilt und ihre Verweigerung nicht zu unmittelbaren Nachteilen für die Person führt.

- Die Anbahnung oder die Erfüllung eines Vertrages erfordert die Verarbeitung personenbezogener Daten, zum Beispiel bei Abschluss eines Kaufvertrags, Arbeitsvertrags oder auch für die Teilnahme an einem Kundenbindungsprogramm.

- Die Verarbeitung personenbezogener Daten ist aufgrund einer rechtlichen Verpflichtung des Verantwortlichen erforderlich. Hierunter fallen nur rechtliche Verpflichtungen, die im Unionsrecht bzw. im Recht eines Mitgliedstaats geregelt sind. So sind zum Beispiel behördliche Anforderungen aus Ländern außerhalb der EU in der Regel nicht als eine solche Verpflichtung anzusehen.

- Die Verarbeitung erfolgt zur Wahrung lebenswichtiger Interessen der betroffenen Person oder einer anderen natürlichen Person, soweit diese Verarbeitung nicht auf eine andere Rechtsgrundlage gestützt werden kann. Hierunter fallen vor allem Verarbeitungen zu humanitären Zwecken, zum Beispiel im Fall von Epidemien oder Naturkatastrophen.

- Die Verarbeitung dient einer Aufgabe, die im öffentlichen Interesse liegt oder die Wahrnehmung einer öffentlichen Gewalt darstellt, die dem Verantwortlichen übertragen wurde. In der Praxis wird diese Rechtsgrundlage für Privatunternehmen sehr wenig relevant sein.

- Die Verarbeitung dient der Wahrung berechtigter Interessen des Verantwortlichen oder eines Dritten. Dabei ist zu berücksichtigen, dass dieser Verarbeitung keine Interessen oder Grundrechte und Grundfreiheiten des Betroffenen entgegenstehen, die den Schutz seiner personenbezogenen Daten erfordern und demzufolge das Interesse des Verantwortlichen überwiegen. Die Interessen von Kindern fallen dabei besonders ins Gewicht. Die Verarbeitung sensibler Daten ist regelmäßig nicht mit dem berechtigten Interesse des Verantwortlichen begründbar. Praxisbeispiele sind die Nutzung von Kundendaten für Werbezwecke oder zur Verhinderung von Betrugsfällen. Die Rechtsgrundlage „berechtigtes Interesse" ist keine Lösung für alle Fälle, sondern bedarf einer sorgfältigen und dokumentierten Interessenabwägung für jeden einzelnen Verarbeitungszweck.

- In Artikel 6 Abs. 4 DSGVO wird als letzte mögliche Erlaubnis zur Verarbeitung personenbezogener Daten, die Verarbeitung zu einem sogenannten kompatiblen Zweck eingeführt. Das bedeutet, dass eine Verarbeitung von personenbezogenen Daten zu einem mit dem ursprünglichen Zweck vereinbaren weiteren Zweck verarbeitet werden dürfen. Die Voraussetzungen für kompatible Zwecke sind hoch, u.a. muss der Betroffene über die Verarbeitung zu einem anderen Zweck vorab informiert werden und ihm steht ein Recht auf Widerspruch gegen die Verarbeitung zu.

Sensible Daten sind nach Artikel 1 DSGVO besondere Kategorien personenbezogener Daten, aus denen die rassische und ethnische Herkunft, politische Meinungen, religiöse oder weltanschauliche Überzeugungen oder die Gewerkschaftszugehörigkeit hervorgehen sowie genetische und biometrische Daten zur eindeutigen Identifizierung einer natürlichen Person, Gesundheitsdaten oder Daten zum Sexualleben, der sexuellen Orientierung.

Die vorgenannten Rechtsgrundlagen sind abschließend. Die DSGVO bietet keine Öffnungsklauseln für die Schaffung weiterer Rechtsgrundlagen.

Profiling

Die DSGVO definiert den Begriff des Profiling (Artikel 4 Ziffer 4) als jede Art der automatisierten Verarbeitung personenbezogener Daten, die einer Bewertung persönlicher Aspekte einer natürlichen Person dienen. Dies können Analysen, Einschätzungen oder Vorhersagen sein, zum Beispiel der Arbeitsleistung, der wirtschaftlichen Lage, der Gesundheit, von persönlichen Vorlieben, Interessen, Zuverlässigkeit, Verhalten, Aufenthaltsort oder Ortswechsel. In der Systematik der DSGVO muss für jede als Profiling charakterisierte Verarbeitung ein definierter Zweck und eine Rechtsgrundlage vorliegen. In vielen Fällen wird eine zulässige, auf Profiling ausgerichtete Verarbeitung aus Sicht des Betroffenen hohe Risiken bergen, zum Beispiel aufgrund von Fehleinschätzungen und deren Folgen für seine Recht und Freiheiten. Deshalb muss der Verantwortliche in solchen Fällen eine ausführliche Risikobewertung aus der Perspektive des Betroffenen durchführen und geeignete technische und organisatorische Maßnahmen treffen, um diese Risiken weitestgehend zu reduzieren.

Automatisierte Entscheidungsfindung

Die DSGVO räumt in Artikel 22 Betroffenen das Recht ein, nicht einer Entscheidung unterworfen zu werden, die ausschließlich auf der automatisierten Verarbeitung personenbezogener Daten beruht. Dies gilt für Entscheidungen, die gegenüber dem Betroffenen eine rechtliche Wirkung entfalten oder ihn in ähnlicher Weise beeinträchtigen. Von dieser Grundregel gibt es Ausnahmen für den Fall, dass die automatisierte Entscheidung für den Abschluss eines Vertrags zwischen dem Betroffenen und dem Verantwortlichen erforderlich ist oder mit der ausdrücklichen Einwilligung des Betroffenen einhergeht. In diesen Fällen hat der Betroffene das Recht, dass eine natürliche Person beim Verantwortlichen in die Entscheidung eingreift und diese überprüft. Er hat das Recht, den eigenen Standpunkt darzulegen und die Entscheidung anzufechten. Weiterhin sind automatisierte Entscheidungen aufgrund bestimmter Rechtsvorschriften erlaubt.

Scoring

Informationen zum Zahlungsverhalten von natürlichen Personen (Konsumenten) dürfen in Deutschland durch Wirtschaftsauskunfteien erhoben und verarbeitet werden. Die Einhaltung des Datenschutzes ist durch eine die DSGVO ergänzende nationale Regelung im § 31 Bundesdatenschutzgesetz bestimmt. Die Vorschrift dient dem Schutz des Wirtschaftsverkehrs und soll explizit die Rechtmäßigkeit von Scoring und Bonitätsauskünften sicherstellen. Der

Score als Wahrscheinlichkeitswert über das zukünftige Verhalten einer natürlichen Person stellt einen Indikator dar, der einen Verantwortlichen in der Entscheidung über die Begründung, Durchführung oder Beendigung eines Vertragsverhältnisses mit einem Betroffenen unterstützen soll. Ein Scoring darf nur unter Einhaltung der Datenschutzvorschriften durchgeführt werden. Das Verfahren an sich muss fachliche Vorgaben erfüllen. Es werden Bedingungen für die einzusetzenden Algorithmen und die zu verwendenden Informationen genannt. Der Betroffene muss über die Verwendung seiner Daten für ein Scoring informiert werden. Insbesondere ist im Detail geregelt, welche Art von Informationen für die Ermittlung der Wahrscheinlichkeit eines zukünftigen Zahlungsverhaltens des Betroffenen verwendet werden darf. Der Betroffene hat ein Recht auf Auskunft, Überprüfung und Anfechtung des Score-Werts.

Umfangreiche Rechte der Betroffenen

Das Kapitel III der DSGVO regelt die Rechte der Betroffenen, wenn Daten zu ihrer Person verarbeitet werden. Dabei ist zu unterscheiden zwischen Informationspflichten des Verantwortlichen und Rechten des Betroffenen, insbesondere auf Auskunft, Berichtigung und Löschung der zu seiner Person verarbeiteten Daten sowie des Widerspruchs gegen eine Verarbeitung, wenn diese auf einem legitimen Interesse des Verantwortlichen beruht.

Die Informationspflichten des Verantwortlichen sind ein wesentlicher Baustein des Grundrechtsschutzes natürlicher Personen, weil damit die Transparenz über die Verarbeitung personenbezogener Daten geschaffen wird. Es ist sicherzustellen, dass den Betroffenen leicht verständliche Informationen über den Zweck, die Art, den Umfang, die Dauer, die Umstände der Verarbeitung seiner Daten zum Zeitpunkt der Erhebung der Daten gegeben werden. Dazu gehören auch Informationen über mögliche Empfänger der Daten. Das können Unternehmen sein, die für den Verantwortlichen Dienstleistungen erbringen oder Behörden aufgrund gesetzlicher Informations-, Berichts- oder Unterstützungspflichten, denen der Verantwortliche unterliegt. Wenn sich solche Empfänger in Drittländern befinden, muss der Verantwortliche angeben, welche Garantien für die datenschutzkonforme Behandlung der Daten im Drittland bestehen.

In der Praxis findet man diese Pflichtangaben in Form von Datenschutzerklärungen oder Datenschutzhinweisen auf den Internetseiten der Verantwortlichen, über die sie mit den Betroffenen agieren. Auch werden entsprechende Informationsblätter zum Beispiel bei Vertragsabschlüssen mit einer Bank oder einer Versicherung in Papierform ausgehändigt.

Die daran anknüpfenden Rechte der Betroffenen sind der zweite wesentliche Baustein zum Schutz personenbezogener Daten. Sie sind die Kontrollinstrumente, mit denen sich jeder über die zu seiner Person verarbeiteten Daten jederzeit informieren kann. Er kann als Folge der erteilten Auskunft die Berichtigung von Fehlern oder auch die Löschung seiner Daten verlangen. Weiterhin kann einer Verarbeitung mit Wirkung für die Zukunft widersprochen werden.

Für die Erfüllung dieser Betroffenenrechte gegenüber Kunden, Mitarbeitern, Geschäftspartnern im privaten Bereich oder gegenüber Bürgern im öffentlichen Bereich müssen Kontaktdaten einer Anlaufstelle für Betroffene veröffentlicht werden. Auskünfte müssen im Regelfall innerhalb von vier Wochen erteilt werden. Bei großen Unternehmen, die im großen Umfang Daten von Kunden verarbeiten, sind diesen Aufgaben in der Regel dezidierte Ressourcen gewidmet und automatisierte Abläufe für die Zusammenstellung der Auskünfte etabliert, um Auskunftsanfragen in der gegebenen Frist zu erledigen.

Geltungsbereich der Datenschutzgrundverordnung
Nach Artikel 2 Abs. 1 DSGVO ist die Verordnung auf jegliche ganz oder teilweise automatisierte Verarbeitung personenbezogener Daten sowie die manuelle Verarbeitung in einem Dateisystem anzuwenden. Abs. 2 nennt dazu Ausnahmen, die einerseits den rein privaten Bereich natürlicher Personen und andererseits die Verarbeitung durch Behörden, die für die Strafverfolgung und Strafvollstreckung sowie für den Schutz und die Abwehr von Gefahren für die öffentliche Sicherheit zuständig sind.

Weiterhin regelt Artikel 3 Abs. 1 DSGVO, dass das Gesetz für alle Verarbeitungen personenbezogener Daten im Rahmen der Tätigkeit von in der EU niedergelassenen Verantwortlichen gilt. Dabei ist es nicht erforderlich, dass die Verarbeitung selbst in der EU stattfindet (sogenanntes Niederlassungsprinzip). Zudem erweitert Abs. 2 den räumlichen Geltungsbereich der DSGVO auf Verarbeitungen von Daten natürlicher Personen, die sich in der EU befinden, auch wenn der Verantwortliche nicht in der EU niedergelassen ist. Dies gilt, wenn die personenbezogenen Daten im Zusammenhang mit dem (auch unentgeltlichen) Angebot von Waren und Dienstleistungen verarbeitet werden oder die Verarbeitung der Beobachtung des Verhaltens von Personen gilt. Dieses sogenannte Marktortprinzip bindet somit jeden außerhalb der EU ansässigen Verantwortlichen, den EU-Datenschutz einzuhalten, sofern er natürlichen Personen, die sich in der EU befinden, Waren und Dienstleistungen anbietet oder deren Verhalten, zum Beispiel auf Social-Media-Plattformen oder durch Auswertungen in Suchmaschinen beobachtet.

Mechanismen zur Durchsetzung der Datenschutzvorschriften
Die DSGVO installiert unabhängige Kontrollinstanzen und verfügt über weitreichende Sanktionsmechanismen. Die Betroffenen erhalten explizite Rechte auf materiellen und immateriellen Schadensersatz bei Schäden durch die Verletzung von Datenschutzregeln.

Unabhängige interne und externe Kontrollinstanzen
Unternehmen,

- deren Kerntätigkeit in der Verarbeitung von personenbezogenen Daten besteht, die eine umfangreiche, regelmäßige und systematische Beobachtung natürlicher Personen erforderlich macht oder
- deren Kerntätigkeit in der umfangreichen Verarbeitung von sensiblen personenbezogenen Daten oder von Daten über strafrechtliche Verurteilungen und Straftaten besteht
- sowie Behörden

müssen eine interne, unabhängige Instanz etablieren, die über Datenschutzvorschriften unterrichtet und berät sowie deren Einhaltung überwacht (Datenschutzbeauftragter nach Artikel 37 DSGVO).

Der Datenschutzbeauftragte ist weisungsfrei und berichtet unmittelbar an die höchste Managementebene des Verantwortlichen und ist frühzeitig in alle für den Datenschutz relevanten Vorgänge einzubeziehen. Jeder Betroffene kann sich in Sachen der Verarbeitung personenbezogener Daten und der Wahrnehmung seiner Rechte an den Datenschutzbeauftragten wenden (Artikel 38 DSGVO).

Der Datenschutzbeauftragte ist der Ansprechpartner für die (unternehmens-)externe Kontroll- und Überwachungsinstanz, die Datenschutzaufsichtsbehörde, die jeder EU-Mitgliedstaat als eine von Weisungen, Steuerung und Kontrolle von außen unabhängige Behörde einrichten muss.

Artikel 57 DSGVO listet insgesamt 22 Aufgaben der Aufsichtsbehörde, die sicherstellen sollen, dass eine umfangreiche Information, Sensibilisierung, Beratung über die Rechte der Betroffenen und Pflichten der Verantwortlichen im privaten und öffentlichen Bereich stattfindet. Dazu gehört auch die Beratung der staatlichen Organe zum Schutz der Rechte und Freiheiten natürlicher Personen in Bezug auf die Verarbeitung ihrer Daten bei allen legislativen und administrativen Maßnahmen. Betroffene sind bei der Wahrnehmung ihrer Rechte zu unterstützen und Beschwerden entgegenzunehmen und zu bearbeiten. Die Aufsichtsbehörden haben auch die Aufgabe, Untersuchungen über die Anwendung der DSGVO durchzuführen. Ergänzend verfügen die Aufsichtsbehörden über umfangreiche Befugnisse zur Kontrolle, Abhilfe gegen unrechtmäßige Verarbeitungen und Sanktionierung von Verstößen durch die Verantwortlichen. Sie erteilt zu dem die nach der DSGVO erforderlichen Genehmigungen, zum Beispiel für Verarbeitungen personenbezogenen Daten, die wegen ihres Umfangs, ihrer Tragweite und Auswirkungen sowie der mit damit verbundenen Risiken für die Betroffenen, einer Vorab-Genehmigung bedürfen.

Rechtsbehelfe für Betroffene

Jeder Betroffene hat das Recht auf Beschwerde bei der zuständigen Aufsichtsbehörde sowie das Recht auf sämtliche zur Verfügung stehenden (außer-) gerichtlichen Rechtsbehelfe. Er hat zudem das Recht auf materiellen und immateriellen Schadensersatz und er kann sich bei einer Klage auch von einer dafür berechtigten Organisation, wie zum Beispiel den Verbraucherschutz vertreten lassen.

Sanktionen und hohe Geldbußen

Die Verstöße gegen die Vorgaben der DSGVO für Verantwortliche sind in zwei Gruppen kategorisiert, die erste Kategorie kann mit Geldbußen von bis zu zehn Millionen Euro oder zwei Prozent des weltweiten Umsatzes des Verantwortlichen belegt werden. Die zweite Kategorie kann mit Bußen von bis zu 20 Millionen Euro oder vier Prozent des Umsatzes bedacht werden. Die höchsten, bisher ausgesprochenen Geldbußen fielen allesamt unter die Kategorie 2 und beliefen sich auf rund 205 Millionen Euro (British Airways), 110 Millionen Euro (Marriot Hotel Gruppe) und 50 Millionen Euro (Google). Das Volumen der seit Inkrafttreten der DSGVO im Mai 2018 verhängten Geldbußen gegen private Unternehmen übersteigt mittlerweile 400 Millionen Euro (Kanzlei CMS, Enforcement Tracker, einsehbar unter: https://www.enforcementtracker.com/).

12.3 Auswirkungen der europäischen Datenschutzvorschriften auf Bonitätssysteme in China

12.3.1 Grundzüge der chinesischen Bonitätssysteme

Das gesellschaftliche Bonitätssystem in China beruht auf einem in 2014 veröffentlichten Plan der Regierung. Es zielt darauf ab, Verhalten zu belohnen, das den finanz-, wirtschafts- und sozialpolitischen Zielen der Regierung entspricht. Entgegenstehendes Verhalten soll hingegen sanktioniert werden (vgl. Ohlberg et al. 2018, S. 4).

Die Systeme zur Bewertung von Unternehmen, Corporate SCS genannt, dienen der Einschätzung, inwieweit die Unternehmen sämtlichen von der Regierung vorgegebenen Compliance-Anforderungen genügen. Dazu werden Daten, die das Unternehmen selbst zur Verfügung stellt, Daten aus Überprüfungen von Regierungsbehörden und zusätzliche Daten aus privatwirtschaftlichen Quellen gesammelt, integriert, ausgewertet und einem Ranking durch das National Internet+ Monitoring System unterworfen (vgl. European Union Chamber of Commerce in China o. J, S. 2).

Weiterhin existieren derzeit unterschiedliche Ansätze zur Bewertung von natürlichen Personen: Landesweite „schwarze Listen", Compliance-Bewertungen in Pilotstädten, und soziale Bonitätssysteme des Finanzbereichs (vgl. Sithig und Siems 2019, S. 12). Es ist offenbar beabsichtigt, die Bewertung von Unternehmen und natürlichen Personen in einer umfassenden Rating-Infrastruktur zu integrieren.

Der Umfang der bereits bestehenden und noch geplanten Erhebung von Daten erscheint immens und erstreckt sich für Unternehmen praktisch auf ihr gesamtes Geschäftsgebaren einschließlich ihrer Geschäftspartner. Für natürliche Personen erstreckt sich die Bewertung auf sämtliche Datenspuren, die in kommerziellen und privaten digitalen Aktivitäten anfallen, bis hin zu Aufzeichnungen aus Video-Überwachungen im öffentlichen Raum.

12.3.2 Gesetze zur Datensicherheit und zum Datenschutz in China

In China wurde per 01.06.2017 das Gesetz „The Cyber Security Law of the People's Republic of China" eingeführt. Dieses Gesetz definiert für Infrastruktur-kritische Unternehmen und für sogenannte Network Operators (Eigentümer von Netzwerken und Netzwerk-Service-Unternehmen) Pflichten zur Umsetzung von Maßnahmen für die Sicherheit der Datenverarbeitung und insbesondere von Datentransfers mittels des Netzwerks. Dieses Gesetz beinhaltet auch einige Vorschriften zu personenbezogenen Daten und „wichtigen Daten" („Measures on the Security Assessment of Cross-border Transfer of Personal

Information and Important Data"). Hierbei steht weniger der Schutz der Rechte und Freiheiten der Betroffenen bei der Verarbeitung personenbezogener Daten im Vordergrund, sondern die sogenannte Lokalisierung der Daten von chinesischen Staatsangehörigen, sprich ihre sichere Speicherung ausschließlich in China, bevor ein Transfer einer Kopie der Daten an Empfänger außerhalb Chinas stattfinden darf. Vor einem Datentransfer müssen die geplanten Sicherheitsmaßnahmen in Bezug auf den Schutzbedarf der zu übermittelnden Daten durch eine kompetente Behörde oder die Aufsichtsbehörde der jeweiligen Industrie bewertet werden. Die Definition personenbezogener Daten entspricht in etwa der der DSGVO. Die Unternehmen im Geltungsbereich dieser Vorschriften müssen bestimmte Vorgaben bei der Verarbeitung von personenbezogenen Daten beachten. So muss die Verarbeitung rechtmäßig, begründet und erforderlich sein. Die Regeln für die Erhebung und Verarbeitung der Daten, der Zweck, der Umfang und die Methoden der Verarbeitung müssen offengelegt werden. Die Betroffenen müssen in die Übermittlung ihrer Daten einwilligen. Die Verarbeitung muss durch technische Maßnahmen abgesichert werden, um den Diebstahl, die unbeabsichtigte Weitergabe oder die Verfälschung und Beschädigung von personenbezogenen Daten zu verhindern. Die betroffenen Unternehmen müssen eine Art Plattform für Beschwerden und für ein Reporting zu ihrer Netzwerk-Sicherheit bereitstellen.

Verglichen mit den europäischen Regeln zum Datenschutz lassen sich Datenschutzregeln in China bestenfalls als fragmentarisch bezeichnen (vgl. Sithig und Siems 2019, S. 19 mit weiteren Quellen).

12.3.3 Reichweite des EU-Datenschutzes bei Bewertungen, die auf Unternehmen ausgerichtet sind

Für in der EU ansässige Unternehmen
Nach Artikel 1 schützt die DSGVO natürliche Personen bei der Verarbeitung personenbezogener Daten. Der Schutz ist unabhängig davon, ob die natürliche Person als Privatperson, als Unternehmer oder als Mitarbeiter einer privaten oder einer öffentlichen Institution die Verarbeitung seiner personenbezogenen Daten erfährt.

In der EU ansässige Unternehmen müssen deren Anforderungen bei der Übermittlung von Informationen an ein Bewertungssystem für Unternehmen in China berücksichtigen. Vereinfacht bedeutet das, dass nur Informationen und Daten zur Verfügung gestellt werden dürfen, die keinen Bezug auf Personen aufweisen, die mit dem Unternehmen in Verbindung stehen, sei es als Mitarbeiter, Organvertreter, Kunden, Lieferanten, Aktionäre oder andere Geschäftspartner. Denn bisher gibt es keine Rechtsgrundlage in Form einer in der EU wirksamen rechtlichen Verpflichtung der Unternehmen, personenbezogene Daten an ein staatliches oder privates Bewertungsinstitut in China zu übermitteln.

Eine weitere Möglichkeit wäre, eine Übermittlung von personenbezogenen Daten an ein chinesisches Bewertungssystem darauf zu stützen, dass das betroffene Unternehmen ein berechtigtes Interesse (Artikel 6 Abs. 1 Buchstabe f) DSGVO) hat sich bewerten zu lassen, um geschäftliche Chancen in China zu vorteilhaften Bedingungen nutzen zu können. Dabei ist zu berücksichtigen, dass das Schutzbedürfnis der Betroffenen, deren Daten in ein Bewertungssystem eingehen, schwerer wiegen kann, als das Interesse des Unternehmens. Zieht man alle Rahmenbedingungen der Verarbeitung durch chinesische Bewertungssysteme in Betracht, so hat der Betroffene in der EU kaum realistische Chancen eingehende Informationen über die Verarbeitung der ihn betreffenden Daten zu erhalten oder seine Rechte auf Auskunft, Berichtigung und Löschung in China auszuüben. Deshalb würde das Interesse des Betroffenen am Schutz seiner Daten immer das Interesse eines Unternehmens an der Teilnahme an einem Bewertungssystem überwiegen.

Selbst wenn ein Unternehmen eine stichhaltige Interessenabwägung zu seinen Gunsten darlegen könnte, wären für eine rechtmäßige Übermittlung weitere Hürden zu überwinden, die nur durch die Empfänger der Daten in China erfüllt werden können. Aus Sicht der DSGVO ist China im Hinblick auf die Verarbeitung personenbezogener Daten ein unsicheres Drittland, weil das Datenschutzniveau in China bisher nicht auf seine Adäquanz mit dem in der EU gültigen Schutzniveau bewertet wurde. Ein nach Artikel 45 DSGVO dafür erforderlicher Angemessenheitsbeschluss der EU-Kommission liegt nicht vor.

Die die Daten empfangende Stelle in China könnte sich auch freiwillig in Form eines Vertrages zur Schaffung und Einhaltung eines den EU-Regeln entsprechenden Datenschutzes verpflichten. Dafür sieht die EU den Abschluss sogenannter Standard Contractual Clauses vor (https://eur-lex.europa.eu/legal-content/DE/TXT/?uri=CELEX%3A32010D0087 zum Beschluss und den Vertragsmustern für die Datenübermittlung in Länder außerhalb der EU ohne anerkanntes Datenschutzniveau). Darin verpflichtet sich der Empfänger der Daten zu weitreichenden Garantien für den Datenschutz. Diese muss er auch gegenüber weiteren Empfängern der Daten gewährleisten. Die Ausnahmeregelungen nach Artikel 49 DSGVO für eine Übermittlung personenbezogener Daten in ein aus Datenschutzsicht unsicheres Drittland bieten insgesamt keine gangbare Lösung für die Übermittlung personenbezogener Daten, weil sie auf die Interessenslage und Willenserklärungen einer betroffenen Person abstellen und deshalb den Sachverhalt hier nicht treffen. Zudem dürfte eine solche soweit zulässige Übermittlung nur im Einzelfall und nicht systematisch und regelmäßig stattfinden (Artikel 49 Abs. 1 Satz 2 DSGVO).

Zusammenfassend ist eine Übermittlung von Unternehmensdaten an ein chinesisches Bonitätssystem für Unternehmen im Geltungsbereich der DSGVO nur dann möglich, wenn diese Unternehmensdaten anonym sind, das heißt keinerlei Rückschlüsse auf natürliche Personen im Kontext der unternehmerischen Tätigkeiten des Unternehmens ermöglichen (vgl. European Union Chamber of Commerce in China o. J, S. 4, zum Erfordernis einer genauen Analyse der Datenflüsse an die Bonitätssysteme mit dem Ziel der Erhaltung integrer Daten). Unternehmensaktivitäten sind immer verbunden mit menschlichem Handeln, erlauben Rückschlüsse auf die Handelnden oder legen das Handeln natürlicher Personen offen, allein schon, weil diese Handlungen mehr und mehr digital unterstützt werden und damit einhergehend zahlreiche Datenspuren des Handelnden hinterlassen. Bei automatischen Übermittlungen von Daten aus unternehmenseigenen Systemen ist eine Trennung der personenbezogenen Bestandteile von Datensätzen oft nur schwer oder gar nicht möglich. Der Datenschutz soll dafür sorgen, dass diese Datenspuren nicht über das erforderliche Maß hinaus erzeugt werden und dass sie vor allem nicht für andere, als die ursprünglich festgelegten Zwecke genutzt werden. Deshalb muss an den relevanten Schnittstellen für die Übermittlung von Daten an staatliche Überwachungssysteme für eine wirksame Anonymisierung personenbezogener Daten gesorgt werden. Wirksamkeit ist dann gegeben, wenn es mit allen objektiv zur Verfügung stehenden Mitteln nicht mehr möglich ist, einen Personenbezug herzustellen. Anonyme Daten unterliegen keinerlei Anforderungen aus dem Datenschutz.

In China ansässige Unternehmen

Chinesische Unternehmen, auch wenn es sich um Tochtergesellschaften europäischer Firmen handelt, müssen an erster Stelle lokales Recht im Land ihres Sitzes befolgen. Für solche Unternehmen gilt aber auch die DSGVO, wenn sie natürlichen Personen in der EU Waren und Dienstleistungen anbieten oder das Verhalten natürlicher Personen in der EU beobachten (sogenanntes Marktortprinzip der DSGVO nach Artikel 3 Abs. 2).

Für chinesische Anbieter von Produkten und Services auf dem europäischen Markt an europäische Endverbraucher gelten sämtliche Verpflichtungen der DSVO unabhängig davon, wo die Datenverarbeitung stattfindet. Das Unternehmen darf in diesem Zusammenhang lediglich aufgrund der gesetzlichen Grundlagen der EU Daten erheben und verarbeiten. Es muss die Kunden über die Datenverarbeitung eingehend informieren, eine Kontaktadresse nennen, an die der Kunde sich wenden kann, um seine Rechte zum Beispiel auf Auskunft, auf Widerspruch und Löschung geltend zu machen. Das chinesische Unternehmen muss einen in der EU ansässigen Ansprechpartner benennen, an den sich die zuständige Datenschutzaufsichtsbehörde wenden kann.

Ebenso gelten die Regeln der DSGVO für chinesische Unternehmen, die auf Weisung von in der EU ansässigen Auftraggebern personenbezogene Daten verarbeiten. Sie sind im Hinblick auf die Verarbeitung von personenbezogenen Daten im Geltungsbereich der DSGVO den Verantwortlichen in vieler Hinsicht gleichgestellt. Sie können nur ordnungsgemäß beauftragt werden, wenn sie die Regeln der DSGVO nachweislich einhalten (Artikel 28 DSGVO). Solche sogenannte Auftragsverarbeitungen können zum Beispiel Software-Entwicklung, Applikationswartung und Rechenzentrumsleistungen, aber auch Produktionsbetriebe und Kundendienstleistungen sein, bei der der Auftragnehmer Zugriff auf personenbezogene Daten im Verantwortungsbereich des Auftraggebers hat oder dieser Zugriff technisch nicht ausgeschlossen werden kann. Es genügt schon, wenn bei einer Software-Wartung der Zugriff auf in den gewarteten Systemen gespeicherte personenbezogene Daten nicht ausgeschlossen werden kann.

Eine Weiterleitung personenbezogener Daten von europäischen Kunden an andere Unternehmen in China ist ebenfalls nur unter den engen Voraussetzungen der DSGVO möglich und nur dann, wenn der Empfänger ein den Vorgaben der EU adäquates Datenschutzniveau garantiert, zum Beispiel durch den Abschluss der EU Standard-Vertragsklauseln. Diese Klauseln verpflichten den Datenempfänger u. a. zur Umsetzung von technischen und organisatorischen Maßnahmen, die den Schutz der personenbezogenen Daten gewährleisten. Außerdem muss der Datenempfänger die Erfüllung der Rechte der Betroffenen sicherstellen und unterstützen.

Für die Weitergabe personenbezogener Daten von natürlichen Personen aus der Geschäftstätigkeit in der EU an staatliche Stellen und deren Beauftragte in China gelten die gleichen Voraussetzungen und Grenzen, die im vorherigen Abschnitt bereits beschrieben wurden. Üblicherweise werden solche Datenübermittlungen in den jeweiligen Ländern auf gesetzliche Vorgaben gestützt, die das Daten verarbeitende Unternehmen verpflichten, bestimmte Daten für gesetzlich definierte Zwecke zur Verfügung zu stellen. Diese gesetzlichen Vorgaben müssen nach Artikel 6 Abs. 1 Buchstabe c sowie Abs. 3 im Recht des Mitgliedsstaats bzw. im Unionsrecht niedergelegt sein. Das bedeutet, die EU und/oder die einzelnen Mitgliedsstaaten müssten Gesetze verabschieden, die eine Übermittlung von personenbezogenen Daten nach China erlauben.

So ist die Übermittlung von Passagierdaten von Reisenden aus der EU in die USA an amerikanische Sicherheitsbehörden in einem Abkommen zwischen der EU und den USA geregelt. Dieses Abkommen stellt für die europäischen Fluggesellschaften eine rechtliche Verpflichtung dar, sodass die Übermittlung der Daten auf der Grundlage des Artikels 6 Abs. 1 Buchstabe c) rechtmäßig ist. Die EU-Kommission hat den amerikanischen Behörden für die Verarbeitung dieser Passagierdaten ein EU-adäquates Datenschutzniveau zuerkannt.

12.3.4 Voraussetzungen für ein EU-adäquates Datenschutzniveau

Würde die chinesische Regierung anstreben, dass Unternehmen im Geltungsbereich der DSGVO rechtmäßig personenbeziehbare Informationen für die Bonitätssysteme in China zur Verfügung stellen können, müsste für diese Datenverarbeitungen ein EU-adäquates Datenschutzniveau hergestellt werden. Die EU-Kommission könnte nach entsprechender Prüfung eine Adäquanzentscheidung nach Artikel 45 DSGVO für den Sektor Bonitätssysteme treffen. Die Anforderungen beziehen sich nicht nur auf die Datenverarbeitung an sich, sondern auf eine ganze Reihe im Artikel 45 Abs. 3 genannte grundlegende Aspekte eines Landes oder einer internationalen Organisation, die bei einer Entscheidung über ein Schutzniveau zu berücksichtigen sind. Die Rechtsstaatlichkeit, die Achtung der Menschenrechte und der Grundfreiheiten, aber auch die geltenden Rechtsvorschriften in Bezug auf öffentliche Sicherheit, Verteidigung, nationale Sicherheit und Strafrecht. Der Zugang von Behörden zu personenbezogenen Daten, die geltenden Datenschutzvorschriften, Berufsregeln, Sicherheitsvorschriften müssen den Schutz des Betroffenen beim Umgang mit seinen Daten berücksichtigen. Das gilt auch für Vorschriften, die eine Weiterübermittlung von Daten an andere Drittländer oder internationale Organisationen regeln. Hinzu kommt die Rechtsprechung, das Vorhandensein von wirksamen und durchsetzbaren Rechten der Betroffenen einschließlich der notwendigen verwaltungsrechtlichen und gerichtlichen Rechtsbehelfe. Die Zuerkennung eines adäquaten Datenschutzniveaus erfordert auch die Einrichtung unabhängiger und durchsetzungsbefugter Aufsichtsbehörden für den Datenschutz, die den

Betroffenen bei der Wahrnehmung seiner Rechte unterstützen und mit den europäischen Aufsichtsbehörden kooperieren. Völkerrechtliche Verpflichtungen die ein Land eingeht, wie die Datenschutzkonvention des Europarates (Sammlung europäischer Verträge Nr. 108: Übereinkommen zum Schutz des Menschen bei der automatischen Verarbeitung personenbezogener Daten abgerufen unter https://rm.coe.int/1680078b38), gelangen ebenfalls in die Waagschale. Die Anerkennung des Datenschutzniveaus erfolgt in einem Durchführungsrechtsakt. Die EU-Kommission ist gesetzlich verpflichtet, mindestens alle vier Jahre zu überprüfen, ob die Angemessenheit des Datenschutzniveaus noch besteht. Alle Entwicklungen, die das Datenschutzniveau beeinträchtigen könnten, sind dabei einzubeziehen. Eine Liste aller Länder mit einem anerkannten Datenschutzniveau ist auf den Internetseiten der EU abrufbar (die aktuelle Liste ist abrufbar unter: https://ec.europa.eu/info/law/law-topic/data-protection/international-dimension-data-protection/adequacy-decisions_en).

12.3.5 (Mindest-)Anforderungen an Bonitätssysteme aus europäischer Sicht

Die Anerkennung eines EU-adäquaten Datenschutzniveaus erfordert nicht die Schaffung genau gleicher Regelungen in einem Drittland. Wenn man die Regelungen der DSGVO sinngemäß anwendet, stellen sich zumindest gleichartige grundlegende Anforderungen an die Verarbeitung von personenbezogenen Daten in Bonitätssystemen.

Die Zwecke der Verarbeitung müssten eindeutig und abschließend benannt werden. Die Legitimität der Verarbeitungszwecke müsste in der EU anerkannt sein. Besonders problematisch erscheint in diesem Zusammenhang die Absicht, privatwirtschaftliche und staatliche Bewertungssysteme zu integrieren. Denn dadurch entstehen aller Wahrscheinlichkeit nach neue Verarbeitungszwecke für in privatwirtschaftlichen Verarbeitungen gespeicherte Daten, über die der Betroffene keine Kenntnis hat. Als weitere Kernanforderung käme die Datenminimierung hinzu. Die Bonitätssysteme müssten also nachweisen können, dass sie nur die für ihre Analysen erforderlichen Daten verwenden.

Die Speicherdauer aller verwendeten Daten muss definiert sein und ein darauf aufbauendes Löschkonzept muss gewährleisten, dass eine physische Löschung oder unumkehrbare Anonymisierung der Daten stattfindet, die den Personenbezug abschließend eliminiert.

Die Systeme müssen nachweislich so gestaltet werden, dass eine Verfälschung der Daten und eine Verletzung ihrer Vertraulichkeit ausgeschlossen ist. Für die Wahrung der Vertraulichkeit müsste allerdings genau definiert werden, wer die berechtigten Nutzer der Informationen sind. Eine Information der Betroffenen darüber ist ebenso sicherzustellen. Angesichts des immensen Umfangs der Datensammlungen aus den unterschiedlichsten Lebensbereichen der Menschen und der möglichen Tragweite der Folgen einer (unzutreffenden) Bewertung, sind die Verarbeitungen in Bonitätssystemen aus europäischer Sicht als hochrisikobehaftet einzustufen. Das heißt, es wäre eine eingehende Bewertung aller Folgen für den Betroffenen aus der Verletzung des Schutzes der ihn betreffenden Informationen durchzuführen. Diese Folgen müssen durch geeignete Maßnahmen auf ein mögliches Mindestmaß reduziert werden. In der EU wäre die zuständige Aufsichtsbehörde vor einer möglicherweise so weitreichenden Verarbeitung zu konsultieren (Art. 35 und 36 DSGVO).

Transparenz ist herzustellen, insbesondere über die Bewertungsalgorithmen und die dafür genutzten Informationen. Stellt man sich die bereits in den Pilotprojekten vorhandene Fülle von Verarbeitungsvorgängen und die hohe Komplexität der zu integrierenden Systeme zu einer Gesamtplattform vor, erscheint es höchst schwierig, in verständlicher Form die Funktionsweise der Bonitätssysteme auf Daten- und Prozessebene zu beschreiben.

Es sind zudem bei allen Gelegenheiten und in allen Vorgängen, bei denen Daten manuell oder automatisch erhoben werden, sei es direkt beim Betroffenen oder durch Nutzung anderer Quellen sowie auch aus physischen Überwachungsvorgängen, darüber zu informieren. Es sind leicht zugängliche und verständliche Informationen anzubieten. Diese müssen die Art der erhobenen Daten und die Verarbeitungsform der durch den Betroffenen bereitgestellten Informationen verständlich erklären. Die Transparenzerfordernisse gelten nicht nur für die Erhebung der Daten, sondern auch für die Erfüllung der Betroffenenrechte auf Auskunft, Berichtigung, Widerspruch und Löschung. Sofern in China ansässige Institutionen in den Bonitätssystemen Informationen von Betroffenen in der EU verarbeiten, müssen sie in der EU einen Beauftragten benennen, der für Anfragen von Betroffenen und Datenschutz-Aufsichtsbehörden zur Verfügung steht.

Technische und organisatorische Maßnahmen, die eine legitime und sichere Verarbeitung der Daten gewährleisten und Datenpannen wirksam verhindern können, müssen nachgewiesen werden und überprüfbar sein, zum Beispiel unabhängige Kontroll- und Überwachungsinstanzen für die Einhaltung der Datenschutzregeln und die Unterstützung der Betroffenen bei der Wahrnehmung ihrer Rechte.

12.4 Fazit

Die Verarbeitung von Daten für die Bonitätsbewertung natürlicher Personen in dem beschriebenen Umfang kann nach Ansicht der Verfasserin aufgrund ihrer Komplexität kaum in verständlicher Weise offengelegt und beschrieben werden, wie es nach den in Artikel 5 der DSGVO festgelegten Grundsätzen für die Verarbeitung personenbezogener Daten in der EU erforderlich ist. Es wäre darzulegen, ob die Datenerhebung für die Zwecke der Bewertungen überhaupt erforderlich ist. Es wäre sicherzustellen, dass die Daten auch tatsächlich nur für die Zwecke der festzulegenden Bewertungsschemata genutzt werden. Schließlich wäre die Korrektheit der Daten und ihre Löschung nachzuweisen, sobald die Daten für den ursprünglichen Zweck nicht mehr benötigt werden. Das Aufheben und Nutzen der Daten zu einem anderen als den ursprünglichen Zweck wäre nicht zulässig.

Die Bewertungssysteme treffen im Grunde automatisierte Entscheidungen, auf die der Betroffene kaum Einfluss hat, zumal, wenn die Erhebung der Daten und die Zwecke ihrer

Verwendung nicht vollständig erkennbar sind. Die Weiterleitung von Daten aus kommerziellen Systemen an die staatlichen Bewertungs- und Monitoring-Systeme stellen eine Weiterverarbeitung zu neuen, fremden Zwecken dar, für die der europäische Datenschutz hohe Hürden aufstellen würde.

In vielen Fällen liegt der Schluss nahe, dass Daten zunächst auf Vorrat, also ohne konkreten Verarbeitungszweck erhoben werden, um für spätere zum Erhebungszeitpunkt noch nicht feststehende Auswertungen genutzt zu werden. Eine solche sogenannte Vorratsdatenspeicherung ist in der EU im Zusammenhang mit der Bekämpfung von Terrorismus und anderen schweren Verbrechen unter engen Vorgaben zulässig, wie die zeitlich beschränkte Speicherung von Telekommunikationsdaten oder die Übermittlung von Passagierdaten an staatliche Datensammelstellen (Richtlinie (EU) 2016/681 des Europäischen Parlaments und des Rates vom 27. April 2016 über die Verwendung von Fluggastdatensätzen (PNR-Daten) zur Verhütung, Aufdeckung, Ermittlung und Verfolgung von terroristischen Straftaten und schwerer Kriminalität). Letzteres ist zum Beispiel in Deutschland derzeit umstritten und Gegenstand einer Klage von Betroffenen gegen die zuständige Behörde (https://www.spiegel.de/netzwelt/netzpolitik/fluggastdaten-speicherung-aktivisten-wollen-eu-richtlinie-mit-klagen-kippen-a-1267331.html, abgerufen am 20.12.2019).

Bewertungssysteme, die unter Einbeziehung personenbezogener Informationen Score-Werte ermitteln, sind vor dem Hintergrund der EU-Datenschutzregeln nur unter engen Voraussetzungen möglich und unterliegen der unabhängigen Datenschutzaufsicht.

Rating-Systeme für Unternehmen sind auch in der EU weit verbreitet. Sie beruhen allerdings auf öffentlich verfügbaren Informationen aus der Finanzberichterstattung und dienen zum Beispiel der Einstufung hinsichtlich des Investitionsrisikos für Anteilseigner. Die Informationen für solche Rating-Systeme werden i.d.R. in Form von Berichten durch die Unternehmen selbst zur Verfügung gestellt und durch die Rating-Agenturen verarbeitet. Bonitätsauskünfte für Unternehmen stützen sich überwiegend auf öffentlich verfügbare Informationen und Register. Private Unternehmen, die Bonitätsauskünfte erstellen, aggregieren üblicherweise öffentlich recherchierbare Informationen zu einem Gesamtbild über die Bonität von Unternehmen, um potenzielle Kunden oder Geschäftspartner gegen Entgelt hinsichtlich der möglichen Risiken einer Geschäftsbeziehung zu beraten.

Eine immer größere Bedeutung gerade für börsenorientierte Unternehmen stellen Ratings über die Nachhaltigkeit der Unternehmenstätigkeit im Sinne einer gesamthaften gesellschaftlichen Verantwortung dar. Solche Indizes werden durch direkte Befragung und Einsammeln von Nachweisen der Unternehmen hinsichtlich der Einhaltung von entsprechenden Vorgaben ermittelt. Aktienrechtliche Nachweispflichten bestehen für große Unternehmen hinsichtlich der Steuerung von Risiken aus ihrer Geschäftstätigkeit für die Gesellschaft. Damit soll eine aktive Auseinandersetzung, zum Beispiel mit internationalen Codes of Conduct oder Umwelteinflüssen durch die Geschäftstätigkeit in den Unternehmen erreicht werden. Die Unternehmen sollen selbst Regeln implementieren, die die Einhaltung diesbezüglicher (globaler) Standards sicherstellen. Die dafür angestellten Bewertungen basieren auf der im Wesentlichen durch das Unternehmen gesteuerten Informationsweitergabe und einer teilweisen Überprüfung durch unabhängige Wirtschaftsprüfer.

Europäische Unternehmen mit Geschäftstätigkeit, Niederlassungen, Tochterunternehmen oder Geschäftspartnern in China sollten auf jeden Fall nicht nur die Anforderungen der Informationsbereitstellung an chinesische Bewertungssysteme im Auge haben. Sie sollten gleichzeitig auch die möglichen Restriktionen für die Bereitstellung von im weitesten Sinne personenbezogenen Daten aus der europäischen Datenschutzgesetzgebung betrachten. Der Geltungsbereich der DSGVO eröffnet möglicherweise die Anwendbarkeit der Regeln auf Datenflüsse von der EU nach China. Die mit der DSGVO möglichen Sanktionen, insbesondere die beträchtlichen Geldbußen und Schadensersatzanforderungen stellen ein hohes Risiko für die Unternehmen dar.

Literatur

Gesetzestexte

Verordnung (EU) 2016/679 des Europäischen Parlaments und des Rates vom 27. April zum Schutz natürlicher Personen bei der Verarbeitung personenbezogener Daten, zum freien Datenverkehr zur Aufhebung der Richtlinie 95/46/EG (Datenschutz-Grundverordnung).

Richtlinie (EU) 2016/681 des Europäischen Parlaments und des Rates vom 27. April 2016 über die Verwendung von Fluggastdatensätzen (PNR-Daten) zur Verhütung, Aufdeckung, Ermittlung und Verfolgung von terroristischen Straftaten und schwerer Kriminalität.

BDSG (neu) 2018, zuletzt angepasst durch das zweite Gesetz zur Anpassung des Datenschutzrechts an die Verordnung (EU) 2016/679 und zur Umsetzung der Richtlinie (EU) 2016/680 (Zweites Datenschutz-Anpassungs- und Umsetzungsgesetz EU – 2. DSAnpUG-EU) geändert, am 26. November 2019 in Kraft getreten.

https://eur-lex.europa.eu/legal-content/DE/TXT/?uri=CELEX%3A32010D0087 zum Beschluss und den Vertragsmustern für die Datenübermittlung in Länder außerhalb der EU ohne anerkanntes Datenschutzniveau.

https://ec.europa.eu/info/law/law-topic/data-protection/international-dimension-data-protection/adequacy-decisions_en

Sammlung europäischer Verträge Nr. 108: Übereinkommen zum Schutz des Menschen bei der automatischen Verarbeitung personenbezogener Daten abgerufen unter https://rm.coe.int/1680078b38

Weitere Quellen

Ohlberg, M., Ahmed, S., Lang, Bertram: Zentrale Planung, lokale Experimente, MERICS China Monitor, 4.4.2018.

European Union Chamber of Commerce in China (Hrsg.): The Digital Hand.

Síthig, D.M. and Siems, M.; The Chinese social credit system: A model for other countries? EUI Working Papers LAW 2019/01 https://www.spiegel.de/netzwelt/netzpolitik/fluggastdaten-speicherung-aktivisten-wollen-eu-richtlinie-mit-klagen-kippen-a-1267331.html, abgerufen am 20.12.2019.

Kanzlei CMS, Enforcement Tracker, einsehbar unter: https://www.enforcementtracker.com/

Dr. Barbara Kirchberg-Lennartz publiziert seit 2009 in Form von Vorträgen, Fachpublikationen und Kommentierungen von Datenschutzgesetzen zu rechtlichen und operativen Fragestellungen des Datenschutzes. Sie ist als Datenschutzbeauftragte eines großen deutschen DAX-Konzerns tätig.

Social Credit Rating im Spannungsfeld zwischen möglichst umfassender Informationsgrundlage für Entscheidungen und dem Schutz der Privatsphäre

13

Steffen Salvenmoser

Zusammenfassung

Der Artikel betrachtet die Frage, ob und unter welchem Umständen das Modell des Social Credit Rating in einem europäischen, insbesondere deutschen Rechtssystem umsetzbar wäre und kommt zu dem Ergebnis, dass bei allem verständlichen Interesse an einer umfassenden Sachverhaltsgrundlage für eine Entscheidung die Idee gegen fundamentale Prinzipien unseres Rechtssystems, wie dem Recht auf informationelle Selbstbestimmung und dem Rechtsstaatsprinzip diametral entgegen läuft.

13.1 Vorbemerkung

Dieser Beitrag erhebt nicht den Anspruch, das chinesische System des Social Credit Rating im Einzelnen zu beschreiben oder zu analysieren und zu bewerten. Vielmehr verfolgt er das Ziel, losgelöst von einer konkreten Umsetzung oder Ausformung eines solchen Systems abstrakt Chancen und Risiken abzuwägen. Dabei liegt im Geschäftsverkehr in der einen Waagschale das Interesse einer Entscheidung (insbesondere einer Prognoseentscheidung), wie zum Beispiel der Frage, ob mit dem Eingehen eines Geschäfts bzw. genauer, dem Eingehen einer Geschäftsbeziehung mit einer konkreten Person oder einem Unternehmen, ein besonderes Risiko verbunden ist. Dagegen zu stellen ist das Interesse insbesondere natürlicher Personen am Schutz ihrer Privatsphäre.

S. Salvenmoser (✉)
Kanzlei Salvenmoser, Erlenbach, Deutschland
E-Mail: stsa@kanzlei-salvenmoser.de

© Springer Fachmedien Wiesbaden GmbH, ein Teil von Springer Nature 2020   249
O. Everling (Hrsg.), *Social Credit Rating*,
https://doi.org/10.1007/978-3-658-29653-7_13

So wie der Verfasser das System versteht, dient dieses u. a. dazu, auf der Basis einer möglichst umfassenden Datenbasis das Verhalten von Personen aber auch Unternehmen zu bewerten und den Zugang zu bestimmten insbesondere staatlichen Einrichtungen und Leistungen zu ermöglichen oder bei einem schlechten Rating auch einzuschränken. Abhängig vom Social Credit Rating könnten aber auch Entscheidungen nicht staatlicher Einrichtungen sein, wie zum Beispiel die Frage der Gewährung eines Darlehens durch eine Bank.

Während es im ersten Fall dann wohl stärker darum ginge sozial unerwünschtes Verhalten zu sanktionieren bzw. sozial erwünschtes Verhalten zu belohnen und damit die Bereitschaft zur Normbefolgung zu erhöhen, geht es im zweiten Beispiel (Darlehen) darüber hinaus auch um eine Risikoeinschätzung. Dabei wird aus gesammelten Daten der Vergangenheit eine Verhaltensprognose für die Zukunft abgeleitet. Neben vielen anderen Frage wirft dies aber etwa auch die Frage auf, ob es ohne weiteres möglich ist, aus Verhalten der Vergangenheit sichere Schlüsse auf die Zukunft zu ziehen. Wir alle kennen alle den Satz, „wer einmal nicht die Wahrheit spricht, dem glaubt man nicht." Er wird häufig, allerdings meistens in dieser unvollständigen Form zitiert. Weiter geht er: „… und wenn er doch die Wahrheit spricht." Schon dieses kleine Beispiel macht deutlich, dass es wohl keinen Mechanismus gibt, dass früheres Fehlverhalten zwangsläufig wiederholt wird.

Mit dem Beitrag soll der Versuch unternommen werden, auszuloten wie weit die soeben abstrakt beschriebene Idee einer umfassenden Informationsbeschaffung in einem Rechtssystem, wie dem Deutschen umsetzbar wäre.

13.2   Entscheidungsfindung und Informationsstand

Die (subjektiv empfunden) richtige Entscheidung zu finden ist häufig nicht einfach. Viele Faktoren beeinflussen den Entscheidungsprozess. Die (subjektive) Bedeutung der zu treffenden Entscheidung, die Zeit, die für die Entscheidungsfindung zur Verfügung steht und natürlich der Informationsstand auf dessen Grundlage die Entscheidung getroffen wird sind maßgebliche Faktoren.

Es ist eine Binsenweisheit, dass eine Entscheidung umso besser wird, je stärker sie auf der Kenntnis relevanter Fakten gründet.

Es ist deswegen ein naheliegender und faszinierender Gedanke, zum Beispiel durch einen erweiterten Zugriff auf vorhandene Daten, das erfassen weiterer Daten und deren effektive, gegebenenfalls maschinengestützte Aufbereitung und Auswertung mittels Algorithmen (Stichwort Big Data) zu besseren, weil informierten Entscheidungen zu kommen.

Im Geschäftsleben und in der Rechtswissenschaft ist die Frage des erforderlichen Informations- und Kenntnisstandes im Zusammenhang mit der sogenannten „Business Judgment Rule" oder dem „breiten unternehmerischen Ermessen" von besonderer Bedeutung.

13.2.1 Die Business Judgement Rule als Beispiel für die Notwendigkeit einer umfassenden Tatsachengrundlage für geschäftliche Entscheidungen

Der Bundesgerichtshof (BGH) hat in seiner sogenannten „ARAG/Garmenbeck Entscheidung", in der es im Kern um die Frage ging, ob es eine Verpflichtung des Aufsichtsrats geht, vermutete Pflichtverletzungen eines Vorstands zu verfolgen und allfällige Schadenersatzansprüche durchzusetzen, formuliert, dass Ermessen nur ausgeübt werden kann, wenn die gegeneinander abzuwägenden Umstände festgestellt werden(BGHZ 135, 244).

Entscheidend ist unter den Aspekten dieses Beitrags zunächst lediglich, dass der Grundgedanke, dass eine Entscheidung nur dann richtig getroffen werden kann, wenn sie auf einem möglichst umfassenden Tatsachengrundlage ruht.

Ohne dass dies in der Entscheidung ausdrücklich gesagt wird, steht aber außer Zweifel, dass die darin zum Ausdruck kommende Pflicht zur Sachverhaltsaufklärung nicht unbegrenzt ist, sondern Ihre Grenze in den geltenden Gesetzen findet.

13.2.2 Grenzen der Aufklärungspflicht

Die soeben abgeleitet Aufklärungspflicht als Grundlage für eine angemessene Ermessensentscheidung ist nicht unbegrenzt. Sie findet eine Vielzahl von Grenzen. Nach geltendem Recht können dies strafrechtliche Grenzen sein (zum Beispiel das Briefgeheimnis, das Steuergeheimnis …). Grenzen können auch aus dem öffentlichen Recht kommen, oder aus dem Zivilrecht. Dabei sind viele Einzelvorschriften, insbesondere solche des Datenschutzrechts, letztlich nichts anderes als die Umsetzung des verfassungsrechtlich geschützten „Rechts auf informationelle Selbstbestimmung" und des allgemeinen Persönlichkeitsrechts.

Selbstverständlich gibt es auch innerhalb dieser Grenzen keine Pflicht zur Sachverhaltsaufklärung um jeden Preis. Vielmehr muss der Aufwand in einem vernünftigen Verhältnis zu dem zu erwartenden Ergebnis stehen.

Und wenn es nicht gerade so spezifische Situationen wie die, die der Entscheidung des BGH zu Grunde lag, dann darf im geschäftlichen Umfeld natürlich auch eine Abwägung zwischen Chancen und Risiken vorgenommen werden und es ist als Bestandteil unternehmerischer Freiheit natürlich nicht verboten auch riskante Geschäfte einzugehen.

13.2.3 Zwischenergebnis

Das geltende Recht beschränkt das unternehmerische Ermessen und die unternehmerische Handlungsfreiheit insoweit, als eine verantwortungsvolle Entscheidung Kenntnis der wesentlichen Fakten betrifft.

Unterstellt man nun, dass bei der Entscheidung eine möglichst umfassende Datengrundlage über den Geschäftspartner vorliegt, die weit über das hinaus geht was heute in Deutschland aus allgemein zugänglichen Quellen über natürliche Personen und Unternehmen in Erfahrung zu bringen ist, dann müsste dadurch an sich die Entscheidung besser und rechtssicherer werden.

Dem stehen indes vielfältige „aber" entgegen, die im Weiteren näher beleuchtet werden sollen. „Aber", die sich zum Beispiel aus unserer deutschen Perspektive auf das Thema Datenschutz und Schutz der Privatsphäre ergeben.

13.3   Datenschutz und das allgemeine Persönlichkeitsrecht

Der Begriff des informationellen Selbstbestimmungsrechts geht zurück auf ein Gutachten von Wilhelm Steinmüller, Bernd Lutterbeck u. a. aus dem Jahr 1971 (vgl. Steinmüller et al. 1971). Im sogenannten „Volkszählungsurteil" des BVerfG hat dieses das Recht auf informationelle Selbstbestimmung zunächst als grundrechtlich geschützt definiert und näher erläutert (BVerfG, Urteil des Ersten Senats vom 15. Dezember 1983, 1 BvR 209/83 u. a. – Volkszählung –, BVerfGE 65, 1).

Im Leitsatz der Entscheidung heißt es:

> „Unter den Bedingungen der modernen Datenverarbeitung wird der Schutz des Einzelnen gegen unbegrenzte Erhebung, Speicherung, Verwendung und Weitergabe seiner persönlichen Daten von dem allgemeinen Persönlichkeitsrecht des GG Art 2 Abs. 1 in Verbindung mit GG Art 1 Abs. 1 umfaßt. Das Grundrecht gewährleistet insoweit die Befugnis des Einzelnen, grundsätzlich selbst über die Preisgabe und Verwendung seiner persönlichen Daten zu bestimmen.
>
> Einschränkungen dieses Rechts auf „informationelle Selbstbestimmung" sind nur im überwiegenden Allgemeininteresse zulässig. Sie bedürfen einer verfassungsgemäßen gesetzlichen Grundlage, die dem rechtsstaatlichen Gebot der Normenklarheit entsprechen muß. Bei seinen Regelungen hat der Gesetzgeber ferner den Grundsatz der Verhältnismäßigkeit zu beachten. Auch hat er organisatorische und verfahrensrechtliche Vorkehrungen zu treffen, welche der Gefahr einer Verletzung des Persönlichkeitsrechts entgegenwirken.
>
> Bei den verfassungsrechtlichen Anforderungen an derartige Einschränkungen ist zu unterscheiden zwischen personenbezogenen Daten, die in individualisierter, nicht anonymer Form erhoben und verarbeitet werden, und solchen, die für statistische Zwecke bestimmt sind."

Das BVerfG hat damit bereits im Jahre 1983 im Hinblick auf die damaligen Möglichkeiten elektronischer Datenverarbeitung formuliert, dass es grundsätzlich in der Befugnis jedes Einzelnen steht, darüber zu entscheiden, wem welche persönlichen Daten zur Verfügung stehen sollen. Einschränkungen bedürfen einer ausdrücklichen gesetzlichen Grundlage und müssen im überwiegenden Interesse der Allgemeinheit liegen.

Im Urteil des BVerfG findet sich hierzu eine Passage, die auch nach 36 Jahren an Aktualität nichts verloren hat, sondern Rückblickend auf die technischen Möglichkeiten der

Achtzigerjahre als nahezu visionär zu bezeichnen ist. Ich zitiere diese Passage deswegen wörtlich:

> Es (das allgemeine Persönlichkeitsrecht; d. Verf.) umfaßt … auch die aus dem Gedanken der Selbstbestimmung folgende Befugnis des Einzelnen, grundsätzlich selbst zu entscheiden, wann und innerhalb welcher Grenzen persönliche Lebenssachverhalte offenbart werden …
>
> Diese Befugnis bedarf unter den heutigen und künftigen Bedingungen der automatischen Datenverarbeitung in besonderem Maße des Schutzes. Sie ist vor allem deshalb gefährdet, weil bei Entscheidungsprozessen nicht mehr wie früher auf manuell zusammengetragene Karteien und Akten zurückgegriffen werden muß, vielmehr heute mit Hilfe der automatischen Datenverarbeitung Einzelangaben über persönliche oder sachliche Verhältnisse einer bestimmten oder bestimmbaren Person (personenbezogene Daten [vgl. § 2 Abs. 1 BDSG]) technisch gesehen unbegrenzt speicherbar und jederzeit ohne Rücksicht auf Entfernungen in Sekundenschnelle abrufbar sind. Sie können darüber hinaus – vor allem beim Aufbau integrierter Informationssysteme – mit anderen Datensammlungen zu einem teilweise oder weitgehend vollständigen Persönlichkeitsbild zusammengefügt werden, ohne daß der Betroffene dessen Richtigkeit und Verwendung zureichend kontrollieren kann. Damit haben sich in einer bisher unbekannten Weise die Möglichkeiten einer Einsicht- und Einflußnahme erweitert, welche auf das Verhalten des Einzelnen schon durch den psychischen Druck öffentlicher Anteilnahme einzuwirken vermögen.
>
> Individuelle Selbstbestimmung setzt aber – auch unter den Bedingungen moderner Informationsverarbeitungstechnologien – voraus, daß dem Einzelnen Entscheidungsfreiheit über vorzunehmende oder zu unterlassende Handlungen einschließlich der Möglichkeit gegeben ist, sich auch entsprechend dieser Entscheidung tatsächlich zu verhalten. Wer nicht mit hinreichender Sicherheit überschauen kann, welche ihn betreffende Informationen in bestimmten Bereichen seiner sozialen Umwelt bekannt sind, und wer das Wissen möglicher Kommunikationspartner nicht einigermaßen abzuschätzen vermag, kann in seiner Freiheit wesentlich gehemmt werden, aus eigener Selbstbestimmung zu planen oder zu entscheiden. Mit dem Recht auf informationelle Selbstbestimmung wären eine Gesellschaftsordnung und eine diese ermöglichende Rechtsordnung nicht vereinbar, in der Bürger nicht mehr wissen können, wer was wann und bei welcher Gelegenheit über sie weiß. Wer unsicher ist, ob abweichende Verhaltensweisen jederzeit notiert und als Information dauerhaft gespeichert, verwendet oder weitergegeben werden, wird versuchen, nicht durch solche Verhaltensweisen aufzufallen. Wer damit rechnet, daß etwa die Teilnahme an einer Versammlung oder einer Bürgerinitiative behördlich registriert wird und daß ihm dadurch Risiken entstehen können, wird möglicherweise auf eine Ausübung seiner entsprechenden Grundrechte (Art. 8, 9 GG) verzichten. Dies würde nicht nur die individuellen Entfaltungschancen des Einzelnen beeinträchtigen, sondern auch das Gemeinwohl, weil Selbstbestimmung eine elementare Funktionsbedingung eines auf Handlungs- und Mitwirkungsfähigkeit seiner Bürger begründeten freiheitlichen demokratischen Gemeinwesens ist.

Das BVerfG setzt mit dieser Entscheidung dem Gedanken des Social Credit Rating und der für jedermann verfügbaren umfassenden Information als Grundlage für Entscheidungen sehr enge Grenzen.

Auf der Grundlage dieser Entscheidung hat sich die häufig als sehr restriktiv empfundene deutsche Rechtslage zum Datenschutz, entwickelt. Durch die im Jahr 2018 in Kraft getretene EU-Datenschutzgrundverordnung sind viele ihrer Leitgedanken auch EU-weit geltendes Recht geworden.

Das Recht auf informationelle Selbstbestimmung fußt, wie es das BVerfG auch darstellt auf der allgemeinen Handlungsfreiheit, die sich aus Art 2 Abs. 1 des Grundgesetzes ergibt und der Menschenwürde (Artikel 1 Grundgesetz). Besonders eindrücklich und für die weiteren Überlegungen von besonderer Bedeutung ist dabei wohl der Hinweis, dass die Sorge, nicht zu wissen, welche Informationen über einen selbst gespeichert sind, ob diese Informationen wirklich zutreffend oder möglicherweise unzutreffend bereits zu einer massiven Einschränkung der Handlungsfreiheit führen kann. Das Gericht weist darauf hin, dass allein schon die Sorge, dass auf der Grundlage zutreffender Daten Schlüsse über eine Person gezogen werden, die unzutreffend sind und für den Betroffenen zu Nachteilen führen die Entschließungsfreiheit stark beinträchtigen kann. weil Menschen aus Angst vor unberechtigten Konsequenzen von ihren Freiheitsrechten keinen Gebrauch mehr machen könnten.

Ein Beispiel für einen solchen falschen Schluss ist in der Vorbemerkung mit dem Sprichwort „Wer einmal lügt, dem glaubt man nicht, …" bereits erwähnt worden.

Er verkennt evident, dass nicht jeder, der bereits einmal gelogen hat, wieder lügen wird. Ebenso wird nicht jeder wieder straffällig der bereits einmal eine Straftat begangen hat.

Aus einem Gutachten des Bundesjustizministeriums (vgl. Jehle et al. 2016) ergibt sich eine durchschnittliche Rückfallquote von lediglich etwa 35 Prozent. Das heißt im Umkehrschluss, dass 65 Prozent der Straftäter*innen nicht wieder straffällig werden. Bei allen Unwägbarkeiten solcher empirischen Untersuchungen liegt man somit aber immer noch richtig, wenn man davon ausgeht, dass im Falle einer rechtskräftigen strafrechtlichen Verurteilung die Wahrscheinlichkeit größer ist, dass jemand sich in Zukunft straffrei führen wird, als die Wahrscheinlichkeit, dass er eine weitere Straftat begeht.

Das Beispiel macht damit anschaulich deutlich, was das BVerfG in seiner Entscheidung deutlich gemacht hat. Nämlich, dass es ein erhebliches Risiko gibt, dass selbst aus zutreffenden Daten unzutreffende, für die Betroffenen nachteilige Schlüsse gezogen werden. Und in einem wie auch immer gestalteten System eines Social Credit Rating würde sich eine frühere Straftat wohl zweifelsohne auf das Scoring negativ auswirken.

Ein solches Modell läuft im Übrigen auch Gefahr, die Idee der Resozialisierung zu unterlaufen. Das Bundeszentralregistergesetz sieht deswegen Löschungsfristen für rechtskräftige Verurteilungen vor und nicht staatlichen Einrichtungen werden Verurteilungen unter drei Monate Freiheitsstrafe bzw. zu einer entsprechen Freiheitsstrafe nicht mitgeteilt.

Aus denselben Gedanken heraus kann der Arbeitgeber nach deutschem Recht nicht ohne weiteres die Vorlage eines Führungszeugnisses verlangen. Der Arbeitgeber darf nur nach Straftaten fragen, die für die Stelle, auf die sich der Bewerber bewirbt, von Bedeutung sind und der Bewerber muss nur dann darauf wahrheitsgemäß antworten. In allen anderen Fällen darf er lügen.

Bei einem Beschäftigten mit Verantwortung für die Kasse oder Verfügungsmacht über Bankkonten könnte dies die Frage nach Verurteilungen wegen Diebstahls, Unterschlagung oder Betrug sein. Fragen nach anderen Delikten dürften zur Not verschwiegen werden.

Entsprechendes gilt dies zum Beispiel bei Tätigkeiten, bei denen der Bewerber Kontakt zu Kindern und Jugendlichen hat, ist die Frage nach Strafen wegen Sexualdelikten und Körperverletzungsdelikten zulässig.

Nach dem zuvor gesagten kann festgehalten werden, dass die Idee einer umfassenden Datensammlung über Personen oder Unternehmen (soweit diese Daten als personenbezogene Daten zu verstehen sind) mit dem Ziel einer umfassenden Beurteilungsgrundlage aller aus einer Geschäftsbeziehung möglicherweise entstehender Risiken aufgrund datenschutzrechtlicher, letztlich aus dem grundgesetzlichen Schutz der informationellen Selbstbestimmung abgeleiteter Gründe sehr enge Grenzen gesetzt sind, bzw. sie schlicht nicht realisierbar ist.

Diese Wertung weder in gleichem Masse noch annähernd in gleichem Masse in ganz Kontinentaleuropa gezogen. Während in Deutschland das Steuergeheimnis aus § 370 Abgabenordnung sogar strafrechtlichen Schutz genießt und damit der Einblick in eine fremde Steuererklärung praktisch undenkbar ist, ist es zum Beispiel in Schweden völlig selbstverständlich, dass die Steuerunterlagen aller Bürger für jeden jederzeit

13.4 Social Credit Rating als Instrument zur Prävention von Straftaten und zur Sanktionierung von unerwünschtem Verhalten

Nach der Rechtsprechung des BVerfG findet der Schutz des Rechts auf informationelle Selbstbestimmung dort seine Grenzen, wo es überwiegende Allgemeininteressen gibt. Es dürfte außer Zweifel stehen, dass die Verhinderung von Straftaten und die Sanktionierung begangenen Unrechts in diesem Sinne im überwiegenden Allgemeininteresse ist, sodass es grundsätzlich in diesem Kontext vorstellbar ist, dass eine im Detail noch näher zu bestimmende Ausformung des Social Credit Rating im Prinzip rechtskonform anwendbar wäre.

Im Weiteren betrachte ich zunächst den Einsatz zu präventiven Zwecken und sodann den Einsatz zu repressiven Zwecken

13.4.1 Social Credit Rating als Instrument der Kriminalitätsprävention

Bei der folgenden Betrachtung unterstelle ich folgende grundsätzliche Wirkmechanismen des Social Rating Systems.

Das System bewertet erwünschtes Verhalten mit einem positiven und unerwünschtes Verhalten mit einem negativen Punktwert. Ausgehend von einem für alle gleichen Punktwert wird eine Vielzahl verschiedener Datenquellen ausgewertet. Von der Bewertung (dem Scorewert) hängt der Zugang zu bestimmten staatlichen oder nicht-staatlichen Leistungen ab.

So formuliert liegen die Bedenken auf der Hand.

Der gleiche Zugang zu allen staatlichen Leistungen, die Chancengleichheit und ähnliche Werte sind fundamentaler Bestandteil unseres Staatsverständnisses. Es fällt uns schwer, vorzustellen, dass jemand keinen Kredit bekommen soll, kein Flugzeug oder nicht die Bahn benutzen darf, weil er beispielsweise wiederholt bei rot über die Ampel gegangen ist.

Das Überqueren der Ampel bei rot ist nicht von der grundgesetzlich garantierten Handlungsfreiheit gedeckt, denn diese findet ihre Grenzen in den allgemeinen Gesetzen. Im konkreten sind es aber wohl vor allem zwei Fragen, die näher beleuchtet werden müssten, nämlich einerseits, die Frage, wie der Weg vom (vermeintlichen) Verstoß zur Sanktionierung durch Punktabzug (Verschlechterung des Scorings) ist und zum zweiten ob es eine Konnexität zwischen Sanktion und Fehlverhalten geben sollte.

13.4.1.1 Social Credit Rating und Rechtsstaatsprinzip

Der Satiriker Marc-Uwe Kling beschreibt in seinem Roman Quality Land (vgl. Kling 2017) eine fiktive Welt, von der man meinen könnte, dass sie die (satirisch zugespitzte) aber im Prinzip nur konsequent zu Ende gedachte Umsetzung der Idee des Social Credit Rating ist. Der Effekt auf den ich vor allem eingehen möchte ist dabei, dass in einer immer stärker Rating orientierten, automatisierten Welt (wie im Roman) das Rating automatisch angepasst wird, wenn aufgrund einer Beobachtung, zum Beispiel einer Kamera an einer Ampel, ein Verkehrsverstoß festgestellt wird. Die Kamera zeichnet das Fehlverhalten auf, mittels zum Beispiel Gesichtserkennung wird der Betroffene identifiziert und ein Algorithmus entscheidet über Schuld und Sanktion, die er in Form des Punktabzugs auch unmittelbar vollstreckt.

Das Beispiel mag auf den ersten Blick übertrieben wirken, es ist aber, wenn heute noch nicht technisch perfekt, in naher Zukunft sicher realisierbar.

Das Rechtsstaatsprinzip bedeutet kurz gefasst ein Willkürverbot. Es beinhaltet darüber hinaus auch das Recht, das staatliches Handeln durch unabhängige Gerichte überprüft werden kann. In Deutschland geht das so weit, dass selbst ein Verwarngeld über zehn Euro wegen eines Parkverstoßes vom Bürger angefochten und gegebenenfalls auch vom Gericht zu überprüfen ist.

Das Rechtsstaatsprinzip bedeutet nicht, dass der Gesetzgeber auf ewig daran gehindert ist die gerichtlichen Zuständigkeitsregelungen, wie sie derzeit gelten, unverändert zu lassen. Es bedeutet aber, dass es eine Möglichkeit geben muss staatliches Handeln durch unabhängige Gerichte prüfen zu lassen.

Der oben beschriebene Automatismus würde ohne jeden Zweifel gegen das Rechtsstaatsprinzip verstoßen, wenn es keine Möglichkeit gibt, die Entscheidungen, die einer (negativen) Veränderung des Scorings zu Grunde liegen, überprüfen zu lassen. Vermutlich würde man auch sagen müssen, dass eine negative Veränderung des Scorings erst umgesetzt werden dürfte, wenn dem Betroffenen die Möglichkeit gegeben wurde, sich zu dem Vorwurf zu äußern. Die entspricht dem Grundsatz des rechtlichen Gehörs.

Die Umsetzung der Idee eines Social Credit Rating würde damit vermutlich einen erheblichen administrativen Aufwand auslösen und zu einer weiteren Flut von Gerichtsverfahren, während die Gerichte schon jetzt über Überlastung klagen.

13.4.1.2   Konnex zwischen Fehlverhalten und Sanktion

Dieser Beitrag kann nicht die Aufgabe erfüllen, die Antwort auf die Frage nach dem Sinn und Zweck von Strafe zu beantworten. Soviel aber kann gesagt werden:

Unser Strafrecht kennt Geldstrafe und Freiheitsstrafe. Darüber hinaus gibt es sogenannte Nebenstrafen. Dies ist nach geltendem Recht in Deutschland ausschließlich das Fahrverbot nach § 44 StGB. Bis zur Änderung des § 44 StGB im Jahre 2017 konnte ein Fahrverbot nur verhängt werden, wenn eine Tat vorlag, die im Zusammenhang mit dem Straßenverkehr stand. Seither ist dies auch in anderen Fällen möglich.

Die Änderung ist kriminalpolitisch hoch umstritten. Mit guten Argumenten wird vertreten, dass die Nebenstrafe ebenso wie die Hauptstrafe tatschuldbezogen und von dieser abhängig sei. Sie kommt neben einer Hauptstrafe nur in Betracht, wenn über die Hauptstrafe hinaus ein besonderer spezialpräventiver Zweck verfolgt wird. Dabei muss es gerade die vom Täter begangene Tat sein, die das besondere spezialpräventive Bedürfnis anzeigt (vgl. Zopfs 2017, S. 401 ff.). Unter diesen Umständen wären im deutschen Rechtsumfeld auch alle solche Sanktionsmechanismen nur schwer oder gar nicht umsetzbar anzusehen, die eine Ahndung vorsehen, die keinen Bezug zur Tat haben. Wenn also zum Beispiel ein Kredit nicht gewährt würde, weil der potenzielle Kreditnehmer zu oft bei Rot über die Ampel gegangen ist.

13.5   Schlussbetrachtung

Die Idee des Social Credit Rating, sei es nun als staatliches Instrument, oder erst recht im Umgang Privater untereinander kollidiert an vielen Stellen mit elementaren Grundregeln unseres Rechtssystems.

Es liegt auch auf der Hand, dass eine derartige Anhäufung von Information und Macht einer besonders intensiven Kontrolle unterworfen sein muss. Selbst in einem Rechtsstaat mit funktionierender Gewaltenteilung, umfassender Transparenz behördlicher Entscheidungen und einer funktionierenden Kontrolle durch unabhängige Medien müsste man besorgt sein, ob ein Missbrauch dieser Macht rechtzeitig (wenn überhaupt) bemerkt und die notwendigen Konsequenzen daraus gezogen würden.

So faszinierend der Grundgedanke also ist, auf der Basis vollständiger und damit besserer Informationen zu besseren Entscheidungen zu kommen, so beängstigend ist dann am Ende die Furcht vor einem umfassenden Überwachungsstaat, in dem bürgerliche Freiheitsrechte nicht mehr garantiert wären.

Literatur

Jehle, J.-M., Albrecht, H.-J., Hohmann-Fricke, S. und Tetal, C. in Kooperation mit dem Bundesamt für Justiz (2016): Legalbewährung nach strafrechtlichen Sanktionen – Eine bundesweite Rückfalluntersuchung 2010 bis 2013 und 2004 bis 2013; .herausgegeben vom Bundesministerium der Justiz und für Verbraucherschutz Berlin 2016.

Kling, M.-U. (2017): Quality Land, Ullstein Verlag 2017.

Steinmüller, W./Lutterbeck, B./Mallmann, C./Harbort, U./Kolb, G./Schneider, J. (1971): Grundfragen des Datenschutzes – Gutachten im Auftrag des Bundesinnenministeriums, Drucksache VI/3826.

Zopfs, J. (2017): Dogmatische Verwerfungen im Sanktionenrecht bei Einführung eines Fahrverbots ohne Straftat; in: SVR Straßenverkehrsrecht 2017.f.

Steffen Salvenmoser ist als Rechtsanwalt mit dem Schwerpunkt Criminal Compliance und interne Untersuchungen in eigener Praxis tätig. Zuvor war er mehr als 20 Jahre Partner bei einer Big 4 Wirtschaftsprüfungsgesellschaft im Bereich Forensic Services. Herr Salvenmoser war darüber hinaus als Richter und als Staatsanwalt an einer Staatsanwaltschaft für Wirtschaftsstrafsachen tätig. Von 2011 bis 2019 hatte er einen Lehrauftrag an der Universität Osnabrück im Postgraduiertenstudiengang Wirtschaftsstrafrecht. Seit dem Wintersemester 2019 ist er Lehrbeauftragter an der Hochschule Fresenius in Wiesbaden im Masterstudiengang Wirtschaftsforensik.

Influencer Marketing und Recht

Scarlett Lüning

Zusammenfassung

Das Influencer Marketing ist in den letzten Jahren in der Welt des Marketings ein wichtiger Bestandteil geworden, insbesondere für Unternehmen. Auch hat es einen vollständig neuen Beruf zu Tage gebracht: der Influencer. Lange Zeit von vielen belächelt, ist mittlerweile ein umfassendes Business daraus entstanden. Ganz klar ist – am Influencer Marketing kommt man so schnell nicht vorbei. Der nachfolgende Inhalt beleuchtet unter anderem die Grundlagen des Influencer Marketings sowie des Influencers an sich, insbesondere aber die rechtlichen Herausforderungen.

14.1 Was ist Influencer Marketing?

Für Influencer gibt es viele Umschreibungen. Sie sind Einflussnehmer, Gestalter, lebende Werbeflächen, Markenbotschafter. Durch sie wurde eine neue Kultur des Mitgestaltens geschaffen. Die Medienlandschaft hat einen radikalen revolutionären Wandel erlebt. Individualität und Ästhetik des Auftritts eines Influencers stehen besonders im Fokus.

Beim Influencer Marketing werden gezielt Meinungsmacher, in vielen Fällen mit einer hohen Reichweite in ihrer Community, für das Online-Marketing von Unternehmen eingesetzt. Sie ist eine Weiterentwicklung der klassischen Testimonial-Werbung, welche insbesondere in die digitale Welt transportiert wird (vgl. Solmecke und Kocatepe 2018, S. 423). Unternehmen setzen diese sogenannten Influencer insbesondere aufgrund ihrer Glaubwürdigkeit innerhalb ihrer Community und des zielgruppenorientierten Einflusses

S. Lüning (✉)
Wilde Beuger Solmecke Rechtsanwälte Partnerschaft mbB, Köln, Deutschland
E-Mail: luening@wbs-law.de

© Springer Fachmedien Wiesbaden GmbH, ein Teil von Springer Nature 2020
O. Everling (Hrsg.), *Social Credit Rating*,
https://doi.org/10.1007/978-3-658-29653-7_14

für die Vermarktung ihrer Produkte und Dienstleistungen ein (vgl. Troge 2018, S. 87). Der klassische Influencer unterhält Profile auf Instagram, Facebook, YouTube und Co. und postet hier zielgruppenorientierten Content. Im Influencer Marketing ist vor allem die Interaktion mit den Followern von großer Bedeutung, sodass der Influencer zu seinen Followern eine Art persönliche Beziehung aufbaut und für sie so quasi zum Vorbild bzw. zu einem Star zum Anfassen wird.

14.1.1 Mega-Influencer

Zu den Mega-Influencern zählen bereits bekannte Personen, wie Prominente und Stars, die schon im klassischen Testimonial-Bereich Erfahrungen haben und dort bekannt sind. Der Aufbau einer Community ist für sie ohne große Anstrengungen möglich. Das Engagement mit den Followern ist allerdings besonders gering bis nicht vorhanden.

14.1.2 Makro-Influencer

Makro-Influencer sind in der Regel durch ihre Social Media Kanäle bekannt geworden und haben sich dort ihre Reichweite erarbeitet. Makro-Influencer zeichnen sich häufig dadurch aus, dass sie eine hohe Reichweite, das heißt eine Followeranzahl zwischen 50.000 und einer Million und sich auf zwei bis drei Themengebiete fokussieren. Makro-Influencer posten in der Regel mehrmals täglich besonders hochwertigen Content, der meist akribisch geplant wurde. Je höher allerdings die Followerschaft, desto geringer die Engagement-Rate, das heißt, die Interaktionen zwischen dem Influencer und den Followern sinkt, da dies aufgrund der Masse an Kommentaren und Likes oftmals nicht mehr zu realisieren ist. In vielen Fällen stellen Makro-Influencer bestimmte Funktionen der Social Media Plattformen, wie zum Beispiel das Kommentieren oder Reagieren auf Stories aus, sodass eine Interaktion nicht mehr möglich ist.

14.1.3 Mikro-Influencer

Mikro-Influencer befinden sich mit der Anzahl ihrer Follower im eher unteren unpopulären Mittelfeld, bis zu 50.000 Follower. Dennoch sind sie für viele Unternehmen vom besonderen Interesse, da Mikro-Influencer in der Regel inhaltlich eher eine Sparte bedienen, die zwar von weniger Personen beachtet wird, aber in dieser Community eine große Aufmerksamkeit hat. Diese Influencer spezialisieren sich im Gegenteil zu Makro-Influencer auf bestimmte wenige Themen. Der Trust-Factor des Influencers ist trotz seiner eher wenigen Follower besonders hoch. Dies liegt vor allem an der hohen Interaktionsrate, welche gleichzeitig die Engagement-Rate steigert.

14.1.4 Nano-Influencer

Nano-Influencer bewegen sich im Bereich von 1000 bis 10.000 Followern. Die Nachfrage nach Nano-Influencer steigt jährlich, insbesondere da die Werbekosten für Makro-Influencer stetig steigen. Dies bedeutet: je höher die Werbekosten, desto geringer die Engagement-Rate, welche für Unternehmen mittlerweile essenziell geworden und der Schlüssel zum Erfolg geworden ist. Nano-Influencer sprechen nur eine sehr kleine Community an. Ihre Follower haben das gleiche Interesse wie der Nano-Influencer, sodass er als einer von ihnen angesehen wird. Die Glaubwürdigkeit des Influencers und das Vertrauen innerhalb der Community sind besonders hoch. Die Vorteile eines Nano-Influencers werden auch von den Unternehmen gesehen. Sie haben zu den Themen, Produkten oder Dienstleistungen eine tatsächlich ernst gemeinte Meinung und können diese authentisch vermitteln. Auch wenn der Influencer nur eine geringe Anzahl an Follower hat, sind dies die, welche angesprochen werden sollen.

14.2 Rechtliche Herausforderungen des Influencer Marketings

Jahrelang nicht im Fokus rechtlicher Auseinandersetzungen konnten sich Influencer und Unternehmen ziemlich frei und offen bei der Einbindung von werblichen Inhalten bewegen. Dies änderte sich allerdings vor ein paar Jahren: sowohl Influencer als auch die Unternehmen merken zunehmend, dass die oft fließende Vermischung von redaktionellen und kommerziellen Beiträgen mit gesetzlichen Kennzeichnungspflichten für Werbung und dem Trennungsgebot kollidiert und von Medienanstalten, Wettbewerbern und Verbänden verfolgt werden kann (vgl. Troge 2018, S. 87).

14.2.1 Rechtliche Rahmenbedingungen

Beiträge von Influencern in den sozialen Medien unterliegen als Telemedien dem Regelungsbereich des Rundfunkstaatsvertrages (RStV) sowie dem Telemediengesetz (TMG). Das Lauterkeitsrecht (UWG) findet insbesondere gegenüber anderen Influencern und Unternehmen Anwendung.

Das Trennungsgebot des RStV besagt, dass Werbung als solches in Telemedien klar und eindeutig erkennbar und vom übrigen Inhalt klar getrennt sein muss, § 58 Abs. 1 RStV. Darüber hinaus müssen kommerzielle Beiträge strikt von journalistisch-redaktionellen Beiträgen getrennt werden, § 6 Abs. 1 Nr. 1 TMG. Das UWG regelt zudem in § 5a Abs. 6 UWG, dass eine unlautere Handlung vorliegt, wenn der kommerzielle Zweck einer geschäftlichen Handlung nicht kenntlich gemacht wird und sich auch nicht aus den Umständen ergibt und dadurch ein Verbraucher zu einer geschäftlichen Handlung, zum Beispiel zum Kauf eines Produkts, verleitet wird, welche er sonst nicht getätigt hätte.

14.2.2 Vorliegen einer kommerziellen Handlung

Oftmals problematisch ist für viele Influencer das Erkennen einer solchen kommerziellen Handlung in ihrem eigenen Verhalten. Besteht eine Kooperation mit einem Unternehmen und erhält der Influencer eine Gegenleistung für seine Tätigkeit, ist unstreitig von einer geschäftlichen Handlung auszugehen. Abgrenzungsschwierigkeiten bereiten in der Regel, wenn der Influencer vermeintlich privaten Content veröffentlicht, hierbei aber Nennungen und Verlinkungen zu kommerziellen Personen oder Unternehmen erfolgen (vgl. Köhler et al. 2019, § 5a, Rn. 7.71).

Das Lauterkeitsrecht definiert in § 2 Abs. 1 Nr. 1 UWG eine geschäftliche Handlung als das Verhalten einer Person zugunsten des eigenen oder eines fremden Unternehmens zur Förderung des Absatzes. Eine geschäftliche Handlung wird im Influencer Marketing insbesondere durch das Setzen von kommerziellen Links angenommen, unabhängig davon, ob der Influencer hierfür eine Gegenleistung erhalten hat oder nicht. Die Rechtsprechung nimmt an, dass durch die Verlinkungen die Follower den Weg zu den jeweiligen Unternehmen sowie deren Produkte oder Dienstleistungen finden und diese auch erwerben. Der Erwerbswille wird insbesondere angenommen, da Influencer als eine Art Vorbild agieren und die Follower ihnen möglichst ähnlich und nahe sein wollen. Dies funktioniert vor allem durch den Kauf der verlinkten Produkte, die der Influencer vermeintlich selbst nutzt und trägt.

Ein Nichtkenntlichmachen des kommerziellen Zwecks kommt insbesondere vor, wenn Äußerungen dem Anschein nach privater Natur sind, in Wahrheit aber der Förderung des Erscheinungsbilds oder des Absatzes eines bestimmten Unternehmens dienen und damit eine geschäftliche Handlung darstellen (vgl. Köhler et al. 2019, § 5a, Rn. 7.75).

Das Kammergericht Berlin (Urteil vom 08.01.2019, Az. 5 U 83/18) urteilte zuletzt, dass nicht jeder Post eines Influencers, der Verlinkungen zu kommerziellen Accounts hat, als Werbung zu charakterisieren ist. Stattdessen muss danach differenziert werden, welchen Informationsgehalt der Post und die dazu gehörigen Verlinkungen hat sowie ob die Verlinkungen in einem redaktionellen Zusammenhang mit dem Inhalt des Postings stehen. Demnach sind Beiträge mit Links auf Seiten anderer Unternehmen nicht generell als kennzeichnungspflichtige Werbung anzusehen. Allein die Tatsache, dass ein Influencer nicht über tagesaktuelle Ereignisse berichtet, führt noch nicht dazu, dass diese Inhalte immer als Werbung zu klassifizieren sind. Es stellt keinen Verstoß gegen das Lauterkeitsrecht dar, wenn ein Influencer nur weltanschauliche, wissenschaftliche, redaktionelle oder verbraucherpolitische Äußerungen von sich geben, ohne damit gezielt den Absatz von Waren oder Dienstleistungen zu fördern. Bei der Beurteilung, ob eine Äußerung der Meinungsfreiheit unterfällt, darf nicht nach dem Gegenstand der Berichterstattung differenziert werden, sondern es muss sowohl der konkrete Inhalt des Posts als auch die besonderen Umstände des Einzelfalls geprüft werden.

Anders hingegen sah es das Landgericht München I (Urteil vom 29.04.2019, Az. 4 HK O 4985/18) und entschied, dass Posts dann nicht als Werbung zu kennzeichnen sind, wenn die Gewerblichkeit des gesamten Accounts eines Influencers ohnehin für jedermann zu erkennen ist, da Influencer praktisch immer gewerblich handeln, aber mit der vermeintlichen Darstellung ihres Privatlebens die Follower beeinflussen. Das Gericht stellte allerdings klar, dass dies nur für besonders große und bekannte Influencer gelten kann, da allgemein bekannt sein dürfte, dass solch ein Influencer nicht mit seiner gesamten Followerschaft befreundet ist. Zudem gelten der blaue Haken bei Instagram und der Umstand, dass es sich um ein öffentliches Profil handelt, als Indiz für die Bekanntheit eines Influencers. Dennoch ist jeder Fall einzeln zu betrachten.

So entschied das Landgericht Karlsruhe (Urteil vom 21.03.2019, Az. 13 O 38/18), dass eine Werbekennzeichnung nicht entbehrlich ist, da keinesfalls alle Follower den werblichen Charakter des Auftretens eines Influencers einschätzen können; dies gilt vor allem für besonders junge Follower.

14.2.3 Kennzeichnungspflicht

Wurde das Vorliegen einer geschäftlichen Handlung festgestellt, besteht die Kennzeichnungspflicht, das heißt, der Influencer muss seinen Beitrag klar und eindeutig als Werbung kennzeichnen. Hierbei ist insbesondere die Art der jeweiligen Kennzeichnung zu beachten, nicht alle Kennzeichnungsmöglichkeiten sind auch rechtssicher. So entschied das Oberlandesgericht Celle (Urteil vom 08.06.2017, Az. 13 U 53/17), dass die Verwendung des Hashtags #ad unter einem Instagram-Post nicht ausreichend ist, um den kommerziellen Zweck des Beitrags zu kennzeichnen. Das Kammergericht Berlin (Urteil vom 11.10.2017, Az. 5 W 221/17) entschied zudem, dass auch eine Kennzeichnung mit #sponsoredby den rechtlichen Anforderungen nicht genügt. Tatsächlich rechtssicher sind bislang nur die Begriffe „Anzeige" und „Werbung". Das Landgericht Hagen (Urteil vom 13.09.2017, Az. 23 O 30/17) wies zudem darauf hin, dass bei besonders jungen Followern ein entsprechend strenger Maßstab an die Erkennbarkeit des werblichen Charakters einzelner Posts gestellt werden muss, da für diese Zielgruppe das Vermischen von werblichen und vermeintlich privaten Posts nicht sofort und deutlich erkennbar sind. Es kristallisierte sich somit heraus, dass eine Werbekennzeichnung an den Anfang eines Posts zu stellen ist und keinesfalls in einem Meer aus Hashtags verschwinden darf. Auch die eigens durch Instagram eingeführte Kennzeichnung „Bezahlte Partnerschaft mit …" ist nicht ausreichend; hier muss zusätzlich gekennzeichnet werden, obwohl man davon ausgehen könnte, dass sich dadurch die geschäftliche Handlung aus dem Umstand ergeben könnte.

14.2.4 Fehlende oder fehlerhafte Kennzeichnung

Unterlässt ein Influencer die entsprechende Kennzeichnung riskiert er neben einer Abmahnung von Wettbewerbern oder Verbänden, Bußgelder von den Landesmedienanstalten.

Nach dem Lauterkeitsrecht können Wettbewerber, das heißt sogar andere Influencer und Unternehmen, sowie Verbände im Rahmen einer Abmahnung einen Unterlassungsanspruch sowie Kostenerstattung für die außergerichtliche Tätigkeit geltend machen. Nicht ganz risikofrei ist die Abgabe einer strafbewehrten Unterlassungserklärung. Der Influencer verpflichtet sich dadurch gegenüber einem Dritten es zu unterlassen, kommerzielle Beiträge nicht als solche zu kennzeichnen. Verstößt er gegen diesen Unterlassungsvertrag, der unter Umständen ein Leben lang Gültigkeit hat, ist er verpflichtet dem Unterlassungsgläubiger eine Vertragsstrafe zu zahlen. Diese liegen in der Regel zwischen 2500 und 5000 Euro pro Verstoß. Dies kann dazu führen, dass der Influencer einer dauerhaften Beobachtung des Unterlassungsgläubigers ausgesetzt ist. Um das Risiko zu verringern verbleibt die Möglichkeit eine einstweilige Verfügung gegen sich wirken zu lassen oder eine Unterlassungsklage. Bei einem Verstoß besteht nur das Risiko eines Ordnungsgeldes, welches allerdings der Staat erhält und nicht der Abmahner.

Bei einem Verstoß gegen die rundfunkstaatsvertraglichen Regelungen können die Landesmedienanstalten Bußgelder in Höhe von bis zu 500.000 Euro verhängen. Ein solcher Verstoß kommt insbesondere bei der Nicht- oder fehlerhaften Kennzeichnung von YouTube Videos, da diese fernsehähnlich sind, in Betracht. Dagegen ist eine Anwendbarkeit des Rundfunkstaatsvertrages auf Stories bei Instagram und Snapchat nicht anzunehmen.

Verstöße gegen das Telemediengesetz, also bei Nichtbeachtung des Trennungsgebotes, können Aufsichtsbehörden Bußgelder in Höhe von bis zu 50.000 Euro verhängen.

Eine weitere rechtliche Problematik ergibt sich für Influencer, wenn die mit Agenturen oder Unternehmen geschlossene Verträge für eine Kooperation zum Gegenstand haben, dass eine Kennzeichnung von werblichen Inhalten zu erfolgen hat, der Influencer sich dieser vertraglichen Verpflichtung allerdings widersetzt. Hat der Vertragspartner dadurch einen Schaden erlitten, besteht ein Schadensersatzanspruch gegen den Influencer. Zudem besteht unter Umständen kein Anspruch auf die vollständige Vergütung aufgrund der Verletzung einer Vertragspflicht.

14.3   Kaufen von Followern und Likes

Die Anzahl der Follower ist für Influencer ein besonders wichtiges Element, da es eine Art Gradmesser ihres Erfolges ist. Der Aufbau einer eigenen Community bedeutet somit viel Arbeit, das heißt, es muss täglich Content kreiert werden, mittels einer eigenen Strategie, bestimmten Zielsetzungen, interessanten Inhalten und kontinuierlicher Interaktion mit den Followern.

Allerdings greifen in einigen Fällen Influencer, aber auch Unternehmen, auf den Kauf von Followern, Likes, Kommentare und Bewertungen zurück, um dadurch ihre eigene Reichweite und Bekanntheit zu steigern. Insbesondere für Neu-Influencer kann dies ein Einstieg ins Influencer-Marketing sein, da für die meisten Unternehmen eine gewisse Reichweite Voraussetzung für eine Kooperation darstellt.

Der Kauf dieser „Waren" ist rechtlich nicht ganz unbedenklich, insbesondere wenn es sich lediglich um Bots handelt, also computergesteuerte Interaktionen. Zusätzlich kann dies auch das Ende der Karriere eines Influencers bedeuten, da durch den Kauf von Followern und Likes oftmals die Glaubwürdigkeit geschmälert wird und sich die „echten Follower" von dem jeweiligen Influencer abwenden.

14.3.1 Rechtliche Betrachtung

Eine irreführende geschäftliche Handlung gemäß § 5 Abs. 1 S. 2 Nr. 3 UWG liegt vor, wenn ein Unternehmen oder ein Influencer durch den Kauf von Likes und/oder Followern einen Umstand vorgibt, der tatsächlich nicht gegeben ist. Der Werbende erweckt somit den Eindruck, dass die Personen, die ein Like gesetzt haben, dies aufgrund des Unternehmens und ihrer Waren oder Dienstleistungen getan haben (LG Stuttgart, Beschluss vom 06.08.2014, AZ: 37 O 34/14). Dem Käufer drohen in diesen Fällen Abmahnungen von Mitbewerbern, Wettbewerbs- oder Verbraucherschutzvereinen. Sie werden dann aufgefordert eine strafbewehrte Unterlassungserklärung abzugeben sowie Aufwendungsersatz oder gar Schadensersatz zu leisten.

Das Kaufen von Followern ist vor allem ein Instagram-Phänomen. Auch hier wird ein Umstand vorgetäuscht, der in Wirklichkeit nicht gegeben ist, sodass eine irreführende Handlung angenommen werden kann. Auf dem Markt erhält man bereits 10.000 neue Follower für ca. fünf Euro, sodass die Hemmschwelle für viele schnell überschritten ist, da es finanziell keine Belastung darstellt. Hierbei sollte man allerdings beachten, dass man auch nur das bekommt, wofür man bezahlt – und das sind in den meisten Fällen Bots, also nicht aktive Accounts, die von Computern gesteuert werden und nicht mit dem eigenen Account interagieren. Kostspieligere Alternativen (ab 1000 Euro aufwärts für 10.000 Follower) interagieren dann tatsächlich mit dem eigenen Account, da es sich um aktive Accounts handelt.

Rechtlich muss unterschieden werden zwischen den Nutzungsbedingungen der jeweiligen Plattform sowie die Auswirkungen auf eine Kooperation des Influencers mit einem Unternehmen.

14.3.2 Nutzungsbedingungen

Die meisten Social Media Plattformen verbieten neben dem Erstellen von Fake-Accounts auch das Kaufen von Fake-Followern, insbesondere Bots. Solche Bots erkennt man vor allem an ihren generischen Namen sowie ihrer Herkunft; die meisten Bots stammen aus dem asiatischen oder russischen Raum.

Instagram löscht in regelmäßigen Abständen diese bot-generierten Follower. Man riskiert zudem die Löschung des Accounts sowie eine Sperre für die Neuerstellung eines Accounts. Für Influencer könnte dies im schlimmsten Fall das Ende ihrer Karriere bedeuten, da die sozialen Medien elementar für das Influencer Marketing sind.

14.3.3 Kooperationen mit Unternehmen

Basiert eine Kooperation dem Grunde nach darauf, dass ein Influencer eine bestimmte Reichweite hat, welche nach den tatsächlichen Gegebenheiten allerdings nicht besteht, kann der Werbetreibende den Vertrag aufgrund arglistiger Täuschung nach § 123 BGB anfechten. Als Rechtsfolge kann der Anfechtende sodann bereits gezahltes Honorar zurückfordern, auch wenn der Influencer seiner vertraglichen Hauptleistung (Erstellen von Content) nachgekommen ist. Darüber hinaus könnte sich der Influencer gegenüber seinem Vertragspartner schadensersatzpflichtig gemacht haben, § 823 BGB, das heißt, der Influencer muss den durch sein Verhalten entstandenen Schaden ersetzen. Mögliche Schäden wären zum Beispiel der Imageverlust des Vertragspartners, Umsatzeinbußen usw.

14.4   Engagement Rate

In den sozialen Medien heißt die Währung „Engagement Rate", das heißt das Zusammenspiel zwischen Followern, Likes und Kommentaren. Mit der Engagement Rate wird die Anzahl an Interaktionen eines Social Media Profils gemessen. Zur Berechnung des Engagements gibt es verschiedene Methoden, die gängigsten werden nachfolgend aufgelistet (vgl. Sehl 2019).

14.4.1 Engagement-Rate nach Reichweite (ERR)

Diese Formel wird in der Regel zur Berechnung des Engagements mit Inhalten genutzt. Die ERR misst den Prozentsatz der Personen, die mit ihrem Content interagieren, nachdem sie diese gesehen haben.

$$ERR = Gesamt - Engagement (alle\ Interaktionen) pro\ Post\ /\ Reichweite\ pro\ Post \times 100$$

Möchte man den Durchschnittswert ermitteln, werden alle ERR der Posts, für die der Durchschnitt berechnet werden soll, addiert und dann durch die Anzahl der Post geteilt.

$$Durchschnittliche\ ERR = Gesamt\ ERR\ /\ Gesamte\ Posts$$

Vorteil: Das Messen der Reichweite ist präziser als das Messen der Follower-Anzahl, da nicht alle Follower den gesamten Content eines Influencers sieht (liegt insbesondere am geänderten Algorithmus der sozialen Medien; der Content wird nicht mehr chronologisch angezeigt). Nicht-Follower können zudem Posts durch Shares, Hashtags, Suchen-Funktion usw. wahrgenommen haben.

Nachteil: Die Reichweite kann aus vielen Gründen schwanken. Auch eine sehr geringe Reichweite kann zu einer hohen Engagement-Rate führen und umgekehrt.

14.4.2 Engagement Rate nach Posts (ER Posts)

Diese Formel misst das Zusammenspiel nach Followern bei einem bestimmten Post, das heißt, es wird die Anzahl der Follower ermittelt, die mit dem jeweiligen Content interagieren.

$$ER\ Post = Gesamt - Engagement\ eines\ Posts\ /\ Gesamte\ Follower \times 100$$

Zur Ermittlung des Durchschnittswerts werden sodann alle ER Posts addiert und durch die Anzahl der der Posts dividiert.

$$Durchschnittlicher\ ER\ pro\ Post = Gesamt - ER\ pro\ Post\ /\ Gesamte\ Posts$$

Vorteil: Während ERR sich besser eignet, um Interaktionen auf Basis der Anzahl der Personen zu messen, die den Beitrag gesehen haben, ersetzt diese Formel die Reichweite durch Follower – generell die stabilere Kennzahl.

Nachteil: Diese Berechnung liefert nicht unbedingt ein vollständiges Bild, da die virale Reichweite nicht berücksichtigt wird. Steigt die Anzahl der Follower steigt, kann die Engagement-Rate leicht rückläufig sein.

14.4.3 Engagement Rate nach Impressions (ER Impressions)

Die Impressions (Seitenaufrufe) zeigt, wie oft die Inhalte eines Influencers auf dem Screen eines Followers angezeigt werden.

$$ER\ Impressions = Gesamt - Engagement, das\ ein\ Post\ erzielt\ /\ Gesamt - Impressions \times 100$$

$$Durchschnittliche\ ER\ Impressions = Gesamt - ER - Impressions\ /\ Gesamte\ Posts$$

Vorteil: Diese Berechnungsmethode bietet sich an, wenn ein Influencer mit Paid Content arbeitet und die Effektivität auf Basis von Impressions evaluieren muss.

Nachteil: Eine Engagement-Rate, die auf Grundlage von Impressions berechnet wird, fällt zwangsläufig niedriger aus, als ERR- und ER-Post-Berechnungen.

14.4.4 Tägliche Engagement-Rate (Daily ER)

Diese Berechnungsmethode legt offen, wie oft die Follower eines Influencers mit seinem Account interagieren.

$$Daily\ ER = Gesamt - Engagement\ an\ einem\ Tag\ /\ Gesamtzahl\ Follower \times 100$$

$$\text{Durchschnittliche Daily ER} = \text{Gesamt} - \text{Engagement in X Tagen} / (\text{X Tage} * \text{Follower}) \times 100$$

Vorteil: Diese Formel bietet sich dafür zu messen, wie oft die Follower mit dem gesamten Account interagieren und nicht nur mit einem bestimmten Post. Zudem wird das Engagement bei alten und neuen Posts mit einbezogen.

Nachteil: Die Berechnung birgt ein hohes Maß an Fehlerpotenzial, da nicht unterschieden wird, ob derselbe Follower mehr als einmal am Tag mit einem Account interagiert hat.

14.4.5 Engagement Rate nach Views (ER Views)

Für viele Influencer gehört YouTube zum alltäglichen Leben. Mit dieser Methode kann die Engagement nach dem Anschauen eines Videos berechnet werden, das heißt, wie viele Viewer danach mit dem Video interagieren, zum Beispiel durch Likes und Kommentare.

$$\text{ER View} = \text{Gesamt} - \text{Engagement bei einem Video} - \text{Post} / \text{Gesamte Video Views} \times 100$$

$$\text{Durchschnittlicher ER View} = \text{Gesamter ER View} / \text{Gesamte Posts}$$

Vorteil: Wie bereits beschrieben, kann hiermit die Interaktion nach dem Anschauen eines Videos gemessen werden.

Nachteil: Ähnlich wie bei der Daily ER wird hier nicht berücksichtigt, dass sich ein Viewer das Video öfter angeschaut hat, aber nicht ebenso oft interagiert hat.

14.5   Wie wichtig sind Likes?

Likes sind für Influencer und Unternehmen von enormer Bedeutung; hieran wird unter anderem ihr Einfluss und Erfolg gemessen. Likes sind ein positives Feedback; Likes können Kunden bringen; Likes bedeuten Vertrauen, Interaktion und eine höhere Sichtbarkeit. Likes entscheiden, ob ein Beitrag relevant ist. Viele Influencer aber auch „normale" Nutzer vergleichen sich anhand der Anzahl ihrer Likes. Für eine große Anzahl an Usern bedeuten viele Likes sogar einen höheren Selbstwert. Oftmals gehen sie davon aus, dass sie nur durch eine gewisse Anzahl an Likes, die sie für ihre Beiträge erhalten, Anerkennung erhalten und wirklich erfolgreich sind. Diesem Phänomen will vor allem ein Unternehmen entgegenwirken: Instagram.

14.5.1 Instagram stellt die Likes ab – und jetzt?

Die Verantwortlichen des Unternehmens Instagram, welches seit ein paar Jahren zum Facebook Konzern gehört, befürchten, dass Likes in naher Zukunft einen noch größeren Stellenwert erhalten werden und ein noch höherer Konkurrenzkampf entsteht. Ziel ist es, den Usern wieder vor Augen zu führen, dass es nicht um die Anzahl der Likes für einen Beitrag geht, sondern der Beitrag an sich wieder mehr Aufmerksamkeit erlangt.

Aus diesem Grund stellt Instagram die Likes ab; die Like-Funktion bleibt allerdings weiterhin bestehen. Dies hat zur Folge, dass andere User nicht mehr sehen können, wie viele Likes ein fremder Beitrag erhalten hat, sondern es kann nur noch die Anzahl der eigens erhaltenen Likes gesehen werden. Dies soll vor allem den Erfolgsdruck immer mehr Likes zu bekommen unter den Influencern eindämmen.

14.5.2 Auswirkungen für das Influencer Marketing

Bislang wurden Likes unter Influencer als eine Art Währung gehandelt, das heißt, je höher die Anzahl an Likes ist, desto höher ist auch die Wahrscheinlichkeit, dass ein Unternehmen auf den Influencer aufmerksam wird und gegebenenfalls eine Kooperation anbietet. In vielen Fällen werden Influencer auch nach der Anzahl der Likes für einen bestimmten Post vergütet. Folglich müssen neue Möglichkeiten geschaffen werden, den Wert eines Influencers festzulegen. Nicht auszuschließen ist, dass die Anzahl der Kommentare dann als digitale Währung unter den Influencern gehandelt wird, wobei nicht außer Acht gelassen werden darf, dass die womöglich nur eine Verschiebung des Problems hervorruft und keine echte Lösung darstellt.

14.6 Fazit

Das Influencer Marketing ist im ständigen Wandel, insbesondere juristisch. Mangels einer einheitlichen Rechtsprechung sind noch einige Fragen offen und viele Influencer stehen vor großen Herausforderungen. Die Notwendigkeit einer höchstrichterlichen Rechtsprechung des Bundesgerichtshofs ist damit immens.

Literatur

Köhler, H., Bornkamm, J., Feddersen, J. (2019): Kommentar zum UWG, 38. Auflage, München.
Sehl, K. (2019): Alle 6 Methoden: So berechnen Sie die Engagement-Rate, unter: https://blog.hootsuite.com/de/engagement-rate/, zuletzt abgerufen am 13. März 2020.
Solmecke, C., Kocatepe, S. (2018): Recht im Online Marketing, Rheinwerk Verlag.
Troge, T. (2018): Herausforderung: Influencer Marketing, in: GRUR-Prax, Heft 4/2018.

Scarlett Lüning ist seit März 2017 bei der Rechtsanwaltskammer Köln zugelassene Rechtsanwältin und in der Kanzlei Wilde Beuger Solmecke Rechtsanwälte in Köln vor allem im Bereich des gewerblichen Rechtsschutzes tätig. In dieser Zeit hat sie den Bereich des Influencer Marketings in der Kanzlei ausgebaut und vertritt hier einige Influencer. Zudem hält sie zu den verschiedensten Themen des Wettbewerbsrechts Vorträge und Workshops.

Teil IV

Ethik im Social Credit Rating

Geschäftsethik in China – ein Praxisbericht

15

Thomas Stewens und Axel Rose

Zusammenfassung

Die herausragende wirtschaftliche Entwicklung der Volksrepublik China hat in den vergangenen Jahrzehnten viele positive Effekte herbeigeführt: Sinkende Armut, wachsende Mittelschicht, innovative High-Tech-Unternehmen. Die Liste ließe sich lange fortführen. Der massive Bedeutungsgewinn materieller Leitmotive hat aber auch eine Ethik des Gewinnens gefördert, in der ein clever agierender Geschäftsmann sich zum Erreichen des Erfolges sehr weitreichender Mittel bedienen kann, während der Verlierer nicht geschickt oder klug genug war, um das Risiko zu erkennen. Ein moralisch ethischer Konsens im Geschäftsleben fehlt. Manipulation, Spekulation sowie ein riesiger grauer Kapitalmarkt sind die Kehrseite des wirtschaftlichen Erfolgs. Folge ist ein gesellschaftlicher Vertrauensverlust, der sich negativ auf die gesamte volkswirtschaftliche Entwicklung auswirkt. Der 2014 veröffentlichten Plan der chinesischen Regierung, im Jahr 2020 ein umfassendes, datenbasiertes Social Credit System einzuführen, ist unter anderem der Versuch, das geringe gesellschaftliche Vertrauen zu adressieren und ein faires, transparentes, und berechenbares Wirtschaftsumfeld zu schaffen. Das Mammutprojekt bezieht auch alle in China tätigen Unternehmen mit ein, die Daten aus rund 30 Bereichen an die zuständigen lokalen und nationalen Behörden weiterreichen müssen. Dort werden die Daten in einer zentralen Datenbank konsolidiert und analysiert. Mittels eines komplexen, intransparenten Algorithmus wird schließlich ein Rating ermittelt, das sowohl Strafen als auch Vergünstigungen zur Folge haben kann. Der vorliegende Beitrag liefert einen Praxisbericht über die chinesische Geschäftsethik und

T. Stewens · A. Rose (✉)
BankM AG, Frankfurt am Main, Deutschland
E-Mail: thomas.stewens@bankm.de; axel.rose@bankm.de

© Springer Fachmedien Wiesbaden GmbH, ein Teil von Springer Nature 2020   273
O. Everling (Hrsg.), *Social Credit Rating*,
https://doi.org/10.1007/978-3-658-29653-7_15

analysiert, inwieweit das Corporate Social Credit System geeignet ist, die Schattenseiten des wirtschaftlichen Erfolgs einzudämmen.

15.1 Die Stellung des lǎo bǎn

Wer die in der Volksrepublik China vorherrschende Geschäftsethik verstehen möchte, muss zunächst die chinesische Geschäftsführungspraxis analysieren. Im Gegensatz zur deutschen Praxis, in welcher sich zunehmend ein System der wechselseitigen Kontrolle durch Funktionsteilung und Aufsichtsgremien etabliert hat, wird das typische chinesische Unternehmen durch den „lǎo bǎn" geführt. Übersetzt bedeutet „lǎo bǎn" der „alte Boss" und bezeichnet in aller Regel den Eigentümer und Geschäftsführer eines Unternehmens. Dieser ist gleichzeitig annähernd immer auch der „gesetzliche Vertreter" der Gesellschaft. Die Rolle des gesetzlichen Vertreters hat in China insbesondere im Bereich der Haftung erhebliche Implikationen. Der gesetzliche Vertreter ist grundsätzlich einzelvertretungsberechtigt und ist so auch gegenüber dem allgegenwärtigen Staat für jegliche Vorkommnisse im Unternehmen persönlich verantwortlich. Dies ist umso bedeutungsvoller, als in China auch für Delikte der Wirtschafts- und Steuerkriminalität drakonische Strafen bis hin zur Todesstrafe verhängt werden können.

Für ausländische Investoren spielt die Bündelung der Verantwortung in den Händen des lǎo bǎn eine erhebliche Rolle. Alle Verhandlungen, Taktiken und wesentlichen Gespräche sind auf die Rolle des lǎo bǎn abgestellt. Alles, was von den Spezialisten in langwierigen Verhandlungen mühsam erarbeitet wurde, kann durch einen Satz des lǎo bǎn wieder auf „Anfang" gestellt werden. Dieses Verständnis wird auch oft taktisch eingesetzt, indem eine intensive, scheinbar finale Verhandlung in Abwesenheit des lǎo bǎn zugespitzt wird, um die Verhandlungsspielräume zu testen. Die hierzu vom Verhandlungsteam eingegangenen Verpflichtungen können dann in der tatsächlich finalen Verhandlung durch den lǎo bǎn wieder aufgelöst werden.

Unbegrenzte Freiräume in der Entscheidungsfindung hat ein lǎo bǎn trotz seiner starken Stellung im Verhältnis zu den Mitarbeitern und Geschäftspartnern aber nicht. Annähernd jedes erfolgreiche Unternehmen ist eng mit den parteilichen Entscheidungs- und Förderungsstrukturen verwoben. Auch die Finanzierung insbesondere von kleineren mittelständischen Unternehmen erfolgt sehr oft über private Netzwerke, die ebenfalls einen hohen Einfluss und Druck auf den lǎo bǎn ausüben können. Die Partei und die Investoren eines Unternehmens treten bei Verhandlungen sehr selten auf, können aber Kooperationen oder Transaktionen unterstützen oder unmöglich machen, ohne selbst jemals am Verhandlungstisch gesessen zu haben.

Für ausländische Investoren ist es hingegen schwer bis unmöglich, einem lǎo bǎn seine Kompetenzen zu entziehen. Denn der dafür notwendige Entzug des Unternehmensstempels ist in der Regel an die Zustimmung des lǎo bǎn geknüpft. Erschwerend kommt hinzu, dass ein Unternehmen durch die Entlassung seines lǎo bǎn auch dessen persönliches Kontaktnetzwerk verliert. Guanxi (das durch unausgesprochene Regeln von Verpflichtungen und Austausch charakterisierte komplexe Beziehungsnetzwerk persönlicher Verbindungen) ist aber ein unerlässlicher Bestandteil jedes Unternehmenserfolges in China. So kommt es, dass selbst bei einem Verkauf von Geschäftsanteilen oder einem kompletten Unternehmensverkauf das alte Rollenverständnis noch lange nach dem Abschluss der Transaktion bestehen bleibt. Für chinesische Geschäftspartner bis hin zu den finanzierenden Banken und nicht zuletzt im Selbstbild des lǎo bǎn, steht der „alte Boss" weiterhin unverrückbar an der Unternehmensspitze.

Dieses fest verankerte Rollenverständnis kann mitunter sogar gesetzliche Regelungen aushebeln. So mussten viele Auslandsgesellschafter chinesischer Tochtergesellschaften erleben, dass die angestellten chinesischen Manager (die ehemaligen lǎo bǎn) in der Lage waren, Gelder der Unternehmen, die sie zuvor veräußert hatten, bei den heimischen Banken als Sicherheit für private Spekulationsgeschäfte zu hinterlegen. Auch nach chinesischem Recht ist das nicht zulässig, insbesondere, wenn ausländische Investoren Anteile an der Gesellschaft halten. Die starke Rolle der ehemaligen Gründer ermöglicht jedoch Verhaltensweisen, die in anderen Kulturen nach einem Unternehmensverkauf schlichtweg unmöglich wären.

In der Quintessenz ist der lǎo bǎn in einem chinesischen Unternehmen nach Innen und in Bezug auf die Mitarbeiter und Geschäftspartner ein annähernd autokratischer Entscheider, während er in Bezug auf die politischen und finanzierenden Stakeholder in einem sehr engmaschigen, für Europäer nicht durchschaubaren Netzwerk persönlicher Beziehungen agiert.

15.2   Bedeutung der Geschäftszahlen und Praxis der dreifachen Buchführung

Die herausgehobene Rolle des lǎo bǎn zeigt sich auch mit Blick auf die Buchführung chinesischer Unternehmen. In der Praxis existieren dort nämlich oft drei Bücher: Eines für die Bank und die Investoren, eines für das Finanzamt und eines für den lǎo bǎn. Letzteres ist in der Regel Grundlage für die Steuerung und Kontrolle des Unternehmens und somit am ehesten mit einer Buchführung nach europäischem Verständnis vergleichbar. Allerdings ist diese Version für fremde Investoren in den allermeisten Fällen nicht zugänglich, sondern kursiert nur intern und wird streng vertraulich behandelt. Für die Bank und die Investoren sowie für das Finanzamt werden jeweils eigene Bücher mit abweichendem Zahlenwerk erstellt, um den geschäftlichen und persönlichen Nutzen auf der Grundlage verschiedener Rechnungslegungsstandards zu maximieren.

Natürlich kennen auch westliche Rechnungslegungssysteme Unterschiede etwa zwischen Handels- und Steuerbilanz und bereinigte Zahlen für Investoren oder Banken sind ebenfalls nicht gänzlich unbekannt. Doch während die Buchführung hierzulande ganz überwiegend als Instrumentarium zur objektivierten Informationsbeschaffung mit dem Ziel, das Unternehmen zu steuern und für Investoren transparent zu machen, dient, werden die Buchhaltung und das Controlling in der chinesischen Geschäftskultur gerade in mittelständischen Unternehmen immer noch oft als Mittel zum Zweck gesehen. Ausgangspunkt ist immer die Frage, für wen die Zahlen erstellt werden sollen und was mit der Darstellung bezweckt wird. Das Ergebnis ist dann die angesprochene mehrfache Buchführung, deren jeweilige Zahlen sich nach dem adressierten Leserkreis richten.

Die chinesische Regierung bemüht sich seit den Neunzigerjahren des letzten Jahrtausends, allgemeingültige Standards zu definieren und bei den Unternehmen durchzusetzen. Ein Hauptaugenmerk liegt dabei gerade auf Unternehmen mit ausländischer Beteiligung. So gibt es seit 2002 das spezifisch entwickelte „Accounting System for Business Enterprises" speziell für sogenannte „Foreign Invested Enterprises" (FIE). Im Jahr 2006 hat China dann ein neues, eng an die International Financial Reporting Standards (IFRS) angelehntes System für Rechnungslegungsstandards eingeführt und allgemeingültige „New Chinese Accounting Standards" entwickelt. Seit 2007 dürfen diese aufgrund der zunehmenden Ausdifferenzierung auch von FIEs alternativ angewendet werden. Im Jahr 2010 veröffentlichte das chinesische Finanzministerium als Reaktion auf eine Initiative der G20, einen weltweit einheitlichen, hochwertigen Rechnungslegungsstandard zu etablieren, schließlich die „Roadmap für die kontinuierliche Konvergenz der chinesischen Rechnungslegungsstandards und der internationalen Rechnungslegungsstandards". Als Folge wurden im Jahr 2014 acht neu formulierte oder grundlegend überarbeitete Unternehmensrechnungslegungsstandards herausgegeben. Darunter die Bewertung des beizulegenden Zeitwerts und die Darstellung des Abschlusses, mit denen die fortlaufende Konvergenz der chinesischen Unternehmensrechnungslegungsstandards mit den internationalen Rechnungslegungsstandards postuliert wurde. Für kleinere Unternehmen gibt es aber nach wie vor spezielle „Chinese Accounting Standards for Small Enterprises" mit erheblichen Erleichterungen und Spielräumen.

15.3 Spielwiese für Manipulation

Allen Bemühungen der Zentralregierung zum Trotz ist der Stellenwert der Buchführung in China jedoch immer noch anders als in der wesentlichen Welt. Besonders für Externe gibt es nach wie vor eine besondere Buchführung unter der Maxime, dass die Zahlen in erster Linie gut auszusehen haben. Besonders die Posten „Forderungen" und „Sonstige Forderungen" sind hierbei häufig als Quelle von Übertreibungen zu identifizieren. Jedes moderne Industrieunternehmen hat eine große Anzahl an Forderungen und Sonstigen Forderungen. Erstere beziehen sich hauptsächlich auf Waren und letztere auf sonstige laufende Konten. Dies ist die bequemste und schnellste Art, falsche Konten zu erstellen. Um den Jahresüberschuss zu steigern, können Unternehmen beispielsweise Warenkreditgeschäfte mit verbundenen Unternehmen oder auch ausländischen „Kunden" abwickeln. Da es sich um kreditbasierte Transaktionen handelt, wird der kostbare Cashflow geschont. Wenn also in der Gewinn- und Verlustrechnung eines börsennotierten Unternehmens ein starker Gewinnanstieg zu sehen ist, die Kapitalflussrechnung jedoch keinen großen Nettomittelzufluss aufweist und die Bilanz gleichzeitig eine große Anzahl an „Forderungen" aufweist, ist höchste Vorsicht geboten.

Der Lebenszyklus von Kreditkontotransaktionen ist allerdings endlich. Im Allgemeinen beträgt auch in China die Toleranzperiode für den Ausgleich offener Forderungen weniger als ein Jahr. Forderungskonten, die länger offenbleiben, werden in die Forderungsausfälle einbezogen und wirken sich mindernd auf die Gewinne des betroffenen Unternehmens aus. Für die gutgläubigen Investoren ist es dann jedoch oft zu spät. Denn um die Umsatzerhöhung aufzulösen und Bilanz sowie Gewinn- und Verlustrechnung des Vorjahres anzupassen, muss zunächst das Ursprungsgeschäft rückabgewickelt werden. Dazu wird der Geschäftspartner aufgefordert, die Ware zurückzugeben und einen Retourenbeleg auszufüllen. Das Beispiel eines börsennotierten Autoherstellers verdeutlicht das Dilemma. Letzterer hatte im Jahr 2004 gemäß der veröffentlichten Zahlen 10.000 Autos verkauft und dabei einen Gewinn von zehn Millionen US-Dollar erzielt. In der Bilanz und Gewinn- und Verlustrechnung wurden natürlich Kreditverkäufe bzw. offene Forderungen ausgewiesen. Der Aktienkurs stieg bis zum Ende des Jahres 2005 aufgrund der guten Zahlen kontinuierlich. Dann gab die Autofirma plötzlich bekannt, dass ein Großteil der Autos zurückgegeben worden seien. Der für das Jahr 2004 veröffentlichte Gewinn von zehn Millionen US-Dollar wurde annulliert und der Aktienkurs sank dramatisch.

Neben dieser Art von Gewinnmanipulationen ist es in den vergangenen Jahren immer wieder auch zu noch drastischeren Betrugsfällen gekommen mit fiktiven Forderungen und gefälschten Verkaufsunterlagen. So wurden nach der Jahrtausendwende unter anderem die Unternehmen „Yin Guangxia" und „Zheng Baiwen" für die Bilanzierung fiktiver Forderungen schwer bestraft. Bei einem weiteren, besonders schwerwiegenden Skandal spielten fiktive Forderungen ebenfalls eine wichtige Rolle. Beschaffung, Produktion und Vertrieb wurden hier zu großen Teilen virtuell durchgeführt. Interessant auch: Das Unternehmen bezahlte tatsächlich „echte" Steuern für die virtuellen Erträge und lebte so vom Börsengang bis hin zur Aufdeckung des Betrugs acht Jahre in dieser virtuellen Zahlenwelt.

In einem relativ jungen Kapitalmarkt wie dem chinesischen, in dem viele Anleger, Analyseinstitute und Wertpapiermedien unerfahren sind und sich wirksame Kontrollinstanzen und -mechanismen erst entwickeln müssen, ist es für Betrüger naturgemäß einfacher als in etablierten Märkten mit gewachsenen Strukturen. Allerdings entwickeln sich nicht nur die Strukturen weiter und die Wirtschaftsprüfer werden mit jedem aufgedeckten Fall wachsamer, sondern auch die Betrüger passen ihre Taktiken an. Wurden bei den sino-amerikanischen Betrugsfällen beispielsweise noch relativ einfache, gefälschte Bilanzen der chinesischen Töchter an die amerikanische Holding übersandt, machte bei den deutschen Betrugsfällen in den Folgejahren eher der „Trick" der Mehrfachverwendung von Gesellschaftsmitteln Schule. Diese Betrugsform konnte auch von erfahrenen, international tätigen Wirtschaftsprüfungsgesellschaften unter Anwendung internationaler Prüfungsstandards nicht aufgedeckt werden.

In den Bilanzen der Gesellschaften wird derzeit noch nach den Tätigkeiten im Rahmen des wesentlichen Unternehmenszwecks, dem sogenannten „Hauptgeschäft" und den unter

„Andere Geschäfte" subsumierten sonstigen Tätigkeiten unterschieden. Müssen alle Einnahmen und Kosten des Hauptgeschäfts in der Gewinn- und Verlustrechnung ordnungsgemäß erfasst werden, ist die Dokumentationserfordernis für die Anderen Geschäfte wesentlich geringer. Dies eröffnet Spielräume für Manipulationen. Undurchsichtige Geschäftsvorfälle zwischen Hauptgeschäft und Anderen Geschäften erschweren es Anlegern zusätzlich, ein transparentes Bild zu gewinnen. Beispielsweise definierte ein börsennotiertes Unternehmen in seinem Geschäftsbericht die Immobilien-, Hotel- und Tourismusbranche, als Hauptgeschäft, berechnete jedoch nur den Immobiliengewinn auch tatsächlich dort, während sowohl die Erträge aus der Hotel- als auch die der Tourismusbranche unter anderen Geschäften klassifiziert wurden. Inkonsistenzen dieser Art lassen sich manchmal bereits im Geschäftsbericht ablesen und sollten Investoren auf jeden Fall misstrauisch machen. Im Extremfall ist es sogar schon vorgekommen, dass Unternehmen den Gewinn eines bestimmten Geschäfts gleichzeitig in den Hauptgeschäftsgewinn und den sonstigen Geschäftsgewinn einbezogen haben und somit den Ertrag kurzerhand verdoppelten.

Die Folgen dieser Praxis der beschönigenden Buchführung sind gravierend. Sie veranlasst die gesamte Gesellschaft, den Unternehmensabschlüssen zu misstrauen. Schon bei ersten Schwierigkeiten zwischen zwei Firmen kommt es deshalb häufig zu harten Reaktionen. In der Folge führt dies immer wieder zu unnötigen Forderungsausfällen, die sich negativ auf die gesamte volkswirtschaftliche und gesellschaftliche Entwicklung auswirken.

15.4 Die Rolle des Immobiliensektors

Einen noch größeren Spielraum als Umsatzmanipulationen im Hauptgeschäft bieten überhöhte Bilanzansätze. Der Immobilienbereich mit seinen langen Bauzyklen und zögerlichen Investitionsfortschritten ist hierfür besonders anfällig. Zumal es für die Regulierungsbehörden eine große Herausforderung darstellt, auffällige Bewertungsansätze zu überprüfen. Selbst wenn der gesunde Menschenverstand eine Abwertung unausweichlich erscheinen lässt, muss oft abgewartet werden, bis ein Projekt tatsächlich abgeschlossen ist. Erst dann können die Bewertungsansätze mit stichhaltigen Argumenten hinterfragt werden. Dies ist auch der Grund, warum bei vielen Immobilienprojekten die Fertigstellung oder auch die Kommerzialisierung (Vermietung/Verkauf) immer wieder hinausgezögert werden. Ist es dann endlich soweit, sind die Verluste der Anleger allerdings längst zementiert und Maßnahmen zur Wertaufholung kaum mehr möglich. Welche Rolle die schwierige Bewertung von Immobilienprojekten dabei spielt, dass die Anlageinvestitionsprojekte im Immobilienbereich bei vielen börsennotierten Unternehmen in China in den vergangenen Jahren stark an Volumen gewonnen haben, darüber kann nur spekuliert werden. Die gebundenen Mittel belaufen sich jedenfalls auf Hunderte von Millionen, teilweise sogar Milliarden US-Dollar.

Insgesamt hat die rasche Entwicklung der Wirtschaft nach der Jahrtausendwende in China zu einer sprunghaften Entwicklung der Immobilienbranche geführt. Viele Unternehmen haben aufgrund der erwarteten hohen Renditen das Geschäftsmodell sukzessive umgestellt, angefangen oft mit privaten Investments des lǎo bǎn. Anfangs wurden diese Immobilien-Geschäfte häufig als kurzfristiger „sonstiger Geschäftsbereich" bilanziert bzw. im Unternehmen selbst überhaupt nicht bilanziert. Im Verlauf haben einige Unternehmen ihr früheres Kerngeschäft dann eingestellt und wurden zu reinen Immobilienunternehmen. Allein zwischen 2005 und 2010 (ein Zeitraum stark steigender Immobilienpreise) wechselten fast 40 börsennotierte Unternehmen von ihrem ursprünglichen Hauptgeschäft zum Immobiliengeschäft. Der Effekt war die wirtschaftliche Aushöhlung des operativen Geschäftes vieler chinesischer Firmen durch Immobiliengeschäfte und Spekulationen.

Ein typisches Beispiel hierfür ist die „Schmuck-Aktie" ZHONG BAO KE KONG, die ab 2002 schrittweise die Schmuckproduktionslinien mit niedrigeren Rohertragsmargen veräußerte und die freigewordenen Mittel sukzessive in Immobilien investierte. Die Performance des Unternehmens und damit der Aktie hat sich durch die Umwandlung in ein Immobilienunternehmen jedoch nicht verbessert: Bis heute sind Eigenkapitalrendite und Kursentwicklung nicht zufriedenstellend. Es zeigt sich, dass der Wechsel vom angestammten Kerngeschäft zu einem scheinbar erfolgversprechenden Immobiliengeschäft mit erheblichen Risiken verbunden ist.

Wenn eine Vielzahl von Unternehmen ihr Stammgeschäft zugunsten des verlockenden Immobiliengeschäftes einstellt, ist das gesamtwirtschaftlich ebenfalls kritisch zu betrachten. Entsprechend versucht die chinesische Regierung auch hier gegenzusteuern und verpflichte die Unternehmen laut § 15 des Wertpapiergesetzes die Mittel, die durch das öffentliche Angebot von Aktien einer börsennotierten Gesellschaft eingeworben werden, nur noch in Übereinstimmung mit dem Zweck der im Prospekt aufgeführten Mittelverwendung zu investieren.

15.5 Spekulation und die Ethik des Misserfolgs

Ähnlich wie andere Großprojekte oder technologische Investitionen wurde die Immobilienentwicklung in China lange Zeit sehr stark staatlich unterstützt. Auch auf Unternehmensebene, wo zum Beispiel der Bau neuer Produktionsstätten bis heute großzügig gefördert wird. In vielen anderen Bereichen fehlt aber gerade mittelständischen Unternehmen ein institutioneller Zugang zu Wachstumskapital. Sowohl der klassische Bankensektor als auch der Kapitalmarkt stehen in der Regel nur wenigen meist deutlich größeren und staatlich geförderten Konzernen offen. Viele kleinere Unternehmen sind zur Wachstumsfinanzierung deshalb gezwungen, sehr stark auf private Kredite zurückzugreifen. So entstand in China ein sehr leistungsfähiger „grauer" Kreditmarkt mit hohen Zinsen. 15 bis 20 Prozent p.a. sind hier mehr Regel als Ausnahme.

Auch in Deutschland und vielen anderen Ländern gibt es einen florierenden grauen Kapitalmarkt, also einen Bereich, der nicht unter direkter staatlicher Kontrolle steht und für den deshalb wenige bis gar keine gesetzlichen Vorgaben gelten. In China weist der graue Kapitalmarkt jedoch eine entscheidende Besonderheit auf: Gläubiger können grundsätzlich jedes Jahr zum Frühlingsfest den Ausgleich der Darlehen verlangen. Schuldner müssen sich dadurch sehr kurzfristig refinanzieren, Vermögenswerte werden aus Mangel an Alternativen vielfach zu niedrigen Preisen veräußert. So entsteht Jahr für Jahr ein enormer Druck am grauen Kapitalmarkt. Geld oder Liquidität werden plötzlich noch wertvoller. Diese Besonderheit führt in China genau wie die annähernd ständig steigenden Aktien- und Grundstückspreise zu einer massiven Spekulationsneigung. Geschickte Spekulanten können durch vorteilhaftes Agieren quasi über Nacht reich werden und das wird natürlich durch Freunde und Bekannte beobachtet, registriert und nachgeahmt. Es ist deshalb nicht verwunderlich, dass Kassenbestände in Unternehmen nicht wie in Deutschland üblich in sichere Anlageformen investiert werden, sondern oft in hochspekulative Investitionen im grauen Markt fließen – zwar hochverzinslich, aber eben auch sehr riskant.

Das typische Spekulationsziel in China sind Immobilien und Aktien. Bei Aktien versprechen besonders vorbörsliche Investments hohe Erträge, und der Preis für Immobilien kennt quasi seit der marktwirtschaftlichen Öffnung vor rund 40 Jahren nur eine Richtung. Die erwarteten Renditen aus solchen Geschäften liegen in der Regel deutlich jenseits von 20 Prozent p.a., oft wird sogar von einer Vervielfachung der Investitionen ausgegangen. Wenn die Chance auf ein Investment in ein Grundstück, ein Hotel oder eben ein aussichtsreiches Börsenunternehmen besteht, ist der Anreiz deshalb groß, sich das Geld trotz der hohen Graumarktzinsen zu leihen.

Die alljährlich drohende Rückzahlung macht den grauen Kapitalmarkt in China jedoch anfällig. So gesehen in den Jahren 2012 und 2013, als das Wachstumstempo in China erstmals seit der Jahrtausendwende unter die Acht-Prozent-Marke rutschte. Immobilienpreise stagnierten und gingen in den B-Städten sogar leicht zurück. Auch die Kapitalmärkte versagten vorübergehend die rauschhaften Gewinne der Vergangenheit. Zeitweise wurden keine Neuemissionen in Mainland China mehr zugelassen. Was nach außen weit von einem Zusammenbruch oder dem Platzen einer Blase entfernt schien, traf den grauen Kapitalmarkt und damit die Spekulanten empfindlich. Kein Wunder, bei Zinsen von 15 Prozent und mehr sowie dramatisch anzupassenden Rendite- und Zeiterwartungen der getätigten Investments.

Genau zu dieser Zeit tauchten sehr viele der alten Bosse in China unter und waren weder für die Belegschaft noch für die Gesellschafter zu erreichen. Der Zusammenhang zwischen dem geplatzten Spekulationshype und dem Verschwinden der alten Bosse erscheint zunächst merkwürdig. Aufklärung brachten erst spätere Analysen und Prüfungen von Firmen wie Youbisheng, einem innovativen Papierhersteller oder von Joyou, einer damaligen Tochter der deutschen Grohe AG und später Bestandteil des japanischen Baukonzerns Lixil. Die Gründer und ehemaligen Hauptgesellschafter, die klassischen lǎo bǎns, hatten die reichlich gefüllten Unternehmenskassen als Sicherheit für private Spekulationsgeschäfte hinterlegt. Solange diese Transaktionen positiv verliefen, entstand bei keinem Beteiligten ein Schaden. Im Gegenteil, die Unternehmen schienen bestens aufgestellt und auch die anderen unternehmerischen Initiativen der Gründer befeuerten das volkswirtschaftliche Wachstum. Als dann aber der graue Kapitalmarkt in Schieflage kam und die massiv durch Fremdkapital gehebelten Investitionen unter Druck gerieten, verwertete die

Hausbank der Unternehmen die Sicherheiten und entzog somit den Unternehmen die Lebensgrundlage.

Für den lǎo bǎn ist diese Situation nach Jahren des stetigen, programmierten Aufschwungs nahezu ausweglos. Er verliert sein Gesicht vor der Belegschaft, den Geschäftspartnern und natürlich auch vor den geprellten Gläubigern und Aktionären. Unter diesem Druck sind viele der betroffenen Bosse dann einfach verschwunden. Einige tauchten später in Gefängnissen wieder auf, andere auch gar nicht mehr.

15.6 Social Scoring als Lösung?

Die chinesische Wirtschaft hat sich in den letzten 30 Jahren rasant entwickelt, aber die Entwicklung des Finanzsystems und des Systems der sozialen Integrität hat nicht mit der wirtschaftlichen Entwicklung Schritt gehalten. Oft herrscht eine Ethik des Gewinnens bzw. des Siegers, das heißt der clever agierende Geschäftsmann kann sich zum Erreichen des Erfolges sehr weitreichender Mittel bedienen. Der Verlierer war eben nicht geschickt oder klug genug, um das Risiko zu erkennen. Ein im weltweiten Vergleich hoher Wert von 66 Punkten im Hofstede-Index für Maskulinität unterstreicht die Bedeutung von Wettbewerb, Gewinnstreben und Erfolg in der chinesischen Gesellschaft.

Es ist durchaus denkbar, dass in China aufgrund des massiven Bedeutungsgewinns materieller Leitmotive in den letzten Jahrzehnten ein moralisch ethischer Konsens im Geschäftsleben fehlt. Zumal China sich zwar durch ein hohes gesamtgesellschaftliches Vertrauen auszeichnet, innerhalb von Familie und Guanxi gleichzeitig aber ein striktes System gegenseitiger Abhängigkeiten existiert. Soziale Netzwerke bleiben zumeist auf diesen engen Kreis beschränkt. Dabei sind Vertrauen und übergreifende Netzwerke entscheidende Merkmale für das Sozialkapital einer Gesellschaft. Ohne die Überzeugung, dass Kooperationspartner sich an getroffene Absprachen halten, bzw. ohne die Gewissheit, dass Verstöße durch andere Gesellschaftsmitglieder sanktioniert werden, ist keine Kooperation möglich. In der wissenschaftlichen Diskussion werden zwei Arten von Vertrauen unterschieden. Interpersonelles Vertrauen meint Vertrauen im zwischenmenschlichen Bereich und verringert die Transaktionskosten unternehmerischen Handelns. Systemisches Vertrauen hingegen wird als Zuversicht in das politische, wirtschaftliche und institutionelle System beschrieben. Mangelt es Wirtschaftsteilnehmern an Vertrauen in das System, werden sie dieses nicht nutzen und auf informelle Sektoren wie den Schwarzmarkt ausweichen.

Der 2014 veröffentlichten Plan der chinesischen Regierung, im Jahr 2020 ein umfassendes, datenbasiertes Social Credit System einzuführen, ist unter anderem der Versuch, enge Vertrauensradien aufzubrechen und ein faires, transparentes und berechenbares Wirtschaftsumfeld zu schaffen. Angestrebt werden neben der Steigerung der „Aufrichtigkeit in Regierungsangelegenheiten", der „sozialen Integrität" und der „gerichtlichen Glaubwürdigkeit" ganz offiziell auch die Erhöhung der „kommerziellen Integrität". Während westliche Medien überwiegend die Auswirkungen auf Privatpersonen beleuchten, geht oftmals ein wenig unter, dass es für Firmen ein eigenes Corporate Social Credit System geben wird. Dafür müssen Unternehmen Daten aus rund 30 Bereichen an die zuständigen lokalen und nationalen Behörden weiterreichen. Dort werden die Daten in einer zentralen Datenbank konsolidiert und analysiert. Mittels eines komplexen Algorithmus wird schließlich ein Rating ermittelt, das sowohl Strafen als auch Vergünstigungen zur Folge haben kann.

Die meisten im Rahmen des Corporate Social Credit System erhobenen Daten sind für die Unternehmen nicht neu, sondern betreffen Kriterien, die sowieso bereits gesetzlich vorgeschrieben sind. Dazu gehören die pünktliche Zahlung von Steuern, der Besitz aller für den Geschäftsbetrieb erforderlichen Lizenzen, die Einhaltung von Produktqualitätsstandards und die Erfüllung von Umweltschutzauflagen. Darüber hinaus unterliegen die Unternehmen je nach Art ihrer Geschäftstätigkeit einer Reihe an branchenspezifischen Anforderungen. Problematisch ist, dass Firmen auch für das Wohlverhalten ihrer Zulieferer und das korrekte Verhalten des Managements und einzelner Mitarbeiter bewertet werden. Auf Beides haben Unternehmen nur bedingt Einfluss, zudem verschwimmen hier unternehmerische und persönliche Ebene.

Wir möchten an dieser Stelle aber weder tiefer in die genaue Rating-Berechnung einsteigen noch eine moralische Diskussion führen. Ziel des Beitrags ist es vielmehr, zu beleuchten, welchen Einfluss das Corporate Social Credit System auf die in den vorangegangenen Kapiteln dargestellten praktischen Probleme im chinesischen Geschäftsalltag haben könnte. Auf dem Papier ist eine ganze Reihe von positiven Wirkungsmechanismen denkbar, angefangen mit der Rolle des lǎo bǎn als autokratischem Entscheider, der in einem engen Netzwerk persönlicher Beziehungen agiert. Ist der lǎo bǎn gegenüber dem Staat bislang für jegliche Vorkommnisse im Unternehmen persönlich verantwortlich, könnte das Corporate Social Credit System dazu beitragen, diese Vereinheitlichung von Unternehmen und Unternehmer aufzuweichen, indem es ein differenzierteres Bild auf die betrieblichen Vorgänge ermöglicht. Ein voll automatisiertes System, basierend auf leidenschaftslosen Algorithmen, sollte zudem die aktuelle Bedeutung möglichst enger Verbindungen zwischen Unternehmenslenkern und lokalen Politikern verringern. Auch das verkleinert die Abhängigkeit vom lǎo bǎn und schränkt zudem Korruptionsanreize ein.

Einer der wichtigsten potenziellen positiven Effekte des Corporate Social Credit System ist aber sicherlich die Verbesserung des Kreditzugangs insbesondere kleinerer mittelständischer Unternehmen. Finanzieren sich mehr Firmen über Banken und institutionelle Investoren, können private Netzwerke weniger Einfluss und Druck auf den lǎo bǎn ausüben. Gleichzeitig, und vielleicht noch wichtiger, dürfte der graue Kreditmarkt an Bedeutung verlieren. Wachstumsunternehmen können sich im besten Fall günstiger und langfristiger finanzieren, der jährliche Refinanzierungsdruck nimmt ab oder entfällt ganz. Indem auch Faktoren wie Produktqualität, Umweltschutz oder soziale Standards bewertet werden, müssten besonders innovative Unternehmen in der Theorie sogar überproportional stark profitieren. Insgesamt sinkt der Anreiz, das Hauptgeschäft auf kurzfristig erfolgversprechende Wirtschaftsbereiche auszurichten oder hochspekulative Geschäfte einzugehen. Dies gilt auch für private Spekulationsgeschäfte auf Ebene des lǎo bǎn, zumal dieser sich darüber bewusst sein muss, dass sein persönliches Verhalten ebenfalls bewertet wird und in das Unternehmensrating einfließt.

Selbst die Flucht nach fehlgeschlagenen Spekulationen wird durch das Social Credit System schwieriger: Laut Medienberichten wurden 2018 während der Pilotphase 17,5 Millionen Flugticket-Anfragen und 5,5 Millionen Buchungsanfragen im Bahn-Fernverkehr zurückgewiesen, weil der Social Score zu niedrig war. Hat dies eher eine komische Note, ist der folgende Gedanke umso wichtiger: Dadurch, dass private Spekulationsgeschäfte erstmals im Unternehmenskontext sichtbar werden, erhöht sich die Transparenz gegenüber Investoren. Müssen diese weniger Angst haben, dass Firmengelder für private Geschäfte zweckentfremdet werden, steigt die Investitionsbereitschaft. Die Transparenz, die ein einzelnes umfassendes System bereitstellt, macht es darüber hinaus schwerer, einzelne Zahlen oder sogar ganze Bilanzen zu manipulieren. In dem Moment, wo Banken, Gesellschafter und Steuerbehörden zuallererst auf den Social Credit Score schauen, wird eine eigene Buchführung für unterschiedliche Adressaten ineffektiv. Natürlich können Zahlen immer noch manipuliert werden, doch das Risiko in China auf eine staatliche Blacklist zu geraten, werden nur wenige Firmen eingehen.

Gerade aus der Sicht ausländischer Firmen, die oft eine willkürliche Benachteiligung gegenüber heimischen Konkurrenten beklagen, könnte das Corporate Social Credit System einen wichtigen Schritt hin zum viel beschworenen „Level Playing Field" sein. Mehr noch, dass sie sich tendenziell stärker als ihre chinesischen Wettbewerber an Umweltauflagen oder Sozialstandards halten, könnte künftig belohnt werden. Wenn bisher beispielsweise an Tagen mit starker Luftverschmutzung verfügt wurde, dass alle produzierenden Unternehmen einen Produktionsstopp einlegen müssen, könnte dies im Corporate Social Credit System differenziert geschehen, je nachdem, wie stark die Firmen die Grenzwerte einhalten. Dass internationale Konzerne grundsätzlich mehr Erfahrung in der Umsetzung von Compliance-Standards und Regulierungsanforderungen haben ist ein weiterer Vorteil. Allerdings ist davon auszugehen, dass einheimische Unternehmen sich besser in den Feinheiten des Systems zurechtzufinden und über einen überlegenen Informationsfluss zu Regierungsbehörden verfügen. Besonders ausländische KMU dürften Schwierigkeiten haben, mit den sich ständig ändernden Vorschriften und Bewertungsstandards Schritt zu halten und die Ressourcen zu heben, die für die Einhaltung der Vorschriften erforderlich sind.

Wie das Corporate Social Credit System die Geschäftspraxis und -ethik in China am Ende tatsächlich verändern wird, ist aktuell noch schwer abzuschätzen. Ob und wieweit die beschriebenen potenziellen positiven Auswirkungen eintreten, wird stark von der finalen Ausgestaltung und Umsetzung abhängen. Zwar wird das System bereits seit Jahren in einigen Provinzen getestet, doch der genaue Algorithmus ist noch unklar, der Datenaustausch fehleranfällig, die Implementierung der Sanktionsmechanismen unvollständig und die Big-Data-Technologien noch nicht ausgereift. Und selbstverständlich besteht eine hohe Gefahr, dass das System missbraucht werden könnte, gerade mit Blick auf die schwelenden Handelskonflikte. Vor allem die schwarze Liste, auf der sogenannte stark verdächtige Unternehmen erfasst werden, öffnet der Willkür Tür und Tor. Einen Klageweg gibt es nicht.

Schon jetzt von einem orwellianischen System zu schreiben ist jedoch zu früh. Standardisierte Compliance-Systeme sind in der westlichen Unternehmenshemisphäre weit verbreitet. So muss jede Bank Firmenkunden und deren wirtschaftlich Berechtigte schon lange einem Compliance-Check unterziehen. Basis ist meist standardisierte Software mit zum Teil fragwürdigen Ergebnissen. Anlass zur Hoffnung gibt der weltweite Erfolg von Plattformen wie Ebay, Amazon, Uber oder Airbnb. Dieser basiert zu einem nicht geringen Anteil auf datenbasierten Bewertungssystemen. Erst diese schaffen das Vertrauen, das für eine Transaktion notwendig ist.

Mit dem Corporate Social Credit System versucht die chinesische Regierung diese Erfolgsmodelle auf staatlicher Ebene umzusetzen. Nicht von ungefähr war Alibaba, das chinesische Pendant zu Amazon, am Aufbau des Ratingsystems beteiligt. Zwar lässt sich nachhaltiges Sozialkapital in der Regel nur durch freiwillige Aktivitäten bilden, doch wenn die Algorithmen transparent und diskriminierungsfrei funktionieren, kann das neue System dazu beitragen, gleiche Wettbewerbsbedingungen zu schaffen. Im besten Fall werden Compliance-Verstöße von Unternehmen erkannt oder verhindert und interpersonelles sowie systemisches Vertrauen werden gestärkt. Im schlechtesten Fall verlagern sich Manipulationen von der betrieblichen auf die systemische Ebene und Sozialkapital wird zerstört. So oder so wird das Corporate Social Credit System einen erheblichen Einfluss auf alle in China tätigen Unternehmen haben.

Thomas Stewens ist bei der BankM AG beschäftigt, einer 2007 gegründeten deutschen Spezialbank, die in der Mittelstandsfinanzierung tätig ist und hier insbesondere als Spezialist für strategische Kapitalbeschaffung im Fremd- und Eigenkapitalbereich eine führende Stellung einnimmt. Dabei unterstützt BankM auch chinesische Unternehmen mit Interessen in Deutschland und deutsche Unternehmer bei Expansionsvorhaben in China und hat in dieser Funktion mehrere Börsengänge, M&A-Transaktionen und Wirtschaftsförderungsprojekte begleitet. Thomas Stewens ist Gründungspartner, Vorstand und Leiter China Desk der BankM sowie Gründungsvorstand des Vereins Kapitalmarkt KMU und Mitglied der Landesfachkommission Finanzmarktpolitik & Vorsorge im Wirtschaftsrat der CDU. Herr Stewens verfügt über langjährige Erfahrung im Investment Banking und hat zahlreiche Transaktionen mit China-Bezug in führender Position begleitet.

Axel Rose ist bei der BankM AG beschäftigt, einer 2007 gegründeten deutschen Spezialbank, die in der Mittelstandsfinanzierung tätig ist und hier insbesondere als Spezialist für strategische Kapitalbeschaffung im Fremd- und Eigenkapitalbereich eine führende Stellung einnimmt. Dabei unterstützt BankM auch chinesische Unternehmen mit Interessen in Deutschland und deutsche Unternehmer bei Expansionsvorhaben in China und hat in dieser Funktion mehrere Börsengänge, M&A-Transaktionen und Wirtschaftsförderungsprojekte begleitet. Axel Rose ist seit 2013 bei der BankM im Projektgeschäft tätig und hat seitdem verschiedene Projekte in China begleitet. Herr Rose ist Spezialist für Unternehmens- und Investorenkommunikation, hat mehrere Jahre für die F.A.Z.-Gruppe als Finanzjournalist gearbeitet und ist Autor des 2009 im Peter Lang Verlag erschienenen Buches „China – Indien: Wettbewerbsfähigkeit im Vergleich".

Das Sozialkreditsystem in China aus ethischer Sicht

16

Eine Diskussion der Chancen und Risiken für die Betroffenen

Oliver Bendel

Zusammenfassung

Das Sozialkreditsystem ist ein im Test bzw. in den Anfängen befindliches Überwachungs-, Erfassungs- und Bewertungssystem zur Angleichung des Verhaltens der Bürger, Behörden und Firmen von China an die moralischen, sozialen, rechtlichen, wirtschaftlichen und politischen Anforderungen der Kommunistischen Partei (KP). Es findet, so der Plan, ein permanentes Rating und Scoring mit Blick auf die Lebenssituation, das Sozialverhalten oder Verwaltungs- und Wirtschaftsaktivitäten statt. Der vorliegende Beitrag skizziert die Diskussion über das Sozialkreditsystem in den westlichen Medien und zwischen den Experten in China. Ausgehend von Grundannahmen, die von der Planung und Projektierung abgeleitet sind, werden Überlegungen aus der Perspektive der Ethik angestellt, und zwar mit Blick auf die betroffenen Bürgerinnen und Bürger. Die Anwendung des Systems auf Unternehmen spielt im vorliegenden Beitrag keine Rolle. Der Befund ist, dass eine bestimmte Umsetzung des Sozialkreditsystems die Lebensqualität heben, aber auch die Persönlichkeitsrechte und die Menschenrechte verletzen kann.

16.1 Einleitung

Das Sozialkreditsystem (engl. „social credit system") ist ein in Planung bzw. in den Anfängen befindliches elektronisches Überwachungs-, Erfassungs- und Bewertungssystem (vgl. Bendel 2019b). Es soll das Verhalten der Bürgerinnen und Bürger, Behörden und Firmen von China gezielt verändern, sodass die moralischen, sozialen, rechtlichen, wirtschaftlichen und

O. Bendel (✉)
Zürich, Schweiz
E-Mail: oliver.bendel@fhnw.ch

© Springer Fachmedien Wiesbaden GmbH, ein Teil von Springer Nature 2020
O. Everling (Hrsg.), *Social Credit Rating*,
https://doi.org/10.1007/978-3-658-29653-7_16

politischen Ansprüche der Kommunistischen Partei (KP) bzw. des Großteils der Bevölkerung erfüllt werden. Es findet, so die Idee, ein permanentes Rating und Scoring („citizen score" bzw. „social scoring") mit Blick auf die Lebenssituation, das Sozialverhalten oder Verwaltungs- und Wirtschaftsaktivitäten statt. Dabei werden vernetzte Datenbanken sowie Bild- und Tonsysteme in Verbindung mit Big-Data-Analysen und Methoden der Künstlichen Intelligenz (KI) eingesetzt. Bei Identifizierung, Quantifizierung, Qualifizierung und Evaluierung in öffentlichen Bereichen, etwa über Kameras und Mikrofone und damit verbundene Sprach-, Stimm- und Gesichtserkennung, einschließlich Emotionserkennung, sind Echtzeitverfahren von Bedeutung.

Nach der mehrjährigen Testphase, die u. a. in Rongcheng stattfand, einer Millionenstadt in der ostchinesischen Provinz Shandong, sollte im Jahre 2020 das „moralische und soziale Bonitätssystem" (Strittmatter 2018) in den Normalbetrieb übergehen. Manche Autoren haben diesen Zeitpunkt als unrealistisch betrachtet.[1] Das Punktekonto wird je nach Bewertung nach oben oder unten korrigiert. In Rongcheng startete man mit 1000 Punkten. Bei über 1050 Punkten galt man als mustergültig, bei weniger als 599 als unehrlich. Es sind im System einerseits Belohnungen vorgesehen, andererseits Bestrafungen wie Karrierebehinderungen, Reiseverbote, Steuererhöhungen oder Betriebsbeschränkungen (Strittmatter 2018). Chinesische Unternehmen wie Baidu, Alibaba und Tencent sind nicht nur (neben Bürgern und Behörden) Ziel, sondern wohl auch Teil der Kontrolle, indem sie technische Mittel, Infrastrukturen und Expertise zur Verfügung stellen (vgl. Bendel 2019b).

Der vorliegende Beitrag fasst die Diskussion über das Sozialkreditsystem in den westlichen Medien und unter Experten in China möglichst knapp und ausgewogen zusammen. Ausgehend von Grundannahmen, die von der Planung und Projektierung abgeleitet sind, werden Überlegungen aus der Perspektive der Ethik angestellt, und zwar mit Blick auf die betroffenen Bürgerinnen und Bürger. Die Unternehmen, auf die das System ebenfalls angewendet werden kann, spielen im vorliegenden Beitrag keine Rolle. Eine Zusammenfassung samt Ausblick rundet den Beitrag ab.

16.2 Darstellungen und Standpunkte

16.2.1 Die Diskussion in den westlichen Medien

Spätestens seit 2017 berichten die deutschsprachigen Medien intensiv über das Sozialkreditsystem in China und die damit verbundene Testphase, etwa in Rongcheng. Es wird immer wieder darauf hingewiesen, dass Überwachung und Bewertung der Aktivitäten sich nicht nur

---

[1] Vgl. dazu Genzsch: „Ursprünglich war eine landesweite Realisierung der Pläne bis 2020 vorgesehen. Dieses Ziel kann zum derzeitigen Zeitplan als unrealistisch betrachtet werden. Derzeit gibt es über 40 Pilotprojekte und Testsysteme, die sich aufgrund der Heterogenität des Landes auf unterschiedliche Schwerpunkte konzentrieren." (Genzsch 2019, S. 136; vgl. ferner Dorloff 2018)

auf den virtuellen, sondern auch auf den realen Raum beziehen. Dabei sollen optische Erkennungs- und Auswertungssysteme und KI-Systeme eine Rolle spielen, also wiederum digitale Möglichkeiten. Die ZEIT bemerkt, die Regierung wolle sich bei der Bewertung mit dem Verhalten ihrer Bürger im Internet nicht zufriedengeben, und führt dann aus:

> In Kombination mit der Gesichtserkennungstechnik moderner Videokameras, die schon bald flächendeckend in den chinesischen Großstädten installiert werden sollen, lässt sich künftig auch das Verhalten der Bürger in der Öffentlichkeit erfassen und in die Bewertung aufnehmen. Dazu gehören dann nicht nur Verstöße im Straßenverkehr, sondern auch das Benehmen etwa beim Anstehen vor der Kasse im Supermarkt. Natürlich muss die Technik dann zuverlässig sein, aber nicht nur in China wird hart daran gearbeitet. Und Gesichtsdatenbanken zum Abgleich hat der Staat längst, denn jeder chinesische Bürger hat einen Personalausweis mit einem biometrischen Foto. (Lee 2017)

Hervorgehoben wird hier, wie in weiteren deutschsprachigen Berichten, dass es sich im Rahmen der Einstufung und Beurteilung nicht nur um rechtliche, sondern auch um moralische und soziale Kategorien handelt:

> Neu in Chinas Sozialkreditsystem ist …, dass gesellschaftliches und moralisches Verhalten der Bürger in die Bewertung mit einfließt. Alle Informationen sollen perspektivisch ein großes Ganzes ergeben. Der gläserne Bürger, über den alles bekannt ist. (Lee 2017)

Das Ziel der Partei sei, „die Menschen zu moralisch einwandfreien Bürgern zu erziehen" (Lee 2017). Verbessert werden sollen die Ordnung des Markts und die Ordnung in der Gesellschaft, zugunsten einer „harmonischen Gesellschaft", wie sie seit 2003 bzw. 2004 erträumt wird, und einer „wissenschaftlichen Entwicklung".[2]

Kai Strittmatter, über viele Jahre Korrespondent in Peking für die Süddeutsche Zeitung, sieht mit dem Sozialkreditsystem, so stellt es einer seiner Kritiker, der Autor Marcus Hernig, in der Republik dar, ein Zeitalter der totalen Überwachung heraufziehen. „Das flächendeckende Kontrollnetz, gefüttert von etlichen Datenquellen und Überwachungskameras, werde noch die privatesten Regungen der Bürgerinnen erfassen – und einen unmündigen, unfreien Menschen hervorbringen." (Hernig 2018) Tatsächlich ist der Chinakenner ein vehementer Gegner des Sozialkreditsystems. Und er wird nicht müde zu betonen, dass es nicht den einen Chinesen gibt, dem die Überwachung nichts ausmacht, der sie begrüßt. Er verweist auf Hongkong, auf Taiwan, aber auch auf die Volksrepublik China selbst, wo durchaus abweichende Meinungen zu finden sind, von Künstlern, Bloggern und Dissidenten, die die Partei beobachten, verfolgen und einsperren lässt. An einer Stelle seines Buchs „Die Neuerfindung der Diktatur" schreibt er:

---

[2] Der SWP-Studie ist zu entnehmen, dass die Kommunistische Partei seit 2003/2004 zwei Leitbilder propagiert, nämlich die wissenschaftliche Entwicklung und die harmonische Gesellschaft. „Damit will die Partei ihre Legitimität und ihr Ansehen in der Gesellschaft stärken. Konkret steht dahinter die Absicht, den identifizierten Fehlentwicklungen entgegenzusteuern und dabei gleichzeitig einen Ersatz für den mittlerweile ausgehöhlten ideologischen Überbau anzubieten." (Wacker und Kaiser 2008, S. 5) Die harmonische Gesellschaft habe eine ideologische Dimension und Funktion.

Zentraler Bestandteil dieses neuen Chinas wird zum Beispiel das „Soziale Bonitätssystem", das von 2020 an jede Handlung eines jeden Chinesen in Echtzeit aufzeichnen und die Summe seines wirtschaftlichen, sozialen und moralischen Verhaltens sodann mit Belohnungen und Strafen vergelten soll. Allgegenwärtige Algorithmen schaffen in dieser Vision den ökonomisch produktiven, sozial harmonierenden und politisch gefügigen Untertanen, der sich am Ende stets vorbeugend selbst zensiert und sanktioniert. (Strittmatter 2018, S. 12)

Hier wird also zum einen wieder auf den technischen Kern der Überwachung hingewiesen, zum anderen darauf, dass die Betroffenen im Laufe der Zeit die äußere Überwachung sozusagen zur inneren machen. Der zitierte Gastautor der Republik äußert sich zunächst zu seiner Person. Er verweist darauf, dass er schon seit 20 Jahren in China lebt, also die Verhältnisse genau kennt (Strittmatter begann seine Tätigkeit etwa zur gleichen Zeit, nämlich 1997). Dann äußert er sich zur Sache:

Ja, das künftige Datensystem wird auch der Kontrolle dienen – vor allem aber soll es Vertrauen unter den Chinesinnen schaffen. Es mag auf den ersten Blick an Orwells Big Brother erinnern – vor allem aber steht es in der Tradition von Konfuzius. Es genügt nicht, westliche Werte eins zu eins auf China zu übertragen. Es genügt auch nicht, lineare Projektionen anzustellen: Wenn der Staat jetzt das und das tut, wird sich China so und so verändern. Wer das Land begreifen will, muss es aus seiner Geschichte und seinen Werten verstehen. Muss einen Schritt zurücktreten und das ganze Bild anschauen. (Hernig 2018)

Er gibt zudem eine Einschätzung ab bezüglich der internationalen Medienlandschaft und unterscheidet englischsprachige und deutschsprachige Autoren respektive Medien. Die einen relativierten die apokalyptischen Szenarien mit Vergleichen und Erzählungen über ähnliche Tendenzen, wie sie etwa in den USA anzutreffen seien. In der deutschsprachigen Welt jedoch steche China heraus, als Reich einer bösen Zukunft. Das gilt indes ebenso für einige Zeitungen und Zeitschriften in Großbritannien. So ist die Grundaussage in einem Artikel des Wired bereits in der Überschrift klar und deutlich ersichtlich: „Big data meets Big Brother as China moves to rate its citizens". Zu lesen ist dann:

You could see China's so-called trust plan as Orwell's *1984* meets Pavlov's dogs. Act like a good citizen, be rewarded and be made to think you're having fun. It's worth remembering, however, that personal scoring systems have been present in the west for decades. (Botsman 2017)

Jeremy Daum, Forscher am Paul Tsai China Center an der Yale Law School, erklärte im Mai 2019 auf der re:publica, es gebe vor allem im Westen zahlreiche falsche Informationen über das Vorhaben (vgl. Krempl 2019). Es gehe bislang kaum um KI, sondern vor allem um altbekannte Scoring-Verfahren, wie sie die Finanzwirtschaft für Bonitätsprüfungen verwende. Dazu kämen, so der Jurist, schwarze Listen, die künftig stärker durchgesetzt werden sollten.[3] Im Kern lege es Peking vor allem auf Propaganda an:

---

[3] Vgl. dazu Genzsch: „In vielen Städten und Kommunen existieren eine Reihe von ‚schwarze Listen' [!], die 2014 zum Joint-Punishment-System zusammengeführt wurden." (Genzsch 2019, S. 138)

Den Bürgern solle beigebracht werden, ehrlich zu sein. Die Rede vom „Citizen Score" habe so vor allem erzieherischen Charakter. In Metropolen wie Schanghai werde das Punktesystem offiziell aber gar nicht erwähnt, potenzielle Sanktionen würden nicht durchgesetzt … Die chinesische Regierung nähre auch den Glauben, dass die Bürger auf der Straße ständig per Videoüberwachung beschattet würden. Die meisten Kameras seien aber gar nicht in Betrieb. (Krempl 2019)

Der Technologiekorrespondent der New York Times in Schanghai, Paul Mozur, schlägt in dieselbe Kerbe, freilich, wie Strittmatter, mit einer anderen Geisteshaltung: Es „könne selbst die Annahme, man werde überwacht, genügen, um die Bevölkerung auf Linie zu halten" (Kormann 2019). Nach Daum projiziert der Westen seine eigenen Bedenken und Ängste vor der Technik auf China. Nach ihm ist die oben skizzierte Berichterstattung offensichtlich irreführend oder sogar inhaltlich falsch.

16.2.2 Die Haltung chinesischer Experten und Betroffener

Der Deutschlandfunk zitiert in einem schriftlichen Beitrag einen der Strategen hinter dem Sozialkreditsystem, Professor Zhang Zheng von der Peking University, die auch Beida genannt wird. Der Wirtschaftsprofessor leitet die Forschungsstelle, die sich mit dem Sozialkreditsystem beschäftigt. Für Zhang Zheng sei dieses ein künftiger Grundpfeiler für die moralische Ordnung der chinesischen Gesellschaft:

Wenn ein junger Mensch heiraten möchte und die Eltern sich über den ausgesuchten Partner unklar sind, können sie dessen Punktestand im Sozialkredit-System erfragen. Es gibt Heirats-Vermittlungen, die das bereits nutzen. Welche Informationen über die Bürger gesammelt werden dürfen, das muss die Politik entscheiden. Ob es zum Beispiel erfasst werden soll, ob man regelmäßig mit seinem Hund spazieren geht oder seine Eltern besucht – dazu gibt es bislang keine Vorgaben. (Dorloff 2018)

Dieser Punkt ist interessant, weil damit deutlich wird, wie verschiedene Interessengruppen ihren Vorteil aus dem Sozialkreditsystem schlagen könnten. Der junge Mensch, der heiraten will, wäre offenbar ein Spielball der Interessen. Am Ende hätten die Eltern den Spielball in der Hand, und sie entschieden dann auf der Grundlage des Systems über die Zukunft ihres Kinds. Hier scheinen alte Traditionen mit neuen Optionen verbunden zu werden. Der Wissenschaftler der Beida lobt die erwähnte Testphase im Osten von China:

Die Stadt Rongcheng hat sehr viel ausprobiert. Vieles mit Erfolg. In Rongcheng herrscht eine hervorragende Ordnung. Die Bewohner, das medizinische und wirtschaftliche Umfeld – alles sehr gut. Wir ziehen daraus den Schluss, dass das Sozialkreditsystem gut für die Atmosphäre in Wirtschaft und Gesellschaft ist. (Dorloff 2018)

Damit wird die angebliche Verbesserung der Lebensqualität angesprochen. Der Zweck heiligt die Mittel, wobei die Mittel hier nicht beanstandet werden. Auch wenn vieles noch im Aufbau sei – China habe mit der Sanktionierung der Bürger bereits begonnen, moniert der Deutschlandfunk. Er stellt den Ausführungen von Zhang Zheng das Wort des Schriftstellers Hao Qun, der unter seinem Pseudonym Murong Xuecun bekannt ist, gegenüber. Sein Debütwerk „Leave Me Alone" von 2006, das in China online verbreitet wurde, machte ihn berühmt.

> Die chinesische Regierung will seine 1,4 Milliarden Bürger künftig besser und effizienter kontrollieren. Die Führung in Peking hat verstanden, dass die alten Werkzeuge der Kontrolle nicht mehr greifen: Aufenthalts-Registrierung, Polizei, Personenspitzel. Das reicht nicht im digitalen Zeitalter der sozialen Medien. Um das System der sozialen Kontrolle entsprechend weiter zu entwickeln, schafft der Staat ein Sozialkreditsystem. Es ist Teil einer totalitären Internet-Gesellschaft des 21. Jahrhunderts. (Dorloff 2018)

Damit wird u. a. Jeremy Daum widersprochen, für den keine technische Revolution im Gange ist. Murong Xuecun beanstandet auch den Begriff der harmonischen Gesellschaft und dessen Vereinnahmung durch den Staatspräsidenten:

> China redet dauernd über die harmonische Gesellschaft. Aber die Harmonie, die Präsident Xi Jinping meint, unterscheidet sich von der Harmonie, wie sie Leute wie ich verstehen. Unter Harmonie versteht Präsident Xi eine strenge Ordnung, in der nur eine Stimme zugelassen ist und keine Opposition. Aber eine Gesellschaft, die so strikt von der Regierung kontrolliert wird, kann weder innovativ noch kreativ sein. (Dorloff 2018)

Murong Xuecuns Beispiel zeigt, dass in einem System der Konformität, mit dem die harmonische Gesellschaft enge Beziehungen hat, zeitweilig Stimmen der Individualität ertönen können. Diese sind in China zu hören und von Exilanten wie Ai Weiwei, der durch seine Kunst und Politik in der Volksrepublik zur unerwünschten Person geworden (und dort zum Schweigen gebracht worden) ist. Damit wird auch Hernigs Einteilung in westliche und östliche Werte relativiert. Das Zitat ist aber auch interessant, weil die Innovationskraft einer kontrollierten Gesellschaft hinterfragt wird und damit das Leitbild der wissenschaftlichen Entwicklung.

Die Zustimmung scheint zu einem guten Teil aus dem Volk zu stammen. Zumindest zeichnet der Deutschlandfunk in seinem Bericht dieses Bild. Er hat die 37-jährige Sui Yuxiang in einem Restaurant getroffen (vgl. Dorloff 2018). Sie ist der Meinung, dass die Moral sich durch die soziale Kontrolle und Bewertung verbessert hat.

> Zu meinem Bereich gehören 260 Familien, wir haben insgesamt 560 Einwohner. Einmal im Monat gibt es einen neuen Sozialkredit-Punktestand, zum 25. jeden Monats. Dann sehen wir, wer welche Bewertung bekommt. Wer anderen hilft oder sich engagiert, bekommt Zusatzpunkte, fünf oder zehn. Und je mehr Punkte jemand hat, als desto ehrlicher gilt die Person. Die gesamte Atmosphäre hier ist besser geworden. Die Menschen sind aktiver und engagierter, die Qualität des Zusammenlebens hat sich verbessert. (Dorloff 2018)

Die Idee und Umsetzung von Gamification, von Spielifizierung, wird hier augenfällig (vgl. Marczewski 2017; Bendel 2013a). Verwiesen wird wiederum auf die Verbesserung der Lebensumstände. Der Radiosender hat sich zudem im Dorf Fulushan umgeschaut und -gehört. An einer Hauswand hänge eine Tafel mit Fotos der Menschen mit Vorbildcharakter, also derjenigen Bewohner mit einem ausnehmend hohen Punktestand. Als besonders integer und moralisch einwandfrei gelte Chen Shengzhang, der mit folgenden Worten zitiert wird:

> Wir haben das Sozialkredit-System in unserem Dorf nun schon seit einigen Jahren. Was immer wir auch tun, wir denken dabei an unsere Kreditpunkte. Wir unterstützen das Dorf, wo es geht. Wir machen sehr oft sauber und fegen die öffentlichen Flächen. Müll oder auch nur Gras vor die eigene Tür zu legen – das ist nicht erlaubt. Wenn einer diese Regeln nicht befolgt, gilt er als unehrlich. Wenn der Dorfvorsteher nach etwas verlangt, folgen wir. Wer alles sauber und in Ordnung hält, gilt als Vorbild. (Dorloff 2018)

Einerseits geht es also um den Erwerb von Kreditpunkten, um das Spielen des Spiels – andererseits um das Verhalten selbst. Vorbild ist man offenbar nicht allein wegen eines gefüllten Punktekontos. Es interessiert durchaus das damit verbundene Handeln, zumindest das Ergebnis davon.

16.3   Diskussion des Sozialkreditsystems

Wie deutlich wurde, gibt es ganz unterschiedliche Darstellungen des politischen bzw. gesellschaftlichen Vorhabens und der technischen Möglichkeiten, außerhalb und innerhalb von China. Eine sozialwissenschaftliche und ethische Bewertung sollte ausgehen von tatsächlichen oder mit hoher Wahrscheinlichkeit erwarteten Begebenheiten. Man kann die zitierten Aussagen herbeiziehen und auf ihren sachlichen Kern reduzieren. Offensichtlich will die Kommunistische Partei 1) die Bürgerinnen und Bürger stärker überwachen, auch und gerade mit informationstechnischen Mitteln, will sie 2) ihnen ihre bzw. verbreitete Moralvorstellungen auferlegen, 3) sie für ein bestimmtes Verhalten belohnen oder bestrafen und 4) ein System der Konformität schaffen. Diese Themen werden im Folgenden vorausgesetzt und behandelt.

16.3.1 Überwachung

Kontrolle und Überwachung finden in China – über die erwähnten Schwarzbücher hinaus – seit langem statt. Mit der „Great Firewall" schränkt die Partei den Zugriff auf das Internet bzw. das WWW stark ein (vgl. Griffiths 2019). Das Projekt Goldener Schild, wie es auch genannt wird, wird vom Ministerium der Volksrepublik China für Staatssicherheit verantwortet. Es begann mit ersten Vorbereitungen bereits im Jahre 1998. Die Systeme wurden im November 2003 in Betrieb genommen. Laut Ministerium ist der Zweck, ein

Kommunikationsnetzwerk und ein Informationssystem für die Polizei zu bauen, um deren Leistungsfähigkeit und Effizienz zu verbessern und „Staatssicherheit, Würde, Entwicklung, internationale Sicherheit und soziale Stabilität" (Ackeret 2014) herzustellen. Zugleich geht es aber um Zensur.

Im realen Raum ist Überwachung ebenfalls durch analoge und digitale Mittel omnipräsent. Peking wird sicherlich eine der Hauptstädte in diesem Zusammenhang sein, die Einwohnerschaft ein bevorzugtes Ziel, ob man sich im öffentlichen oder halböffentlichen Bereich aufhält:

> Beobachtet wird man dabei von mehreren Millionen Überwachungskameras. Bis 2020 sollen 626 Millionen Kameras in Peking installiert sein, Gesichtserkennung inklusive. Sie hängen vor Gebäuden, an Kreuzungen und selbst in Parks. Auch in den Gängen unseres Hotels sind alle zehn Meter 360-Grad-Kameras angebracht. (Borchardt 2019)

Damit wird es zunehmend schwieriger, sich Freiräume zu schaffen. Man wird im virtuellen wie im realen Raum ständig beobachtet. Dabei handelt es sich nicht einfach um zusätzliche Anwesende, sondern es besteht ein Ungleichgewicht zwischen Beobachteten und Beobachter. Die einen wissen nicht, wer sie observiert, wann und warum. Der Beobachter bleibt unsichtbar und anonym. Man kann das Ganze als Übergriff interpretieren, zumal die Kameras auch persönliche Details freilegen können, nicht zuletzt zusammen mit Gesichts- und Emotionserkennung. Dies hat potenziell eine sexuelle Dimension. Man weiß von einzelnen Mitarbeitern der Polizei und der Geheimdienste, dass sie ihre Macht in diesem Sinne missbrauchen. Edward Snowden schildert in seinem Buch „Permanent Record", dass ein auf elektronischem Wege gestohlenes Nacktbild von Mädchen oder Frauen unter CIA-Mitarbeitern als informelle Währung galt (vgl. Snowden 2019, S. 354).[4] Wer weiß, wo überall die Kameras angebracht sind und was sie erfassen können? Und wer weiß, wie die Männer (oder Frauen) hinter den Systemen agieren und reagieren? Grundsätzlich entsteht ein digitaler Graben der besonderen Art, zwischen Beobachtern und Beobachteten, zwischen Mächtigen und Ohnmächtigen. Er kann kaum überwunden werden, es sei denn durch Hacker, die in gewisser Weise selbst Mächtige sind und die Mächtigen angreifen, die von Beobachteten selbst zu Beobachtern (auch der Beobachter) werden.

Das Ungleichgewicht ist noch in anderer Hinsicht vorhanden. Es ist unwahrscheinlich, dass der Staatspräsident und führende Funktionäre selbst einer solchen Überwachung und Kontrolle ausgesetzt sind. Sie werden sich eher selten im (halb-)öffentlichen Raum der Kameras bewegen, eher abgeschirmt sein in Hochsicherheitsgebäuden und Spezialfahrzeugen. Vermutlich müssen sie auch keine Punkte erwerben, werden weder belohnt noch

---

[4] Was die technische Reife der Überwachungssysteme im Internet angeht, spricht er übrigens eine deutliche Sprache: „Was ich über die technischen Details der Überwachung privater Kommunikation durch China zu lesen bekam – in einem umfassenden, detaillierten Bericht über die Mechanismen und Geräte, die für die permanente Sammlung, Speicherung und Auswertung der unzähligen täglichen Anrufe und Internetverbindungen von über einer Milliarde Menschen gebraucht wurden –[,] war schlicht unfassbar." (Snowden 2019, S. 218) Und er fügt hinzu, im Vergleich mit den USA: „Immerhin war China ein explizit antidemokratischer Einparteienstaat." (Snowden 2019, S. 218)

bestraft. Sie spielen das Spiel, das sie für ihr Volk erfunden haben, nicht mit. Es würde natürlich wenig Sinn ergeben: Soll der Staatspräsident, wenn er unter einen Punktestand rutscht, dann nicht mehr fliegen oder Kredite aufnehmen dürfen? Sich als Politiker aus dem Recht herausnehmen zu können, ist keine Zierde für einen Rechtsstaat, für den man die USA im Jahre 2020 noch halten könnte, und ein Indiz für einen Unrechtsstaat.

Die technischen Möglichkeiten der Überwachung werden von den zitierten Autoren ganz unterschiedlich beurteilt. Generell engagiert man sich stark im Bereich Big Data, auch mit Bezug zum Sozialkreditsystem (vgl. Chen und Cheung 2017; Shi und Lechner 2019). Es ist zudem kein Geheimnis, dass China massiv in Künstliche Intelligenz investiert (vgl. Lauterbach 2019) und enorme Fortschritte im Bereich der Gesichts- und Objekterkennung stattfinden. Dabei wendet man sich am Rande der Pseudowissenschaft der Physiognomik zu (vgl. Bendel 2018). Xiaolin Wu und Xi Zhang von der Jiao Tong University in Shanghai haben 2016 angeblich einer Software beigebracht, Kriminelle von Nichtkriminellen mit einer Genauigkeit von 89,5 Prozent zu unterscheiden (vgl. Wu und Zhang 2016). Den Forschern zufolge gibt es drei verschiedene Gesichtsmerkmale und -züge, die darauf hindeuten, dass jemand kriminell ist. Das ist nicht nur konzeptionell und technisch fragwürdig. Es werden auch moralische und rechtliche Kategorien vermischt. Im Zuge des Leaks, den die New York Times im November 2019 verwertet hat, wurde nicht zuletzt bekannt, dass Xi Jinping in internen Reden mit Blick auf die Uiguren persönlich „den Aufbau eines lückenlosen Überwachungsapparats" forderte, „der sowohl modernste Technik als auch klassische Spitzelnetzwerke umfassen solle" (Böge 2019).

Grundsätzlich muss man die Beobachtung und Überwachung ohne Grund bzw. ohne Einverständnis in Frage stellen. Womöglich werden grundlegende Persönlichkeitsrechte verletzt und zentrale Menschenrechte missachtet. China ratifizierte die wichtigsten Menschenrechtskonventionen der Vereinten Nationen. Aber Massenüberwachung mit Hilfe von Kameras und Gesichts- und Emotionserkennung sowie weitergehenden Mitteln kann, ausgehend von der Allgemeinen Erklärung der Menschenrechte (vgl. Vereinte Nationen 1948), gegen Artikel 1 (Freiheit, Gleichheit, Brüderlichkeit), Artikel 2 (Verbot der Diskriminierung), Artikel 3 (Recht auf Leben und Freiheit), Artikel 12 (Freiheitssphäre des Einzelnen), Artikel 18 (Gedanken-, Gewissens-, Religionsfreiheit), Artikel 19 (Meinungs- und Informationsfreiheit) und Artikel 22 (Recht auf soziale Sicherheit) verstoßen. Folgeaktivitäten wie Verhaftungen und Festsetzungen sind bezüglich Artikel 7 (Gleichheit vor dem Gesetz), Artikel 8 (Anspruch auf Rechtsschutz), Artikel 9 (Schutz vor Verhaftung und Ausweisung), Artikel 10 (Anspruch auf faires Gerichtsverfahren) und Artikel 11 (Unschuldsvermutung) zu prüfen. Mit den Begriffen der Informationsethik kann gesagt werden, dass die informationelle Autonomie verletzt wird, also die Möglichkeit, selbstständig auf Informationen zuzugreifen, über die Verbreitung von eigenen Äußerungen und Abbildungen selbst zu bestimmen sowie die Daten zur eigenen Person einzusehen und gegebenenfalls anzupassen (vgl. Bendel 2019a). Damit sind eben bestimmte Persönlichkeitsrechte tangiert sowie mehrere der oben aufgeführten Menschenrechte, etwa Artikel 12 und 19. Der juristische Begriff der informationellen Selbstbestimmung kann ebenfalls herangezogen werden.

Unklar ist, ob nur die Bürgerinnen und Bürger der Volksrepublik China oder auch zum Beispiel von Hongkong oder Taiwan vom Sozialkreditsystem abgedeckt werden. Zudem ist die Frage, was man ausländischen Arbeitskräften und Touristen zumutet. Auch sie werden von Überwachungssystemen erfasst und ihre Daten verarbeitet. Aber werden Profile angelegt und gespeichert? Nicht zuletzt ist die Frage, was mit Chinesinnen und Chinesen passiert, die zeitweilig oder dauerhaft im Ausland unterwegs sind. Erfasst das Smartphone bzw. eine App ihre Aktivitäten, und wenn ja, in welcher Weise?

16.3.2 Die Herstellung von Moral

Immer wieder wird, von Gegnern wie Befürwortern des Sozialkreditsystems, betont, dass es sowohl um rechtliche als auch um moralische Aspekte geht. Eine Moral, die verloren gegangen ist, soll wiedergewonnen werden. Es soll eine Moral entstehen, die vorbildlich und verbindlich ist. Ein geläutertes Denken (bzw. das Denken an Sanktionen) soll unmittelbar zu einem entsprechenden Handeln führen, letztlich zu einer harmonischen Gesellschaft, wie sie der Bevölkerung und vor allem dem Staatspräsidenten anscheinend ein Anliegen ist (vgl. Genzsch 2019). Hier sind zahlreiche Widersprüche vorhanden.

Wenn die Moral in China an einem Tiefpunkt ist, muss der Niedergang ja auch in den Jahrzehnten erfolgt sein, in denen die Partei das Sagen hatte. Das erinnert an das Klischee von manchen westlichen Politikern, die einen Zustand beklagen, den sie selbst zu verantworten haben. Sie wenden sich vorwurfsvoll an andere Parteien oder Individuen und stellen keinen Zusammenhang mit ihren eigenen Aktivitäten her. Sie entziehen sich der Verantwortung, indem sie keine wahrnehmen und keine erkennen. Man kann es im Falle der Volksrepublik freilich ganz anders interpretieren, die Kolonialzeit in Erinnerung rufen und die Verheißungen der Moderne und des Westens:

> Neben verheerenden ökologischen Schattenseiten konnten wichtige strukturelle Aspekte des politischen, wirtschaftlichen und sozialen Lebens dem Tempo nicht standhalten. Darüber hinaus brachte der „amerikanische Traum" den Hunger nach Status, schürte Ungleichheit und Rücksichtslosigkeit in der Bevölkerung, mit folgenreichen gesellschaftlichen Fehlentwicklungen. (Genzsch 2019, S. 129)

Auf jeden Fall werden, wie bereits dargestellt, seit 2003/2004 zwei Leitbilder propagiert, nämlich die wissenschaftliche Entwicklung und die harmonische Gesellschaft, um den identifizierten Fehlentwicklungen entgegenzusteuern. Ob das ein Schuldeingeständnis ist, müsste genauer untersucht werden.

Aus ethischer Perspektive kann problematisiert werden, dass Moral wiedergewonnen werden soll, indem man sie extrinsisch vorgibt. Es gibt verschiedene Möglichkeiten, eine Handlung moralisch zu begründen (vgl. Pieper 2007, S. 189 ff.). Man kann sich auf Fakten beziehen, auf Folgen, auf Gefühle, auf sein Gewissen. Man kann sich ebenso auf einen Moralkodex berufen oder auf eine moralische Autorität. Dabei fehlt indes die innere Überzeugung, der innere Antrieb. Ein stures Festhalten an festgeschriebenen Regeln scheint nicht für eine reife Moral zu sprechen, und es ist auch die Frage, ob ein solches Konstrukt nachhaltig ist oder es nicht zwangsläufig in der modernen Welt in sich zusammenfallen muss. Das ist keineswegs sicher. Zumindest sind widersprüchliche Entwicklungen und unterschiedliche Kulturen zu berücksichtigen. Auf jeden Fall kann das System, wie das der Religion, nur unter permanentem Druck funktionieren. Es verträgt sich weder mit Aufklärung noch mit Eigenständigkeit und Erwachsensein.[5]

Dass die Moral von der Regierung bzw. der Partei selbst (nach Vorbildern) entworfen und vorgegeben wird, nicht etwa von einer unabhängigen Kommission oder als Ergebnis einer Abstimmung, ist ein weiterer Problempunkt.[6] Es ist grundsätzlich erstaunlich, wie die Stärkung der Moral so herausgehoben wird, obwohl klar sein dürfte, dass es keine einheitliche Moral gibt, man allenfalls mit Grundzügen argumentieren kann, die in den Menschenrechten ihren Niederschlag gefunden haben, mit denen das Sozialkreditsystem gerade nicht im Einklang steht. Es ist jedoch vor allem erstaunlich, weil die KP nach Meinung vieler Beobachter ein totalitäres Regime ist, das seine tatsächlichen Gegner oder vermeintliche Feinde verfolgt und einsperrt bzw. verschwinden lässt. Die Verantwortlichen würden dies entweder zurückweisen oder womöglich der Meinung sein, dass die Moral der Herrschenden historisch immer wieder die herrschende Moral war und man nicht zimperlich sein durfte, um sie durchzusetzen.[7]

Recht und Moral werden in der Diskussion immer wieder vermischt. Es hat indes gute Gründe, warum man beide zuweilen scharf trennen sollte. Die Moral, das ist ein Setting aus allgemeinen Handlungsregeln und Wertmaßstäben bzw. persönlichen Überzeugungen in Bezug auf das, was gut und böse ist (vgl. Höffe 2013; Bendel 2018). Moralfähigkeit ist, wie Sprachfähigkeit, dem Menschen angeboren. Die Moral selbst wird dann in der Familie, Gruppe und Gesellschaft im sozialen Austausch geschaffen, wie ja auch die Sprache, und schließlich, je nachdem, wie dies zugelassen wird, individuell adaptiert. Das Recht fußt durchaus zum Teil auf der Moral, und entsprechend wird es immer wieder an den gesellschaftlichen Wandel angepasst. Es wäre aber fatal, wenn es sich einfach nach der herrschenden Moral richten würde. Vielmehr sind allgemeine Maßstäbe herauszuarbeiten, nicht zuletzt in der Rechtsethik, die dann herangezogen werden können. Es

---

[5] Immer wieder wird das politische System Chinas als atheistisches dargestellt, das religiöse Bewegungen bekämpft. Allerdings erwächst aus der atheistischen Idee zunächst einmal gar nichts. Es wird lediglich der Glaube an einen Gott abgelehnt bzw. dessen Existenz in Frage gestellt. Es müssen weitere Ideen hinzukommen, um ein Durchgreifen zu legitimieren. Es kann sogar vermutet werden, dass das politische System, um das es hier geht, selbst ein religiöses ist, ohne Anbetung vielleicht, aber mit dem Hang zur Unterwerfung, nicht zuletzt eben unter eine extrinsische, diktierte Moral.

[6] Damit soll nicht gesagt werden, dass eine unabhängige Kommission oder eine Abstimmung eine gültige Moral vorschlagen soll oder erzeugen kann. Es soll vor allem darauf hingewiesen werden, dass die staatliche Gewalt der Urheber ist, in erstaunlicher Eigenermächtigung.

[7] Genzsch, die es verneint, dass China eine Diktatur ist, bemerkt immerhin: „Auch muss gewährleistet sein, dass die Bewertungsparameter, die einen ‚guten Bürger' ausmachen, in verantwortungsvollen Händen liegen, da sonst die Gefahr einer destruktiven, machtzentristischen Manipulation besteht." (Genzsch 2019, S. 139)

darf indes durchaus eine Absicht darin vermutet werden, wenn Recht und Moral in eins gesetzt werden: Recht wird scheinbar fundiert, Moral legitimiert, die Moral der Herrschenden zwangsläufig zur herrschenden Moral.

Das Sozialkreditsystem wird angeblich von der überwiegenden Mehrheit der Bürgerinnen und Bürger der Volksrepublik befürwortet (vgl. Genzsch 2019).[8] Es würde also der allgemeinen Moral entsprechen. Allerdings ist es in einem Rechtsstaat wichtig, wie bereits angedeutet, dass auch Minderheiten geschützt werden, welche Moral diese auch haben mögen. Wie Murong Xuecun betont hat, teilt er keineswegs die Meinung des Regimes, und er dürfte nicht allein sein. Immer wieder wird behauptet, die Chinesen seien halt so und so. Offensichtlich sind sie das nicht, und wenn man noch die Chinesen von Hongkong und Taiwan einbezieht, ergibt sich ein heterogenes Bild. Man fühlt sich erinnert an die Behauptung arabischer Staaten, die Bevölkerung sei muslimisch, allenfalls noch zu einem winzigen Prozentsatz christlich. Dass es auch atheistische Bewegungen gibt, wird in der Regel geleugnet, und wenn nicht, werden sie als von westlichen Akteuren gesteuert angesehen. Jedes Volk ist aber zum Denken und Zweifeln bereit, und seine Angehörigen sind nie vollkommen gleich.

Wie dem auch sei: Ist nicht das Ergebnis erfreulich? Die Sauberkeit, die an manchen Orten nun zu finden ist, die Ruhe und Gelassenheit im Verkehr, die Ehrlichkeit, die sich ihren Weg zu bahnen scheint (vgl. Genzsch 2019, S. 137; Dorloff 2018)? Ein Urteil ist hier ausgesprochen schwierig. Wie oben deutlich wurde, scheint ein gefülltes Punktekonto nicht auszureichen, um als Vorbild zu gelten. Es geht vielleicht nicht um den inneren Antrieb, das veränderte Denken, aber um das sichtbare Handeln und vor allem das Resultat. Und dieses scheint für viele Menschen eine Verbesserung darzustellen. Die äußeren Umstände ändern sich zum Guten, könnte man sagen, ohne dass das Gute dafür verantwortlich wäre. Freilich müsste man sich genauer anschauen, wie diese neuen Lebensumstände im Detail sind. Sind sie die richtigen für alle? Sind sie mehr Schein als Sein? Sind sie dauerhaft? Und wie hoch ist der Preis, um sie zu gewinnen und zu erhalten? Letztlich geht es wohl nicht darum, dass die Einzelnen oder die Gesellschaft gewinnen, sondern die Verantwortlichen der Partei.[9]

---

[8] „Ein Großteil der Bevölkerung (rund 80 Prozent) steht hinter den Plänen der Regierung und befürwortet das SCS. In erster Linie ist diese Reaktion darauf zurückzuführen, dass Chinesen kulturell bedingt ein anderes Verständnis davon haben, wie Gesellschaft funktioniert und welche Rolle Politik dabei einnimmt. Darüber hinaus genießt die Regierung um Xi Jinping derzeit großes Vertrauen in der Gesellschaft, denn sie verdankt ihr den Wohlstand, den die Entwicklungen der vergangenen Jahrzehnte gebracht haben. Zudem erhoffen sich viele Chinesen einen persönlichen Nutzen vom SCS, denn mit dem rasanten Tempo der vergangenen Jahre konnten wichtige strukturelle Aspekte nicht mithalten und so gibt es starke Defizite im Rechtssystem, in finanzwirtschaftlichen und lokalpolitischen Strukturen sowie im gesellschaftlichen Miteinander." (Snowden 2019, S. 131)

[9] In dieser Richtung argumentiert auch Marczewski (2017): „It is worth noting that for some the idea of a social credit system is positive. But the potential outcomes for the user are less than ideal, and arguably the intention of the design is not to benefit the users but to benefit the state. Therefore, it fails our definition of being ethical."

16.3.3 Die Idee von Belohnung und Bestrafung

Die Idee von Belohnung und Bestrafung, um bestimmte Ziele zu erreichen, ist keine chinesische Erfindung. Sie ist zunächst einmal in jeder Erziehung auf der Welt anzutreffen. Eltern schenken oder entziehen Ressourcen und Zuneigung, um ihre Vorstellungen durchzusetzen. Das Kind befreit sich von diesen Zwängen in der Regel einfach, indem es erwachsen wird, seine eigenen Überzeugungen formt und hinaus in die Welt geht (vgl. Mayer und Schulte 2007). Das ist nicht überall einfach, und manche Gesellschaften erwarten, dass man in der Familie verbleibt und sich den Eltern bis zum Ende widmet. Damit bleibt man in gewisser Weise zeitlebens Kind. In anderen Gesellschaften gehört es zum Erwachsensein, dass man sich dem System von Belohnung und Bestrafung innerhalb der Familie verweigert, es allenfalls tauscht gegen eines im Betrieb oder gegenüber Firmen, deren Kunde man ist, oder es gegenüber dem eigenen Nachwuchs (oder dem Partner bzw. der Partnerin) anwendet.

Das System von Belohnung und Bestrafung findet sich natürlich auch seit alters her im Staat. Der Rechtsstaat ist aber gerade dadurch ausgezeichnet, dass er dem Einzelnen Freiheit lässt (vgl. Bielefeldt 2004). In monotheistischen Religionen und autoritären bis totalitären Staatsformen ist dies anders. Dort soll der Einzelne in seinem Handlungsspielraum, ja sogar in seinem Gedankenspielraum, beschränkt werden, soll er sich dem Führer oder der Gruppe unterwerfen und es hinnehmen, dass diese ihn mit Lobpreisungen und Drohungen überhäuft. Da es keine Götter gibt und diese nicht direkt Macht ausüben können, ermächtigen sich die Fürsprecher und Stellvertreter, predigen das Wort der Götter, das zu ihrem eigenen wird, predigen ihr eigenes Wort, das sie für das der Götter ausgeben. Die Verwandtschaft zwischen religiösen und politischen Strukturen totalitärer Ausprägung ist nicht zu übersehen: Es sind wenige, die auserwählt sind, diese lassen niemanden neben sich zu, sie setzen ihre Macht unbedingt ein, und sie benutzen ein System von Belohnung und Strafe, um ihre Macht zu sichern und die Bevölkerung an diese glauben zu lassen.

Gamification ist dabei die moderne Form, Belohnung und Bestrafung umzusetzen. Es handelt sich um die Übertragung von spieltypischen Elementen und Vorgängen in spielfremde Zusammenhänge (vgl. Bendel 2013b). Ziele von Gamification sind Motivationssteigerung und Verhaltensänderung bei Anwenderinnen und Anwendern oder eben Bürgerinnen und Bürgern.[10] Dabei geht es vielfach um Manipulation (vgl. Marczewski 2017). Zu den spieltypischen Elementen gehören Beschreibungen (Ziele, Beteiligte, Regeln, Möglichkeiten), Punkte, Preise und Vergleiche. Zu den spieltypischen Vorgängen zählt die Bewältigung von Aufgaben durch individuelle oder kollaborative Leistungen (vgl. Bendel 2013b). Ein großes Spiel also, das die gesamte Gesellschaft umfasst, das Gemeinschaften in Stadtvierteln und auf dem Dorfe in Bann hält. Dies wird von den erwähnten Bewohnern anschaulich geschildert. Das Ergebnis ist nicht nur, wie suggeriert wird, eine höhere Le-

---

[10] Das Sozialkreditsystem wurde u. a. in Zusammenarbeit mit dem auf Onlinespiele spezialisierten Unternehmen Tencent sowie mit Alibaba entwickelt (vgl. Söffner 2018). Es hat eine größere Nähe zu bestimmten kommerziellen Anwendungen und Lernanwendungen als etwa zum Schufa-System, bei dem nicht die Verhaltensänderung im Zentrum steht.

bensqualität, mehr Sauberkeit, mehr Ehrlichkeit, weniger Korruption, sondern auch die Erlaubnis oder das Verbot, in andere Länder zu reisen oder einen Kredit zu bekommen.[11] Genzsch weist auf die lange Tradition von Bewertungssystemen in China hin:

> Die modernen, auf innovativer Technologie basierenden Bewertungssysteme kennen viele historische Vorbilder, die sich in den verschiedenen Phasen chinesischer Entwicklung in den Dörfern und Gemeinden etabliert haben. Das historische Baojia-System fasste beispielsweise zehn Familien zu einer Bao und mehrere Bao zu einer Jia zusammen. Innerhalb der Baojias wurde gegenseitig für Recht und Ordnung, sowie die Einhaltung moralischer Standards gesorgt. Auch wurde bei Fehlverhalten nicht die Einzelperson, sondern die gesamte Baojia bestraft. (Genzsch 2019, S. 137)

Auch das Sozialkreditsystem soll anscheinend eine Gruppe sanktionieren, etwa indem der Punktestand sinkt, wenn man mit den falschen Personen befreundet ist (vgl. Strittmatter 2018; Botsman 2018; Söffner 2018). Allerdings erhält man wohl auch die Gelegenheit, diesen zu helfen, und kann sich so im Ranking verbessern – und man kann sich verschlechtern, wenn man sie vorschnell im Stich lässt (vgl. Söffner 2018).

16.3.4 Das System der Konformität

Die westliche Welt kennt das System der Konformität durchaus. Die Individualität, die man als ihr Gegenüber ansehen kann, hat sich erst in der neueren Zeit herausgebildet, und dies vor allem in den Städten (vgl. Hahn und Willems 1996). Auf dem Lande ist man nach wie vor zum Teil erheblichem Druck ausgesetzt, wegen der großen Nähe, in der man lebt, aber auch aufgrund von politischen und religiösen Strukturen, die oft ihre Starrheit bewahrt haben. Die Aufklärung hat den Siegeszug der Individualität begünstigt, in unterschiedlichem Tempo und in unterschiedlicher Verbreitung, doch im Prinzip hat sie die relative Freiheit von Konventionen und Sanktionen ermöglicht (vgl. Ricken 2014, S. 559 ff.). Dies gilt eben dort, wo die Aufklärung stattfand – woanders nicht unbedingt.

Ein System der Konformität hat eine gewisse Trägheit. Es ändert sich nicht von einem Tag zum anderen. Es ist die pure Masse an Gleichgesinnten, die es trägt und nährt (vgl. Asch 1951). Das Sozialkreditsystem wird, wie gesagt, angeblich von der überwiegenden Mehrheit der Bürgerinnen und Bürger der Volksrepublik befürwortet. In einem perfekten System der Konformität würde sich dies kaum so schnell ändern. Man würde auch achtgeben, dass man nicht nur selbst, sondern jeder Mensch in der Umgebung ihm entspricht. Konformität wird durch gegenseitige Überwachung und Kontrolle ermöglicht. Ein wesentliches Element des Sozialkreditsystems ist, dass es nicht nur von oben, sondern auch von unten getragen wird. Es kombiniert in verschiedener Weise George Orwell und Aldous Huxley. Gerade in Huxleys „Brave New World" wird die Begeisterung offenbar, die

---

[11] Ein weiteres Problem könnte sein, dass durchaus nicht alle Menschen gerne spielen – obwohl dies gern behauptet wird. Vielmehr empfinden manche Gamification als unpassend oder kindisch (vgl. Bendel 2013a).

der Betroffene entfalten kann. Damit könnte man das Gesagte wie folgt erweitern: Das Sozialkreditsystem wird von oben, von unten und von innen getragen. Und damit ist es das perfekte neue Rückgrat der Konformität.

Das System der Konformität hängt eng mit dem Konzept der harmonischen Gesellschaft zusammen. Dieses knüpft nach Wacker und Kaiser (2008) zum einen an die marxistisch-maoistische Denktradition an, zum anderen an die konfuzianische Philosophie.[12] Die Rehabilitation der Tradition greife dabei populäre Strömungen in der Gesellschaft auf und lasse sie in den Aufbau der harmonischen Gesellschaft einfließen. „Dies dient dem obersten Ziel der Kommunistischen Partei, nämlich dem Erhalt der Stabilität (im Sinne von Regimestabilität)." (Wacker und Kaiser 2008) Populäre Strömungen also, die freilich mit technologischen Neuerungen verknüpft sind. Die Menschen gehen auf der ganzen Welt seit Jahren mit sozialen Medien und mit Suchmaschinen um, sie begeistern sich für innovative Entwicklungen, sie benutzen Computer und Smartphones, und viele von ihnen lieben Spiele. Individuelle Gewohnheiten gehen zusammen mit staatlichen Vorgaben und Umsetzungen, die mit Geschick und Gespür und in Zusammenarbeit mit Profis ausgedacht sind (vgl. Söffner 2018).

Wenn Konformität nicht gänzlich hergestellt werden kann, droht ein Bruch in der Gesellschaft. Diese Gefahr sieht selbst Genzsch:

> Neben der Problematik um den Datenschutz und die Sicherung der Privatsphäre (eine Debatte, die hauptsächlich im westlichen Kulturkreis geführt wird) weisen Kritiker mahnend darauf hin, dass ein Vergehen in einem Bereich zu Konsequenzen in sämtlichen Lebensbereichen des Einzelnen führen kann. Das kann zu Ausgrenzung und damit verbunden zu einem Zweiklassensystem führen, eine den Visionen der kommunistischen Partei gegenläufige Entwicklung. (Genzsch 2019, S. 139)

Das ist in gewisser Weise durchaus richtig. Die KP will eben eine Einklassengesellschaft (einmal abgesehen von ihrer eigenen Klasse), sprich die totale Konformität. Es ist zu befürchten, dass diejenigen, die die Wünsche der Partei nicht erfüllen, in vielfältiger Weise ohnmächtig werden und unsichtbar gemacht werden.

16.4 Zusammenfassung

Das Sozialkreditsystem wird in Medien und Wissenschaft ganz unterschiedlich dargestellt. Es lassen sich jedoch einige Grundzüge erkennen. Die Kommunistische Partei von China will die Bürgerinnen und Bürger stärker überwachen, auch und gerade mit informa-

---

[12] Die konfuzianische Ethik, die keine wissenschaftliche philosophische Ethik ist, geht davon aus, dass die Welt von einer Ordnung regiert wird, die im Kern moralischer Natur ist (vgl. Steininger 1988). Der Mensch als Mitglied der Gesellschaft soll sich in moralischer Hinsicht perfektionieren. Maßstab und Leitbild sind dabei die sogenannten fünf Konstanten.

tionstechnischen Mitteln, ihnen ihre Moralvorstellungen auferlegen, für ein bestimmtes Verhalten belohnen oder bestrafen und ein System der Konformität schaffen. Es konnten Widersprüche aller Art ausgemacht werden, die keineswegs nur im Verhältnis zum Westen entstehen, sondern auch innerhalb der chinesischen Kultur und Gesellschaft.

Die Ethik widmet sich der fragwürdigen Idee einer von oben verordneten und von unten freiwillig oder unfreiwillig sowie unkritisch gestützten Moral von Personen und Einrichtungen, die Wirtschaftsethik, die hier vernachlässigt wurde, der zweifelhaften Rolle der beteiligten Internet- und IT-Firmen. Deren Entwicklungen wendet sich die Informationsethik zu. Sie wurde hier wiederholt einbezogen, wobei sie nach dem digitalen Graben und der informationellen Autonomie und nicht zuletzt nach den Möglichkeiten des Hackens und Manipulierens bzw. Modifizierens fragt.

Die Chancen und Risiken für die Betroffenen zu untersuchen, war das erklärte Ziel dieses Beitrags. Es wurden vor allem Risiken gefunden, was der Verfasser nicht allein mit seinem kulturellen und sprachlichen Hintergrund und seinen Vorbehalten erklären kann. Er gesteht durchaus zu, dass viele Bürgerinnen und Bürger einer Verbesserung der Lebensumstände durch das Sozialkreditsystem applaudieren. Er fragt sich freilich, ob diese Verbesserung nicht anderweitig erzielt werden könnte, und ohne den Begriff der Moral zu strapazieren und zu korrumpieren. Letztlich ist die Frage, was von einer Moral zu halten ist, die von oben vorgegeben und von oben und unten erzwungen ist.

Die Anhänger von monotheistischen Religionen und totalitären Staaten müssen darauf antworten, dass von einer solchen Moral viel zu halten ist, und man kann sie lediglich mit zahlreichen Widersprüchen konfrontieren, die sich auch innerhalb ihrer Systeme ergeben. Sie müssten immerhin zugestehen, dass Angehörige der Systeme unterschiedliche Werte haben können. Natürlich kann man sich dafür entscheiden, diese (abweichende Werte wie bestimmte Angehörige) mit allen Mitteln zu bekämpfen und ihnen keinerlei Möglichkeit der Verbreitung und Entfaltung zu geben. Aber genau dies zeichnet eben fundamentalistische und totalitäre Systeme aus.

Eine Gefahr, die im vorliegenden Beitrag nicht behandelt wurde, besteht im Export von technischen Mitteln. Auch der Westen kennt zahlreiche Rating- und Scoring-Systeme. Kaum eines davon will jedoch Moralvorstellungen grundsätzlich und einheitlich verändern.[13] Und keines davon ist derart an Überwachungssysteme gekoppelt, wie sie in China zur Verfügung stehen und zur Anwendung kommen. Wohl aber wird durch kommerzielle und behördliche Anwendungen der Boden für weitergehende Möglichkeiten bereitet. Hier kommen wieder das System von Belohnung und Bestrafung und der Ansatz der Gamification ins Spiel. Wer diese verinnerlicht und erprobt hat, könnte anfällig sein für das Sozialkreditsystem, über das er oder sie sich weitere Vorteile verschaffen kann. Die Hoffnung ist,

---

[13] In die Kritik kamen das Anytime Feedback Tool, re:Work von Google und Zonar von Zalando, die zur Bewertung im betrieblichen Kontext eingesetzt werden (vgl. Staab und Geschke 2019).

dass Aufklärung immer noch wirkt, wie schon einmal vor Jahrhunderten und während des Aufbruchs der 1960er- und 1970er-Jahre.[14]

Literatur

Ackeret A (2014) Die neue chinesische Mauer. NZZ, 22. November 2014. https://www.nzz.ch/international/asien-und-pazifik/die-neue-chinesische-mauer-1.18429859

Asch S E (1951) Effects of group pressure upon the modification and distortion of judgements. In: Guetzkow, H. Groups, Leadership, and Men. Carnegie Press, Pittsburgh.

Botsman R (2017) Big data meets Big Brother as China moves to rate its citizens. Wired, 21. Oktober 2017. https://www.wired.co.uk/article/chinese-government-social-credit-score-privacy-invasion

Bendel O (2019a) 400 Keywords Informationsethik: Grundwissen aus Computer-, Netz- und Neue-Medien-Ethik sowie Maschinenethik. 2. Aufl. Springer Gabler. Wiesbaden.

Bendel O (2019b) Sozialkreditsystem. Gabler Wirtschaftslexikon. Springer Gabler. Wiesbaden. https://wirtschaftslexikon.gabler.de/definition/sozialkreditsystem-100567

Bendel O (2018) The Uncanny Return of Physiognomy. In: The 2018 AAAI Spring Symposium Series. AAAI Press, Palo Alto. 10–17.

Bendel O (2013a) Schluss mit lustig. Netzwoche, (2013) 16. 40–41.

Bendel O (2013b) Gamification. Gabler Wirtschaftslexikon. Springer Gabler. Wiesbaden. http://wirtschaftslexikon.gabler.de/Definition/gamification.html

Bielefeldt H (2004) Freiheit und Sicherheit im demokratischen Rechtsstaat. https://www.ssoar.info/ssoar/bitstream/handle/document/31609/ssoar-2004-bielefeldt-Freiheit_und_Sicherheit_im_demokratischen.pdf?sequence=1&isAllowed=y&lnkname=ssoar-2004-bielefeldt-Freiheit_und_Sicherheit_im_demokratischen.pdf

Böge F (2019) Die undichte Stelle in Chinas Machtapparat. FAZ, 17. November 2019. https://www.faz.net/aktuell/politik/ausland/umerziehungslager-in-xinjiang-die-undichte-stelle-in-chinas-machtapparat-16489775.html

Borchardt K (2019) Was ist erlaubt und was verboten? Die Räume für chinesische Verlage sind eng, Nischen aber bleiben. NZZ, 15. Juli 2019. https://www.nzz.ch/feuilleton/china-seine-verlage-die-zensur-was-ist-erlaubt-was-verboten-ld.1493420

Botsman R (2017) Big data meets Big Brother as China moves to rate its citizens. Wired, 21. Oktober 2017. https://www.wired.co.uk/article/chinese-government-social-credit-score-privacy-invasion

Chen Y, Cheung A S Y (2017) The Transparent Self Under Big Data Profiling: Privacy and Chinese Legislation on the Social Credit System (June 26, 2017). https://ssrn.com/abstract=2992537

Dorloff A (2018) China auf dem Weg in die IT-Diktatur. Deutschlandfunk, 23. Juni 2018. https://www.deutschlandfunk.de/sozialkredit-system-china-auf-dem-weg-in-die-it-diktatur.724.de.html?dram:article_id=421115

Genzsch M (2019) Harmonie durch Kontrolle? Chinas Sozialkreditsystem. In: Loitsch T (Hrsg.) China im Blickpunkt des 21. Jahrhunderts. Springer Gabler. Berlin, Heidelberg. 129–142.

---

[14]Erinnert sei an die Aussage von Kant: „Aufklärung ist der Ausgang des Menschen aus seiner selbst verschuldeten Unmündigkeit. Unmündigkeit ist das Unvermögen, sich seines Verstandes ohne Leitung eines anderen zu bedienen. Selbstverschuldet ist diese Unmündigkeit, wenn die Ursache derselben nicht am Mangel des Verstandes, sondern der Entschließung und des Muthes liegt, sich seiner ohne Leitung eines anderen zu bedienen. Sapere aude! Habe Muth, dich deines eigenen Verstandes zu bedienen! ist also der Wahlspruch der Aufklärung." (Kant 1784)

Griffiths J (2019) The Great Firewall of China: How to Build and Control an Alternative Version of the Internet. Zed Books. London.

Hahn A, Willems H (1996) Wurzeln moderner Subjektivität und Individualität. Aufklärung, vol. 9, no. 2, 1996, 7–37. JSTOR, www.jstor.org/stable/24361335

Hernig M (2018) Errichtet China eine Big-Data-Diktatur? Nein. Republik, 4. Oktober 2018. https://www.republik.ch/2018/10/04/errichtet-china-die-erste-big-data-diktatur-des-21-jahrhunderts-nein

Höffe O (2013) Ethik: Eine Einführung. C. H. Beck. München.

Kant I (1784) Beantwortung der Frage: Was ist Aufklärung? Berlinische Monatsschrift 4 (1784), 481–494.

Kormann J (2019) Chinas beunruhigendes Zukunftsmodell: Wenn Technologien zu Feinden der Demokratie werden. NZZ, 6. August 2019. https://www.nzz.ch/meinung/chinas-zukunft-technologien-werden-zum-feind-der-demokratie-ld.1499377

Krempl S (2019) re:publica: US-Forscher hält Chinas Social-Credit-System für Propaganda. Heise News, 7. Mai 2019. https://www.heise.de/newsticker/meldung/re-publica-US-Forscher-haelt-Chinas-Social-Credit-System-fuer-Propaganda-4415221.html

Lauterbach A (2019) Trojanische Verhältnisse? Der Wettbewerb um die Marktdominanz Künstlicher Intelligenz ist noch längst nicht entschieden In: Loitsch T. (eds) China im Blickpunkt des 21. Jahrhunderts. Springer Gabler, Berlin, Heidelberg. 1–17.

Lee F (2017) Die AAA-Bürger. Zeit Online, 30. November 2017. https://www.zeit.de/digital/datenschutz/2017-11/china-social-credit-system-buergerbewertung/komplettansicht

Marczewski A (2017) The Ethics of Gamification. XRDS, Fall 2017, Vol. 24, No. 1. https://www.researchgate.net/profile/Andrzej_Marczewski/publication/319855897_The_ethics_of_gamification/links/5bc26314458515a7a9e72be1/The-ethics-of-gamification.pdf

Mayer S, Schulte D (2007) Die Zukunft der Familie. Wilhelm Fink Verlag. München.

Pieper A (2007) Einführung in die Ethik. 6., überarb. u. akt. Auflage. A. Francke. Tübingen.

Ricken N (2014) Individualität. In: Wulf C, Zirfas J (Hrsg.) Handbuch Pädagogische Anthropologie. Springer VS. Wiesbaden. 559–566.

Shi C M, Lechner M (2019) Big-Data in China: Ein Überblick. Industrie Management, Nr. 2, 2019, 61–65.

Snowden E (2019) Permanent Record. S. Fischer. Frankfurt am Main.

Söffner J (2018) Bewerten wir uns zu Tode? Die totalitären Züge des digitalen Kapitalismus. NZZ, 29. September 2018. https://www.nzz.ch/meinung/die-totalitaeren-zuege-des-digitalen-kapitalismus-ld.1418802

Staab P, Geschke S-C (2019) Ratings als arbeitspolitisches Konfliktfeld: Das Beispiel Zalando. Study 429, September 2019. https://www.boeckler.de/pdf/p_study_hbs_429.pdf

Steininger H (1988) Das fernöstliche Bildungsverständnis und sein Verfall in der Neuzeit. In: Böhm W, Lindauer M (Hrsg.) „Nicht Vielwissen sättigt die Seele": Wissen, Erkennen, Bildung, Ausbildung heute. 3. Symposium der Universität Würzburg. Ernst Klett. 107–128.

Strittmatter K (2018) Die Neuerfindung der Diktatur: Wie China den digitalen Überwachungsstaat aufbaut und uns damit herausfordert. Piper. München.

Vereinte Nationen (1948) Resolution der Generalversammlung 217 A (III). Allgemeine Erklärung der Menschenrechte. https://www.un.org/depts/german/menschenrechte/aemr.pdf

Wacker G, Kaiser M (2008) Nachhaltigkeit auf chinesische Art: Das Konzept der „harmonischen Gesellschaft". SWP-Studie. https://www.swp-berlin.org/fileadmin/contents/products/studien/2008_S18_wkr_ks.pdf

Wu X, Zhang X (2016) Automated Inference on Criminality Using Face Images. arXiv, November 13, 2016. https://arxiv.org/abs/1611.04135v1

Prof. Dr. Oliver Bendel hat an der Universität Konstanz sowohl Philosophie und Germanistik als auch Informationswissenschaft studiert und an der Universität St. Gallen in der Wirtschaftsinformatik promoviert. Er lebt und arbeitet in der Schweiz. Seine Forschungsschwerpunkte sind Informationsethik und Maschinenethik. Seit 1998 sind ca. 400 Fachpublikationen entstanden, darunter verschiedene Bücher und Buchbeiträge sowie zahlreiche Artikel in Praktiker- und Fachzeitschriften.

Social Scoring als Mensch-System-Interaktion

17

Ulrich Hoffrage und Julian N. Marewski

17.1 Einleitung und Überblick

Derzeit wird in China das sogenannte *Social Credit Rating* vorbereitet. Dieses Verfahren sammelt diverse Daten über die Bürger dort und errechnet daraus dann eine Zahl. Dafür bedarf es vielerlei Absprachen, Entscheidungen und Normierungen, aber auch praktisch-technischer Dinge, wie Infrastruktur für Informationssammlung, -speicherung, und -integration. Beim Social Credit Rating handelt es sich in der Tat um ein komplexes System, bei dem viele Institutionen, Personen und Technik zusammenwirken müssen. Dies soll in diesem Kapitel künftig als ‚kleines' System bezeichnet werden.

Auch der gesellschaftliche Rahmen, in den dieses Verfahren eingebettet ist, lässt sich als System bezeichnen. In China ist ein solches Ratingverfahren von der Regierung initiiert, um Bürger zu beurteilen. In der westlichen Hemisphäre hingegen werden derartige Systeme derzeit vornehmlich von Firmen implementiert, um Eigenschaften oder Verhalten von Kunden oder Konsumenten vorherzusagen. Dieser Unterschied lässt erahnen, dass ein Ratingverfahren nicht isoliert zu sehen ist von dem übergeordneten, gesellschaftlichen ‚großen' System, in dem es eingesetzt wird. Insbesondere die Ausgestaltung und Nutzung, aber auch die Legitimation des Ratingverfahrens sowie die Durchsetzung der Konsequenzen, die sich für den Einzelnen aus einem bestimmten Rating ergeben, dürften wohl maßgeblich von den übergeordneten politischen und wirtschaftlichen Ideologien, Strukturen und Machtverhältnissen abhängen.

Das vorliegende Kapitel möchte einen Beitrag zu der Frage leisten, wie Menschen überhaupt dazu kommen, derartige Systeme zu schaffen (und damit sind jetzt sowohl die ‚großen' als auch die ‚kleinen' gemeint) und wie diese Systeme auf den Menschen

U. Hoffrage (✉) · J. N. Marewski
University of Lausanne, Lausanne, Schweiz
E-Mail: ulrich.hoffrage@unil.ch; julian.marewski@unil.ch

© Springer Fachmedien Wiesbaden GmbH, ein Teil von Springer Nature 2020
O. Everling (Hrsg.), *Social Credit Rating*,
https://doi.org/10.1007/978-3-658-29653-7_17

zurückwirken. Es geht also weder um Details noch um den gegenwärtigen Stand der Entwicklung in China. Auch stellt sich uns das Social Credit Rating in China nur als eines von vielen Beispielen dar, Menschen in irgendeiner Form zu ‚vermessen'. Entsprechend wird in diesem Kapitel generell von *Social Scoring* die Rede sein. Unter Social Scoring verstehen wir ein Verfahren, welches darauf abzielt, bestimmte Aspekte eines einzelnen Menschen, die sich jeweils aus der Interaktion mit seinem sozialen Umfeld ergeben, auf eine einzige Zahl (eben den *Social Score*) abzubilden (und in der Tat wird im Zusammenhang des chinesischen *Social Credit* Rating auch vom Social Score oder vom *Citizen Score* gesprochen; zum Beispiel Helbing et al. 2019). Schwingt im Wort *Rating* noch etwas von einer subjektiven Einschätzung bzw. einem Konstrukt mit, so verschiebt sich mit der Verwendung des Wortes *Scoring* die Bedeutung subtil in Richtung einer objektiven Größe, denn der aus dem Scoring resultierende Score ist ein Maß, in der Regel eine Anzahl von Punkten (im Englischen bezeichnet Score auch die Anzahl der erzielten Tore oder Punkte bei Ballspielen).

Der Mensch schafft diese ‚kleinen' Systeme, gestaltet und betreibt sie, liefert Daten (wobei die meisten wohl maschinengeneriert sein dürften), und ist obendrein der Träger der Merkmale und Variablen, die in einen Score einfließen. Man kann also sagen, der Mensch ist zunächst Schöpfer diverser Scoringsysteme, dann interagiert er mit ihnen, und schlussendlich ist er das zu vermessende Objekt. Und genau dieser Messvorgang verändert ihn: Ein Mensch mag sich, wenn er gemessen wird und das weiß, anders verhalten, als wenn er nicht gemessen wird oder wenn er es nicht weiß (vgl. auch *Hawthorne-Effekt*). Hinzukommt, dass ein Score Dinge ermöglicht, die vorher nicht möglich waren. So gestaltet der Mensch die Systeme, aber er wird auch von ihnen verändert.

Diese wechselseitige Beeinflussung ist ein zentraler Gegenstand des vorliegenden Kapitels und kann durch folgendes Bild veranschaulicht werden: Einerseits bestimmt das Flussbett den Weg, den das Wasser zum Meer nehmen wird; andererseits ist es aber auch das fließende Wasser, das das Flussbett ausformt. Wie das Wasser des Flusses durch das Flussbett, so geht auch ein einzelner Mensch durch (‚große') Systeme hindurch, die vieles in seinem Leben bestimmen. Aber umgekehrt formen die vielen Zeitgenossen auch das System, in dem sie dahinschwimmen, und wirken auf dieses ein – in der Regel durch viele kleine Änderungen, aber mitunter auch durch Ausgestaltung von abrupten Durchbrüchen und Revolutionen, zu denen es kommt, wenn zuvor eine natürliche Entwicklung gestaut wurde (sofern in diesen Zeiten des Tumults überhaupt noch von Gestaltung gesprochen werden kann).

Dieses Bild von Fluss und Flussbett kann für vieles, was im hier vorliegenden Kapitel gesagt wird, hilfreich sein. Doch wir gehen noch einen Schritt darüber hinaus und versuchen, im Abschnitt 17.2, aufzuzeigen, wie ein Blick auf die Konstitution und die Bedürfnisse des Menschen helfen kann, die ‚großen' Systeme, die für unsere Gegenwart in Betracht kommen, zu verstehen als Ausstülpung von etwas, das im Menschen gefunden werden kann – wie also, um im Bilde zu bleiben, das Wasser sein Bett hervorbringt und formt. Abschnitt 17.3 geht der Frage nach, wie diese Systeme dann wiederum auf den Menschen zurückwirken, also wie das Bett des Flusses das Fließen des Wassers bedingt.

Doch nicht nur Mensch und ‚großes' System gehen wechselseitig auseinander hervor und beeinflussen sich, sondern sie haben in dem Verbund, in dem sie wirken, auch einen Einfluss auf ein Drittes. Gemäß der Ausrichtung dieses Buches betrachten wir hier vornehmlich Social Scoring. Gefragt werden soll, im Abschnitt 17.4, wie Social Scoring (als ‚kleines' System) seine Ausgestaltung, seine Nutzung und seine Legitimation durch die Interaktion des Menschen mit dem überordneten ‚großen' gesellschaftlichen System erhält. Konkret: Welche Typen von Dimensionen gibt es, die durch ein Social Scoring auf einen Social Scores abgebildet werden könnten und inwieweit lassen sich die Unterschiede zwischen diesen Typen durch den jeweiligen gesellschaftlichen, politischen und wirtschaftlichen Rahmen und die darin jeweils wirkenden Ideen verstehen? Im Abschnitt 17.5 soll wieder die Perspektive umgedreht und gefragt werden, welchen Einfluss die Verwendung dieser Dimensionen im Rahmen von (‚kleinen') Scoringsystemen auf die Menschen ausübt, die damit vermessen und beurteilt werden. Abschnitt 17.6 wendet den Blick von der hier dargelegten (und offensichtlich stark europazentrierten) Sichtweise auf die gegenwärtige Entwicklung in China, aber fragt auch, wie wir uns hier dazu positionieren könnten.

17.2 Der Mensch als Ursprung von Systemen

Inwiefern urständet Social Scoring als ein ‚kleines' System, aber auch Kapitalismus oder Kommunismus als Beispiele für ‚große' Systeme, im Menschen? Welche Bedürfnisse, Ideen und Ideale standen bzw. stehen an der Wiege solcher Systeme? Zur Beantwortung dieser Frage erscheint es sinnvoll, einen Blick auf die menschliche Konstitution zu werfen.

17.2.1 Der Mensch in der Subjekt-Objekt-Spaltung

Seit Beginn des 15. Jahrhunderts ist die Bewusstseinsverfassung der Menschen des europäischen Raumes zunehmend durch eine Subjekt-Objekt-Spaltung charakterisiert (der Text hierfür ist weitgehend unverändert entnommen aus Hoffrage 2019, S. 24). Ich als Subjekt stehe im Mittelpunkt der Welt (meiner Welt) und bin umgeben von Objekten, die mir gegenüberstehen und fremd erscheinen (*subiectum* ist das Unterworfene, das Zugrundeliegende; *obiectum* hingegen das Entgegengeworfene). Beide Pole dieser Spaltung gehören zusammen. Der eine ist ohne den anderen nicht denkbar. Dennoch wurden sie auseinandergerissen und wurden so jeweils zum Dreh- und Angelpunkt vieler philosophischer Strömungen wie zum Beispiel des subjektorientierten Idealismus oder Rationalismus einerseits und des objektorientierten Empirismus oder Materialismus andererseits.

Im Menschen wirken diese Pole auch als Triebe. In seinen Briefen über die ästhetische Erziehung des Menschen nennt Friedrich Schiller (1793/1794, insbesondere 12.–15.

Brief) sie den „auf Behauptung der Persönlichkeit" drängenden *Formtrieb* und den „auf die Realität des Daseins" drängenden *Stofftrieb*. Sie seien hier der Bemächtigungstrieb und der Unterwerfungstrieb genannt. Als Subjekt stehe ich im Zentrum, alles dreht sich um mich, ich muss und will mich behaupten. Ich bin mir selbst der Nächste und bemächtige mich der Dinge um mich herum. Aber angesichts einer als objektiv wahrgenommenen Außenwelt voller Sachzwänge und Konkurrenten ist dies manchmal gar nicht so leicht. Man ist ihr oft mehr oder weniger machtlos ausgeliefert, kommt sich vor wie ein Nichts oder ein Rädchen in einer großen Maschinerie und muss sich unterwerfen.

Der eine Zeitgenosse neigt mehr zu Selbstbehauptung und Kampf, ein anderer mehr zu Wegducken und Anpassung an die Verhältnisse. Trotzdem wirken in jedem Menschen immer beide Triebe – einfach, weil er ab dem dritten Lebensjahr, sobald sich das Kind als Ich zu erleben beginnt, durch die Subjekt-Objekt-Spaltung konstituiert ist. Man kann ihr Zusammenspiel sehen in dem bekannten Bild des Radfahrers, der nach unten tritt und nach oben buckelt, und der so als Chimäre zwischen Egoist und Duckmäuser durch die Welt radelt.

17.2.2 Der Mensch als soziales Wesen

Nun gibt es in der Umwelt aber eine ganz besondere Sorte von Objekten, nämlich solche, die zu sich ebenfalls „ich" sagen. Wir fühlen uns mit ihnen verbunden, schon allein deshalb, weil wir davon ausgehen, dass sie sich ebenfalls als Subjekt erleben und eine Erste-Person-Perspektive haben. Aber genau dieses Ich-Sagen trennt uns auch von ihnen – denn Ich-Sagen kann ein jeder nur für sich selber – und ist die Grundlage für unterschiedliche Standpunkte und für Interessenkonflikte.

Wie nun wird der Mensch unterschiedlich durch den Objektpol affiziert, je nachdem, ob er einem Stück Natur oder einem anderen Menschen gegenübersteht – alles jeweils betrachtet unter der Annahme, dass er sich als Subjekt ausschaltet und sich vollständig vom Objekt bestimmen lässt? Reine Hingabe oder auch Unterwerfung an ein Naturobjekt kann ein sich-Hingeben in der reinen Wahrnehmung eines Steines bedeuten, wie es in manchen Meditationsformen geübt wird, es kann die liebevolle Pflege eines Gartens bedeuten, es kann aber auch ein sich-Anpassen an Naturgewalten bedeuten, die einen auszulöschen vermögen. Reine Hingabe an oder auch Unterwerfung unter einen anderen Menschen kann reichen von selbstloser Liebe oder aufopfernder Pflege eines Menschen bis hin zu blindem Gehorsam, sei es aus Angst, Verblendung oder glühender Verehrung.

Was aber ergibt sich, wenn der Mensch angesichts eines Anderen weder ganz im Subjektpol verbleibt, noch sich ganz in den Objektpol hineinstellt? Wenn er die Position der Mitte einnimmt, wenn er sich dem Anderen prinzipiell gleichgestellt sieht? Dann ergibt sich eine Balance zwischen Subjektpol und Objektpol, ein Austausch, ein Geben und Nehmen. Was in einer Dyade oder kleineren Gruppe noch recht überschaubar ist, wird sehr schnell recht komplex, wenn man eine (größere) Gruppe ins Auge fasst. Wir lernen, zu koordinieren sowie uns zu spezialisieren, Arbeitsteilung einzuführen und Handel zu betreiben. Dieses Geben und Nehmen hat zwei Seiten. Vom Subjektpol aus gesehen nehmen wir: Als Gegenleistung für unsere Arbeit erhalten wir Lohn, Waren oder Dienstleistungen. Zum Objektpol hin geben wir: Durch unsere Arbeit befriedigen wir die Bedürfnisse der Anderen. Was wir tun, ist also letztlich durch diese Anderen bestimmt. Wenn jeder sich spezialisiert und den Anderen genau das gibt, was diese brauchen (aber nicht gut für sich selber erbringen könnten, weil sie sich ja anderweitig spezialisiert haben), ist füreinander gesorgt.

Dieses Vermittelnde, dieses Geben und Nehmen, das Halten der Balance fängt bereits in frühen Kindertagen in der Familie an. In der Regel achten Geschwister peinlichst darauf, dass auch keines der anderen bevorteilt wird und alle gleich viel abbekommen vom Nachtisch. Jeder Mensch hat ein Bedürfnis, gerecht behandelt zu werden, und das heißt zunächst als Geschwister vor den Eltern, als Schüler vor dem Lehrer und letztlich als Bürger vor dem Gesetz gleich dazustehen und nicht aufgrund irgendwelcher Merkmale oder Eigenschaften benachteiligt zu werden. In der Tat kann das Bedürfnis nach Gerechtigkeit oft mit einer Gleichbehandlung befriedigt werden.

17.2.3 Freiheit – Gleichheit – Brüderlichkeit

Bisher haben wir Bedürfnisse betrachtet, die – indem sie elementar wirken – auch als Triebe bezeichnet werden können. Werden diese Bedürfnisse, bzw. Triebe nicht nur gefühlt und empfunden, sondern gedacht, so treten sie auf der Bühne des menschlichen Bewusstseins als Idee auf. Angesichts des Ursprunges nimmt es dann nicht Wunder, wenn der Mensch darin etwas Natürliches und Erstrebenswertes erblickt. So werden ihm diese Ideen zu Idealen. Dabei ergibt die Zusammenschau der drei oben besprochenen Bedürfnisse – dem nach Entfaltung (Subjektpol), nach Gerechtigkeit und Gleichbehandlung (Mitte), und nach Unterordnung, Einordnung, Hingabe und Füreinander-Sorgen (Objektpol) – einen vielleicht nicht ganz uninteressanten historischen Bezug. In der Tat sind die Ideale, die aus diesen Bedürfnissen hervorgehen, genau die der Französischen Revolution: Freiheit, Gleichheit, Brüderlichkeit.

Ein jedes dieser Ideale hat gemäß der hier eingenommenen Betrachtungsweise also ein Woher (und, wie wir unten noch sehen werden, ein Wohin). Damit erscheint ein jedes Ideal dieses Trios als eine Stufe auf einem Weg: vom Bedürfnis zur Idee, zum Ideal und schließlich hin zur Ideologie und zum System. Ferner lässt sich für jeden dieser Wege ein wirksames Prinzip postulieren, das von innen nach außen drängt. Das lateinische *principio* könnte man mit Urbeginn übersetzen, und diese Urbeginne haben wir, wie oben ausgeführt, im Menschen bzw. in seiner Konstitution gesehen. So wie das Wasser sich sein Flussbett schafft, so schafft der Mensch sich seine Systeme und formt sie.

Wir werden unten noch einen Schritt weitergehen und den Systemen, beziehungsweise den ihnen jeweils korrespondierenden Prinzipien, Orte zuweisen: West, Mitte und Ost. Bekanntermaßen haben sich zu verschiedenen Zeiten und an verschiedenen Orten verschiedene gesellschaftliche Systeme herausgebildet. So war noch vor wenigen Jahrzehnten die Welt, stark vereinfacht gesagt, in zwei Blöcke eingeteilt: Westlich und östlich des sogenannten Eisernen Vorhangs gab es die beiden großen Systeme des Kapitalismus und Kommunismus. Sie unterschieden sich grundstürzend hinsichtlich ihrer Antworten auf die gleichen Fragen, nämlich wie Ich und Welt, wie der Einzelne und die Gemeinschaft in eine Balance zu bringen sind und wie ihr Verhältnis geregelt sein soll. Die Unterschiedlichkeit zwischen Systemen rührt unseres Erachtens daher, dass bei der Ausstülpung der Prinzipien von innen nach außen und bei der Ausformung der Systeme jeweils verschiedene Ideale dominierten.

Selbstverständlich sind wir uns der Tatsache bewusst, dass konkrete Systeme in einem bestimmten Raum zu einer bestimmten Zeit viele Facetten und Varianten haben. Dies ist nicht weiter verwunderlich, zumal nicht ein Prinzip ausschließlich alleine wirkt, sondern sie immer zusammenspielen, vielleicht vergleichbar einem Orchester. Wenn wir im Folgenden zum Beispiel von „dem Kapitalismus" oder „dem Osten" sprechen, so meinen wir damit nicht eine konkrete historische Ausgestaltung, die ja immer von einem Orchester hervorgebracht ist, sondern wollen damit auf das in all diesen Ausgestaltungen wirksame Urbild in Reinform deuten, also gewissermaßen den Beitrag einzelner Instrumente. Dadurch wollen wir dem Leser einen für wissenschaftliche Schriften wohl eher ungewöhnlichen Zugang zu Systemen ermöglichen. Wir glauben, dass unser Blick auf die menschliche Konstitution und das Bild vom Wasser und vom Flussbett hilfreich sein kann, ein Verständnis der Französischen Revolution, des 20. Jahrhunderts, aber auch der Gegenwart zu bekommen, das nicht im Widerspruch zum chronologischen Zugang eines Historikers steht, wohl aber dessen Analysen sinnvoll ergänzen kann. In diesem Sinne empfehlen wir, von uns etwaig gemachte einzelne Aussagen zu historischen Details *cum grano salis* zu nehmen und all dessen unbeschadet die Gestalt beziehungsweise das Ganze zu suchen, das mehr ist als die Summe der Teile. Beginnen wir, in diesem Sinne, nunmehr wieder bei den Polen und stellen die Mitte ans Ende der Betrachtung.

17.2.4 Liberalismus und Kapitalismus

Der nach Freiheit strebende Einzelne merkt bald, dass er mit seinem Freiheitsstreben mit anderen in Konflikt gerät. In milden Fällen erscheinen die lieben Mitmenschen lediglich als Verkehrshindernisse auf dem Weg der eigenen Selbstverwirklichung, aber das kann auch eskalieren und dann werden diese Hindernisse potenziell als bedrohlich für die eigene Freiheit, wenn nicht gar als Unterdrücker oder als Feinde gesehen. Wird dieses Streben des Menschen nach Freiheit, Wachstum und Entfaltung erst gedacht und dann in ein System gebracht, so springt dabei der Liberalismus heraus. Der Liberalismus ist eine Position in der politischen Philosophie, die aber durchaus sehr praktische Konsequenzen hatte. Liberale Bürgerbewegungen versuchten im Zeitalter des Absolutismus ihren Königen und Fürsten einige Rechte abzutrotzen.

Der Liberalismus geht eine natürliche Allianz mit dem Kapitalismus ein. Das lateinische *caput* ist der Kopf (Plural: *capita*), der oft auch als Sitz des Denkens und der

Ideen gesehen wird. Und wenn Karl Marx sein Hauptwerk *Das Kapital* nennt, so meint er damit nicht das Münzgeld im Portemonnaie, sondern große Summen von Geld, die in die Umsetzung von Ideen investiert werden. So braucht es Kapital, um Fabriken zu bauen, in denen wiederum Erfindungen – Produkte des menschlichen Geistes – hergenommen werden, um Produkte des täglichen Lebens herzustellen. Aber auch Hollywood oder Walt Disney sind letztlich Fabriken – Traumfabriken – in denen Romane und Drehbücher hergenommen werden, um Filme und andere Unterhaltungsgüter zu produzieren. Der Kapitalismus ist ein System, das darauf hin organisiert ist, Menschen, die Ideen haben und diese als Unternehmer wirtschaftlich umsetzen wollen, Geld zu verschaffen, wobei die Beschaffer dieses Geldes am Unternehmergewinn mitprofitieren. Dabei kam den Banken zunehmend eine Schlüsselposition zu, sie wurden zu eigenständigen Spielern im Wirtschaftsleben, es entstand der Begriff der Finanzindustrie (als ob dort etwas produziert würde), und in der Tat wurde Geld zu einer Ware, die man ‚machen‘ kann. (Nota bene hat das englische „to make money" eine Doppelbödigkeit: Die meisten meinen damit wohl „Geld verdienen", was letztlich auf ein Verschieben hinausläuft, aber es kann auch verstanden werden als Geldschöpfung von sogenanntem *Fiat money* aus dem Nichts – durch profitorientierte Banken und mittels Bilanzverlängerung; Creutz 2012.)

Im Neoliberalismus[1] schließlich wurde die Freiheit auch zum dominierenden Ideal der Akteure des Wirtschaftslebens (inklusive Banken): Diese sollten möglichst ohne Grenzen nach maximalem Profit streben dürfen. Dabei gibt es zwar Verlierer, aber die von Adam Smith formulierte unsichtbare Hand des Marktes sorgt für maximale Effizienz und die Verlierer können sich ja neue Betätigungsfelder suchen, auf denen es mehr zu gewinnen gibt. Diese Argumentation mit der Effizienz des Marktes wurde auch zur Legitimierung dieser ökonomischen Schule herangezogen. So wie vor einigen hundert Jahren der Liberalismus die Köpfe von (überwiegend englischen) Philosophen verließ und politische Bewegungen in ganz Europa befeuerte, wurden in den 1970ern die Ideen, die jeweils im Liberalismus und mit Adam Smith ihren Ausgang nahmen, an den Schreibtischen von Ökonomen der sogenannten Chicagoer Schule (mit Milton Friedman als deren vielleicht bekanntestem Vertreter) zusammengeführt und anschließend unter den Regierungen Ronald Reagan und Margot Thatcher in den 1980ern politisches Programm. Im Zuge der Deregulierungen, die mit diesem Programm einherging, wurde auch die Leine, an denen Banken zuvor noch hingen, länger und sie durften zunehmend die Finanzmärkte in ein globales Kasino verwandeln, um dort Profite zu machen (vgl. Gonin et al. 2012; Niederhausen 2011).

---

[1] Der Begriff „Neoliberalismus" kann wohl als „essentially contested concept" betrachtet werden, das obendrein auch noch von verschiedenen ökonomischen Denkschulen und zu verschiedenen Zeiten unterschiedlich verwendet wurde. Wir meinen hier nicht den Ordoliberalismus der Freiburger Schule um Walter Eucken, sondern die Gruppe der angelsächsisch geprägten Varianten.

17.2.5 Kommunismus und Sozialismus

Der wohl bedeutendste Gegenspieler des Kapitalismus ist der Kommunismus. Das Ideal, das diesem System zugrunde liegt, ist heute bereits ein gutes Stück weit verwirklicht: In der Tat kommen in unserer modernen, arbeitsteiligen Welt die Erzeugnisse unserer Arbeit letztlich anderen zugute. Dies wird offenbar, wenn man ausschließlich die Warenströme und die Erbringung von Dienstleistungen betrachtet. Sieht man nur diese, so muss man sagen, dass wir alle, weltweit, zu einer Gemeinschaft gehören, in der alle füreinander sorgen. Dies macht auch verständlich, warum die Vordenker dieses Systems eine internationale Ausrichtung hatten und diese sich auch in der Propaganda niedergeschlagen hat („*Völker, hört die Signale! Auf zum letzten Gefecht! Die Internationale erkämpft das Menschenrecht*").

Nimmt man die Geldströme hinzu und betrachtet das Wirtschaftsleben durch die Brille der angelsächsischen Philosophen und Denker, ergibt sich ein anderes Bild. Danach arbeitet ein jeder nämlich nur für sich bzw. für Lohn. In der sogenannten freien Marktwirtschaft versuchen nun diese rationalen Maximierer des eigenen Nutzens (*homo oeconomicus*) ständig, einen möglichst großen Teil vom Kuchen, den wohlgemerkt andere gebacken haben, zu ergattern: Die unsichtbare Hand des Marktes mit Preisen als Steuerinstrumente treibt sie an genau die Plätze, an denen sie am meisten Geld für ihre Leistungen erhalten – Geld, von dem sie sich anschließend eben jene Kuchenstücke kaufen können. So wird genau das produziert, was gebraucht wird, und sollten irgendwo Bedürfnisse ungestillt bleiben, werden dort wiederum die Preise steigen, was wiederum mehr Menschen dazu antreibt, diesen Mangel zu beheben und für genau diese Nischen zu produzieren. Es lässt sich unschwer erkennen, dass (zumindest dem Anspruch nach) der Mechanismus freier Märkte letztlich darauf hinzielt, das Ideal der Brüderlichkeit (heute würde man wohl besser Solidarität sagen), bzw. das des Kommunismus zu verwirklichen: Menschen, die sich frei entfalten und ihre Ideen umsetzen können (Liberalismus und Kapitalismus), sorgen maximal effizient füreinander (Kommunismus und Sozialismus).

17.2.6 Recht und Demokratie

Der Liberalismus hat noch einen anderen Gegenspieler, und der besteht in der Einsicht, dass grenzenloses Freiheitsstreben und Selbstentfaltung in Anarchie und Chaos und letztlich in den Kampf Aller gegen Alle führt – so bezeichnet der englische Staatsphilosoph Thomas Hobbes in seinem Hauptwerk *Leviathan* das Untergangsszenario, welches durch den Liberalismus heraufbeschworen wird. Und um genau das zu verhindern, bedarf es eines starken Staats, der für Ordnung, Recht und Schranken sorgt. Darin wirken sich die Ideale Gleichheit, Gerechtigkeit und Fairness aus. Diese Ideale sind es auch, die sich in Demokratie und Parlamentarismus geltend machen – bis hinein in eine schwerfällige Bürokratie, in der jedes auch noch so kleine Detail nach einem langwierigen Kräfteringen diverser Interessengruppen (zumindest dem Anspruch nach fair und manchmal vielleicht auch zur Zufriedenheit aller) festgelegt wurde.

Sagt der Westen, salopp gesprochen „Soll doch jeder nach seiner eigenen Façon glücklich werden und dabei auch die Ellbogen einsetzen dürfen", und sagt der Osten „Jeder hat seinen Dienst an der Gemeinschaft zu erbringen", so sagt die Mitte „Wir müssen alle irgendwie miteinander klarkommen und aufeinander Rücksicht nehmen – und dabei müssen hemmungslose Egoisten eingeschränkt und faule Drückeberger eingespannt werden". Entsprechend sehen wir im kapitalistischen Westen eine tendenziell neodarwinistisch angehauchte Marktwirtschaft, die als Plattform für freie Unternehmen und freie Konsumenten dienen soll, und im sozialistischen Osten eine Planwirtschaft, die meint, freien Individuen misstrauen und stattdessen alles überwachen und kontrollieren zu müssen. Und in der europäischen Mitte sehen wir, zumindest dem Anspruch nach, eine soziale Marktwirtschaft, die mit dem Markt zum Westen und mit dem Sozialen zum Osten hinschielt.

Das Janusgesicht der Mitte soll noch durch ein weiteres Beispiel illustriert werden. Es ist interessant zu sehen, wie die Ideale der Mitte, Gleichheit und Gerechtigkeit, nach Westen und Osten hin unterschiedlich akzentuiert werden. Im Westen wird Chancengerechtigkeit hochgehalten: Jeder soll die gleiche Chance auf Wohlstand haben und vom Tellerwäscher zum Millionär aufsteigen können – hier schimmert das Ideal der Freiheit durch. Im (ehemaligen) Kommunismus lag der Akzent eher auf der Einordnung in die Gemeinschaft: Jeder sollte seinen Anteil zum Gemeinwohl liefern, die Schere zwischen Arm und Reich sollte nicht allzu sehr auseinanderklaffen, und die Uniformen der Werktätigen unter Mao hatten alle die gleiche Farbe.

17.3 Systeme als Rahmen für den Menschen

Wie ist es nun umgekehrt: Wie wirken diese Systeme auf den Menschen zurück?

17.3.1 Der Mensch im Kapitalismus

Der Kapitalismus gründet auf und appelliert an den Subjektpol und das Freiheitsstreben. Genau das spiegelt sich auch im amerikanischen Traum wieder. Der verspricht, dass jeder Fleißige und Rechtschaffene es zu Glück und Wohlstand bringen kann. Dieses Glücksversprechen breitete sich über die ganze Welt aus, und ging schließlich im Wettstreit der Systeme nach dem Fall des Eisernen Vorhangs als der vermeintliche Sieger hervor, sodass Fukuyama (2006/1992) meinte, das „Ende der Geschichte" ausrufen zu können. Seither wurde von vielen Kanzeln der Wirtschaftsuniversitäten weltweit das Streben nach Profit als normal und legitim erklärt und hat das Denken von Abermillionen geprägt. Mensch und System: Im Fall von Selbstentfaltung als menschlichem Bedürfnis und Kapitalismus als ‚großem' System, so scheint es, haben sich hier zwei gesucht und gefunden, die zusammengehören wie Huhn und Ei, bzw. wie fließendes Wasser und Flussbett.

17.3.2 Der Mensch im Kommunismus

Beim Kommunismus sieht das anders aus. Genaugenommen ist das Bedürfnis nach Geborgenheit im Subjektpol verankert, aber um es zu befriedigen, braucht es eben die Anderen, die etwas von sich hergeben. Bei der Frage, weshalb sie das tun, scheiden sich die
Geister. Gemäß Adam Smith tun sie dies aus egoistischem Kalkül, aber dies entspricht
nicht der kommunistischen Idee. Auch wenn es etwas antiquiert oder religiös klingt, aber
letztlich ist, wie oben bereits gesagt, das dem Kommunismus zugrunde liegende Ideal die
Selbstlosigkeit bzw. die Liebe (Niederhausen 2011). Das Problem: Die Liebe schenkt aus
freien Stücken, aber was in der Familie funktioniert, lässt sich offenbar nicht ohne weiteres (und auch nicht durch Propaganda) zu dem Klebstoff erheben, der eine ganze Gesellschaft zusammenhält. Eine Lösung: Zwang und Kontrolle. So führte von den philosophischen Texten Marx und Engels, über die glühenden Reden Lenins und Trotzkis, ein fast
schon zwangsläufiger Weg zu den Umerziehungslagern und Terrorregimen Stalins und
Maos. Im Fall des Kommunismus drängt sich der Eindruck auf, dass hier Mensch und
System anscheinend eben doch nicht so zusammengepasst haben, wie dies beim Kapitalismus der Fall ist. Zumindest deutet der Zusammenbruch vieler kommunistischer Staaten
in den 1990ern darauf hin (wobei für den Kapitalismus das letzte Wort noch nicht gesprochen scheint, und in der Tat gemahnen *Club of Rome* und *Fridays for Future*, dass auch
dieses Modell nicht nachhaltig und damit nicht überlebensfähig ist).

17.3.3 Der Mensch als Bürger

Die Wiegen von Demokratie und Bürgerrechten werden oft in der Antike verortet, beispielsweise in Athen oder Rom (unbeschadet der Tatsache, dass zum Beispiel auf der
Agora in Athen weder Frauen noch Sklaven wählen durften). Nach dem Untergang der
antiken und spätantiken Staatengebilde knüpften die großen Reiche Europas jedoch nicht
an diese Traditionen an, sondern hatten an ihrer Spitze Könige und Kaiser. Doch deren
Legitimation war schon nicht mehr die aus früheren Zeiten: Wurden die Pharaonen im
alten Ägypten noch als gottgleich angesehen, so wirkten die Könige des europäischen
Mittelalters nur noch von „Gottes Gnaden". Und spätestens mit der Renaissance und dem
Humanismus schien auch der göttliche Atem, der jene Herrscher angehaucht hatte, nicht
mehr tragfähig zu sein. Im Jahre 1587 bestieg mit Maria Stuart erstmals eine gesalbte
Königin das Schafott, und ihre Halbschwester Elisabeth begründete, zusammen mit dem
House of Lords, einen Zwitter aus Monarchie und Parlamentarismus – und markierte damit den Beginn der angelsächsischen Weltherrschaft. Im Jahre 1793 fielen im Zuge der
Französischen Revolution auch in Paris zwei gekrönte Häupter, und 1918 wurde die russische Zarenfamilie ermordet und es dankte der deutsche Kaiser ab. In all diesen Fällen
folgten Zeiten des Tumults, die dann jeweils von einem starken Mann bzw. diktatorischem
Herrschaftssystem abgelöst wurden (so beendete zum Beispiel Napoleon die Schreckensherrschaft der Guillotine, zu der es zuvor gekommen war, nachdem sich die „Göttin Ver-

nunft" nicht als regierungsfähig erwiesen hatte; die Kämpfe zwischen Roter und Weißer Armee mündeten in die Diktatur Stalins; und auf die Instabilität der Weimarer Republik folgte Hitler). Heute sind in Europa starke Führer diskreditiert: überall gibt es Demokratien, Gewaltenteilung, Parteien, Kommissionen und eine EU-Bürokratie, die es unter anderem als ihre Aufgabe ansah, die maximal erlaubte Krümmung von Gurken festzulegen (Enzensberger 2011). Hier scheint das dominante Ideal das der Gleichheit zu sein, in der jeder sprechen darf und jede Stimme gleich viel zählt.

17.3.4 Perversionen und Schatten

Die Ideale der Französischen Revolution klingen groß und hehr. Doch löst der Mensch die Ideale aus sich und aus dem Verbund, in dem sie stehen, heraus und macht sie zum Leitstern für Systeme, so können diese leicht eine Art Eigenleben entfalten, das dem Menschen entgleitet. Dazu Rudolf Steiner (1918): „Man muss sich der Idee erlebend gegenüberstellen können; *sonst* gerät man unter ihre Knechtschaft" (S. 282). Ausgehend von Ideen und Idealen gelangt man oft unversehens zu Ideologien, Systemen und -ismen, denen offenbar die Tendenz innewohnt, Grenzen und Knechte zu schaffen. Was als Auseinandersetzung Andersdenkender beginnt, endet oft, und sei es auch erst nach Jahrzehnten oder Jahrhunderten, als Krieg zwischen Armeen. In der Tat können wir in vielen Kriegen, insbesondere auch in vielen des 20. Jahrhundert, die Fratzen der Ideale der Französischen Revolution erkennen (vgl. Swassjan 2018). In diesen Fratzen zeigt sich, wie ein System, das aus dem Menschen herausgesetzt wurde, auf die Menschen zurückwirkt und sie in den Griff nimmt.

Der Westen ist getragen von einer Begeisterung für den amerikanischen Traum, aber nicht wenige (zum Beispiel Perkins 2005; Roberts 2015; Wenders 1986/1968) beschreiben auch seinen Schatten und das Umschlagen in einen Albtraum. Einige der hässlichen Fratzen, die einem hier (ähnlich wie bei einem Karnevalsumzug) entgegenspringen und erschrecken lassen, sind: Neodarwinismus, Raubtierkapitalismus, Raubbau an der Erde um den American Way of Life mit seinem riesigen ökologischen Footprint durchzusetzen, Umweltskandale mitverschuldet durch Herabsetzung oder Umgehung kostspieliger Sicherheitsstandards, und ein Auseinanderklaffen der Gesellschaft in Arm und Reich mit einer hohen Obdachlosigkeit auf der einen und einer *Greed-is-good* Mentalität auf der anderen Seite.

In der Mitte, insbesondere in Deutschland, sahen wir die Perversion der Gleichheit. Ihre hässliche Fratze hieß Gleichschaltung und praktiziert wurde sie von den Nationalsozialisten. Pluralismus und ein freier Wettstreit von Ideen war nicht erlaubt. Überhaupt scheint die Ausschaltung des Individuums ein Wesensmerkmal faschistischer Systeme zu sein – was vielleicht variiert, das ist, zugunsten wessen dies geschieht. In diesem Fall: „Du bist nichts, Dein Volk ist alles."

Im Osten sahen wir die Gulags und die Umerziehungslager Stalins und Maos. Auch hier galt es, das Individuum auszuschalten, es also gewissermaßen zwangsweise ganz in den Objektpol zu stellen – diesmal allerdings nicht zugunsten nationaler Dominanz, sondern, wie oben bereits gesagt, mit einer internationalen Ausrichtung. Hier geht es also nicht um nationalen Stolz, sondern um Klassenkampf, und da ist man gemeinsam stärker: „Werktätige aller Länder vereinigt euch!" Zwang und Terror waren aus Sicht des Regimes nötig, weil die Einsicht in die Richtigkeit dieses Systems bei vielen noch nicht hinreichend weit gereift war. Der Verrat des praktizierten Bolschewismus am Ideal des Kommunismus?

Diese Perversionen sind extrem und jedes der Systeme hat wohl Millionen von Toten produziert – nicht nur auf den Schlachtfeldern. Es ist tragisch und frappierend zu bemerken, wie gutgemeinte Intenonen umschlagen können. In den Worten Friedrich Hölderlins: „Immerhin hat das den Staat zur Hölle gemacht, dass ihn der Mensch zu seinem Himmel machen wollte." Und vielleicht noch als Anregung zum Nachdenken, wenn man nicht gleich auf die Extreme schauen will: Wie sehen die Schäden aus, die von gemäßigten Varianten dieser Einseigkeiten, die diese Systeme nun einmal darstellen, hervorgerufen werden?

Nachdem bislang einige ‚große' Systeme besprochen wurden (ihr Hervorgehen aus dem Menschen und ihre Rückwirkung auf den Menschen), soll die Betrachtungsweise nunmehr auf ‚kleine' Systeme ausgeweitet werden.

17.4 Vom Homo Socialis zum Social Scoring

In unserer sozialen Welt werden tagtäglich Menschen von Menschen beurteilt. Mehr noch, es gibt eine Fülle von mehr oder weniger automatischen Beurteilungssystemen. Diskutieren wir auch hier wieder zunächst, woher derartige ‚kleine' Systeme stammen, das heißt, inwieweit sie im Menschen urständen. Anschließend soll betrachtet werden, wie sie sich innerhalb der jeweils ‚großen' Systeme ausgestalten und wie in diese Ausgestaltungen die unterschiedlichen Bedürfnisse und Ideale hineinwirken.

17.4.1 Vom Leben in der sozialen Welt

Wir leben in einer sozialen Welt, und die hat einen Namen: Tina – There is no alternative. In der Tat ist ein Leben außerhalb dieser sozialen Welt praktisch unmöglich: Wer könnte als Eremit in einer Höhle ganz allein überleben? Und selbst wenn, wer will das schon? Die Abhängigkeit von Anderen beginnt denkbar früh: Im Gegensatz zu vielen Tieren ist ein menschliches Neugeborenes ohne Fürsorge lange Zeit lebensunfähig. In dieser ersten Zeit müssen einige Mitmenschen, in der Regel die Eltern, sehr viel leisten. Umgekehrt begegnet der Säugling den Menschen (und) seiner Umgebung mit Vertrauen, Offenheit und Neugierde. Ungefähr im dritten Lebensjahr erwacht das Kleinkind dann schließlich zu einem Selbstbewusstsein und sieht sich dieser Welt mit ihren Mitmenschen und anderen Objekten als Subjekt gegenübergestellt. Irgendwann folgen Enttäuschungen: das Kind wird mit Lüge konfrontiert, es wird ermahnt, nicht mit Fremden mitzugehen, und es stellen sich Fragen, von wem es was zu halten hat. Es erfährt, dass es dabei zu Fehleinschätzungen kommen kann und dass es in der sozialen Welt die vielfältigsten Unsicherheiten gibt. Es lernt den Wert von Informationen, die andere über dritte weitergeben, zu schätzen. Dies kann Unsicherheit reduzieren. Aber es kann auch neue schaffen, denn wer sagt, dass man diesen Anderen mehr vertrauen kann als jenen, vor denen sie warnen?

Das Kind erfährt, dass es von außen gesehen wird und sich andere auch ein Bild von ihm machen. Und es lernt, dass es wichtig und vorteilhaft ist, in einem guten Licht dazustehen. Es hört Sprichwörter wie *„Wer einmal lügt, dem glaubt man nicht – und wenn er auch die Wahrheit spricht"*. Kurzum, der Begriff der Reputation wird erworben und mit Erfahrung gefüllt. Eine gute Reputation erweist sich als vorteilhaft für ein Leben in der Gruppe, während eine schlechte Reputation schädlich ist und tunlichst vermieden werden sollte. Andererseits kann man unter Umständen aber auch die Erfahrung machen, dass man sich Vorteile erschleichen kann, wenn man Gebote übertritt, zum Beispiel indem man lügt oder stiehlt. Zugleich weiß man, dass dergleichen unvereinbar ist mit einer guten Reputation. So wird man mit komplexen Abwägungen konfrontiert. Wiegt der kurzfristige Vorteil einer Lüge den langfristigen Nachteil einer schlechten Reputation auf? Wie spielen hier Vergessen und Verzeihen seitens der Mitmenschen hinein? Wie ist es um das eigenen Gewissen bestellt?

Doch nicht nur andere denken über uns nach – und wir sind uns dessen wohl bewusst – sondern auch wir machen uns ein Bild von Anderen. Ungeachtet der Tatsache, dass wir dabei tausenderlei Irrtümern ausgesetzt sind, verwenden wir viel Zeit und Aufmerksamkeit darauf, Informationen über andere zu verarbeiten, zu hinterfragen, zu bewerten und zu Urteilen zu gelangen. Was wir über andere wissen und denken, hilft uns bei der Frage, wie wir uns ihnen gegenüber verhalten sollen. Und oft sind wir da auch recht mitteilsam. Dieser sogenannte Klatsch hat eine wichtige Funktion im sozialen Gefüge. Dabei wird sowohl Positives als auch Negatives weitergetragen – einige Mitmenschen werden auf ein Podest gehoben, andere an den Pranger gestellt. So spricht sich einerseits soziales Verhalten herum, was wiederum für viele ein positives Beispiel setzt und auch dem nützt, der es gezeigt hat, denn eine gute Reputation kann unter Umständen sehr viel wert sein. Aber auch unsoziales Verhalten macht sehr schnell die Runde. Das wird dann in der Regel nicht der Situation, sondern der Person zugeschrieben (*fundamentaler Attributionsfehler*). In Folge werden diese sogenannten Egoisten dann oft stigmatisiert, und Vorteile, die sich aus der Kooperation ergeben, werden ihnen vorenthalten. Oft werden sie erst gar nicht als Kooperationspartner akzeptiert. Mindestens ebenso wichtig wie diese Sanktionsfunktion des Klatsches ist die Abschreckungsfunktion: Allein das Wissen, dass sich allzu egoistisches Verhalten herumsprechen wird, wird manche davon abhalten, es zu zeigen. Es ist unschwer zu erkennen, dass das Social Scoring, psychologisch gesehen, hier seinen Ursprung hat. Der Klatsch der Waschweiber am Fluss vor tausend Jahren, das Getuschel in weltweit abertausenden Betriebskantinen und Pausenhöfen heutzutage („Hast Du schon gehört, dass Peter und Sandra …") und das Social Scoring im digitalen Zeitalter befriedigen das gleiche Bedürfnis und haben die gleiche Funktion: Teilhabe des Einzelnen am sozialen Leben, Dazugehörigkeit, Integration, aber auch Schutz der Gemeinschaft vor Ausbeutung durch sogenannte Egoisten.

17.4.2 Das Messproblem

Wie kommt es nun von sozialen Urteilen zu Social Scores? Neben den zahlreichen Beurteilungen, die Menschen über andere Menschen abgeben, haben sich nämlich auch Beurteilungssysteme herausgebildet. Bei diesem Übergang vom beurteilenden Mensch zum System findet sich interessanterweise die oben bereits angesprochene Dreiheit wieder. Der Ausgangspunkt bzw. das Urphänomen: Jemand urteilt über jemanden. (1) Das urteilende Subjekt bemächtigt sich eines Objekts, eines Mitmenschen, und hat dabei alle Freiheiten, bei dessen Beurteilung zu verfahren, wie es will. Nicht von ungefähr hört man in diesem Kontext oft, dass persönliche Beurteilungen oft willkürlich seien. (2) Idealerweise sollte sich das beurteilende Subjekt dabei jedoch völlig auslöschen und in der Beurteilung sollte ausschließlich das Objekt zum Vorschein kommen, so wie es „an sich" ist, also ohne von einem Subjekt gesehen zu werden. (3) Das geht aber nicht, denn jede Beurteilung stützt sich auf Beobachtungen, und schon deren Auswahl ist subjektiv. In der Regel sind auch die Interpretationen der Beobachtungen subjektiv. In dieser aussichtslos erscheinenden Situation hat man sich dann damit beholfen, sich in einem sozialen Prozess auf ein Messverfahren (oder, wenn es komplexer wird, auf ein bestimmtes System) zu einigen. Man hat also gewissermaßen die Willkür eines Subjekts durch den Konsens zwischen vielen Subjekten aufgefangen. Und diese vielen Subjekte, wie auch die, die mit einem solchen Messfahren bzw. System zu Beurteilungen gelangen, sollten im Prinzip alle zum gleichen Ergebnis gelangen. Diese Ergebnisse werden dann als intersubjektiv gültig betrachtet. So wie vor dem Gesetz alle Bürger gleich sind, so sind vor dem Messverfahren bzw. dem ‚kleinen' System alle zu beurteilenden Objekte gleich (was natürlich nicht heißt, dass am Ende alle die gleiche Beurteilung bzw. den gleichen Score bekommen). So lassen sich also Messverfahren (und Social Scoring Systeme sind hier explizit miteingeschlossen) beschreiben als sozial akzeptable Kompromisse, die der Mensch bei seinen Versuchen, objektiv zu sein, eingegangen ist (vgl. auch Hoffrage und Marewski 2015, Abschn. 2.1: „The culture of objectivity and the fight against subjectivity").

Welche Ausgestaltung können Social Scoring Systeme nun annehmen? Wir sagten bereits, dass derartige Systeme bestimmte Aspekte eines Menschen, die sich jeweils aus der Interaktion mit seinem sozialen Umfeld ergeben, auf eine einzige Zahl abbilden. Diese Aspekte dienen somit als Inputvariablen für einen wie auch immer gearteten Integrationsmechanismus. Dabei müssen diese Inputvariablen nicht selbst Scores sein, können es aber. Sicherlich gibt es eine Vielzahl an Möglichkeiten, diese Inputvariablen zu kategorisieren. Die Typologie, die wir hier vorschlagen wollen, ergibt sich aus dem oben Entwickelten.

17.4.3 Input vom Subjektpol: Wettbewerb und Selbstbespiegelung

Menschen suchen Anerkennung und Bestätigung. Dabei stellen sie gerne zur Schau, was sie so „drauf" haben: Tabellenplätze und Pokale bei Fußballvereinen, Medaillen bei Olympiamannschaften, Elozahlen bei Schachspielern – die Liste ließe sich leicht fortsetzen. Im

Grunde genommen sind all dies Zahlen, die reflektieren, wie jemand im Vergleich zu einer bestimmten Referenzgruppe, die sein soziales Umfeld bildet, dasteht. Konkreter: wie oft und gegen wen er in welchen Wettbewerben gewonnen hat.

Hier ist eine ähnliche Liste: Zitationen von wissenschaftlichen Werken, Oscars in der Medienbranche, Likes und Dislikes von auf YouTube hochgeladenen Liedern, Followers, Downloads, und Views, um nur einige zu nennen. Auch dies sind Zahlen. Sie erfassen, wie das, was jemand geleistet hat, bei anderen ankommt – sie können als Ergebnis eines Abstimmungsprozesses angesehen werden. Konkreter: als Aggregationen subjektiver Einzelbeurteilungen in einem standardisierten Rahmen.

Beide Gruppen von Zahlen können dem Subjektpol zugeordnet werden. Sie erfassen, wie weit es jemand gebracht hat, der in der (vermeintlich?) freien Entfaltung seiner Persönlichkeit womöglich sogar das Äußerste aus sich herausholen wollte. Manchmal ist oberstes Ziel, besser zu sein als andere, zum Beispiel bei sportlichen Wettbewerben – wobei es in der Regel egal ist, wie lang jemand braucht, um im Kreis herumzufahren, solange er dafür nur weniger Zeit braucht als andere. Und manchmal will man es einfach nur gut machen, zum Beispiel beim Schreiben eines Buches oder Drehen eines Films – und dabei dann Ranglisten und Preise für viele (wenngleich auch nicht für alle) eher im Hintergrund stehen.

Viele dieser Maße, die dem Subjektpol zugeordnet werden können, befriedigen nicht nur die Egos, die letztlich hier vermessen werden, sondern haben durchaus auch geldwerte Vorteile, sprich Werbeeinnahmen. Es gibt sogar Scores, die, sofern sie hinreichend hoch sind, in den Augen vieler heutzutage fast schon für einen neuen Beruf qualifizieren, nämlich den des „influencers". Bezeichnenderweise sind es bei diesem Typus von Score Individuen, die mit ihrem Score, so sie damit gut dastehen, für sich werben – und große Anstrengungen unternehmen, um einen noch besseren zu erhalten.

17.4.4 Input aus der Mitte: Regelkonformität und Linientreue

Ein anderer Typus von Inputvariablen segelt unter der Flagge von Thomas Hobbes' *Leviathan*. Ein Beispiel wäre der Punktestand eines deutschen Bürgers beim Kraftfahrtbundesamt in Flensburg. Ein weiteres Beispiel wäre das polizeiliche Führungszeugnis. Kann man den Punktestand in Flensburg direkt als Score bezeichnen, so enthält das Führungszeugnis entweder keine Einträge oder spezifischen Text, aber eben nichts, was einem Score gleichkäme. Doch wir erinnern daran, dass wir hier keine Typen von Scores besprechen, sondern Typen von Inputvariablen und es wäre natürlich ein leichtes, die im polizeilichen Führungszeugnis verfügbaren Kategorien durch Gewichtungen und Additionen auf einen Score abzubilden.

Dergleichen Variablen erfassen, inwieweit sich jemand an Recht und Ordnung hält, und bezeichnenderweise sind es hier auch in der Regel nicht Privatpersonen oder Firmen, die diese ‚kleinen' Systeme betreiben, sondern staatliche Kontrollorgane. Vor dem Gesetz sind alle gleich, und um Chaos zu vermeiden, muss sich jeder an die Regeln halten, die in einer Gemeinschaft gelten. Im Gegensatz zum erstgenannten Typus von Inputvariablen (vgl. Abschn. 17.4.3) bleibt man hier am liebsten unauffällig und hat keine Einträge.

17.4.5 Input vom Objektpol: Hingabe und Gemeingüter

Dieser dritte Typus von Variablen quantifiziert, wie viel jemand für Andere leistet. Hier steht der Mensch ganz im Objektpol und unterwirft sich, idealerweise, ganz diesen Anderen bzw. gibt sich ihnen hin. Die hier zu nennenden Ideale wären die Liebe, die Hingabe und die Brüderlichkeit bzw. Solidarität. Und damit liegt das Problem auch schon auf dem Tisch: Lässt sich Liebe quantifizieren und messen? Handwerker schreiben Rechnungen aus, aber keine Mutter legt ihrem Säugling eine Rechnung fürs Stillen vor. Und wenn Eheleute beginnen, eine Buchhaltung darüber zu führen, wer wie viel für den anderen geleistet hat, so ahnt man schon, wo das endet. Andererseits kann man aber auch sagen: Je weiter man von engsten persönlichen Beziehungen zu größeren Kollektiven fortschreitet, desto legitimer werden externe Buchhaltung und Sanktionsinstrumente – zumindest aus der Perspektive des Westens. Dabei wird dann zumeist auch irgendwann einmal genau festgelegt, welchen Beitrag jeder zum Gemeinwohl zu leisten hat. Vielleicht könnte man diese Beiträge „erzwungene Liebestaten" nennen. Beispiele im Kleinen wären: Putzen des Treppenhauses in einem Mehrfamilienhaus bzw. Entrichtung eines Entgelts an die Hausverwaltung, die dann jemanden dafür einstellt. Beispiele im Großen wären: Staatsabgaben in Form von Steuern oder, in vielen Ländern, Militärdienst bzw. soziales Jahr. Und nachdem dergleichen Beiträge genau festgelegt sind, oft durch das Gleichheitsprinzip, fällt jedwedes Maß, das man aus der Erbringung dieser Beiträge ermitteln könnte, eigentlich auch schon in die zweite Kategorie, also in die im letzten Abschnitt besprochene Regelkonformität. Doch es gibt auch Beiträge zum Allgemeinwohl, die in die Freiwilligkeit gestellt sind, etwa Spenden an gemeinnützige Organisationen oder ehrenamtliche Mitarbeit in entsprechenden Einrichtungen. Diese könnten dann dem dritten Typus in Reinform zugeschlagen werden.

Elinor Ostrom (1990), die für ihre Forschung zur Verwaltung von Gemeingütern mit dem Nobelpreis für Ökonomie ausgezeichnet wurde, zeigt, wie schwer es ist, diese Güter zu schützen, denn hier liegt ein Dilemma vor. Egoisten haben (per definitionem) ein Interesse daran, sich so viel wie möglich aus dem gemeinsamen Topf zu holen und so wenig wie möglich in dessen Pflege zu investieren. Aber wenn alle Egoisten wären und alle genau das machen würden, wäre der Topf bald leer. Beispiele reichen vom Klopapier einer Wohngemeinschaft zu Fischbeständen in internationalen Gewässern. Das Ergebnis ihrer Forschung: In kleinen Gemeinschaften, wo jeder jeden kennt, funktioniert die Pflege von Gemeingütern sehr gut, aber auch hier haben sich strenge, oft auch ungeschriebene Regeln, herausgebildet – mit abgestuften und zum Teil drakonischen Strafen für egoistisches Verhalten. Bei größeren Kollektiven werden zumeist Details verhandelt und verschriftlicht, aber auch hier braucht es die gleichen Prinzipien: zum Beispiel Fairness, Transparenz, Regeln und Sanktionen bei Regelverstößen (für verwandte Laborexperimente zur

Verhaltensökonomie, vgl. zum Beispiel Fehr und Gächter 2000). Und genau für die Pflege und Verwaltung von Gemeingütern können Social Scores ein schrecklich wichtiges Überwachungs-, Steuer- und Gleichschaltungsinstrument auf dem Wege in neue Diktaturen werden.

Um in diesem Sinne einen kleinen, spekulativen Blick in die Zukunft zu riskieren: Im Zentrum von Saci Lloyds (2008) viel beachteten Science-fiction Roman *The Carbon Diaries* steht die Jugendliche Laura. So wie jeder Bürger dieser (aus Sicht des Erscheinungsjahres) zukünftigen Welt hat auch sie ein CO_2-Kontingent und jede ihrer Aktivitäten wird daran bemessen, wie viel CO_2 dafür ausgestoßen wird. Damit wird erfasst, wie stark ein bestimmtes Privileg (zum Beispiel 100 km mit dem Auto zu fahren) die Allgemeinheit belastet. So ergibt sich, in der Aggregation über verschiedene Aktivitäten, Lauras individueller ökologischer Fußabdruck. Und der hat eine soziale Dimension. Sie liegt darin, dass wir alle auf dem gleichen begrenzten Planeten wohnen. Dort verringern die Privilegien, die der eine für sich in Anspruch nimmt, die Ressourcen, die anderen noch zur Verfügung stehen. Verteilungsprobleme sind soziale Probleme und insofern könnte man leicht auf die Idee kommen, den individuellen ökologischen Fußabdruck in einen Social Score einfließen zu lassen.

Dicke und Helbing (2017) spielen in ihrem Science-fiction Roman *iGod* einmal durch, wie eine Welt aussehen könnte, in der die Kontingente, die dabei jedem Bürger zustehen, von seinem *Citizen Score* abhängen. Wer eine Straftat begeht, dem wird weniger Wasser zugeteilt, was in den *Smart Homes* der Zukunft, die von einer weltumspannenden Künstlichen Intelligenz gesteuert werden, technisch leicht umzusetzen wäre. Man ahnt, worauf das hinauslaufen kann: Auf einen Ökofaschismus in einer Zeit, die von Verteilungskämpfen um knapper werdende Ressourcen geprägt sein wird und in der die Frage virulent wird, wer überleben darf und wer nicht (vgl. auch Amery 1998).

17.4.6 Integration und Mischformen

Betrachten wir abschließend zwei Beispiele für Social Scoring Systeme, die verschiedene Typen von Inputvariablen integrieren. Somit kann der resultierende Score auch nicht einem dieser Typen in Reinform zugeordnet werden, was ja per se kein Problem darstellen muss. So erfasst der von der deutschen SCHUFA vergebene Credit Score, verkürzt gesagt, die Bonität und Zahlungsmoral potenzieller Kreditnehmer. Kann die Zahlungsmoral noch unter Regelkonformität subsumiert werden (Abschn. 17.4.4), ist dies bei der erstgenannten Komponente, der Bonität, nicht der Fall. Diese ist damit korreliert, wie weit es jemand in der Gesellschaft gebracht hat, und ist daher dem ersten Typus (Abschn. 17.4.3) zuzuordnen. Auch was die Verwendung dieses Maßes anbelangt, ist der Credit Score sehr verschieden zu den Punkten, die man in Flensburg hat. Der SCHUFA-Eintrag wird nicht von einer Regierung genutzt, die Bürger sanktionieren bzw. erziehen will, sondern von Banken, die selber Profit machen und dabei Risiken vermeiden wollen, also zum Beispiel das Risiko, eine Hypothek an eine Person zu vergeben, die später dann ihre Raten nicht zahlt.

Das zweite Beispiel knüpft an das oben bereits erwähnte polizeiliche Führungszeugnis an. Von dort ist es wohl nur noch ein kleiner gedanklicher Schritt zu dem, was man vielleicht „staatliches Führungszeugnis" nennen könnte. Fokussiert ersteres auf Ordnungswidrigkeiten und Straftaten, die von der Polizei erfasst und ermittelt werden (und das Flensburger Zentralregister für Verkehrsdelikte kann als kleiner Ausschnitt dessen gesehen werden), so könnte in einem staatlichen Führungszeugnis noch viel mehr stehen. Immerhin unterhält der Staat neben der Polizei noch viel mehr Organe und Institutionen, wie zum Beispiel Finanzamt, Innenministerium, Verfassungsschutz oder Schulen, um nur einige zu nennen.

An dieser Stelle bietet sich eine Bemerkung über die Stasi an, das Ministerium für Staatssicherheit in der ehemaligen Deutschen Demokratischen Republik (DDR). „Dabei denken viele vielleicht an George Orwells Roman *1984* mit seinem Slogan *Big Brother is watching you!* Dieser 1948 fertiggestellte Roman stellt eine Fiktion dar und fokussiert auf psychologische Aspekte der Einführung und Aufrechterhaltung von Überwachung und Gedankenkontrolle durch ein totalitäres politisches System. Orwells Big Brother war seinerzeit an Stalin angelehnt, und in der Tat hatten zur Zeit des Kalten Kriegs viele totalitäre Staaten aus dem ehemaligen Sowjetblock Überwachungssysteme installiert" (Hoffrage 2019, S. 25; vgl. auch Martignon und Hoffrage 2019, Kap. 9). In einem Einparteiensystem, in welchem Pluralität unterdrückt wird, gilt es, die Bürger linientreu zu machen und ihre Treue auch zu überwachen. Dabei beachte man, dass der Begriff der politischen Gesinnung deutlich weiter gefasst ist als das Einhalten von Regeln. Man kann in einem derartigen Überwachungssystem die oben bereits erwähnte Perversion des Ideals der Gleichheit, nämlich die Fratze der Gleichschaltung erblicken (wenngleich deren Züge doch deutlich anders aussahen als bei den Nazis). Das Ideal der Gleichheit, welches sich auf den Menschen als Bürger beziehen und alle vor dem Gesetz gleichstellen sollte, wurde in der ehemaligen DDR aus diesem Bereich herausgerissen und dafür missbraucht, eine Diskussion über Ideen zur Sozialgestaltung zu unterdrücken: Alle hatten diesbezüglich der gleichen Meinung zu sein wie das Regime. Dies führte dann zu Ereignissen, auf die das im ersten Abschnitt erwähnte Bild vom angestauten Fluss passen würde. Zur Stasi sollte noch hinzugefügt werden, dass sie ein Spitzelsystem installiert hat und die Menschen einander prinzipiell misstrauten, denn sie wussten nicht, wer Agent oder inoffizieller Mitarbeiter der Stasi war und Informationen an diese weitergetragen würde. Ebenfalls ist erwähnenswert, dass die Stasi keine Scores berechnet hat: Ähnlich wie beim polizeilichen Führungszeugnis, so bestanden auch die Stasi-Akten überwiegend aus Text (wobei es prinzipiell natürlich kein Problem gewesen wäre, die zahlreichen Informationen, die die Stasi gesammelt hat, auf eine Zahl abzubilden).

17.5 Auswirkung von Social Scoring auf den Menschen

Ein Social Score ist eine Zahl, die erfasst, wie ein einzelner Mensch mit seiner sozialen Umwelt interagiert. Um diese zu erhalten, werden oft zunächst viele Dimensionen durch viele Operationalisierungen und Messungen auf viele Zahlen abgebildet und diese werden dann irgendwie aggregiert. Die Vereinfachung, die mit einer derartigen Reduktion einhergeht, eröffnet Möglichkeiten, die es sonst so nicht gegeben hätte, denn man kann mit einer Zahl ganz anders umgehen als mit einem langen Text. Aber es geht dabei auch viel verloren. Hinzukommt, dass die Integration vieler Dimensionen in eine einzige Zahl stark subjektive Züge trägt, was angesichts einer Zahl, die am Ende dann ‚objektiv' dasteht, oft und gerne in den Hintergrund rutscht. Will man sagen, welcher von zwei Bewerbern auf eine Stelle der bessere ist, so ist das nach dem Durchlesen der beiden jeweils zehnseitigen Dossiers unter Umständen alles andere als leicht, und viele werden sagen, die Präferenz eines Entscheiders sei subjektiv. Aber wenn ein Algorithmus die jeweils zehn Seiten in zwei Scores überführt, so kann leicht der Eindruck entstehen, dass man objektiv sei, wenn man den Bewerber mit dem höheren Score einstellt.

Dabei zeigt sich ein ähnliches Muster, wie es bereits oben festgestellt wurde: Im Fall der ‚großen' Systeme ging der Weg von Bedürfnissen zu Ideen, zu Idealen bis hin zu Systemen. Hier, im Fall der ‚kleinen' Scoring Systeme, ist es der Weg von subjektiven Beurteilungen im sozialen Kontext zu Operationalisierungen, zu statistischen Modellen und letztlich hin zu Algorithmen.

Der Mensch und seine Systeme: Bei den ‚großen' wie bei den ‚kleinen' ist es beide Male der Weg des Menschen vom Schöpfer zum Objekt. Am Ende ist er in die (von ihm geschaffenen) ‚großen' Systeme eingebettet und wird von den Algorithmen seiner ‚kleinen' Systeme bewertet. Noch sind diese Algorithmen menschengemacht, doch dies kann sich schnell ändern! Wie für die oben besprochenen Ideologien, die sich eben auch in Perversionen und Fratzen offenbaren, so gilt auch für diese ‚kleinen' Scoring Systeme, dass sie eine Eigendynamik entfalten können. Sie tragen maßgeblich zum Entstehen einer kalten und technisch-bürokratischen Welt bei (Nezig 2019) – und dies schleichend und kaum merklich, aber gerade dadurch umso gefährlicher (Hoffrage 2011). Aber gerade an diesen Fratzen kann der Mensch auch aufwachen und die Qualität seines eigenen, menschlichen und moralischen Urteils entdecken. Sie können ihn also knechten, aber ihn gerade dadurch auch herausfordern und gegebenenfalls stärken. Ausgang ungewiss. So schweben beständig die Fragen im Raum, was sie im Menschen zutage fördern und wer hier eigentlich wen beherrscht. Nicht nur Ideologien, sondern auch Social Scores haben ein Janusgesicht. Schauen wir uns für die drei Typen von Inputvariablen, die ein mögliches Social Scoring System benutzen könnte, getrennt an, wie sie auf den Menschen zurückwirken.

17.5.1 Auswirkungen am Subjektpol

Wettbewerb stimuliert und spornt an. Das Wissen, dass wir von Anderen gesehen werden, kann unsere Leistungen steigern. Messverfahren und Zahlen können dies befeuern, ja, manchmal auch erst ermöglichen, denn oft wird erst durch sie feststellbar, wer der Gewinner ist. Scores dieses Typs können uns also helfen, zu wachsen und uns zu entfalten. Gleichzeitig befeuern sie aber auch das im Menschen, was man wohl klar als Schattenseiten bezeichnen muss. Schädigung von Konkurrenten (durch Sabotage, Mobbing oder wie auch immer), Betrug und Bestechung bei Messverfahren, oder Doping im Sport mögen als Beispiele genügen. Auch kann man sich fragen, ob die Ziele, die sich Menschen setzen und die durch derartige Scores noch mehr aufgewertet werden, wirklich so erstrebenswert sind. Da gibt es zum Beispiel erwachsene Menschen, die ihren Sinn und Lebenszweck darin zu sehen vermeinen, schneller als andere mit Maschinen im Kreis herumzufahren, wobei sie obendrein auch noch ihr Leben riskieren. Ohne die durch Scores mitbegünstigte Wettbewerbsatmosphäre käme mancher vielleicht noch auf ganz andere Ideen, wofür er seine Zeit und Energie nutzen möchte.

17.5.2 Auswirkungen in der Mitte

Scores, die aus Inputvariablen des zweiten Typus resultieren, können eine Art Erziehungshilfe darstellen, um gewünschtes Verhalten zu erzeugen und unerwünschtes zu unterdrücken. Dazu braucht man nur einen hohen Score mit Privilegien und einen niedrigen mit Einschränkungen verbinden. Kurzum: Zuckerbrot und Peitsche. So werden die Methoden der Lernpsychologie behavioristischer Provenienz, die überwiegend mit Ratten und Tauben experimentierte, zur Steuerung des Verhaltens einer Bevölkerung eingesetzt. Einige Menschen können mit Hilfe von Anreizen und Strafen ihren eigenen Schweinehund in den Griff bekommen und mögen um diese Wirkung von Scores vielleicht auch froh sein. Und selbst wenn sich ein Autofahrer in einer Spielstraße nicht aus Einsicht oder Rücksicht auf Kinder, sondern nur aus Rücksicht auf seinen Social Score (und den daran gekoppelten Geldbeutel) an die Geschwindigkeitsbegrenzung hält – die Anwohner werden den Unterschied nicht bemerken.

Aber es wird Adaptation und Auswüchse geben, die man nicht sehen möchte und die es ohne diese Regeln und ohne diese Scores, die auf Regelkonformität basieren, so in dieser Form nicht geben würde. Denn da sind immer auch diejenigen, die keine Einsicht in die Notwendigkeit von Hobbes' *Leviathan* haben und die ihr Drang nach Selbstentfaltung dazu treiben wird, nach Schlupflöchern im Regelwerk und dessen Durchsetzung zu suchen. Beispiele reichen von Rasern, die vor der Radarfalle scharf abbremsen, hin zu Experten und Beratern, die sich darauf spezialisiert haben, die Steuerlast von milliardenschweren Unternehmen auf ein absolutes Minimum zu drücken – und zwar ganz legal.

In diesem Zusammenhang eine Bemerkung zum Thema Steuern. Diese machen unter anderem genau das, was das Verbum besagt: sie steuern. Sie setzen Anreize für das Tun und Lassen der Besteuerten und machen sie so zu Gesteuerten. In Hinblick auf klar definiertes Verhalten, bei dem sich klar sagen lässt, ob es erwünscht oder unerwünscht ist, macht es also Sinn, dies klar zu benennen und genau dadurch auch eine Steuerwirkung zu erzielen. In Hinblick auf diesen Effekt sollten die Algorithmen, die die Scores bestimmen, offengelegt werden. Damit können dann auch Social Scoring Verfahren das Verhalten von Menschen steuern. Doch sollten sich die Architekten von Social Scoring Systemen genau überlegen, was transparent sein sollte und was nicht, denn was transparent ist, wird eine Wirkung entfalten (vgl. auch SVRV 2018). So gab es in einigen europäischen Ländern früher eine Fenstersteuer, mit dem Effekt, dass zum Beispiel Mietskasernen, die in England zur Zeit der Industrialisierung gebaut wurden, kaum Fenster hatten (im Jahre 1851 wurde diese nicht nur ungeliebte, sondern auch dysfunktionale Steuer dort wieder abgeschafft; Brockhaus Enzyklopädie, 19. Aufl. 1988, Bd. 7, S. 194, Stichwort „Fenstersteuer").

17.5.3 Auswirkungen am Objektpol

Wie passt sich jemand an ein System an, das seinen Beitrag zum Gemeinwohl misst und gegebenenfalls auch kommuniziert? An unserer Fakultät, einer Institution, die mit Lehre und Forschung zum Gemeinwohl beitragen möchte, gibt es im Eingangsbereich eine Wand, in deren Fliesen die Namen von Sponsoren, meist Firmen, eingraviert sind. Je größer der Beitrag, desto größer die Fliese. Mancher möchte schenken und dabei ungenannt bleiben. Genau das ist ja auch das Wesen der selbstlosen Hingabe am Objektpol und deshalb sei dieser Fall hier als erster genannt. Aber für manchen spielt bei der Entscheidung, ob und wie viel er spendet, vielleicht doch ein wenig aus dem Subjektpol mit hinein (nach dem Motto: *Tue Gutes und rede darüber).* Kurzum, die Offenlegung von Social Scores, die aus Inputvariablen des dritten Typus resultieren, kann altruistisches Verhalten stimulieren, und die Möglichkeit, dass dies mit einem subtilen Hinschielen auf die eigene Reputation verbunden sein kann (aber auch nicht muss), ändert nichts am Ergebnis: Spende ist Spende.

Das Erfassen von Beiträgen zum Gemeinwohl, die nicht freiwillig erfolgen, sondern die durch Regelung und als ‚Zwangsschenkungen' erbracht werden, führt zu Inputvariablen des zweiten Typs (Regelkonformität). Auswirkungen darauf aufbauender Scores auf den Menschen wurden bereits in Abschn. 17.5.2 besprochen.

17.6   Der Mensch und seine Systeme

In diesem Kapitel haben wir versucht aufzuzeigen, dass Mensch und Systeme auseinander hervorgehen wie Huhn und Ei, und – um das andere Bild zu wiederholen – aufeinander wirken wie Wasser und Flussbett. So sind wir, ausgehend von der Subjekt-Objekt-Spaltung und unter Hinzunahme der ausgleichenden Mitte, die wir im Sozialen gefunden haben, zu menschlichen Urbedürfnissen gekommen: Selbstentfaltung (des Subjekts), Gerechtigkeit (ausgleichende Mitte) und Hingabe (an ein Objekt). Wir haben diese Bedürfnisse wiedergefunden in den Idealen der Französischen Revolution – Freiheit, Gleichheit, Brüderlichkeit – und wir haben gesehen, wie daraus Ideologien werden können, die Großartiges im Menschen hervorbringen aber auch viel Unheil anrichten können. Schließlich war es vielleicht nicht ganz uninteressant zu sehen, wie sich das, was sich hier im Großen zeigen konnte, auf einer kleineren Ebene widerspiegelt. So haben wir drei Typen von Inputvariablen für Social Scoring Systeme identifizieren können: Solche, die anzeigen, was jemand für sich errungen hat, inwieweit sich jemand an die Regeln hält, und was jemand zum Gemeinwohl beiträgt.

Bei den bisherigen Betrachtungen sind wir noch nicht explizit auf die gegenwärtige Entwicklung in China eingegangen. Dies zum einen, weil wir uns weder kompetent noch berufen fühlen, dies zu tun (es sei auf andere Kapitel dieses Buches verwiesen). Hinzukommt, dass sich dort alles noch in der Aufbau- und Testphase befindet und es derzeit noch unklar ist, wie sich die vorläufigen Ergebnisse aus Pilotprojekten generalisieren lassen. Ferner ist das chinesische System ohnehin dynamisch angelegt und wird laufender Anpassung und Änderung ausgesetzt sein, nicht zuletzt auch aufgrund der sich rasch ändernden technologischen Möglichkeiten und ihrer Akzeptanz. Trotz all dieser Unsicherheit vermeinen wir, als Laien, doch sehen zu können, dass das chinesische Social Credit Scoring darauf hinauslaufen wird, Elemente eines jeden der hier besprochenen Typen von Inputvariablen in einen einzigen Superscore zu integrieren. Diese Mischung innerhalb dieses ‚kleinen‘ Systems scheint in eigentümlicher Weise damit zu korrespondieren, dass auch das ‚große‘ System in China eine weltweit vielleicht einzigartige Mischung der Elemente darstellt, die wir oben besprochen haben. Welches Land sonst ist (1) in einem derartigen Ausmaß eingebunden in den Prozess der Globalisierung, der von der unsichtbaren Hand des Marktes westlicher Provenienz geprägt ist, und hat (2) mit seinem Ein-Parteien-System einen derart starken Staat implementiert, der im Sinne des Thomas Hobbes' *Leviathan* für Ordnung im Inneren sorgt, und ist (3) derart stark in einer kommunistischen Tradition verhaftet?

Der dominierende Pol in Chinas ‚großem‘ System scheint uns der starke Staat zu sein. Und genau der ist es auch, der das Social Credit Rating ausgestalten und betreiben, und der den resultierenden Superscore benutzen wird. Das wird zweifelsfrei seine Macht stärken, insbesondere, wenn man sich vergegenwärtigt, dass die Reduktion vieler Dimensionen auf einen einzigen Score mit Handhabbarkeit einhergeht (für eine Übertragung des vom Wahrnehmungspsychologen James Gibson geprägten Begriff der *Affordance* auf Scoring, siehe Marewski und Hoffrage 2021). Diese Handhabbarkeit wiederum ist nicht zu trennen von digitalen Technologien, deren Einsatz hier eine Art Quantensprung darstellt, sowohl bei der Berechnung von Scores als auch bei der Durchsetzung daran gekoppelter Konsequenzen. Mit diesen Technologien ist Überwachung und Steuerung in einem völlig neuen Ausmaß möglich und ‚alte‘ Science-Fiktion Romane zu diesem Thema wie Georg Orwells *1984*, Aldois Huxleys *Brave New World* und B. F. Skinners *Futurum II (Walden Two)* müssten heute noch einmal ganz neu geschrieben werden (wir zitierten oben bereits Dicke und Helbings *iGod*). Standen im Zentrum des amerikanischen Traums Werte wie Freiheit, Selbstverwirklichung und Wohlstand, so scheint die chinesische Führung mit ihrem Social Credit Rating in einer digitalen Welt Ziele wie Kontrolle, Überwachung und Steuerung zu verfolgen. So wie wir oben von den Schatten und Perversionen des amerikanischen Traums sprachen, könnte man auch hier fragen, inwieweit das schnell in einen Albtraum umschlagen könnte. Bekanntermaßen stehen die Technologien, die hier zum Einsatz kommen, weltweit zur Verfügung und allein von daher sollten die Fragen, die man daran haben kann, weltweit gestellt werden.

Datensammeln im Internetzeitalter ist, so scheint es, aufs Engste verzahnt mit dem ‚großen' System. Bude und Staab (2019) kontrastieren das chinesische mit dem amerikanischen Modell. Zu China schreiben sie, dass die Bevölkerung das dortige Sozialkreditsystem „nicht als skandalöse Überwachungsmaschinerie" erlebt, „sondern offenbar vor allem als Chance für jeden einzelnen, sich durch aktive gesellschaftliche Teilhabe Privilegien für die private Lebensgestaltung zu sichern" (Bude und Staab 2019, S. 4). Dies klingt sehr nach dem, was wir im Subjektpol lokalisieren würden. Aber sie schreiben auch, dass dabei die „Privilegien für den Einzelnen … an das Vorangehen der ganzen Gesellschaft gebunden" werden, was im Grunde bedeutet, dass das Subjekt an die Leine des Objekts und der Gemeinschaft gelegt wird. Für den Westen stellen sie, konsistent mit unseren Ausführungen, fest: „Das Gegenmodell dazu stellen die USA dar. Eine zentrale Vergabestelle für Privilegien ist in dem Land unvorstellbar, in dem jede Person das verfassungsmäßige Recht hat, ihr Glück zu suchen" (Bude und Staab 2019, S. 4). Wird in China „Massenloyalität durch die Vergabe von Privilegien" gesichert, so geschieht das in den USA „durch die Gratisversorgung mit digitalen Lebenshilfen" – wobei die Menschen im Westen die persönliche Datensouveränität bereitwillig zugunsten des Zugewinns an Bequemlichkeit opfern.

Das Bemerkenswerteste an Bude und Staabs (2019) Essay liegt unseres Erachtens aber in ihrer These, die auch den Titel hergibt: „Da passt noch was dazwischen." Damit steht Europa wieder einmal in der Mitte, diesmal allerdings nicht zwischen den USA und den UDSSR, sondern zwischen Google, Apple, Facebook, Amazon einerseits und Huawei und SESAME Credit andererseits. Lassen wir die Autoren wieder selbst zu Wort kommen: „Die europäische Erfahrung sozialer Teilhabe und gesellschaftlichen Reichtums beruht auf der Idee sozialer Rechte. Man gewinnt die Hingabe der Einzelnen für die gemeinsame Zukunft weder dadurch, dass man individuelle Entfaltung an Gehorsam bindet, noch dadurch, dass man bei ihnen immer wieder die *frontier*-Erwartung schürt, ein neues und unbekanntes Land zu erobern und sich untertan zu machen. Europa macht die Einzelnen zu Trägern unveräußerlicher Rechte, die ihrem Stolz genügen, ihren Einsatz fordern und sich dabei gemeinsam an einer Idee des guten Lebens orientieren. Ich fühle mich nicht groß, weil …, sondern weil ich Teil einer Solidargemeinschaft bin, in der man füreinander aufkommt und so die gemeinsame Zukunft gestaltet" (S. 4). Man beachte, dass der Mensch hier weder als jemand gesehen und angesprochen wird, der sich selber verwirklichen möchte, noch als jemand, der in ein System eingespannt ist, und schon gar nicht als jemand, den man durch eine Zahl repräsentieren könnte. Vielmehr als jemand, der als Einzelner Verantwortung für andere übernehmen kann und möchte. Gemäß diesem Traum der (europäischen) Mitte werden zur Verteilung von Ressourcen keine Algorithmen (bzw. daraus resultierende Scores) eingesetzt, vielmehr bleibt der Mensch als Mensch gefordert.

Doch wer ist das: der Mensch? Eine unbehaarte Ratte und ungefiederte Taube? Und was sind die von ihm geschaffenen Scoringsysteme? Skinnerkäfige? Beurteilungsinstrumente, mit deren Hilfe entschieden werden kann, wann es das Zuckerbrot und wann es die Peitsche gibt? Die Reduktion der Vielfalt von Aspekten eines Menschen auf einen einzigen Superscore kann, wie oben bereits gesagt, als Verarmung und dessen Handhabbarkeit als Gefahr gesehen werden. Paradoxerweise liegt gerade darin aber auch eine Chance: Wird ein Mensch nur noch als Zahl repräsentiert, so geht viel verloren, doch eben das provoziert Fragen. Inwieweit unterscheidet sich der Mensch von seinem jeweiligen „Datendoppelgänger", der in Form eines Vektors einer Künstlichen Intelligenz als Input dient? Vielleicht kann er auf seiner Suche nach Antworten Gebiete entdecken, die einem Algorithmus auf immer und ewig verschlossen bleiben müssen. Und inwieweit kann er so handeln lernen, dass er eine Lernpsychologie und ein „big nudging" (Helbing et al. 2019) durch sich selbst widerlegt? Können Systeme, die der Mensch aus sich herausgesetzt hat und die ihn zu Knechten machen, ihm dazu verhelfen, für sein ureigen-Menschliches aufzuwachen? Dazu Friedrich Schiller (Die Worte des Glaubens, 1797): „Der Mensch ist frei geschaffen, ist frei, und würd' er in Ketten geboren. ... Vor dem Sklaven, wenn er die Kette bricht, vor dem freien Menschen erzittert nicht!" Braucht es die Ketten, damit er dies merkt?

Literatur

Amery, C. (1998). *Hitler als Vorläufer: Ausschwitz – der Beginn des 21. Jahrhunderts?* München: Luchterhand.

Bude, H., Staab, P. (2019). Da passt noch was dazwischen. Die USA und China wetteifern um die digitale Weltherrschaft. Europa könnte mit einem eigenen Modell dagegenhalten. *DIE ZEIT* (14. November), 47, 4.

Creutz, H. (2012). *Das Geld-Syndrom. Wege zu einer krisenfreien Wirtschaftsordnung.* Verlag Mainz.

Dicke, W., Helbing, D. (2017). *iGod.* Wroclaw, Poland: Amazon Fullfillment.

Enzensberger, H. M. (2011). *Sanftes Monster Brüssel oder Die Entmündigung Europas.* Suhrkamp.

Fehr, E., Gächter, S. (2000). Cooperation and punishment in public goods experiments. *American Economic Review*, 90(4), 980–994.

Gonin, M., Palazzo, G., & Hoffrage, U. (2012). Neither bad apple nor bad barrel: How the societal context impacts unethical behavior in organizations. Journal of Business *Ethics: A European Review*, 21(1), 31–46.

Helbing, D., Frey, B. S., Gigerenzer, G., Hafen, E., Hagner, M., Hofstetter, Y., ... Zwitter, A. (2019). Will democracy survive big data and artificial intelligence? In D. Helbing (Ed.) *Towards digital enlightenment: Essays on the dark and light sides of the digital revolution* (pp. 73–98). Cham, Switzerland: Springer.

Hoffrage, U. (2011). How people can behave irresponsibly and unethically without noticing it. In G. Palazzo, M. Wentland (Ed.) *Responsible Management Practices for the 21st century* (p. 173–182). Paris: Pearson Education France.

Hoffrage, U. (2019). Digitalisierung: Verheissung und Verhängnis. *AGORA – in geänderter Zeitlage, 5/6/7*, 25–29.

Hoffrage, U., Marewski, J. N. (2015). Unveiling the Lady in Black: Modeling and aiding intuition. *Journal of Applied Research in Memory and Cognition, 4*, 145–163.

Lloyd, S. (2008). *The Carbon Diaries: 2015*. London: Hodder Children's.

Marewski, J. N., & Hoffrage, U. (2021). The winds of change: The Sioux, Silicon Valley, society and simple heuristics. In R. Viale (Ed.), Routledge Handbook of Bounded Rationality (pp. 280–312). Abingdon, Oxon: Routledge.

Martignon, L., Hoffrage, U. (2019). *Wer wagt, gewinnt? Wie Sie die Risikokompetenz von Kindern und Jugendlichen fördern können*. Bern: Hogrefe Verlag.

Nezig, A.-K. (2019). Wenn Maschinen kalt entscheiden. *DIE ZEIT* (24. Oktober), 44, 21.

Niederhausen, H. (2011). *Zeit der Entscheidung. Die „Finanzkrise" und neue Begriffe für eine grundlegend menschliche Gesellschaft*. Niederhausen-Verlag.

Ostrom, E. (1990). *Governing the commons: The evolution of institutions for collective action*. Cambridge University Press.

Perkins, J. (2005). *Bekenntnisse eines Economic Hit Man*. München: Riemann.

Roberts, P. C. (2015). *Amerikas Krieg gegen die Welt … und gegen seine eigenen Ideale*. Rottenburg: Kopp Verlag.

Steiner, R. (1918). *Die Philosophie der Freiheit*. Berlin: Philosophisch-Anthroposophischer Verlag.

SVRV (2018). *Verbrauchergerechtes Scoring. Gutachten des Sachverständigenrats für Verbraucherfragen [Consumer-friendly scoring. Expert report of the Advisory Council for Consumer Affairs]*. Berlin: Sachverständigenrat für Verbraucherfragen. Heruntergeladen am 13. Dezember 2019 von: http://www.svr-verbraucherfragen.de/wp-content/uploads/SVRV_Verbrauchergerechtes_Scoring.pdf.

Swassjan, K. (2018). *Das Böse im Zeitalter des Goetheanismus: Bolschewismus, Nationalsozialismus, Amerikanismus*. Tagung vom 2.–5.08.2018 in Hertenstein/Schweiz.

Wenders, W. (1986). Der amerikanische Traum. In: *Emotion Pictures: Essays und Filmkritiken* (S. 141–170). Frankfurt: Verlag der Autoren.

Ulrich Hoffrage ist seit 2004 Professor für Entscheidungstheorie an der wirtschaftswissenschaftlichen Fakultät der Universität Lausanne (Schweiz). Zuvor, von 1995 bis 2004, war er Mitglied der Arbeitsgruppe Adaptives Verhalten und Kognition in Berlin. Seine Forschungsgebiete sind Urteilen und Entscheiden, einfache Heuristiken als Modelle begrenzter Rationalität, Risikokommunikation, kognitive Täuschungen, (un)ethisches Verhalten und Entscheiden, Konsumentenentscheidungen, evolutionäre Psychologie und Gruppenentscheidungen. Seit kurzem interessiert er sich auch für die Auswirkungen der Digitalisierung. Er zählt zu den meist zitierten lebenden Psychologen in der Schweiz.

Julian N. Marewski ist ebenfalls Professor an der wirtschaftswissenschaftlichen Fakultät der Universität Lausanne. Seine Forschungsgebiete sind Urteilen und Entscheiden, einfache Heuristiken als Modelle begrenzter Rationalität sowie das Zusammenspiel der statistischen Struktur der Umwelt mit Entscheidungs- und Gedächtnisprozessen. Er interessiert sich ferner für Auswirkungen der Digitalisierung auf individuelles Entscheidungsverhalten sowie die Veränderungen, die die Digitalisierung für Firmen, öffentliche Einrichtungen und andere Institutionen hervorbringt und unterrichtet Manager und andere Teilnehmer von Executive Education Programmen u.a. in *data analytics, data risks,* und *data ethics*.

Diskriminierung im Sozialkreditsystem

18

Dirk Schlotböller

18.1 Einleitung

Jeder Mensch diskriminiert – er unterscheidet verschiedene Verhaltensweisen und Merkmale seiner Mitmenschen und behandelt sie entsprechend unterschiedlich. Diese Aussage gilt jedenfalls in der Bedeutung des Begriffs „Diskriminierung" als „Unterscheidung". Unterscheidungen sind üblich und allgemein akzeptiert, selbst wenn daraus eine Benachteiligung folgt: Beispielsweise wählen Kunden das Produkt des Anbieters mit dem besten Preis-/Leistungsverhältnis – und entscheiden sich gegen dessen Konkurrenten. Im Alltagsleben ist Diskriminierung sogar unvermeidlich: „Knappheit der Ressourcen, Wettbewerb und Diskriminierung sind daher untrennbar miteinander verbunden" (Kirchgässner 1997, S. 9, oder ähnlich Block und Walker 1981, S. 6). In diesem Sinne diskriminiert auch jeder Staat: Er bestraft Gesetzesübertretungen, während er bestimmte Tätigkeiten fördert und unterscheidet damit ebenfalls zwischen verschiedenen Verhaltensweisen.

Der Begriff der Diskriminierung wird in der Regel aber als „Herabwürdigung" oder „ungerechtfertigte Behandlung" verstanden. So kann eine Benachteiligung zum einen unverhältnismäßig stark ausfallen, zum anderen können als ungerecht empfundene Kriterien herangezogen werden. Vor allem an der Frage, welche Kriterien denn legitimerweise zur Unterscheidung herangezogen werden dürfen, scheiden sich die Geister – das zeigt sich auch in der Debatte um das staatliche Sozialkreditsystem Chinas (SKS). Hierzulande werden beispielsweise folgende Fälle diskutiert:

- Der Staat überwacht und erfasst weite Teile des Privatlebens der Bürger, ohne dass sich diese entziehen können, zum Beispiel durch „Gesichtsscanner".

D. Schlotböller (✉)
Münster, Deutschland
E-Mail: Dirk.Schotboeller@web.de

- Einzelpersonen werden wegen „Fehlverhaltens" von bestimmten Dienstleistungen wie der Nutzung von Schnellzügen und Flügen, Auslandsvisa oder den in China besonders bedeutsamen Singlebörsen ausgeschlossen.
- „Wohlverhalten" ermöglicht umgekehrt einen besseren Kreditzugang oder eine kautionsfreie Miete.
- Familienangehörige werden sanktioniert, beispielsweise durch einen eingeschränkten Zugang zu Privatschulen für Kinder.
- Unternehmen werden abhängig vom Verhalten ihrer Manager und Lieferanten im Berufs- und Privatleben bewertet. Eine Sanktionierung erfolgt in Form eines schlechteren Zugangs zu Krediten, öffentlichen Aufträgen, häufigeren und umfangreicheren Inspektionen und Zollkontrollen, oder der Aussicht auf Steuernachlässe im Falle eines besonders „wünschenswerten" Verhaltens.
- Individuelles Verhalten wird transparent gemacht („naming and shaming"), mit (legitimierten) positiven und negativen Konsequenzen zum Beispiel für die berufliche Entwicklung.

Beispiele dieser Art sollen im Folgenden diskriminierungstheoretisch und -politisch untersucht werden. Zwar können aus ökonomischer Perspektive Fragen der Legitimität „diskriminierender" Bewertungskriterien nicht beantwortet werden. Wohl aber können sie beschrieben und anhand von Plausibilitätsüberlegungen mit unseren verglichen werden. Insbesondere lassen ökonomische Theorie und Empirie zuverlässige Erkenntnisse zu, wie sich staatliche Normen und Eingriffe ins Privat- und Wirtschaftsleben voraussichtlich auswirken.

Angesichts der unklaren Definition und der teilweise vorschnell wertenden Nutzung des Begriffs „Diskriminierung" soll in Abschn. 18.2 zunächst für begriffliche Klarheit gesorgt und potenzielle Diskriminierungsarten im SKS ermittelt werden. Anschließend werden diese in Abschn. 18.3 näher betrachtet sowie mögliche Effizienz- und Verteilungswirkungen abgeschätzt.

Dabei geht es nicht um eine Beschreibung der tatsächlichen Ausgestaltung. Das können andere Beiträge in diesem Band besser leisten (oder beispielsweise Creemers 2018). Auch zur Marktstruktur in China können hier nur grundsätzliche Überlegungen angestellt werden. Im Mittelpunkt stehen vielmehr eine Veranschaulichung sowie wahrscheinlich Konsequenzen für das Verhalten, die Allokation und die Verteilung. Es wird sich zeigen, dass sich die oben aufgeworfenen Fragen ebenfalls in diese Systematik einordnen lassen.

18.2 Begriffsdefinition und -systematik

Von Diskriminierung wird häufig gesprochen, wenn eine Person oder eine Gruppe anhand von Kriterien, die aus normativen Gründen abgelehnt werden, unterschiedlich behandelt wird (vgl. ausführlich zu den Darstellungen dieses Abschnitts Schlotböller 2008). Als abgelehnte Diskriminierungskriterien gelten gemeinhin die ethnische Herkunft (in den

Varianten Rasse, Hautfarbe oder Sprache), die Nationalität, das Geschlecht (einschließlich Schwangerschaften), Behinderungen, die Religion/Weltanschauung (nicht: politische Überzeugung), das Alter, die sexuelle Orientierung/Ausrichtung und der Gesundheitszustand. Die konkrete Festlegung der abgelehnten Kriterien bleibt freilich umstritten, insbesondere international, eindeutig abgrenzbar ist sie jedenfalls nicht. Zudem bestehen selbst hierzulande Abstufungen: Im Wirtschaftsleben sind bestimmte Kriterien verboten, bei der Wahl des Lebenspartners nicht. Im SKS steht jedoch ohnehin keines dieser Kriterien im Mittelpunkt, jedenfalls nach Kenntnisstand des Autors dieses Beitrags. Die obige Abgrenzung des Begriffs der Diskriminierung hilft hier also kaum weiter.

In der ökonomischen Literatur findet sich daher die Differenzierung zwischen Unterscheidungen entweder aufgrund von Wettbewerbsparametern oder aufgrund als sachfremd eingestufter Kriterien, zum Beispiel das „Geschlecht" statt der „Leistungsfähigkeit" von Arbeitnehmern. Was „sachfremd" bedeutet, hängt wiederum ebenfalls von einer normativen Bewertung ab. Gary S. Becker, Pionier der ökonomischen Diskriminierungsforschung, vermeidet daher konsequenterweise Definitionen von Diskriminierung anhand von „zulässiger" oder „unzulässiger" Kriterien, da diesen immer eine subjektive Komponente zugrunde liegt (vgl. Becker 1971, S. 13 f.). Er hält daher eine exakte allgemeine Definition gar nicht für notwendig. Stattdessen geht Becker davon aus, dass für jede persönliche Affinität eine Zahlungsbereitschaft am Markt offenbart werden muss, also auch für „Diskriminierung". In dieser mittlerweile gemeinhin akzeptierten Lesart geht Diskriminierung somit unabhängig von den Kriterien als eine Art „Geschmack" in anderen Präferenzen auf, da diese empirisch nicht messbar und voneinander abgrenzbar sind. Unmittelbar anschaulich ist dies bei vielen persönlichen Dienstleistungen, bei denen sich das Gut und die erstellende Person nicht voneinander trennen lassen, und dementsprechend die jeweiligen Vorlieben ebenso wenig. Einige Aspekte des SKS lassen sich in dieser Auslegung hinreichend einordnen: Die dort herangezogenen Kriterien sind vielfach schlicht eine staatliche Präferenz für bestimmtes Verhalten seiner Bürger.

Eine Erweiterung des Begriffs der Diskriminierung besteht in der Ausnutzung von Marktmacht, wenn die Verwendung bestimmter Kriterien zur Preisdifferenzierung der Gewinnmaximierung dient. Beispielsweise kann eine unterschiedliche Lohnelastizität der Geschlechter bei Nachfragemacht zu Diskriminierung aufgrund eines inakzeptablen Kriteriums führen. Der diskriminierende Arbeitgeber hat in diesem Fall nichts gegen die Opfergruppe, macht sich aber ihre ungünstige Situation zunutze. Stabil können solche Konstellationen jedoch nur sein, wenn der Staat selbst über Marktmacht verfügt oder wenn er private Marktmacht zulässt. Tarifkartelle, Mindestlöhne, restriktive Kündigungsregeln und Elemente der Mitbestimmung schränken den Wettbewerb ein und erleichtern Diskriminierung. Im chinesischen System stellen die Wettbewerbsbeschränkungen daher eine ernste Gefahr dar, zumal wenn geschützte Unternehmen auf viele Informationen über ihre Kunden zurückgreifen und zur Preisdifferenzierung nutzen können.

Berücksichtigt der Kreditzins oder die Miete die Wahrscheinlichkeit eines Zahlungsausfalls oder richtet sich der Lohn nach der Produktivität, ist dies üblicherweise allgemein akzeptiert und führt auch zu einer effizienten Allokation, wie sich zeigen lässt. In der

realen Welt sind diese Informationen aber nicht immer bekannt – entweder beiden Markt-
seiten oder zumindest nur einer (sogenannte Informationsasymmetrie, vgl. grundlegend
Akerlof 1970). Der Marktteilnehmer mit dem geringeren Informationsstand (in den Beispielen der Kreditgeber, Vermieter bzw. Arbeitgeber) kann dann Behelfsmaßstäbe für ein
anerkanntes Kriterium nutzen, wie etwa die Herkunft als Indikator für die Leistungsfähigkeit. Dies kann zu einer systematischen unterschiedlichen Behandlung von Menschen führen (sogenannte strukturelle Diskriminierung). Ist die Korrelation mit der Leistungsfähigkeit lediglich vorgeschoben, dient also nur der Kaschierung des tatsächlichen Kriteriums,
so handelt es sich hier ebenfalls um eine Diskriminierung aufgrund einer Präferenz (siehe
oben). Darüber hinaus können der Einschätzung *fehlerhafte* Informationen über Gruppen
und dementsprechend durchschnittlich falsche Rückschlüsse auf einzelne Personen zugrunde liegen (sogenannte Irrtumsdiskriminierung, vgl. Petersen 2005, S. 677 ff.). Besteht
die Korrelation jedoch, kann das Verhalten rational und die Verwendung auch volkswirt-
schaftlich effizient sein (sogenannte statistische Diskriminierung). Gemessen an der Leistungsfähigkeit, erfährt in diesem Fall nicht jedes Mitglied der Gruppe einen Nachteil,
sondern nur diejenigen „Sonderfälle", die überdurchschnittlich leistungsfähig sind, denen
also diese ungewöhnlichen Fähigkeiten nicht zugetraut werden. Dies gilt ebenso für Korrelationen über erwartetes Verhalten, beispielsweise Vermutungen über die Kreditwürdigkeit im Zusammenhang mit dem Wohnort.

Strukturelle Diskriminierung dürfte im täglichen Leben die wichtigste Diskriminie-
rungsart sein, da die meisten Entscheidungen in Unkenntnis getroffen werden müssen:
Arbeitgeber kennen die Leistungsfähigkeit von Bewerbern nicht, Kreditgeber orientieren
sich über Wahrscheinlichkeiten am Zahlungsausfallrisiko. Für polizeiliche Ermittler liegt
die Unsicherheit sogar in der Natur der Sache, da Täter in der Regel ihre Spuren zu verwischen versuchen.

Wenn Diskriminierung auf Basis des SKS erfolgt, handelt es sich offensichtlich um
institutionelle Diskriminierung, das heißt, um eine unterschiedliche Behandlung von
Menschen durch Institutionen im sozialwissenschaftlichen Sinne, begründet oder korreliert mit Gruppenmerkmalen.

Des Weiteren kann zwischen absichtlicher und unbewusster Diskriminierung unterschieden werden. Dies soll hier jedoch nicht weiter verfolgt werden, da beide Varianten
von den bereits genannten erfasst sind. Vor allem in den USA gängig ist der Begriff der
„positiven Diskriminierung". Dabei handelt es sich nicht um eine Bevorteilung (dies wäre
ja nur spiegelbildlich zur Benachteiligung). Vielmehr meint sie den Ausgleich von Be-
nachteiligungen durch negative Diskriminierung (also zum Beispiel die Förderung von
Schwarzen in den USA). Dieser Ansatz spielt im SKS aber offenbar keine Rolle.

Diskriminierung kann sich in vielen, nicht nur materiellen Formen zeigen: Dies können
höhere Preise, schlechtere Leistungen, niedrigere Einkommen oder ein eingeschränkter
Zugang zu Gütern jeglicher Art sein (Ausgrenzung). Eine Rufschädigung kann ebenfalls
eine Diskriminierung sein, zum einen, weil sie zu den zuvor genannten Formen führen
kann. Zum anderen messen die allermeisten Menschen ihrem Ansehen auch einen Wert an

sich zu. Der Staat kann außerdem dadurch diskriminieren, dass er (Bürger)Rechte unterschiedlich zuteilt.

Für die empirische Feststellung von Diskriminierung greifen im Übrigen bloße Korrelationen zu kurz. Unterschiede zwischen Gruppen können verschiedenste Ursachen haben, wie zum Beispiel unterschiedliche Präferenzen. Auf Diskriminierung als Ursache lässt dies allerdings keineswegs schließen.

Wie sich gezeigt hat, sind vor allem vier Arten der Diskriminierung im SKS relevant:

- staatliche Präferenzen für das Verhalten seiner Bürger
- das Ausnutzen von Marktmacht durch Preisdifferenzierung
- Irrtumsdiskriminierung
- statistische Diskriminierung

Sie sollen im Folgenden näher betrachtet werden, außerdem einige besondere Aspekte des SKS.

18.3 Diskriminierung im SKS

18.3.1 Staatlich geäußerte Präferenzen

Jeder Staat macht bestimmte Vorgaben, wie sich seine Bürger verhalten sollen – und wie nicht. Diese Vorgaben sollten sich aus den Wünschen der Bevölkerung ableiten. Beispiele sind der Verzicht auf Gewalt der Bürger untereinander, die Achtung privaten Eigentums, allgemeine Beiträge zu den Staatseinnahmen und redliches Handeln staatlicher Mitarbeiter. Dementsprechend ahndet der Staat Körperverletzungen, Diebstahl, Steuerhinterziehung und Korruption. Umgekehrt honoriert der Staat gemeinnützige Spenden oder ehrenamtliches Engagement. Die Identifikation einiger dieser Verhaltensweisen wird für den chinesischen Staat durch das SKS offensichtlich erleichtert.

Die Legitimität staatlicher Regeln kann die Rechtswissenschaft besser bewerten als eine ökonomische Arbeit. Daher werden an dieser Stelle nur knappe grundsätzliche Überlegungen angestellt. In einem demokratisch legitimierten System mit einem Minderheitenschutz sollte von akzeptablen Regeln ausgegangen werden (Zwar neigt auch ein demokratischer Staat dazu, diskriminierenden Neigungen seiner Wähler zu folgen. Dieses Phänomen wird hier aber nicht weiter betrachtet.). In Demokratien soll der „Wettbewerb als Entdeckungsverfahren" im Hayekschen Sinne die Präferenzen der Bürger bei der Wahl ihrer Vertreter möglichst gut zum Ausdruck bringen. In nicht-demokratischen Systemen gelingt dies deutlich schwieriger. Sie verfügen über geringere Anreize, den Wählerwillen in Erfahrung zu bringen – und ihn auch noch umzusetzen. Hier liegt grundsätzliche ein ernstes Problem Chinas. Für das SKS bedeutet dies, dass die dort definierten Maßgaben nicht unbedingt den Vorlieben der Bevölkerung entsprechen. Dies ist jedoch keine diskriminierungspolitische Fragestellung, weswegen sie in diesem Beitrag nicht weiter betrachtet wird.

Die Strafwürdigkeit und die Verhältnismäßigkeit von Strafen können an dieser Stelle nicht geklärt werden. Am Beispiel von Sanktionen gegen Familienangehörige können aber zumindest Plausibilitätsüberlegungen durchgespielt werden. Unserem Rechts- und Individualitätsverständnis widersprechen solche Maßnahmen. Sicherlich trifft die Strafe auch denjenigen, der bestraft werden soll, zumal er sich innerfamiliärem Druck auseinander gesetzt sehen wird. Die größte Last wird allerdings jemandem anders aufgebürdet, der allenfalls über die familiäre soziale Kontrolle Einfluss nehmen kann.

Die Konsequenzen und insbesondere die Verteilungswirkungen von diskriminierenden Präferenzen auf Märkten im weitesten Sinne – also Güter-, Kapital-, Arbeitsmärkte oder auf den Märkten um Wählerstimmen – hat Gary Becker (1971) umfassend dargelegt. Demnach muss für besondere Präferenzen eine entsprechende Zahlungsbereitschaft offenbart werden – dies gilt gerade für diskriminierende. Ein Kunde, der bestimmte Geschäftspartner bevorzugt und sich anderen verweigert, muss höhere Preise zahlen. Ein Arbeitgeber mit persönlichen Vorlieben für Mitarbeiter muss diesen höhere Löhne bieten, oder für den gleichen Lohn eine geringere Produktivität akzeptieren. Ein Mitarbeiter, der bestimmte Arbeitgeber oder Vorgesetzte ablehnt, schmälert sein Einkommen und seine Beschäftigungschancen. Dies gilt jedenfalls in wettbewerblichen Systemen. Der Wettbewerb verhindert also nicht unbedingt Diskriminierung, bestraft sie aber finanziell.

Dies gilt grundsätzlich auch für Monopolisten bzw. Monopsonisten: Sie haben Spielräume, Preise zu setzen und ggf. zu differenzieren. Ihr Gewinn fällt am höchsten aus, wenn sie sich an den Preiselastizitäten orientieren. Nehmen sie stattdessen persönliche Vorlieben als Maßstab, sinkt ihr Gewinn. Allerdings sind sie vor Wettbewerben geschützt, die sie dank höherer Gewinne vom Markt verdrängen könnten.

Marktbeschränkungen wie beispielsweise Mietendeckel, Mindestlöhne, Lohnkartelle oder Zutrittsbeschränkungen schränken den wettbewerblichen Disziplinierungsmechanismus ein. Sie erhöhen damit den Spielraum für Diskriminierung.

Diese Überlegungen lassen sich grundsätzlich auch auf den Staat und seine Präferenzen übertragen: Gibt der Staat bei öffentlichen Ausschreibungen zusätzliche Kriterien vor, die über den Preis und die Leistung hinausgehen, wird sich dies in höheren Preisen oder schlechteren Leistungen niederschlagen. Verlangt er von seinen Mitarbeitern zusätzliche Eigenschaften, wird er ihnen mehr bezahlen oder bei anderen Qualifikationen Abstriche machen müssen. Dies gilt sowohl für Fälle, in denen der Staat Präferenzen seiner Bürger zum Ausdruck bringt, als auch wenn seine Akteure im eigenen Interesse handeln. Allerdings muss die Allgemeinheit stets die Kosten tragen – auch im zweiten Fall.

Da der Staat – und gerade der chinesische – häufig als Akteur nur in einem eingeschränkten Wettbewerbsumfeld handelt, wirken die disziplinierenden Kräfte nur sehr eingeschränkt. So verfügt er bei der inneren und äußeren Sicherheit über ein weitgehendes (Gewalt)Monopol, ebenso bei der Gesetzgebung und Besteuerung. Auf vielen Gütermärkten verfügen staatliche Unternehmen ebenfalls über Monopole oder profitieren von abgeschotteten Märkten (zum Beispiel Verkehr, Medien, Telekommunikation). Dadurch besteht gerade im öffentlichen Sektor Diskriminierungsspielraum (im Sinne des Prinzipals und der Agenten), häufig ist der Gewinn nicht einmal das Hauptziel (vgl. Cain 1992, S. 720 f.; Farron 2000, S. 180).

Zusammenfassen lässt sich an dieser Stelle, dass das SKS eine höhere Effizienz für das Handeln staatlicher Akteure sicherstellt. Als Monopolist in vielen Bereichen stehen dem chinesischen Staat besondere Diskriminierungsmöglichkeiten offen. Dabei gibt es immer auch etliche Profiteure, nämlich diejenigen, die die gewünschten Kriterien erfüllen. Durch die sinkenden Kosten der Informationsgewinnung dürfte die Eingriffsintensität steigen. Dazu zählen auch wachsende Differenzierungsmöglichkeiten, wenn nicht mehr nur binär zwischen „erlaubt" und „unerlaubt" unterschieden, sondern Verhalten auf einer Skala eingeordnet wird. Völlig einzigartig ist das chinesische System freilich nicht: Hierzulande wird beispielsweise zunehmend über staatliches „Nudging" diskutiert, also das subtile „Anstupsen" der Bürger in Richtung erwünschter Verhaltensweisen. Das SKS lenkt in dieser Hinsicht zumindest transparent.

18.3.2 Staatliche Prinzipale und ihre Agenten

Problematisch kann eine unzureichende demokratische Kontrolle sein, wenn staatliche Akteure („Agenten") Informationen und Handlungsspielraum im eigenen Sinne missbrauchen (sogenannte Prinzipal-Agenten-Konstellationen als „moralisches Risiko", vgl. grundlegend Williamson 1990). Dabei handeln sie zumindest teilweise gegen die Interessen ihres Auftraggebers (Prinzipals, im Falle von Staaten also das Volk). Dies kann eine persönliche Bereicherung sein, zum Beispiel durch Korruption, oder die Begünstigung nahestehender Personen, und somit auch das Ausleben diskriminierender Präferenzen.

Die hohe staatliche Eingriffsintensität und der fehlende demokratische Wettbewerb in China geben Anlass zu großen Bedenken. So können Spitzenvertreter der Regierung persönliche Ziele – insbesondere den eigenen Machterhalt – als gesellschaftliche Wünsche deklarieren. Beispielsweise können sie „Ruhe im Land" als gesellschaftlichen Wunsch vorgeben, damit tatsächlich aber politische Kritik im Keim ersticken. Für erwünschtes Verhalten gilt dies entsprechend umgekehrt. In einem demokratischen System mit gesicherten Grundrechten wäre dies deutlich schwieriger. Das SKS erleichtert somit die Diskriminierung politisch unerwünschter Verhaltensweisen.

Prinzipal-Agenten-Konstellationen sind zumindest in gewissem Umfang allgegenwärtig, sowohl in staatlichen Einheiten als auch in Unternehmen. Die Schwierigkeit, problematische Entwicklung zu erkennen und einzuschränken, wächst mit der Größe der Institution. In jedem Fall hat der Prinzipal aber ein hohes Eigeninteresse daran, solche Praktiken zu verhindern. Daher führt er zum Beispiel Regeln, die den Ermessensspielraum definieren, Dokumentationspflichten oder ein Vieraugenprinzip ein. Neben den erwähnten Missbrauchsspielräumen kann das SKS seinerseits auch Missbrauch aufdecken, indem Korruption oder diskriminierendes Verhalten von Akteuren aufgedeckt wird.

Schafft nun also der chinesische Staat ein Anreizsystem aus Strafen und Belohnungen für das Verhalten seiner Bürger, um sie entsprechend zu beeinflussen, so lässt sich aus diskriminierungspolitischer Sicht zu unterscheiden zwischen

- der Durchsetzung von Rechtstreue im Interesse seiner Bürger, beispielsweise der Schutz geistigen Eigentums, Lebensmittelsicherheit oder die Ahndung von Umweltverschmutzung,
- der Kontrolle seiner eigenen „Agenten" im Interesse des „Prinzipals", also der Bürger, wie zum Beispiel die Korruptionsprävention,
- dem Missbrauch von Macht durch „Agenten", insbesondere von Machthabern, wie die Einschränkung demokratischer und freiheitlicher Bestrebungen und des Zugangs zu Informationen (einschl. der Instrumentalisierung anderer Agenten für diese Zwecke).

Die ersten beiden Punkte sind diskriminierungspolitisch unproblematisch, abgesehen von der fraglichen Abbildung der Präferenzen der Bürger durch die Regierung. Dies gilt für Freiheits- und Geldstrafen, ebenso wie für unkonventionelle Strafen wie zum Beispiel die Verweigerung von Auslandsvisa. Zudem sollte der Nutzen der Regeleinhaltung die Kosten wie insbesondere Eingriffe in die Privatsphäre übersteigen (vgl. 3.6). Diese Aussagen gelten auch für die Förderung bestimmter Tugenden wie „Zuverlässigkeit". Hier werden Anbieter für Liefertreue begünstigt (bzw. im umgekehrten Fall bestraft) – mit anderen Worten: Der Staat offenbart eine Zahlungsbereitschaft für Leistungen, die ein Produkt aus seiner Sicht attraktiver macht.

Dagegen stellt der dritte Punkt einen Machtmissbrauch dar, indem eigene Vorlieben diskriminierend ausgenutzt werden, ohne dass die disziplinierenden Kräfte des Wettbewerbs ausreichend wirken könnten.

Die Digitalisierung vereinfacht die Erfassung individuellen Verhaltens und Rückschlüsse auf politische Positionen – und damit natürlich auch den politischen Missbrauch. Insofern sorgt das SKS in den ersten beiden Fällen für eine effiziente Sicherung der Präferenzen der Bürger (die erwähnten Einschränkungen nun außer Acht gelassen), im dritten Fall erleichtert es Missbrauch. Hier spielt auch strukturelle Diskriminierung eine Rolle: So können in Unkenntnis der Gesinnung der Bürger pauschale Einschränkungen vorgenommen werden, etwa eine nächtliche Ausgangssperre (an dieser Stelle ist es unerheblich, ob es sich um Irrtums- oder statistische Diskriminierung handelt). Identifiziert der Staat „aufmüpfige" Bürger nun durch das SKS treffsicherer, könnten die pauschalen Maßnahmen zurückgeführt werden, zur Erleichterung der „linientreuen" Bürger (vgl. Abschn. 18.3.4).

Bei Sanktionen gegen Unternehmen wegen des Verhaltens ihrer Mitarbeiter greifen die bereits beschriebenen Mechanismen. Dass Entscheidungen eines Mitarbeiters sich auf sein Unternehmen auswirken, ist im Arbeits- und Geschäftsleben nicht ungewöhnlich. Dies gilt auch für illegales Verhalten. Das Unternehmen ist in diesem Fall der Prinzipal, und kann sich über Schadenersatzregeln gegen Fehlverhalten seiner Agenten absichern. Dies gilt beispielsweise für verweigerte Auslandsvisa für dienstliche Reisen. Sie zwingen Mitarbeiter zu Inlandstätigkeiten und schmälert dadurch Produktivität und Lohn oder kann Schadenersatzzahlungen an das Unternehmen auslösen. Die Strafwürdigkeit und die Verhältnismäßigkeit von Strafen lassen sich diskutieren, dies ist aber keine diskriminierungspolitische Frage.

Beachtung verdient der Aspekt des SKS, dass Mitarbeiter ausländischer Firmen ebenfalls einbezogen werden sollen. Zwar gelten hier zunächst die bisherigen Überlegungen. Doch lässt sich veranschaulichen, dass sich auch der chinesische Staat mit seinen Regeln einem Wettbewerb stellen muss. Denn diskriminierende Regeln zulasten von Unternehmen machen eine Ansiedlung in China weniger attraktiv. Außerdem werden Mitarbeiter diskriminierende Einschränkungen etwa beim Zugang zu Informationen so unattraktiv finden, dass sie zumindest eine Zulage von ihrem Arbeitgeber verlangen (zudem eine Segregation ist zu erwarten, dass also besonders freiheitlich orientierte Arbeitskräfte mit höheren Gehaltsforderungen seltener nach China gehen). Dies gilt besonders, wenn auch das Verhalten ausländischer Unternehmen bzw. seiner Mitarbeiter im Ausland in China sanktioniert. Angesichts der hohen wirtschaftlichen Bedeutung werden die meisten ausländischen Unternehmen diese Mehrkosten in Kauf nehmen, sicherlich allerdings nicht ausnahmslos alle. China zahlt somit einen Preis für diskriminierende Präferenzen. Der Standortwettbewerb sorgt auch bei der Diskriminierung inländischer Unternehmen für Mäßigung, selbst wenn hier die Elastizität deutlich geringer sein dürfte. Nutzt China diskriminierende Regeln protektionistisch, drohen Gegenmaßnahmen. Insgesamt werden handelspolitische Überlegungen aber nicht weiter betrachtet.

18.3.3 Marktmacht staatlicher Unternehmen

In diesem Abschnitt sollen staatliche Unternehmen betrachtet werden, die dank einer gewissen Marktmacht Preise differenzieren können, um ihren Gewinn zu maximieren (Preise wiederum im weitesten Sinne, also einschl. Löhnen, Mieten oder Zinsen). Gesamtwirtschaftlich ist solches Verhalten jedoch ineffizient, hinzu können problematische Verteilungswirkungen zulasten einzelner Gruppen kommen (vgl. zu den theoretischen Grundlagen und verschiedenen Varianten der Preisdifferenzierung Fritsch et al. 2005). Preisdifferenzierung setzt die Möglichkeit voraus, verschiedene Gruppen anhand bestimmter Kriterien voneinander abzugrenzen und Arbitrage zu verhindern. Die Preissetzung erfolgt dann anhand der jeweiligen Preiselastizität. Dieses Verhalten kann als mittelbare Diskriminierung bezeichnet werden. In der Praxis kann sich das vor allem auf dem Arbeitsmarkt (wenn ein Bewerber aufgrund familiärer Bindungen wenig flexibel ist) und auf dem Wohnungsmarkt (wenn bestimmte Mieter nicht mobil sind) zeigen. Außen vor bleibt an dieser Stelle Preisdifferenzierung aus tatsächlichen oder vorgeblichen sozialpolitischen Motiven wie Rabatte für Schüler oder Senioren.

In China verfügen der Staat bzw. seine Unternehmen grundsätzlich über diese Möglichkeiten, und prinzipiell auch private Unternehmen mit einer starken Marktstellung. Mit einem besseren Zugang zu persönlichen Daten lässt sich die Reaktion auf Preisänderungen einfacher schätzen und die Preispolitik entsprechend gestalten. Gefragt wäre hier die

Wettbewerbspolitik, um wohlfahrtsschädliche Preisdifferenzierung zu unterbinden oder zumindest einzuschränken. Dies ist jedoch ein Problem des chinesischen Wirtschaftssystems insgesamt, unabhängig davon ob Personengruppen systematisch diskriminiert werden.

18.3.4 Abbau von Informationsasymmetrien

In diesem Abschnitt werden nun ebenfalls Unternehmen betrachtet, die strukturell diskriminieren (in der Variante der Irrtums- oder der statistischen Diskriminierung). An dieser Stelle ist zunächst unerheblich, ob es sich um staatliche oder private Unternehmen handelt. Typische Beispiele sind Kreditvergaberegeln einer Bank, Vermietungskonditionen einer Wohnungsgesellschaft oder Einstellungsregeln eines Betriebes.

Wie erwähnt, ist es grundsätzlich ebenso effizient wie allgemein akzeptiert, wenn der Kreditzins oder die Miete die Wahrscheinlichkeit eines Zahlungsausfalls berücksichtigt oder wenn sich der Lohn nach der Produktivität richtet. Die genannten Beispiele können auch im SKS Anwendung finden. Wichtig ist nun, ob die genutzten Kriterien tatsächlich eine Aussagekraft über das gesuchte Merkmal haben (Irrtums- oder statistische Diskriminierung). Grundsätzlich ermöglicht ein solch umfassendes System bessere Prognosen, selbstverständlich ist dies aber nicht:

- Sagt beispielsweise die Rechtstreue eines Bürgers etwas über seine Solvenz oder seine Produktivität aus, empfiehlt sich unter Effizienzgesichtspunkten eine Nutzung dieses Kriterium (dementsprechend verlangt auch der deutsche Staat von Mitarbeitern ein polizeiliches Führungszeugnis).
- Beim Zugang zu Flugzeugen oder Schnellzügen hingegen lässt sich die Zahlung grundsätzlich im Vorfeld sicherstellen (andernfalls handelt sich wiederum um eine Form eines Kredits). Schließt der Staat also Bürger von der Nutzung dieser Verkehrsmittel aus, kann man dies als die Nutzung eines irrelevanten Kriteriums bezeichnen (Irrtumsdiskriminierung), oder wohl eher als eine Form der Sanktion aufgrund bestimmter Präferenzen. In beiden Fällen schädigt er sich selbst.
- Der Ausschluss von Singlebörsen könnte durch eine solche Form der Qualitätssicherung zur Attraktivität für andere Nutzer beitragen.

Ein leistungsfähiges SKS im Sinne eines „Ratings" sollte also die Prognosefähigkeit erhöhen, das heißt, weder „gute" Kunden ausschließen, noch „schlechte" Kunden zu günstig zum Zuge kommen lassen. Dann sorgt es sogar für eine „faire" Behandlung und für Nicht-Diskriminierung in dem Sinne, dass die Entlohnung der Produktivität entspricht bzw. der Zins oder die Miete der Ausfallwahrscheinlichkeit.

Zur Gewinnung relevanter Informationen liefert der Markt die besten Anreize. Im Wettbewerb maximieren dezentrale Akteure ihren Gewinn, wenn sie Informationen so lange sammeln und auswerten, bis die Kosten hierfür den Nutzen übersteigen – und die Digitalisierung senkt diese Kosten massiv. Dabei lässt sich der Sanktions- und Selektionsmechanismus des

Marktes (Kane 1998) an folgendem Beispiel veranschaulichen: Korreliert beispielsweise die Ethnie eines Kreditnehmers für sich genommen mit der Ausfallwahrscheinlichkeit, kann dieses Kriterium die Kreditvergabe einer Bank effizienter gestalten. Nimmt jedoch ein Wettbewerber das Einkommen als Indikator hinzu, und stellt sich hier ein Zusammenhang heraus, der die Ethnie als Korrelation insignifikant macht, ist dieses Kriterium überlegen. Die zweite Bank hat einen Wettbewerbsvorteil gegenüber der ersten. Je schärfer der Wettbewerb, umso eher werden ineffiziente Praktiken aussortiert. Umgekehrt schirmt ein eingeschränkter Wettbewerb um Erkenntnisse suboptimale Unterscheidungskriterien und damit Irrtumsdiskriminierung vom Realitätscheck ab – die Konstellationen auf dem Markt für Ratings zur Kreditwürdigkeit mit wenigen dominierenden Agenturen mag ein Beispiel sein. Wettbewerbspolitisch kann es aufgrund der fehlenden Nutzungsrivalität der Daten effizient sein, diese (anonymisiert) öffentlich zur Verfügung zu stellen, um Irrtumsdiskriminierung zu minimieren (vgl. Schweitzer et al. 2018, S. 128 ff.). Insgesamt behindert der unzureichende Wettbewerb auf vielen chinesischen Märkten also den Wettbewerb um die beste Informationsverarbeitung.

Effizienzsteigerungen lassen sich auch dort realisieren, wo der Staat ein Gewaltmonopol hat, insbesondere als Dienstleister für innere Sicherheit. So dürften strengere Kontrollen von Personen oder Unternehmen, die bereits durch Verstöße gegen Zoll-, Steuer- oder andere Bestimmungen aufgefallen sind, zu einer effizienteren Regeleinhaltung beitragen. Gleichzeitig liegt in diesem Bereich aber der beschriebene Effizienznachteil der Informationsverarbeitung in der Natur der Sache: Die Polizei nutzt mutmaßliche Korrelationen für ihre Ermittlungen. Lässt beispielsweise das Anklicken bestimmter Internetseiten Rückschlüsse auf das Planen von Terroranschlägen zu (oder im Falle von „moralischem Risiko" der Machthaber: von demokratischen Umtrieben?)? Der Wettbewerb als Entdeckungsverfahren fällt allerdings aus, sodass ihn interne Organisationsverfahren ersetzen müssen. Das SKS erleichtert also den Informationszugang und die -nutzung (strukturelle Diskriminierung). Es filtert zwar Irrtumsdiskriminierung nicht so effizient aus wie auf Wettbewerbsmärkten, gleichwohl dürfte es auch hier es zu allokativen Verbesserungen beitragen.

Das Problem von Informationsasymmetrien besteht zwar zunächst häufig, allerdings hat jemand mit „gewünschten" Eigenschaften ein Interesse daran, diese kundzutun. Ist also ein Kreditnehmer besonders solvent, wird es dies über Einkommens- und Vermögensnachweise so gut wie möglich nachweisen wollen. Weniger solvente Kandidaten werden diese nicht tun, was Kreditgeber (zurecht) als unterdurchschnittliche Kreditwürdigkeit interpretieren. Auch mittelprächtige Kandidaten haben dadurch einen Anreiz, sich zumindest etwas abzuheben. Auf diese Weise entsteht im Falle nachweisbarer oder glaubwürdiger Indikatoren letztlich doch eine hohe Transparenz (Akerlof 1970).

Die Verwendung mancher Kriterien erscheint fragwürdig, etwa die von Vorerkrankungen für den Abschluss einer Krankenversicherung. Angesichts der gerade beschriebenen Anreize wird jedoch deutlich, dass ein bloßes Verbot der Abfrage dieses Kriteriums wirkungslos ist. „Gute Risiken" werden freiwillig Auskunft erteilen. Gibt hingegen jemand keine Auskunft, erweckt er Misstrauen und wird entsprechend schlecht eingestuft. In solchen Sonderfällen erscheint ein sozialer Ausgleich die bessere Alternative als Vorgaben der Informationsnutzung.

In einem System mit diskriminierenden Präferenzen gewinnen wie erwähnt jene, die die gewünschten Kriterien erfüllen. In Falle von Irrtumsdiskriminierung bzw. unzureichender statistischer Diskriminierung profitieren diejenigen, die scheinbar hilfreiche Kriterien erfüllen oder vorgaukeln.

Das SKS erhöht somit den Informationsstand und damit die Chancen auf eine effiziente Informationsverarbeitung und Entscheidungen. Ein Wettbewerbsumfeld wäre durch bessere Anreize leistungsfähiger, Irrtumsdiskriminierung durch statistische Diskriminierung zu verdrängen. Die unvollständige Konkurrenz erleichtert es dem chinesischen Staat also in vielen Bereichen nicht nur, offen gemäß seinen Präferenzen (Abschn. 18.3.1) zu diskriminieren, sondern auch Fehleinschätzung vorzunehmen, ohne wie in einem scharfen Wettbewerb bestraft zu werden. Gleichwohl muss er selbst dort, wo er als Monopolist einer Irrtumsdiskriminierung unterliegt, Einbußen hinnehmen.

18.3.5 Staatliches „naming and shaming"

Ein wichtiger Aspekt des SKS ist die Veröffentlichung von Informationen über Bürger. Die Konsequenzen und die Bewertung lassen sich aus den bereits diskutierten Argumenten ableiten.

So können Akteure diese Informationen über andere nutzen, um diese anders zu behandeln – ihn also zu diskriminieren. Beispielsweise könnten sie Menschen, die ihre Steuererklärung später abgeben, schlechter oder großzügige Spender besser behandeln. In diesem Fall handeln sie entsprechend ihrer Präferenzen optimal und offenbaren dafür eine Zahlungsbereitschaft. Ihr Gegenüber muss im Falle unerwünschten Verhaltens Einbußen hinnehmen, umgekehrt profitiert es. Die Folgen können unangenehm sein, entsprechen aber den Wünschen der Menschen. Auch hierzulande kann ein Lebensmittelskandal für einen Gastronomen oder Produzenten existenzgefährdend sein.

Nutzen Unternehmen mit Marktmacht die Informationen missbräuchlich, um ihren Gewinn zu erhöhen, ist dieser Problematik eher wettbewerbspolitisch beizukommen.

Verwenden schließlich Unternehmen Informationen gewinnmaximierend, indem sie statistisch diskriminieren, steigt die Wohlfahrt, während Irrtumsdiskriminierung unwahrscheinlicher wird. Umgekehrt entsteht im Falle eines begrenzten Zugangs zu Informationen ein Anreiz, andere Indikatoren zu Rate zu ziehen, mit dem Risiko ineffizienter und womöglich diskriminierender Ergebnisse. Beispielsweise können Einschränkungen der Preisdifferenzierung sogar dazu führen, dass Kontrakte mit „schlechten Risiken" wie „Problemmietern" gar nicht zustande kommen (Im Extremfall kommen ganze Märkte nicht mehr zustande.). Dies ist eine typische Ausweichreaktion der Vermietungspraxis, die im Vergleich zu Preisdifferenzierungen kaum feststellbar ist.

18.3.6 Ausmaß der Informationsgewinnung

In allen dargestellten Varianten liefern zusätzliche Informationen des SKS staatlichen und privaten Unternehmen eine bessere Entscheidungsgrundlage (sei es zur Diskriminierung oder zur Gewinnmaximierung), wie sich gezeigt hat. Die Informationsgewinnung führt also zweifellos zu einem Nutzen für viele Beteiligte. Allerdings ist das „moralische Risiko" des Missbrauchs nicht zu vernachlässigen – und damit die gesellschaftlichen Zusatzkosten. Außerdem entstehen staatlichen und privaten Datensammlern durch das SKS monetäre Überwachungskosten. Zudem dürften den überwachten Bürgern zum Beispiel eine ständige Beobachtung sowie eine Identifikation über Gesichtsscanner als unangenehm empfinden. Theoretisch gesprochen sollten diese gesamten Kosten unter dem Nutzen der Informationsgewinnung und -verarbeitung bleiben (juristisch entspricht dies der Verhältnismäßigkeit der Mittel). Hierzulande wird mittlerweile ein Überwachungsniveau gesellschaftlich akzeptiert, das zwar vor einigen Jahren noch undenkbar erschien, allerdings deutlich grobmaschiger ausfällt als in China. Das bedeutet zwar noch nicht, dass es wohlfahrtsökonomisch optimal austariert wäre. Die demokratische Beteiligung sollte aber eine bessere Balance sicherstellen als ein Einparteiensystem. Nicht eingegangen werden kann hier auf Aspekte wie die Datensicherheit und das Risiko von Datenmanipulationen.

18.4 Zusammenfassung

Im SKS spielen vor allem vier Arten von Diskriminierung eine Rolle: staatliche Präferenzen für das Verhalten seiner Bürger, das Ausnutzen von Marktmacht durch Preisdifferenzierung, Irrtumsdiskriminierung und statistische Diskriminierung.

Generell erleichtert das SKS dem chinesischen Staat, Informationen über das Verhalten seiner Bürger zu gewinnen. Damit ermöglicht es Effizienzsteigerungen für staatliches Handeln. Dementsprechend kann eine Veröffentlichungspflicht anonymisierter Daten die Effizienz weiter erhöhen, damit andere Marktteilnehmer, diese ohne Zusatzkosten mutzen können.

Grundsätzlich positiv ist zunächst, dass das SKS die Durchsetzung von Rechtstreue im Interesse der Bürger erleichtert (Schutz geistigen Eigentum, Lebensmittelsicherheit, Erhalt der Umwelt). Es kann außerdem dazu beitragen, Missbrauch im Staatswesen wie Korruption und Diskriminierung aufzudecken.

Als problematisch erweist sich dabei allerdings der Verzicht auf den „Wettbewerb als Entdeckungsverfahren". Unzureichender politischer Wettbewerb erschwert staatliches Handeln entsprechend den Vorlieben der Bevölkerung. Gleichzeitig steigt die Gefahr von Missbrauch durch Regierungsvertreter, insbesondere die Diskriminierung politisch unerwünschter Verhaltensweisen.

Auf anderen Märkten vergrößert eingeschränkter Wettbewerb den Diskriminierungsspielraum ebenfalls – das gilt auch und gerade in der chinesischen Wirtschaft. Verfügt der Staat bzw. staatliche Unternehmen über ein Monopol, hat er besonders große Möglichkeiten, zu diskriminieren, ohne die sonst übliche Prämie für seine Präferenz zahlen zu müssen. In solchen Konstellationen erleichtert das SKS auch eine gewinnmaximierende Preisdifferenzierung. Ein leistungsfähiges SKS bietet grundsätzlich großes Potenzial, Ausfallwahrscheinlichkeiten oder Produktivitäten besser abzuschätzen (und damit Irrtums- durch statistische Diskriminierung zu verdrängen). Ein eingeschränkter Wettbewerb lähmt jedoch die Effizienz der Erkenntnisgewinnung und -verarbeitung.

Das SKS birgt also nicht nur neue Missbrauchsmöglichkeiten wie jede andere Innovation auch. Durch den eingeschränkten politischen und wirtschaftlichen Wettbewerb entstehen systematische Fehlanreize. Gerade im Umgang mit ausländischen Unternehmen kann China trotz seiner wirtschaftlichen Bedeutung aber nicht agieren – der Standortwettbewerb schränkt den Diskriminierungsspielraum ein.

Literatur

Akerlof, George A. (1970): The Market for „Lemons" – Quality Uncertainty and the Market Mechanism, in: Quarterly Journal of Economics, Vol. 84, No. 3, S. 488–500.

Becker, Gary S. (1971): The Economics of Discrimination, in: Milton Friedman (Hrsg.). Economics Research Studies of the Economics Research Center of the University of Chicago, Studies in the Quantity Theory of Money, Chicago/London.

Block, Walter/Michael Walker (1981): Introduction, in: Walter Block und Michael Walker (Hrsg.), Discrimination, Affirmative Action, and Equal Opportunity. An Economic and Social Perspective, The Fraser Institute, S. 5–31, Vancouver.

Cain, Glen G. (1992): The Economic Analysis of Labor Market Discrimination: A Survey, in: Orley Ashenfelter und Richard Layard (Hrsg.), Handbook of Labor Economics, 2. Ed., Vol. 1, Chapter 13, S. 693–785, Amsterdam.

Creemers, Rogier (2018): China's Social Credit System: An Evolving Practice of Control, May 9, 2018, https://papers.ssrn.com/sol3/papers.cfm?abstract_id=3175792 (letzter Zugriff am 29.01.2020).

Farron, Steven (2000): Prejudice is free, but discrimination has costs, in: Journal of Libertarian Studies, Vol. 14, No. 2, S. 179–245. https://mises.org/library/prejudice-free-discrimination-has-costs (letzter Zugriff am 29.01.2020).

Fritsch, Michael/Thomas Wein/Hans-Jürgen Ewers (2005): Marktversagen und Wirtschaftspolitik, 6. Aufl., München.

Kane, Thomas (1998): Racial and Ethnic Preferences in College Admissions, in: Christopher Jencks und Meredith Phillips (Hrsg.), The Black-White Test Score Gap, S. 431–456, Washington, D. C.

Kirchgässner, Gebhard (1997): Auf der Suche nach dem Gespenst des Ökonomismus: Einige Bemerkungen über Tausch, Märkte, und die Ökonomisierung der Lebensverhältnisse, in: Universität St. Gallen, Volkswirtschaftliche Abteilung (Hrsg.), Diskussionspapier Nr. 9703, Januar 1997, St. Gallen.

Petersen, Trond (2005): Discrimination, Measurement, in: Kimberly Kempf-Leonard (Hrsg.), Encyclopedia of Social Measurement, Bd. 1, S. 677–684, Amsterdam.

Schlotböller, Dirk (2008): Diskriminierung – eine kritische Analyse der Arten, Ursachen und Handlungsansätze, Berlin.

Schweitzer, Heike/Justus Haucap/Wolfgang Kerber/Robert Welker (2018): Modernisierung der Missbrauchsaufsicht für marktmächtige Unternehmen, Endbericht, Projekt im Auftrag des Bundesministeriums für Wirtschaft und Energie, Projekt Nr. 66/17.

Williamson, Oliver E. (1990): Die ökonomischen Institutionen des Kapitalismus: Unternehmen, Märkte, Kooperationen, Die Einheit der Gesellschaftswissenschaften, Bd. 64, Tübingen.

Dr. Dirk Schlotböller hat seit dem Jahr 2001 am Centrum für angewandte Wirtschaftsforschung in Münster zum Thema Diskriminierung geforscht und 2008 seine Promotion zu diesem Thema abgeschlossen. Von 2008 bis 2018 war er beim Deutschen Industrie- und Handelskammertag in verschiedenen Funktionen der Abteilungen Wirtschaftspolitik, Außenwirtschaft und Innovation/Industrie tätig. 2018 wechselte er ins Wirtschaftsministerium des Landes Nordrhein-Westfalen. Der Artikel gibt seine private Einschätzung wieder.

Veränderung der relationalen Vorstellungen, des Sozialverhaltens und der Moralvorstellungen durch Social Credit Systeme

19

Bernhard Streicher und Johannes F. W. Arendt

Zusammenfassung

Menschen haben ein grundlegendes Bedürfnis nach Zugehörigkeit zu Gruppen und Gemeinschaften. Wenn sie sich mit der Gemeinschaft identifizieren, verhalten sie sich freiwillig kooperativ und engagieren sich für Gemeinschaftsziele. Zur Steuerung der sozialen Interaktion mit anderen verwenden Menschen unterschiedliche kognitive Modelle, die ihr Verhältnis zum Interaktionspartner und ein situativ angemessenes Verhalten bestimmen. Wir argumentieren, dass die Einführung eines allgemeingültigen Social Credit Systems nachhaltig die soziale Wahrnehmung und das soziale Denken so verändert, dass die Betroffenen sich in sozialen Interaktionen nur noch dann kooperativ verhalten, wenn dies für sie von Vorteil ist und vom Beurteilungssystem honoriert wird. Freiwilliges, aus innerem Antrieb motiviertes, kooperatives Verhalten und Engagement wird abnehmen. Dies wird sich mittelbar nicht nur negativ auf die soziale Interaktion zwischen einzelnen Personen, sondern auch auf Unternehmen auswirken, weil die Kreativität und damit die Innovationsfähigkeit abnehmen. Zusätzlich sind flächendeckende Social Credit Systeme aus unserer Sicht unvereinbar mit demokratischen Grundprinzipien.

B. Streicher (✉) · J. F. W. Arendt
UMIT Tirol, Hall, Österreich
E-Mail: bernhard.streicher@umit.at; johannes.arendt@umit.at

© Springer Fachmedien Wiesbaden GmbH, ein Teil von Springer Nature 2020
O. Everling (Hrsg.), *Social Credit Rating*,
https://doi.org/10.1007/978-3-658-29653-7_19

19.1 Einleitung

Die allgemeingültige, flächendeckende und verpflichtende Einführung eines Social Credit Systems wird, zumindest außerhalb Chinas, meist mit Skepsis betrachtet. Die Kritik richtet sich dabei in erster Linie auf das Missbrauchspotenzial eines derartigen Systems in Hinblick auf einen Überwachungsstaat, in dem non-konformes Verhalten in allen Lebensbereichen überwacht und sanktioniert werden kann. Eine mögliche Folge eines Social Credit Systems, der bisher vergleichsweise wenig Beachtung zuteilwurde, betrifft die Auswirkungen auf die psychische Regulation von sozialen Beziehungen und die substanzielle Veränderung der sozialen Wahrnehmungen und des Verhaltens. In diesem Beitrag stellen wir aus sozialpsychologischer Perspektive zunächst die Grundlagen und Merkmale menschlicher Gesellschaften dar. Wir zeigen auf, warum die Zugehörigkeit zu Gruppen für Menschen existenzielle Bedeutung hat und warum Menschen in ihrem Denken und Verhalten stark von sozialen Normen beeinflusst sind. Schließlich führen wir aus, welche Effekte ein allgegenwärtiges Punktesystem auf die sozialen Beziehungen und die sozialen Interaktionen zwischen Individuen innerhalb einer Gesellschaft haben kann und wie es sich auf die Grundlagen sozialer Interaktion sowie auf Moral- und Fairnessvorstellungen auswirken kann.

19.2 Kooperation als evolutionäres Erfolgsrezept

Sobald Menschen keinen umfänglichen Zugriff auf notwendige Ressourcen haben, sind sie zur Befriedigung ihrer Bedürfnisse auf Kooperation mit anderen angewiesen. Dies beginnt bei einfachen Arbeitsteilungen in Stammesgesellschaften und endet derzeit in den hochkomplexen Abläufen und Interdependenzen globaler Wirtschaftsprozesse. Neben dem Zugriff auf materielle Ressourcen und der damit ursprünglich verbundenen Sicherung des physischen Überlebens, beinhaltet Kooperation aber auch die für Menschen äußerst bedeutsamen Aspekte des sozialen Kontaktes und der Gruppenzugehörigkeit. Menschen sind soziale Wesen und die Zugehörigkeit zu Gruppen hat für sie existenzielle Bedeutung. Die gesamte hominide Evolution ist vom Leben in Gruppen und sozialen Kontexten geprägt. So können sich Kinder nicht normal entwickeln, wenn sie in sozialer Isolation oder Deprivation aufwachsen (Morrison 2004). Der tatsächliche oder angedrohte Ausschluss aus einer Gruppe und der damit einhergehende reale oder symbolische Verlust der materiellen, emotionalen, kognitiven und verhaltensbezogenen Vorteile wird von Menschen als existenziell bedrohlich erlebt (Williams et al. 2013). Soziale Exklusion, beispielsweise in Form von Mobbing, gehört, neben den Formen körperlicher Verletzung, zu den effektivsten und schädlichsten Formen destruktiven Verhaltens gegenüber anderen Personen.

Unsere Fähigkeit zu komplexen sozialen Beziehungen ermöglicht uns, auch komplexe, arbeitsteilige Prozesse zu organisieren. Anthropologen gehen davon aus, dass sich das menschliche Gehirn durch die zunehmende Komplexität sozialer Interaktionen und Kooperationen entwickelt hat (Tomasello 2010). Dabei ist menschliche Kommunikation insofern einzigartig, als wir uns in unseren Wahrnehmungen, Denken, Gesten und sonstigem Verhalten sehr stark auf andere beziehen. Menschliche Kommunikation ist nach Kooperationsprinzipien strukturiert. Menschliche Wahrnehmung, Denken, Fühlen und Handeln findet im Regelfall nicht isoliert statt, sondern berücksichtigt immer auch andere. Selbst wenn Menschen egozentrisch an die Erreichung eigener Ziele denken, findet dieses Denken in einem sozialen Kontext und in realen oder gedachten Beziehungen zu anderen statt.

19.3 Ausbeutung oder Vorteil: das fundamentale soziale Dilemma

Ein großer Vorteil komplexer, kooperationsbezogener sozialer Interaktionen ist, dass nicht mehr alle Beteiligten über alle Fähigkeiten verfügen müssen, die zur Erreichung eines Ergebnisses notwendig sind. Wenn ein großes Beutetier gejagt wird, ist es nicht notwendig, dass alle Jäger sich gleich gut anschleichen, gleich schnell laufen oder gleich zielgenau Speere werfen können. Entscheidend ist die Koordination dieser Fähigkeiten. In noch komplexeren Gesellschaften ist es für einzelne auch nicht mehr notwendig, an der Jagd teilzunehmen oder überhaupt über jagdrelevante Kompetenzen zu verfügen. Aber es ist notwendig, sich auf den gegenseitigen Austausch von Ressourcen und der gegenseitigen Bereitstellung von Fähigkeiten zu verlassen. Dies führt in Kontexten sozialer Interaktion zu einem fundamentalen Dilemma für alle Beteiligten: Woher können sie wissen, dass die soziale Interaktion mit anderen oder die Zugehörigkeit zu einer Gruppe für sie von Vorteil ist oder aber von Nachteil, weil sie übervorteilt oder ausgebeutet werden?

Eine Möglichkeit, dieses fundamentale soziale Dilemma zu lösen ist, über einen längeren Zeitraum und über unterschiedliche Situationen Erfahrungen mit dem Interaktionspartner zu sammeln. Den Verlauf und das Ergebnis dieser Erfahrungen nehmen Menschen als Grundlage für die Beurteilung der Vertrauenswürdigkeit einer Person, Gruppe oder Organisation. Genau dies geschieht in langjährigen familiären, freundschaftlichen, partnerschaftlichen oder geschäftlichen Beziehungen: „Wir kennen uns schon so lange und haben schon so viel gemeinsam erlebt. Ich weiß, dass ich dieser Person (nicht) vertrauen kann." Dennoch schützen auch langjährige positive Interaktionen nicht vor Übervorteilung. Ehepartner können fremdgehen, gute Freunde stellen sich gegen einen oder verlässliche Geschäftspartner werden wortbrüchig. Dieses Dilemma sozialer Interaktionen ist deswegen fundamental, weil es nicht endgültig gelöst werden kann.

Ein evolutionär sehr erfolgreicher Lösungsansatz liegt in der kooperativen Ausrichtung unserer Kommunikation und sozialen Interaktion. Je mehr Mitglieder einer Gruppe sich kooperativ verhalten, desto geringer ist die Wahrscheinlichkeit für ein einzelnes Mitglied übervorteilt zu werden. Kooperation ist das Prinzip, dass uns evolutionär betrachtet als soziale Gruppen das Überleben in Mangelumwelten ermöglichte. Darüber hinaus ermöglichte das Kooperationsprinzip Gruppen sich dauerhaft stabil arbeitsteilig zu organisieren. Dieses Prinzip ist in unseren sozialen Interaktionen stark verankert: Kooperation im Sinne der Ziele und des Wohlergehens der Gemeinschaft ist die zentrale implizite soziale Norm für alle Mitglieder einer Gruppe. Schwerwiegende Verstöße gegen diese Norm wurden in Stammesgesellschaften mit dem sozialen Tod durch dauerhaften Ausschluss aus der Gruppe oder dem physischen Tod bestraft (vgl. Junger 2016). Auch wenn wir heute überwiegend nicht mehr in Stammesgesellschaften, sondern in großen sozialen Gebilden mit vielfältigen, zum Teil unklaren und sich überschneidenden, Gruppengrenzen leben, ist unser Interaktionsverhalten immer noch vom Kooperationsprinzip geprägt. So erwarten Menschen nach dem Reziprozitätsprinzip, dass sich ihre Interaktionspartner auch kooperativ verhalten, wenn sie selbst kooperativ waren. Diesen Effekt nutzen beispielsweise Unternehmen oder Parteien, wenn sie kleine Geschenke an Passanten verteilen, weil sich diese dann gegenüber den Unternehmen bzw. Parteien verpflichtet fühlen.

Neben dem Kooperationsprinzip ist die Möglichkeit der Sanktion von Normverletzungen das zweite zentrale Prinzip zur Gestaltung komplexer, arbeitsteiliger Organisationsformen. Je komplexer die sozialen Interaktionen und Strukturen während der evolutionären Entwicklung wurden, desto mehr mittelbare Interaktionen und Abhängigkeiten entstanden. Personen können die Verlässlichkeit mittelbarer Interaktionspartner aber nicht unmittelbar beurteilen, weil sie keinen direkten Kontakt zu ihnen haben. Zur Lösung des fundamentalen sozialen Dilemmas und um den Vorteil des Kooperationsprinzips für die Gesamtgruppe aufrecht zu erhalten, müssen sich alle Gruppenmitglieder darauf verlassen können, dass sich Einzelne zur Erreichung eigener Vorteile nicht unkooperativ verhalten. Dies wird durch die Entwicklung von Regeln des Miteinanders und unterschiedliche Sanktionsmöglichkeiten bei Regelverstößen gewährleistet. Je komplexer soziale Systeme werden, desto differenzierter werden die allgemeingültigen Regelwerke. Diese Regelwerke finden heute ihren Ausdruck in Gesetzen und internationalen Abkommen. Voraussetzung für die Umsetzung von Sanktionen ist die Sichtbarkeit des individuellen Verhaltens. Um das Ausmaß gruppenkonformen und damit kooperativen Verhaltens beurteilen zu können, muss das Verhalten auch beobachtbar oder messbar sein. Zahlreiche Studien zeigen, dass die Möglichkeit anonymen Verhaltens egoistisches und unkooperatives Verhalten fördert. Haben Menschen dagegen nur den Eindruck, dass ihr Verhalten identifizierbar ist, verhalten sie sich konformer, kooperativer und sind motivierter, sich für Gruppenziele einzusetzen.

Ein fundamentaler Ausdruck regulativer Vorstellungen, wie Menschen miteinander interagieren sollen, sind Fairnessprinzipien. Die empirische Fairnessforschung konnte zeigen, dass Menschen im Wesentlichen zwischen vier Dimensionen unterscheiden (vgl. Gollwitzer et al. 2013). Erstens, nach welchen Prinzipien Ressourcen verteilt werden (*Verteilungsgerechtigkeit*); zweitens, nach welchen Prinzipien Entscheidungen getroffen werden (*Prozedurale Gerechtigkeit*); drittens, nach welchen Prinzipien Informationen weitergeben werden (*Informationale Gerechtigkeit*); und viertens, nach welchen Prinzipien Menschen miteinander umgehen (*Interpersonale Gerechtigkeit*). Diese vier Dimensionen finden sich in unterschiedlichsten Kulturen, was auf eine evolutionäre Entwicklung der Prinzipien hindeutet. Es finden sich aber kulturelle Unterschiede in der Ausgestaltung der

Umsetzung. So bevorzugen beispielsweise Kollektivisten zur Herstellung prozeduraler Gerechtigkeit eher Konsistenz bei den Entscheidungsträgern während es Individualisten eher wichtig ist, ihre Meinung äußern zu können (vgl. Summereder et al. 2014). Auch wenn sich identische Dimensionen von Fairness über viele Kulturen hinweg finden, können das Verständnis, was fair oder unfair ist, und die entsprechenden Regeln zwischen sozialen Gruppen völlig unterschiedlich sein. Dabei muss sich die subjektive Übereinkunft innerhalb einer sozialen Gruppe darüber, was fair und was unfair ist, auch nicht an allgemeinen normativen Prinzipien oder an unveräußerlichen Werten orientieren, sondern kann auch diskriminierend sein. Regeln, die beispielsweise Nicht-Mitgliedern den Zugriff auf Ressourcen verwehren oder Mitglieder zu Nicht-Mitgliedern erklären, wenn diese Gruppennormen verletzen, können von der Mehrheit der Gruppenmitglieder durchaus als fair erlebt werden. Ein Ziel fundamentaler Rechte wie der Grundrechte oder der Menschenrechte ist es, genau diese Form von diskriminierender Unfairness durch Macht oder Mehrheit zu verhindern. Subjektiv können Regeln innerhalb einer Gruppe also als fair empfunden werden, die bei normativer, wertebasierter Betrachtung diskriminierend und ungerecht sind.

Neben dem Kooperationsprinzip ermöglichten Fairnessprinzipien gepaart mit Sanktionsmöglichkeiten die Entwicklung komplexer sozialer Systeme. Gleichzeitig eröffnen Fairnessprinzipien für den Einzelnen einen weiteren Ansatz zur Lösung des fundamentalen sozialen Dilemmas: Ob ich einem mir nicht bekannten Interaktionspartner vertrauen kann oder nicht, kann ich danach beurteilen, wie fair ich von dieser Person behandelt werde. Einerseits konnten die positiven Auswirkungen der Erfüllung von Fairnessprinzipien auf Emotionen, Einstellungen und Verhalten vielfach in unterschiedlichsten Kontexten nachgewiesen werden (vgl. zum Beispiel Greenberg und Colquitt 2005): Wenn Menschen fair behandelt werden, dann fühlen sie sich als wertvolles Mitglied dieser Gruppe, identifizieren sie sich stärker mit der Gruppe und den Entscheidungsträgern und sie engagieren sich mehr für die Gruppenziele. Andererseits sind Menschen auch stark bemüht, sich an die Regeln von Gruppen zu halten, wenn diese Gruppe bzw. die Mitgliedschaft in dieser Gruppe für sie eine hohe Attraktivität hat. Eine Belohnung für regelkonformes Verhalten kann dann eine faire Behandlung durch die Gruppenautoritäten sein. So entsteht eine Wechselwirkung: Ich werde fair behandelt, weil ich mich im Sinne der Gruppe verhalte; ich verhalte mich konform, weil ich mich fair behandelt fühle. Hierin liegt auch ein Grund, warum sich Menschen üblicherweise in Übereinstimmung mit den tradierten gesellschaftlichen Werten ihrer Kultur verhalten. Dadurch vermitteln sie unter anderen die Botschaft, Mitglied dieser Gesellschaft zu sein, und dürfen erwarten, nach den gesellschaftlichen Regeln fair behandelt zu werden. Prinzipiell sind Menschen umso bemühter, Mitglied einer Gruppe zu sein und sich gruppenkonform zu verhalten, je attraktiver die Gruppe für sie ist. Insofern hat die Gesellschaft als Gruppe eine Sonderstellung und besonders hohe Attraktivität, weil ein Ausschluss aus dieser Gruppe gleichbedeutend mit dem Verlust der kulturellen Zugehörigkeit ist.

19.4   Existenzielle Bedeutung der Gruppenzugehörigkeit

Da in Stammesgesellschaften der Ausschluss aus der Gesellschaft potenziell lebensbedrohlich war, hatten diejenigen Personen einen evolutionären Vorteil, die sich kooperativer verhielten und sensibler gegenüber Signalen eines möglichen Ausschlusses waren. Eine bessere Passung (*fit*) an diese sozialen Bedingungen sollte wiederum zu mehr Nachkommen führen, die über ähnliche Merkmale verfügen sollten. Gemeinschaften, deren Mitglieder sich untereinander kooperativer verhielten als die Mitglieder anderer Gemeinschaften hatten als Gruppe wiederum einen Vorteil: sie konnten sich komplexer und arbeitsteiliger organisieren und erweiterten damit ihre Handlungsoptionen und ihr kreatives Potenzial für Innovationen. Die Auswirkungen dieser Entwicklung einer kooperativen Grundhaltung, die hohe Bedeutsamkeit von Gruppenmitgliedschaft, die Ausrichtung des eigenen Verhaltens nach Gruppennormen und die hohe Sensitivität gegenüber möglichen Ausschlusssignalen findet sich in modernen Menschen als zentrales und grundlegendes menschliches Motiv nach Zugehörigkeit wieder (*Need to belong*; vgl. Baumeister und Leary 1995). Dieses Motiv ist darauf ausgerichtet, von anderen Menschen akzeptiert zu werden und zu sozialen Gruppen zu gehören, und stellt einen der mächtigsten und universellsten Antriebe des Menschen dar. Die Entwicklung und Aufrechterhaltung eines positiven Selbstwertes sind maßgeblich von positiven, selbstwertrelevanten Informationen bedeutsamer Anderer abhängig. Für die Aufrechterhaltung ihres Selbstwertes versuchen Menschen Teil sozialer Gruppen zu sein, die ihnen vermitteln, dass sie ein wertvoller und bedeutsamer Mensch sind. Diese hoch selbstwertrelevante Rückmeldung bekommen sie von anderen Gruppenmitgliedern insbesondere dann, wenn sie sich gruppenkonform verhalten. Soziale Isolation dagegen birgt auch heute noch erhebliche psychische und physische Risiken. Sozial isolierte Menschen haben beispielsweise eine deutlich geringere Lebenserwartung als Menschen mit positiven sozialen Beziehungen. Damit kann auch erklärt werden, warum Menschen ihr Verhalten sehr stark an den wahrgenommenen sozialen Normen ausrichten.

Zusammengefasst hat die Mitgliedschaft in Gruppen und in einer Gesellschaft für Menschen existenzielle Bedeutung. Sie sind motiviert, Mitglied von für sie bedeutsamen Gruppen zu bleiben, indem sie die wahrgenommene soziale Norm erfüllen und sich gruppenkonform verhalten. Diese kognitiven Prozesse sind üblicherweise hoch automatisiert und unbewusst. Leitfragen zur Beurteilung einer sozialen Situation und zur Verhaltenssteuerung sind dabei:

- Was ist hier die Norm?
- Wie muss ich mich verhalten, um Gruppenmitglied zu bleiben?
- Wie verhalten sich die anderen?
- Was denken die anderen über mich?
- Wie kann ich mögliche Sanktionen durch mein Verhalten vermeiden?

19.5   Social Credit Systeme zur Herstellung von Vertrauen

Je stärker Personen von anderen Personen abhängig werden, mit denen sie nicht mehr direkt interagieren und die sich auch nicht mehr persönlich kennen, desto stärker müssen sie sich auf ein funktionierendes Regel- und Sanktionssystem verlassen können und diesem System vertrauen können. Vertrauen in soziale Systeme entsteht, wenn Fairnessprinzipien umgesetzt werden, die keine Gruppenmitglieder systematisch bevorzugen oder benachteiligen. Dies ist beispielsweise dann der Fall, wenn Regeln verlässlich sind, alle berechtigten Interessen angemessen berücksichtigt werden und relevante Informationen verfügbar sind oder zeitnah weitergeben werden. Besonders wichtig ist das Vorhandensein verlässlicher Kontroll- und Sanktionsmechanismen wie zum Beispiel ein funktionierendes Rechtssystem. Eine wichtige Frage in diesem Kontext ist, wie gegenseitiges Vertrauen in Interaktionspartner wiederhergestellt werden kann, wenn dies beispielsweise aufgrund dramatischer historischer Ereignisse wie dem Nationalsozialismus in Europa, den Militärdiktaturen in Südamerika, der Apartheid in Südafrika, der Sklaverei und Rassentrennung in den USA oder der Kulturrevolution in China zerstört wurde. Diese Ereignisse zeichnen sich unter anderem dadurch aus, dass gegenseitiges Vertrauen systematisch ersetzt wurde durch Kontrolle, Bespitzelung und Misstrauen (vgl. Wang 2017). Die genannten Beispiele zeigen, wie langwierig und schwierig es ist, gegenseitiges Vertrauen (wieder) herzustellen. Neben Wiedergutmachung, Schuldanerkennung verantwortlicher Täter oder Anerkennung und Benennen des Geschehenen (zum Beispiel in Form von Wahrheitskommissionen wie in Südafrika) scheint zentral die Etablierung eines allgemein gültigen Kontroll- und Sanktionssystems zu sein, auf das sich alle verlassen können und das für alle gleichermaßen gilt. In den westlichen Staaten geschieht dies durch das Rechtssystem. Das westliche Rechtssystem macht keine Aussagen zur Vertrauenswürdigkeit einzelner Personen (im Gegenteil: Persönlichkeitsrechte schützen sogar die Veröffentlichung von berechtigtem Misstrauen gegenüber einer Person), sondern etabliert Vertrauen durch allgemeingültige Regeln in Form von Rechtsvorschriften und Gesetzen (das heißt Verlässlichkeit) und deren Einklagbarkeit (das heißt Sanktionsmöglichkeiten). In so einem System muss sich ein Interaktionspartner – theoretisch – a-priori keine Gedanken darüber machen, ob er einem anderen Interaktionspartner vertrauen kann, weil er/sie Fehlverhalten a posteriori einklagen und Schadensersatzansprüche geltend machen kann (aufgrund möglicher Folgekosten wird man sich in der Praxis über die Vertrauenswürdigkeit natürlich meist trotzdem Gedanken machen).

Social Credit Systeme würden dagegen an der Vertrauensbildung auf der Individualebene anknüpfen und eine Aussage über die Vertrauenswürdigkeit einer einzelnen Person, eines Unternehmens oder einer Institution machen. Dies versetzt den einzelnen Interaktionspartner in die Lage, a priori das Risiko einer Interaktion zu beurteilen und sich für oder gegen die Kooperation zu entscheiden. Dies sollte die Anzahl von Fehlentscheidungen im Sinne des sozialen Dilemmas reduzieren. Das heißt, Interaktionspartner sollten keine oder nur wenige Interaktionen eingehen, bei denen sie benachteiligt werden, weil sie durch den

Rang im oder die Beurteilung durch das Social Credit System wissen, ob sich die Person in der Vergangenheit regelkonform verhalten hat und daher vertrauenswürdig ist. Theoretisch sollte ein ideales und perfekt funktionierendes Social Credit System ohne nachträgliche Regulationen wie der rechtlichen Regelung von Schadensersatzansprüchen auskommen. Alle Interaktionspartner haben in so einem System (theoretisch) einen hohen Anreiz, sich im Sinne der aufgestellten Regeln kooperativ und konform zu verhalten, weil sonst ihre Vertrauensbeurteilung und damit ihr Wert im System fallen. Voraussetzung hierfür ist, dass der Beurteilungsalgorithmus des Social Credit Systems transparent ist und die Beurteilungsregeln für alle Beteiligten gleichermaßen gelten. Das Grundprinzip solcher Systeme findet sich beispielsweise in den Kundenbewertungen und Ratings einzelner Produkte oder ganzer Unternehmen auf digitalen Plattformen. Hier beurteilen Personen nach mehr oder minder objektivierbaren und strukturierten Kategorien ihre Interaktionserfahrungen. Diese Bewertungen dienen wiederum anderen Personen als Grundlage für die Abschätzung der Vertrauenswürdigkeit. Die Wirkung von Social Credit Systemen ist aber in der skizzierten Form übersimplifiziert, da zahlreiche (sozial)psychologische Mechanismen außer Acht gelassen werden, deren negative Auswirkungen die theoretische, idealtypische Wirkung verringern bzw. übersteigen. Neben den oben ausgeführten Mechanismen für die zentrale Bedeutsamkeit von sozialen Systemen wie Gruppen und Gesellschaften für Menschen, ist es daher wichtig näher zu verstehen, auf welcher Grundlage Menschen ihre Interaktion mit anderen Menschen steuern.

19.6 Steuerung sozialer Beziehungen: relationale Rollenvorstellungen

Eines der Phänomene menschlichen Sozialverhaltens besteht darin, dass Interaktionspartner (selbst wenn sie ihr Gegenüber noch nicht lange kennen) außerordentlich schnell und ohne bewusste Denkprozesse eine Vorstellung davon haben, wie sie sich selbst in Relation zu ihrem Gegenüber verorten und welches Verhalten dieser Person gegenüber in einer bestimmten Situation *angemessen* oder *unangemessen* ist. Bedenkt man die Vielzahl an möglichen Handlungsoptionen und die sich daraus ergebenden möglichen inkompatiblen Verhaltensweisen von Interaktionspartnern, so ist es eigentlich erstaunlich, wie reibungslos menschliche Kommunikation meist verläuft. Dies spricht für die Existenz von elementaren, automatisierten Wahrnehmungs-, Denk- und Handlungsskripten (sogenannten kognitiven Schemata), auf die wir bei der Regulation unserer sozialen Interaktionen zurückgreifen.

Ein theoretisches Rahmenmodell zur Beschreibung der kognitiven Schemata, die menschlicher Interaktion zugrunde liegen, ist die *Relational Models Theory*, die von dem US-Anthropologen Alan Fiske (1992) formuliert wurde. Im Kern besagt die Theorie, dass Menschen vier elementare, mental repräsentierte Beziehungsmodelle nutzen, um die Beziehungen zu ihren Mitmenschen zu regulieren. Diese sogenannten *relationalen Modelle* bestimmen, wie Menschen zueinander stehen und wie sie sich *in Relation zueinander* sehen. Welches Verhalten in welchen sozialen Interaktionen als angemessen oder unangemessen, als gerecht oder ungerecht wahrgenommen und erwartet wird, hängt davon ab, welches relationale Modell in der gegebenen Situation der sozialen Interaktion zugrunde gelegt wird. Betroffen sind hiervon alle Aspekte sozialer Interaktion wie beispielsweise die Verteilung von Ressourcen, das gemeinsame Treffen von Entscheidungen oder wahrgenommene Verpflichtungen und Verbindlichkeiten gegenüber Interaktionspartnern. Aus dem Blickwinkel der Relational Models Theory ist auch moralisches Urteilen und Handeln eng mit den in einer Situation aktiven relationalen Modellen verknüpft, da unterschiedliche Modelle mit unterschiedlichen moralischen Grundmotiven einhergehen. In anderen Worten: Was als moralisch richtig oder gar in höchsten Maßen verwerflich wahrgenommen wird, hängt maßgeblich von dem als gültig erlebten relationalem Modell und dem, diesem innewohnenden, moralischen Motiv ab. Handlungen, die vor dem Hintergrund eines bestimmten relationalen Modells vollkommen angemessen und passend erlebt werden, können vor dem Hintergrund eines anderen relationalen Modells in höchsten Maßen als unpassend oder unmoralisch erlebt werden. Relationale Modelle fungieren also als handlungsleitende Prinzipien, mittels derer fremdes und eigenes Verhalten beurteilt, antizipiert, interpretiert und gegebenenfalls sanktioniert wird.

Die Relational Models Theory unterscheidet vier Modelle: das *Communal Sharing Modell*, das *Authority Ranking Modell*, das *Equality Matching Modell* und das *Market Pricing Modell*. Wenn Interaktionspartner das *Communal Sharing* (CS) Modell anwenden, nehmen sie sich als Gemeinschaft mit einer gemeinsamen Identität wahr. Interaktionen zeichnen sich durch Zugehörigkeitsgefühl, Solidarität und Selbstlosigkeit aus. Werden Entscheidungen getroffen, wird ein gemeinsamer Konsens angestrebt; werden Ressourcen geteilt, geschieht dies nach dem Bedürfnisprinzip wobei eine Aufrechnung von geleisteten und empfangenen Gütern und Leistungen nicht vorgenommen und gar als moralisch verwerflich wahrgenommen wird.

Wenn Interaktionspartner das *Authority Ranking* (AR) Modell anwenden, nehmen sie einander in einer Hierarchie wahr, die durch unterschiedliche Charakteristika wie formaler Rang, Expertise oder Seniorität gebildet wird. Interaktionen zeichnen sich dabei durch Gefühle von Macht, Unterordnung, Loyalität und Respekt aus. Werden Entscheidungen getroffen, hat die Stimme ranghöherer Gruppenmitglieder mehr Gewicht als derer mit niedrigerem Rang; werden Ressourcen verteilt ist es sozial akzeptiert, dass eine ranghöhere Person mehr erhält oder diese nach eigenem Gutdünken verteilt.

Wenn Interaktionspartner das *Equality Matching* (EM) Modell anwenden, nehmen sich als gleichberechtigte, aber distinkte Individuen mit exakt denselben Rechten und Pflichten war. Interaktionen zeichnen sich durch Reziprozität und Balance aus. Werden Entscheidungen getroffen, zählt jede Stimme gleich viel, werden Ressourcen verteilt, erhalten alle Gruppenmitglieder einen gleich großen Anteil. Interaktionspartner behalten individuelle Beiträge im Blick und versuchen Ungleichheiten wie einseitige Gefallen ohne entsprechende Rückzahlung zu vermeiden.

Wenn Interaktionspartner das *Market Pricing* (MP) Modell anwenden, sehen sie einander als unabhängige Individuen in einer von rationalen Abwägungen geprägten

Austauschbeziehung. Interaktionen sind gekennzeichnet durch Kosten-Nutzen-Rechnungen und eine konsequenten Nachverfolgung individueller Beiträge. Sowohl das Treffen von Entscheidungen als auch die Verteilung von Gütern erfolgt unter Berücksichtigung individueller Beiträge der einzelnen Gruppenmitglieder.

Diese vier relationalen Modelle bilden die kognitiven Bausteine sozialer Interaktion. Sie steuern unser Denken, Handeln und Erleben im sozialen Kontext indem sie uns eine Antwort auf die Frage „Wie sehe ich mich selbst in Relation zu meinem Gegenüber" erlauben. Aus dieser Antwort wiederum ergibt sich, welches Verhalten für einen selbst und das Gegenüber als passend, angemessen und moralisch richtig (oder falsch) wahrgenommen wird. Relationale Modelle sind dabei in allen Ebenen sozialer Beziehungen handlungsleitend: Auf dyadischer Ebene, auf Gruppenebene, in Organisationen und der Gesellschaft als Ganzes. Sie beinhalten nicht nur ein Selbstkonzept und ein Bild des Gegenübers, sondern auch Motive, Bedürfnisse sowie moralische Emotionen (wie zum Beispiel Empörung oder Schuld).

Die Unterschiede zwischen den vier relationalen Modellen werden deutlich, wenn man beispielsweise die Motive für Hilfeverhalten betrachtet: Werden die eigenen Handlungen von einem Communal Sharing Modell geleitet, hilft man einem anderen Menschen aus Solidarität und Altruismus ohne dafür eine direkte Gegenleistung zu verlangen oder zu erwarten. Ist man hingegen von einem Market Pricing Modell geleitet, erfolgt Hilfe auf Grundlage einer Kosten-Nutzen-Rechnung, in Erwartung einer Gegenleistung und nur in dem Ausmaß, wie es sich für einen selbst lohnt. Dieses Beispiel verdeutlicht zum einen, dass viele soziale Handlungen prinzipiell in allen relationalen Modellen stattfinden können. Zum anderen zeigt es allerdings auch, dass die einer Handlung zugrunde liegenden Motive, das Ausmaß sowie auch die sozialen Konsequenzen einer bestimmten Verhaltensweise stark von dem relationalen Modell abhängen, vor dessen Hintergrund die Handlung stattfindet. So kann beispielsweise Hilfeverhalten durch alle Modelle erklärt werden, allerdings sollten Menschen, die von einem Market Pricing Modell geleitet sind, tendenziell weniger Hilfeverhalten zeigen (da dieses nur dann geleistet wird, wenn es sich auch für einen selbst auszahlt) im Vergleich zu Menschen, deren Verhalten von Solidarität und Fürsorge durch das Communal Sharing Modell motiviert ist (vgl. Bridoux und Stoelhorst 2016; Mossholder et al. 2011). Ebenso hat sich gezeigt, dass die emotionale Reaktion der *Empfänger* von Hilfeverhalten stark vom wahrgenommenen Motiv des Helfenden beeinflusst wird (vgl. Ames et al. 2004).

Relationale Modelle sind nicht als starres Korsett zu verstehen, in das Beziehungen zur Gänze eingeordnet werden, sondern vielmehr als eine Art Grundbausteine sozialer Interaktion, eine *Grammatik sozialer Beziehungen* (Hupfeld-Heinemann 2005), derer sich Interaktionspartner in sozialen Situationen in unterschiedlicher Weise bedienen. Vor diesem Hintergrund stellt sich die Frage nach den Einflussfaktoren, die entscheiden, welches Modell in welcher Situation als angemessen oder unangemessen erlebt wird und welches der kognitiven Schemata in einer sozialen Situation aktiviert wird. Hierbei können verschiedene Aspekte von Person und Situation eine Rolle spielen: Erstens können Menschen individuelle Tendenzen aufweisen, bestimmte relationale Modelle situations- und beziehungsübergreifend anzuwenden (vgl. Bridoux et al. 2016). Zweitens können Personen aufgrund ihrer physischen und/oder psychischen Merkmale *bei ihren Interaktionspartnern* ein bestimmtes Modell aktivieren (vgl. Reh et al. 2017). Derartige individuelle Dispositionen können allerdings nur zu einem Teil die Anwendung bestimmter relationaler Modelle in verschiedenen Situationen erklären. Der dritte, ausgesprochen bedeutsame Einflussfaktor, sind situative Hinweisreize. In der Sozialpsychologie gibt es eine Vielzahl an Studien, die die starke, oft unbewusste Beeinflussung von Verhalten durch situative Hinweisreize belegen. So konnte auch gezeigt werden, dass sich relationale Modelle durch bewusst und unbewusst wahrgenommene Hinweisreize aktivieren lassen. Beispielsweise konnten Brodbeck, Kugler, Reif und Maier (2013) in einer Reihe von Experimenten durch bestimmte Hinweiswörter in einer Teilnehmerinstruktion sowie mittels Hinweisreizen in Form kurzer Sätze mit bestimmten Schlagwörtern unterhalb der Wahrnehmungsschwelle in einer Versuchsbedingung Communal Sharing Modelle, in einer anderen Market Pricing Modelle aktivieren und so das Verhalten ihrer Versuchspersonen beeinflussen.

Wenn sich nun bereits derartige kleine Hinweisreize auf die Aktivierung relationaler Modelle auswirken, kann davon ausgegangen werden, dass die Ausgestaltung des sozialen Systems als Ganzes, vor dessen Hintergrund sich soziale Interaktionen abspielen, einen ausgesprochen bedeutsamen Einfluss darauf ausübt, welche relationalen Modelle sozialen Beziehungen zugrunde gelegt werden. Diesem Gedanken folgend beschäftigen sich Autoren im Bereich der Arbeits- und Organisationspsychologie zunehmend mit der Frage, wie sich soziale Systeme innerhalb von Organisationen auf die handlungsleitenden relationalen Modelle auswirken. So diskutieren beispielsweise Mossholder et al. (2011), wie sich das aus organisationalen Regeln und Praktiken gebildete Human-Ressource-System (HR-System) eines Unternehmens auf die relationalen Modelle, welche den Beziehungen zwischen Mitarbeitern zugrunde liegen, auswirkt und welche Konsequenzen dies für das Hilfeverhalten am Arbeitsplatz hat. Die Autoren postulieren hierbei, dass die durch das HR-System geschaffene soziale Umwelt einer Organisation ein bestimmtes *relationales Klima* schafft und sich hierdurch auf die sozialen Beziehungen zwischen Mitarbeitern auswirkt, indem bestimmte relationale Modelle aktiviert werden. So wird beispielsweise angenommen, dass Mitarbeiter eines Unternehmens mit einem bewusst kompetitiv gestalteten HR-System eher dazu tendieren, ihre sozialen Beziehungen auf Grundlage eines Market Pricing Models zu regulieren als Mitarbeiter eines Unternehmens, dessen HR-System auf ein solidarisches Miteinander und das Wohlbefinden der Mitarbeiter ausgerichtet ist. Das Unternehmen schafft demnach also durch die Regeln und Praktiken, welche die Mitarbeiter betreffen, eine soziale Umwelt. Da Menschen sich, wie oben ausgeführt, stark an den expliziten und impliziten sozialen Regeln der Gruppen, in denen sie Mitglied sind, orientieren, beeinflussen Unternehmen gewollt oder ungewollt so das Erleben und Verhalten der Mitarbeiter in sozialen Situationen innerhalb des Systems. Die Überlegungen, die Mossholder et al. (2011) auf organisationaler Ebene aufstellen, lassen sich, wie wir im Weiteren anführen werden, auch auf die Gesellschaftsebene übertragen.

19.7 Der lange Schatten von Social Credit Systemen: Die Veränderung sozialer Interaktionen und Kognitionen

Ziel eines gesellschaftsweiten Social Credit Systems ist es, das Verhalten der Menschen in allen Bereichen des Lebens zu kontrollieren und zu beeinflussen. Der „gute" Bürger zeigt sich nicht allein in seinem Verhalten als Geschäftspartner oder gegenüber dem Staat, sondern in allen Bereichen des Lebens: am Arbeitsplatz, im Bildungswesen oder in seiner Freizeit. Oben haben wir ausgeführt, dass Menschen sich insbesondere dann freiwillig prosozial verhalten und kooperativ sind, wenn sie sich als wertvolles, wertgeschätztes Mitglied einer attraktiven Gruppe wahrnehmen, das individuelle Verhalten für die anderen Gruppenglieder sichtbar ist und/oder es Sanktionsmöglichkeiten für Fehlverhalten gibt. Genau an diesen Punkten setzen allgemeingültige Social Credit Systeme an: Die Mitgliedschaft in der Gesellschaft ist für Menschen existenziell bedeutsam; durch die umfassende Beurteilung Aller, wird individuelles Verhalten für alle Gruppenmitglieder sichtbar; und bei Fehlverhalten werden bestimmte Leistungen verweigert bzw. konformes Verhalten wird mit Zugang zu Ressourcen belohnt. Damit sind Social Credit Systeme zur Verhaltensbeeinflussung voraussichtlich extrem effektiv, weil sie das grundlegende Motiv nach Zugehörigkeit ansprechen und die Handlungsoptionen von Personen, die sich nicht systemkonform verhalten, stark einschränken, was einer sozialen Exklusion gleichkommt.

Befürworter eines Social Credit Systems mögen anführen, dass dieses als eine Art „verhaltenstherapeutischer Maßnahme" angesehen werden kann, um positives Verhalten zu verstärken und negatives Verhalten zu verringern. In der Tat erinnert es an Belohnungs- und Verstärkungssysteme, die in der Verhaltenstherapie eingesetzt werden, um im Alltag erwünschte Verhaltensweisen zu verstärken. Was könnte also daran auszusetzen sein, „gutes" Verhalten zu belohnen und „schlechtes" Verhalten zu bestrafen? Ist es nicht angemessen und im Sinne der Gemeinschaft, dass beispielsweise Personen, die ein gutes Verhältnis zu ihren Mitbürgern pflegen und etwa prosoziales Verhalten zeigen, hierfür belohnt werden und im Alltag Vorteile genießen, während Personen, die sich unsozial verhalten und wenig zum Gemeinwohl beitragen, mit negativen Konsequenzen rechnen müssen? Hierbei wird allerdings außer Acht gelassen, dass ein derartiges Regelsystem, insbesondere wenn es alle Bereiche des Lebens durchdringt, einen tief greifenden Eingriff in die Psyche der Menschen darstellt und damit einen nicht zu unterschätzenden Einfluss auf die Art der sozialen Beziehungen zwischen Individuen haben kann. Ein solches System, bei dem durch extrinsische Anreize bestimmte Verhaltensweisen gefördert werden sollen, wirkt sich massiv auf die sozialen Regeln und Normen aus, vor deren Hintergrund soziales Verhalten stattfindet.

Wenn sich soziale Beziehungen auf das eigene Punktekonto auswirken, werden diese Beziehungen selbst leicht einer Kosten-Nutzen-Analyse unterzogen. Mit welchen Vor- und Nachteilen geht eine bestimmte Beziehung oder eine bestimmte soziale Handlung einher? Lohnt es sich mit einer bestimmten Person zu interagieren? Sind positive Auswirkungen auf das eigene Punktekonto zu erwarten oder muss gar mit einer Herabstufung gerechnet werden, da die Person einen niedrigeren Score hat? Aus dem Blickwinkel der

Theorie der relationalen Modelle liegt es nahe, dass soziale Beziehungen in einem solchen System zunehmend von einem Market Pricing Modell geleitet werden: Man sieht sich nicht mehr als Teil eines solidarischen Miteinanders (wie es bei einem Communal Sharing Modell der Fall wäre), sondern als Teil in einer von rationalen Kosten-Nutzen-Rechnungen geprägten Austauschbeziehung. Durch die Dominanz des Market Pricing Modells sollte es auch zu Folgewirkungen auf das Erleben und Verhalten der beteiligten Personen kommen. Am Beispiel von Hilfeverhalten könnte dies folgendes bedeuten: Einerseits kann, isoliert betrachtet, ein Social Scoring System durchaus zu einem erwünschten Verhalten führen, wenn Menschen sich, in Erwartung positiver Auswirkungen auf den eigenen Score, in positiver Weise verhalten. Eine Belohnung durch Scoring Punkte könnte beispielsweise auch solche Menschen dazu motivieren, ihren Mitmenschen gegenüber Hilfeverhalten zu zeigen, die dies ohne extrinsischen Anreiz vielleicht nicht oder in geringerem Ausmaß getan hätten. Andererseits muss dabei berücksichtigt werden, dass eine Hilfeleistung vor dem Hintergrund eines Social Scoring Systems sowohl von Seiten des Helfenden als auch von Seiten des Hilfempfängers auf Grundlage eines Market Pricing Modells wahrgenommen und beurteilt wird. Für den Helfenden bedeutet dies, dass Verhalten nur dann und nur solange gezeigt wird, wie auch positive Auswirkungen für ihn selbst zu erwarten sind. Ist dies nicht (mehr) garantiert, ist es nach der Logik eines Market Pricing Modells nicht mehr angebracht das Verhalten aufrechtzuerhalten. Im Extremfall kann ein solches System, das auf extrinsischen Anreizen basiert, gar dazu führen, dass Personen, die unter anderen Umständen bedingungslos und motiviert durch Solidarität und Gemeinschaftsgefühl geholfen hätten, in ein Market Pricing Modell und eine Kosten-Nutzen-Rechnung gedrängt werden, wodurch das eigene Hilfeverhalten reduziert wird. Dies bedeutet, dass die ursprüngliche intrinsische Motivation für ein bestimmtes Verhalten, aufgrund externer Anreize (eine Vergütung durch Social Scoring) durch eine extrinsische Motivation ersetzt wird – ein Mechanismus, der in der sozialpsychologischen Forschung unter dem Namen *Korrumpierungseffekt* bekannt ist (vgl. Deci et al. 1999, 2001) und seit Jahrzehnten erforscht wird. Auf Seite des Hilfempfängers ist wiederum zu erwarten, dass die Hilfeleistung vor dem Hintergrund eines Market Pricing Modells (unabhängig von der tatsächlichen Motivation des Helfenden) nicht als Akt der Solidarität und des Mitgefühls, sondern als Ergebnis einer Kosten-Nutzen-Rechnung erlebt wird und dementsprechend die positiven affektiven Reaktionen, die üblicherweise von Hilfeverhalten hervorgerufen werden (Ames et al. 2004), ausbleiben, weil dem Hilfegeber egoistische Motive unterstellt werden. Dies könnte im Extremfall dazu führen, dass Hilfeverhalten abgelehnt wird oder negativ besetzt ist, weil es aus Kalkül geschieht. Hilfeverhalten könnte dann auch in der unmittelbaren sozialen Interaktion als Referenz für berechtigtes Misstrauen gegenüber dem Hilfegebenden verstanden werden: „Die Person hilft nur, um ihren Score zu verbessern. Daher sollte man ihr misstrauen." Durch solche Folgewirkungen würden intendierte positive Effekte von Social Credit Systemen konterkariert und ins Gegenteil verkehrt werden.

Das Beispiel Hilfeverhalten verdeutlicht bereits, dass die Einflussnahme auf Sozialverhalten durch Social Scoring leicht unerwartete und möglicherweise ungewollte Nebeneffekte in Hinblick auf das soziale Denken und Handeln mit sich bringen kann. Wir haben dieses Beispiel zur Veranschaulichung auf der kleinsten sozialen Betrachtungsebene, der Dyade angesiedelt. Die beschriebenen Mechanismen lassen sich aber auch auf die Gesellschaft als Ganzes übertragen. Wie zu Beginn dieses Kapitels beschrieben, hängt das Gemeinwohl einer Gesellschaft von unzähligen Akten der Kooperation ab. Dabei ist es oftmals weder möglich noch erforderlich, alle Beiträge der einzelnen Individuen nachzuverfolgen. Menschen verhalten sich freiwillig gemeinschaftsdienlich und kooperativ, wenn sie sich als respektiertes und wertgeschätztes Mitglied der Gemeinschaft erleben und so ihr Motiv nach Zugehörigkeit befriedigt wird. Wenngleich aus dem Blickwinkel der Relational Models Theory kooperatives Verhalten prinzipiell aus jedem relationalen Modell heraus möglich ist, wird davon ausgegangen, dass bestimmte relationale Modelle (insb. Communal Sharing) mit einem deutlich höheren Ausmaß kooperativen Verhaltens einhergehen als andere (insb. Market Pricing) (Bridoux und Stoelhorst 2016; Mossholder et al. 2011). Dies hängt nicht allein damit zusammen, dass Kooperation ohne Abwägung von Aufwand und persönlichem Nutzen „unkomplizierter" vonstattengeht, sondern auch damit, dass sich Interaktionspartner bei aktiviertem Communal Sharing Modell als Teil der Gemeinschaft mit gemeinsamen Zielen und Interessen wahrnehmen und von Solidarität und Fürsorge geleitet sind, während sie sich bei einem Market Pricing Modell als unabhängige Individuen in einer Austauschbeziehung verstehen. In sozialen Beziehungen, die von einem Communal Sharing Modell reguliert werden, erfolgt Kooperation um ihrer selbst willen und nicht aus direktem Eigeninteresse und einer Kosten-Nutzen-Abwägung wie es bei einem Market Pricing Modell der Fall wäre.

Relationale Modelle regulieren nicht allein Austauschprozesse zwischen einer Person und ihrer sozialen Umwelt, sondern erzeugen sowohl fundamental unterschiedliche mentale Repräsentanzen einer Person von sich selbst und anderen als auch fundamental unterschiedliche moralische Grundmotive und Fairnessprinzipien. Diese Repräsentanzen, Grundmotive und Fairnessvorstellungen bilden die regulative Grundlage für das Zusammenleben in der Gemeinschaft. Eine sukzessive Verschiebung sozialer Kognitionen in Richtung eines dominierenden Market Pricing Modells durch Spezifizierung und Nachverfolgung individueller Beiträge in Form eines Social Credit Systems beeinflusst somit nicht nur die Bedingungen und das Ausmaß kooperativen Verhaltens, sondern auch die Art und Weise, wie eine Person sich selbst und ihr Umfeld wahrnimmt, wie sie moralisch urteilt und handelt und welche selbst- und fremdbezogenen Emotionen hierbei empfunden werden. In dem Ausmaß, in dem durch das Communal Sharing Modell regulierte Interaktionen seltener werden, reduziert sich auch die Identifikation des einzelnen mit der Gemeinschaft sowie die psychologische Nähe zu deren Mitgliedern. Handlungsleitende Motive wie Solidarität und Fürsorge werden ersetzt durch transaktionales Kalkül und Kosten-Nutzen-Abrechnungen. Ebenso ändern sich die mit dem moralischen Grundmotiv einhergehenden Gerechtigkeitsprinzipien anhand derer Handlungen des Einzelnen und der Gemeinschaft als Ganzes beurteilt werden. In sozialen Interaktionen, die von einem Communal Sharing Modell reguliert werden, gilt eine Güterverteilung nach dem Bedürfnisprinzip als moralisch richtig und gerecht. Dagegen gilt es in sozialen Interaktionen, die von einem Market Pricing Modell reguliert werden, als moralisch angemessen und gerecht, Güter abhängig vom individuellen (spezifizierbaren) Beiträgen zu verteilen. Dies bedeutet, dass Personen, die nicht gewillt oder nicht imstande sind, einen spezifizierbaren Beitrag zu leisten (oder auch solche, die einen zwar vorhandenen, aber nicht direkt messbaren Beitrag leisten), bei der Verteilung von Ressourcen nicht berücksichtigt oder bewusst ausgeschlossen werden und dass dies von der Gemeinschaft als moralisch angemessen wahrgenommen wird. Die in China bereits gefassten Pläne, Personen mit einem niedrigen Punktestand den Zugang zu Teilen der Infrastruktur wie dem öffentlichen Verkehrswesen zu erschweren oder zu verwehren, sind bereits ein Beispiel für eine derartige Verteilung von Gütern und Ressourcen, die in einem fundamentalen Widerspruch zu demokratischen Prinzipien staatlicher Fürsorgepflichten und Gleichbehandlungsgeboten steht. Die Konsequenz für die einzelnen Bürgerinnen und Bürger ist, dass das eigene Denken und Verhalten zunehmend darauf ausgerichtet ist, sich konform zu den Regeln des Social Credit Systems zu verhalten, Punkte zu erlangen und einen Punkteverlust zu vermeiden, wobei für Solidarität und Mitgefühl gegenüber Mitmenschen, die nicht imstande sind belohnenswertes (respektive systemkonformes) Verhalten zu zeigen, zunehmend kein Platz mehr ist.

Social Credit Systeme fördern über die beschriebenen Mechanismen eine Anpassung individuellen Verhaltens an die Vorgaben des Systems und die bestmögliche Ausnutzung der Schwächen des Systems inklusive gemeinschaftsschädigender Handlungen, solange diese vom System nicht aufgedeckt werden. Social Credit Systeme entfalten also eine korrumpierende Wirkung, indem die Betroffenen ihr Denken und Verhalten auf die Algorithmen des Systems ausrichten und nicht mehr auf die Grundlagen und Fähigkeiten menschlicher Kooperation. Auf zwischenmenschlicher Ebene wird die spontane, evolutionär geprägte Kooperationsbereitschaft von Menschen ersetzt durch die maschinellen Vorgaben eines Systems. Dies sollte mittelfristig zu mehr Anpassung, aber weniger spontanen sozialen kooperativen Handlungen führen. Warum sollte man jemandem helfen, wenn man keine Punkte dafür bekommt? Durch Social Credit Systeme wird intrinsische Motivation, also freiwilliges Verhalten, durch extrinsische Motivatoren, also Belohnung und Bestrafung, ersetzt. Diese Verschiebung des Anreizes zum kooperativen Verhalten von innen nach außen hat fatale Auswirkungen auf alle Formen von Engagement und ist in seiner destruktiven Wirkung in unterschiedlichen Kontexten und Verhaltensweisen wie der Mitarbeitermotivation oder der Kreativität empirisch gut belegt (vgl. Deci et al. 1999; Amabile 1996)

Aber auch (volks-)wirtschaftlich könnten sich mittelfristig negative Auswirkungen durch die Einführung von Social Credit Systemen ergeben, weil Denken und Verhalten in eine bestimmte Richtung gelenkt werden und dadurch Diversität verloren geht. Zusätzlich findet, wie ausgeführt, ein Verlust an intrinsischer Motivation für freiwilliges Engagement statt. Diversität und intrinsische Motivation sind aber essenzielle Voraussetzungen für Kreativität und Innovation (Maier et al. 2007). Für wissensbasierte, dienstleistungs- und entwicklungsgetriebene, rohstoffarme Volkswirtschaften, wie die meisten europäischen Länder sie haben, sind Innovationen aber der unabdingbare Motor für Fortschritt und Wohlstand. Zusammenfassend argumentieren wir, dass allgemeingültige Social Credit

Systeme neben der offensichtlichen Unvereinbarkeit mit demokratischen Grundprinzipien wie staatlichen Fürsorgepflichten, Persönlichkeitsrechten oder Freizügigkeit, durch eine nachhaltige Veränderung der sozialen Wahrnehmungen destruktive Wirkungen entfalten, die sowohl nachteilig für die sozialen Interaktionen zwischen einzelnen Personen als auch für die Gesellschaft und Wirtschaft als Ganzes sind.

Literatur

Amabile, T. M. (1996). *Creativity in context. Update to the social psychology of creativity.* Boulder, CO: Westview.

Ames, D. R., Flynn, F. J., Weber, E. U. (2004). It's the thought that counts: On perceiving how helpers decide to lend a hand. *Personality and Social Psychology Bulletin, 30*(4), 461–474. https://doi.org/10.1177/0146167203261890

Baumeister, R.F., Leary, M.R. (1995). The need to belong: Desire for interpersonal attachments as a fundamental human motivation. *Psychological Bulletin, 117*, 497–529.

Brodbeck, F. C., Kugler, K. G., Reif, J. M., Maier, M. A. (2013). Morals matter in economic games. *PLoS ONE, 8*(12). https://doi.org/10.1371/journal.pone.0081558

Bridoux, F., Stoelhorst, J. W. (2016). Stakeholder relationships and social welfare: A behavioral theory of contributions to joint value creation. *Academy of Management Review, 41*(2), 229–251. https://doi.org/10.5465/amr.2013.0475

Deci, E. L., Koestner, R., Ryan, R. M. (1999). A meta-analytic review of experiments examining the effects of extrinsic rewards on intrinsic motivation. *Psychological Bulletin, 125*(6), 627–668. https://doi.org/10.1037/0033-2909.125.6.627

Deci, E. L., Koestner, R., Ryan, R. M. (2001). Extrinsic rewards and intrinsic motivation in education: Reconsidered once again. *Review of Educational Research, 71*(1), 1–27. https://doi.org/10.3102/00346543071001001

Fiske, A. P. (1992). The four elementary forms of sociality: Framework for a unified theory of social relations. *Psychological Review, 99*(4), 689–723.

Gollwitzer, M., Lotz, S., Schlosser, T., Streicher, B. (2013). *Soziale Gerechtigkeit.* Göttingen: Hogrefe.

Greenberg, J., Colquitt, J. (2005). *Handbook of organizational justice.* Mahwah, NJ: Lawrence Erlbaum Associates.

Hupfeld-Heinemann, J. (2005). *Die Grammatik sozialer Beziehungen.* (Habilitation Thesis), Universität Bern, Bern.

Junger, S. (2016). *Tribe: On homecoming and belonging.* Toronto: Harper Collins.

Maier, G. W., Streicher, B., Jonas, E., Frey, D. (2007). Kreativität und Innovation. In D. Frey, L. von Rosenstiel (Eds.), *Enzyklopädie der Psychologie: Wirtschaftspsychologie* (S. 809–855). Göttingen: Hogrefe.

Morrison, L. (2004). Ceausescu's legacy: Family struggles and institutionalization of children in Romania. *Journal of Family History, 29*(2), 168–182.

Mossholder, K. W., Richardson, H. A., Settoon, R. P. (2011). Human resource systems and helping in organizations: A relational perspective. *The Academy of Management Review, 36*(1), 33–52. https://doi.org/10.5465/amr.2009.0402

Reh, S., Van Quaquebeke, N., Giessner, S. R. (2017). The aura of charisma: A review on the embodiment perspective as signaling. *The Leadership Quarterly, 28*(4), 486–507. https://doi.org/10.1016/j.leaqua.2017.01.001

Summereder, S., Streicher, B., Batinic, B. (2014). *Voice or consistency? What you perceive as procedurally fair depends on your level of power distance Journal of Cross-Cultural Psychology, 45*, 192–212. https://doi.org/10.1177/0022022113505356.

Tomasello, M. (2010). *Origins of human communication.* Cambridge, MA: MIT Press.

Wang, Y. (2017). For whom the bell tolls: The political legacy of China's Cultural Revolution. Retrieved from https://scholar.harvard.edu/files/yuhuawang/files/cultural_revolution_0.pdf

Williams, K. D., Forgas, J. P., Von Hippel, W. (2013). *The social outcast: Ostracism, social exclusion, rejection, and bullying.* New York, NY: Psychology Press.

Bernhard Streicher ist Universitätsprofessor für Sozial- und Persönlichkeitspsychologie und Leiter des Departments für Psychologie und Sportmedizin an der Privatuniversität Hall in Tirol. Seine Forschungsinteressen umfassen die psychologischen Mechanismen von Entscheidungen unter Risiko und Unsicherheit; die Risikokultur von Unternehmen und Gruppen; und die Verbesserung der Risikokompetenz. Zusätzlich zu seiner wissenschaftlichen Forschung arbeitet er als Redner und Berater zum Thema Risiko für profit und non-profit Unternehmen.

Johannes F. W. Arendt promovierte an der LMU München zu relationalen Modellen am Arbeitsplatz und ist seit 2019 wissenschaftlicher Mitarbeiter am Department für Psychologie und Sportmedizin an der Privatuniversität Hall in Tirol. In seiner Forschung beschäftigt er sich unter anderem mit verschiedenen Aspekten sozialer Beziehungen wie Gerechtigkeitswahrnehmung, Hilfeverhalten und interpersonellen Risiken im Arbeitskontext.

Teil V

Nachhaltigkeit im Rating

Green Economy, Green Deal und Sustainable Finance – Die zentrale Rolle von Nachhaltigkeitsratings

Henry Schäfer

20.1 Green Economy und Green Finance – neue Programmatiken für drängende Probleme aufgrund mangelnder Nachhaltigkeit

Gibt man heutzutage den Begriff „Nachhaltigkeit" in eine Internet-Suchmaschine ein, erhält man mit ziemlich hoher Wahrscheinlichkeit Suchergebnisse im zweistelligen Millionenbereich. Diese hohe Zahl offenbart auch, dass weltweit eine Vielfalt von Verständnissen, Definitionen und Programmatiken zu Nachhaltigkeit kursiert. Lange Zeit hingegen hatte es den Anschein, dass es einen gewissen inhaltlichen Konsens zu Nachhaltigkeit gibt. Historische Bezugsgrundlage bildete das 1713 vom kurfürstlich-sächsischen Oberberghauptmann Hans Carl von Carlowitz veröffentlichte forstwirtschaftliche Werk „Sylvicultura Oeconomica". Hierin wurde nicht nur die Idee und der Begriff der Nachhaltigkeit erstmals explizit behandelt, sondern auch in klaren Umrissen die Dreidimensionalität der Nachhaltigkeit formuliert (vgl. Höltermann und Oesten 2001). In der Nachhaltigkeitsvorstellung wie sie 1987 von der von den Vereinten Nationen eingesetzten „World Commission on Environment and Development" (sogenannte Brundtland-Kommission) formuliert wurde, spiegeln sich die damaligen Grundelemente der Nachhaltigkeit wieder: „(…) eine Entwicklung, die die Bedürfnisse der Gegenwart befriedigt, ohne zu riskieren, dass künftige Generationen ihre eigenen Bedürfnisse nicht befriedigen können" (Hauff 1987, S. 46).

Ihre aktuellste „Modernisierung" erhielt das Konzept der nachhaltigen Entwicklung mit dem einstimmigen Beschluss über die 17 Sustainable Development Goals (SDGs) im Jahr 2015 durch die 193 Mitgliedsstaaten des UN-Nachhaltigkeitsgipfels in New York im

H. Schäfer (✉)
Universität Stuttgart, Stuttgart, Deutschland
E-Mail: henry.schaefer@bwl.uni-stuttgart.de

© Springer Fachmedien Wiesbaden GmbH, ein Teil von Springer Nature 2020
O. Everling (Hrsg.), *Social Credit Rating*,
https://doi.org/10.1007/978-3-658-29653-7_20

Rahmen der 2030-Agenda. Sie lösen die seit dem Jahrtausendwechsel geltenden Millennium Development Goals der Vereinten Nationen ab.

Die 17 SDGs mit ihren 169 Unterzielen streben eine ganzheitliche nachhaltige Entwicklung an. Viel eindringlicher als die SDGs haben sich aber über die letzten Jahre die Auseinandersetzungen um die Folgen des Klimawandels, verursacht durch den anthropogenen Treibhausgasausstoß der vergangenen Jahrzehnte und deren negativen Folgen für Umwelt, Gesellschaft und Wirtschaft in der öffentlichen Aufmerksamkeit festgesetzt. Aufgeschreckt durch die Krisenszenarien der zahlreichen Studien des Weltklimarates wurde auf der Weltklimakonferenz im Dezember 2015 mit dem sogenannten „Paris Alignment" ein neuer Meilenstein für die internationalen Bemühungen zur Reduktion des weltweiten Treibhausgasausstoßes gesetzt. Es stellt eine von vielen als längst überfällig erachtete Fortführung der 1992 in Rio de Janeiro formulierten UN-Klimarahmenkonvention dar. Sie mündete 1997 in das Kyoto-Protokoll, mit dem erstmals eine internationale, weitreichende Vereinbarung bestand, um konkrete klimaschützende Maßnahmen und Institutionen wie Emissionshandelssysteme auf der Basis internationaler Konventionen einzurichten. Mit dem Paris Alignment zur UN-Klimarahmenkonvention wurden erstmals verbindliche Ziele zu Treibhausgasemission gesetzt, welche auf nachfolgenden Conferences of the Parties fortwährend zu spezifizieren sind (UNFCCC 1998).

Seither ist es in vielen Industrie- und Schwellenländern zur Formulierung und Verabschiedung von Klimaschutzzielen gekommen, begleitet von einer intensiver werdenden Auseinandersetzung um Sinn und Zweck des Wirtschaftens. Ordnungspolitische Aufmerksamkeit erzielte in jüngster Zeit das Leitbild einer Green Economy wie es erstmals vom Umweltprogramm der Vereinten Nationen (UNEP) 2010 formuliert wurde: „A green economy (GE) can be defined as one that results in improved human well-being and social equity, while significantly reducing environmental risks and ecological scarcities." (UNEP FI 2010, S. 4). Die Vorstellung einer Green Economy entstand im Nachgang zu den weltweiten Folgen der Finanzkrise und soll im Rahmen eines Global Green New Deal eine wachstumsorientierte und gesellschaftlich verträgliche Wirtschaftspolitik ermöglichen. Aus den Grundüberlegungen entwickelte sich über mehrere Arbeitsschritte in den Vereinten Nationen das Leitbild einer Kohlenstoff reduzierten, Ressourcen effizienten und sozial inklusiven Wirtschaft: „In a green economy, growth in income and employment are driven by public and private investments that reduce carbon emissions and pollution, enhance energy and resource efficiency, and prevent the loss of biodiversity and ecosystem services" (UNEP 2011, S. 16). Die im November 2019 neu gebildete EU-Kommission hat dies zu ihrer ordnungs- und wirtschaftspolitischen Programmatik erkoren, indem sie für ihre Amtsperiode die Umsetzung einer Strategie des „European Green Deal" proklamiert: „The European Green Deal is (…) a new growth strategy that aims to transform the EU into a fair and prosperous society, with a modern, resource-efficient and competitive economy where there are no net emissions of greenhouse gases in 2050 and where economic growth is decoupled from resource use." (EU-Commission 2019, S. 2). Dadurch erweitert sie ihren 2018 proklamierten Aktionsplan für nachhaltiges Wachstum (EU Kommission 2018).

Basierend auf den Nachhaltigkeitsvorstellungen der Brundtland-Kommission wurde mit der Verkündung der sechs Principles for Responsible Investment im Jahr 2006 durch den damaligen UN-Generalsekretär, Kofi Annan, dem Finanzsektor eine Potenzialrolle für die Erreichung nachhaltiger Entwicklungsziele beigemessen. Aber erst die Empfehlungen der High-Level Expert Group on Sustainable Finance Ende 2017 und der „EU Action Plan for a Sustainable Economy" vom Frühjahr 2018 haben breite vorregulatorische Aktivitäten einer radikalen Instrumentalisierung von Finanzmärkten und – instituten für Nachhaltigkeitsziele in Gang gesetzt. Es setzt sich seither zunehmend in politischen und zivilgesellschaftlichen Kreisen die Überzeugung fest, dass Finanzmärkte mit ihren Akteuren einen gewichtigen Beitrag zur Erreichung der Ziele einer Green Economy zu leisten haben (vgl. OECD 2016; UNCTAD 2015; Sikken 2011). Hierzu wurde der Begriff „Green Finance" geprägt. Er zielt auf eine ganzheitliche Betrachtung der ökologisch nachhaltigen Entwicklung ab: „A ‚green finance system' refers to a series of policies, institutional arrangements and related infrastructure building that, through loans, private equity, issuance of bonds and stocks, insurance and other financial services, steer private funds toward green industry" (Green Finance Task Force 2015, S. 6).

Green Finance umfasst die Finanzierung sowohl von öffentlichen als auch privaten Investitionen für Innovationen, Unternehmen, Geschäftsmodelle, Großprojekte und Infrastrukturen, die positive ökologische Wirkungen (Impacts) erzielen. Diese sind vielfältig und reichen von einer Reduzierung der Luft-, Wasser- oder Landverschmutzung über eine Reduzierung von Treibhausgasemissionen, einer verbesserter Energieeffizienz und Verwendung natürlicher und nachwachsender Ressourcen bis hin zu Mitigations- und Adaptionsbemühungen in Wirtschaft und Gesellschaft (vgl. Stern 2009, S. 79–81). Die Finanzierung von Aktivitäten zur Mitigation oder von Anpassungsmaßnahmen wird unter dem Begriff „Klimafinanzierung" (Climate Finance) zusammengefasst. Green Finance ist als Querschnittsthema zu verstehen, denn es kann alle Geschäftsmodelle im Finanzsektor durchziehen und ist nicht beschränkt auf bestimmte Finanzdienstleister und Finanzprodukte. Dies macht Green Finance zu einem universellen Leitbild für den Finanzsektor – korrespondierend zum Leitbild der Green Economy im Realsektor.[1]

20.2   ESG-Informationen als zentrale Voraussetzung für eine nachhaltige Entwicklung

Dem Finanzsektor wird aufgrund dieser Entwicklungen zukünftig eine zentrale Scharnierfunktion als Risikomanager, Kapitalvermittler und Liquiditätsanbieter für Nachhaltigkeit und insbesondere die Dekarbonisierung zuteil. Es ist festzustellen, dass der Einfluss des

---

[1] Derzeit kursiert neben Green Finance auch der Begriff „Sustainable Finance". Die Trennlinie ist unscharf, da beiden Begriffen der systemische Gedanke, vor allem die Verbindung zwischen einzelwirtschaftlichen Finanzentscheidungen und Wirkungen auf Umwelt, Soziales und Gesellschaft zu eigen ist. Stellenweise finden sich Vorstellungen, dass mit dem vorangestellten Adjektiv „grün" und dem Substantiv „Klima" eine Spezialisierung des Blickwinkels auf die Finanzsphäre aus Umweltsicht erfolgt, während der Nachhaltigkeitsblick den Anspruch umfassender, zum Beispiel auf die 17 Sustainable Development Goals gerichtete Vorstellung beansprucht.

Klimawandels zunehmend in das Bewusstsein von Finanzsektor und -aufsicht rückt. Bemerkenswert ist jedoch an der gegenwärtigen Diskussion um Green und Sustainable Finance der Aspekt des Risikos, das aufgrund mangelnder Nachhaltigkeit entsteht. Es sind vor allem klimabedingte Risiken, die derzeit von Aufsichtsbehörden diskutiert und als regulierungs bedürftig angesehen werden (vgl. Network for Greening the Financial System 2019). Solche Risiken haben oft die spezielle Eigenschaft sogenannter „Tail Risks", das heißt, im Gegensatz zu üblichen normalverteilten Marktpreisrisiken können Nachhaltigkeits- und Klimarisiken zwar eine geringe Eintrittswahrscheinlichkeit, aber enorme negative finanzielle Auswirkungen haben (zum Beispiel Vermögensverluste erzeugt durch plötzliche klimabedingte Kursrückgänge bei Aktien). Sie sind mit den bei Finanzinstituten am häufigsten verwendeten Modellen der Risikoquantifizierung (Value at Risk-Modelle) nicht effizient genug erfassbar (vgl. Schäfer 2016, S. 151 ff.). Risiken aufgrund mangelnder Nachhaltigkeit erfordern im Gegensatz dazu eine intensive Prüfung der sie verursachenden Ereignisse (zum Beispiel durch Stresstests). Ferner sind etliche dieser Risiken systemischer Natur, das heißt, sie können nicht durch die übliche Diversifikation einer „guten Mischung" von Kapitalanlagen gemindert oder gar ausgeschaltet werden (vgl. TCFD 2016).

Die Erreichung von Nachhaltigkeitszielen und die Umsetzung der Prinzipien einer Green Economy bedürfen eines funktionsfähigen Zusammenspiels von Kapitalgebern und -nehmern sowie begleitender privatwirtschaftlicher und öffentlicher Ermöglicher. Abb. 20.1 veranschaulicht die Strukturkomponenten und Akteure eines solchen „nachhaltigen" Eco-Systems.[2]

Mit einem grünen Eco-System einher geht ein sprunghaft steigender Bedarf an Informationen zu Umwelt, Sozialem und guter Unternehmensführung, was im Englischen mit „ESG" abgekürzt wird. Vor allem um Klimadaten zu erfassen, bedarf es nicht nur spezifischer Messkonzepte, sondern auch entsprechender Produktionstechnologien, die die sogenannten nicht-monetäre Intermediäre als Bindeglieder zwischen Kapitalgeber und -nehmerseite bereits seit den 1980er-Jahren anwenden.[3]

Kennzeichnend für solche (nicht-monetären) Ratingeinrichtungen, bankinternen Analystenabteilungen und Betreibern von Wertpapierindizes ist das Fehlen eines einheitlichen bzw. genau definierten Nachhaltigkeitsparadigmas und Kriterienkataloges (vgl. Schäfer et al. 2006). Trotz wiederkehrender Elemente verbleibt in den derzeit bekannten Ratingkonzepten von Nachhaltigkeit eine hohe Individualität. Dies ist jedoch kaum nachvollziehbar, weil die einzelnen Konzepte sehr unterschiedlichen, individuellen Motiven der Anbieter und unter Umständen sehr unterschiedlichen Vorstellungen von Nachhaltigkeit bzw. Ethik entspringen.

---

[2] In den folgenden Ausführungen wird ausschließlich auf die Rolle von Unternehmen für die Erzielung von Nachhaltigkeit Bezug genommen. Die Ausführungen zu Nachhaltigkeitsratings beziehen sich demzufolge ebenfalls ausschließlich auf Unternehmen. Auf das ebenfalls anzutreffende Nachhaltigkeitsrating von Staaten sei auf Schäfer und Sauter 2016 verwiesen.

[3] Gleichwohl stellen die klimabezogene Berichterstattung und die Ermittlung von klimarelevanten Indikatoren bislang noch eine zentrale Herausforderung dar. Entsprechende Erhebungsmodelle und Datenqualitäten stehen bislang noch eher am Anfang (vgl. Busch et al. 2018).

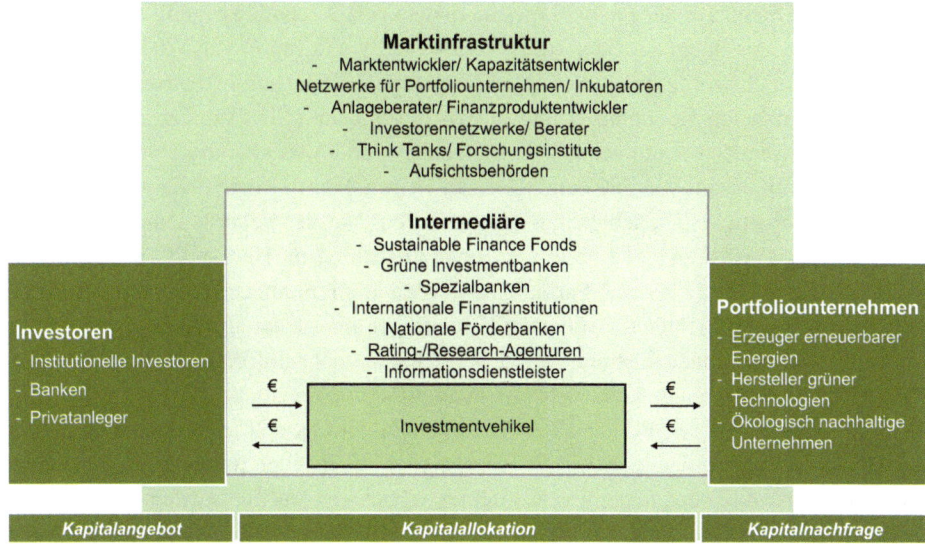

Abb. 20.1 Strukturkomponenten eines nachhaltigen Eco-Systems im Finanzbereich. (Quelle: eigene Darstellung)

Seit den 1970er-Jahren wurde im Zuge von Weiterentwicklungen der staatlichen Umweltpolitik immer mehr auch die Frage nach einer Messung von Wertschöpfung nicht nur auf der Makroebene einer Gesamtwirtschaft, sondern auch auf der Mikroebene von Unternehmen thematisiert.[4]

Diese einstmals sehr umweltorientierte Diskussion mündete ein in Systeme zur Umweltberichterstattung, die mittlerweile in Großunternehmen beinahe überall Einzug gehalten haben (vgl. Schäfer und Lindenmayer 2004). Seit den 1990er-Jahren fordern zudem einzelne Stakeholder und Nichtregierungsorganisationen (NGOs) immer stärker sozial verantwortliches Verhalten von Unternehmen ein. Unternehmen werden damit nicht länger nur hinsichtlich ihres finanziellen Erfolgs für die Kapitalgeber, sondern auch bezüglich ihres Beitrags für die Gesellschaft und von für Unternehmen kritischen Stakeholderkreisen beurteilt. Damit entstand ein völlig neuer Informationsbedarf, da jetzt Unternehmen nach dem Prinzip der Triple Bottom Line zu beurteilen sind: Gemessen wird nicht nur der finanzielle Erfolg (die Single Bottom Line einer finanziellen Rechnungslegung), sondern ergänzend oder integriert die ökologische und soziale Leistung eines Unternehmens (vgl. Elkington 1998). Weltweit dokumentieren heutzutage bereits zahlreiche Unternehmen ihre Leistungen und Erfolge durch eine ökologische und soziale Rechnungslegung. Im Gegensatz jedoch zur finanziellen Rechnungslegung, die oft an international weitgehend einheitlichen Standards etwa des International Financial Reporting Systems (IFRS) oder der US Generally Accepted Accounting Principles (US-GAAP) ausgerichtet ist, liegen

---

[4] Beispielhaft für die auf gesamtwirtschaftlicher Ebene seinerzeit entwickelten Messkonzepte sei auf Sozialindikatorensysteme verwiesen (vgl. zum Beispiel Nordhaus und Tobin 1972).

vergleichbare Standards für die ökologische und soziale Rechnungslegung derzeit allenfalls im Ansatz des Integrated Reportings vor (McElroy und van Enjelen 2012, S. 136).

In jüngster Zeit wurde durch die Umsetzung der EU-Reporting Directive 2014/95/EU zur nicht-finanziellen Berichterstattung in den EU-Staaten (EU Commission 2018) für kapitalmarktorientierte Unternehmen sowie Banken und Versicherungsunternehmen ab einer Beschäftigtenzahl von 500 eine gesetzliche Grundlage zur regelmäßigen und strukturierten Nachhaltigkeitsberichterstattung geschaffen. Mit der Verabschiedung des Non-Financial Disclosure Amendment für Finanzinstitute im Herbst 2019 durch die EU-Kommission und das EU-Parlament werden dem Finanzsektor weitergehende Berichtserstattungspflichten u. a. über die Klimafolgen ihrer Kapitalanlagen auferlegt. Im Gegenzug werden Finanzinstitute von Unternehmen mehr Informationen über deren Treibhausgasauswirkungen einfordern (EU Kommission 2019).

In Verbindung mit den Berichterstattungspflichten wird aktuell auch wieder der Blick auf diejenigen Intermediäre gerichtet, die solche nicht-finanziellen Informationen erheben, verdichten, auswerten und in den Finanzmärkten verbreiten. Das Ziel solcher Anbieter von Nachhaltigkeitsratings ist es, durch die Auswertung von unternehmensspezifischen ESG-Daten sowohl eine absolute, unternehmensindividuelle Nachhaltigkeitsbewertung, als auch eine relative Nachhaltigkeitsposition von Unternehmen etwa im Branchen- oder Länderkontext zu bestimmen. Anbieter von Nachhaltigkeitsratings entwickelten hierfür eigenständige Erhebungssysteme, das heißt auf Paradigmen der Nachhaltigkeit oder der Corporate Social Responsibility (CSR) ausgerichtete Messkonzepte und Beurteilungsverfahren. Sie übernehmen damit primär auf Finanzmärkten eine intermediäre Rolle in einem arbeitsteiligen Prozess der Allokation von Kapital wie es aus Abb. 20.2 zu ersehen ist. Damit ist auch beschrieben, dass die primäre Nachfrage nach nicht-finanziellen Ratings traditionell von Kapitalmarktteilnehmern, insbesondere Anlegern und Finanzinstituten ausgeht.

Wirtschaftstheoretisch machen Ratings ökonomisch Sinn, wenn Wirtschaftssubjekte nur begrenzte Kapazitäten haben, um die für optimale Entscheidungen benötigten Infor-

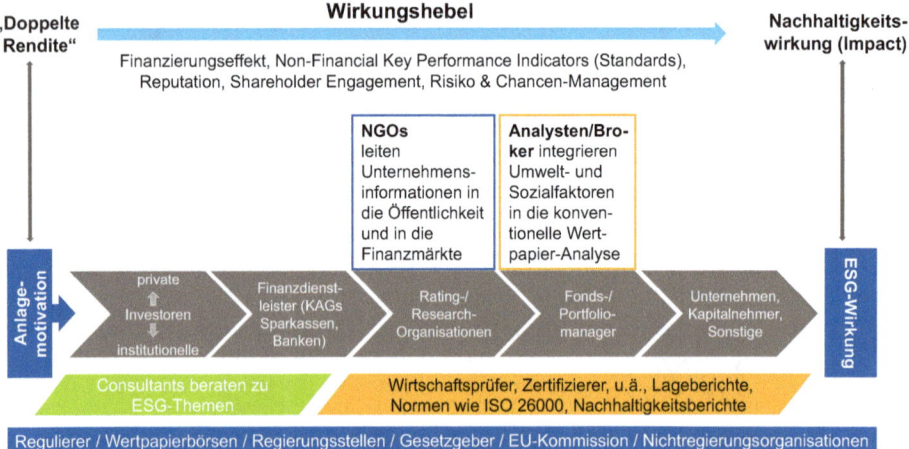

Abb. 20.2 Wertschöpfungskette in einem „grünen" Eco-System. (Quelle: eigene Darstellung)

mationen zu erheben und zu verarbeiten. Wie die Neo-Institutionenökonomik lehrt liegen in solchen Fällen asymmetrische Informationsverteilungen vor, die zu opportunistischen Verhaltensweisen führen, das Entstehen von Märkten behindern oder vorhandene Märkte ineffizient machen können. Sofern sich technologisch Wege finden, diese Informationslücken zu füllen und sich dafür ein Markt etablieren lässt, können Intermediäre Informationsleistungen wie Ratings anbieten, mit denen die Informationsprobleme beseitigt oder gelindert werden.

Existieren auf Märkten Anbieter, die aufgrund von Economies of Scale und Economies of Specialization die Transaktionskosten für zum Beispiel Anlageentscheidungen senken, so werden Wirtschaftssubjekte unter Abwägung des Kosten/Nutzen-Verhältnisses anstelle der eigenen Informations- und Kommunikationsaktivitäten auf die (marktmäßig) angebotenen Informationsdienstleistungen der Intermediäre zurückzugreifen. Abb. 20.3 liefert einen Überblick über die wichtigsten Gruppierungen von Anbietern für Nachhaltigkeitsinformationen auf Finanzmärkten und ihren Beziehungen untereinander.

Begründet wird die Existenzberechtigung von Informationsintermediären wie Ratinganbieter mit ihrem komparativen Vorteil in der Technologie ihrer Informationsproduktion, also nach Erhebung von direkt in den Märkten befindlichen ESG-Informationen diese in neue Informationsformate umzuwandeln (zum Beispiel in Form eines Ratingberichts oder Ratingurteils) (vgl. Schäfer 2005a). Die Transformation der ursprünglichen Informationen basiert auf der Selektion und Anwendung geeigneter Erhebungstechniken sowie der Technologien zur Informationsbeschaffung, – verarbeitung und – strukturierung. Die individuelle Produktion einer Ratingeinrichtung ist bestimmt durch ihre Kapa-

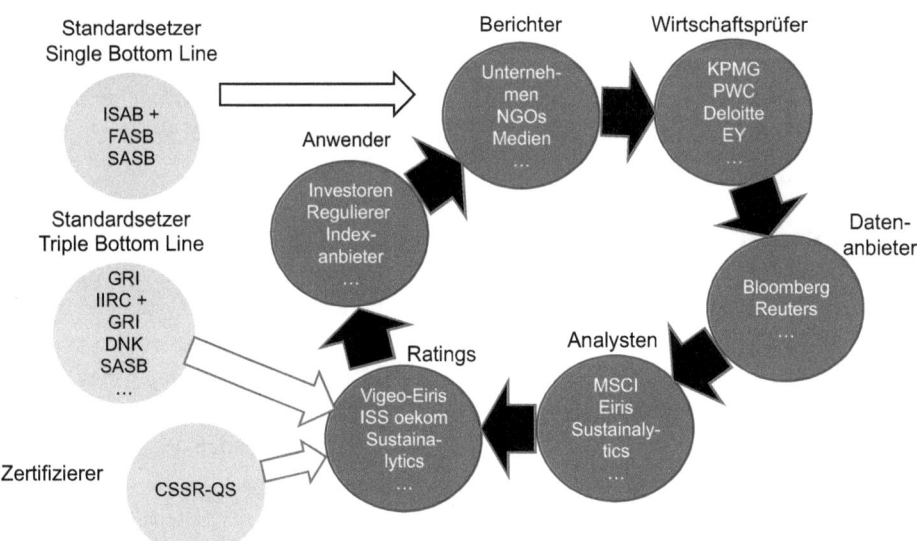

Abb. 20.3 Beziehungsgeflecht am Markt für Nachhaltigkeitsinformationen. (Quelle: White 2012, S. 6)

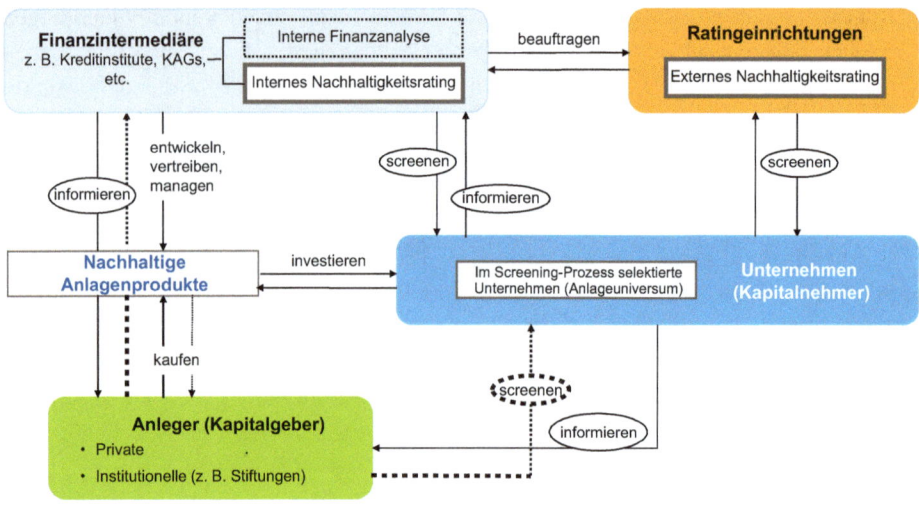

Abb. 20.4 Netzwerkbeziehungen bei SRI. (Quelle: eigene Darstellung)

zitäten, angewandten Technologien und den Zwecken der Informationsproduktion. Ratinginformationen zu ESG stellen somit ökonomische Güter dar, die produziert und auf speziellen Märkten (Märkte für Ratingdienstleistungen) gehandelt werden. Bislang dominiert hier noch das Auftragsverhältnis des sogenannten Unsolicited Rating, das heißt, die Ratingeinrichtung führt ein Nachhaltigkeitsrating auf eigene Veranlassung durch. Mittlerweile geht immer öfter von Unternehmen die Initiative zur Erstellung von Nachhaltigkeitsratings aus. Dies ähnelt dann dem Solicitated Rating, bei dem wie im marktmäßigen Credit Rating ein explizites Auftragsverhältnis vorliegt (vgl. Schäfer 2001). Da die Hauptabnehmer Finanzmarktakteure sind, sind Einrichtungen des Nachhaltigkeitsratings auf Finanzmärkten eingebettet in einen arbeitsteiligen Prozess im Bereich von Sustainable and Responsible Investments (SRI) wie er in Abb. 20.4 dargestellt ist.[5]

20.3   Strukturmodell des CSR-Ratings

Unabhängig von individuellen Ausgestaltungen und methodischen Analysetechnologien lässt sich ein allgemeines Strukturmodell des Nachhaltigkeitsratings feststellen (vgl. im Folgenden Schäfer 2009). Dabei ist relevant, für welche Adressaten ein solches Rating erstellt wird. Derzeit werden sie fast überwiegend für Kapitalmarktzwecke angeboten. In Abb. 20.5 wird ein Strukturmodell vorgestellt, das einen stilisierten Regelprozess des

---

[5] „Sustainable and responsible investment (‚SRI') is a long-term oriented investment approach which integrates ESG factors in the research, analysis and selection process of securities within an investment portfolio. It combines fundamental analysis and engagement with an evaluation of ESG factors in order to better capture long term returns for investors, and to benefit society by influencing the behaviour of companies" (Eurosif 2018, S. 12.).

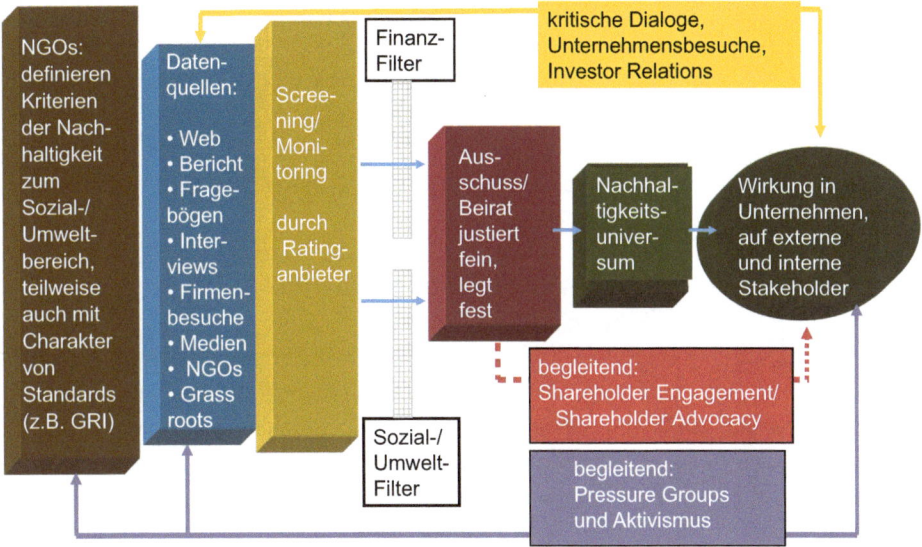

Abb. 20.5 Strukturmodell des CSR-Ratings für Kapitalmarktakteure. (Quelle: eigene Darstellung)

Screening und Monitoring von Unternehmen im Hinblick auf deren Nachhaltigkeitsleistung.

Ausgangspunkt ist die Spezifikation, Qualifizierung und Quantifizierung relevanter Fi-
nanz-, Sozial- und Umweltkriterien. Hierbei finden bereits relevante Prozesse statt, die das
spätere Ratingurteil prägen. Die Analysebasis für die nicht-finanziellen Kriterien liefern in
fast allen Nachhaltigkeitsratings mittelbar oder unmittelbar Vorstellungen von NGOs.
Teilweise haben einzelne NGOs und supranationale Organisationen hier Standards gesetzt
(zum Beispiel OECD Guidelines for Multinational Enterprises) (vgl. zum Überblick
KPMG 2014, S. 36 ff.). Grundlage für diese Kriterienformulierungen, oftmals gerade auch
mit normativem Anspruch, sind die Erfahrungen von NGOs im Umgang mit Unternehmen
auf ganz spezifischen Konfliktfeldern zwischen Gesellschaft und Unternehmen.

Ratingeinrichtungen verdichten die von ihnen vor allem aus veröffentlichten Nachhal-
tigkeitsberichten erhobenen ESG-Daten von Unternehmen zu einem Gesamturteil und
liefern eine relative Vergleichbarkeit von Unternehmen über deren Nachhaltigkeitsleistung. Je nach Ausgestaltung der Ratingsysteme kann die Auswahl von Unternehmen auf
Filtern basieren, denen

- entweder ausschließlich Sozial- und Umweltkriterien (oder auch nur eine Kriterien-
  gruppe) zugrunde gelegt werden oder
- zusätzlich Finanzkriterien integriert werden (wobei hier noch unterschieden werden
  kann, ob der Finanzfilter dem Sozial- und Umweltfilter vor- oder nachgelagert ist bzw.,
  ob eine gleichrangige Berücksichtigung stattfindet).

Das endgültige Urteil hinsichtlich der Erfüllung der von einer einzelnen Ratingeinrichtung gesetzten Anforderungen an Unternehmensnachhaltigkeit wird bei vielen Ratinganbietern durch ein Gremium erledigt, das als Beirat oder Ausschuss installiert ist. Resultat dieses Prozessschrittes ist ein Universum verantwortungsvoller Unternehmen (zum Beispiel einer Branche), auf denen dann Adressaten des Ratings ihre Entscheidungen und Handlungen (wie die Bestückung eines Anlageportfolios) basieren können.

20.4   Was zählt: Von Materialität und Indikatoren

Von jeher kann wie in Abschn. 20.3 dargelegt ein gewisses Grundmodell des Nachhaltigkeitsratings im Markt festgestellt werden und die angebotenen Ratingleistungen der mittlerweile sechs global agierenden Ratingeinrichtungen (ISS ESG (vormals oekom research), Morningstar (Sustainalytics), MSCI (ehemals KLD), Moody's (Vigeo-Eiris), Thomson Reuters (Ethical Corp.) und Standard & Poor's (SAM ESG)) scheinen auf den ersten Blick vergleichsweise ähnlich zu sein. Dies wird u. a. auch durch die in den letzten Jahren stark zugenommene Konzentration unter den Ratingeinrichtungen deutlich (vgl. Wong und Petroy 2020, S. 7).

Bei genauerem Hinsehen zeigen sich jedoch methodische und analytische Unterschiede und Datenquellen, die nicht unerheblich sind und historische Ursprünge haben. Dies zeigt sich insbesondere im Hinblick auf die interne Konzeptualisierung von „Nachhaltigkeit", der Definition von Wesentlichkeit (Materialität) und die thematische Spezialisierung bis hin zum Leistungsschwerpunkt eines jeden Anbieters (vgl. Schäfer et al. 2006, S. 16).[6]

Evolutorisch gesehen haben sich Anbieter von Nachhaltigkeitsratings aus ganz unterschiedlichen Gründen etabliert und weiterentwickelt. Es entstanden Agenturen als Non Profit-Organisationen aus dem kirchlichen Bereich (zum Beispiel die britische Eiris), durch zivilgesellschaftliche Initiativen (zum Beispiel die französische Vigeo), aus dem Medienbereich (zum Beispiel die deutsche oekom research), durch Mitwirkungen von Alternativbanken (zum Beispiel Sustainalytics) oder durch Gründungen von ehemaligen Investmentbankern (zum Beispiel die amerikanische KLD) (vgl. Stroehle 2019, S. 39).

Grob unterscheiden lassen sich die Stoßrichtungen der heute dominierenden Anbieter von Nachhaltigkeitsratings als einerseits normativ, werteorientierte Gruppierung und andererseits privatwirtschaftlich ertragsorientiert (vgl. Stroehle 2019, S. 38).[7]

Ein zentrales Merkmal der Anbieter von nicht-finanziellen Daten aus dem Nachhaltigkeitsbereich ist, dass sie mit Nachhaltigkeit als einem Diskurthema zurechtkommen müssen (vgl. grundsätzlich hierzu Schäfer 2019, S. 27 ff.). Im Gegensatz zu finanziellen

---

[6] „The concept of extra-financial materiality expands reporting to include those factors that have not yet manifested in market opportunities or risks, or that have not yet resulted in news products, let alone in financial transactions that can be recorded in financial accounts" (Garz und Volk 2007, S. 8). Das Konzept der Wesentlichkeit ist mittlerweile kennzeichnend für weite Bereiche der Nachhaltigkeitsmessung und -berichterstattung.

[7] Frühe Autoren des Nachhaltigkeitsratings betonten noch den ethisch-ökologischen Anspruch (vgl. Homolka und Nguyen-Khac 1996).

Analysen, bei denen in Wissenschaft und Praxis ein breiter Konsens über bestimmte Parameter und deren Interpretation sowie Bedeutung besteht (sogenannter Key Financial Performance Indicator), vermisst man im Nachhaltigkeitsrating einen solchen Konsens (vgl. Hawken 2004, S. 18 und 31). Individuelle Vorstellungen über Nachhaltigkeit äußern sich dann operational in Gestalt der sogenannten „Wesentlichkeit" oder Materialität, also dessen, was der Anbieter hinsichtlich Nachhaltigkeit für wichtig erachtet. Es können grob zwei Unterschiede von am Markt operierenden Ratingeinrichtungen festgestellt werden (vgl. Novethic 2013, S. 5 ff.):

- Datenanbieter mit Werteorientierung tendieren dazu, den gesamten Einfluss eines Unternehmens auf Wirtschaft, Gesellschaft und Umwelt zu erfassen und im Ratingurteil abzubilden. Es geht vornehmlich um die Erfassung der externen Effekte unternehmerischer Wertschöpfungsprozesse und deren Output. Hierzu werden relativ häufig qualitative Daten erhoben, die sich auf die Unternehmenswertschöpfung verschiedener Stufen erstrecken. Internationale Konventionen wie der Global Compact der Vereinten Nationen, Konzepte von Standarsetzungseinrichtungen wie der International Standardisation Organisation (ISO) oder Richtlinien von international anerkannten Institutionen wie zum Beispiel der Internationalen Arbeitsorganisation, dem Dachverband nationaler Gewerkschaften, bilden die Grundlage solcher Ratings.
- Die Gruppe der ertragsorientierten Datenanbieter konzentriert sich demgegenüber bei der Bearbeitung von ESG-Daten auf die finanzielle Leistung und Rendite, die in Verbindung mit ESG-Faktoren gesehen wird. Verwendung finden in den Analysen vor allem quantitative, leistungsbasierte Indikatoren, die intern entwickelt werden oder es werden marktorientierte Benchmarks eingesetzt.

20.5 Kritische Betrachtungen

Mit der Wahl ihres Bewertungsansatzes legt eine Ratingeinrichtung ein individuelles Analysemodell zugrunde, das im Wesentlichen aus den Analysekategorien, -indikatoren und – gewichten, der Berechnungsweisen sowie den Grenzwerten, ab denen ein Unternehmen als nachhaltig eingestuft wird, besteht (vgl. Delmas und Blass 2010, S. 254 ff.). Dadurch ergeben sich signifikante Einflüsse auf das finale Ratingergebnis und institutsindividuelle Ergebnisunterschiede. Dadurch kann ein Unternehmen bei unterschiedlichen Ratingeinrichtungen zu voneinander abweichenden Beurteilungen seiner Nachhaltigkeitsleistungen gelangen. Abb. 20.6 gibt eine beispielhafte Vorstellung unterschiedlicher Nachhaltigkeitsratingergebnisse ausgewählter US-Unternehmen und Ratingeinrichtungen.

Will ein beurteiltes Unternehmen das Zustandekommen des Ratingergebnisses bei einem Anbieter ergründen, so wird es häufig keinen vollständigen Einblick in das System erhalten, da es für die Anbieter keinen Anreiz gibt, sich mit der Veröffentlichung ihrer Methodik der Nachahmung durch Wettbewerber oder neue Anbieter auszusetzen. So äußern Unternehmen häufig Kritikpunkte wie Komplexität, Intransparenz, mangelnde Sinn-

Ranking	Newsweek Green Rankings					CR's Best Corporate Citizens (US)				DJSI
	US–500	US–500	Global–100	US–500	Global–500	Top 100	Top 100	Top 100	Top 100	Index
Year	2009	2010	2010	2011	2011	2009	2010	2011	2012	2011
Pepsi Co.	119	135	87	182	296	85	13	25	22	YES
Coca-Cola	58	141	NO	289	399	56	90	14	14	NO
Haliburton	169	222	NO	277	389	NO	NO	NO	NO	YES
PG&E	66	20	NO	330	NO	28	25	22	38	YES
Nike	7	10	NO	243	355	26	23	10	9	YES
BP	N/A	N/A	92	N/A	376	N/A	N/A	N/A	N/A	NO

Abb. 20.6 Beispiel unterschiedlicher Ratings und Rankings ausgewählter US-Unternehmen und Ratingeinrichtungen. (Quelle: White 2012, S. 10); DJSI: Dow Jones Sustainability Index, N/A: „not available"

haftigkeit der Fragen und unzureichende Auseinandersetzung mit dem Unternehmen im Bewertungsprozess (vgl. GlobeScan und SustainAbility 2012, S. 1 ff.). Neben diesem „Black Box"-Problem (vgl. Windolph 2011, S. 69 f.), ist die Art und Weise sowie das Ausmaß wie Ratinganbieter Transparenz über das Zustandekommen ihrer Ratings im Einzelfall an Unternehmen weitergeben, uneinheitlich: „Corporations are given mixed messages about what is being measured" (Dillenburg et al. 2003, S. 169).

Andererseits ist festzustellen, dass Unternehmen bei der Art ihrer ESG-Berichterstattung (als die zentrale Grundlage für Nachhaltigkeitsratings) teilweise opportunistisch vorgehen. Dies liegt vor allem daran, dass der Zusammenhang zwischen berichteter und tatsächlicher Nachhaltigkeitsleistung zum gegenwärtigen Zeitpunkt weder theoretisch noch empirisch eindeutig belegt ist und dadurch Spielräume für opportunistische Verhaltensweisen in der Berichterstattung ermöglicht:

- So postuliert die Legitimacy Theory, dass Nachhaltigkeitsberichte von Unternehmen mit schwacher Nachhaltigkeitsleistung zur positiven Beeinflussung der öffentlichen Wahrnehmung strategisch eingesetzt werden (vgl. Deegan 2002). Kennzeichnend für derartige Berichte ist deren hoher qualitativer, verbaler Anteil und eine geringe Festlegung auf quantitative überprüfbare Unternehmensleistungen. Im äußersten Fall kann durch geschickte Publikation und Kommunikation ein Greenwashing geringer Nachhaltigkeitsleistung entstehen (vgl. Weber 2014, S. 104).

- Im Gegensatz dazu erklärt die Voluntary Disclosure Theory, dass Unternehmen mit hoher Nachhaltigkeitsleistung diese gezielt mit quantitativen Daten speisen, damit vor allem auf den Finanzmärkten nachhaltig ausgerichtete Investoren einen direkten Zugang zum nachhaltigen Unternehmen als Anlageobjekt erhalten (vgl. Clarkson et al. 2008).

Hummel und Schlick (2013) haben in ihrer Studie mit 50 Großunternehmen aus Deutschland und der Schweiz nachgewiesen, dass beide Theorien in der Lage sind, jeweils einen Teil der Nachhaltigkeitsberichterstattung zu erklären.[8]

Bei der Berichtsform wurde ein signifikant positiver Zusammenhang zwischen dem quantitativen Ausmaß des Berichts und der Nachhaltigkeitsleistung nachgewiesen. Zwischen dem Ausmaß der verbalen Berichterstattung und der Nachhaltigkeitsleistung besteht dagegen ein signifikant negativer Zusammenhang.

20.5.1 Kosten und Nutzen von Nachhaltigkeitsratings für Unternehmen

Für die Aufnahme und Auswertung von Informationen über die Ratingobjekte „Unternehmen" ist eine effiziente Kommunikation zwischen Ratingeinrichtungen und Unternehmen unabdingbar. Dies erfordert nicht nur seitens des Ratinganbieters entsprechende Kernkompetenzen und Kapazitäten, sondern auch für Unternehmen das Vorhalten und Pflegen von für die Datenlieferungen nutzbaren Aufbau- und Ablauforganisationen sowie Kommunikationsplattformen. Die damit verbundenen Maßnahmen können derart weitreichend sein, dass bestehende Organisationsstrukturen neu eingerichtet werden müssen und es zu Um- und Neuverteilungen von Kompetenzen innerhalb von Unternehmen kommt. Dabei ist besonders herausfordernd, dass die Anfragen und Kommunikationen einzelner Ratingeinrichtungen im Rahmen ihrer jeweiligen Methodologien und Datenformate erfolgen, die sich wie bereits erwähnt zwischen den Anbietern unterscheiden.

Für Unternehmen ergibt sich daraus oft die Notwendigkeit, unterschiedliche Datenformate vorzuhalten, zu aktualisieren und zu kommunizieren. Mittlerweile haben sich mit den Konzepten internationaler Organisationen wie der Global Reporting Initiative (GRI – G4), dem Sustainability Accounting Standards Board (SASB), der International Integrated Reporting Council (IIRC) oder der International Standards Organization (ISO 26000) universell einsetzbare Rechnungslegungswerke zur Nachhaltigkeitsberichterstattung global etabliert, auf denen die Ratingeinrichtungen ihre Datenerhebung weitestgehend stützen können (vgl. McElroy und van Enjelen 2012, S. 136). Laermann (2016, S. 12) stellt hierzu aber einschränkend fest: „At this stage, there are no universally accepted ESG reporting,

---

[8] In der Grundgesamtheit waren sämtliche Unternehmen, die zum 31.12.2011 im Deutschen Aktienindex (DAX) oder im Swiss Leader Index (SLI) abgebildet wurden, enthalten.

auditing, or accounting standards like there are for financial data". Teilweise bestehen parallel dazu noch nationale Rechnungslegungswerke wie zum Beispiel in Deutschland der Deutsche Nachhaltigkeitskodex des Rates für Nachhaltige Entwicklung (vgl. DNK 2019).

Neben solcherart Kritik an bestehenden Systemen und Praktiken von Nachhaltigkeitsratings wird ihnen unternehmensseitig durchaus ein gewisser Nutzen für die Unternehmensführung zugebilligt. Zunehmende Bedeutung erhalten Nachhaltigkeitsratings bezüglich der Erfassung von systemischen Risiken und Megarisiken in den Wertschöpfungsketten und Stakeholder-Netzwerken (vgl. Schäfer 2005b, S. 57). Vielversprechend scheint außerdem die mögliche Nutzung von Nachhaltigkeitsratings als Grundlage von SWOT-Analysen zu sein, aus welchen das bewertete Unternehmen Erkenntnisse für eigene Produkt-, Produktions- und Marktstrategien ziehen kann (vgl. Reisch 1998, S. 200). So ergab eine im Jahr 2013 durchgeführte Befragung von knapp 200 durch die Ratingagentur oekom research beurteilte internationale Großunternehmen, dass mehr als 84 Prozent angaben, Nachhaltigkeitsratings unternehmensintern für eigene Stärken/Schwächen-Analysen und als Trendradar zur frühzeitigen Erkennung neuer sozialer und umweltbezogener Themen und Entwicklungen einzusetzen (vgl. Häßler und Markmiller 2013, S. 22 ff). Ferner werden bei einigen Unternehmen Nachhaltigkeitsratings und die dazugehörigen Report zu Wettbewerbs- und Best Practise-Analysen verwendet (vgl. Kahlenborn et al. 2010, S. 9 ff.).

Da Nachhaltigkeitsratings primär von Anlegern und Finanzmarktteilnehmern genutzt werden, müssten solche Ratings Kauf- und Verkaufstransaktionen an Wertpapierbörsen mit beeinflussen und Auswirkungen auf Aktien- und Anleiherenditen sowie spiegelbildlich auf die Kapitalkosten von Unternehmen haben. Auch könnte sich durch Erschließung neuer oder Intensivierung vorhandener Anlegerkreise aus dem Bereich nachhaltiger Investoren für Unternehmen ein erweiterter Zugang zu Finanzierungsquellen ergeben (vgl. Beckmann und Schaltegger 2014, S. 351). Eine dadurch bewirkte Senkung von Eigen- und Fremdkapitalkostensätzen würde dann zur Steigerung des Unternehmenswertes führen (Nachhaltigkeit als sogenannter Investment Case) (vgl. Schäfer 2007, S. 21 ff.). Diesen Fragen gehen bereits seit mehreren Dekaden empirische wissenschaftliche Untersuchungen nach. Mehrere Meta-Studien zeigen, dass Unternehmen, die eine überdurchschnittlich gute Nachhaltigkeitsleistung aufweisen, oft auch ein niedrigeres Anlagerisiko oder eine höhere Rendite kennzeichnen (vgl. zum Beispiel Friede et al. 2015).

Dass Unternehmensleitungen von vor allem Großunternehmen den finanziellen Wertbeitrag von Nachhaltigkeitsratings erkannt haben, zeigt u. a. eine Befragung von Unternehmen zu ihrer Nachhaltigkeitskommunikation. Arnold (vgl. Arnold 2012, S. 230 ff.) ermittelte für deutsche Unternehmen, dass Anbietern von Nachhaltigkeitsratings, gefolgt von nachhaltig ausgerichteten institutionellen Investoren und Fondsgesellschaften die höchste Bedeutung beigemessen wird. Auch die Erschließung nachhaltig ausgerichteter Anlegergruppen als Kapitalgeber zum Beispiel durch die Aufnahme in einen der zahlreichen Nachhaltigkeits-Wertpapierindizes stellt mittlerweile ein Ziel der Unternehmenskommunikation dar.

20.5.2 Bewirken Nachhaltigkeitsratings Verhaltensänderungen zu mehr Nachhaltigkeit?

In einer 2010 durchgeführten Befragung von DAX- und MDAX-Unternehmen gab die überwiegende Mehrheit an, ihr Nachhaltigkeitsmanagement gar nicht oder nur teilweise an der Platzierung von Nachhaltigkeitsratings auszurichten (vgl. Kahlenborn et al. 2010, S. 19). Auch ein spätere Befragung von 22 europäischen Aktiengesellschaften ergab, dass nur für drei der befragten Unternehmen Nachhaltigkeitsratings einen Hauptantriebsfaktor ihrer Unternehmensführung darstellen, etwa um in einen der Wertpapiernachhaltigkeitsindizes aufgenommen zu werden oder um dort verbleiben zu können (vgl. Südwind 2014, S. 22 ff.). Die gleiche Erhebung förderte zu Tage, dass lediglich zwei der 22 Unternehmen konkrete Veränderungen in ihrer Unternehmensführung und Wertschöpfung aufgrund von Nachhaltigkeitsratings benennen können (zum Beispiel Formulierung unternehmenseigener Klimaziele). Ferner ergaben Erhebungen, dass die Selbsteinschätzung von Unternehmen zur Bedeutung ihres Nachhaltigkeitsleitbildes und damit verbundener Aktivitäten von der externen Beurteilung durch Ratingeinrichtungen mitunter abweicht (vgl. Häßler und Markmiller 2013, S. 19).

Wenn die Erhebungsergebnisse auch nicht repräsentativ sind, so geben sie doch Anlass zu einer kritischen Reflektion, inwiefern von Nachhaltigkeitsratings Verhaltensänderungen auf Unternehmen hin zu erhöhten Nachhaltigkeitsanstregungen und – leistungen ausgehen. Die bereits in Abschn. 20.5.1 dargelegte Ambivalenz vieler Unternehmen gegenüber dem Kosten/Nutzen-Verhältnis von Nachhaltigkeitsratings scheint demnach in der Praxis ein gewisses Beharrungsvermögen zu haben. Damit erscheint aber die wohlfahrtsökonomische Wirkung von Nachhaltigkeitsratings mit Blick auf Verbesserungen von Umwelt- und Sozialbedingungen sowie eine gute Unternehmensführung derzeit doch eher vage.

20.5.3 Methodische Kritik am Nachhaltigkeitsrating und Verbesserungsansätze

Der Grundgedanke von konventionellen Finanzratings als auch Nachhaltigkeitsratings ist eine einheitliche und objektive Beurteilung der finanziellen bzw. nicht-finanziellen Leistung von Kapitalnehmern. Während bei Finanzratings und hier insbesondere Credit Ratings weitgehender Konsens über deren Parametrisierungen und Methoden sowie den ökonomischen Folgewirkungen vorliegen (vgl. Schäfer 2010, S. 450), bestehen bei Nachhaltigkeitsratings schon wegen der unterschiedlichen Vorstellungen über die Wesentlichkeit von ESG-Informationen Abweichungen (vgl. Chatterji und Levine 2006, S. 33–35). So deckte eine aktuelle Studie durch Vergleich der Ratingergebnisse von fünf unterschiedlichen Ratingeinrichtungen teilweise erhebliche Abweichungen in den Ratingergebnissen für die gleichen Unternehmen auf. Sie beruhen insbesondere auf der Wahl der Indikatoren, deren Gewichtung und der Aggregationsmethodik (vgl. Berg et al. 2019).

20.5.3.1 Output- versus Inputbezug von Nachhaltigkeitsratings

Die Erfassung und Analyse der Nachhaltigkeitsleistung eines Unternehmens kann methodisch in eine Input oder Output orientierte Vorgehensweise ausgerichtet werden. Output basierte Ratings fokussieren auf die Ergebnisse von Unternehmensmaßnahmen hinsichtlich Umwelt-, Sozial- und Unternehmensbeziehungen (vgl. Chen und Delmas 2010). Dagegen stellt die Input orientierte Vorgehensweise auf unternehmensinterne Maßnahmen wie Managementsystem und -qualität, Produktangebot, Umweltschutzmaßnahmen etc. ab. Allen heutigen Systemen ist zu Eigen, dass sie unabhängig von Input- oder Outputorientierung weitgehend vergangenheitsorientiert sind. Damit reduziert sich die Brauchbarkeit von Nachhaltigkeitsratings für zukunftsorientierte Unternehmensentscheidungen.

Ein Defizit besteht auch immer noch darin, dass vor allem prozessbasierte Kriterien benötigt würden, aus denen aktuelle Managementpraktiken und -entscheidungen eine Prognose über die zukünftige Nachhaltigkeitsleistung eines Unternehmens zulassen, da aus dem vergangenen Output an Nachhaltigkeitsleistungen nicht zuverlässig auf den zukünftigen geschlossen werden kann (vgl. Chatterji et al. 2009, S. 162–165; vgl. Delmas et al. 2013, S. 258).

Interessant ist in diesem Kontext der Bezug zwischen Nachhaltigkeits- und Finanzleistung. So scheint die Methodik von Nachhaltigkeitsratings vor allem wegen der Auswahl der Ratingkriterien, -indikatoren und -gewichte aufgrund daraus resultierender unterschiedlicher Einschätzungen der Nachhaltigkeitsperformance unter Ratingeinrichtungen nur bedingt in einem nachweisbaren Einklang von nachhaltiger und finanzieller Performance zu stehen (vgl. Griffin und Mahon 1997, S. 20–26). Ein Einklang lässt sich dagegen nachweisen, wenn die Nachhaltigkeit auf Grundlage outputorientierter Kriterien gemessen wird (vgl. Busch und Hoffmann 2011, S. 252).

20.5.3.2 Stakeholderbezüge von Nachhaltigkeitsratings

Ein entscheidender Aspekt für die Aussagekraft von Nachhaltigkeitsratings ist das individuelle Verständnis der Ratingeinrichtungen wie Nachhaltigkeit sowohl aus der Prozess- als auch aus der Output-Perspektive heraus in den Unternehmen zu verstehen und zu messen ist (vgl. Delmas und Blass 2010, S. 246; Döpfner 2016, S. 59; Windolph 2011, S. 50 f.). Nachhaltigkeitsratings bilden damit eine Art Prisma, indem sie die individuellen Vorstellungen von Stakeholdern selektieren, bündeln und parametrisieren. Damit geht naturgemäß eine Durchschnittsbildung von Nachhaltigkeitspräferenzen verschiedener Stakeholder einher. Da Stakeholderbeziehungen teilweise auch konfliktär sein können, kann ein einheitliches Nachhaltigkeitsrating für unterschiedliche Stakeholder und deren Präferenzen Verzerrungen aufweisen.

Ein Ausweg wäre, durch Stakeholderbefragungen implizit die für ein Nachhaltigkeitsrating maßgeblichen Kriterien, Indikatoren, Gewichte, Schwellenwerte etc. zu ermitteln. In einer diesbezüglichen Studie wurden sowohl Unternehmen als auch NGOs zu deren Nachhaltigkeitsverständnis befragt. Es wurden Unterschiede in der den Kriterien zugeordneten Gewichten aufgedeckt. Anschließend wurden Stakeholder zu ihren präferierten Nachhaltigkeitsindikatoren und -gewichten befragt. Es ergaben sich dadurch Änderungen in den ursprünglichen Nachhaltigkeitsratings. Allerdings änderte sich die Reihenfolge der Unternehmen im untersuchten Unternehmenssamples hinsichtlich deren Nachhaltigkeitsleistungen nur geringfügig (vgl. Graafland et al. 2004, S. 147–151). Demnach scheint der Unterschied in den Präferenzen der Stakeholder für die relative Einordnung von Unternehmen bezüglich deren Nachhaltigkeit weniger kritisch zu sein. Dies würde auch das Problem eigentlich, dass bei fehlendem universellem und einheitlichem Nachhaltigkeitsverständnis individuelle Nachhaltigkeitsratings erforderlich wären.

20.5.3.3 Offset-Effekt von Nachhaltigkeitsratings

Um ein möglichst hochtransparentes Abbild der Nachhaltigkeitsleistung eines Unternehmens zu erhalten, ist eine differenzierte und vollständige Abbildung seiner ESG-Ausprägungen erforderlich. Nachhaltigkeitsratings sind allerdings bisher nicht oder nur sehr eingeschränkt in der Lage, eine solche Differenzierung zu leisten (vgl. Strike et al. 2006, S. 856–860). Die Operationalisierung von ESG-Kriterien, -Indikatoren und -Gewichten unabhängig von den Präferenzen unterschiedlicher Stakeholder eines Unternehmens und die Verdichtung der teilweise bis zu mehreren Hundert Einzeldaten zu einem Gesamtrating auf einen einzigen Wert führt neben der Ignoranz individueller Präferenzen zu einem Verlust von Detailinformationen. Es kann zudem zu einer Kompensation von positiven und negativen Ausprägungen der Nachhaltigkeitsleistung eines Unternehmens auf der Indikatoren- und Kriterienbasis kommen. Dieser Offset-Effekt wird aufgrund zunehmender Unternehmensgröße, fortschreitender Globalisierung und Diversifikation von Geschäftsmodellen innerhalb von Unternehmens verstärkt (vgl. Escrig-Olmedo et al. 2014, S. 561–563, und Escrig-Olmedo et al. 2017, S. 142–143).[9]

Ein methodisch überzeugender Lösungsansatz für den Offset-Effekt stammt von Escrig-Olmedo et al. (2014). Sie integrieren mittels einer Fuzzy-Logik die Gewichtung und Aggregation der Kriterien. Die Autoren wenden in ihrer Studie diese Methodik an, um Ratingergebnisse des Anbieters „Accountability Ratings" mittels eines Fuzzy-Inferenz-Systems zu simulieren. Im Vergleich zum Ranking des Accountability Ratings 2008 ergaben die Neubewertungen in einem Fuzzy-Nachhaltigkeitsranking, dass sich für einige der bewerteten Unternehmen eine Veränderung ihrer Platzierung um mehr als zehn Positionen gegenüber der Ausgangsplatzierung im Accountability Ranking ergab. Durch die Fuzzy-Methodik könnten also durchaus Offset-Effekte verringert und subjektive Nachhaltigkeitsverständnisse von Investoren in Form einer Experteneinschätzung in ein Nachhaltigkeitsrating Rating integriert werden (vgl. Escrig-Olmedo et al. 2014, S. 570–572).

---

[9] Ein diesbezüglich prägnantes Beispiel ist die Nachhaltigkeitsbewertung des Siemens Konzerns aus dem Jahr 2008. Die Vermögensverwaltungsgesellschaft Sustainable Asset Management (SAM), auf deren Rating der Dow Jones Sustainability Index (DJSI) basiert, beurteilte das damals durch Korruption gebeutelte Unternehmen zum achten Mal als qualifiziert für den Nachhaltigkeitsindex. Zwar führte das damalige Fehlverhalten von Siemens zu einigen Punkteverlusten im SAM-Rating. Es begründete aber keinen Ausschlussgrund, da SAM keinen ethischen Ansatz seinen Ratings zugrunde legt. Ein solcher wirkte bei den Ratings der oekom research (jetzt ISS ESG) und führte zu einer zu SAMs Meinung entgegengesetzten Einschätzung der Nachhaltigkeitsleistung von Siemens (vgl. Döpfner und Schneider 2012, S. 13).

20.6   Fazit und Ausblick für Nachhaltigkeitsratings

Nachhaltigkeitsratings werden von sogenannten nicht-monetären Finanzintermediären auf einem internationalen Markt für spezialisierte Informationsleistungen angeboten. Mittlerweile sind durch Fusionen und Übernahmen oligopolistische Marktstrukturen entstanden. Nach wie vor wendet sich aber die überwiegende Zahl der verbliebenen Ratinganbieter an Finanzmarktakteure; übrige Stakeholder spielen so gut wie keine Rolle. Da sich die Auswahl der zu beurteilenden Unternehmen meist an der Zusammensetzung von Wertpapierindizes ausrichtet, werden in den Ratingsystemen überwiegend börsennotierte Großunternehmen behandelt. Mittlerweile wurde allerdings der Kreis der gerateten Emittenten auf Staaten, supranationale Organisationen und Small Cap-Unternehmen sowie etliche Schwellenländer ausgedehnt. Auch Nachhaltigkeitsratings für spezielle Assetklassen wie Pfandbriefe sind am Markt vorzufinden.

Nachhaltigkeitsratings dienen dazu, Informationslücken zwischen Unternehmen und Stakeholderkreisen kostengünstig zu schließen bzw. bestehende Informationslücken zu verkleinern. Gut funktionierende Einrichtungen des Nachhaltigkeitsratings erfüllen damit eine wichtige ökonomische Bindegliedfunktion, ohne die Investoren und Finanzinstitute kaum in der Lage wären, ihre Nachhaltigkeitsziele und – vorstellungen durch Anlageentscheidungen gegenüber Unternehmen zu kommunizieren und sie entsprechend zu sanktionieren. Insofern können Nachhaltigkeitsratings auch als Social Accountings verstanden werden.

Mit der Übertragung von Intermediärsfunktionen zur Lösung eines ansonsten informationsökonomisch ineffizienten ESG-Informationsmarktes entstehen wiederum neue Unsicherheitsquellen: die Qualität der Intermediärsleistung mag schwanken, überraschende Methoden- oder Kriterienwechsel können eintreten (sogenannte Hold-up-Situation) und am Ende des Tages mag es für Außenstehende verborgen bleiben, welche konkreten Absichten und Maßnahmen sich im Ratingmodell bei der Einrichtung niedergeschlagen haben (sogenanntes Moral Hazard-Problem). Ratingeinrichtungen im Nachhaltigkeitsbereich haben dies durchaus erkannt und nach eigenen Wegen der Qualitätssicherung gesucht. Im Jahr 2004 wurde dazu die „Association for Independent Corporate Sustainability and Responsibility Research" (AI CSRR) als Dachverband europäischer Ratingagenturen gegründet, um gemeinsam den freiwilligen Qualitätsstandard Qualitätsstandard ARISTA (bis 2012 CSRR-QS) zu entwickeln. Eine marktweite Durchsetzung des Qualitätsstandards scheint aber bislang noch nicht erfolgt zu sein (vgl. Dietsche 2014, S. 22). Um die Transparenz von Nachhaltigkeitsratings zu erhöhen, taucht immer wieder die Idee eines „Rating der Rater" durch externe Stakeholder, wissenschaftliche Einrichtungen oder Beratungsinstituten auf (vgl. GlobeScan und SustainAbility 2011, S. 7).

Die hauptsächlichen Herausforderungen für Nachhaltigkeitsratings und deren Anbieter dürften in naher Zukunft sein (vgl. auch weiterführend Laermann 2016, S. 22 ff.):

- ergänzender Fokus auf den Treibhausgasausstoß von Wertpapieremittenten,
- breite Ausweitung des Ratinguniversums auf Schwellenländer und deren Wertpapieremittenten,

- Erfüllung regulatorischer Anforderungen der EU-Kommission im Rahmen der laufenden Arbeiten der Technischen Arbeitsgruppe zu Sustainable Finance,
- höhere Anpassungen an kundenindividuelle Bedarfe wie zum Beispiel Auftragsratings vergleichbar mit denen im Credit Rating-Bereich,
- neue Wettbewerber aus dem Bereich der Credit Ratings durch Integration nicht-finanzieller ESG-Parameter in deren bestehenden Systeme.

Literatur

Arnold, J. (2012): CSR-Kommunikation am nachhaltigen Kapitalmarkt, in: Umwelt-Wirtschafts-Forum (uwf), Vol. 19, S. 229–236.

Beckmann, M./Schaltegger, S. (2014): Unternehmerische Nachhaltigkeit, in: Heinrichs, H., Michelsen, G. (Hrsg.) „Nachhaltigkeitswissenschaften", S. 321–367, Berlin, Heidelberg.

Berg, F./Kölbel, J./Rigobon, R. (2019): Aggregate Confusion: The Divergence of ESG Ratings (August 17, 2019). MIT Sloan Research Paper No. 5822-19, https://doi.org/10.2139/ssrn.3438533, abgerufen am 05.01.2020.

Busch, T./Johnson, M./Pioch, Th./Kopp, M. (2018), Consistency of Corporate Carbon Emission Data, Hamburg, University of Hamburg/WWF Deutschland, https://www.wiso.uni-hamburg.de/forschung/forschungsschwerpunkte/sustainable-finance/03-topics/consistency-of-corporate-carbon-data.pdf, abgerufen am 05.01.2020.

Busch, T./Hoffmann, V. H. (2011): How Hot Is Your Bottom Line?, in: Business & Society, Vol. 50, No 2, S. 233–265.

Chatterji, A./Levine, D. (2006): Breaking down the Wall of Codes, in: California Management Review, Vol. 48, No 2, S. 29–51.

Chatterji, A. K./Levine, D. I./Toffel, M. W. (2009): How Well Do Social Ratings Actually Measure Corporate Social Responsibility?, in: Journal of Economics & Management Strategy, Vol. 18, No 1, S. 125–169.

Chen, C.M./Delmas, M. (2010): Measuring Corporate Social Performance: An Efficiency Perspective, in: Production and Operations Management, Vol. 20, No 6, S. 789–804.

Clarkson P.M./Li, Y./Richardson, G.D./Vasvari, F.P. (2008): Revisiting the Relation Between Environmental Performance and Environmental Disclosure: An Empirical Analysis. Accounting, Organizations and Society, Vol. 33, No 4, S. 303–327.

Deegan, C, (2002): Introduction: The Legitimizing Effect of Social and Environmental Disclosures – A Theoretical Foundation, in: Accounting, Auditing & Accountability Journal, Vol. 15, No 3, S. 282–311.

Delmas, M./Blass, V. D. (2010): Measuring Corporate Environmental Performance, in: Business Strategy and the Environment, Vol. 19, No 4, S. 245–260.

Delmas, M. A./Etzion, D./Nairn-Birch, N. (2013): Triangulating Environmental Performance, in: Academy of Management Perspectives, Vol. 27, No 3, S. 255–267.

Dietsche, C. (2014): Nachhaltigkeitsratings. Basisinformationen für Arbeitnehmervertretungen in Aufsichtsräten, Düsseldorf, https://www.boeckler.de/pdf/pb_mbf_dietsche_nr_2014.pdf, abgerufen am 05.01.2020.

Dillenburg, S./Greene, T./Erekson, O. H. (2003): Approaching Socially Responsible Investment with a Comprehensive Ratings Scheme: Total Social Impact, in: Journal of Business Ethics, Vol. 43, No 3, S. 167–177.

DNK (2019): Leitfaden zum Deutschen Nachhaltigkeitskodex, https://www.deutscher-nachhaltigkeitskodex.de/de-DE/Documents/PDFs/Sustainability-Code/Leitfaden-zum-Deutschen-Nachhaltigkeitskodex-Orien, Berlin, abgerufen am 05.01.2020.

Döpfner, C. (2016): Wie nützlich sind Nachhaltigkeitsratings für eine nachhaltige Entwicklung von Unternehmen?, in: Kopp, H. E. (Hrsg.): CSR und Finanzratings, Berlin, Heidelberg, S. 55–63.

Döpfner, C./Schneider, H.-A. (2012): Nachhaltigkeitsratings auf dem Prüfstand, Frankfurt am Main. https://www.cric-onlne.org/images/individual_upload/publikationen/nachhaltigkeitsstudie2012.pdf, abgerufen am 05.01.2020.

Elkington, J. (1998): Cannibals With Forks: The Triple Bottom Line of 21st Century Business, Oxford.

Escrig-Olmedo, E./Muñoz-Torres, M. J./Fernández-Izquierdo, M. Á./Rivera-Lirio, J. M. (2017): Measuring Corporate Environmental Performance, in: Business Strategy and the Environment, Vol. 26, No 2, S. 142–162.

Escrig-Olmedo, E./Muñoz-Torres, M. J./Fernández-Izquierdo, M. Á./Rivera-Lirio, J. M. (2014): Lights & Shadows on Sustainability Rating Scoring, in: Review of Managerial Science, October 2014, Vol. 8, Issue 4, S. 559–574.

EU Commission (2018): Non Financial Reporting, Bruxelles, https://ec.europa.eu/info/business-economy-euro/company-reporting-and-auditing/company-reporting/non-financial-reporting_en, abgerufen am 05.01.2020.

EU-Commission (2019): The European Green Deal, Bruxelles, https://ec.europa.eu/info/sites/info/files/european-green-deal-communication_en.pdf, abgerufen am 05.01.2020.

EU-Kommission (2019): Verordnung (EU) 2019/2088 des Europäischen Parlaments und des Rates vom 27. November 2019 über nachhaltigkeitsbezogene Offenlegungspflichten im Finanzdienstleistungssektor, https://eur-lex.europa.eu/legal-content/DE/TXT/PDF/?uri=CELEX:32019R2088&from=EN, abgerufen am 10.01.2020.

EU Kommission (2018): Aktionsplan: Finanzierung nachhaltigen Wachstums, Brüssel, https://eur-lex.europa.eu/legal-content/DE/TXT/PDF/?uri=CELEX:52018DC0097&from=EN, abgerufen am 05.01.2020.

Eurosif (2018): European SRI Study 2018. URL: http://www.eurosif.org/wp-content/uploads/2018/11/European-SRI-2018-Study-LR.pdf, abgerufen am 05.01.2020.

Friede, G./Busch, T./Bassen, A. (2015): ESG and Financial Performance. Aggregated Evidence from more than 2000 Empirical Studies, in: Journal of Sustainable Finance & Investment, Vol. 5, No 4, S. 210–233.

Garz, H./Volk, C. (2007): What Really Counts. The Materiality of Extra-financial factors, WestLB, Düsseldorf.

GlobeScan/SustainAbility (2012): Rate the Raters 2012, Polling the Experts, https://globescan.com/rate-the-raters-2012-polling-the-experts/, abgerufen am 05.01.2020.

GlobeScan/SustainAbility (2011): Rate the Raters Phase Three, Uncovering Best Practices, http://www.sustainability.com/library/rate-the-raters-phase-three#.VLksxme1mSo, abgerufen am 05.01.2020.

Graafland, J. J./Eijffinger, S./SmidJohan, H. (2004): Benchmarking of Corporate Social Responsibility, in: Journal of Business Ethics, Vol. 53, No 1/2, S. 137–152.

Green Finance Task Force (2015): Establishing China's Green Financial System. Final Report of the Green Finance Task Force, Beijing, https://www.cbd.int/financial/privatesector/chinaGreen%20Task%20Force%20Report.pdf, abgerufen am 05.01.2020.

Griffin, J.J./Mahon, J.F. (1997): The Corporate Social Performance and Corporate Financial Performance Debate: Twenty-five Years of Incomparable Research, in: Business and Society, Vol. 36, S. 5–31.

Häßler, R./Markmiller, I. (2013): Der Einfluss nachhaltiger Kapitalanlagen auf Unternehmen. Eine empirische Analyse von oekom research, München, http://www.oekom-research.com/homepage/german/oekom_Impact-Studie_DE.pdf, abgerufen am 05.01.2020.

Hauff, V. (1987): Unsere gemeinsame Zukunft. Der Brundtland-Bericht der Weltkommission für Umwelt und Entwicklung, Greven.

Hawken, P. (2004): Socially Responsible Investing. How the SRI industry has failed to respond to people who want to invest with conscience and what can be done to change it, Sausalito Cal., https://www.baldwinbrothersinc.com/wp-content/uploads/2018/03/report-harkin.pdf, abgerufen am 05.01.2020.

Höltermann, A./Oesten, G. (2001): Forstliche Nachhaltigkeit; in: Der deutsche Wald, H. 1, S. 39–45.

Homolka, W./Nguyen-Khac, T.-Q. (1996): Ethisch-ökologisches Rating; in: Büschgen, H./Everling, O. (Hrsg.), Handbuch Rating, Wiesbaden; S. 675–699.

Hummel, K./Schlick, C. (2013): Zusammenhang zwischen Nachhaltigkeitsperformance und Nachhaltigkeitsberichterstattung – Legitimität oder finanzielle Überlegungen?, in: Die Unternehmung, 67. Jg., Nr. 1, S. 36–61.

Kahlenborn, W./Dierks, H./Wendler, D./Keitel, M. (2010): Klimaschutz durch Kapitalanlagen. Wirkung von Klima- und Nachhaltigkeitsfonds auf deutsche Aktienunternehmen, Berlin, http://www.adelphi.de/files/uploads/andere/pdf/application/pdf/klimaschutz_durch_kapitalanlagen.pdf, abgerufen am 05.01.2020.

KPMG (2014) A New Vision of Value. Connecting Corporate and Societal Value Creation, http://www.upj.de/nachrichten_detail.81.0.html?&tx_ttnews[tt_news]=2848&tx_ttnews[backPid]=20&cHash=c0c79d2f9a, abgerufen am 05.01.2020.

Laermann, M. (2016): The Significance of ESG Ratings for Socially Responsible Investment Decisions: An Examination from a Market Perspective, ohne Ort, https://doi.org/10.2139/ssrn.2873126, abgerufen am 05.01.2020, abgerufen am 05.01.2020.

McElroy M.W./van Enjelen, J.M.L. (2012): Corporate Sustainability Management. The Art and Science of Managing Non-Financial Performance, London.

Network for Greening the Financial System (2019), A Call for Action: Climate Change as a Source of Financial Risk – First Comprehensive Report, https://www.ngfs.net/sites/default/files/medias/documents/synthese_ngfs-2019_-_17042019_0.pdf, abgerufen am 11.01.2020.

Nordhaus, W.D./Tobin J. (1972): Is Growth Obsolete? National Bureau of Economic Research, 96, New York, http://www.nber.org/chapters/c3621, abgerufen am 05.01.2020.

Novethic (2013): Overview of ESG Research Agencies, https://www.novethic.com/fileadmin/user_upload/tx_ausynovethicetudes/pdf_complets/2013_overview_ESG_rating_agencies.pdf, abgerufen am 05.01.2020.

OECD (2016): Development Co-operation Report 2016: The Sustainable Development Goals as Business Opportunities, https://doi.org/10.1787/dcr-2016-en, abgerufen am 05.01.2020.

Reisch, L. A. (1998): „Triple A" für die Moral? Ethisch-ökologische Bonität von Unternehmen auf dem Prüfstand, in: Neuner, M./Reisch, L. A. (Hrsg.), Konsumperspektiven. Verhaltensaspekte und Infrastruktur. Beiträge zur Verhaltensforschung, H. 33, S. 187–205.

Schäfer, H. (2019): On Values in Finance and Ethics. Forgotten Trails and Promising Pathways, Cham.

Schäfer, H. (2016): Der Einfluss ökosozialer Risiken auf die Bewertung von Assets im Investmentbanking – Forschungsstand und Anknüpfungspunkte für die Praxis, in: Wendt, K. (Hrsg.), CSR und Investment Banking. Investment und Banking zwischen Krise und Positive Impact, Berlin u.a., S. 149–169.

Schäfer, H. (2010): Corporate Social Responsibility Rating, in: Aras, G./Crowther, D. Aras, G., Crowther, D. (Eds.), A Handbook of Corporate Governance and Corporate Social Responsibility, 2009, Surrey, S. 449–465.

Schäfer H (2009): Selbstbindung von Unternehmen in der Corporate Social Responsibility (CSR) – ein neo-institutionenökonomischer Erklärungsansatz mittels Intermediären des Ratings, in: Schaal, G. (Hrsg.), Techniken rationaler Selbstbindung, Berlin, S. 165–194.

Schäfer, H. (2007): Lohnt sich verantwortungsbewusstes Handeln? Corporate Social Responsibility im Kontext der wertorientierten Unternehmensführung, in: Controlling. Zeitschrift für erfolgsorientierte Unternehmenssteuerung, 19. Jg., S. 21–26.

Schäfer, H. (2005a): International Corporate Social Responsibility Rating Systems – Conceptual Outline and Empirical Results, in: Journal of Corporate Citizenship, Vol. 20, S. 107–120.

Schäfer, H. (2005b): CSR-Rating. Ökonomisches Bindeglied zwischen Investoren und Unternehmen, in: RATINGaktuell, H. 3, 2005, S. 52–57.

Schäfer, H. (2001): Triple Bottom Line Investing – Ethik, Rendite und Risiko in der Kapitalanlage, in: Zeitschrift für das gesamte Kreditwesen, 54. Jg., H. 13, S. 740–744.

Schäfer, H, Sauter, F. (2016): Das Rating der Nachhaltigkeit von Staaten – Analyse und Bewertung existierender Ratingmethoden, in: Kopp, H. (Hrsg.), CSR und Finanzratings, Berlin, Heidelberg, S. 109–133.

Schäfer, H./Beer, J./Zenker, J./Fernandes, P. (2006): Who is Who in Corporate Social Responsibility Rating, Forschungsbericht, Bertelsmann Stiftung, Gütersloh, https://www.upj.de/fileadmin/user_upload/MAIN-dateien/Infopool/Forschung/bs_stiftung_whoiswho_2006.pdf.pdf, abgerufen am 05.01.2020.

Schäfer, H./Lindenmayer, P. (2004): Sozialkriterien im Nachhaltigkeitsrating, Düsseldorf, https://www.boeckler.de/pdf/p_edition_hbs_104.pdf, abgerufen am 05.01.2020.

Sikken, B.J. (2011): Accelerating the Transition towards Sustainable Investing – Strategic Options for Investors, Corporations and other Key Stakeholders, Geneva, https://papers.ssrn.com/sol3/papers.cfm?abstract_id=1891834, abgerufen am 05.01.2020.

Stern, N.H. (2009): The Global Deal – Climate Change and the Creation of a New Era of Progress and Prosperity, New York.

Strike, V. M./Gao, J./Bansal, P. (2006): Being Good While Being Bad, in: Journal of International Business Studies, Vol. 37, No 6, S. 850–862.

Stroehle, J. (2019): Die Ursprünge unterschiedlicher Ratingansätze, in: Absolut impact, Nr. 4, S. 36–41.

Südwind (2014): Klassenziel erreicht? Der Beitrag von „Best-in-Class"-Ratings zur Einhaltung von Menschenrechten im Verantwortungsbereich von Unternehmen, Siegburg, https://institutional.union-investment.de/dam/jcr:060862d9-4f3c-41d0-9816-d82c0720ea16/2014-01_Klassenziel_erreicht__Beitrag_von_Best-in-Class-Ratings.pdf?download=true, abgerufen am 05.01.2020.

TCFD (2016): Recommendations of the Task Force on Climate-related Financial Disclosures, https://www.fsb-tcfd.org/publications/recommendations-report/, abgerufen am 05.01.2020.

UNCTAD (2015): Investing in Sustainable Development Goals. Action Plan for Private Investments in SDGs, http://unctad.org/en/PublicationsLibrary/osg2015d3_en.pdf, abgerufen am 05.01.2020.

UNEP (2011): Towards a Green Economy: Pathways to Sustainable Development and Poverty Eradication – A Synthesis for Policy Makers, URL: https://sustainabledevelopment.un.org/index.php?page=view&type=400&nr=126&menu=1515, abgerufen am 05.01.2020.

UNEP FI (2010): Universal Ownership – Why Environmental Externalities Matter to Institutional Investors, Geneva, https://www.unepfi.org/fileadmin/documents/universal_ownership_full.pdf, abgerufen am 05.01.2020.

UNFCCC (1998): Kyoto Protocol to the United Nations Framework Convention on Climate Change, New York

Weber, T. (2014): Nachhaltigkeitsberichterstattung als Bestandteil marketingbasierter CSR-Kommunikation, in: Fifka, M. S. (Hrsg.), CSR und Reporting, Heidelberg, S. 95–106.

White, A. (2012): Redefining Value: The Future of Corporate Sustainability Ratings, in: Private Sector Opinion, Issue 29, https://www.ifc.org/wps/wcm/connect/d0bf7a6b-4a35-4a9d-8b3b-7f9bff131ee4/IFC_PSO_29.pdf?MOD=AJPERES&CVID=jJyvvWO, abgerufen am 05.01.2020

Windolph, S. E. (2011): Assessing Corporate Sustainability Through Ratings: Challenges and Their Causes, in: Journal of Environmental Sustainability, Vol. 1, No. 1, S. 61–80.

Wong, Chr./Petroy, E. (2020): Rate The Raters 2020: Investor Survey and Interview Results, https://sustainability.com/wp-content/uploads/2020/03/sustainability-ratetheraters2020-report.pdf, abgerufen am 12.08.2020.

Prof. Dr. Henry Schäfer war bis 2019 Ordinarius der Universität Stuttgart und Inhaber des Lehrstuhls „Allgemeine Betriebswirtschaftslehre und Finanzwirtschaft" sowie Leiter der Abteilung III des Betriebswirtschaftlichen Instituts der Universität Stuttgart. Eine besondere Bedeutung hat bis heute der Forschungsbereich „Sustainability & Finance". Seit 2007 ist er geschäftsführender Gesellschafter der von ihm gegründeten EccoWorks GmbH, eine Beratungsgesellschaft für Sustainable Finance und Werte orientierte Unternehmensführung.

ESG-Risiken und ihre Quantifizierung

Werner Gleißner und Frank Romeike

Zusammenfassung

Das Akronym ESG ist eine englischsprachige Abkürzung und steht für „Environment", „Social" und „(Corporate) Governance" (Umwelt, Soziales/Gesellschaft und Unternehmensführung/-struktur).

Allgemeine Beispiele für den Bereich „Environment" sind Höhe des Energieeinsatzes, Anteil erneuerbarer Energieträger, Strategie rund um das Thema Klimawandel, Emissionsausstoß. Unter „Social" sind Aspekte wie beispielsweise Achtung der Menschenrechte, Verbot von Kinder- und Zwangsarbeit, Chancengleichheit und Diversität, Arbeitsplatzgestaltung, Weiterentwicklung zu verstehen. Das Kriterium „Governance" zielt darauf ab, inwieweit Nachhaltigkeit strukturell im Unternehmen verankert ist. Darunter fallen beispielsweise Themen wie Nachhaltigkeitsmanagement, Maßnahmen zur Korruptionsbekämpfung, Umwelt- & Qualitätsmanagementsysteme, finanzielle Nachhaltigkeit und Risikomanagementsysteme.

ESG-Risiken und -Chancen haben als Ursache eine hohe Relevanz, da sie hinsichtlich Wirkung beispielsweise die Reputation oder die immateriellen Werte eines Unternehmens erheblich beeinflussen können.

Sammle deinen Reichtum, ohne seine Quellen zu zerstören, dann wird er beständig zunehmen. (Siddhartha Gautama)

W. Gleißner (✉)
FutureValue Group, Leinfelden-Echterdingen und Technische Universität Dresden, Deutschland
E-Mail: w.gleissner@futurevalue.de

F. Romeike
RiskNET – The Risk Management Network, Brannenburg, Deutschland
E-Mail: romeike@risknet.de

© Springer Fachmedien Wiesbaden GmbH, ein Teil von Springer Nature 2020
O. Everling (Hrsg.), *Social Credit Rating*,
https://doi.org/10.1007/978-3-658-29653-7_21

Es wird dabei aufgezeigt, dass auch ESG-Risiken oft finanzielle Auswirkungen haben, die als solche zu quantifizieren und bei der Beurteilung des Gesamtrisikoumfangs (Eigenkapitalbedarfs), der Kapitalkosten und des Grads der „Bestandsgefährdungen" eines Unternehmens (u. a. auch beispielsweise im Sinne von § 91 des deutschen Aktiengesetzes [AktG, eingeführt durch durch Kontroll- und Transparenzgesetz, KonTraG]) als solche zu berücksichtigen sind.

Der nachfolgende Beitrag setzt sich mit der Relevanz und Bewertung von ESG-Risiken in der Praxis auseinander. Hierbei werden auch Parallelen zum „Social Credit Rating" aufgezeigt und diskutiert. Ein besonderer Schwerpunkt liegt hierbei auf der Erläuterung der Bedeutung von Simulationsverfahren, die eine quantitative Bewertung komplexer Systeme, wie beispielsweise von Umweltsystemen oder sozialer Systeme, ermöglichen.

21.1 Überblick ESG und CSR

Vor dem Hintergrund der zunehmenden Relevanz von Klima- und Umweltrisiken (vgl. Abschn. 21.2) und sozialer Risiken, sind diese als Risiken aus der Unternehmensführung zu interpretieren, die mit der Abkürzung ESG (Environmental, Social and Governance) zusammenfasst werden.

ESG-bezogene Regularien und Offenlegungspflichten haben in den vergangenen Jahren stark zugenommen. Bereits im Jahr 2014 wurde die EU-Richtlinie 2014/95/EU verabschiedet, welche die Angabe nichtfinanzieller und Diversität betreffender Informationen durch bestimmte große Unternehmen und Unternehmensgruppen vorschreibt.

Richtlinie 2014/95/EU des Europäischen Parlaments und des Rates vom 22. Oktober 2014 zur Änderung der Richtlinie 2013/34/EU im Hinblick auf die Angabe nichtfinanzieller und die Diversität betreffender Informationen durch bestimmte große Unternehmen und Gruppen.

Die Umsetzung in nationales Recht erfolgte in Deutschland bereits im Jahr 2017 durch das CSR-Richtlinie-Umsetzungsgesetz (CSR-RUG) zur Stärkung der nichtfinanziellen Berichterstattung der Unternehmen in ihren Lage- und Konzernlageberichten. So müssen Unternehmen in ihrer Berichterstattung Aspekte wie Umwelt-, Sozial- und Arbeitnehmerbelange, Menschenrechte und Korruption berücksichtigen.

Als weiterer wichtiger Aspekt von „Governance" wird die finanzielle Nachhaltigkeit und Stabilität des Unternehmens angesehen, also der „Grad der Bestandsgefährdung" (vgl. Gleißner 2018a; Romeike 2018 sowie Romeike und Hager 2020), den man als Spitzenkennzahl des Risikofrüherkennungssystems nach KonTraG auffassen kann. Die zentrale Anforderung des KonTraG besteht gerade darin, mögliche „bestandsgefährdende Entwicklungen" früh zu erkennen (vgl. Gleißner 2018b; Romeike 2008 sowie Romeike und Hager 2020). Die Messung der finanziellen Nachhaltigkeit basiert daher auf einem, oder vielleicht besser mehreren finanziellen Indikatoren, wie im einfachsten Fall einer Insolvenzwahrscheinlichkeit (Ratingnote), die abhängig ist vom Risikodeckungspotenzial und dem aggregierten Gesamtrisikoumfang (und selbst ein Einflussfaktor für den Unternehmenswert darstellt, siehe zum Zusammenhang Gleißner 2019f sowie Gleißner und Ernst 2019). Hohe finanzielle Nachhaltigkeit hat einen positiven Einfluss auf die Performance von Unternehmen, auch an der Börse (siehe die Studie von Günther/Gleißner/Walkshäusl, 2020).

ESG-Kriterien spielen seit langer Zeit eine große Rolle, etwa im Underwriting-Prozess bei Erst- und Rückversicherungsunternehmen. In den letzten Jahren steigen jedoch vor allem ESG-Kriterien bei der Auswahl von Kapitalanlagen. Auch Ratingagenturen berücksichtigen im Rahmen der Kreditwürdigkeitsprüfung und Risikoanalyse bereits seit langem ESG-Faktoren (vgl. hierzu Verordnung (EG) Nr. 1060/2009, ABl. EU L 302/1).

Das IOSCO Growth and Emerging Markets Committee (GEMC) – als Einheit der internationalen Organisation der Wertpapieraufsichtsbehörden IOSCO (International Organisation of Securities Commissions) – hat im Jahr 2019 ein Papier mit dem Titel „Sustainable finance in emerging markets and the role of securities regulators" (vgl. IOSCO 2019) veröffentlicht. So werden in der Publikation nachhaltige Kapitalmarktprodukte, wie beispielsweise grüne und Nachhaltigkeitsfonds, sozialethische Fonds und Investitionen in erneuerbare Energien analysiert.

In elf Empfehlungen beschreibt die IOSCO ihre Erwartungen zu Nachhaltigkeitsaspekten (für eine umfassendere Darstellung des Themas Nachhaltigkeit sei verwiesen auf Günther und Günther 2017) an Aufsichtsbehörden, Unternehmen und Produkte:

- Empfehlung 1: Emittenten und beaufsichtigte Unternehmen sollen ESG-spezifische Aspekte in ihrem Risikoappetit/Risikoakzeptanz und ihre Unternehmensführung berücksichtigen;
- Empfehlung 2: ESG-spezifische Offenlegungs- und Berichtspflichten. Die Regulierungsbehörden sollten eine Offenlegung in Bezug auf wesentliche ESG-spezifische Risiken und Chancen in Bezug auf Governance, Strategie und Risikomanagement eines Emittenten oder CIS verlangen;
- Empfehlung 3: Datenqualität. Stellen die Regulierungsbehörden fest, dass eine zusätzliche ESG-spezifische Berichterstattung erforderlich ist (gemäß Empfehlung 2), sollten sie eine angemessene Datenqualität für die ESG-spezifische Berichterstattung anstreben, u. a. durch die Aktualisierung der Börsennotierungsvorschriften, die Verwendung externer Überprüfungen und durch die Tätigkeit anderer Informationsdienstleister, zum Beispiel Ratingagenturen, Benchmarks und Wirtschaftsprüfer.
- Empfehlung 4: Definition und Taxonomie nachhaltiger Instrumente. Nachhaltige Instrumente sollten klar definiert werden und sich auf die Kategorien der förderfähigen Projekte und Vermögenswerte beziehen, für die die durch ihre Emission aufgenommenen Mittel verwendet werden können;
- Empfehlung 5: Förderfähige Projekte und Aktivitäten. Mittel, die durch nachhaltige Instrumente aufgebracht werden, sollten für Projekte und Aktivitäten verwendet werden, die unter eine oder eine Kombination der unten aufgeführten allgemeinen ESG-Kategorien fallen: 1. Umwelt (erneuerbare Ressourcen, Bekämpfung des Klimawandels, Umweltverschmutzung und Abfall sowie andere Umweltmöglichkeiten); 2.

Soziales (Humankapital, Produkthaftung und andere soziale Chancen); 3. Governance (Unternehmensführung; Unternehmensverhalten).

- Empfehlung 6: Anforderungen an die Angebotsunterlagen. Die Regulierungsbehörden sollten Anforderungen für das Angebot nachhaltiger Instrumente festlegen, die unter anderem die Verwendung und Verwaltung der durch die Emission solcher Instrumente aufgenommenen Mittel sowie die von den Emittenten für die Bewertung und Auswahl der Projekte angewandten Verfahren umfassen.
- Empfehlung 7: Laufende Offenlegungsanforderungen. Die Aufsichtsbehörden sollten laufende Offenlegungspflichten hinsichtlich der Verwendung der durch die Emission nachhaltiger Instrumente aufgenommenen Mittel einschließlich des Umfangs der nicht verwendeten Mittel, falls vorhanden, festlegen.
- Empfehlung 8: Ordnungsgemäße Verwendung der Mittel. Die Regulierung sollte Maßnahmen zur Verhinderung, Aufdeckung und Sanktionierung des Missbrauchs der durch die Ausgabe nachhaltiger Instrumente aufgenommenen Mittel vorsehen.
- Empfehlung 9: Externe Überprüfungen. Die Emittenten sollten den Einsatz externer Überprüfungen in Erwägung ziehen, um die Konsistenz mit der Definition der nachhaltigen Instrumente gemäß Empfehlung 4 sicherzustellen.
- Empfehlung 10: Integration von ESG-spezifischen Aspekten in die Analyse und Strategien der Investments und die gesamte Unternehmensführung bei institutionellen Investoren und
- Empfehlung 11: Aufbau von Kapazität und Expertise für ESG-Belange.

Eine Umsetzung der oben aufgeführten Kriterien bedingt eine Methodik zur methodisch fundierten Quantifizierung der ESG-Risiken und -Chancen.

Aufgrund ihrer Nähe werden nachfolgend ESG- und CSR-Risiken als Begriffe synonym verwendet.

Das „Committee of Sponsoring Organizations of the Treadway Commission" (COSO) und das „World Business Council for Sustainable Development" (WBCSD) empfiehlt in ihrer Veröffentlichung „Enterprise Risk Management: Applying enterprise risk management to environmental, social and governance-related risks" quantitative Ansätze zur Bewertung von ESG-Risiken (vgl. The Committee of Sponsoring Organizations of the Treadway Commission (COSO) and World Business Council for Sustainable Development (WBCSD) 2018).

Das COSO (Committee of Sponsoring Organizations of the Treadway Commission) ist eine privatwirtschaftliche Organisation in den USA, die helfen soll, Finanzberichterstattungen durch ethisches Handeln, wirksame interne Kontrollen und gute Unternehmensführung qualitativ zu verbessern. COSO wurde im Jahr 1985 als Plattform für die National Commission on Fraudulent Financial Reporting (Treadway Commission) gegründet und wird durch die fünf bedeutendsten US-Organisationen für Kontrolle im Finanz- und Rechnungswesen unterstützt: Institute of Internal Auditors (IIA), American Institute of Certified Public Accountants (AICPA), Financial Executives International (FEI), Institute of Management Accountants (IMA) und American Accounting Association (AAA).

Das WBCSD ist ein Zusammenschluss von mehr als 200 Mitgliedsunternehmen, die zusammen mehr als 8,5 Billionen Dollar Umsatz und 19 Millionen Mitarbeiter repräsentieren. Das WBCSD hat sich unter anderem zum Ziel gesetzt Einfluss auf die internationale Politik zur Förderung nachhaltiger Entwicklungen zu nehmen, seine Mitglieder bei eigenen Initiativen zu unterstützen und nachhaltige Projektentwicklungen zu fördern.

In Abb. 21.1 ist der von COSO/WBCSD empfohlene Risikokatalog für ESG-Risiken wiedergegeben.

In Abb. 21.2 sind die Methoden zur Bewertung von ESG-Risiken in Form einer semiquantitativen Bewertung zusammengefasst.

COSO/WBCSD empfehlen eine quantitative Bewertung ESG-bezogene Risiken auf eine relevante Ziel- bzw. Wirkungsgröße. Für viele Unternehmen bedeutet dies, dass Risikomanagement- und Nachhaltigkeitsexperten die Wirkung eines ESG-bezogenen Risikos in Bezug auf Umsatz, Kosten oder EBITDA bewerten müssen.

Die Autoren weisen aber auch darauf hin, dass der Bedarf an quantitativen Bewertungen in der Praxis einige Herausforderungen mit sich bringen kann. Insbesondere die Wechselwirkungen vieler Unternehmen mit ESG-Themen (beispielsweise in einer globalen Supply Chain) haben vielfältige Einflüsse auf den Marktwert oder den Preis von Produkten, Materialien oder Cashflows.

Bei ausgewählten ESG-bezogenen Risiken empfiehlt COSO/WBCSD die Aufnahme einer nicht-finanziellen Messgröße in den Priorisierungskriterien zu berücksichtigen. Als Alternative wird somit auch eine rein qualitative Bewertung empfohlen, unabhängig davon, ob sich eine finanzielle Auswirkung quantifizieren lässt.

Eine Auswahl an qualitativen, semi-quantitativen und quantitativen Ansätzen ist in Abb. 21.3 zusammengefasst.

Der weitere Beitrag gliedert wie folgt. In Abschn. 21.2 wird zunächst beispielhaft das aktuell besonders intensiv diskutierte Thema der Risiken in Bezug auf Klimawandel und Nachhaltigkeit – einer wesentlichen Komponente der ESG-Risiken – vorgestellt. Neben

Strategic	Operational	Financial	Compliance
• Vision and core values	• Research and development	• Interest rate volatility	• Fraud
• Corporate governance	• New products	• Foreign currency volatility	• Bribery
• Organizational structure	• Marketing	• Cash management	• Conflicts of interest
• Strategic planning	• Budgeting and forecasting	• Credit risk	• Country/state/local regulation
• Mergers and acquisitions valuation and pricing	• Raw material availability	• Accounting policies	• Tax regulation
• Investor relations	• Suppliers	• Accounting estimates	• Trade regulation
• Competition	• Production management	• Internal control	• IP management and protection
• Changing customer preferences or lifestyles	• Product stewardship	• Tax strategy and planning	• Greenhouse gas emissions
• Growing middle class	• Inventory management		• Water treatment
• Urbanization/growing population	• Employee engagement		• Health and safety
• Emerging markets	• Labor relations		
	• Human rights		
	• IT investment		
	• Cybersecurity		
	• Business continuity		
	• Pandemic		
	• Physical impacts of climate change		

Abb. 21.1 Risikoinventar für ESG-Risiken. (Quelle: COSO/WBCSD)

Risk rating	Definition
Catastrophic	• Financial loss: [ ]% of earnings before interest, taxes, depreciation and amortization (EBITDA) or more than [ ]% impact on share price • International negative media coverage for more than six months that results in at least [ ]% revenue loss • More than [ ]% employee turnover • Prosecution, fines and litigation greater than [ ]% of expenses • Threatened or actual loss of [ ]% or more strategic customers
High	• Financial loss: [ ]% of EBITDA or share price • Reputation damage from media coverage that persists for one to six months and results in [ ]% nonrecurring revenue loss • Results from employee survey showing staff morale more than [ ]% less than peer organizations • Threatened or actual loss of [ ]% strategic customers
Medium	• Financial loss: [ ]% of EBITDA or share price • Reputation damage from media coverage that persists for less than one month and results in [ ]% nonrecurring revenue loss • Results from employee survey showing morale [ ]% less than peer organizations • Threatened or actual loss of [ ]% strategic customers
Low	• Financial loss: less than [ ]% of EBITDA or share price • Local reputation damage from NGO or media resulting in less than [ ]% revenue loss • Individual feedback from employees on low staff morale • Customer complaints from less than [ ]% of strategic customers

Abb. 21.2 Bewertungsmethoden für ESG-Risiken. (Quelle: COSO/WBCSD)

Measure	Example risk severity metrics
Quantitative (monetary) Quantitative (non-monetary) Qualitative	**Revenue:** Projected or identified impact on revenue or expenditures **Expenditures:** Projected or identified impact on expenditures or costs **EBITDA:** Projected or identified impact on EBITDA **Assets and liabilities:** Write-off, asset impairment and early retirement of existing assets **Capital and financing:** Impact to cost of capital or access to capital, operating losses **Share price:** Impact (%) in share price[c] **Customer/reputation:** Reduction in customer confidence (%) (may also be measured in revenue) **Safety:** Lost time due to injuries **Social media coverage:** Number of viewers of the entity's video **Business continuity:** Maximum allowable outage **Greenhouse gas emissions:** Total emissions by type of greenhouse gas (GHG); carbon intensity (GHG/USD $ million) **Energy/fuel:** Total energy consumption in megawatt hours **Water:** Total freshwater withdrawn in cubic meters from water-stressed regions **Land use:** Percentage change in land cover type (e.g., grassland, forest, cultivated, pasture, urban) **Location:** Number of locations within a designated flood zone **Capital and financing:** Increase or decrease in ability to raise capital **Reputation:** Type of complaints received from stakeholders[d] **Staff morale/turnover:** Engagement survey results/level of engagement

Abb. 21.3 Qualitative, semi-quantitative und quantitative Bewertung von ESG-Risiken. (Quelle: COSO/WBCSD)

der Erläuterung des Themas wird hier auch schon gezeigt, welche Bedeutung der Einsatz von Simulationsverfahren bei der adäquaten Beschreibung und Quantifizierung von ESG-Risiken haben. Im anschließenden Abschn. 21.3 werden die verschiedenen Wirkungsweisen von ESG-Risiken, finanzielle und nicht-finanzielle, erläutert. Anschließend wird in Abschn. 21.4 das Feld der ESG-Risiken und die hier verwendeten Kriterien mit insbesondere aus China stammenden Konzept des „Social Credit Ratings", in der Anwendung für Unternehmen, verglichen. Der anschließende Abschn. 21.5 beschäftigt sich mit der Quantifizierung von ESG bzw. CSR-Risiken, den hier bestehenden Herausforderungen und Lösungsstrategien (auch basierend auf einem Fallbeispiel). Der abschließende Abschn. 21.6 fasst die wichtigsten Aussagen knapp zusammen und bietet einen kurzen Ausblick.

21.2 Beispiel: Nachhaltigkeit und Klima

21.2.1 Grundlagen

Die Diskussion um die Zukunft der Menschheit ist nicht erst ein Phänomen unserer heutigen Zeit. Im Jahr 1972 wurde unter dem Titel „The Limits to Growth" (vgl. Meadows et al. 1972) die Ergebnisse eines systemdynamischen Simulationsmodells veröffentlicht. Die Ergebnisse dieser wissenschaftlichen Studie wurden häufig emotionalisiert (vgl. vertiefend Romeike 1994). Das Massachusetts Institute of Technology (MIT) wurde vom Club of Rome, eine im Jahr 1968 gegründete gemeinnützige Organisation, die sich für eine nachhaltige Zukunft der Menschheit einsetzt, beauftragt, eine methodisch fundierte Studie zu erstellen, die die Komplexität der Fragestellung adäquat abbilden kann. Donella und Dennis Meadows sowie deren wissenschaftlichen Mitarbeiter der Sloan School of Management am MIT führten dazu eine Systemanalyse und Computersimulationen verschiedener Szenarien durch.

Grundlage hierfür bildete die Simulationsmethodik System Dynamics. Das verwendete Simulationsmodell untersuchte fünf Szenarien von Parametern mit globaler Wirkung: Industrialisierung, Bevölkerungswachstum, Unterernährung, Ausbeutung von Rohstoffreserven und Zerstörung von Lebensraum. Im Jahr 1973 wurde dem Club of Rome der Friedenspreis des Deutschen Buchhandels verliehen.

Die Kernaussage der Studie lautete, dass „die Ausbeutung der wichtigsten Rohstoffe und die wachsende Belastung durch Umwelt- und Luftverschmutzung zunehmend steigende Risiken für die Weltwirtschaft in der Zukunft verursachen" würde. Viele verstanden den Bericht so, als würde die Weltwirtschaft innerhalb weniger Jahrzehnte zum Stillstand kommen. Dies war allerdings nicht die Aussage der „Grenzen des Wachstums". Der Bericht stützte sich auf eine Perspektive von 50 bis 100 Jahren und legte darüber hinaus seinen Fokus auf die steigenden, physischen Auswirkungen des Wirtschaftswachstums (den ökologischen Fußabdruck), nicht auf Wachstum an sich.

Die Studie des Club of Rome hat für einige wesentlichen Aspekte sensibilisiert, wenngleich heute viele der damals getroffenen Aussagen so nicht mehr ohne weiteres akzeptiert werden können.

Auch bei dieser kritischen Bestandsaufnahme bleibt jedoch unbestritten: ESG-Risiken, die Unternehmen eingehen (müssen), bleiben auch aus volkswirtschaftlicher Perspektive relevant, wie auch insgesamt Risiken hohe volkswirtschaftliche Bedeutung haben (vgl. Gleißner 2018c; Romeike 2018 sowie Romeike und Hager 2020). Auch wenn von einer „sicheren" Wachstumsgrenze (zum Beispiel durch die Limitierung von Rohstoffen) aufgrund der skizzierten Möglichkeiten von Substitutionen und Recycling nicht ausgegangen werden muss, besteht eindeutig das Risiko, dass beispielsweise bestimmte Rohstoffe (zu) knapp werden könnten (beispielsweise, weil Substitutionsmöglichkeiten zu spät entwickelt wurden). Heute wird insbesondere der Klimawandel als besonders hohes Risiko bewertet (vgl. exemplarisch hierzu World Economic Forum 2020).

Auf dem Weltgipfel für Umwelt und Entwicklung im Jahr 1992 in Rio de Janeiro haben zunächst 154 Staaten die Klimarahmenkonvention der Vereinten Nationen (United Nations Framework Convention on Climate Change, UNFCCC) unterzeichnet. Das multilaterale Übereinkommen trat im Jahr 1994 in Kraft. Aktuell haben 196 Vertragsparteien sowie die EU als regionale Wirtschaftsorganisation die Klimarahmenkonvention ratifiziert. Mit dieser UN-Klimarahmenkonvention erkennt die internationale Staatengemeinschaft globale Klimaveränderungen als wichtiges und relevantes Problem an und verpflichtet sich zum gemeinsamen Handeln.

Das Ziel der Klimarahmenkonvention ist die Stabilisierung der Treibhausgaskonzentrationen auf einem Niveau, bei dem eine gefährliche vom Menschen verursachte Störung des Klimasystems verhindert wird. Dies soll in einem Zeitraum geschehen, der es den komplexen Ökosystemen erlaubt, sich auf natürliche Weise an die Klimaänderungen anzupassen (siehe Artikel 2 UNFCCC).

Die Klimarahmenkonvention ist eher als Prozess zu verstehen und wird permanent durch ergänzende Beschlüsse der einzelnen COPs erweitern und modifiziert (die Vertragsstaaten treffen sich jährlich zu Vertragsstaatenkonferenzen: Conference of the Parties, COP). Hierbei ist insbesondere zu nennen:

- Berlin, Deutschland (COP 1, 1995): Berliner Mandat zur Erarbeitung eines Protokolls mit rechtlich verbindlichen nationalen Minderungszielen; Einrichtung des Sekretariats der Klimarahmenkonvention in Bonn.
- Kyoto, Japan (COP 3, 1997): Verabschiedung des Kyoto-Protokolls als Zusatz zur Klimarahmenkonvention, es enthält erstmals rechtsverbindliche Minderungsverpflichtungen für die Industrieländer.
- Marrakesch, Marokko (COP 7, 2001): Festlegung der Durchführungsbestimmungen des Kyoto-Protokolls („Marrakesh Accords").
- Montreal, Kanada (COP 11, zugleich 1. Vertragsstaatenkonferenz unter dem Kyoto-Protokoll, 2005): Beschluss zur Aufnahme von Verhandlungen über neue Emissionsreduktionsziele für Industriestaaten ab 2013.
- Bali, Indonesien (COP 13, 2007): Beschluss über Verhandlungen zu einem umfassenden Klimaschutzabkommen ab 2013, an dem sich alle Staaten beteiligen (Bali-Roadmap).
- Kopenhagen, Dänemark (COP 15, 2009): Verhandlungen scheitern, Vertragsstaaten finden zudem keinen Konsens über den „Kopenhagen-Accord".
- Cancún, Mexiko (COP 16, 2010): Wiederaufnahme der Verhandlungen; Einigung auf das zentrale Ziel der internationalen Klimaschutzanstrengungen die Treibhausgasemissionen so zu mindern, dass die globale Temperaturerhöhung die Zwei-Grad-Obergrenze nicht überschreitet.
- Durban, Südafrika (COP 17, 2011): Einigung auf die sofortige Aufnahme von Verhandlungen über ein umfassendes Klimaschutzabkommen („Durban Plattform"), das 2015 verabschiedet und bis spätestens 2020 wirksam werden soll.
- Doha, Katar (COP 18, 2012): Einigung auf eine zweite Verpflichtungsperiode unter dem Kyoto-Protokoll, Diskussion über Verhandlungsfahrplan zum neuen Klimaschutzabkommen.
- Warschau, Polen (COP 19, 2013): Einigung auf wesentliche Eckpunkte zur Finanzierung von Klimaschutzmaßnahmen; Errichtung eines Mechanismus für Verluste und Schäden; Durchbruch beim Waldschutz.

- Lima, Peru (COP 20, 2014): Grundlage für neuen Weltklimavertrag in Paris gelegt; Aufruf nationale Klimaschutzbeiträge vorzulegen; Einzahlungen von über zehn Milliarden US-Dollar in den Grünen Klimafonds (Green Climate Fonds, GCF).
- Paris, Frankreich (COP 21, 2015): Übereinkommen von Paris als zweite Ergänzung zur Klimarahmenkonvention verabschiedet; erstmals legen nahezu alle Staaten nationale Klimaschutzbeiträge (Nationally Determined Contribution, NDC) vor; 1,5 °C-Obergrenze soll angestrebt werden.
- Marrakesch, Marokko (COP 22, zugleich 1. Konferenz unter dem Übereinkommen von Paris, 2016): Verhandlungen über die technische Ausgestaltung des Übereinkommens von Paris; Start der weltweiten NDC-Partnerschaft unter deutscher und marokkanischer Führung.
- Bonn, Deutschland (COP 23, 2017): 13. Treffen zum Kyoto-Protokoll sowie als 2. Treffen der Conference of the Parties serving as the meeting of the Parties to the Paris Agreement (CMA 1–2). Ziel war ein „Regelbuch" zur Umsetzung des Paris-Abkommens, was auf der nächsten Klimakonferenz COP 24 verabschiedet werden soll.
- Katowice, Polen (COP 24, 2018): Auf der COP 24 in Katowice sollte festgelegt werden, welche Rechte und Pflichten die einzelnen Staaten haben, um das in Paris anvisierte 1,5-Grad-Ziel zu erreichen. Dabei spielen einheitliche oder zumindest vergleichbare Methoden für die Messung von Treibhausgasen eine wesentliche Rolle – diese existieren bislang lediglich für die „Industrieländer"; die „Entwicklungsländer" müssen in diesem Zusammenhang noch Berichtssysteme aufbauen.
- Madrid, Spanien (COP 25, 2019): Schlüsselthema der Vertragsstaaten des Übereinkommens von Paris war die Fertigstellung des ergänzenden Regelwerks. Der größte Teil des Regelwerks wurde während der COP 24 / CMA 1-3 in Katowice vereinbart, Regeln zum Artikel 6 des Paris-Übereinkommens wurden dort aber ausgeklammert. So sollen insbesondere marktbasierte Mechanismen den teilnehmenden Staaten einen Anreiz bieten, Emissionsminderungen in anderen Staaten zu finanzieren, wo dies kostengünstiger möglich ist. Von der Kosteneffizienz verspricht man sich höhere Klimaschutz-Ambitionen und damit höhere Minderungsbeiträge (Nationally Determined Contributions, NDCs) der Mitgliedsländer in den Folgejahren. Die Ergebnisse der COP25 wurde von Natur-, Umwelt- und Klimaschutzorganisationen scharf kritisiert.

Das Folgetreffen COP 26 soll Ende November 2020 im schottischen Glasgow stattfinden. Die internationale Gemeinschaft kann die in der Klimarahmenkonvention definierten Ziele nur erreichen, wenn alle Länder, Unternehmen sowie Teile der Gesellschaft einen Beitrag leisten.

21.2.2 Einsatz von Simulationsverfahren am Beispiel World3

21.2.2.1 Hintergrund des Simulationsmodells

Im Jahre 1968, als viele Industrieländer seit Jahren durch rasches Wirtschaftswachstum geprägt waren, wurde, wie erwähnt, in der Accademia dei Uncei (Rom) der Club of Rome als informelle Vereinigung von Wissenschaftlern und Praktikern verschiedenster Disziplinen gegründet (die nachfolgenden Ausführungen basieren auf Romeike 1994). Anlass waren die zunehmenden globalen Krisenerscheinungen, u. a.

- die vermehrten internationaler Konflikte,
- das exponentielle Bevölkerungswachstum,
- der Ressourcenabbau,
- die zunehmende Asymmetrie zwischen Industrie- und Entwicklungsländern,
- die Atmosphärenveränderungen durch Spurengase.

Ziel des Club of Rome ist nicht die Verfolgung bestimmter politischer Ziele oder Ideologien, sondern die Erforschung von Ursachen und Interdependenzen der wirtschaftlichen, politischen, ökologischen, sozialen und demografischen Situation der Menschheit.

Der Club of Rome will nicht nur eine Institution für wissenschaftliche Forschung sein, sondern auch weltweite Diskussionen anstoßen, was innerhalb offizieller Strukturen oft schwieriger ist, und die erforderlichen Strategien vorschlagen. Hierbei sollten die globalen Probleme von einem interdisziplinären, übernationalen und langzeitlichen Gesichtspunkt aus analysiert werden.

Als Forschungsansatz hatte man sich zunächst für die Delphi-Analyse (vgl. hierzu Romeike 2018) als Kreativitätsmethodik entschieden. Um der Komplexität der untersuchten Systeme gerecht zu werden, schlug Eduard Christian Kurt Pestel im Jahr 1969 vor, ein kybernetisches Computermodell basierend auf einer Simulationsmethodik zu erstellen, in welchem die Interdependenzen zwischen wirtschaftlichem Wachstum, Umwelt und Gesellschaft untersucht werden.

Pestel war ein deutscher Ingenieur und Ökonom sowie Professor für Mechanik und Regelungstechnik. Im Jahr 1975 gründete er zusammen mit sechs weiteren Wissenschaftlern das Institut für angewandte Systemforschung und Prognose (ISP) in Hannover.

Seit einiger Zeit befasste sich bereits der Informatiker Jay Wright Forrester am Massachusetts Institute of Technology (MIT) mit einer kybernetischen Modellmethode zur Analyse komplexer Systeme: Industrial Dynamics (später System Dynamics genannt). Forrester hatte mit diesem Ansatz bereits im Jahr 1969 versucht, die Entwicklung und die Probleme der Städte darzustellen (vgl. Forrester 1969).

Diese Simulationsmethodik erschien dem Club of Rome für ihr Projekt geeignet und nach einem ersten Modellentwurf (World1) wurde ein modifizierter Entwurf (World2) der Stiftung Volkswagenwerk vorgelegt, die daraufhin die Finanzierung des Projekts übernahm.

Leiter des Projektes wurde Dennis L. Meadows vom MIT, unterstützt von 16 Wissenschaftlern aus verschiedenen Disziplinen. Im März 1971 veröffentlichte Forrester sein auf dem World2-Modell basierendes Buch World Dynamics, welches allerdings nur geringe Aufmerksamkeit auf sich zog.

Meadow übernahm das System-Dynamics-Verfahren und modifizierte das World2-Modell durch eine größere Anzahl von Variablen und Verknüpfungen. Außerdem stützt sich das World3-Modell auf eine größere empirische Basis als Forresters Ausgangsmodelle (vgl. Pestel 1988, S. 34 f.).

Insbesondere zwei Problemkomplexe sollten analysiert werden (vgl. Pestel 1980, S. 115):

- Die mögliche Diskrepanz zwischen Bevölkerungs- und Wirtschaftswachstum sowie der Begrenztheit der Erde hinsichtlich Ressourcen und Senken sollte aufgezeigt werden.
- Die Interdependenz und die Einwirkungen wesentlicher Faktoren, die das physische Verhalten des globalen Systems bestimmen, sollten analysiert werden.

Im März 1972 wurde der Bericht unter dem Titel „The Limits to Growth" (Meadows et al. 1972) nach einer 18monatigen Studienarbeit in einer populär-wissenschaftlichen Form der Öffentlichkeit vorgestellt und ist mittlerweile in 30 Sprachen übersetzt worden. Im Jahre 1973 wurde ergänzend eine Sammlung von 13 Einzelberichten über den Aufbau der Subsysteme (vgl. Meadows und Meadows 1973) und ein Jahr später der „technische Bericht" (vgl. Meadows et al. 1974) – Darstellungen zur Methodologie, zu Gleichungssystemen und zur Datenbasis – publiziert.

The Limits to Growth hat weltweit Kontroverse und Kritik hervorgerufen, die Modellbauer wurden als Fortschrittspessimisten und Untergangspropheten verspottet.

21.2.2.2 Aufbau, Grundannahmen und Prämissen des Simulationsmodells

„No single element of this world model is new to human thought. What is new is the synthesis of many isolated, incomplete perceptions into a more complete picture, an attempt to comprehend the whole system rather than just its single parts." (Meadows et al. 1974, S. 3)

Mit Hilfe des World3-Modells haben die MIT-Wissenschaftler (basierend auf der Methodik System Dynamics) die Abhängigkeiten und die Komplexität eines Systems beschrieben und simuliert. Das MIT-Team bezeichnet World3 als ein formales mathematisches Modell eines komplexen sozialen Systems. Es wird versucht, das langfristige Wachstumsverhalten der Weltwirtschaft für 70 Jahre ex ante zu analysieren und Szenarien über einen Zeitraum von 130 Jahren zu simulieren. Ziel war nicht eine exakte Prognose, sondern vielmehr das „Lernen aus der Zukunft" zu ermöglichen (vgl. vertiefend Romeike und Spitzner 2013; vgl. Romeike 2015a, b).

Basierend auf der System-Dynamics-Methode ging das MIT-Team bei der Konstruktion des World3-Modells in folgenden neun Phasen vor (vgl. Meadows et al. 1974, S. 5):

1. Allgemeine verbale Beschreibung des komplexen Systems.
2. Genaue Angabe der Zielsetzung und des Modellzwecks.
3. Definition des Zeithorizonts.
4. Bestimmung der wichtigsten Elemente, die die relevanten Aspekte des Systems abbilden.
5. Strukturierung des Modells unter Berücksichtigung der Kausalstrukturen.
6. Schätzung der Parameter sowie Festlegung der Anfangswerte der Bestandsgrößen (durch Expertenurteile, historische Zeitreihenanalysen, Schätzungen et cetera).
7. Evaluierung der Parametersensitivität durch Computersimulation.
8. Experimente durch Simulationen, um das dynamische Modellverhalten zu analysieren.
9. Diskussion und kritische Analysen der Ergebnisse.

Meadows et al. weisen darauf hin, dass die Zielsetzung des World3-Modells nicht in einer präzisen Prognose im engeren statistischen Sinne liegt, sondern das mittels einer Projektion (unpräzise Prognose im weiteren Sinne) typische Verhaltensmuster aufgezeigt werden sollen. Und exakt hier liegt die Wurzel für die Fehlinterpretation vieler Kritiker an dem Modellansatz. Viele Kritiker haben den Unterschied zwischen Prognose und Szenarien nicht beachtet (vgl. zu den Missverständnissen im Zusammenhang mit Simulationen Romeike und Spitzner 2013, S. 50–54).

> „We had to limit ourselves to conditional and imprecise questions, rather than precise predictions, for two reasons. First social systems are by their nature unpredictable in the absolute sense. Since any prediction made about the future of a social system becomes an influence on social policy, the prediction itself may change the system's behavior. Second, the incomplete and inaccurate world data base currently available does not permit precision, even for conditional long-term prediction of social systems." (Meadows et al. 1974, S. 7 f.).

Den langen Modellzeitraum von 200 Jahren begründen Meadows at al. mit den Langzeitwirkungen einiger Modellvariablen (insbesondere Umwelteinwirkungen, aber auch die technologische Entwicklung). Im Zentrum des World3-Modells, das aus insgesamt 280 Gleichungen (im Modell sind 146 Differenzen- und Hilfsgrößengleichungen und 139 zeitinvariante Gleichungen enthalten) besteht, stehen fünf Variablen, wobei diese wiederum in verschiedene Teilgrößen aufgespalten wurden:

1. Bevölkerung (Population)
2. Kapital- und Industrieoutput (Capital)
3. Nahrungsmittel (Agriculture)
4. Sich nicht regenerierende Rohstoffe (Nonrenewable Resources)
5. Umweltverschmutzung (Pollution)

Ein Blick auf diese Teilgrößen zeigt, dass es im World3-Modell bereits um die Simulation von ESG-Kriterien bzw. -Risiken ging. Diese fünf Modellsektoren wurden als hinreichend repräsentativ angesehen, um das charakteristische Verhalten des globalen Wachstums abzubilden.

Nach Meadows könnten andere Sektoren hinzugefügt werden, beispielsweise ökonomische Faktoren (Preissteuerung), regenerierende Rohstoffe oder gesellschaftlicher Wandel (vgl. Meadows et al. 1974, S. 10). Auf der anderen Seite wäre das Modell dadurch zu komplex und unüberschaubar geworden.

Ein weiteres Problem ergab sich bei der Suche nach dem Aggregationsgrad. Möglich wäre beispielsweise eine Aufteilung der Bevölkerung nach Nationalität oder Alter. Ökonomische Variablen könnten nach Wirtschaftssystemen, Regionen et cetera unterteilt werden. Doch World3 ist ein hoch aggregiertes Modell. Meadows et al. führen folgende Begründung auf:

„A highly disaggregated model with much detail may leave out important relationships that could alter the behavior of the model and the conclusions drawn from it." (Meadows et al. 1974, S. 12).

World3 wurde nicht in industrialisierte und nicht-industrialisierte Regionen aufgeteilt, da das MIT-Team der Ansicht ist, dass das physikalische System strukturell in jeder geografischen Region gleich ist. Für andere Modellzwecke wäre es evtl. notwendig, eine Disaggregation der entsprechenden Größen vorzunehmen.

Nach der Definition der relevanten Modellvariablen wurden in einem nächsten Schritt die jeweiligen Kausalbeziehungen festgelegt. Die Beziehungen zwischen den jeweiligen Systemgrößen wurden zunächst grafisch dargestellt und später über Gleichungssysteme beschrieben.

Hierbei bilden der Bevölkerungs- und der Kapital- und Industrieoutputsektor zwei große, wachstumsinduzierende positive Regelkreise. Die ökologischen Grenzen werden durch drei negative Regelkreise induziert: den Nahrungsmittelsektor, den Sektor der nicht regenerierbaren Rohstoffe sowie den „Umweltsektor".

Folgende Prämissen, die durchaus kritisch diskutierbar sind, wurden bei diesen drei Sektoren unterstellt (vgl. Meadows et al. 1974, S. 15 f.):

- Das potenziell kultivierbare Land ist begrenzt. Die Grenzkosten für eine zusätzliche Landerschließung steigen.
- Die Nahrungsmittelproduktion weist eine obere Grenze auf, wobei die Erträge zwar durch agrartechnologische Innovationen zunehmen, aber insgesamt durch die zunehmende Umweltverschmutzung mit sinkenden Grenzerträgen zu rechnen ist.
- Der Vorrat an sich nicht regenerierenden Rohstoffen ist begrenzt. Bei der Erschließung und Förderung neuer Vorkommen steigen die Grenzkosten durch den vermehrten Energie- und Kapitaleinsatz.
- Die Absorptionsfähigkeit der natürlichen Umwelt ist für Schadstoffemissionen begrenzt. Bei zunehmenden Emissionen nimmt die Grenzleistungsfähigkeit des in Umwelttechnologie investierten Kapitals ab.

Der interdependente Aufbau des World3-Modells ist in Abb. 21.4 wiedergegeben, aus dem die Verbindungen zwischen den einzelnen interagierenden rückgekoppelten Regelkreisen ersichtlich werden. Auch die Vielzahl der ins Modell (zwischen Ursache und Wirkung) eingebauten Verzögerungsglieder, die zu einem instabilen Verhalten führen können, ist abgebildet. Auch wenn aus heutiger Sicht einige der oben genannten Prämissen zu relativieren sind, ist wesentlich, dass die von Meadows et al. verwendete Methodik der Simulation notwendig ist, um die komplexen Interdependenzen adäquat zu erfassen (entgegen den Vorstellungen des Modells ist bis heute z. B. durch technischen Fortschritt die Versorgung der Menschen mit Lebensmitteln immer einfacher geworden, was man an einem sinkenden Anteil der Landwirtschaft am globalen BIP erkennen kann).

Meadows führen für den Empiriebezug basierend auf interdisziplinären Expertenschätzungen die folgenden Gründe an (Meadows et al. 1974, S. 21):

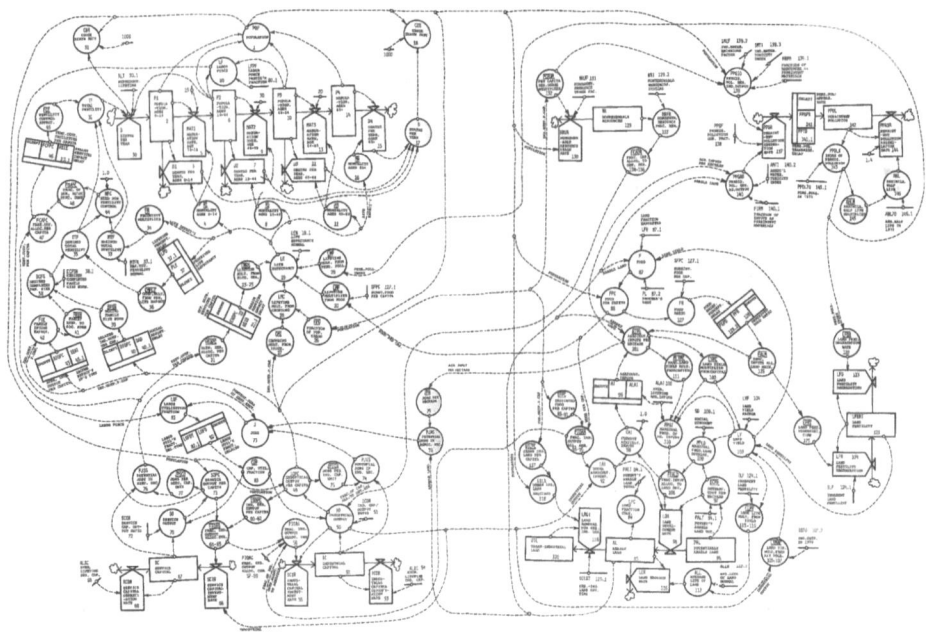

Abb. 21.4 Aufbau des World3-Modells (Stock and flow diagram). (Quelle: Meadows et al. 1974)

„[…] experience in modeling feedback systems rapidly demonstrates that even the most sophisticated numerical estimation techniques will not produce useful conclusions from a faulty
or incomplete model structure."

„[…] system dynamics models are usually concerned with imprecise questions about the
general behavioral tendencies of social systems."

„[…] a correct causal structure generally produces realistic model behavior, even with
only approximate numerical parameters."

In einem nächsten Schritt erfolgte die Sensitivitätsanalyse. Hierbei werden die Reaktionen des Outputs durch Variation des Inputs sowie das Stabilitätsverhalten des Modells
analysiert. Außerdem wurde mittels Sensitivitätsanalyse untersucht, ob auf eine empirische Untermauerung verzichtet werden kann, falls die Outputvariable von bestimmten
Variablen/Parametern nicht signifikant sensitiv ist.

Diese Überprüfung führte im World3-Modell zu einer Anpassung einiger Konstanten,
wobei sowohl die einzelnen Subsektoren als auch das Globalmodell analysiert und adjus-
tiert wurden.

Aufbauend auf einem Standardlauf (Szenario 1 des World3-Modells) hat das MIT-
Team versucht, die Anwendung verschiedener politischer Strategien (beispielsweise durch
Regulierung in den Bereichen Soziales sowie Technologie und Umwelt) durch Datenvari-
ationen zu simulieren.

Das heißt, die MIT-Wissenschaftler haben sich im World3-Modell bereits mit den Einflussfaktoren „Environment", „Social" und „(Corporate) Governance" (Umwelt, Soziales/
Gesellschaft und Unternehmensführung/-struktur) auseinandergesetzt und die Ursache-/

Wirkungszusammenhänge in Form eines Simulationsmodells nicht nur beschrieben, sondern potenzielle Wirkungseffekte aufgezeigt und diskutiert.

21.2.2.3 Nutzen und Bewertung des SD-Modells

Mit dem World3-Modell, basierend auf der Simulationsmethodik System Dynamics, haben Meadows et al. untersucht, ob das Wachstum der Wirtschaft beziehungsweise der Bevölkerung zu

- „[…] a smooth transition to a steady state,
- oscillation around an equilibrium position, or
- overshoot and decline" (Meadows et al. 1974, S. 561)

führt.

Hierbei ist das MIT-Team von dem Status quo des gesellschaftlichen Wertesystems ausgegangen. Unter diesen Prämissen ist World3 auch zu interpretieren.

„Die Kritik hat unseren Bericht oft missverstanden als eine normative Studie, die besagt, wie die Zukunftsentwicklung aussehen soll. Ich sehe sie dagegen als eine Art Radar, welcher die Zukunft abtastet und die Gefahrenstellen zeigt. Der Radarlotse bestimmt aber nicht, wohin die Fahrt geht, und man sollte ihn auch nicht für schlechte Wetteraussichten verantwortlich machen. Die langfristige Simulation von globalen Entwicklungsprozessen basiert auf dem bisherigen Verhalten der Menschheit und wird dann unrichtig, wenn politische Entscheidungen das Verhalten ändern. Als soziales Frühwarnsystem ist aber auch unsere Studie nur ein erster Schritt." (Zitat von D.L Meadows in: Tuchtfeld 1973, S. 129 ff.).

Ein großer Verdienst der MIT-Wissenschaftler ist darin zu sehen, dass sie die globalen Interdependenzen von sozio-ökonomischen und komplexen Systemen abgebildet haben. Und hierbei wurden sowohl Umweltaspekte, als auch soziale Aspekte und Governancethemen berücksichtigt.

Trotz einiger methodologischer Schwächen des World3-Modells (eine vertiefende und wissenschaftliche Analyse enthält Romeike 1994) kann man resümieren, dass Meadows et al. – basierend auf einer fundierten Simulationsmethodik – potenzielle Szenarien über die Zukunft aufgezeigt haben (vgl. Tuchtfeld 1973 S. 129 ff.; im dem Bericht wird D.L Meadows zitiert). Ziel des Forschungsprojektes war es nicht, den „Untergang der Menschheit" (Handelsblatt vom 15.03.1972) zu prognostizieren oder einen „Kassandraruf" (Frankfurter Allgemeine Zeitung vom 17.07.1972) heraufzubeschwören.

Bekanntermaßen verhallten die Prophezeiungen Kassandras, Tochter des trojanischen Königs, ungehört. Der Gott Apollon gab Kassandra wegen ihrer Schönheit die Gabe der Weissagung. Kassandra hatte damit alle Fähigkeiten, über die eine perfekte Risikomanagerin verfügen sollte. Doch als sie die Verführungsversuche von Apollon zurückwies, verfluchte er sie und ihre Nachkommenschaft, auf dass niemand ihren Weissagungen Glauben schenken werde.

Ziel der Analyse war es vielmehr, Verhaltensänderungen auszulösen, damit die skizzierten Szenarien nicht Realität werden.

Dies haben viele Rezensenten der World3-Studie nicht erkannt. Die Gründe hiervon sind vielfältig und reichen von Dogmatismus bis zu methodologischer Unkenntnis von systemdynamischen Modellen oder Simulationsmethoden insgesamt (siehe zur Unkenntnis von Simulationsmethoden die empirische Studie: Meyer et al. 2012).

Um die Erkenntnis des interdisziplinären World3-Modells im vollen Umfang werten zu können, sind zudem fundierte Kenntnisse der ökologischen (physikalischen, chemischen, biologischen) und ökonomischen Zusammenhänge notwendig. Ein solches kybernetisches Denken in mehrfach rückgekoppelten vernetzten Zusammenhängen ist genetisch unterentwickelt. Vielleicht ist auch dies ein Grund, warum die Kritiken oft im Dilettantischen steckenbleiben.

Viele Kritiker sehen in der World3-Studie lediglich eine Trendextrapolation, da sie sich bei der Analyse ausschließlich auf den Standardlauf konzentriert haben (gleicher Ansicht: Heck 1992, S. 54). Eberwein führte hierzu aus: „Über den vielen Unsinn, der meist aus nachweisbarer Inkompetenz zum Thema Weltmodelle geschrieben wird, lohnt es sich kaum nachzudenken, geschweige denn darauf einzugehen." (Eberwein 1990, S. 1).

Meadows at al. haben versucht, durch die Variation der Annahmen, alternative Szenarien aufzuzeigen – dies wurde seitens der Kritiker oft nicht zur Kenntnis genommen. So ist das MIT-Team beispielsweise von der Annahme jährlich konstanter Zuwachsraten für bestimmte Technologien ausgegangen (exponentielle Zunahme), was zu einer vorübergehenden Stabilisierung geführt hat.

An dieser Stelle muss jedoch auch auf eine große Schwäche der Methodik System Dynamics sowie des World3-Modells hingewiesen werden. Die Simulationsergebnisse des World3-Modells basierten auf verschiedenen deterministischen Szenarien, die durch eine Expertenschätzung zustande gekommen sind. Dies führt im Ergebnis dazu, dass es kein „richtiges" Szenario geben kann und die einzelnen Szenarien auch nicht mit Wahrscheinlichkeiten unterlegt wurden (zu den Grenzen einer deterministischen Szenarioanalyse vgl. Romeike und Spitzner 2013, S. 94 ff.).

Die Ergebnisse deterministischer Szenarien sind – je nach Stärke der subjektiven Beeinflussung durch die Teilnehmer des Szenarioteams – nicht wertfrei und sind damit auch stets angreifbar.

21.2.2.4 Einordnung: System Dynamics, Grenzen und Perspektiven

System Dynamics Methoden kann man als Vorstufe einer stochastischen Simulation (Monte-Carlo-Simulation) auffassen. Sie ersetzen diese aber nicht. System Dynamics-Modelle sind hilfreich, um wesentliche Zusammenhänge zu analysieren und abzubilden und mögliche „Einzelszenarien" zu durchdenken. Zusätzlich erforderlich ist jedoch Information über die Bandbreite bestimmter Entwicklungen, und damit auch die Wahrscheinlichkeit von Szenarien (Wahrscheinlichkeitsdichte). Ein denkbarer Übergang ist möglich, wenn man im Modell gesetzte unsichere Annahmen durch Wahrscheinlichkeitsverteilungen beschreibt.

Folglich bietet es sich an, das SD-Modell durch eine stochastische Simulation zu ergänzen. Mit Hilfe einer stochastischen Simulation werden für sämtliche in ein Modell einbezogene Zufallsvariablen Realisationen erzeugt (basierend auf Zufallszahlen gezogen), die asymptotisch einer zuvor spezifizierten Verteilungsannahme (je Variable) gehorchen. Im nachfolgenden Kapitel gehen wir auf den Einsatz von stochastischen Simulationsmodellen zur Bewertung von ESG-Risiken ein.

21.3   ESG-Kriterien als Ursache für vielfältige Wirkungsmechanismen

Die ESG-Kriterien und die aus diesen resultierenden ESG-Risiken haben vielfältige Wirkungen. Zunächst einmal sind Wirkungen – direkte wie indirekte – auf das Unternehmen von solchen zu unterscheiden, die Gesellschaft oder Umwelt betreffen. Für das Unternehmen relevant sind zunächst Auswirkungen finanzieller Art, also Veränderungen bei (1) erwarteter Höhe oder (2) Volatilität von Zahlungen, weil diese zusammen den fundamentalen Ertragswert eines Unternehmens bestimmen (die Ertragsrisiken, beispielsweise ausgedrückt im Variationskoeffizient der Erträge, bestimmt den Kapitalkostensatz und beeinflussen, neben dem Risikodeckungspotenzial, auch das Insolvenzrisiko, siehe beispielsweise Gleißner 2011; Gleißner 2019g; Gleißner und Ernst 2019 sowie Gleißner 2019e). Die Erfassung der finanziellen Auswirkungen von ESG-Wirkungen ist auch heute schon aufgrund der gesetzlichen Vorgaben an ein Risikofrüherkennungssystem (§ 91 AktG) notwendig. Wesentlich ist zu beachten, dass die finanziellen Auswirkungen von ESG-Risiken oft „indirekt" sind. So wird die erwartete Höhe und das Risiko der Cashflows zum Beispiel dadurch beeinflusst, dass zunächst die Reputation eines Unternehmens beeinträchtigt wird, was wiederum eine Vielzahl finanzieller Auswirkungen zur Konsequenz haben kann (zum Beispiel Umsatzverluste durch Verlust von Kunden).

Es ist eine Binsenweisheit, dass eine gute Unternehmensreputation der wesentliche und dominante immaterielle Vermögensgegenstand eines Unternehmens ist. Der Aufbau und die Weiterentwicklung des „guten Rufs" dauern oft Jahre oder Jahrzehnte. Umgekehrt kann jedoch die Reputation in Windeseile beschädigt oder gar gänzlich zerstört werden. Wenn die Gerüchteküche brodelt, ist es für Unternehmen höchste Zeit einzugreifen, bevor Themen in der Öffentlichkeit ihre eigene Dynamik entfalten (vgl. hierzu Bauer et al. 2012 sowie Weißensteiner 2014).

Die Verknüpfung von Marke und Reputation schafft eine besondere Form symbiotischer Abhängigkeit. Das fragile und facettenreiche Gebilde Reputation kann innerhalb weniger Augenblicke zerstört werden. Daher muss es das Ziel jeden Unternehmens sein, Reputationsbedrohungen rechtzeitig zu erkennen und die Reputation durch Prävention langfristig zu erhalten. Denn die „Dominorallye" beim Eintritt von Reputationsrisiken kann rasend schnell verlaufen.

Heute wird der zukünftige wirtschaftliche Erfolg oder Misserfolg eines Unternehmens nicht nur vom realen Sachkapital bestimmt, sondern vielmehr auch durch seine immateriellen Vermögensgegenstände. Hierbei zählt die Unternehmensreputation zu einem der wichtigsten immateriellen Vermögenswerte. Reputation eignet sich ideal zum Aufbau und Ausbau strategischer Wettbewerbsvorteile.

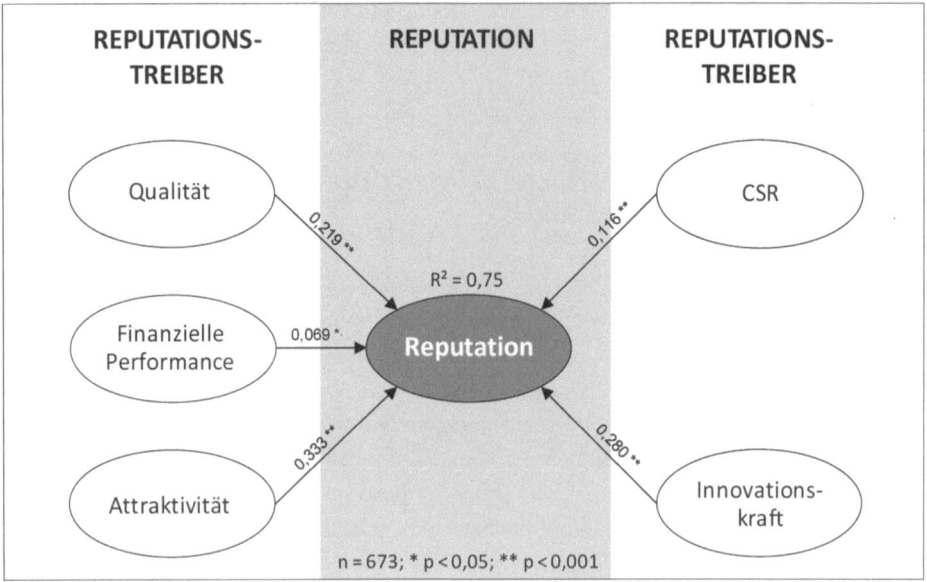

Abb. 21.5 Reputations-Treibermodell. (Quelle: Weißensteiner 2014)

Im Jahr 2012 hat das Kompetenzportal RiskNET in Kooperation mit der Technischen Universität Graz eine Studie zu den Ursachen und Treibern für Reputationsverluste-/ gewinne durchgeführt. An der Studie haben 430 Personen teilgenommen. Ziel der wissenschaftlichen Analyse war die Bestimmung der wesentlichen Reputationstreiber. Zur Bestimmung des Impacts der einzelnen Treiber auf das fragile Konstrukt der Unternehmensreputation stellen die Pfadkoeffizienten, im Speziellen deren Stärken und Signifikanzen, das wesentliche Beurteilungskriterium dar.

Wie das Ergebnis der Regressionsanalyse zeigt (vgl. Abb. 21.5), weist die Wahrnehmung der Unternehmensattraktivität mit einem Regressionskoeffizienten von 0,333 den größten positiven Einfluss auf die Unternehmensreputation auf. Die von der Öffentlichkeit wahrgenommene Innovationskraft des Unternehmens ($\beta = 0{,}280$) und die empfundene Qualität der Produkte und Dienstleistungen ($\beta = 0{,}219$) belegen den zweit- bzw. dritthöchsten Koeffizienten. Hinsichtlich Corporate Social Responsibility (CSR) lässt sich das empirisch erhobene Ergebnis so interpretieren, dass die wahrgenommene Attraktivität als Arbeitgeber eines Unternehmens dreimal bedeutender hinsichtlich der Unternehmensreputation ist, als die wahrgenommen CSR-Aktivitäten eines Unternehmens ($\beta = 0{,}116$). Den geringsten Einfluss auf die Unternehmensreputation übt die wahrgenommene finanzielle Performance eines Unternehmens aus ($\beta = 0{,}069$) (vgl. Weißensteiner 2014, S. 173).

Das aggregierte Ergebnis der Treiberanalyse ist in Abb. 21.6 dargestellt. Die beiden Reputationskonstrukte Sympathie und Kompetenz können im aufgestellten Modell zu 58 Prozent bzw. 71 Prozent durch ihre fünf Treiber erklärt werden. Die Sympathie-

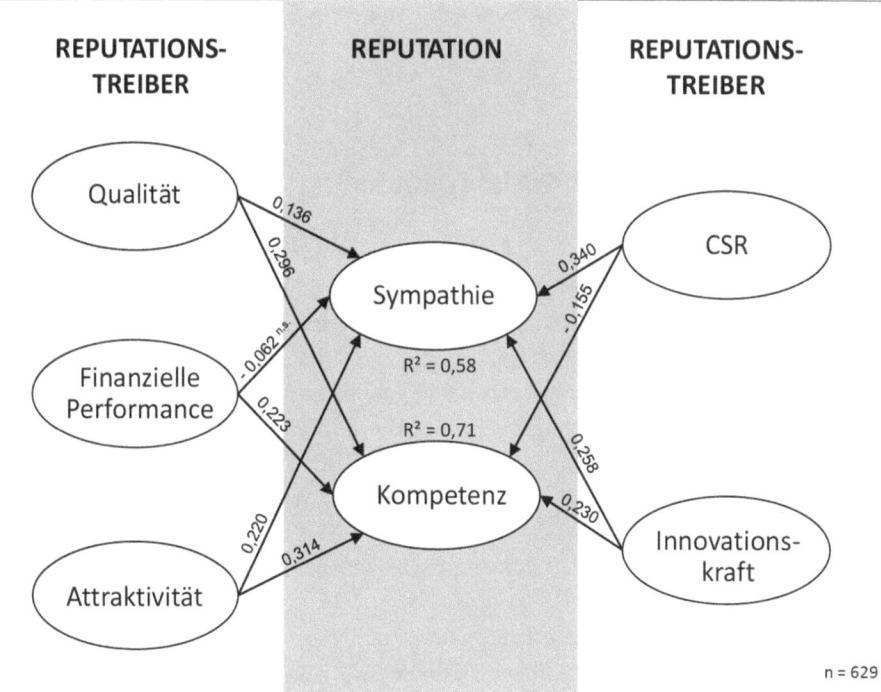

Abb. 21.6 Wirkung von CSR-Kriterien im Reputations-Treibermodell. (Quelle: Weißensteiner 2014)

Dimension erfährt einen positiven Einfluss über die Treiber Qualität, Attraktivität, Innovationskraft und Corporate Social Responsibility (CSR).

Corporate Social Responsibility (CSR) umschreibt die unternehmerische Gesellschaftsverantwortung oder auch Sozialverantwortung. CSR steht für verantwortliches unternehmerisches Handeln in der eigentlichen Geschäftstätigkeit (Markt), über ökologisch relevante Aspekte (Umwelt) bis hin zu den Beziehungen mit Mitarbeitern (Arbeitsplatz) und dem Austausch mit den relevanten Anspruchs- bzw. Interessengruppen (Stakeholdern).

Die Überlappung der beiden Begriffe CSR und ESG ist offensichtlich. Der wesentliche Unterschied zwischen ESG und CSR liegt darin begründet, dass vor allem institutionellen Investoren sich stärker an ESG-Kriterien orientieren, um den Zustand eines Vermögenswertes zu beurteilen.

Abb. 21.6 zeigt auf, dass die Sympathie negativ von der finanziellen Performance beeinträchtigt wird, wobei dieser Beziehungszusammenhang als einziger im Modell als nicht signifikant zu bezeichnen ist. Der Kompetenz-Dimension können positive Wirkungen von den Treibern Finanzielle Performance, Qualität, Attraktivität und Innovationskraft zugewiesen werden. Der negative Einfluss über den Treiber CSR erweist sich als signifikant.

ESG-Kriterien sollten daher in einem Treibermodell und als Subsystem eines Gesamtsystems betrachtet werden. Wichtig ist in diesem Kontext vor allem, dass die Wirkung auf die Finanzperformance in einer quantitativen Form abgebildet wird.

21.4 ESG-Kriterien und Social Credit Rating

Wo liegt die Verbindung zwischen den ESG-Kriterien und einem „Social Credit Rating", wie es beispielsweise die Regierung der Volksrepublik China in Form eines „Social Scoring"-Systems eingeführt hat. Basierend auf einem Scoring-/Punktesystem wird das soziale und politische Verhalten von Privatpersonen, Unternehmen und anderen Organisationen (wie beispielsweise Non-governmental organization, NGO) zur Ermittlung ihrer „sozialen Reputation" bewertet und analysiert. Hiermit verfolgt die Kommunistische Partei Chinas eine umfassende Überwachung zu mehr „Aufrichtigkeit" im sozialen Verhalten. Hierbei werden u. a. die folgenden Ziele verfolgt (vgl. Creemers 2014):

- Steigerung der „Aufrichtigkeit in Regierungsangelegenheiten";
- Steigerung der „kommerziellen Integrität";
- Steigerung der „sozialen Integrität" sowie
- Steigerung der „gerichtlichen Glaubwürdigkeit".

Der Fokus der Betrachtung hier ist die Anwendung auf Unternehmen, um Bezug zu ESG-Ansätze herzustellen.

Die kontroverse Diskussion der Anwendung auf Privatpersonen wird hier nicht aufgegriffen. Viele Analysen, insbesondere aus einer nicht-chinesischen Perspektivem, neigen dazu, das „Social Credit Rating" in China, speziell in Anwendung auf einzelne Bürger, als Orwell'schen Überwachungsstaat zu betrachten. Chinesische Politiker und auch viele Wissenschaftler bewerten das System anders. Als Begründung wird angeführt, dass sich das System nicht sehr stark von einem regulären Kreditratingsystem unterscheidet, das heißt der Einstufung der Bonität eines Wirtschaftssubjekts (Einzelperson, Unternehmen, Staat) oder eines Finanzinstruments. Dem entgegnet wird, dass gerade die umfassende Verfügbarkeit von Daten der Bürger beim Staat Machtmissbrauch ermöglicht

Das chinesische Wort für „Kredit", „信用" (xinyong), stammt aus der konfuzianischen Ethik und Moral und ist in Wirklichkeit ein Moralbegriff, der die moralische Würde und Vertrauenswürdigkeit eines Menschen anzeigt. So wurden während der Pilotphase des Social Credit Ratings verschiedene Handlungen von Bürgern und Unternehmen als „Vergehen" betrachtet, wie beispielsweise das Nicht-Zurückzahlen von Schulden, das Überqueren des Straßenverkehrs bei roter Ampel, das Wegwerfen von Müll in der Natur, das intensive Spielen von Online-Spielen oder das Nichteinhalten von Regeln im Straßenverkehr. Auf der anderen Seite erhalten Bürger, die sich an die vom Staat definierten „Ethik-Kriterien" halten, Punkte. Beispielsweise erhöhen Spenden an Wohltätigkeitsorganisationen, Blutspenden oder die Teilnahme an ehrenamtlicher und sozialer Arbeit, die Scoringpunkte. Dies führt in der Konsequenz beispielsweise zu günstigeren Wohnungsbaudarlehen, einem Verzicht auf Bibliotheksgebühren oder einem vergünstigten Zugang zu öffentlichen Verkehrsmitteln oder kürzeren Wartezeiten in Krankenhäusern.

Neben Bürgern werden auch Unternehmen über einen eigenen Social-Credit-Score verfügen. So plant die chinesische Regierung allerdings keinen zentralen Credit-Score, sondern einen diversifizierten und dezentralisierten Markt für Sozialkredit-Ratings. So werden verschiedene Credit-Rating-Systeme existieren, mit unterschiedlichen Bewertungskriterien, die jeweils für einen anderen Zweck eingesetzt werden. So soll ein System beispielsweise ESG-Ratings automatisieren. Diese Ratings decken Bereiche wie Steuern, Zollauthentifizierung, Umweltschutz, Produktqualität, Arbeitssicherheit, E-Commerce und Cybersicherheit ab.

Wenn ein Unternehmen sich beispielsweise nicht an Gesetze hält oder Compliance nicht ernst nimmt, wird das „Social Credit Rating" des Unternehmens in Echtzeit gesenkt; im Umkehrschluss steigt das Credit Rating, wenn sich das Unternehmen ethisch konform verhält. Wenn ein Unternehmen beispielsweise Kredite nicht rechtzeitig zurückzahlt oder wenn es die Emissionsziele, die Sicherheitsstandards der Arbeitnehmer oder die staatlichen Investitionsauflagen nicht einhält, so wird sich dies im Scoring widerspiegeln.

Und was wären die Konsequenzen? Mögliche Wirkungen eines Downgrades wären beispielsweise ungünstigere Kreditkonditionen, ein Verbot zur Ausgabe von Anleihen oder eine reduzierte Chance zur Teilnahme an öffentlich finanzierten Projekten. Somit haben Unternehmen eine große Motivation ihr Social-Credit-Rating bzw. ihre ESG-Performance zu verbessern.

Insgesamt ist davon auszugehen, dass in das „Social Credit Rating" auch Daten der chinesischen Internetunternehmen Alibaba Group, Tencent sowie Baidu in die Bewertung einfließen. Bei Ant Financial, einer Tochtergesellschaft der Alibaba Group, heißt das System „Sesame Credit".

So hatte in einem Interview der Manager Min Wanli der Alibaba Group gegenüber der Redaktion des Handelsblatt bestätigt, dass Alibaba ein eigenes Bonitätssystem aufgebaut hat, das als Vorlage für das staatliche System dienen könnte: „Wir sind überzeugt, dass unser Punktesystem eine gute Hilfe für die Regierung sein kann. Der Staat überlegt sogar, unser Punktesystem zu übernehmen. Falls er das möchte, unterstützen wir gerne". Vgl. Interview „Es gibt Firmen, die Zeit in Brettspiele investieren – wir machen Krankenwagen schneller" mit dem Chefdatenwissenschaftler von Alibaba, Min Wanli, Handelsblatt vom 27.10.2018. Im Interview hat Min Wanli auch darauf hingewiesen, dass die Daten zur Kreditwürdigkeitsprüfung genutzt werden, und keinesfalls für andere Zwecke: „Als Technologieunternehmen kann ich nicht erfassen, ob jemand der Regierung gegenüber kritisch ist. Diese Daten gibt es nicht und sie lassen sich auch nicht generieren. Deshalb werden sie nicht Teil des Punktewertes sein. Selbst wenn das geplant sein sollte, ließe es sich praktisch nicht umsetzen."

Was in den Diskussionen um das Thema „Social Credit Rating" häufig ausgeblendet wird, ist die Tatsache, dass nicht nur die chinesischen Internetgiganten Verhaltensdaten sammeln und diese auswerten, sondern auch die GAFA-Konzerne (Google, Apple, Facebook, Amazon) und staatliche Stellen und privatwirtschaftliche Unternehmen diese Informationen für ihre Zwecke nutzen (vgl. hierzu Zuboff 2018).

In der Zwischenzeit wissen wir, dass die Daten der GAFA-Konzerne und weiterer Datensammler auch an Geheimdienste weitergeleitet und für deren Zwecke verwendet werden. So enthüllte Anfang 2013 der US-amerikanische Whistleblower und ehemalige Geheimdienstmitarbeiter Edward Snowden, wie die Vereinigten Staaten von Amerika und das Vereinigte Königreich seit spätestens dem Jahr 2007 in großem Umfang die Telekommunikation und insbesondere das Internet global und verdachtsunabhängig überwachen und hierbei auch mit einigen Internet konzernen zusammenarbeiten (vgl. vertiefend Beckedahl und Meister 2013). Nach Angaben von Edward Snowden betreiben NSA-Abhörspezialisten auf dem Gelände der Mangfall-Kaserne in Bad Aibling eine eigene Kommunikationszentrale und eine direkte elektronische Verbindung zum Datennetz der NSA.

Die NSA hat sich zudem Zugriff auf Netzwerke von Google und anderen Internetkonzernen verschafft und zapft die Daten in deren Rechenzentren an (vgl. vertiefend die Analyse bei Zuboff 2018). Außerdem ist die Zusammenarbeit mit dem Bundesnachrichtendienst (BND) beim Datenaustausch nachgewiesen.

So weist Shoshana Zuboff, eine US-amerikanische Wirtschaftswissenschaftlerin und emeritierte Professorin für Betriebswirtschaftslehre der Harvard Business School, darauf hin, dass die Menschheit an einem Scheideweg steht. Sie zeichnet in ihren Publikationen ein unmissverständliches Bild der neuen digitalen Märkte, auf denen Menschen nur noch Quelle eines kostenlosen Rohstoffs sind: Lieferanten von Verhaltensdaten. Sie beschreibt einen Totalitarismus einer „Dritten Moderne" und skizziert ein Zeitalter des Überwachungskapitalismus (vgl. Zuboff 2018).

In dieses Bild passen die von der Politik, beispielsweise der EU-Kommission und dem EU-Parlament, angedachten Ansätze zur Regulierung der Finanzströme basierend auf ESG-Kriterien. Um die Klimaziele zu erreichen, sollen (gesteuert durch planwirtschaftliche Ansätze) das Kapital von Anlegern gezielt in die entsprechenden Branchen und Unternehmen geleitet werden, die ihren Anteil zu einer ethisch, ökologisch und sozial besseren Welt beitragen. Sowohl private als auch institutionelle Anleger sollen mit regulatorischem Druck motiviert werden, künftig verstärkt nach ESG-Kriterien zu investieren.

Dass die Kapitalmärkte hierfür keine planwirtschaftlichen Ansätze aus der Politik und Regulierung benötigen, zeigt ein Blick auf einen der ältesten Fonds überhaupt, den Pioneer Fund, der bereits im Jahr 1928 als Investmentprodukt für die streng religiösen Gemeinschaften der Quäker und Methodisten in Boston aufgelegt wurde und alle heutigen ESG-Kriterien erfüllen würde. So verzichtet der Ethikfonds seit seinem Start auf Investitionen in Glücksspiel sowie die Alkohol- und Tabakindustrie. Für diese Art des Investments hat sich im angelsächsischen Raum der Begriff „ethical" oder „social responsible investment" herausgebildet.

Unabhängig von der Sinnhaftigkeit einer zentral- und planstaatlichen Lenkung von ESG-Investments, stellt sich die wichtige Frage nach einer trennscharfen Definition der ESG-Kriterien sowie einer methodisch fundierten Bewertung. ESG-Kriterien sind Teile eines komplexen Systems und lassen sich daher nicht mit einfachen qualitativen Kriterien bewerten. Daher sind Methoden erforderlich, die fundiert die einzelnen ESG-Kriterien

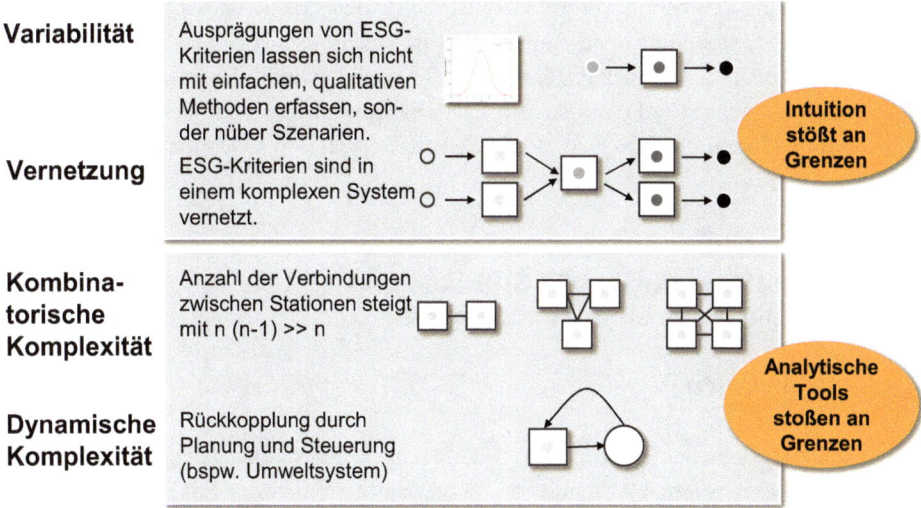

Abb. 21.7 Gründe für den Einsatz von quantitativen Methoden zur Bewertung von ESG-Risiken. (Quelle: Romeike und Hager 2020)

sowie deren Abhängigkeiten bewerten. In Abb. 21.7 sind einige wesentliche Gründe für den Einsatz von quantitativen Methoden zur Bewertung von ESG-Risiken zusammengefasst.

In Abschn. 21.2.2 wurde bereits eine quantitative Simulationsmethode zur Bewertung von ESG-Risiken vorgestellt. Bereits Anfang der 1950er-Jahre hatte Jay Wright Forrester an der Sloan School of Management des Massachusetts Institute of Technology „System Dynamics" (SD) als Methodik zur ganzheitlichen Analyse und Simulation komplexer und dynamischer Systeme entwickelt. System Dynamics war auch die grundlegende Methodik zur Simulation des Weltmodells World3, einer Studie zur Zukunft der Weltwirtschaft, die der Club of Rome in Auftrag gegeben hatte. In diesem Modell spielten sowohl Umweltkriterien, als auch soziale und Governance-Aspekte eine wichtige Rolle.

System Dynamics ist eine Methodik zur Modellierung, Simulation, Analyse und Gestaltung von dynamisch-komplexen Sachverhalten in sozioökonomischen Systemen (vgl. zur Vertiefung Romeike und Spitzner 2013, vgl. Sterman 1989, S. 321–339, vgl. Heij et al. 1997). Dynamische und komplexe Systeme zeichnen sich unter anderem sowohl durch verzögerte Ursache-/Wirkungseffekte als auch durch Rückkopplungsbeziehungen zwischen einzelnen Variablen aus. Dies gilt sowohl für Unternehmen als Systeme, als auch für Umweltsysteme und soziale Systeme.

System Dynamics beschäftigt sich mit dem Verhalten von gelenkten Systemen im Zeitablauf. Es verfolgt das Ziel, Systeme mit Hilfe qualitativer und quantitativer Modelle nicht nur zu beschreiben, sondern auch zu verstehen, wie Rückkopplungsstrukturen das Systemverhalten determinieren.

Aufbauend auf den bereits vorgestellten SD-Systemen skizzieren wir im anschließenden Kapitel die Erweiterung der Systeme mit Hilfe stochastischer Simulationsmethoden.

Diese Methoden bilden auch die Grundlage für viele methodische Ansätze im Bereich „Artificial Intelligence" (AI), in dem stochastische Simulationsmethoden (Monte Carlo Simulation) mit neuronalen Netzen, die die Arbeitsweise unseres Gehirns abbilden, kombiniert werden.

21.5 Quantifizierung von ESG-Risiken und stochastische Simulationsmodelle

21.5.1 Grundlagen

Ausgehend von den eher „qualitativen" Überlegungen der vorangegangenen Abschnitte befasst sich dieser zentrale Abschnitt mit der Quantifizierung von ESG-Risiken. Dabei wird auf die Notwendigkeit der Quantifizierung von ESG-Risiken, speziell ihrer finanziellen Auswirkungen, die hier bestehenden Herausforderungen (beispielsweise durch Defizite bei den verfügbaren Daten) und Lösungsstrategien eingegangen. Insbesondere wird auch auf die Bedeutung von Simulationsverfahren (stochastische Simulation bzw. Monte-Carlo-Simulation) in diesem Zusammenhang verwiesen.

Mögliche Anhaltspunkte für eine Quantifizierung von ESG-Risiken bieten das „Natural Capital Protocol" (vgl. Natural Capital Coalition 2016) und das „Social & Human Capital Protocol" (vgl. Social & Human Capital Coalition 2016).

Als konkrete Methoden empfehlen COSO und WBCSD beispielsweise die Durchführung einer Delphi-Analyse, einer deterministischen Szenarioanalyse, einer stochastischen Simulation sowie ESG-spezifischer Methoden.

Als weitere Methoden werden in der Dokumentation aufgeführt (vgl. The Committee of Sponsoring Organizations of the Treadway Commission (COSO) and World Business Council for Sustainable Development (WBCSD) 2018, S. 60):

- Greenhouse Gas Protocol: Der „Greenhouse Gas Protocol Corporate Accounting and Reporting Standard" bietet Unternehmen eine Anleitung für die Berechnung von Treibhausgasinventaren.
- WBCSD Water Tool: Das „WBCSD Water Tool" ist eine multifunktionale Ressource zur Identifizierung und Berechnung von Wasserrisiken und -chancen eines Unternehmens, einschließlich einer Arbeitsmappe, (für Standortinvestoren, wichtige Berichtsindikatoren und Metriken), einer Kartierungsfunktionalität und Google-Earth-Kompatibilität.
- InVEST: Das „Integrated Valuation of Ecosystem Services and Trade-offs" (InVEST) ist eine Suite von Open-Source-Softwaremodellen. InVEST ermöglicht es Entscheidungsträgern, die Auswirkungen von Management-Entscheidungen auf das zukünftige

Klima zu bewerten und zu erkennen, wo Investitionen die menschliche Entwicklung und die Ökosysteme nachhaltig fördern können.

- WRI Aqueduct: WRI Aqueduct ist ein Werkzeug zur Risikokartierung, das Unternehmen dabei unterstützt zu verstehen, wo und wie Wasserrisiken und -chancen weltweit entstehen. Der Atlas verwendet eine Peer-Review-Methodik, um anpassbare globale Karten der Wasserrisiken zu erstellen.
- World Bank Climate Change Knowledge Portal: Das „Climate Change Knowledge Portal" ist eine zentrale Drehscheibe für Informationen, Daten und Berichte über den Klimawandel weltweit. Es ermöglicht den Nutzern, wichtige klima- und klimarelevante Informationen abzufragen, zu kartografieren, zu vergleichen, darzustellen und zusammenzufassen.
- B Analytics, Global Impact Investment Rating System (GIIRS): Die GIIRS verwendet die Methode der B-Folgenabschätzung, um die Auswirkungen eines Anlageportfolios auf Arbeitnehmer, Kunden, Gruppen und die Umwelt zu erfassen.
- Impact Measurement Framework: Diese Sammlung von sektorspezifischen Rahmenwerken identifiziert relevante sozioökonomische Auswirkungen, Indikatoren und Metriken.
- Organisation for Economic Co-operation and Development (OECD) Guidelines on Measuring Subjective Well-being: Diese Richtlinien geben Ratschläge für die Erhebung und Verwendung von Maßen des subjektiven Wohlbefindens. Sie sollen die nationalen statistischen Ämter und andere Interessengruppen bei der Gestaltung, Erhebung und Veröffentlichung von Messungen des subjektiven Wohlbefindens unterstützen.

Im Dokument „Enterprise Risk Management: Applying enterprise risk management to environmental, social and governance-related risks" wird auch auf Bewertungsfehler (Availability bias, Confirmation bias, Groupthink bias, Illusion of control, Overconfidence effect, Status quo bias) kritisch hingewiesen (vgl hierzu vertiefend Gleißner und Romeike 2012, S. 43–46, vgl. Romeike 2013, S. 25–29).

Der häufig zu lesende Hinweis, manche Risiken seien nicht quantifizierbar, trifft nicht zu (vgl. hierzu auch Gleißner 2019d; Holton 2004 sowie Sinn 1980). Wenn man nicht von einem traditionellen „frequenzialistischen" Ansatz ausgeht, und sinnvollerweise die Quantifizierung von Risiken basierend auf den besten verfügbaren Informationen zulässt, ist jedes Risiko quantifizierbar und die Unterscheidung der Unsicherheit von Knight (vgl. Knight 1921) in Ungewissheit und einem quantifizierbaren Risiko obsolet. Mit den besten verfügbaren Informationen lässt sich jedes Risiko durch eine Expertenschätzung quantifizieren, die aber transparent zu erläutern ist.

21.5.2 Stochastische Methode

Die Risikoquantifizierung ist aus folgenden Gründen nützlich und wichtig (in Anlehnung an Gleißner 2017a):

1. Die Quantifizierung einzelner Risiken ermöglicht deren Priorisierung und den Vergleich mit anderen Risiken eines Unternehmens. Hierzu ist es notwendig, ein Risikomaß zu definieren und/oder die Konsequenz eines Risikos für den Erfolgsmaßstab des Unternehmens (oberstes Ziel, zum Beispiel Unternehmenswert) zu berechnen. Falls Risiken außerdem in mehreren Wirkungsdimensionen gemessen werden, muss eine Verrechnung zwischen den Dimensionen erfolgen (Zeit, Geld, Reputation, menschliche Gesundheit etc.); erst dann ist der Vergleich möglich. Bei Unternehmen zählt letztlich die Wirkung auf Gewinn, Ertrag, Cashflow bzw. Unternehmenswert.
2. Die quantitative Beschreibung von Einzelrisiken ist zudem eine unverzichtbare Grundlage, um anschließend mittels einer Risikoaggregation eine Gesamtrisikoposition zu berechnen und die Wirkungsmechanismen durch die kombinierte Wirkung mehrerer Einzelrisiken zu erkennen.
3. Erst durch die Risikoquantifizierung kann das Risikomanagement in den Kontext von Planung und Controlling gestellt werden, um die Planungssicherheit zu beurteilen.
4. Mit einer Risikoaggregation, die eine Risikoquantifizierung erfordert, sind Aussagen hinsichtlich der nötigen Bemessung von Eigenkapital (Eigenkapitalbedarf) oder Liquiditätsreserven möglich. Auch Aussagen zum angemessenen Rating – also der Insolvenz- bzw. Überlebenswahrscheinlichkeit – sind dann direkt aus der Unternehmensplanung in Verbindung mit den quantifizierten Risiken ableitbar. Zudem können die Konsequenzen der Risiken auch als „kalkulatorische Eigenkapitalkosten" leicht verständlich dargestellt werden.

Auch im Risikomanagement gilt, wie obige Beispiele zeigen, der bekannte Grundsatz *„If you can meassure it, you can manage it"*. Die Notwendigkeit einer klaren quantitativen Beschreibung von Risiken wird daran deutlich, dass eine alleinige verbale Umschreibung ein sehr breites Interpretationsspektrum zur Folge hat (vgl. Hillson 2005a, b). Einer Befragung zur Folge hat beispielsweise die Wahrscheinlichkeitsaussage „almost certain" eine korrespondierende Eintrittswahrscheinlichkeit von knapp 80 Prozent. „Likely" liegt bei rund 60 Prozent, und „impossible" bei immer noch acht Prozent. Auffällig ist, dass die meisten verbalen Wahrscheinlichkeitsangaben zwischen den Befragten eine Spannweite der zuordenbaren Wahrscheinlichkeiten von zehn Prozent und mehr aufweisen. Die Interpretation einer verbalen Wahrscheinlichkeitsaussage ist zudem stark kontextabhängig.

Dass Risiken dennoch häufig nicht quantifiziert werden, hat verschiedene Ursachen. Zu nennen sind insbesondere Probleme mit verfügbaren Daten über Risiken, Kenntnisdefizite hinsichtlich der Methodik zur Risikoquantifizierung und die Aversion vieler Menschen, mit Zahlen und Mathematik umzugehen (und sich damit nachvollziehbar und klar festzulegen, vgl. hierzu die empirischen Untersuchungen zur Risikoeinstellung von Managern bei March und Shapira 1987; Kesten 2007 sowie Günther und Detzner 2012). Als häufigste Begründung hört man in Unternehmen, dass auf eine quantitative Beschreibung des Risikos verzichtet wird, weil über die quantitativen Auswirkungen und die Eintrittswahrscheinlichkeit bzw. Häufigkeit eines Risikos keine adäquaten (historischen) Daten vorlagen. Das Risiko wird dann nicht quantifiziert und nur als „verbale Merkposition" im

Risikomanagement „verwaltet". Es fließt entsprechend nicht ein in die Beurteilung der Bestandsgefahrdung des Unternehmens, in die Berechnung des Eigenkapitalbedarfs mittels Risikoaggregation oder in die Ableitung risikogerechter Kapitalkostensatze für die Unternehmenssteuerung.

Rechtfertigt eine schlechte Datenqualität einen derartigen Umgang mit einem Risiko? Sicher nicht. Entscheidend ist vor allem, dass mit der hier beschriebenen Vernachlässigung eines Risikos eine „Nicht-Quantifizierung" überhaupt nicht erreicht wird. Tatsächlich wird das Risiko in allen genannten Berechnungen nicht berücksichtigt, das heißt, es wurde faktisch mit Null quantifiziert (das heißt mit einer Eintrittswahrscheinlichkeit/Häufigkeit und einer Schadenshöhe von Null).

Hieraus wird deutlich: Eine Nicht-Quantifizierung von Risiken gibt es nicht; Nicht-Quantifizierung bedeutet Quantifizierung mit Null. Und dies ist sicherlich häufig nicht die beste Abschätzung eines Risikos. Statt einer derartigen „Null-Quantifizierung" eines Risikos bietet es sich an, eine Quantifizierung mit den besten verfügbaren Informationen vorzunehmen und dies können, wenn weder historische Daten noch Vergleichswerte oder andere Informationen vorliegen, selbst subjektive Schätzungen der quantitativen Höhe des Risikos durch „Experten" des Unternehmens oder externe „Experten" sein. Eine akzeptable Qualität solcher Schätzungen lässt sich durch geeignete Verfahren, beispielsweise eine Verpflichtung zu einer nachvollziehbaren Herleitung, durchaus sicherstellen. Auch die Verwendung subjektiv geschätzter Risiken und deren Verwendung im Risikomanagement ist methodisch zulässig und notwendig, was Sinn bereits im Jahr 1980 im Rahmen seiner Dissertation „Ökonomische Entscheidungen bei Unsicherheit" aufgezeigt hat (vgl. vertiefend Sinn 1980). Auch subjektiv geschätzte Risiken können genauso verarbeitet werden, wie (vermeintlich) objektiv quantifizierte. Man muss sich hier immer über die Alternativen klar sein: Die quantitativen Auswirkungen eines Risikos mit den besten verfügbaren Kenntnissen (notfalls subjektiv) zu schätzen, oder die quantitativen Auswirkungen implizit auf null zu setzen und damit den Risikoumfang zu unterschätzen. Insgesamt ist damit klar: Nur die Quantifizierung von Risiken schafft einen erheblichen Teil des ökonomischen Nutzens des Risikomanagements zur Unterstützung von Entscheidungen unter Unsicherheit. Die scheinbare Alternative einer Nicht-Quantifizierung von Risiken existiert, wie schon erwähnt, nicht, da nicht quantifizierte Risiken nichts anderes sind als mit „Null" quantifizierte Risiken. Ein wirksames Risikomanagement bedingt eine Quantifizierung aller relevanten Risiken (vgl. hierzu vertiefend Romeike 2018, Gleißner 2019d sowie Romeike und Hager 2020).

Nach dem Prozessschritt der Risikoidentifikation sind alle wesentlichen Risiken zu quantifizieren. Dies gilt auch für ESG-Risiken, zumindest deren finanzielle Wirkungen. Nur mit quantifizierten Risiken kann man rechnen, sie vergleichen, und beispielsweise im Hinblick auf die Konsequenzen für Rating oder Unternehmenswert beurteilen. Die Risikobewertung umfasst, wie erwähnt, die quantitative Beschreibung eines Risikos durch eine geeignete Wahrscheinlichkeitsverteilung und die Berechnung von Risikomaßen. Da die Bestimmung einer geeigneten quantitativen Beschreibung für ein Risiko durchaus mit erheblichem Arbeitsaufwand, beispielsweise statistischen Analysen, verbunden sein kann, wird man sich hier in der Praxis meist nur auf die für das Unternehmen wichtigen Risiken beschränken. Um eine derartige Fokussierung vornehmen zu können, ist jedoch zumindest eine Grobeinschätzung der quantitativen Höhe eines Risikos erforderlich.

Zur quantitativen Beschreibung eines Risikos kann eine Wahrscheinlichkeitsverteilung genutzt werden, die die Ergebnisauswirkungen eines Risikos in einer Periode (etwas bezogen auf ein Jahr) beschreibt (in enger Anlehnung an Gleißner 2017b). Eine differenziertere Betrachtung ist möglich, wenn man ein Risiko beschreibt durch (1) eine Wahrscheinlichkeitsverteilung für die Häufigkeit des Risikoeintritts in einer Periode und (2) eine Wahrscheinlichkeitsverteilung für die Schadenshöhe je eingetretenen Risikofall.

Dabei ist zwischen „Bruttowirkungen" und „Nettowirkungen" eines Risikos zu unterscheiden. Für die Risikoquantifizierung sind letztlich die Nettowirkungen relevant, bei denen sämtliche momentan realisierte Risikobewältigungsverfahren (zum Beispiel Versicherungen) bereits berücksichtigt sind. Statt von „Bruttowirkungen" und „Nettowirkungen" wäre es angemessener von einem Status-quo-Risiko und einem Ziel-Risiko (Target Risk) zu sprechen (vgl. hierzu vertiefend Romeike und Hager 2020). Bei der Status-quo-Analyse werden alle bereits in der Vergangenheit umgesetzten Maßnahmen berücksichtigt. Das Ziel-Risiko hingegen definiert das angestrebte Niveau nach Umsetzung weiterer und neuer Maßnahmen zur Risikosteuerung. Die Berechnung eines „echten Bruttorisikos" wird in der Praxis nicht möglich sein, da in der Regel keine Informationen über alle Maßnahmen vorliegen, die in der Vergangenheit bereits umgesetzt wurden.

Die wichtigsten **Verteilungsfunktionen** im Rahmen des Risikomanagements sind Binomialverteilung, Normalverteilung, Dreieckverteilung, Poissoverteilung sowie die Compound-Verteilung (vgl. hierzu vertiefend Romeike und Hager 2020; Cottin und Döhler 2013 sowie Stampfer 2019). Diese Verteilungen beschreiben entweder die Häufigkeit oder die monetären Auswirkungen eines Risikos. Oder sie integrieren die Häufigkeit des Eintretens und die Höhe der Auswirkungen des Risikos.

Traditionell häufig Verwendung findet in der Praxis die einfachste **Binomialverteilung**, die ein Risiko nur durch Schadenshöhe und Eintrittswahrscheinlichkeit beschreibt. Diese ist angemessen, wenn man „ereignisorientierte Risiken" betrachtet. Bei diesen kann man näherungsweise davon ausgehen, dass das entsprechende Risiko genau einmal in einem Jahr mit der Wahrscheinlichkeit p eintritt und dann einen Schaden zur Konsequenz hat. Typische Anwendungsfälle sind der Verlust eines Schlüsselkunden, der Brand in einer Fabrik oder der Ausfall einer kritischen Maschine. Ereignisorientierte Risiken sind damit entweder „Chance" oder „Gefahr", aber nicht beides zugleich. Kann ein Ereignis mehr als einmal innerhalb eines Jahres eintreten, benötigt man dagegen die Poissonverteilung oder eine allgemeine Binomial-Verteilung ($n > 1$).

Risiken, die Chance und Gefahr zugleich darstellen, kann man beispielsweise durch die **Normalverteilung** beschreiben. Für ihre Spezifikation benötigt man den Erwartungswert, der als Lageparameter aussagt, was „im Mittel" passiert, und die Standardabweichung, die den Umfang „üblicher" positiver oder negativer Abweichungen spezifiziert. Die Normalverteilung findet insbesondere zur Beschreibung von Risiken Anwendung, die man als Verdichtung vieler einzelner kleiner (und unabhängiger) Einzelereignisse auffassen kann, wie beispielsweise für Nachfrageschwankungen, Umsatzschwankungen, Zinsänderungs- und Währungsrisiken, Aktienrenditen sowie Rohstoffpreisänderungen (speziell also für „marktbezogene" Risiken).

Für die Beschreibung von asymmetrischen Risiken, die entweder einen Chancen- oder einen Gefahrenüberhang aufweisen, kann man im einfachsten Fall die sogenannte **Dreiecksverteilung** verwenden. Bei dieser wird eine betrachtete risikobehaftete Größe (beispielsweise die Kosten eines Projektes) beschrieben durch (a) Mindestwert, (b) wahrscheinlichsten Wert und (c) Maximalwert. Beispiele: risikobedingt mögliche Bandbreite des Marktanteils, der Personalkosten oder der Höhe der Investitionen.

Häufigkeiten können sehr pragmatisch und fundiert mit einer **Poisson-Verteilung** beschrieben werden. Die Poisson-Verteilung wird vor allem dort eingesetzt, wo die Häufigkeit eines Ereignisses über eine gewisse Zeit betrachtet wird. Die Poisson-Verteilung wird auch manchmal als „Verteilung der seltenen Ereignisse" bezeichnet. Die verallgemeinerte Poisson-Verteilung und die gemischte Poisson-Verteilung werden vor allem im Bereich der Versicherungsmathematik angewendet, wo es auch um die Schätzung der Häufigkeit von Schadensereignissen geht. Ist eine Zufallsvariable X Poisson-verteilt, so ist λ zugleich Erwartungswert und Varianz.

Die **Compound-Verteilung** ergibt sich auf natürliche Weise aus Anwendungen in der Praxis, wo sich eine zufällige Zahl von Schadenfällen mit je für sich zufälliger Höhe zu einem Gesamtschaden addieren (vgl. vertiefend Romeike und Hager 2020). In Abb. 21.8 ist exemplarisch die Bewertung mit Hilfe einer Compound-Verteilung wiedergegeben (vgl. hierzu vertiefend Romeike und Hager 2020). Im Beispiel wurde das Schadensszenario basierend auf den Parametern „best case", „realistic case" und „worst case" in Form einer PERT-Verteilung modelliert. Die PERT-Verteilung basiert auf einer Transformation der Vierparameter-Beta-Verteilung mit der Annahme, dass der erwartete Wert sich als gewichtetes Mittel aus dem Minimum, dem Maximum und dem wahrscheinlichsten Wert resultiert. In der Standard-PERT-Verteilung wird dabei Vierfache des Gewichts auf den wahrscheinlichsten Wert angewendet. Durch eine Anpassung des Shape-Parameters lässt sich die Unsicherheit der Expertenschätzungen abbilden. Insbesondere für die Bewertung von ESG-Risiken und eine „seriöse" Berücksichtigung von Unsicherheit bietet die Compound-Verteilung eine solide Basis.

Oft ermöglicht nur eine Kombination von Wahrscheinlichkeitsverteilungen eine adäquate Beschreibung eines Risikos. Eine solche Kombination bildet die Compound-Verteilung ab. Man denke zum Beispiel an den Fall, dass zwar einem ereignisorientierten Risiko eine bestimmte Eintrittswahrscheinlichkeit bzw. Häufigkeit zugeordnet werden kann und auch die Schadenshöhe selbst unsicher ist und nur durch eine Bandbreite beschrieben werden kann (Mindestwert, wahrscheinlichster Wert bzw. Maximalwert).

Beispiel: Der Schaden S tritt zum Beispiel mit $p = 10$-prozentiger Wahrscheinlichkeit ein und der unsichere Schaden ist dann durch $a = 10$ (Mindestwert), $b = 20$ (wahrscheinlichster Wert) und $c = 60$ (Maximalwert) charakterisiert, was zum Beispiel eine Dreiecksverteilung zeigt.

Der **Erwartungswert** des Schadens (S) beträgt dann

1. Schritt:
Beschreibung der Häufigkeit
(bspw. 5 x p.a.)

Beschreibung mit einer Poissonverteilung mit $\lambda = 5$

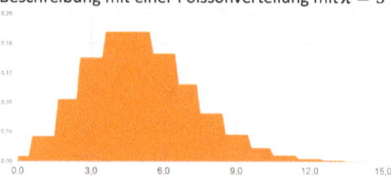

2. Schritt:
Beschreibung des Schadensausmaßes je
Risikoeintritt
(beispielsweise worst case= 100 Mio. EUR;
realistic case= 20 Mio. EUR; best case= 0,25 Mio. EUR)

Beschreibung mit einer PERT-Verteilung mit
wc = 100; rc = 20 und bc = 0,25

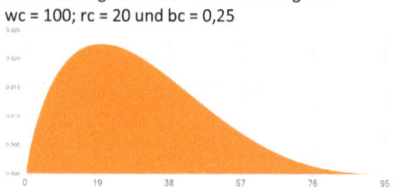

3. Schritt: Simulation und Analyse
potenzieller Schadensszenarien
resultierend aus Häufigkeit und
Schadensausmaßszenario

In mehreren 100.000en von Simulationsläufen
werden unterschiedliche Kombinationen simuliert.
Im nachfolgenden Simulationslauf sind bspw. 4
Ereignisse aufgetreten, für die jeweils
Schadensausmaßverteilungen simuliert wird.

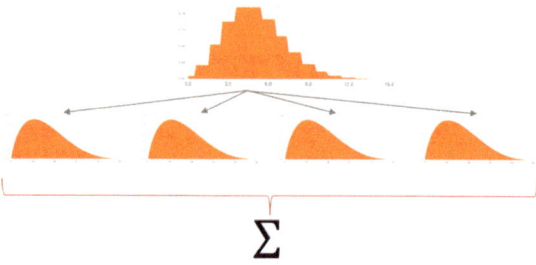

Abb. 21.8 Anwendung der Compoundfunktion in der Praxis. (Quelle: Romeike und Hager 2020)

$$E(S) = p \cdot \frac{(a+b+c)}{3} = 0,1 \cdot \frac{10+20+60}{3} = 3$$

Stochastische Prozesse dienen dazu den Verlauf und Risiken über mehrere Perioden zu beschreiben und ergänzen bisher erläuterte einfache Verteilungen.

Gerade bei der Quantifizierung von ESG- bzw. CSR-Risiken steht man oft vor dem Problem, dass die verfügbaren Daten als unzureichend erscheinen (und man ganz offensichtlich auf Expertenschätzungen angewiesen ist).

Sehr häufig steht man vor der Herausforderung, dass ein ESG-Risiko aus vielen „Facetten" besteht. Die verschiedenen Teilaspekte oder Einzelrisiken sind entsprechend zu aggregieren. Die Aggregation von Risiken erfordert im Allgemeinen eine stochastische Simulation (Monte-Carlo-Simulation); sofern man von einigen wenig Spezialfällen absieht (beispielsweise wenn sämtliche Risiken normalverteilt sind, vgl. hierzu Romeike und Hager 2020 bzw. Romeike und Hager 2010). Die Methode der stochastischen Simulation wird entsprechend im nachfolgenden Unterabschnitt etwas ausführlicher erläutert.

21.5.3 Stochastische Simulation

Die Stochastische Szenarioanalyse (in der Praxis häufig auch als Monte-Carlo-Simulation bezeichnet) basiert auf der Idee, die Eingangsparameter einer Simulation als Zufallsgrößen zu betrachten (die nachfolgenden Ausführungen basieren auf: Romeike 2018, S. 175 ff.). So können analytisch nicht oder nur aufwendig lösbare Probleme mit Hilfe der Wahrscheinlichkeitstheorie (die Teil der Stochastik ist, die Wahrscheinlichkeitstheorie und Statistik zusammenfasst) numerisch gelöst werden. Generell lassen sich zwei Problemgruppen unterscheiden, bei denen die Stochastische Szenarioanalyse angewendet werden kann. Mit ihrer Hilfe können einerseits Problemstellungen deterministischer Natur, die eine eindeutige Lösung besitzen, bearbeitet werden. Auf der anderen Seite sind aber auch Fragen, die sich der Gruppe stochastischer Problemstellungen zuordnen lassen, für eine stochastische Simulation ein geeignetes Anwendungsfeld (vgl. Romeike und Spitzner 2013, S. 104). Die Basis für die Simulation bildet eine sehr große Zahl gleichartiger Zufallsexperimente.

Aus einer betriebswirtschaftlichen Sicht können alle Fragen untersucht werden, die

- entweder aufgrund der Vielzahl ihrer Einflussgrößen nicht mehr exakt analysiert werden (können) und bei denen daher auf eine Stichprobe für die Analyse zurückgegriffen wird;
- oder bei denen die Eingangsparameter Zufallsgrößen sind (Auch die Optimierung von Prozessen oder Entscheidungen bei nicht exakt bekannten Parametern gehören zu dieser Gruppe).

Die Anwendung der Stochastischen Szenarioanalyse ist breit gefächert und reicht unter anderem von der Stabilitätsanalyse von Algorithmen und Systemen, der Aggregation von Einzelrisiken eines Unternehmens zu einem unternehmerischen Gesamtrisiko, der Vorhersage von Entwicklungen, die selbst durch zufällige Ereignisse beeinflusst werden (stochastische Prozesse), der Optimierung von Entscheidungen, die auf unsicheren Annahmen beruhen bis zur Modellierung komplexer Prozesse (Wetter/Klima, Produktionsprozesse, Supply-Chain-Prozesse, Rekonstruktionsverfahren in der Nuklearmedizin) oder der Schätzung von Verteilungsparametern.

Vor diesem Hintergrund ist die Stochastische Simulation auch geeignet, um die Unsicherheit im Bereich von ESG-Risiken abzubilden.

Die Entwicklung der Methode ist eng verbunden mit den Namen der beiden Mathematiker Stanislaw Ulam und John von Neumann. Sie sollen während ihrer Arbeit im Rahmen des Manhattan-Projekts am Los Alamos Scientific Laboratory diese Methode verwendet haben, um hochkomplexe physikalische Probleme nummerisch mit Hilfe einer Simulation zu lösen. Der Anekdote nach wurde als Codename „Monte Carlo" verwendet. Die ersten wissenschaftlichen Publikationen zu diesem Verfahren erschienen Ende der 1940er-Jahre. Mit dem zur damaligen Zeit parallelen Aufkommen elektronischer Computer fand die Monte-Carlo-Simulation zunächst in der Wissenschaft, später auch in der Wirtschaft ihre Verbreitung. Heute ist die Stochastische Simulation eine etablierte Methode in vielen Themengebieten und zur Lösung vielfältiger Fragestellungen.

Die grundlegende Idee der Stochastischen Simulation ist es, für zufällig gewählte Ausprägungen der Parameter über die entsprechenden Zusammenhänge (Ursache-Wirkungs-Geflecht) die zugehörigen Ergebnis- oder Zielgrößen zu ermitteln. Das zur Ermittlung der Zielgrößen verwendete Modell ist in der Regel deterministischer Natur, das heißt, mit dem Festlegen der Parameter sind die Zielgrößen eindeutig bestimmt. Allerdings sind die Zielgrößen durch den Zufallscharakter der Parameter im Prinzip wiederum zufällige Größen. Jedoch kann im Allgemeinen davon ausgegangen werden, dass eine hinreichend große Anzahl so ermittelter Zielgrößen einen guten Näherungswert für die tatsächlichen Werte dieser Zielgrößen darstellt (genau genommen sind nicht die tatsächlichen Werte, sondern die Erwartungswerte der Zielgrößen gemeint). Mathematisches Fundament dieses Vorgehens sind das Gesetz der großen Zahlen, der Fundamentalsatz der Statistik (Satz von Gliwenko-Cantelli) sowie der zentrale Grenzwertsatz. Die Methode ist damit ein Stichprobenverfahren. Aufgrund der zufälligen Auswahl der Parameter hat sich für die Monte-Carlo-Simulation ebenfalls der Begriff der Stochastischen Simulation bzw. Stochastischen Szenarioanalyse etabliert.

Das Vorgehen bei einer Monte-Carlo-Simulation wurde von Metropolis und Ulam in einem Artikel beschrieben, der im Jahre 1949 im Journal of the American Statistical Association erschienen ist. Darin beschreiben beide Wissenschaftler das Vorgehen bei der Monte-Carlo-Methode durch zwei Schritte: „(1) production of ‚random' values with their frequency distribution equal to those which govern the change of each parameter, (2) calculation of the values of those parameters which are deterministic, i.e., obtained algebraically from the others." (Metropolis und Ulam 1949).

An diesem durch Metropolis und Ulam beschriebenen Vorgehen hat sich in den letzten 60 Jahren nicht viel geändert. Seit 20 Jahren ist die Monte-Carlo-Simulation eine unverzichtbare Methode im Risikomanagement und dient der Risikoaggregation.

21.5.4 Besonderheiten der Quantifizierung von ESG/CSR-Risiken: finanzielle und nicht-finanzielle Wirkungen

Den Zusammenhang zwischen den sogenannten „nicht-finanziellen" Risiken und den finanziellen Risiken soll nachfolgend etwas genauer beachtet werden. Hier ist nämlich zu beachten, dass die sogenannten „nicht-finanziellen" Risiken, wie speziell die ESG- oder CSR-Risiken, eben sehr wohl finanzielle Auswirkungen haben können, die im Risikomanagement zu beachten sind.

Infolge der Global Reporting Initiative (GRI) enthalten die Lageberichte der Unternehmen seit 2017 auch eine Nachhaltigkeitsberichterstattung (Corporate Social Responsibi-

Abb. 21.9 Wirkungsbereich von ESG- bzw. CSR-Risiken. (Quelle: Gleißner 2019c, S. 95)

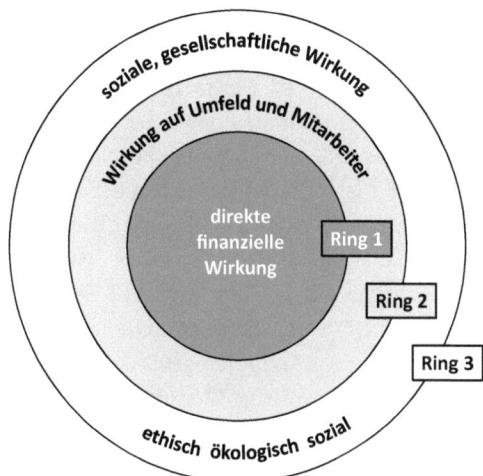

lity, CSR), vgl. auch ESG-Ansätze. Eine Beziehung zum Risikomanagement ergibt sich dadurch, dass hier auch auf wesentliche nicht finanzielle Risiken einzugehen ist (vgl. Abschn. 21.1) (in enger Anlehnung an Gleißner 2019c).

Anzugeben sind im Geschäftsbericht allerdings nur Risiken, die (unter Berücksichtigung von Risikobewältigungsmaßnahmen) sehr wahrscheinlich sind und schwerwiegende negative Auswirkungen haben (beispielsweise auf das Unternehmen, Mitarbeiter, Kunden, Natur oder die Gesellschaft). Die Wesentlichkeitsschwelle ist so hoch, dass bisher noch wenig über ESG- oder CSR-Risiken berichtet wird. Dennoch sind „intern" auch diese CSR-Risiken ein Thema für das Risikomanagement. Empfehlenswert, diese zunächst einmal zu strukturieren (vgl. Abb. 21.9).

Direkte finanzielle Auswirkungen haben etwa „CO_2-Emissionsrisiken", die zukünftig einen teuren Kauf von CO_2-Zertifikaten erfordern können. Beispielsweise fordert in Deutschland § 91 Abs. 2 AktG, dass mögliche „bestandsgefährdende Entwicklungen", auch aus Kombinationseffekten von Einzelrisiken, früh erkannt werden. Von einer Bestandsgefährdung ist jedoch nur auszugehen, wenn ein „CSR-Risiko" auch finanzielle Auswirkungen hat und so zur Illiquidität führen kann (also zu Ring 1 oder 2 gehört, vgl. Abb. 21.9). Solche Risiken sind auch im Risikotragfähigkeitskonzept zwingend zu erfassen (bei anderen ist die Art einer möglichen Einbeziehung zu diskutieren).

Bei den Risiken zu Ring 3 sind gesetzliche Vorgaben einzuhalten; ein Thema für „Compliance". Ob ein Unternehmen mehr als dies tun sollte, ist strittig. Milton Friedman hat argumentiert, Unternehmen sollten nachhaltig (unter Beachtung von Gesetzen) möglichst hohe Gewinne für ihre Eigentümer erwirtschaften und es diesen überlassen, ob sie Erträge für soziale, ökologische oder sonstige Ziele einsetzen möchten (Shareholder-Value-Ansatz).

Im Ergebnis ist festzuhalten, dass sich das Risikomanagement auch mit nicht finanziellen Risiken, speziell den „CSR-Risiken", befassen muss, also Methoden für die Identifikation, Quantifizierung und Überwachung von Risiken entwickeln muss, die primär Auswirkungen haben auf Mitarbeiter, Kunden, Natur oder die Gesellschaft. Man benötigt Messkonzepte, auch für die nichtfinanziellen Auswirkungen (beispielsweise wie den DALY (Disease Adjusted Life Years) zur Erfassung möglicher negativer gesundheitlicher Auswirkungen). Darüber hinaus ist bei jedem CSR-Risiko, wie bei jedem anderen Risiko, immer auch die finanzielle Auswirkung auf das Unternehmen zu erfassen (inklusive indirekter Auswirkungen beispielsweise durch eine negative Reputationsauswirkung). Wie immer ist zu beachten: neben der Häufigkeit/Eintrittswahrscheinlichkeit ist auch die Unsicherheit der Auswirkungen zu quantifizieren. Dies bedeutet, dass die Wirkungen mit Hilfe einer geeigneten statistischen Wahrscheinlichkeitsverteilung zu beschreiben sind, und nicht etwa durch eine „sichere Schadenshöhe". Es geht also zur Vermeidung von Scheingenauigkeiten um Bandbreiten. Solche Überlegungen sind auch für die Modelle zur Messung von Risikotragfähigkeit und Risikotoleranz relevant.

Ein Beispiel für die Möglichkeit der Quantifizierung eines ESG-Risikos seien im nächsten Abschnitt die Risiken durch die CO_2-Emission eines Industriebetriebs betrachtet.

21.5.5 Fallbeispiel zur quantitativen Bewertung von ESG-Risiken am Beispiel CO_2-Emissionen

Die Quantifizierung der finanziellen Auswirkungen eines Umweltrisikos („E-Komponente" in ESG) sei nachfolgend am besonders wesentlichen Thema CO_2-Emissionen verdeutlicht.

Angenommen ein Unternehmen emittiert momentan pro Produktionseinheit, direkt und indirekt, 1000 Tonnen CO_2.

Die heterogenen Produkte werden dabei in eine einheitliche Maßgröße umgerechnet, wobei hierfür vereinfachend sogar der (preisänderungsbereinigte) Umsatz gesetzt werden kann.

Im Jahr 2020 werden 1000 Produktionseinheiten geplant. Darüber hinaus wird geplant, dass die Produktion im betrachteten Planungszeitraum von fünf Jahren mit einer Rate von fünf Prozent pro Jahr (real) gesteigert werden soll (Unsicherheit: vier Prozent).

Es wird eine Normalverteilung mit einer Standardabweichung der Wachstumsrate von vier Prozent unterstellt. Zudem wird Martingaleigenschaft angenommen, das heißt, die in einem Jahr t eingetretenen Planabweichungen führen zu einer Anpassung der Planung in entsprechender Höhe in den folgenden Jahren.

Die CO_2-Intensität der Produktion soll um zehn Prozent pro Jahr reduziert werden (Unsicherheit: fünf Prozent pro Jahr). Unsicherheit bezüglich der geplanten Kosten der CO_2-Emissionen reduzierenden Maßnahmen sind im didaktischen Beispiel vereinfachend vernachlässigt worden. Ebenso vernachlässigt werden weitere hier noch relevante Teilaspekte des Risikos, zum Beispiel aus der Unsicherheit des politischen Umfelds und der Klimapolitik. Wie stellt sich bei diesen Daten das finanzielle Risiko des Unternehmens dar, das auch bei der Risikoaggregation und der Bestimmung des Gesamtrisikoumfangs (Eigenkapitalbedarf) zu berücksichtigen ist? Und wie stellt sich das ESG-Risiko aus Per-

	2020	2021	2022	2023	2024
Produktionseinheiten (geplant)	1000	1050	1103	1158	1216
Geplante Steigerung p.a.		5%	5%	5%	5%
Standardabweichung (Unsicherheit)	4%	4%	4%	4%	4%
Simulierte Produktionseinheiten	1011	1035	1106	1177	1307
CO2/Produktionseinheit (in Tonnen)	1000	1000	1000	1000	1000
CO2 insgesamt (in Tonnen)	1011208	1034665	1106148	1177360	1307474
Geplante Reduktion CO2 p.a.	-10%	-10%	-10%	-10%	-10%
Standardabweichung (Unsicherheit)	5%	5%	5%	5%	5%
CO2-Reduktion simuliert (in Tonnen)	-99683	-104080	-112239	-119327	-126497
CO2 insgesamt nach Reduktion (in Tonnen)	911525	930585	993909	1058034	1180977
Kosten CO2-Zertifikate / Tonne (realistic case)	20 €	20 €	20 €	20 €	20 €
worst case (PERT-Distribution)	23 €	24 €	26 €	28 €	30 €
best case (PERT-Distribution)	16 €	17 €	17 €	18 €	18 €
Simulierte Kosten CO2-Zertifikate / Tonne	21 €	21 €	19 €	22 €	21 €
Kosten CO2-Zertifikate insgesamt (Simulation)	18.824.592 €	19.078.845 €	19.337.614 €	22.845.399 €	25.306.378 €
Erwartungswert p.a.	17.871.421 €	19.081.426 €	20.352.336 €	21.884.509 €	23.339.266 €
VaR 99%	20.971.914 €	23.092.915 €	25.837.378 €	28.536.611 €	31.600.413 €
Expected Shortfall 99 %	21.343.149 €	23.613.091 €	26.769.732 €	29.841.663 €	32.975.643 €
					Plot

Abb. 21.10 Aufbau des stochastischen Simulationsmodells.

spektive des Stakeholders „Gesellschaft" dar, das heißt, wie relevant sind also die negativen Auswirkungen der CO_2-Emission des Unternehmens infolge der damit einhergehenden Temperaturerhöhung?

In Abb. 21.10 sind die grundsätzlichen Parameter sowie der Aufbau des stochastischen Simulationsmodells wiedergegeben. Sowohl die gewählten Parameter als auch der Aufbau des Modells sollte hierbei vor dem Hintergrund eines didaktischen Beispiels interpretiert werden. Auf die Berücksichtigung von stochastischen Prozessen wurde beispielsweise im Beispiel verzichtet.

Zunächst werden die finanziellen Risiken des Unternehmens berechnet. Ausgangspunkt ist dabei die Messung der Zusatzkosten, die dem Unternehmen durch eine unplanmäßige Entwicklung der CO_2-Emission entstehen können.

Die Kosten für die planmäßige CO$_2$-Emission sind ebenso wie andere Kosten, speziell auch für die CO$_2$-reduzierenden Maßnahmen durch Investition und Technologieveränderung, natürlich in der „erwartungstreuen" Planung berücksichtigt.

Wird angenommen, dass (real oder zumindest fiktiv) für die CO$_2$-Emission CO$_2$-Zertifikate gekauft werden, die momentan einen Preis von 20 Euro je Tonne aufweisen.

Die zukünftige Preisentwicklung ist unsicher. Diese Unsicherheit wird im Simulationsmodell mit Hilfe einer PERT-Verteilung abgebildet (vgl. Abb. 21.13). Die unsicheren Kosten der CO$_2$-Emission ergeben sich daher aus dem unsicheren Preis des CO$_2$-Zertifikats einerseits und der unsicheren CO$_2$-Emissionsmenge andererseits. Mit einer stochastischen Simulation lässt sich leicht der in Abb. 21.10 gezeigte Korridor für die zukünftigen Kosten „CO$_2$-Emissionsrisiko" angeben.

Der Erwartungswert der Kosten ist in der Planung, wie gesagt, berücksichtigt. Also Risikomaß wird zusätzlich der Value-at-Risk berechnet, und zwar zunächst für jedes einzelne Jahr. Darüber hinaus wird das „CO$_2$-Emissionskostenrisiko" für den gesamten Planungszeitraum von fünf Jahren angegeben. Konkret wird ermittelt, dass beispielsweise mit einer Sicherheit von 99 Prozent aus Sicht des Unternehmens die „CO$_2$-Emissionskosten" von rund 31,6 Millionen Euro im Jahr 2024 nicht überschritten werden.

99-Prozent-Quantil, also Value-at-Risk zum entsprechenden Wahrscheinlichkeitsniveau (VaR$_p$). Die konkrete Definition des Sicherheitsniveaus ist abhängig vom definierten Risikoappetit bzw. Risikoakzeptanz.

In Abb. 21.11 (Histogramm) sowie in Abb. 21.12 (Kumulierte Dichtefunktion) können sowohl die Erwartungswerte als auch der Value at Risk sowie der Expected Shortfall (vgl. vertiefend Romeike und Hager 2020 sowie Albrecht und Maurer 2002) abgelesen werden.

Man hat hier eine Quantifizierung des CO$_2$-Emissionsrisikos „Stand alone". Im Kontext des Risikomanagements werden die von der Planung abweichenden CO$_2$-Emissionskosten unmittelbar bei der Risikoaggregation berücksichtigt, um die Wechselwirkung

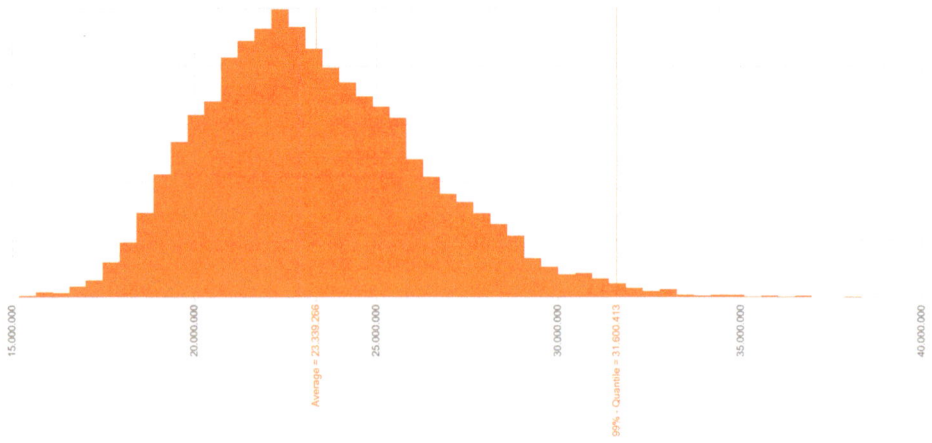

Abb. 21.11 Histogramm der Simulationsergebnisse.

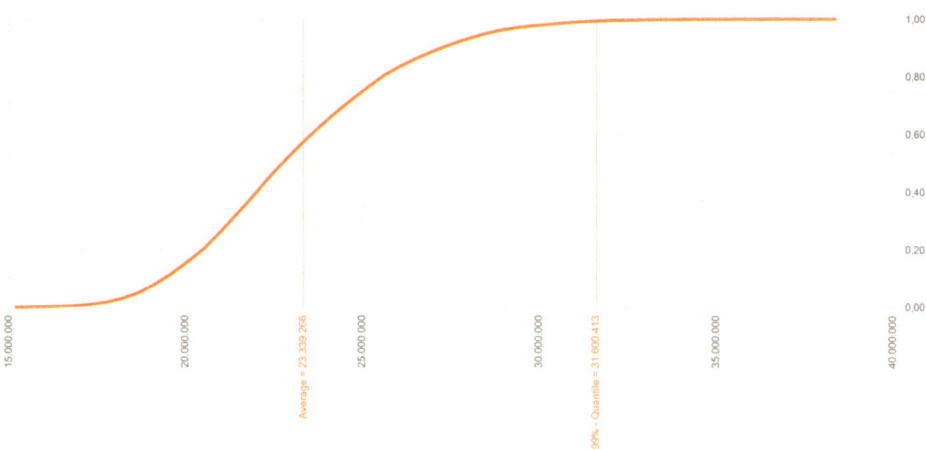

Abb. 21.12 Kumulierte Dichtefunktionen (Cumulative distribution function, CDF) der Simulationsregebnisse.

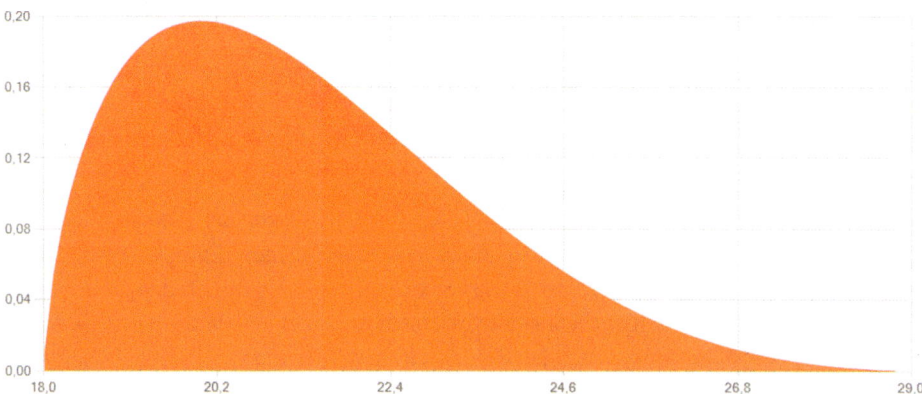

Abb. 21.13 Unsicherheit der Preise für CO_2-Zertifikate wurde mit einer PERT-Verteilung berücksichtigt.

zwischen unsicherer Produktionsmenge und unsicheren CO_2-Emissionskosten adäquat zu berücksichtigen.

Wie stellt sich die Konstellation aus Ebene der Gesellschaft dar? Natürlich kann man zunächst einmal das CO_2-Emissionsrisiko als ein „nicht-finanzielles" Risiko auffassen und die CO_2-Emissionsdaten, zum Beispiel „in Tonnen" angeben. Ist aber in vielen Fällen gar nicht nötig, auf dieser Ebene zu verbleiben. Man kann auch hier die finanziellen Auswirkungen angeben, was es ermöglicht, auch unterschiedliche Risiken (in gewissen Grenzen) miteinander zu vergleichen. So sind beispielsweise die Schäden durch CO_2-Emissionen in verschiedenen wissenschaftlichen Studien durchaus schon quantifiziert worden (vgl. zusammenfassend zum Beispiel Nordhaus 2018).

Nehmen wir an, die globalen Schäden durch die Temperaturerhöhung infolge der CO_2-Emission lägen bei ca. 30 Euro pro Tonne (und sind unsicher). Durch die Kombination der unsicheren CO_2-Emissionsmenge des Unternehmens und den unsicheren Schäden durch die CO_2-Emission kann man nun das ESG-Risiko „CO_2-Emission" aus globaler Perspektive wiederum mit einem Simulationsmodell quantifizieren. Zu beachten ist dabei, dass ein Teil der Schäden, die durch die CO_2-Emission ausgelöst werden, das Unternehmen bereits durch den Kauf von CO_2-Zertifikaten „kompensiert" (unter der Annahme, dass die Einnahmen durch die CO_2-Zertifikate der Start auch adäquat klimapolitisch einsetzt; vgl. Abb. 21.13). Ohne die Details des Modells hier weiter erläutern zu wollen, können auch die volkswirtschaftlichen Kosten (in Form von Erwartungswerten über die Zeit sowie als „Bandbreite") quantifiziert werden.

21.6 Zusammenfassung und Ausblick

Der Beitrag hat sich mit der Relevanz und Bewertung von ESG-Risiken in der Praxis auseinandergesetzt und zeigt Ähnlichkeiten zu Social Credit Systemen auf. Wir haben die „Verwandtschaft" zwischen ESG-Ratings und SCR-Ratings („Social Credit Ratings") aufgezeigt und uns hierbei auch mit dem potenziellen Missbrauch der Daten sowie der Konzentration in Datenmonopolen (sei es staatlich oder privatrechtliche Unternehmen) auseinandergesetzt. Hierbei müssen sowohl die gesamtgesellschaftlichen als auch individuellen negativen Folgen berücksichtigt und kritisch diskutiert werden. Viele Publikationen, insbesondere aus einer nicht-chinesischen Perspektive, neigen dazu, das „Social Credit Rating" in China als Orwell'schen Überwachungsstaat zu betrachten. Chinesische Politiker und auch viele Wissenschaftler bewerten das System anders. Als Begründung wird angeführt, dass sich das System nicht sehr stark von einem regulären Kreditratingsystem unterscheidet, das heißt der Einstufung der Bonität eines Wirtschaftssubjekts (Einzelperson, Unternehmen, Staat) oder eines Finanzinstruments.

In unserem Beitrag haben wir diese grundsätzliche Diskussion weitestgehend ausgeblendet und haben uns im Schwerpunkt auf die Möglichkeiten einer Quantifizierung mit Hilfe von Simulationsverfahren konzentriert. ESG-Risiken werden dabei in ihrer Relevanz für Unternehmen *und* Umfeld erläutert und die Notwendigkeit sowie die Möglichkeit der Quantifizierung gezeigt.

Es ist wesentlich zu beachten, dass auch ESG-Risiken finanzielle Auswirkungen haben, die auf jeden Fall zu quantifizieren sind. Würde man die finanziellen Komponenten von ESG-Risiken für das Unternehmen ignorieren, würde man den „Grad der Bestandsgefährdung" des Unternehmens und den aggregierten Gesamtrisikoumfang (Eigenkapital und Liquiditätsbedarf) unterschätzen. Die Notwendigkeit der Quantifizierung von ESG-Risiken (finanzielle Auswirkungen) ergibt sich aus den gesetzlichen Anforderungen in Deutschland (§ 91 AktG) sowie vielen weiteren gesetzlichen Anforderungen für bestimmte Branchen und auch in verschiedenen Ländern.

Die Möglichkeiten für die Quantifizierung solcher Risiken sind vorhanden, wenn man sinnvollerweise eine Quantifizierung aus den jeweils bestverfügbaren Informationen zulässt (also auch die transparente Quantifizierung basierend auf Expertenschätzungen). Inwieweit man die nichtfinanziellen Aspekte, die Wirkungen für die Gesellschaft und die Umwelt, quantifiziert, ist eigenständig zu diskutieren.

Das im Beitrag skizzierte und didaktische Beispiel der CO_2-Emmisionen zeigt, dass eine solche Quantifizierung durchaus möglich ist. Hilfreich ist hier zunächst eine Strukturierung der Risiken entsprechend der vorgeschlagenen „drei Kreise" (vgl. Abb. 21.9). Grundsätzlich ist festzuhalten, dass auch die Konsequenz von ESG-Risiken für Gesellschaft und Umwelt entsprechend den allgemeinen Erläuterungen über die Quantifizierungsmöglichkeiten von Risiken quantifizierbar sind (vgl. hierzu Abschn. 21.5.5).

Literatur

Albrecht, P., R. Maurer (2002): Investment- und Risikomanagement: Modelle, Methoden, Anwendungen, Schäffer-Poeschel Verlag, Stuttgart 2002.

Bauer, U./Romeike, F./Weißensteiner, Chr. (2012): Der gute Ruf als nachhaltiger Erfolgsfaktor – Management und Controlling von Reputationsrisiken, Studienergebnisse, RiskNET GmbH und Technische Universität Graz, Brannenburg/Wendelstein und Graz 2012.

Beckedahl, M./Meister, A. (Hrsg.): Überwachtes Netz: Edward Snowden und der größte Überwachungsskandal der Geschichte, epubli, Berlin 2013.

Berg, T./Kreft, M. (2020): Digitale Fußabdrücke im Kredit-Scoring – Ein Vergleich von traditionellen statistischen und Machine Learning Verfahren, in: FIRM Jahrbuch 2020, Frankfurt am Main 2020.

Blum, U./Gleißner, W./Xiao, X.: How Does Market Structure Determine Business Models in a Circular Economy? A Theory-Based Analysis for Plastic-Waste and High-Performance Magnet Markets (noch nicht erschienen).

Cottin, C./Döhler, S. (2013): Risikoanalyse – Modellierung, Beurteilung und Management von Risiken mit Praxisbeispielen, 2. Auflage, Springer Verlag, Wiesbaden 2013.

Creemers, R. (2014): Planning Outline for the Construction of a Social Credit System (2014–2020), Internet: https://chinacopyrightandmedia.wordpress.com/2014/06/14/planning-outline-for-the-construction-of-a-social-credit-system-2014-2020/ [Abruf am 19.8.2020].

Eberwein, W.-D. (1990): Globale Trends und Strukturbrüche – Weltmodelle als Forschungsinstrumente, Wissenschaftszentrum Berlin (WZB) für Sozialforschung, Paper 90-307, 10/1990.

Economist Intelligence Unit (2014): Retail banks and big data. Big data as the key to better risk management; Internet: http://www.eiuperspectives.economist.com/financial-services/retail-banks-and-big-data/white-paper/retail-banks-and-big-data-big-data-key-better-risk-management [Abruf am 19.8.2020].

Financial Stability Board (2017): Artificial intelligence and machine learning in financial services - Market developments and financial stability implications, November 2017.

Ford, M. (2018): Architects of Intelligence – The truth about AI from the people building it, Packt Publishing, Birmingham 2018.

Forrester, J.W. (1969): Urban Dynamics, Cambridge/Mass. 1969.

Füser, K./Gleißner, W./Meier, G. (1999): Risikomanagement (KonTraG) – Erfahrungen aus der Praxis, in: Der Betrieb, Heft 15/1999, S. 753–758.

Gleißner, W. (2006): Risikomaße und Bewertung, dreiteilige Serie, in: Risikomanager, Teil 1 – Grundlagen 12/2006, S. 1–11; Teil 2 – Downside-Risikomaße 13/2006, S. 17–23; Teil 3 – Kapitalmarktmodelle 14/2006, S. 14–20.

Gleißner, W. (2011): Risikoanalyse und Replikation für Unternehmensbewertung und wertorientierte Unternehmenssteuerung, in: WiSt, Heft 7/2011, S. 345–352.

Gleißner, W. (2017a): Grundlagen des Risikomanagements, 3. Aufl., Vahlen Verlag, München 2017.

Gleißner, W. (2017b): Risikoanalyse, Risikoquantifizierung und Risikoaggregation, in: WiSt, Heft 9, 2017, S. 4–11.

Gleißner, W. (2018a): Insolvenzrisiko: Top-Kennzahlen für Controlling, Balanced Scorecard und Risikomanagement, in: Controller Magazin, Heft 4, Juli/August 2018, S. 10–15.

Gleißner, W. (2018b): Risikomanagement 20 Jahre nach KonTraG: Auf dem Weg zum entscheidungsorientierten Risikomanagement, in: Der Betrieb vom 16.11.2018, Heft 46, S. 2769–2774.

Gleißner, W. (2018c): Risiko, Volkswirtschaft und Wohlstand, in: Growitsch, C./Loose, S./Wehrspohn, R. B. (Hrsg.): Beiträge zu Wirtschaftspolitik und -forschung - Festschrift anlässlich der Emeritierung von Prof. Dr. Dr. h.c. Ulrich Blum, Center for Economics of Materials CEM, Halle (Saale), S. 55–68.

Gleißner, W. (2019a): Risikoanalyse: Grundlagen der Risikoquantifizierung (Teil 1), in: Controller Magazin, Heft 2 (März/April 2019), S. 42–46.

Gleißner, W. (2019b): Risikoanalyse: Ein strukturierter Leitfaden zur Risikoquantifizierung (Teil 2), in: Controller Magazin, Heft 3, Mai/Juni 2019, S. 31–35.

Gleißner, W. (2019c): Nachhaltigkeit, CSR-Risiken und Risikomanagement. Vom CSR-Risiko zum finanziellen Risiko, in: Controller Magazin, Heft 4, Juli/August 2019, S. 95.

Gleißner, W. (2019d): Cost of capital and probability of default in value-based risk management, in: Management Research Review, Vol. 42, No. 11, S. 1243–1258.

Gleißner, W. (2019e): Wertorientierte Unternehmensführung, Strategie und Risiko, eBook (amazon kindle).

Gleißner, W. (2019f): Insolvenzrisiko, Rating und Unternehmenswert, in: WISU, Heft 6/19, S. 692–698.

Gleißner, W. (2019g): The real dark side of Valuation. Ertragsrisiken und Insolvenzrisiken, in: BOARD, Heft 6/2019, S. 215–219.

Gleißner, W./Ernst, D. (2019): Company valuation as result of risk analysis: replication approach as an alternative to the CAPM, in: Business Valuation OIV Journal, Vol. 1, No. 1 (Frühjahr 2019), S. 3–18.

Gleißner, W./Füser, K. (2000): Innovative Prognoseverfahren für Unternehmensplanung auf Basis des Risikomanagements, in: Der Betrieb, Heft 19/2000, S. 933–941.

Gleißner, W./Romeike, F. (2012): Psychologische Aspekte im Risikomanagement – Bauchmenschen, Herzmenschen und Kopfmenschen, in: Risk, Compliance & Audit (RC&A), 06/2012, S. 43–46.

Gleißner, W./Wolfrum, M. (2015): Problemfelder der Risikoquantifizierung, Datenprobleme und Lösungsstrategien, in: Gleißner, W./Romeike, F. (Hrsg): Praxishandbuch Risikomanagement, ESV, Berlin, S. 274–263.

Günther, T./Detzner, M. (2012): Das Risiko-Entscheidungsverhalten von Managern – Ergebnisse einer empirischen Studie, in: Altenburger, O. A. (Hrsg.): Instrumente und Aufgaben des Controlling, Linde Verlag, Wien, S. 9–52.

Günther, T./Günther, E. (2017): Finanzielle Nachhaltigkeit – Messung, finanzielle Steuerung und Herausforderungen, in: Hoffjan, A./Knauer, T./Wöhrmann, A.: Controlling – Konzeptionen, Instrumente, Anwendungen, Schäffer Poeschel, Stuttgart 2017, S. 79–90.

Günther, T./Gleißner, W./Walkshäusl, C. (2020): What happened to financially sustainable firms in the Corona crisis?, auf: springer.com NachhaltigkeitsManagementForum, https://link.springer.com/article/10.1007/s00550-020-00503-3 [Abruf am 19.8.2020].

Harari, Y. N. (2018): Homo Deus: Eine Geschichte von Morgen, C.H. Beck Verlag, München 2018.

Hassani, H., Silva, E. S. (2015): Forecasting with Big Data: A Review. Annals of Data Science 2 (1).

Hausmann, M. (2019): Artifical Intelligence: Künstliche Intelligenz und datengetriebene Geschäftsmodelle, in: RISIKO MANAGER 03/2019, S. 24–32.

Heck, H.-D. (1992): Die neuen Grenzen des Wachstums, in: Bild der Wissenschaft, Heft 6/1992, S. 54

Heij, C./Schumacher, H./Hanzon, B./Praagman, K. (1997) [Hrsg.]: System dynamics in economic and financial models, John Wiley and Sons, Chichester/New York 1997.

Hillson, D. (2005a): Describing Probability: The Limitations of natural Language, PMI Global Congress.

Hillson, D. (2005b): Understanding and Managing Risk Attitudes, Aldershot.

Holton, G.A. (2004): Defining risk, in: Financial Analysts Journal, Vol. 60, No. 6, S. 19–25.

IOSCO (2019): Sustainable finance in emerging markets and the role of securities regulators – Consultation report, Internet: https://www.iosco.org/library/pubdocs/pdf/IOSCOPD621.pdf [Abruf am 19.8.2020].

Intergovernmental Panel on Climate Change (IPCC) (2014): Climate Change 2014: Synthesis Report. Contribution of Working Groups I, II and III to the Fifth Assessment Report of the Intergovernmental Panel on Climate Change, Geneva 2014.

Kesten, R. (2007): Unternehmensbewertung und Performancemessung mit Robichek/Myers-Sicherheitsäquivalent, in: Finanz Betrieb, Heft 2/2007, S. 88–98.

Knight, F. H. (1921): Risk, Uncertainty and Profit, Houghton Mifflin, Boston/New York 1921.

Kurzweil, R. (2012): How to Create a Mind, New York 2012.

Mainzer, K. (2014): Die Berechnung der Welt – Von der Weltformel zu Big Data, C.H. Beck Verlag, München 2014, S. 27.

March, J./Shapira Z. (1987): Managerial Perspectives on risk and risk taking, in: Management Science, Vol. 33, No. 11, S. 1404–1418.

Martin, M. R. W./Quell, P./Wehn, C. S. (Hrsg.) (2013): Modellrisiko und Validierung von Risikomodellen – Regulatorische Anforderungen, Verfahren, Methoden und Prozesse, Bank-Verlag, Köln 2013.

Meadows, D.L. et al. (1972): The Limits to Growth. A Report for the Club of Rome's Project on the Predicament of Mankind, New York 1972.

Meadows, D.L./Meadows, D.H. (1973): Toward Global Equilibrium – Collected Papers, Cambridge/ Mass. 1973.

Meadows, D.L./Behrens III, W.W./Meadows, D.H./Naill, R.F./Randers, J./Zahn, E.K.O. (1974): Dynamics of Growth in a Finite World, Cambridge/Mass. 1974.

Metropolis, N. C./Ulam, S. (1949): The Monte Carlo Method, Journal of the American Statistical Association, Vol. 44, No. 247, (Sep. 1949), S. 335–341.

Meyer, M./Romeike, F./Spitzner, J. (2012): Simulationen in der Unternehmenssteuerung – Studienergebnisse, RiskNET GmbH, Brannenburg/Wendelstein 2012.

MSCI (2020): ESG Ratings Methodology: Executive Summary; Internet: https://www.msci.com/ documents/10199/123a2b2b-1395-4aa2-a121-ea14de6d708a [Abruf am 19.8.2020].

Natural Capital Coalition (2016): Natural Capital Protocol. Internet: www.naturalcapitalcoalition. org/protocol, [Abruf am 19.8.2020].

Nordhaus, W. D. (2018): Climate Change: The Ultimate Challenge for Economics.

Pestel, E. (1980): Unsere Chance heißt Vernunft, Westermann Verlag, Braunschweig 1980.

Pestel, E. (1988): Jenseits der Grenzen des Wachstums – Bericht an den Club of Rome, Deutsche Verlags-Anstalt, Stuttgart 1988.

RiskNET/Bearing Point (2018): Experten-Studie: Standardmodelle im Spannungsfeld von Risk Analytics und Big Data, Brannenburg/Frankfurt am Main 2018.

Romeike, F. (1994): Zum Wechsel von einem traditionellen zu einem interdisziplinären Wachstumsmodell, Köln 1994.

Romeike, F. (2008): Rechtliche Grundlagen des Risikomanagements, Berlin.

Romeike, F. (2013): Fooled by Randomness, in: FIRM Yearbook 2013, Frankfurt/Main 2013, S. 25–29.

Romeike, F. (2015a): Szenarioanalyse: Lernen aus der Zukunft, in: FIRM Jahrbuch 2015, Frankfurt/Main 2015, S. 118–120.

Romeike, F. (2015b): Scenario analysis: Learning from the future, in: FIRM Yearbook 2015, Frankfurt/Main 2015, S. 14–16.

Romeike, F. (2018): Risikomanagement, Springer Verlag, Wiesbaden 2018.

Romeike, F. (2019): Risk Analytics und Artificial Intelligence im Risikomanagement, in: Rethinking Finance, Juni 2019, 03/2019, S. 45–52.

Romeike, F./Hager, P. (2010): Was ist das Varianz-Kovarianz-Modell?, in: Risk, Compliance & Audit (RC&A), 05/2010, S. 10–11.

Romeike, F./Spitzner, J. (2013): Von Szenarioanalyse bis Wargaming – Betriebswirtschaftliche Simulationen im Praxiseinsatz, Wiley Verlag, Weinheim 2013.

Romeike, F./Eicher, A. (2016): Predictive Analytics: Looking into the future, in: FIRM Yearbook 2016, S. 169–171.

Romeike, F./Hager, P. (2020): Erfolgsfaktor Risiko-Management 4.0, Springer Verlag, Wiesbaden 2020.

Schellnhuber, H.J. (2015): Selbstverbrennung. Die fatale Dreiecksbeziehung zwischen Klima, Mensch und Kohlenstoff, C. Bertelsmann Verlag, München 2015.

Schöning, S./Mendel, V./Köse, A. (2020): Mit neuen Controller-Kompetenzen in die Zukunft, in: Controlling & Management Review, Vol. 64, No. 1 (Januar 2020) S. 58–63.

Schöning, S./Sumer Gogus, H./Pernsteiner, H. (Hrsg.): Risikomanagement in Unternehmen. Interkulturelle Betrachtungen zwischen Deutschland, Österreich und der Türkei, Springer Gabler, Wiesbaden, 2017.

Shad, M. K./Lai, F.-W./Fatt, C. L./Klemeš, J. J./Bokhari, A. (2019): Integrating sustainability reporting into enterprise risk management and its relationship with business performance: A conceptual framework, in: Journal of Cleaner Production, Vol. 208 (20. Januar 2019), S. 415–425.

Silver, N. (2013): The Signal and the Noise: The Art and Science of Prediction, Penguin Books, New York 2013.

Sinn, H.-W. (1980): Ökonomische Entscheidungen bei Unsicherheit, Tübingen 1980.

Social & Human Capital Coalition (2016): Social Capital Protocol. Internet: https://www.wbcsd.org/Programs/People/Social-Impact/Social-and-Human-Capital-Protocol, [Abruf am 19.8.2020].

Stampfer, E. (2019): Risikosteuerung in der Industrie, Konzepte, Methoden und Verfahren für projektorientierte Unternehmen, Linde Verlag, Wien 2019.

Sterman, J. D. (1989): Modeling Managerial Behavior: Misperceptions of Feedback in a Dynamic Decision Making Experiment, in: Management Science, 35(3), 321–339.

The Committee of Sponsoring Organizations of the Treadway Commission (COSO) and World Business Council for Sustainable Development (WBCSD) (2018): Enterprise Risk Management: Applying enterprise risk management to environmental, social and governance-related risks, October 2018.

Tuchtfeld, E. (1973): Die Grenzen des Wachstums – Zwischenbilanz einer Diskussion, in: Zeitschrift für Wirtschaftspolitik, April, 1973, S. 129–144.

Velte, P. (2020): Corporate Social Responsibility (CSR) and Earnings Management: a literature review, in: Corporate Ownership and Control, Vol. 17, No. 2, S. 8–19.

Weißensteiner, Chr. (2014): Reputation als Risikofaktor in technologieorientierten Unternehmen, Springer Verlag, Wiesbaden 2014.

Wengrzik, D./Demski, C. (2019): Machine Learning: Nachrichtenbasierte Frühwarnung im Kontext Kreditrisiko, in: RISIKO MANAGER 01/2019, S. 4–7.
World Economic Forum (2020): The Global Risks Report 2020, Davos 2020.
Zuboff, S. (2018): Das Zeitalter des Überwachungskapitalismus, campus Verlag, Frankfurt am Main 2018.

Prof. Dr. Werner Gleißner ist Vorstand der FutureValue Group AG und Honorarprofessor für Betriebswirtschaft, insb. Risikomanagement, an der Technischen Universität Dresden. Seine Forschungs- und Tätigkeitsschwerpunkte liegen im Bereich Risikomanagement, Bewertung & Rating und Unternehmensstrategie sowie der Entwicklung von Methoden für eine simulationsbasierte Risikoaggregation.

Frank Romeike ist Gründer des Kompetenzzentrums RiskNET GmbH – The Risk Management Network sowie der Risk Academy. Er unterstützt Unternehmen und öffentliche Institutionen seit rund 25 Jahren beim Aufbau und der Weiterentwicklung von wirksamen Risikomanagement-Systemen, insbesondere basierend auf Simulationsmethoden.

Methoden und Modelle zum Social Credit Rating

An Economic Approach to China's Social Credit System

Theresa Krause and Doris Fischer

Abstract

In an effort to increase trustworthiness across society, the Chinese government has been building its Social Credit System since 2014. This system targets all natural and legal persons in China and consists of four major elements: a central data platform, a rating system for commercial creditworthiness, a propaganda system for educative purposes and a publicly available listing system with black- and redlists (for negative or positive behavior) as well as consequential joint punishments and rewards. While most of the related academic discourse has focused on the system's political implications, this paper provides an economic perspective on the Chinese government's rationale for setting up such a system. Transaction cost economics has shown that trust is an important factor for business transactions and economic growth. However, China's rapid economic development and modernization has weakened societal trust, including the traditional trust-building approach via guanxi (interpersonal relationships). Hence, the Chinese government is using the Social Credit System as an alternative approach for trust-building. The system is supposed to strengthen institutional mechanisms and incentivize trustworthy behavior. It can be regarded as an add-on to the currently rather weak legal system and fragmented government enforcement apparatus.

T. Krause
München, Deutschland

D. Fischer (✉)
Würzburg, Deutschland
E-Mail: doris.fischer@uni-wuerzburg.de

© Springer Fachmedien Wiesbaden GmbH, ein Teil von Springer Nature 2020   437
O. Everling (Hrsg.), *Social Credit Rating*,
https://doi.org/10.1007/978-3-658-29653-7_22

22.1   Introduction

China's rapid economic development over the past decades has not only lifted more than 850 million Chinese people out of poverty (1981–2013; The World Bank 2018), but has also created new challenges. For example, Pei (1999) estimated that corruption alone might cost China about four percent of its annual GDP and according to Lin (2010) the cost of environmental pollution accounted for around 3 percent of the Chinese GDP in 2004. Furthermore, social systems and values have eroded leading to a discourse about the "moral crisis" (道德信仰危机) in China. The Chinese Communist Party (CCP) has acknowledged those problems and aims at establishing a more rules-based society and this way ensure more sustainable economic growth. The set-up of the Chinese Social Credit System (SCS) is one effort in a line of measures to achieve this goal. While most of the academic literature has looked at the SCS from a political perspective, this paper provides an economic approach to the question why the Chinese government is developing the SCS. An overview of the SCS and its elements in Sect. 22.2 is followed by a short literature review (Sect. 22.3). Section 22.4 then discusses the SCS from an economic perspective in three sub-chapters. It starts by elaborating on the concept of trust in economic literature, continues by explaining the traditional Chinese approach to trust, namely guanxi, and finishes off by highlighting how the SCS is conceptualized as a new approach to achieve trust. Section 22.5 concludes by summarizing the main theses of this paper and expresses some caveats that need to be considered regarding the SCS.

22.2   The Chinese Social Credit System

The conceptualization of the SCS originates from the intention of building a financial credit rating system after the reform and opening in 1978 and culminated in the State Council's Plan for Establishing a Social Credit System (2014–2020) (国务院关于印发社会信用体系建设规划纲要(2014–2020年)). It is the main policy document outlining the Chinese central government's goals, scope and implementation plan for building the SCS. Since its publication in June 2014, the SCS has been forcefully implemented and in 2020 the first phase of this implementation is scheduled to be completed by the nationwide roll-out of one of the current pilot schemes in 41 Chinese cities.

The State Council's plan defines the SCS's main goal in its first section as "to increase the awareness for integrity and honesty, as well as the trustworthiness level of the entire society" (提高全社会的诚信意识和信用水平). The terms "integrity and honesty" (诚信) and "trustworthiness" (信用) are central to understanding the Chinese government's aim. While trustworthiness refers to the way an actor behaves (Scrivens and Smith 2013) and shows that the SCS aims at changing social behavior, the term "integrity and honesty" adds a moral component. Ultimately, the SCS targets a high level of trustworthiness based on an honest and integer moral orientation, not just in financial transactions, but in any type of social interactions.

To achieve the proclaimed goal, the SCS addresses four realms (government, justice, business and society) and consists of four key elements:

1. The central data platform
2. The rating system for commercial creditworthiness
3. The propaganda system for educative purposes
4. The publicly accessible listing-system with blacklist and redlists, as well as consequential punishments and rewards

▶ **Blacklists** provide records of actors who have lost social credit due to violation of SCS requirements.
Redlists provide information about actors who have adopted praiseworthy behavior according to the SCS requirements.

22.2.1 The Central Data Platform

One of the first initiatives for developing the SCS was the built-up of central data platforms that serve as the basis for all social credit initiatives. Since October 2015 the National Credit Information Sharing Platform (NCISP, 全国信用信息共享平台) aggregates data from ministries, state administrations and industry associations. Already in January 2020, the platform had connected 46 government/administrative departments and all provincial-level credit platforms, as well as collected over 50 billion pieces of credit information (信用中国官网 2020). On this national platform, data about natural persons (in total about 500 million pieces of information) and legal persons (in total information about more than 27 million enterprises) is consolidated and stored under one unified social credit code which is uniquely attributed to each actor (信用中国官网 2020). This government platform is not directly accessible for the public, but about 75 percent of the data is available via the Credit China-website (信用中国) with its sub-pages as well as the National Enterprise Credit Information Publicity System (NECIPS, 国家企业信用信息公示系统) (Meissner 2017). The key information published on those data platforms are basic information about natural and legal persons, their listings and their ensuing punishments and rewards. Even though the data-gathering process is still ongoing, the information provided about businesses is already rather comprehensive and consistent.

Furthermore, a market for credit information providers is developing. The biggest and most popular websites that also have official licenses to use the above-mentioned government data are Qichacha (企查查) and Tianyancha (天眼查). More recently, in September 2019, the first English version of a credit information platform was launched, Xinhua Credit (新华信用). It provides credit information for international users and was set up by China Economic Information Services, an affiliate of Xinhua News Agency. It also links to the Chinese government's One Belt, One Road initiative by allowing searches not only for companies operating in China, but also in countries along the new silk road. Additionally, it provides country ratings. While rather big data gaps for those kind of ratings do still prevail, this hints at China's ambitions to become a global player in the credit rating realm.

▶ The SCS applies to all natural and legal persons in China. This includes foreigners residing in China and any type of business organization with an operating license in China regardless of their size, industry sector, shareholders (both private companies and SOEs) and location of the mother company (both domestic and international companies), etc. (Chen et al. 2018).

22.2.2 The Rating System for Commercial Creditworthiness

The comprehensive nationwide commercial credit system is still under development. An initial attempt to build such a system by the People's Bank of China from 2004 to 2014 was unsuccessful: After ten years of operation, it had only managed to cover 25 percent of the Chinese population (Yang 2017). Hence, the central government picked eight private companies to run pilots for commercial credit schemes. However, discord between the private operators and the central government about criteria of creditworthiness and data sharing prevented all those schemes from receiving an official license to operate the nationwide commercial credit system (von Blomberg 2018; What's on Weibo 2018). Instead, the government coopted the schemes in 2018 by setting up Baihang Credit (百行征信). This company has a license to run for three years and is jointly owned by the National Internet Finance Association, an initiative by the People's Bank of China (36 percent of the shares), and the eight pilot companies (each eight percent of the shares) (信用中国官网 2019). Since Baihang Credit has only been set up as the new commercial credit rating system recently, no detailed information, e.g., about its coverage and credit scoring mechanisms, are publicly available yet.

▶ The eight private companies, that were running the initial pilots for a commercial credit system and are now shareholders of Baihang Credit, are: Alibaba's Sesame Credit (芝麻信用), Tencent Credit (腾讯征信), Intellicredit (中智诚征信), Lakara Financial Services Company's Kaola Credit (考拉征信), Pengyuan Credit (鹏元征信), Sinoway Credit (华道征信), Pingan's Qianhai Credit Services (深圳前海征信), China Chengxin Credit (中诚信征信)

22.2.3 The Propaganda System

The propaganda system is already fully established and is working in full swing. Its core aim is to raise awareness for trustworthiness and integrity by placing them prominently in the public discourse. Examples for such propaganda are a popular song about integrity featuring famous Chinese stars (What's on Weibo 2019), roadshows about integrity (Cre-

Abb. 22.1 Educational poster in the Beijing subway warning people to abstain from certain behaviors in the subway (like playing video music loudly or eating) in order to not lose credit

dit China 2018) and publicly displayed posters reminding citizens about acting trustworthily (see Fig. 22.1). However, the SCS as an instrument itself is to date propagated less dominantly.

22.2.4 The Publicly Accessible Listing System

The listing system has two major purposes: to regulate behavior by publicly displaying positive and negative behavior and to incentivize certain behaviors through a deterrence-based approach of punishments and rewards. So far, the level of detail of blacklists is much higher compared to the redlists (Engelmann et al. 2019). This indicates that the current focus of the CCP is on legal compliance (trustworthiness) rather than moral standards beyond legal compliance (honesty and integrity).

Punishments and rewards are administered based on published black- and redlists generated by different authorities. The guiding principle for punishments is the cross-departmental collaboration following the motto 'trust-breaking here, restrictions everywhere' (处失信, 处处受限) (Chen et al. 2018). To substantiate this principle, the different agencies have signed memoranda of understanding (MoU) defining which consequences certain untrustworthy or extremely trustworthy behaviors have in other areas. Common punishments for businesses are increased government monitoring intensity and frequency, denial of government subsidies and exclusion from government sponsored projects. The punishment for misconduct of a business does in certain cases not only affect the company as a legal entity, but also the company's legal representative. In such a case, the latter faces punishments like restricted consumption (e.g., for vacations, cars or real estate). A typical reward related to the corporate sector is faster access to certain administrative procedures such as tax refunds. Like the punishments, the rewards can be connected to behavior of individuals in the company, for example if businesses are run by highly-rewarded volunteers (Chen et al. 2018).

The most important SCS websites at a glance:

NCISP	http://app.gjzwfw.gov.cn/jmopen/webapp/html5/fgwxyxxpc/
Credit China	https://www.creditchina.gov.cn/
NECIPS	http://www.gsxt.gov.cn/
Baihang Credit	http://www.baihangcredit.com/
Qichacha	https://www.qichacha.com/
Tianyancha	https://www.tianyancha.com/
Xinhua Credit	https://www.credit100.com/

22.3 Literature Review

Since the SCS is still in the pilot phase, the academic discourse has only recently picked up on this topic. Thus, academic literature is still relatively scarce. Most of the early literature explains the SCS's general set-up, scope and mechanisms (e.g., Knight 2018; Chen and Cheung 2017; Ohlberg et al. 2017; Creemers 2018). Much of the interest there revolves around the SCS's elements targeting individuals, especially the pilots by the eight private companies that have been stopped by the government in 2018 (e.g., Botsman 2017; Kostka 2018; Kostka and Antoine 2018). The focus of this thread of literature is the discussion of the SCS's potential political and social implications: Chen, Lin and Liu (2018), for example, elaborate on how the SCS erodes the CCP's rule by law. Dai (2018) argues that the SCS marks China's transition towards a reputational state and discusses the political implications of this transition. Liang et al. (2018) explain the SCS's development and mechanisms from a state surveillance perspective and focus on technical aspects such as data collection, aggregation and analytics.

▶ A reputational state is defined as a state where the "government authorities seek to use reputation mechanisms in law, regulation, and governance" (Dai 2018, p. 2).

Arguably, the SCS in its essence does not constitute a drastic shift in political and social governance in China. The governance mechanisms for social and political control applied via the SCS have been adopted by the CCP in different shapes and forms before, e.g., the punishments and rewards of the one-child-policy or the customs rating system for companies. From this perspective, the SCS is no disruptive innovation, but simply an upgrade of previously applied approaches. The technological advances implemented via the SCS (e.g., central digital data base, potential use of AI-technology) allow the CCP to adopt a more comprehensive and stringent governance.

However, we argue that the SCS aims to fundamentally change the economic system. So far, there are only very few authors who discuss the economic implications of the system. An early publication by Meissner (2017) provides insights into the set-up of the SCS targeting the corporate sector and provides initial hypotheses on how this system could change market regulation in China. A joint report by Sinolytics and the European Chamber of Commerce (Sinolytics and EUCCC 2019) lays out the potential repercussions of the system for foreign companies doing business in China. But so far, no comprehensive discussion of the economic rationale for setting up the SCS in the Chinese context exists. This paper aims to provide such a perspective by applying an economic approach to the question why the Chinese government is developing this nationwide SCS.

22.4 An Economic Approach to China's Social Credit System

22.4.1 The Concept of Trust in Economics

While many different definitions of trust exist, most of them share two common elements of what constitutes trust: vulnerability and expectation (Evans and Krueger 2009). In line with this, we adopt Dasgupta's (2000, p. 330) definition of trust as an individual's "expectations about actions of others that have a bearing on this individual's choice of action, when that action must be chosen before he or she can observe the actions of those others. Trust is important because its presence or absence can have a bearing on what we choose to do, and in many cases what we can do." This definition focuses on the individual and highlights the fundamental impact of trust on their behavior, rather than the origins of trust. As compared to narrow definitions of economic trust (e.g., Zaheer et al. 1998), which focus on the firm as an actor, we here want to adopt a comprehensive understanding of the role of trust in economics, not just interfirm transactions. In one threat of academic literature, trust and social capital are conflated (e.g., Paldam and Svendsen 2000; Beugelsdijk et al. 2004). However, it is important to note that these two concepts cannot be equated. Trust constitutes one amongst several components of social capital (e.g., as Putnam 2000 defines it). Nevertheless, social capital literature that specifically focuses on the impact of trust can still provide useful insights into the role of trust in economics, as the notion of 'capital' in the concept underscores its value as an asset, potentially a competitive advantage.

In the economic discourse trust is mostly discussed in the context of transaction cost economics. There the choice to invest trust in a relationship is based on the notion of cost reduction and opportunism (Bachmann 2001).

▶ Transaction costs are friction costs that appear in economic relationships, which economic actors should reduce to a minimum while still defending themselves against opportunism of the counterpart (den Butter and Mosch 2003).

While trust is not included into the mainstream model of transaction cost economics, most scholars agree that it is a vital element of every transaction (e.g., Arrow 1972; Williamson 1993; Chiles and McMackin 1996; Dasgupta 2000). The fundamental problem of trust in exchanges is illustrated in the famous prisoner's dilemma, also known as the game of trust (Greif 2000): If actors decide to cooperate based on trust, they are better off than without this cooperation. Trust in this case becomes a mechanism to mitigate the risk of opportunism.

Generally, there is a broad consensus on the benefits of trust in the economic literature. The role of trust and its benefits have been investigated on three levels: the firm, the national economy and the international economy.

On the firm-level both intra-firm and inter-firm trust have been studied. In the inter-firm context, transaction costs can be reduced through trust in multiple ways. Trust can decrease the amount of negotiation and the complexity of contracts drafted (Chiles and Mcmackin 1996; Zaheer et al. 1998; Dyer and Chu 2003). Furthermore, it can also reduce monitoring efforts between counterparts (Chiles and Mcmackin 1996; Dyer and Chu 2003). Consequently, a reputation for being trustworthy lowers transaction costs and translates into better firm performance (Dyer and Chu 2003). Both on the intra-firm and inter-firm level, trust has been shown to improve cooperation and information sharing, and hence is beneficial for cooperation (La Porta et al. 1997; Beccerra and Gupta 1999; Pollitt 2002; Cai et al. 2010). This indicates that besides the cost-minimizing effects, trust also affects value creation positively (Dyer and Chu 2003).

The benefits of trust in the firm-to-firm relationship in form of reduced transaction costs can add up to significant benefits on the national level. Improved efficiency of key national economic players can contribute to a stronger local economy and improved international competitiveness. And indeed, several studies have proven exactly this link of trust between individual actors in a society and the improved economic development of nations (e.g., Knack and Keefer 1997; Zak and Knack 2001; Beugelsdijk et al. 2004). They show that countries with high levels of trust, as measured by the World Values Survey, exhibit faster economic growth than comparable other countries. Similarly, Dasgupta (2000) identifies a low-trust environment as a key cause for underdevelopment and economic stagnation in certain countries. This is mainly explained by the lack of reliable institutions, as, for example, La Porta et al. (1997) show a positive correlation between trust and bureaucratic quality, judicial efficiency, tax compliance, and civic participation amongst others as well as a negative correlation with government corruption.

On the international level the correlations between trust and economic efficiencies mainly manifest in trade flows. Den Butter and Mosch's (2003) research indicates that trust problems are a significant source of trade barriers and therefore increase transaction costs of trade. Consequently, trust majorly influences international trade flows. Arguably, due to the increased distance of cooperative economic activities in a highly connected global economy, its importance has been increasing (Chiles and Mcmackin 1996).

Trust, however, is not a uniform concept. There are different ways to classify it and theories of its origins. We follow den Butter and Mosch's (2003) classification into formal and informal trust, which is based on the origins of trust. Formal trust relates to the trust built through formal mechanisms. It includes trust in the formal institutional setting ('system trust') and thereof derives trust in institutional tools (e.g., formal legal contracts, legal procedures, government enforcement). Importantly, it is not the enforcement of the regulations per se, which constitutes the basis for formal trust; rather the looming danger of punishment makes formal trust a rather effective guarantee in relationships with strangers. Therefore, reliable legal institutions (such as commercial law) serve as a reassurance function that can also foster further trust building (Bachmann 2001). Mainstream economic literature related to trust is mostly applying this form of trust as it often implicitly refers to Western democratic countries with longstanding legal and constitutional traditions.

Informal trust, on the other hand, is rather personal and comprises trust towards other individuals or specific collectives (e.g., ethnic group or village) (den Butter and Mosch 2003). While it also constitutes a valuable asset in many relationships and is essential for the functioning of a society, this intrinsic type of trust does not necessary follow rational logics. Therefore, from an economic perspective, sometimes it can even be detrimental, e.g., when people prioritize personal relationships over rational decisions or when this type of trust leads to too close communities that hinder exchanges with people from the outside (e.g., Leenders 1999).

22.4.2 Guanxi as the Traditional Chinese Approach for Trust-Building

Traditionally, informal trust has been dominating in China in the form of guanxi (关系), a type of interpersonal relationships (Wang 2007).

▶ While many definitions of guanxi exist (e.g., Alston 1989; Yang 1994; Lee et al. 2001; Ordóñez de Pablos 2005; Wang and Wang 2007), the term is generally attributed to social connections built on the basis of reciprocal interests and benefits in China. Similar phenomena can be found in other countries with shortcomings and limited independence of the legal systems such as Russia (Zhang and Pimpa 2010).

Guanxi have been a core feature of Chinese society, which has been characterized as 'acquaintance society' (熟人社会) by famous Chinese sociologist Fei Xiaotong (1992). Within a clearly defined hierarchical social structure, guanxi and family ties serve as the glue which holds Chinese society together. According to Fei, Western societies emphasize ties to abstract collectives (organizational mode of association; 团体格局), while Chinese society focuses on personal relationships between concrete people to build a network (differential mode of association; 差序格局). Therefore, networks in China are built in concentric circles around the individual depending on the relational distance (Fei 1992).

While this social comparison has some essentialist characteristics, it still offers insights into the Chinese social and institutional structures. Guanxi, rather than abstract instituti-ons, engender trust in Chinese society. They serve as an insurance in a risky business en-vironment with limited legal protection and low faith in the legal system (Cai et al. 2010; Gallagher 2006; Pia 2016).

▶ While the CCP has, over the past decades, established more of a formalist legal system with courts and a comprehensive set of laws, no separation of power exists (Minzner 2012). Instead, the Chinese legal system aims at rule by law (依法治国) (instead of rule of law) where the CCP stands above the law.

Furthermore, guanxi facilitate transactions and can lead to competitive advantages for businesses (Abramson and Ai 1997; Wang 2007). This can, to a certain extent, improve firms' performance (e.g., Yeung and Tung 1996; Luo and Chen 1997; Peng and Luo 2000, 2001; Nie et al. 2011). Therefore, not just existing guanxi are important for a business, but guanxi is also viewed as a strategic asset that can be built.

As has been analyzed by many scholars, guanxi exhibit substantial differences from what is understood as relationships in Western developed nations (e.g., Abramson and Ai 1997; Lee et al. 2001; Wang 2007). The key features of guanxi with focus on the business context are:

- **Reciprocity**: Reciprocity is vital for guanxi. However, in contrast to Western unders-tandings of reciprocal relationships in business, favors (such as preferential treatment in dealings, increased accessibility to controlled information and preferential access to limited resources) in the Chinese context do not necessarily need to be equal in value and the timeframe of reciprocity is rather long-term (Park and Luo 2001; Lee et al. 2001). This means that trust levels within the guanxi-network are extremely high since the return of a favor is expected in the indeterminate future in an uncertain institutional environment (Abramson and Ai 1997).
- **Particularization**: Guanxi are network-specific and cannot be generalized to abstract entities or groups, they are always attached to specific individuals (Lee et al. 2001). Fur-thermore, individuals are categorized as in-group or out-group, where trust and treat-ment differ significantly between those categories (Wang and Wang 2007).
- **Affective value**: Despite providing business value, building and maintaining guanxi goes beyond that goal. Guanxi are not merely seen in the light of transaction costs, but instead there is also affective value attached to guanxi (Lee et al. 2001; Wang and Wang 2007). Personal and professional activity here are not clearly separated, they go hand in hand.
- **Morality and social norms**: As compared to Western developed nations where the basis for business relationships are rules and legality, in China the basis of guanxi are morality and social norms (Arias 1998). Therefore, a key motivation for adequate com-portment in guanxi is boosting (or saving) face (Lee et al. 2001).

Traditionally, trust in China follows those same logics as guanxi, e.g., it is measured by the ability to return favors rather than legal integrity, and is highly particularized (Wang and Wang 2007). This China-specific trust pattern explains the measurements of trust-levels in Chinese society by different studies. According to the World Value Survey, only eleven percent of the surveyed Chinese believe that people they do not know can be completely or somewhat trusted. Accordingly, China is often called a low-trust society. At the same time the level of particularized trust in China within an actor's own network (informal trust) is high: 72 percent of the surveyed Chinese trust people they know personally at least somewhat, 78 percent show that level of trust to their neighborhood and 94 percent to their family (World Value Survey Association 2014). Other studies underscore these results: A survey of 410 villages throughout China in 2011 showed that only 5.8 percent of villagers believe that outsiders can be trusted (Xia 2011).

While guanxi and the ensuing trust patterns have served China's development in an institutional environment characterized by high uncertainty, it can be problematic for further advancement. Negative side-effects of guanxi are, for example, corruption and nepotism which can damage economic competition and market dynamics (Lee et al. 2001). Furthermore, low levels of trust towards strangers, as they manifest in a guanxi system, have become problematic due to China's transition from a face-to-face society to a stranger society in the context of China's economic development and internationalization. The rapid economic development since the reform and opening in 1978 has eroded the previous social norms and institutions while new ones fitting to the changed social set up (e.g., reliable trusted anonymous institutions) have not been installed yet. This has led to Chinese society's moral crisis as morality is traditionally maintained within personal relationships, but not in impersonal transactions. In a recent study by Kostka (2018) 76 percent of the Chinese respondents stated that mistrust between citizens is a problem in China. It manifests in all realms of society, including the corporate sector. There, for example, many corporate scandals in recent years have harmed the Chinese consumers' trust in businesses, e.g., the melamine-tainted milk-powder problem in 2008, the financial frauds by Wanfu Biotech and the vaccine scandal in Shandong in 2016 (Qiu et al. 2016; Dong et al. 2018). In addition, in the course of global expansion, Chinese businesses are at a disadvantage in terms of transaction costs, if they (have to) rely on guanxi to establish trust.

22.4.3 The Social Credit System as an Alternative Approach for Trust-Building

The CCP is building the SCS to counteract exactly these developments by building a reliable and strongly enforced rules-based system that fosters trustworthy behavior in society and in turn generates stronger formal trust in institutions. With higher formal trust levels, transaction costs in the business sector can be reduced since risk in engaging with strangers is mitigated. This way, transactions become more efficient, which potentially spurs further economic growth. In specific, the SCS targets trust-building on two levels, the national and the international.

On the national level, the SCS's core mechanisms of information transparency, punishments and rewards as well as education can foster trustworthy behavior in a company's core relationships. Internally, businesses have more transparency over their employees' legal conduct and own business activities and that way can avoid conflicts of interest or hiring of untrustworthy employees. At the same time, employees and management have more transparency about the company's compliance performance as well as its legal representative's actions. This information can impact employer branding and affect recruiting success. The government's financial incentives for whistleblowing of major non-compliance incidents by employees also raise companies' costs of non-compliance and make trustworthy behavior the financially more beneficial option. Additionally, those reporting channels increase awareness for compliance as well as legal requirements amongst the population.

In the business-to-consumer relationship, the SCS aims to establish trust from both sides. The transparency about corporate actions will allow consumers to include compliance levels in their consumption choices. In return, the consumer credit ratings will provide businesses with a mechanism to evaluate consumer creditworthiness and thus reduce their risk in business transactions.

The SCS aims at increasing trust in the business-to-business relationship, as well as relationships with other relevant stakeholders (such as investors), in two ways. Firstly, the new information transparency reduces the risk inherent in choosing business partners. Secondly, the joint punishments and rewards incentivize trustworthy behavior by increasing the costs of non-compliance not only of the business itself but also affiliated companies. This is especially the case since the MoU's connection of punishments and rewards between departments and actors leads to more drastic consequences for companies. For government tenders, for example, one selection criterium is not only the compliance performance of one's own company, but also the one of your business partners (their blacklisting will have a negative effect on one's own evaluation).

In the relationship between businesses and the government, trust can be increased, and transaction costs reduced by reducing ambiguity related to regulatory requirements as well as minimizing the window of opportunity for corruption on the local level. The goal is to establish more direct data collection channels (e.g., automatic real-time monitoring of CO_2-emissions via emission detectors) to avoid collusion, human error and personal biases during the data collection process on the local level. An increase of trust between different government departments and levels can also positively affect companies' transaction costs: Data sharing and collaboration between departments will smoothen companies' interaction with the authorities, as they previously often had to provide the same information several times when interacting with different departments. Lastly, on the national level the SCS does not only provide opportunities to reduce transaction costs but has also fostered the growth of the previously underdeveloped credit market: credit rating agencies, consultancies, training agencies, etc. are now emerging and competing for the newly created business opportunities.

On the international level, there are three areas where the SCS can possibly create economic impact. First, it can help Chinese companies compete on the international market. Chinese companies that want to export to other countries often face high compliance requirements, especially when doing business with developed countries. The SCS provides transparency and forces Chinese companies to improve their compliance records. This not only helps them to comply to international standards when doing business abroad, but it could also develop into a quality standard for Chinese businesses accepted internationally. This way, Chinese businesses might be generally regarded as more trustworthy internationally in the future which would reduce their transaction costs. Second, a more rules-based Chinese market is also more attractive for international businesses. If the Chinese government achieves to reduce ambiguity in market regulation and government enforcement through the SCS and provides a level playing field for international and domestic businesses, the Chinese market might be able to increase its attractiveness which has been declining recently (e.g., due to increasing labor costs). Third, if successful, the SCS itself could open new international business opportunities for Chinese enterprises. The SCS's scope could be extended to other countries and Chinese businesses in the credit sector could provide their services internationally. This could be especially attractive for developing countries that lack a well-functioning credit system and struggle with a weak legal system, just like China.

22.5 Conclusions

The explanations above show that from an economic vantage point building the SCS is a sensible step to take for the CCP. It has the potential to lower transaction costs for businesses nationally and increase competitiveness internationally. The SCS can be regarded as an institutional instrument to make up for the shortcomings in the legal system and the weak enforcement of laws. However, the SCS also bears risks. The international reaction to the SCS illustrates how such a measure can have unintended consequences. Instead of cultivating more trust internationally, the SCS has been target of major international criticism. In addition, the SCS could also easily lead to an atmosphere of fear and unpredictability if abused. This would not only be detrimental from an economic standpoint. Moreover, while certain transaction costs can be reduced thanks to the SCS, new transaction costs (managing the new requirements emerging from the SCS) will arise for businesses. Hence, the success of the SCS also depends on whether transaction costs can be truly reduced overall. Furthermore, the implementation of such a mammoth project is challenging and will certainly continue beyond the initially-set completion date of 2020. Some key issues to be solved are improving data quality, implementing the system consistently across regions and establishing a solid legal basis for the mechanisms ("Social Credit Law"). Therefore, it is to be seen whether the SCS will be able to achieve the goals the Chinese government is aspiring to.

References

Abramson, N. R. and Ai, J. X. (1997) 'Using Guanxi-Style Buyer-Seller Relationships in China: Reducing Uncertainty and Improving Performance Outcomes', The International Executive, 39(6), pp. 765–804.

Alston, J. P. (1989) 'Wa, Guanxi, and Inhwa: Managerial principles in Japan, China, and Korea', Business Horizons, 32(2), pp. 26–31.

Arias, J. T. G. (1998) 'A relationship marketing approach to guanxi', European Journal of Marketing, 32(1/2), pp. 145–156.

Arrow, K. J. (1972) 'Gifts and Exchanges', Philosophy & Public Affairs, 1(4), pp. 343–362.

Bachmann, R. (2001) 'Trust, Power and Control in Trans-Organizational Relations', Organization Studies, 22(2), pp. 337–365.

Beccerra, M. and Gupta, A. K. (1999) 'Trust Within The Organization: Integrating The Trust Literature With Agency Theory And Transaction Cost Economics', Public Administration Quarterly, 23(2), pp. 177–203.

Beugelsdijk, S., de Groot, H. L. F. and von Schaik, A. B. T. . (2004) 'Trust and economic growth: a robustness analysis', Oxford Economic Papers, 56, pp. 118–134.

von Blomberg, M. (2018) 'The Social Credit System's Greatest Leap Goes Unnoticed', Mapping China, 18 April. Available at: https://mappingchina.org/2018/04/the-social-credit-systems-greatest-leap-goes-unnoticed-2/ (Accessed: 21 January 2020).

Botsman, R. (2017) 'China Social Credit: Big data meets Big Brother as China moves to rate its citizens', Wired UK, 21 October. Available at: https://www.wired.co.uk/article/chinese-government-social-credit-score-privacy-invasion (Accessed: 21 January 2020)

den Butter, F. A. G. and Mosch, R. H. J. (2003) Trade, Trust and Transaction Cost. 03–082/3. Amsterdam and Rotterdam.

Cai, S., Jun, M. and Yang, Z. (2010) 'Implementing supply chain information integration in China: The role of institutional forces and trust', Journal of Operations Management, 28, pp. 257–268.

Chen, Y.-J., Lin, C.-F. and Liu, H.-W. (2018) '"Rule of Trust": The Power and Perils of China's Social Credit Megaproject', Columbia Journal of Asian Law, 32(1), pp. 1–36.

Chen, Y. and Cheung, A. S. Y. (2017) 'The Transparent Self under Big Data Profiling: Privacy and Chinese Legislation on the Social Credit System', Journal of Comparative Law, XII(2), pp. 356–378.

Chiles, T. H. and McMackin, J. F. (1996) 'Integrating variable risk preferences, trust, and transaction cost economics', Academy of Management Review, 21(1), pp. 73–99.

Credit China (2018) 信用中国官网. Available at: https://www.creditchina.gov.cn/ (Accessed: 20 November 2018).

Creemers, R. (2018) 'China's Social Credit System: An Evolving Practice of Control', SSRN Electronic Journal.

Dai, X. (2018) 'Toward A Reputation State: The Social Credit System Project of China', SSRN Electronic Journal.

Dasgupta, P. (2000) 'Economic Progress and the Idea of Social Capital', in Dasgupta, P. and Serageldin, I. (eds) Social Capital: A Multifaceted Perspectives. Washington D.C.: The International Bank for Reconstruction and Development, pp. 325–424.

Dong, W. et al. (2018) 'Social Trust and Corporate Misconduct: Evidence from China', Journal of Business Ethics, 151, pp. 539–562.

Dyer, J. H. and Chu, W. (2003) 'The role of trustworthiness in reducing transaction costs and improving performance: Empirical evidence from the United States, Japan, and Korea', Organization Science, 14(1), pp. 57–68.

Engelmann, S. et al. (2019) 'Clear Sanctions, Vague Rewards: How China's Social Credit System Currently Defines "Good" and "Bad" Behavior', in Conference on Fairness, Accountability, and Transaparency. Atlanta.

Evans, A. M. and Krueger, J. I. (2009) 'The Psychology (and Economics) of Trust', Social and Personality Psychology Compass, 3(6), pp. 1003–1017.

Fei, X. (1992) From the soil, the foundations of Chinese society. Edited by G. Translated by Hamilton. Berkeley: University of California Press.

Gallagher, M. (2006) 'Mobilizing the Law in China: "Informed Disenchantment" and the Development of Legal Consciousness', Law & Society Review, 40(4), pp. 783–816.

Greif, A. (2000) 'The fundamental problem of exchange: A research agenda in Historical Institutional Analysis', European Review of Economic History. Bayerische Staatsbibliothek, 4, pp. 251–284.

Knack, S. and Keefer, P. (1997) 'Does Social Capital Have an Economic Payoff? A Cross-Country Investigation', The Quarterly Journal of Economics, 112(4), pp. 1251–1288.

Knight, A. (2018) In Laotianye We Trust: Credit, Morality and Governance in China's Exemplary Society. University of Oxford.

Kostka, G. (2018) 'China's Social Credit Systems and Public Opinion: Explaining High Levels of Approval', SSRN Electronic Journal.

Kostka, G. and Antoine, L. (2018) 'Fostering Model Citizenship: Behavioral Change in China's Emerging Social Credit Systems', To be published.

Lee, D., Pae, J. H. and Wong, Y. H. (2001) 'A model of close business relationships in China (guanxi)', European Journal of Marketing, 35(1/2), pp. 51–69.

Leenders, R. (1999) Corporate Social Capital and Liability. Edited by R. T. A. J. Leenders and S. M. Gabbay. Boston: Kluwer Academic Publishers.

Liang, F. et al. (2018) 'Constructing a Data-Driven Society: China's Social Credit System as a State Surveillance Infrastructure', Policy and Internet, pp. 1–39.

Lin, L.-W. (2010) 'Corporate Social Responsibility in China: Window Dressing or Structural Change?', Berkely Journal of International Law, 28(1), pp. 64–100.

Luo, Y. and Chen, M. (1997) 'Does guanxi influence firm performance?', Asia Pacific Journal of Management, 14, pp. 1–16.

Meissner, M. (2017) 'China's Social Credit System: A Big-data Enabled Approach to Market Regulation With Broad Implications for Doing Business in China', MERICS China Monitor.

Minzner, C. (2012) 'What Direction for Legal Reform Under Xi Jinping?', China Brief, 12(24), pp. 6–9.

Nie, R. et al. (2011) 'A bittersweet phenomenon: The internal structure, functional mechanism, and effect of guanxi on firm performance', Industrial Marketing Management. Elsevier B.V., 40, pp. 540–549.

Ohlberg, M., Ahmed, S. and Lang, B. (2017) 'Central Planning, Local Experiments: The complex implementation of China's Social Credit System', MERICS China Monitor.

Ordóñez de Pablos, P. (2005) 'Western and Eastern views on social networks', The Learning Organization, 12(5), pp. 436–456.

Paldam, M. and Svendsen, G. T. (2000) 'An essay on social capital: looking for the fire behind the smoke', European Journal of Political Economy, 16, pp. 339–366.

Park, S. H. and Luo, Y. (2001) 'Guanxi and Organizational Dynamics: Organizational Networking in Chinese Firms', Strategic Management Journal, 22(5), pp. 455–477.

Pei, M. (1999) 'Will China Become Another Indonesia?', Foreign Policy, 116, pp. 94–109.

Peng, M. W. and Luo, Y. (2000) 'Managerial Ties and Firm Performance in a Transition Economy: The Nature of a Micro Link', The Academy of Management Journal, 43(3), pp. 486–501.

Pia, A. E. (2016) '"We Follow Reason, Not the Law:" Disavowing the Law in Rural China', Political and Legal Anthropology Review, 39(2), pp. 276–293.

Pollitt, M. (2002) 'The economics of trust, norms and networks', Business Ethics: A European Review, 11(2), pp. 119–128.

La Porta, R. et al. (1997) 'Trust in Large Organizations', American Economic Review, 87(2), pp. 333–338.

Putnam, R. (2000) Bowling Alone: The Collapse and Revival of American Community. New York: Simon and Schuster.

Qiu, J. et al. (2016) 'Vaccine scandal and crisis in public confidence in China', The Lancet, 387, p. 2382.

Scrivens, K. and Smith, C. (2013) Four Interpretations of Social Capital: An Agenda for Measurment, OECD Statistics Working Papers.

Sinolytics and EUCCC (2019) The Digital Hand: How China's Corporate Social Credit System Conditions Market Actors.

The World Bank (2018) China - Systematic Country Diagnostic: towards a more inclusive and sustainable development.

Wang, C. L. (2007) 'Guanxi vs. relationship marketing: Exploring underlying differences', Industrial Marketing Management, 36, pp. 81–86.

Wang, J. and Wang, Z. (2007) 'The political symbolism of business: Exploring consumer nationalism and its implications for corporate reputation management in China', Journal of Communication Management, 11(2), pp. 134–149.

What's on Weibo (2018) Baihang and the Eight Personal Credit Programmes: A Credit Leap Forward. Available at: https://www.whatsonweibo.com/baihang-and-the-eight-personal-credit-programmes-a-credit-leap-forward/ (Accessed: 7 May 2019).

What's on Weibo (2019) 'Be as Good as Your Word': The Chinese Social Credit Song is Here. Available at: https://www.whatsonweibo.com/be-as-good-as-your-word-the-chinese-social-credit-song-is-here/ (Accessed: 5 November 2019).

Williamson, O. E. (1993) 'Calculativeness, Trust, and Economic Organization', The Journal of Law and Economics, 36(1, Part 2), pp. 453–486.

World Value Survey Association (2014) World Value Survey Database - Wave 7 2010–2014. Available at: http://www.worldvaluessurvey.org/WVSOnline.jsp (Accessed: 6 December 2018).

Xia, M. (2011) 'Social Capital and Rural Grassroots Governance in China', Journal of Current Chinese Affairs, 40(2), pp. 135–163.

Yang, F. (2017) Is Xinlian the answer to the Individual Credit Checking System in China? Available at: https://www.kapronasia.com/china-banking-research-category/is-xinlian-the-answer-to-the-individual-credit-checking-system-in-china.html (Accessed: 7 May 2019).

Yang, M. M. (1994) Gifts, Favors and Banquets: The Art of Social Relationship in China. Ithaka: Cornell University Press.

Yeung, I. Y. M. and Tung, R. L. (1996) 'Achieving business success in Confucian societies: The importance of guanxi (connections)', Organizational Dynamics, 25(2), pp. 54–65.

Zaheer, A., McEvily, B. and Perrone, V. (1998) 'Does Trust Matter? Exploring the Effects of Interorganizational and Interpersonal Trust on Performance', Organization Science, 9(2), pp. 141–159.

Zak, P. J. and Knack, S. (2001) 'Trust and Growth', The Economic Journal, 111(470), pp. 295–321.

Zhang, J. and Pimpa, N. (2010) 'Embracing Guanxi: The Literature Review', International Journal of Asian Business and Information Management, 1(1), pp. 23–31.

信用中国官网 (2019) 百行征信入场 8家股东剥离个人征信业务. Available at: https://www.creditchina.gov.cn/gerenxinyong/gerenxinyongliebiao/201806/t20180604_117132.html (Accessed: 7 May 2019).

信用中国官网 (2020) 国家发展和改革委员会体制改革综合司司长徐善长:全国信用信息共享平台归集信息已超500亿条. Available at: https://www.creditchina.gov.cn/home/zhuantizhuanlan/xinyongdashuju/xinyongdashujuqianyan/202001/t20200119_182705.html (Accessed: 21 January 2020).

Theresa Krause ist Doktorandin am Lehrstuhl China Business and Economics der Julius-Maximilians-Universität Würzburg und forscht zum Thema „Corporate Compliance und das Sozialkreditsystem in China". Sie arbeitete 2010/11 im Rahmen des weltwärts-Programms des BMZ in NGOs in Shanghai und studierte anschließend in Karlsruhe, Taipeh und London mit Fokus auf Wirtschaft, Politik und China. Vor ihrer Promotion war Theresa Krause bei einer internationalen Beratungsfirma tätig.

Doris Fischer ist Inhaberin des Lehrstuhls China Business and Economics der Julius-Maximilians-Universität Würzburg. Sie ist Vorsitzende des Vorstands der Deutschen Gesellschaft für Asienkunde und war 2017–2019 im Auftrag des BMBF Vorsitzende der deutschen Expertengruppe für die Deutsch-Chinesische Plattform Innovation. Aktuell kooperiert sie mit KollegInnen der Technischen Universität München in einem vom bidt geförderten Projekt zu den Auswirkungen des chinesischen Sozialkreditsystems auf Unternehmen.

Daten, Verhalten, Persönlichkeit

Wie man datenbasiert Persönlichkeit analysiert und Verhalten prognostiziert – ein empirisches Beispiel

Alexander Schlegel

Zusammenfassung

Verhalten folgt der Persönlichkeit des Handelnden. Wir zeigen mit unserem täglichen Verhalten, wer wir sind, welche Persönlichkeitseigenschaften bei uns wie stark ausgeprägt sind – im privaten wie beruflichen Handeln und in den Spuren, die *unser Verhalten in sozialen Medien und im Internet hinterlässt. Die Analyse der Persönlichkeit des Handelnden auf Basis der* digitalen Spuren steht noch am Anfang – existieren die Verhaltensspuren im Internet doch gerade erst seit einer Dekade. Umso mehr lohnt der Blick auf den Zusammenhang von in diesen Daten gezeigtem Verhalten und Persönlichkeit des Handelnden, um künftiges Verhalten mit einer gewissen Wahrscheinlichkeit zu prognostizieren. So ist der Ansatz des Social Credit Rating grundsätzlich valide.

Wie eng Daten, Verhalten und Persönlichkeit miteinander verbunden sind, wird anhand einer Studie gezeigt, in der mehr als 100 verurteilte Wirtschaftskriminelle in deutschen Justizvollzugsanstalten hinsichtlich Ihrer Persönlichkeit untersucht und mit einer Kontrollgruppe verglichen wurden. Dabei konnten anhand nur weniger gemessener Persönlichkeitseigenschaften 86 Prozent der Wirtschaftsdelinquenten korrekt als solche identifiziert werden.

23.1 Kreditnehmerrating, Persönlichkeit und Verhalten

Die Beurteilung der Bonität von Kreditnehmern beruht in Deutschland auf den regulatorischen Vorgaben vor allem des § 18 Kreditwesengesetz und fokussiert sich in der Praxis auf Einkommen, Ausgaben, bestehende Schuldnerverhältnisse, Zahlungsschwierigkeiten oder

A. Schlegel (✉)
Nieder-Olm, Deutschland

© Springer Fachmedien Wiesbaden GmbH, ein Teil von Springer Nature 2020
O. Everling (Hrsg.), *Social Credit Rating*,
https://doi.org/10.1007/978-3-658-29653-7_23

bereits erfolgte Kreditausfälle, Wohnort, Familienstand etc. Diese Faktoren gießen Banken in ein Modell mittels dessen die Wahrscheinlichkeit eines künftigen Kreditausfalls prognostiziert werden soll. Die notwendigen Daten dafür erhebt die Bank ohnehin im Rahmen der Kontoeröffnung und des Know Your Customer-Prozesses oder der Finanzierungsanfrage.

Gemeinsam ist diesen Daten, dass man sie bereits vor 100 Jahren (wenn auch mit deutlich höherem manuellen Aufwand) hätte erheben können bzw. teilweise auch erhoben hat. In der letzten Dekade ist jedoch etwas passiert. *Potenzielle Kreditnehmer hinterlassen nicht nur bonitätsrelevante „Spuren" auf ihren Konten, sondern auch und vor allem im Internet.* Surfverhalten im Internet insgesamt, Likes und Dislikes sowie selbst produzierter textlicher, videografischer, fotografischer Content auf einer großen Bandbreite von Plattformen. Verfügbar sind diese Daten vorwiegend für Plattformbetreiber und BigTechs, für kreditvergebende Institute oder Ratinganbieter jedoch kaum. Dennoch verfügen auch Banken und Finanzdienstleister sowie (seit der PSD2-Regulierung zahlreiche FinTechs-Zustimmung des Kunden vorausgesetzt) über umfassende Daten der Privatkunden wie beispielsweise monatliche Ausgaben, Einkaufs- und Konsumverhalten, Hobbies, Immobilienbesitz, etc.

Was haben all diese Daten jedoch mit der Prognose von Kreditausfällen zu tun?

Die Verfügbarkeit dieser Datenart ist relativ neu und die Anzahl der Studien dazu entsprechend gering, die Forschung steht hier noch am Anfang. Dennoch konnten beispielsweise Berg et al. (2018) und auch die Bank for International Settlements (BIS 2019) eindrucksvoll zeigen: „a model that uses only digital variables equals or exceeds the information content of the credit bureau score" (Berg et al. 2018, S. 4). Behalten die Autoren Recht, hieße das (etwas zugespitzt), dass *Google, amazon und Alibaba eine mindestens ebenso valide oder validere Probability of Default errechnen können als es auf Basis von creditreform oder Schufa-Daten möglich ist.* Bei der Generalisierung dieser Erkenntnisse ist jedoch Vorsicht geboten, die Studien dazu sind noch jung, gerade die statistische Qualität und Anwendungsbreite für das Kreditgeschäft unklar (Gefahr von Overfitting bei zu vielen Risikofaktoren oder Scheinkorrelationen). Dennoch: Die BigTechs zur Verfügung stehenden Daten können eine ernst zu nehmende Konkurrenz oder *wertvolle Ergänzung zu bestehenden Ratingverfahren* liefern. Auf datenschutzrechtliche, Verbraucherschutz- und ethische Aspekte kann an dieser Stelle nicht näher eingegangen werden. Klar sollte nur sein: Wäre amazon Kreditgeber und registriert beim nächsten Einkauf/Gebühreneinzug einen Zahlungsrückläufer, hätte das für das Kreditverhältnis des Konsumenten vermutlich schnellere und umfassendere und für ihn weniger transparente Konsequenzen im Vergleich zur Bank als Kreditgeber. Die spannende Frage: Wie kann diese neue Art der Daten eine solche prognostische Kraft haben?

Ein Kreditausfall kommt durch das Verhalten einer Person zustande. Beeinflusst wird dieses *Verhalten von der Persönlichkeit des Handelnden als auch der Situation in der die Handlung steht.* Folgerichtig sollte sich die Wahrscheinlichkeit eines Kreditausfalls auf Basis der beiden Faktoren Persönlichkeit und Situation mit einer gewissen Validität bestimmen lassen. Ist das nicht zu komplex? Ja. Denn die Anzahl möglicher Situationen ist unendlich und daher in einem Modell kaum abbildbar. Die Bandbreite grundlegender Persönlichkeitseigenschaften hingegen ist überschaubar und wissenschaftlich, vor allem durch die Psychologie, gut erforscht.

Dabei ist der Ansatz, ausgehend von der Persönlichkeit des Handelnden dessen Verhalten zu prognostizieren weder neu noch kontraintuitiv. Dass die Persönlichkeit des Kreditnehmers und deren Analyse über das von ihm gezeigte Verhalten bei der Bonitätsbeurteilung eine Rolle spielt, zeigt sich bisweilen an der Berücksichtigung so genannter „weicher oder qualitativer Faktoren" im Kreditvergabeprozess. Mit dem Kreditvergabeprozess vertraute Bankern sind Aussagen wie „ich kenne den Kreditnehmer seit Jahren, wir haben schon viele gemeinsame Geschäfte sehr erfolgreich und zuverlässig durchgeführt" bis zum einfachen „ich habe ein gutes Bauchgefühl", „meine Erfahrung sagt mir" nicht fremd. Solche individuellen Einschätzungen sind jedoch nicht objektiv erhoben oder bewertet und schon gar nicht hinreichend valide, um in einem Risikomodell eine gewichtige Rolle zu spielen. Dennoch illustrieren sie die Idee, dass die Persönlichkeit des Kreditnehmers sein Verhalten beeinflusst und Persönlichkeit durchaus bonitätsrelevant ist.

Nota bene: Mit solchen Zusammenhängen beschäftigen sich nicht nur Psychologen. Auch der *Regulator* hat in den letzten Jahren zunehmend die Bedeutung der Persönlichkeit für das Verhalten im Blick. So werden beispielsweise bei der Eignungsprüfung für Vorstände „honesty and integrity" (sogenannte „fit and proper" Guideline, ESMA und EBA 2017, S. 35–37) als notwendige Persönlichkeitseigenschaften und „indepence of mind" inklusive des Verhaltens beschrieben, an dem sich diese Persönlichkeitseigenschaft konkret manifestiere.

Kurzum: Wenn sich Persönlichkeit in Verhalten zeigt, lässt gezeigtes Verhalten Rückschlüsse auf die Persönlichkeit des Handelnden zu. Die psychologische Forschung ist ein Instrument, diesen Zusammenhang wissenschaftlich fundiert zu untersuchen und zu validen Erkenntnissen weit jenseits von „Bauchgefühl" zu kommen.

So ist es beispielsweise, möglich anhand der Persönlichkeitseigenschaften einer Person die Wahrscheinlichkeit abzuschätzen, dass diese Person wirtschaftskriminelles Verhalten zeigt. In einer Studie, in der mehr als 100 wirtschaftskriminelle Straftäter hinsichtlich ihrer Persönlichkeit untersucht und mit einer Kontrollgruppe von Managern verglichen wurden, konnte gezeigt werden, dass anhand eines nur kurzen Fragebogens 86 Prozent der wirtschaftskriminellen Straftäter korrekt als solche identifiziert werden konnten – eine für eine Feldstudie immens hohe Treffsicherheit. Und ein klarer Beleg für die Erklärung von Verhalten mit Hilfe der Ausprägung von Persönlichkeitseigenschaften. Es lohnt also der nähere Blick hinter die Daten, die uns Informationen über das Verhalten von Menschen geben, auf das „underlying", i.e. die Persönlichkeit. Denn mithilfe dieser lässt sich weiteres Verhalten mit einer gewissen Validität prognostizieren. Wir erinnern uns an die einfache Grundhypothese „Verhalten ist eine Funktion von Persönlichkeit und Situation". In der hier darzustellenden Studie führte allein die datenbasierte Analyse der Persönlichkeit zu 86 Prozent korrekten Zuordnungen – völlig unter außer Acht lassen der Situation.

23.2   Zur Persönlichkeit wirtschaftskrimineller Straftäter

Über wirtschaftskriminelle Straftäter gibt es eine Fülle von Literatur, die sich auf die Analyse von Gerichtsakten und Gutachten über diese Personengruppe stützt. In der darzustellenden Studie wurde erstmals in Europa die Persönlichkeit verurteilter wirtschaftskrimineller Straftäter mit Hilfe von ihnen selbst ausgefüllten, persönlichkeitsbezogenen Fragebögen untersucht.

23.2.1   Untersuchte Persönlichkeitseigenschaften

Erhoben wurden vor allem Werthaltungen (aber auch Narzissmus, Kontrollüberzeugung, Gewissenhaftigkeit, soziale Erwünschtheit). Zum besseren Verständnis dieser Persönlichkeitsaspekte eine kurze Einführung.

Werthaltungen
Weshalb nicht die klassischen „Big Five" der Persönlichkeitsforschung (Offenheit für Erfahrungen, Gewissenhaftigkeit, Extraversion, Verträglichkeit und Neurotizismus, vgl. Asendorpf und Neyer 2018, S. 5–19) untersuchen? Warum der nähere Blick auf die Werthaltungen Wirtschaftskrimineller?

> Values are (a) concepts or beliefs, (b) are about desirable end states or behaviors, (c) transcend specific situations, (d) guide selection or evaluation of behaviour and events, and (e) are ordered by relative importance. (Bilsky und Schwartz 1994, S. 164)

Werte unterscheiden sich von klassischen Persönlichkeitseigenschaften durch ihre motivationale Kraft, sie sind *kognitive Repräsentationen der wichtigen Ziele und Motivation einer Person*, entlang derer sich menschliches Verhalten orientiert (vgl. Bilsky und Schwartz 1994). Aufgrund ihrer verhaltenssteuernden Funktion sind Werte als zentraler Bestandteil der Persönlichkeit zu verstehen.

Um Werthaltungen in der Studie konkret zu erheben wurde der in über 20 Ländern validierte Schwartz Values Survey (vgl. Schwartz 1992) verwendet. Dieser ist in zehn motivationale Werttypen gegliedert:

1. Power: Sozialer Status, Prestige, Kontrolle/Dominanz über Personen und Ressourcen (Soziale Macht, Reichtum, Autorität, In der Öffentlichkeit Ansehen bewahren)
2. Achievement: Persönlicher Erfolg durch Demonstration von Kompetenz, entsprechend den sozialen Standards (Erfolgreich, Fähig, Ehrgeizig, Einflussreich)
3. Hedonism: „Lust" im weiten Sinne, sinnliche Freude (Vergnügen, Das Leben genießen)
4. Self-Direction: Unabhängig in Denken und Handeln, auswählend, schaffend, erforschend (Kreativität, Freiheit, Unabhängig, Eigene Ziele wählen, Neugierig)

5. Stimulation: Herausforderung, Neuheit, Abwechslungsreichtum (Ein anregendes Leben, Wagemutig, Ein abwechslungsreiches Leben)

6. Universalism: Verständnis, Toleranz, Schutz des Gemeinwohls, der Natur, Anerkennung, Verständnis (Soziale Gerechtigkeit, Eine Welt in Frieden, Weisheit, Eine Welt der Schönheit, Tolerant, Einheit mit der Natur, Umwelt schützen, Gleichheit)

7. Benevolence: Erhaltung und Weiterentwicklung der Wohlfahrt der Menschen, mit denen man in häufigem persönlichem Kontakt steht (wahre Freundschaft, Loyal, Ehrlich, Hilfsbereit, Vergeben)

8. Tradition: Respekt, Bindung, Akzeptanz der Gebräuche und Ideale von Traditionen (Kultur, Religion), die einem auferlegt werden (Achtung vor der Tradition, Gemäßigt, Demütig, Alle Seiten des Lebens akzeptieren, Fromm)

9. Conformity: Zurückhaltung vor Handlungen, Neigungen und Impulsen, die andere oder soziale Normen empören, verletzen oder schaden können (Gehorsam, Selbstdisziplin, Höflichkeit, Ehrerbietung gegenüber Eltern und älteren Menschen)

10. Security: Sicherheit, Harmonie und Stabilität der Gesellschaft, von Beziehungen und einem selbst (Soziale Ordnung, Nationale Sicherheit, Ausgleich von Gefälligkeiten, Familiäre Sicherheit)

Schwartz gruppiert die zehn Typen in vier Dimensionen, dabei werden Nummer 1–3 der Dimension „Self-Enhancement", 3–5 „Openness to Change", 6–7 „Self-Transcendence" und 8–10 „Conservation" zugeordnet.

Narzissmus

> The most pervasive problem among people who exercise power over others is that of narcissism in its many forms. (Levinson 1998, S. 237)

Gerade in jüngerer Zeit wird häufig über Narzissten und Psychopathen/Soziopathen in Management-Positionen publiziert, beispielsweise die viel beachtete Monografie Snakes in Suits (Babiak und Hare 2006; vgl. auch Dutton 2012), auch ist von der „Dark Triad of Personality" (Narcissism, Machiavellianism, and Psychopathy) die Rede (Paulhus und Williams 2002) sowie jüngeren Adaptionen (Chamorro-Premuzic 2019).

Narzissmus im Sinne der Narzisstischen Persönlichkeitsstörung (World Health Organisation 2019, ICD-10 F60.8) zeichnet sich durch einen *Mangel an Empathie, Selbstüberschätzung und großes Verlangen nach Anerkennung* aus. Die autonome Regulierung des Selbstwertgefühls ist gestört, sodass Selbstwert regelmäßig über die Herabsetzung anderer generiert wird. Der Narzisst nimmt sich als anderen überlegen, geradezu grandios und genial wahr, Angriffe auf den schlecht regulierten Selbstwert wie Kritik werden häufig durch extrem aggressives, verletztes Verhalten bei subjektiv empfundener großer Ungerechtigkeit/Uneinsichtigkeit des Kritisierenden beantwortet.

Hinweise zu *Narzissmus bei Wirtschaftskriminellen* finden sich Schwartz (1942, 1994); Levinson (1998) und Friedrichsen (2001). Levinson beschreibt mit dem „con man"

(Levinson 1998) eine narzisstische Persönlichkeit, die sich im beruflichen Umfeld großer Beliebtheit erfreut und als vertrauenswürdig eingeschätzt wird. Durch ein gutes Netz von Verbindungen im Unternehmen ist er mikropolitisch geschickt, sehr aktiv und erfolgreich. Gerade vom gehobenen Management wird ihm großes Vertrauen geschenkt.

Solche Rahmenbedingungen sowie eigene Fähigkeiten ermöglichen es dem Narzissten, günstige Gelegenheiten für wirtschaftskriminelle Handlungen herzustellen bzw. in solchen Gelegenheiten zuzugreifen und delinquent zu werden – bei gleichzeitig subjektiv empfundener „reiner Weste" – denn man selbst wird nicht kriminell und tut höchstens etwas nicht Legales, weil die eigentlichen Regeln der Welt und die Gerechtigkeit und der eigene, anderen überlegene Überblick über die großen Zusammenhänge die Tat in einem ganz anderen Licht erscheinen lassen, als sie von den „einfachen Menschen" (oder der Justiz) gesehen wird. Folgendes Zitat von Nietzsches Zarathustra ist eine literarische Illustration genau dieser narzisstischen Grundhaltung.

> Ihr höheren Menschen, dies lernt von mir: auf dem Markt glaubt niemand an höhere Menschen. Und wollt ihr dort reden, wohlan! Der Pöbel aber blinzelt „wir sind alle gleich".
> (Nietzsche 1891, Vierter Teil, Vom Höheren Menschen 1)

Erhoben wurde Narzissmus mit einer auf Basis eines strukturierten, klinischen Interviews (SKID-II, vgl. Wittchen et al. o. J.) zusammengestellten Fragenreihe.

Kontrollüberzeugung

Über seine zahlreichen Lernerfahrungen und der eigenen Rolle in Situationen lernt das Individuum, ob der Ausgang einer Situation mit ihm selbst viel oder eher wenig zu tun hat. Diese Erfahrungen generalisieren über die Zeit hinweg zur Kontrollüberzeugung, diese kann als „Kontrollort", sogenannter „locus of control" internal oder external sein. Bei *internaler* Kontrollüberzeugung geht die Person davon aus, dass der gute oder schlechte Ausgang einer Situation von ihm selbst abhängen, bei *externaler* Kontrollüberzeugung glaubt die Person, dass der Ausgang einer Situation von Schicksal, dem Handeln anderer oder Zufall, in jedem Fall von Faktoren, die sich seiner Kontrolle entziehen, bestimmt ist.

Die Erhebung dieser Persönlichkeitseigenschaft ist bedeutsam, weil sie etwas über die *Wirksamkeit meiner eigenen Werthaltungen* aussagt. Wenn Werthaltungen gewünschte Zielzustände repräsentieren und menschliches Verhalten zur Herbeiführung dieser Zielzustände gesteuert wird (vgl. Six 1998), wird ein Individuum mit internaler Kontrollüberzeugung davon überzeugt sein, dass es diese Zustände auch herstellen kann und seine Werthaltungen werden sein Verhalten daher mehr beeinflussen als bei einem Individuum, das durch externale Kontrollüberzeugung geprägt ist.

> *Kontrollpsychologisch* ist zu fragen warum und wie man selbst handeln kann: Ursachen, Möglichkeiten, Chancen und Barrieren für eigenes Handeln sind ergänzbar durch Einfluss auf die Folgen. *Die moralpsychologische Frage* zielt darauf, ob, warum und wie man selbst handeln soll, muss, darf. Dies wird wesentlich von individuellen Werthaltungen bestimmt.
> (Schlegel 2003, S. 116)

Erhoben wurde Kontrollüberzeugung mit Hilfe des „Fragebogens zu Kompetenz- und Kontrollüberzeugungen" (FKK) von Krampen (1991).

Gewissenhaftigkeit

Gewissenhaftigkeit der einzige der „big five" (siehe oben), die in der Studie erhoben wurde. Sie ist ein *Aspekt der Selbstkontrolle*, der zuverlässige, anspruchsvolle, planerisch und gut organisiert vorgehende Individuen von eher nachlässigen, planfrei agierenden unzuverlässigen und wenig engagierten Personen unterscheidet (vgl. Borkenau und Ostendorf 1993, S. 5).

Vorangegangene Studien legten eine eher geringe Ausprägung dieser Eigenschaft bei Wirtschaftskriminellen nahe (vgl. Collins und Schmidt 1993). So könnte eine geringe Gewissenhaftigkeit als ein Aspekt der Selbstkontrolle in einer günstigen Situation eher dazu führen, dass es zur kriminellen Handlungen kommt, gerade weil kein großes Maß an Planung und Organisation als Vorbedingung gesehen oder als notwendig erachtet wird (vgl. General Theory of Crime, Gottfredson und Hirschi 1990). Entsprechend ist dieser Fragebogenteil ein Test, ob sich die Erkenntnisse von Collins und Schmidt durch Analyse der Persönlichkeit der Delinquenten selbst bestätigen.

Erhoben wurde Gewissenhaftigkeit mit dem von Borkenau und Ostendorf übersetzten NEO-Fünf-Faktoren-Inventar (NEO-FFI, Borkenau und Ostendorf 1993).

Soziale Erwünschtheit

Mittels der für die Untersuchung adaptierten Skala von Lück und Timaeus (1969) wurde Soziale Erwünschtheit erhoben. Sie bezeichnet die Antworttendenz, vom Individuum vermutet sozial erwünschte Antworten zu zeigen. Als solche ist sie ein Aspekt der Persönlichkeit, wurde jedoch auch als Kontrollskala genutzt (sogenannte „Lügenskala").

Soziobiografische und deliktspezifische Daten

Zur näheren Charakterisierung sowohl der Stichprobe der Wirtschaftskriminellen als auch der Kontrollgruppe wurden noch wenige soziobiografische und deliktspezifische Daten erhoben: Alter, Geschlecht, Brutto-Jahresgehalt, Art der verurteilten Handlung, geschätzter Vermögensschaden, Gesamthaftstrafe in Jahren, bisherige Dauer des Gefängnisaufenthaltes. Bei der Erhebung detaillierterer Daten verbreitern sich zwar die Datenbasis und die statistisch möglicherweise nachweisbaren Zusammenhänge, jedoch verringert sich die „Ausfüllbereitschaft" der Probanden. Daher galt auch hier, was für Big Data-Themen gilt: Zuerst genau definieren und eingrenzen, was gemessen werden soll und genau das dann auch erheben.

23.2.2 Untersuchte Personenkreise

Die Stichprobe umfasste *102 verurteilte wirtschaftskriminelle Straftäter* aus 14 Justizvollzugsanstalten in vier deutschen Bundesländern. Voraussetzung war dabei die rechtsgültige

Verurteilung aufgrund eines Deliktes, das unter die Definition der Wirtschaftskriminalität nach Bock (2000) fällt, und das Begehen der Straftat in einem Arbeitsverhältnis in einer Organisation.

Die 102 Personen waren zwischen 25 und 64 Jahre als (im Mittel 44 Jahre), darunter neun Frauen und 93 Männer. Das vor der Verurteilung bezogene letzte Brutto-Jahresgehalt lag zwischen 9.000 und 750.000 Euro (im Mittel 79.000 Euro, elf Personen machten dazu keine Angaben).

Die Delikte waren vorwiegend Betrug und Untreue (63 Prozent) sowie Steuerdelikte (9 Prozent), die restlichen 28 Prozent verteilen sich auf Anlagebetrug, Konkursvergehen u. a.

Die Strafmaße bewegen sich zwischen 10 und 131 Monaten (im Mittel 46), die davon bereits verbüßte Zeit zwischen 2 und 71 Monaten (im Mittel 20 Monate). Der verursachte Vermögensschadenlag bei den Delikten lag zwischen 5.000 und 19.000.000 Euro (im Mittel 1.706.000 Euro, 14 Personen machten hier keine Angabe).

Die *Kontrollgruppe* umfasste 88 Personen zwischen 26 und 73 Jahre (im Mittel 48 Jahre) alte Personen, darunter 24 Frauen und 64 Männer. Das Bruttojahresgehalt bewegte sich zwischen 14.000 und 500.000 Euro (im Mittel 84.000 Euro, acht Personen machten hierzu keine Angabe).

23.2.3 Persönlichkeitsunterschiede von wirtschaftskriminellen Straftätern und Managern

Werthaltungen – durchgängig stärker ausgeprägte Werttypen bei Wirtschaftskriminellen

Wirtschaftskriminelle unterscheiden sich gegenüber der Kontrollgruppe in folgenden Werttypen: Power, Achievement, Hedonism, Benevolence, Security und Conformity. In all diesen Werttypen sind die Scores der Wirtschaftskriminellen signifikant höher als die der Kontrollgruppe.

Dass wirtschaftskriminellen Straftätern die Werttypen *Power, Achievement und Hedonism* bedeutsamer sind als der Kontrollgruppe, ist erwartungsgemäß. Je wichtiger einem Individuum Macht, Vorankommen und Genuss sind, desto eher wird es in einer günstigen Situation dazu neigen, auch illegale Mittel zur Erreichung dieser Zielzustände zu nutzen (vgl. Blickle et al. 2006, S. 228).

Dass jedoch gerade *Security und Conformity* für Wirtschaftskriminelle höhere Bedeutsamkeit aufweisen als für die Kontrollgruppe, ist überraschend.

Conformity ist definiert als große Zurückhaltung vor Handlungen, die andere Personen oder soziale Normen verletzen oder schaden können – wie kriminelles und speziell wirtschaftskriminelles Verhalten. Im eingesetzten Schwartz Values Survey wird Conformity durch die Werte Höflichkeit (gute Umgangsformen), Selbstdisziplin (Selbstbeherrschung), Widerstand gegen Versuchung), Ehrerbietung gegenüber Eltern und älteren Menschen (respektvoll), Gehorsam (Pflichten erfüllen) und Pünktlich (verabredete Zeiten genau einhalten) operationalisiert.

Security definiert sich mit Sicherheit, Harmonie und Stabilität der Gesellschaft, der Beziehungen zu anderen und sich selbst, im Schwartz Values Survey mit folgenden Werten operationalisiert: Zugehörigkeitsgefühl (das Gefühl, dass sich andere um mich kümmern), Soziale Ordnung (Stabilität der Gesellschaft), Ausgleich von Gefälligkeiten (Vermeiden von Dankesschuld), Nationale Sicherheit (Schutz meiner Nation gegen Feinde), Ein Privatleben (das Recht auf eine Privatsphäre), Familiäre Sicherheit (Sicherheit für die geliebten Personen), Gesund (physisch und geistig nicht krank sein), Sauber (ordentlich).

Beiden Werttypen ist ein zentrales Motiv gemein: Die *Bewahrung des Status Quo*. Doch scheinen gerade diese Werte mit wirtschaftskriminellen Handlungen nicht vereinbar. Objektiv gesehen. Da der Schwartz Values Survey jedoch abfragt, welche Werte dem Ausfüllenden wichtig sind, obliegt es dessen Deutung, ober er mit seinem wirtschaftskriminellem Verhalten gegen seine Werte verstoßen hat oder ob er diese Werte hochhält und subjektiv keinerlei kognitive Dissonanz empfindet. Zu ähnlichem Ergebnis kommen Weisburd und Waring (vgl. Weisburd und Waring 2001), die bei einem Teil ihrer Wirtschaftskriminellen dem Werttyp Conformity entsprechende Einstellungen feststellten, obwohl sie selbst illegal handelten. Wirtschaftskriminelle verstehen „sich selbst als Ehrenmänner, nicht als Kriminelle" (Schneider 1981, S. 161). Ähnlich Bock „man kann sogar fast den Eindruck gewinnen, daß es (insbesondere bei Gefangenen) zum geäußerten Selbstverständnis gehört, sich von ‚dem Kriminellen' und ‚seine' Werten abzugrenzen" (Bock 2000, S. 224). So ergibt sich *ein in der individuellen Selbstwahrnehmung des Wirtschaftskriminellen kongruentes Bild*: Der Ehrenmann vertritt durchaus konservative Werte und empfindet keine kognitive Dissonanz zu seiner kriminellen Handlung. Insbesondere die deutlich erhöhten Narzissmus-Werte der Wirtschaftskriminellen vereinfachen hier eine dissonanzfreie Selbstwahrnehmung.

Narzissmus – deutlich höhere Werte für verurteilte Wirtschaftskriminelle

Erwartungsgemäß zeigten Wirtschaftskriminelle signifikant höhere Werte in Narzissmus als die Kontrollgruppe. Dies konnte auch in einer Varianzanalyse bestätigt werden.

Zusätzlich zeigte die Varianzanalyse folgende Zusammenhänge: Signifikant höhere Narzissmus-Werte für *Männer gegenüber Frauen* (was bisherige Forschung zu Narzissmus bestätigt) und positiver Zusammenhang mit dem *Jahresgehalt* (je höher je narzisstischer). Diese Effekte zeigten sich über beide Gruppen hinweg – bei Wirtschaftskriminellen wie auch Kontrollgruppe.

Mit dem Selbstbild des Ehrenmanns im Sinn, zeichnet sich ein plausibles Bild: Man hat keinen einzelnen Personen mit seiner Handlung geschadet, im Gegenteil hat man eine Art Ausgleich im Weltgewicht hergestellt, da man aufgrund des eigenen superioren Überblicks über „das große Ganze" die Tat in einem anderen Licht sieht und sie keineswegs den o. a. konservativen Werten widerspricht. Dass andere dies nicht sehen, liegt an deren limitiertem Blick auf die Welt und mangelndem Verständnis ökonomischer Zusammenhänge.

Kontrollüberzeugung – klar internal für Wirtschaftskriminelle und Kontrollgruppe
Keine Überraschung – sowohl die Kontrollgruppe als auch die Wirtschaftskriminellen
sind klar durch eine internale Kontrollüberzeugung gekennzeichnet. Sowohl T-Test als
auch Varianzanalyse zeigen für beide eine klare internale Kontrollüberzeugung, jedoch
keinen signifikanten Unterschied in der Ausprägung der Internalität zwischen den beiden
Personengruppen.

Gewissenhaftigkeit – keine Unterschiede zwischen den Gruppen
Im Mittel zeigen sich zwar höhere Gewissenhaftigkeitswerte für die Wirtschaftskriminellen, jedoch kann eine Varianzanalyse diesen Gruppenunterschied nicht bestätigen. Interessant jedoch: *Je länger die bislang verbüßte Haftdauer, umso höher die Ausprägung der
Gewissenhaftigkeit* bei den Wirtschaftskriminellen. Hier kann Gewissenhaftigkeit der o. a.
(vgl. Bock 2000) Abgrenzung des Wirtschaftskriminellen von „einfachen oder wirkli-
chen" Kriminellen dienen: Zumeist erhalten wirtschaftskriminelle Straftäter entweder
sehr hohe Geldstrafen oder lange Haftstrafen, kurze Haftstrafen sind hingegen selten. Da
die Justizvollzugsanstalten in Deutschland meist nach Haftdauer gegliedert sind, finden
sich Wirtschaftskriminelle nach Ihrer Verurteilung also nicht in der Nachbarzelle des
Betäubungsmittel-Delinquenten, sondern des Mörders, schweren Räubers, Vergewalti-
gers. Dieses herausfordernde Umfeld kann der Beförderung von Verhaltens- und Beschäf-
tigungsweisen dienen, die an die vorherige Tätigkeit in der Wirtschaft/Büroumfeld erinnern, wie Gewissenhaftigkeit – gerade in Abgrenzung gegenüber den „echten" Kriminellen
in gleicher Justizvollzugsanstalt.

23.2.4 Die Abschätzung des wirtschaftskriminellen Potenzials

Die gefundenen Unterschiede zwischen Kontrollgruppe und verurteilten Wirtschaftskriminellen fanden sich vor allem bei Werthaltungen und Narzissmus. Man kann mit Hilfe
eines forschungsstatistischen Verfahrens jedoch zu einem noch weit kürzeren Fragebogen
kommen, der die beiden Gruppen noch deutlicher unterscheidet. Dazu konzentriert man
sich auf diejenigen Variablen, die den Unterschied zwischen den Gruppen maximieren
und (mit Gewichtskoeffizienten versehen) ihren Anteil „am Zustandekommen des Ge-
samtunterschieds" (Bortz 1999, S. 585, sogenannte Diskriminanzanalyse) ausweisen.

Als Ergebnis dieses statistischen Verfahrens erhält man für Wirtschaftskriminelle und
Kontrollgruppe eine Gleichung (Diskriminanzfunktion) mit sieben Variablen und Gewich-
tungskoeffizienten, aus der sich ein kurzer Fragebogen zusammenstellen lässt. Dieser besteht lediglich aus einem Teil des Fragebogens zu Kompetenz- und Kontrollüberzeugungen (FKK), der vom Autor modifizierten Sozialen Erwünschtheitsskala und Teilen des
Wertefragebogens.

Auf dieser Basis lässt sich das wirtschaftskriminelle Potenzial einer Person abschätzen.
Wie treffgenau diese Einschätzung ist, lässt sich statistisch ermitteln. Dabei „stellt sich
das statistische Verfahren blind" gegenüber der Information, ob die Daten von einem

Wirtschaftsdelinquenten oder eines Angehörigen der Kontrollgruppe stammen und ordnet die Fragebögen aufgrund der Ausprägung der Variablen der Diskriminanzfunktion den beiden Gruppen Wirtschaftskrimineller oder Kontrollgruppe zu. Die Klassifizierungsstatistik zeigt dann, wie treffgenau die Zuordnung aufgrund der Diskriminanzfunktion ist. Dabei kann es zu zwei Fehlerarten kommen: Falsch Positiven (Kontrollgruppen-Person wird für Wirtschaftsdelinquenten gehalten) oder zu falsch Negativen (Wirtschaftskrimineller wird für Kontrollgruppen-Person gehalten).

Tatsächlich zeigte sich, dass die Treffgenauigkeit bei den Wirtschaftskriminellen weit höher ist als bei der Kontrollgruppe:

Über beide Gruppen hinweg können insgesamt 83 Prozent der Probanden richtig zugeordnet werden. Der kurze Fragebogen identifiziert 86,3 Prozent der Wirtschaftskriminellen korrekt als solche, die Kontrollgruppe mit etwas geringeren 79,1 Prozent korrekten Zuordnungen, i.e. mit der Diskriminanzfunktion werden mehr Angehörige der Kontrollgruppe als wirtschaftskriminell eingestuft (20,9 Prozent), als Wirtschaftskriminelle der Kontrollgruppe zugehörig erachtet (13,7 Prozent). Das heißt einerseits, dass einige „Unschuldige" für Wirtschaftskriminelle gehalten werden, dafür aber relativ wenige Wirtschaftskriminelle (13,7 Prozent) durch den Fragebogen unentdeckt bleiben.

23.3 Persönlichkeit prägt Verhalten und hinterlässt Daten

Wenn man verhaltensbezogene Daten analysiert, lässt sich der Weg von gezeigtem Verhalten zur Persönlichkeit zurückgehen und auf dieser Basis künftiges Verhalten erklären bzw. prognostizieren. Die dargestellte Studie hat gezeigt, dass dieser Weg möglich ist und sich auf einer breiten Datenbasis Persönlichkeit mittels psychologischer Forschung und statistischer Methoden analysieren und bisheriges Verhalten erklären lässt. Nach vorne gewandt lässt sich daraus eine Prognosekraft ableiten, die auch einen Blick auf künftiges Verhalten freigibt. Im Kontext des Social Credit Ratings liegen nun große, relativ neuartige Datenmengen vor uns, die zu strukturieren und deren Zusammenhänge untereinander sowie mit gezeigtem und künftigem Verhalten aufschlussreiche Erkenntnisse erwarten lassen.

Die Forschungsarbeit wurde in Schlegel (vgl. Schlegel 2003) inklusive aller forschungsstatistischer Details ausführlich dargestellt und die psychologisch wichtigsten Aspekte in Blickle et al. 2006 publiziert.

Literatur

Asendorpf, J., Neyer, F.J. (2018): Psychologie der Persönlichkeit. 6. Auflage. Berlin.
Babiak P., Hare, R. P. (2006): Snakes in Suits. When Psychopaths go to Work. New York.
Bank for International Settlements (BIS) (2019): III. Big Tech in Finance. Annual Report, S. 55–79. Basel.
Berg, T., Burg, V., Gombovic, A., Puri, M. (2018): On the Rise of FinTechs – Credit Scoring using Digital Footprints. Michael J. Brennan Irish Finance Working Paper Series. Dublin.

Bilsky, W., & Schwartz, S. H. (1994): Values and personality. European Journal of Personality, 8(3), 163–181.

Blickle, G., Schlegel, A., Fassbender, P., Klein, U. (2006): Some Personality Correlates of Business White-Collar Crime. Applied Psychology: An International Review, 55 (2), S. 220–233.

Bock, M. (2000): Kriminologie. München.

Borkenau, P., Ostendorf, F. (1993): NEO-Fünf-Faktoren Inventar (NEO-FFI) nach Costa und Mc-Crae. Göttingen.

Bortz, J. (1999): Statistik für Sozialwissenschaftler (5. vollst. überarb. Aufl.). Berlin.

Chamorro-Premuzic, T. (2019): Warum so viele inkompetente Männer in Führungspositionen sind (und was man dagegen tun kann). Berlin.

Collins, J. M., Schmidt, F. L. (1993): Personality, integrity, and white collar crime. A construct validity study. Personnel Psychology, 46, S. 295–312.

Dutton, K. (2012): The wisdom of psychopaths. What Saints, Spies, and Serial Killers Can Teach Us about Success. London.

European Securities and Markets Authoritiy (ESMA), European Banking Authority (EBA) (2017): Joint ESMA and EBA Guidelines on the assessment of the suitability of members of the management body and key function holders under Directive 2013/36/EU and Directive 2014/65/EU. EBA/GL/2017/12.

Friedrichsen, G. (2001): Wie bitte? Die Justiz entdeckt ein neues Krankheitssymptom bei Wirtschaftskriminellen: Größenwahn. Der Spiegel, 36, S. 40.

Gottfredson, M., Hirschi, T. (1990): A general theory of crime. Stanford, CA.

Krampen, G. (1991): Fragebogen zu Kompetenz- und Kontrollüberzeugungen (FKK). Göttingen.

Levinson, H. (1998): A clinical approach to executive selection. In. R. Jeanneret, R. Silzer (Eds.), Individual psychological assessment. Predicting behavior in organizational settings (pp. 228–242). San Francisco.

Lück, H. E., Timaeus, E. (1969): Skalen zur Messung manifester Angst (MAS) und sozialer Wünschbarkeit (SD-E und SD-CM). Diagnostica, 15, S. 134–141.

Nietzsche, F. (1891): Also sprach Zarathustra. Ein Buch für Alle und Keinen. Vierter und Letzter Teil. Leipzig.

Paulhus, D. L, Williams, K. M. (2002): The Dark Triad of personality: Narcissism, Machiavellianism, and psychopathy. Journal of Research in Personality. 36 (6), p. 556–563.

Schlegel, A. (2003): Werthaltungen inhaftierter Wirtschaftsdelinquenten. In A. Schlegel (Hrsg.), Werthaltungen inhaftierter Wirtschaftsdelinquenten (S. 113–173). Niedernhausen.

Schneider, H. J. (1981): Wirtschaftskriminalität. In H.J. Schneider (Hrsg.), Die Psychologie des 20. Jahrhunderts, Bd. 14. Auswirkungen auf die Kriminologie. Delinquenz und Gesellschaft (S. 141–164). Zürich.

Schwartz, H. S. (1942): Narcisstic process and corporate decay. The theory of the organization ideal. New York.

Schwartz, H. S. (1994): Narcissism project and corporate decay. The case of General Motors. In M. Hofmann, M. List (Eds.), Psychoanalysis and Management (pp. 303–324). Heidelberg.

Schwartz, S. H. (1992): Universals in the content and structure of values. Theoretical advances and empirical tests in 20 countries. In M. P. Zanna (Ed.), Advances in Experimental Social Psychology, 25 (pp. 1–65). New York.

Six, B. (1998). Werte. In H. Häcker, K. Stapf (Hrsg.), Dorsch Psychologisches Wörterbuch (13. Aufl., S. 947). Bern.

Weisburd, D., Waring, E. (2001). White-collar crime and criminal careers. Cambridge.

Wittchen, H.-U., Schramm, E., Zaudig, M., Unland, H. (o. J.). SKID-II. Strukturiertes Klinisches Interview für DSM-III-R. Achse II (Persönlichkeitsstörungen). Interviewheft. Basel.

World Health Organisation (2019). Internationale statistische Klassifikation der Krankheiten und verwandter Gesundheitsprobleme.10. Revision (ICD-10). Online-Version. Köln.

Dr. Alexander Schlegel ist Bereichsleiter des Geschäftsbereichs Marketing & Services beim Deutschen Genossenschafts-Verlag eG in Wiesbaden, einem Dienstleister für Genossenschaftsbanken sowie Waren- und Dienstleistungsgenossenschaften. Zuvor war er als Director im Firmenkundengeschäft der BHF-BANK AG in Frankfurt sowie als Unternehmensberater für verschiedene Strategieberatungen tätig. Schlegel ist Diplom-Theologe, Diplom-Psychologie und hat in Moralphilosophie promoviert.

Menschen bewerten und entwickeln?

Andreas Fornefett, Gerd Rupprecht und Uwe Schacher

Zusammenfassung

Der Mensch ist intelligent, er kann täuschen und lügen, vieles kopieren, alles instrumentalisieren und er kommuniziert bereits als Baby niemals ohne eine Absicht. Dabei ist ihm seine Absicht nicht immer bewusst. Während sich danach rein absichtsloses Handeln in Aktivitäten ausdrücken würde, die sich selbst genügen, könnten alle anderen, vom Menschen als Subjekt mit freiem Willen beabsichtigte Aktivitäten also durchaus einer Bewertung unterliegen.

Als Träger seiner Aktivitäten in Form von bewusstem oder unbewusstem Entscheiden und Handeln (Tun oder Unterlassen) ist der Mensch Sender und direkter oder indirekter Empfänger auch seiner eigenen Signale. Will man Einfluss nehmen auf sein Verhalten und eine gewünschte soziale Entwicklung über Bewusstseins-, Einstellungs- oder Haltungskontrolle erreichen, muss man ihn als Subjekt betrachten und behandeln. Das bedeutet unter anderem ihn so zu fragen, dass er sich fragt, sprich: dass er sein Wissen hinterfragt und Bewusstsein darüber erlangt. So kann er seine Einstellung und darüber auf längere Sicht

A. Fornefett (✉)
i.EPC Solutions AG, Kelkheim, Deutschland
E-Mail: andreas.fornefett@prcsm.com

G. Rupprecht
i.EPC Solutions AG, Leipzig, Deutschland
E-Mail: info@rupprechtpartner.de

U. Schacher
U.S. Research (Markt- und Sozialforschung) Höchstadt/Aisch,
Oberursel, Deutschland
E-Mail: uwe.schacher@ujs.vision

© Springer Fachmedien Wiesbaden GmbH, ein Teil von Springer Nature 2020
O. Everling (Hrsg.), *Social Credit Rating*,
https://doi.org/10.1007/978-3-658-29653-7_24

seine Haltung nachhaltig ändern oder aber sich in seinem Denken und Handeln bestätigt fühlen. Nur mittels Vorgaben, Bewertungen, Strafen oder Belohnungen lässt sich sein vom freien Willen getragenes Verhalten weder gesichert steuern noch entwickeln.

Jede Arithmetik, Logik, Methode, jedes Regelwerk, kurz: jede Moral, besitzt einen jeweils definierten Geltungsbereich, innerhalb dessen richtige oder falsche Aussagen ‚existieren' bzw. gutes oder böses Verhalten, Glück oder Unglück, aber eben nicht darüber hinaus. Deshalb helfen Bewertungen im Rahmen nur eines Bezugssystems, wie eben nur einer Moral, in komplexen Welten für sich kaum weiter. Um einer ethischen Gesinnung des Einzelnen im Sinne des Hinterfragens diverser Moralen unterschiedlicher Bezugssysteme, denen er angehört, gerecht zu werden, ist es als Anspruchsteller an ein gewünschtes Verhalten notwendig, zunächst selbst in den verschiedenen Bezugssystemen zu denken. Denn jedes organisierte System besitzt oder folgt seiner eigenen, teils sehr speziellen Moral im Sinne eines Regelwerks und hält für seine Mitglieder gerne die passenden Antworten parat. Fragen dekonstruieren aber nicht etwa solche Antworten. Sie binden rück, indem sie uns eigene Antworten oder die von Anderen besser verstehen lassen. Dafür stellen Wissenschaft und Praxis verschiedene Methoden zur Disposition.

Die meisten heutigen Gesellschaften und Gemeinschaften lassen sich hinsichtlich ihrer politischen Verfassung und wirtschaftlichen Organisation grob charakterisieren als: global, dynamisch, instabil und vernetzt. Um den mit diesen vier Attributen einhergehenden Herausforderungen zu begegnen, verstärken sie als politische oder wirtschaftliche Akteure diese immer weiter mit Maßnahmen wie: mehr internationale oder nationale Arbeitsteilung, vertikale Segmentprozessoptimierung oder Lean Management, immer flachere Hierarchien, weitreichende Dezentralisierung oder temporäre Projekt- und Teamorganisation. Und sie versuchen mit immer mehr Automatisierung und inzwischen auch Digitalisierung die Voraussetzungen hierfür zu schaffen oder den Folgen daraus zu entgehen.

Den erkannten Erfolgspotenzialen, die in der Verstärkung von globaler, dynamischer, instabiler Vernetzung der Akteure schlummerten, stehen jedoch wachsende Gefahren gegenüber. Letztere resultieren beispielsweise aus der hochgradigen Segmentierung von Institutionen, Unternehmen, Insellösungen für deren Probleme und damit verbunden: vielfältigen Schnittstellenproblemen. Sei dies aus Gründen von Reibungsverlusten bei der Koordination von Aufgaben, vor allem aber aus den bekannten Schwierigkeiten bei der Kommunikation immer häufiger wechselnder Mitglieder einer Organisation bzw. ständig neu zusammengesetzter Teams. Dies verhindert ein notwendiges einheitliches Verständnis für die jeweilige Moral des Teams einer oder mehrerer Organisationen. Deshalb bleibt trotz angestrebter Automatisierung und Digitalisierung die Qualität der Mitglieder bzw. Mitarbeiter einer Organisation als Mensch bzw. Subjekt entscheidend für deren Erfolg hinsichtlich der Verfolgung ihrer jeweiligen Ziele. Denn alle technischen und organisatorischen Lösungen bzw. Vorgaben für Aktivitäten müssen von den Mitarbeitern an einer wachsenden Zahl von Schnittstellen interpretiert, umgesetzt und innerhalb der jeweiligen Segmente ständig an ein sich rasch änderndes Umfeld angepasst werden.

Konnten die Arbeitsschritte früher von einem zentralen Plan, der alle notwendigen Aktivitäten determinierte, innerhalb einer hierarchischen Struktur für einen langen Zeitraum konstant vorgegeben werden, so findet heute zwangsläufig in vielen Bereichen eine Entwicklung zu mehr Eigenverantwortlichkeit und Planungssouveränität in eher „heterarchischen" Strukturen statt. Wir dürfen damit aber nicht nur von einem Extrem ins andere fallen. Völlig neue Strategien und Strukturen würden sonst geradezu nach neuen Menschen, Mitgliedern oder Mitarbeitern verlangen, die dann idealerweise alle intrinsisch motiviert, kooperativ und mit verschiedenen besonderen Schlüsselkompetenzen ausgestattet sein sollten. Wodurch der Anspruch an Arbeitsqualität, Selbststeuerungskompetenz und gleichzeitigem Beitrag zur Prozessperformance der Mitarbeiter stark ansteigt. Der Mensch entwickelt sich aber nur langsam fort, wie uns die vergangenen Jahrtausende lehren. Der Mensch liebt eher das, was seiner Bequemlichkeit dient und meidet gerne alles, was ihn aus der ihm sicher scheinenden Position reißt oder ihn belästigt. Und so fügen sich die meisten Mitglieder bereits aus diesem Grund in die Moral ihrer jeweiligen Organisation und hinterfragen diese grundsätzlich nur ungern.

Trotzdem genügt die Forderung nach immer mehr „Compliance", Kontrolle, Bewertung oder einer einseitigen moralischen Entwicklung nicht per se dem neuen Anspruch an Mitglieder und Mitarbeiter. Denn die Dynamik und Vielfalt der Dimensionen moralischer Urteils- und Handlungsfähigkeit und der gleichzeitigen ethischen Gesinnung der Menschen, diese zu reflektieren, werden dabei völlig vernachlässigt. Beide Schlüsselkompetenzen stehen in Zusammenhang mit einer Vielzahl an erwünschten Eigenschaften wie Loyalität, Pflichtbewusstsein, Selbstständigkeit oder kognitiv-strukturelle Entwicklung. Hat man diese erst einmal verstanden und kann sie korrekt einsetzen, werden sie zu einem starken Hebel für den Erfolg. Dazu ist allerdings die Ausrichtung der Arbeit unter den Menschen bzw. der Menschen miteinander auch an ethischen Prinzipien und deren Implementierung im Sinne von Governance und Organisationskultur notwendig. Die Ausrichtung an ethischen Prinzipien wirkt sich erst danach unmittelbar auf das Fundament der Organisation aus und beeinflusst deren Risiko- und Qualitätsmanagement. Die Auftretenswahrscheinlichkeit von unerwünschtem Verhalten sinkt, während die Wahrscheinlichkeit für erwünschtes Verhalten steigt.

Der vorliegende Beitrag bietet hierfür einen neuen Denkansatz bzw. eine universell über Raum und Zeit gültige Methode für die Bewertung und Entwicklung einer Organisation, hoch integriert in deren Aufbau und Abläufen. Der Blick richtet sich dabei gerade nicht ausschließlich auf die Bewertung der Einhaltung spezieller Normen, Gebote, Gesetze etc. beispielsweise durch ein Social Credit Rating oder System. Ergänzend finden das strukturelle Niveau der moralisch-ethischen Urteilskompetenz sowie die ethische Gesinnung von Mitgliedern oder Mitarbeitern Berücksichtigung als auch eine mit dem jeweiligen Gruppen- oder Organisationsniveau abgestimmte Umgebung sowie deren Neigungen wiederum hinsichtlich ihrer Ziele und Werte. Die Abstimmung dieser verschiedenen Niveaus und Umgebungen bedarf eines besonderen Modells, wie dem im Beitrag vorgestellten mehrdimensionalen System aus vier miteinander verknüpften Koordinatensystemen, das zugleich die Grundlage für eine wirksame Organisations- und Mitgliederentwicklung durch Maßnahmen bildet, die im Beitrag beispielhaft aufgezeigt werden.

Das Management von Risiken kommt danach eher ohne die sonst propagierten Maßnahmen aus, wie eine Erhöhung der Kontrolldichte, Sanktionsandrohungen oder die Verstärkung einer Anreizstruktur. Denn der Mensch ist nach Nietzsche das „nicht festgestellte
Tier". Gerade weil er sich neben sich selbst stellen kann, ist er in der Lage, die Kräfte, die
ihn treiben und die er selbst besitzt und ausübt, zielgerichtet zu nutzen. Spätestens die
Neuzeit hat uns auf diesen Stand der Reflexion gehoben. Lassen wir uns und unsere Mitmenschen diese individuellen schöpferischen Kräfte jedes einzelnen Menschen nutzen,
die uns und die Dynamik der Neuzeit weiter antreiben. Aber ohne dabei die jeweils gewünschte oder erwünschte Moral aus den Augen zu verlieren.

24.1 Einführung

Früher wurden Arbeitsschritte von einem zentralen Plan, der alle Aktivitäten determinierte, innerhalb einer hierarchischen Struktur für einen langen Zeitraum konstant vorgegeben. Heute schreitet eine Entwicklung zu mehr Eigenverantwortlichkeit und Planungssouveränität in eher „heterarchischen" Strukturen immer stärker voran. Dabei besteht die
Gefahr mit radikal veränderten Strategien oder Strukturen von einem Extrem ins andere
zu fallen.

Der Mensch entwickelt sich eher langsam und stufenweise fort. Er präferiert von Natur
aus schlicht das, was weniger Energie verbraucht. So halten viele Menschen schon aus
diesem Grund an ihrer Organisation fest und hinterfragen deren Grundsätze, Prinzipien
und Moral nur ungern. Deshalb genügen Forderungen allein nach mehr Regeltreue oder
einer einseitig moralischen Entwicklung nicht dem neuen strategischen Anspruch an uns
Menschen.

Dynamik und Vielfalt der Dimensionen einer „moralischen Urteilsfähigkeit" und „ethischen Gesinnung" des Menschen werden dabei nicht ausreichend berücksichtigt. Diese
beiden Schlüsselkompetenzen stehen in einem engen Zusammenhang mit einer Vielzahl
an erwünschten Eigenschaften. Hat man sie verstanden und kann sie gezielt einsetzen,
verfügt man über einen starken Ansatz zu nachhaltigem Erfolg. Dazu ist die Ausrichtung
von Politik und Wirtschaft auch an ethischen Prinzipien und deren Implementierung im
Sinne von „Governance" und Organisationskultur nicht nur nützlich, sondern notwendig.
Strukturen müssen die gewünschten Aktivitäten ermöglichen und diese klar definieren.
Erst danach können Moral (die Dinge richtig tun) und Ethik (die richtigen Dinge tun) sich
unmittelbar auf das Fundament der Organisation auswirken und deren Risiko- und Qualitätsmanagement beeinflussen: Die Wahrscheinlichkeit von unerwünschtem Verhalten
sinkt, während die Wahrscheinlichkeit erwünschten Verhaltens steigt. – Moral, hier verstanden im Sinne eines Regelwerks, Ethik im Sinne ihrer Reflexion und Berücksichtigung
weiterer Regelwerke oder Moralen darüber hinaus.

Der vorliegende Beitrag bietet einen interdisziplinären Ansatz zur Bewertung und Entwicklung jeder Art von Organisation, seien es Unternehmen, Staaten, Clans oder NGOs.
Der Blick richtet sich dabei nicht ausschließlich auf die Bewertung der Einhaltung spezi-

eller Regelwerke, Normen, Gebote, Gesetze etc., wie beispielsweise durch ein „Social Credit Rating" oder ein entsprechendes System zur Stützung einer einzigen Moral.

Vielmehr berücksichtigt der Ansatz auch das Niveau der moralischen Urteilskompetenz und das der ethischen Gesinnung der beteiligten Menschen und Organisationen sowie mit dem Umfeld korrespondierende Neigungen, Ziele und Werte.

Das im Folgenden skizzierte universell über Raum und Zeit gültige Modell soll insbesondere zeigen, wie Veränderungen auf jeweils einem Gebiet die Stabilität und Entwicklung einer Organisation im Ganzen beeinflusst.

24.2 Modellentwicklung

All unsere Wahrnehmungen von der Welt basieren auf unseren Vorstellungen von „Raum" und „Zeit", meint Immanuel Kant. Alle Empfindungen und Erkenntnisse, der reinen Vernunft nach, sind für ihn nur unter den Bedingungen der individuellen räumlichen oder zeitlichen Anschauung eines Menschen als Subjekt möglich (vgl. Kant 1787, S. 113 ff.).

Dabei folgt Kant den Vorstellungen Newtons von absolutem und relativem Raum sowie absoluter und relativer Zeit (vgl. Newton 1686; vgl. Abb. 24.1).

Nach Arthur Schopenhauer kann das Individuum, können wir uns als Mensch darüber hinaus durch Selbstbeobachtung gewiss werden, wie wir uns als Subjekt im „Willen" selbst erfahren (Schopenhauer 1859, S. 33 ff.). So erkennen wir neben dem Willen zum Leben auch einen tendenziellen Willen zur Macht oder zur Veränderung.

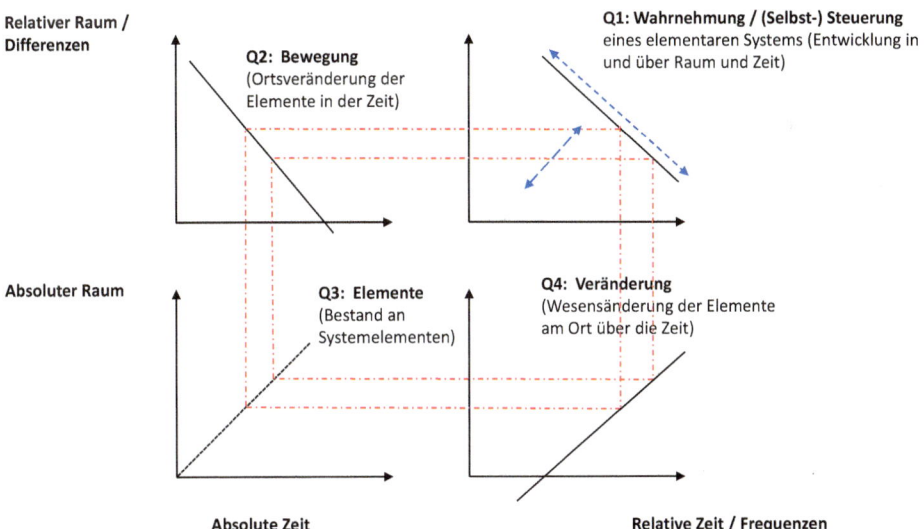

Abb. 24.1 Vier-Quadranten-Modell eines geschlossenen (Sub-)Systems. (Quelle: eigene Darstellung)

Eine Entscheidung zu fällen bedeutet danach für den Menschen als Subjekt, bewusst und willentlich zwischen Alternativen im jeweiligen Kontext zu wählen und die Wahl in Form von Aktivitäten umzusetzen oder wirklich werden zu lassen (vgl. Abb. 24.2).

Eine Aktivität sei hier definiert als „eine von menschlicher Entscheidung getragene Handlung, sei es Tun oder Unterlassen".

Das hier verwendete Modell befasst sich vornehmlich mit den „Aktivitäten" innerhalb einer Organisation und nicht mit dem Menschen als deren Träger. Diese Unterscheidung zwischen den Menschen oder auch Organisationen und deren Aktivitäten ist wichtig.

Denn eine Organisation entsteht und besteht eben nicht per se aus Menschen. Vielmehr wird diese erst über deren Aktivitäten zu einem lebendigen Organismus. Vergleichbar dem Menschen, der ebenso wenig aus der reinen Ansammlung von Zellen zum Menschen wird. Auch er entsteht aus den Aktivitäten seiner Zellen, die interagieren, sich erneuern, während die Aktivitäten als solche fortbestehen und -bestehen müssen, wenn er leben will.

In der Ausrichtung einer Organisation kann so beispielsweise der Wille oder die Neigung (vgl. Abb. 24.2, Q1) vorherrschen, ein Ziel eher im Rahmen von Regularien, Prinzipien oder einer bestimmten Moral zu verfolgen, sprich: Dinge richtig zu tun bzw. tun zu lassen (vgl. Abb. 24.2, Q2). Alternativ kann der Wille oder die Neigung dahin tendieren, die eigenen Ziele mit einem größeren Maß an Offenheit gegenüber anderen Organisationen, der Umwelt oder unter Berücksichtigung verschiedener Moralen ein Ziel zu erreichen. Sprich: die richtigen Dinge zu tun (vgl. Abb. 24.2, Q4).

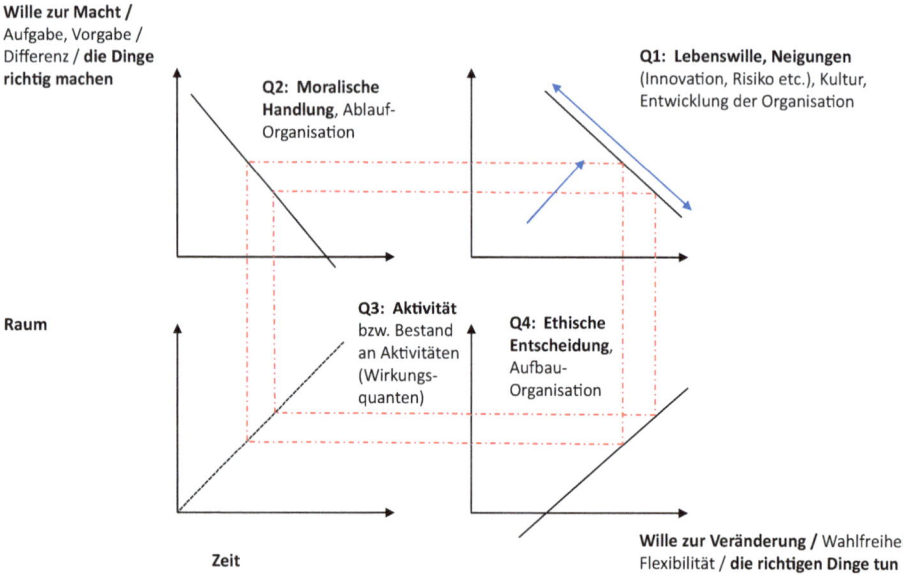

Abb. 24.2 Vier-Quadranten-Modell einer Organisation bzw. eines organisierten (Sub-)Systems. (Quelle: eigene Darstellung)

Auf der Basis solcher Aktivitäten von Menschen, als deren körperlicher oder geistiger Träger, soll im Folgenden die Frage nach einer Bewertung und Entwicklung von Menschen als Mitglieder einer Organisation diskutiert werden.

24.3 Menschliche Aktivitäten als Elemente einer Organisation

Die kleinsten „Wirkelemente" einer Organisation bilden die Aktivitäten ihrer Mitglieder (vgl. Abb. 24.2, Q3). Diesen Aktivitätsträgern werden ihre Aufgaben innerhalb eines bestimmten Rahmens oder einer gewissen Moral zur Erledigung übertragen (vgl. Abb. 24.2, Q2, Ablauf- oder Prozessorganisation). Auf der Basis der Aktivitäten kann man nun die Erledigung der Aufgaben konsistent verknüpfen mit jeglichem Anpassungs- oder Änderungsbedarf im Verlauf ihrer Erledigung und darüber hinaus. Erwünschte oder unerwünschte Anpassungen oder Veränderungen können je nach Grad der Offenheit gegenüber Nachbarn, Kunden oder sonst anderen Mitgliedern oder Organisationen und deren Rahmen und Moralen oder der Wahl alternativer Aktivitäten heraus folgen (vgl. Abb. 24.2, Q4, Aufbau-/Portfolioorganisation). Wichtig für eine jede Bewertung in diesem moralischen (Bezugs-)Raum (Q4) ist unabdingbar der Kontext, in dem eine zu bewertende Aktivität oder ein Prozess aus Aktivitäten (Q2) steht bzw. in den wir diese zur Bewertung stellen.

Wie im Folgenden noch gezeigt werden soll, lässt sich auf diese Weise jedenfalls jedwede Aktivität (Q3) eines zu bewertenden Mitglieds einer Organisation im jeweiligen Kontext der Organisation und deren Umgebung konsistent bewerten. Dies gilt auch für deren Ziele und Werte (Q1) im Kontext ihrer Umgebung und so weiter. Für eine Beurteilung und Bewertung einer Aktivität müssen in jedem Fall moralisches Handeln (Q2) und ethisches Entscheiden (Q4) transparent und nachvollziehbar miteinander verknüpft werden.

Eine solche Verknüpfung im Hinblick auf Regeln oder Moral und Vielfalt aller im Kontext zu beachtenden Dimensionen (Ethik) kann konsistent nur über Raum und Zeit erfolgen. Unter ethischem Entscheiden bzw. Ethik sei dabei das Reflektieren über Moralen verstanden, über Werte, Normen oder Haltungen. Jede Ethik soll sich dabei im Handeln bewähren, und alles Handeln ist von ethischer Relevanz. Die menschliche Handlung umfasst dazu drei Komponenten: die Person mit ihren Haltungen, ihrer ethischen Gesinnung (Q4), das Handeln mit seinen moralischen Normen (Q2) und das Ziel, das durch Werte bestimmt wird (Q1). Moralisches Handeln liegt dann vor, wenn jemand aus bewusster und (selbst-)verpflichteter Grundhaltung heraus, durch beständiges und normengeleitetes Verhalten wertschöpfende und kongruente Ziele anstrebt und verwirklicht. (vgl. Berkel und Herzog 1997, Band 27)

Vielleicht hilft als Erklärung für die Forderung einer konsistenten Betrachtungsweise die Formulierung des Kant-Verehrers Richard Wagner, der seinen Parsifal sagen lässt: „Ich schreite kaum, doch wähn' ich mich schon weit." Und Gurnemanz antwortet: „Du siehst, mein Sohn, zum Raum wird hier die Zeit." (Wagner 2016, Ende der Szene 2 von Aufzug 1; vgl. Abb. 24.2, Q4) Zeit und Raum, Raum und Zeit, bilden also nicht nur nach Kant zwei miteinander verwobene Seins-Ebenen, die auf einander einwirken. Ihre Dimensionen bilden zugleich Basis unseres Verstehens und Grundlage unserer Existenz. Zur Diskussion verwenden wir eine in der Wissenschaft und volkswirtschaftlichen Praxis als „Hicks'sches Modell" bekannte Methode mit vier Quadranten in vier miteinander verknüpften Koordinatensystemen. (siehe dazu näher Assenmacher 1998, S. 89; Woll 2008, S. 343; vgl. Abb. 24.1, 24.2 und 24.3)

Zu diesen Dimensionen von Raum und Zeit tritt im Leben der Wille des Subjekts hinzu (dazu ausführlich Fornefett 2007, S. 243 ff.).

Der Wille tritt quasi neben das Leben selbst, das in seiner jeweiligen Gesamtheit, wie im Falle der Menschheit, in ständiger Erneuerung einen tief greifenden Wandel erfährt. Zeit entsteht überall durch Bewegung, die wir über Veränderungen wahrnehmen oder wenn wir daran beteiligt sind. Zum Beispiel, wenn etwas Neues, Ungewohntes passiert.

Der Mensch ist intelligent, er kann täuschen und lügen, kopieren und instrumentalisieren, und er kommuniziert bereits als Baby niemals ohne eine Absicht. Dabei ist ihm seine Absicht nicht immer bewusst. Als Träger seiner Aktivitäten, sei es in Form von bewusstem oder unbewusstem Entscheiden und Handeln, ist der Mensch Sender und direkter oder indirekter Empfänger auch seiner eigenen Signale (dazu ausführlich Fornefett 2012, S. 13 ff.).

Will man also Einfluss nehmen auf sein Verhalten und dessen erwünschte soziale Entwicklung über Bewusstseins-, Einstellungs- oder Haltungskontrolle erreichen, muss man ihn als Subjekt betrachten und behandeln.

Doch: „Steigen will das Leben und steigend sich überwinden", lässt Friedrich Nietzsche seinen Zarathustra vortragen (vgl. Nietzsche 1883/1999, S. 130). Welchen Wert können wir einer Sache, einem Recht, einer Person gar oder einer Aktivität überhaupt nachhaltig ‚zurecht' beimessen?

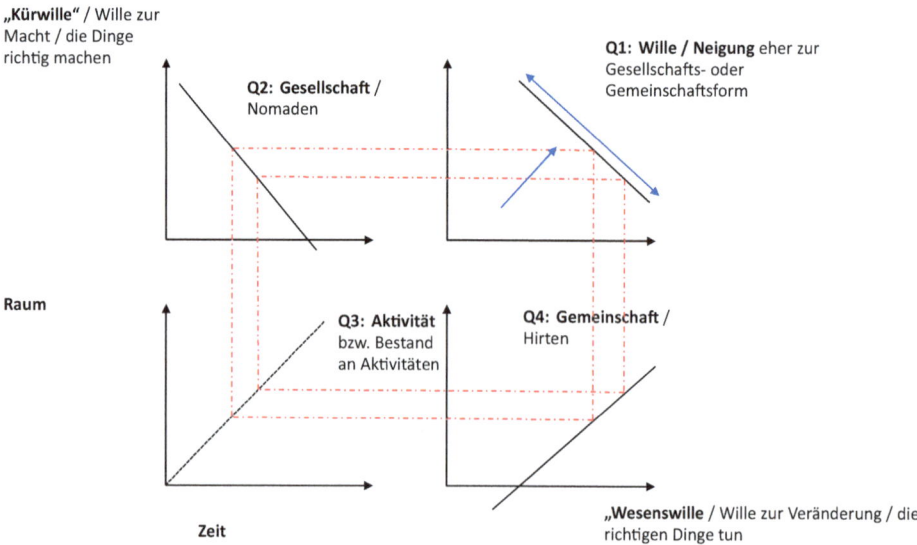

Abb. 24.3 Vier-Quadranten-Modell einer Organisation aus Gesellschaft und Gemeinschaft. (Quelle: eigene Darstellung)

Der Lebenswille des Menschen, der sich im Willen zur Macht und im Willen zur Freiheit bzw. zur Veränderung ausdrückt, wird in Q1 der Abb. 24.2 geometrisch dargestellt. Die Kurve der Moral im zweiten Quadranten und die Ethik-Kurve im vierten Quadranten werden dafür hinzugezogen. Der dritte Quadrant zeigt für jede einzelne Aktivität im Bestand eines betrachteten organisierten Systems (oder Subsystems) die Bedingung für ein Gleichgewicht von moralischem Verhalten und ethischer Gesinnung. Diese ist auf der Winkelhalbierenden, mathematisch betrachtet, immer erfüllt. In der Praxis ist sie erfüllt, wenn beispielsweise eine nicht beabsichtigte Aktivität sich quasi selbst genügt und nicht instrumentalisiert wurde.

Im hier zur Darstellung verwendeten 4-Quadranten-Modell sind die x-Achsen von unten nach oben und die y-Achsen von links nach rechts stets gleich zu lesen. Die verwendeten Kurven oder Funktionslinien sollen an dieser Stelle aber lediglich eine grobe Vorstellung der grundlegenden Verhältnisse ermöglichen. Die rot gestrichelten Beispiele lassen so erkennen, wie im geschlossenen System ein Mehr an Moral stets ein weniger an Ethik und umgekehrt erlauben. Wir sehen aber geradeso, dass Moral und Ethik einander bedingen.

Ein Zuviel des einen oder anderen führt zwangsläufig zu einer Erstarrung, einem Festfahren in sich selbst (Q2) bzw. zu einer Auflösung der Organisation in Beliebigkeit (Q4). Eine willens- oder absichtsfreie Aktivität, die sich selbst genügt, also nicht in der einen oder anderen Weise instrumentalisiert wird, würde hinsichtlich ihrer Darstellung im Modell auf ihre Raum- gleich Zeitdimension in Q3 beschränkt bleiben.

Neben der Berücksichtigung eines freien Willens (Q1) kann nun mit Hilfe einer Aktivitäten-basierten Betrachtungsweise zudem die im Zeitalter von Automatisierung und Digitalisierung wachsende Konsistenzlücke geschlossen werden: zwischen Management als einem eher linearen Ansatz und Führung als einem eher zyklischen Ansatz.

Das Modell in Abb. 24.1 macht deutlich: Die Dimension „die Dinge richtig tun" (Q2) steht über die jeweilige Aktivität (Q1) in Verbindung mit der Dimension „die richtigen Dinge tun" (Q4). Wenn wir also unseren Neigungen weder einseitig dem Willen zur Macht noch dem zur Freiheit bzw. der Veränderung im Sinne eines „Entweder-oder" nachgeben, sondern unserem Lebenswillen mehr Nahrung und Raum gewähren, erst dann kann beides im Sinne eines „Sowohl-als-auch" wachsen. Das bedeutet zugleich: Trägheit (gegen den 0-Punkt y-Achse Q1, Q2) und Angst (gegen den 0-Punkt x-Achse Q1, Q4) sind zu überwinden.

Die erwünschten Aktivitäten eines Mitglieds der jeweiligen Organisation (vgl. Abb. 24.2, Q2) können im Kontext als auch unabhängig von anderen Entscheidungen und Handlungen derselben Person oder anderer Mitglieder derselben Organisation geschehen (vgl. Abb. 24.2, Q4). Gegebenenfalls auch im Sinne oder Kontext von Entscheidungen und Handlungen anderer Entscheidungs- bzw. Aktivitätsträger (Stakeholder oder Dritte), die ihrerseits individuell oder gemeinsam in Form von Aktivitäten umgesetzt werden.

Entgegen der gefühlsmäßig eher positiven Erwartung beispielsweise einer Verbesserung der Effizienz durch eine Automatisierung oder strenge Regulierung (Moral), verringert eine auf Effizienz bzw. Moral fokussierende Kultur nachhaltig den erwarteten Erfolg.

Den Grund dafür bilden die gleichzeitig steigenden Risiken durch eine rigidere, direkte Steuerung oder einen unflexiblen Rahmen bzw. eine strikte Moral. Siehe dazu Q2 der Abb. 24.2.

Für eine operationelle Dimension gilt nämlich, dass Risiken aus Erfolgs- oder eben auch aus Verlusterwartungen erwachsen: „ohne Rendite-Erwartung kein Risiko". Wenn wir hier den Begriff „Risiko" verwenden, so meinen wir damit stets das Risiko im weiteren Sinne von „Chance" oder „Gefahr" und nicht im engeren Sinne, ausschließlich einer Gefahr. (Erben et al. 2010, S. 22)

Auch das umgekehrte Phänomen, dass ich mehr Macht über einen Menschen bzw. seine Aktivität gewinne, je weniger ich diese aktiv steuere, erscheint intuitiv zunächst einmal unglaubwürdig. Was ein wesentlicher Grund mit dafür war, hier das aus der Volkswirtschaft stammende 4-Quadranten Modell zu verwenden. Dieses Modell ermöglicht die Integration psychologischer Aspekte und erlaubt so unserem Verstand, ein scheinbares Paradoxon mit Hilfe der Vernunft leichter aufzulösen.

Wenn nun Hierarchie und Machtwille stärker werden, kann sich dies in Form klarer Regeln positiv auf die Moral auswirken. Demgegenüber gibt es bei einer flexibleren Organisation zwar häufig prozessuale Reibungspunkte und Ineffizienzen, aber der Veränderungswille wirkt innovativ. Wofür in letzterem Fall strategisch Risiken in Kauf genommen werden. Dabei gilt: „ein Erfolg in der Zukunft folgt aus dem Risiko bzw. dessen Erwartung" (Q4). Ohne jede Veränderung oder Anpassung einer Aktivität wäre kein mehr an Erfolg zu erzielen. Es bliebe quasi immer alles beim Alten. Zumindest solange keine Störung oder ein sonstiger einseitiger Impuls von außen auf die Organisation einwirkt. (Erben et al. 2020, S. 114 f.)

Genauso lassen sich Gefühle berücksichtigen: Empfinde ich inneren Widerstand gegen eine gewählte Aktivität (ich möchte die richtigen Dinge tun), erlaubt mir meine damit wachsende Trägheit nicht, die Dinge richtig zu machen. Unabhängig von der wachsenden Spannung aus der Differenz von Trägheit und Vorgaben sinkt die Leistung (der Arbeitsstrom pro Zeit). Was bedeutet, dass die Aufgabe entweder weitgehend unerledigt bleibt oder deren Erledigung länger dauert.

Wie beim Ohm'schen Gesetz der elektromagnetische Strom, so verringert sich hier wegen des wachsenden Widerstands bei steigender Spannung der ‚Fluss'.

Die von Georg Simon Ohm gefundene Gesetzmäßigkeit ist in diesem Zusammenhang ebenfalls interessant: Je größer die elektrische Spannung U bzw. je kleiner der elektrische Widerstand R ist, umso größer ist die Stromstärke I ($I = U/R$). Wird danach an ein Objekt eine veränderliche elektrische Spannung angelegt, so verändert sich der hindurchfließende elektrische Strom in seiner Stärke proportional zur Spannung. Mit anderen Worten: Der als Quotient aus Spannung und Stromstärke definierte elektrische Widerstand ist unabhängig von der Spannung und der Stromstärke (vgl. Wikipedia 2020, dort unter „elektrischer Widerstand" m. w. Nw.).

Ohm suchte damals nach einem mathematischen Zusammenhang – einer Formel – zur Berechnung der „Wirkung fließender Elektrizität", heute genannt: Stromstärke (vergleichbar der Bewältigung von Aufgaben unter der Einhaltung von Vorgaben in Q2). Dies in der

Abhängigkeit von Material und den Dimensionen eines Drahtes (vergleichbar unserer Forderung nach Berücksichtigung einer ethischen Gesinnung in Q4).

Übertragen auf die Bewertung von Aktivitäten bedeutet dies, dass ein Rating nur des moralischen bzw. moralkonformen Verhaltens eines Mitglieds einer Organisation allein schon dem Anspruch an eine objektive Messung von inneren Widerständen gegen einen Moralindex nicht genügen kann.

Dies selbstverständlich unter Berücksichtigung menschlicher Attribute. Das heißt, solcher von Subjekten gegenüber materiellen Attributen von in der Physik untersuchten Objekten.

Gilt dies für alle Organisationen in Form von Gesellschaften oder Gemeinschaften gleichermaßen? Oder lassen sich vielleicht weitere Besonderheiten finden, die bei einem Rating des sozialen Verhaltens ihrer Mitglieder zu berücksichtigen wären?

24.4 Offene versus geschlossene organisierte Systeme bzw. Organisationen

Jede Organisation, sei es im Rahmen einer Gesellschaft oder einer Gemeinschaft, besitzt Regeln: Ge- oder Verbote, Rechte oder Pflichten, explizit oder implizit, niedergeschrieben oder gelebt. Sie richten sich grundsätzlich an die eigenen Mitglieder, an Dritte oder die Organisation selbst, als Ganzes. Um die Legitimität oder die Gültigkeit solcher Regeln („Normen") oder deren Ziele oder Werte zu begründen, werden zugleich Leitlinien der Moral dieser Gesellschaft oder Gemeinschaft kodifiziert. Denn eine Organisation besteht im Allgemeinen aus vielen Subjekten bzw. Aktivitätsträgern.

Diese Mitglieder der Organisation sind direkt oder indirekt, aktiv oder passiv, freiwillig oder unfreiwillig, etabliert oder beziehungslos Mitgestalter dieser sozialen Ordnung. Sie füllen die jeweilige Kultur mit ihrem Leben. Nach Albert Schweitzer ist jedes Mitglied bzw. Subjekt: „Leben, das leben will, inmitten von Leben, das leben will", weshalb eine möglichst breite Zustimmung der Mitglieder zu Normen, Werten und Moral ihrer jeweiligen Organisation nicht nur erwünscht, sondern (zumindest nach außen hin) auch unbedingt notwendig erscheint.

Echte Zustimmung setzt allerdings eine Interpretationsfähigkeit der Werte, des Weiteren eine zumindest rudimentär reflektierte Akzeptanz und drittens eine kontextgerechte Durch- und Umsetzung bzw. Anwendung der Normen durch die Mitglieder voraus. Auf diese Weise können beispielsweise Steuerungsdefizite ausgefüllt werden, die regelmäßig durch abnehmenden Grenznutzen von Einkommen oder Vermögen entstehen. Bzw. durch ein wachsendes Risiko hinsichtlich eines erwarteten Erfolgs aus einer direktiven Aufgabe und deren rigiden Kontrolle in organisierten Systemen, wie Gesellschaften oder Gemeinschaften.

Nach Ferdinand Tönnies (vgl. Tönnies 2019, Bd 2, S. 13 ff.) existieren theoretisch nur zwei kollektive Grundformen willentlicher Bejahung der jeweiligen anderen Menschen durch den Einzelnen: „Gesellschaft" und „Gemeinschaft". Der Einzelne oder die Organisation als Ganzes orientieren sich am höheren Zweck des jeweils größeren Kollektivs („Wesenswille", Neigung zur Gemeinschaft). Oder der Einzelne oder eine Organisation bedienen sich des Kollektivs auf instrumentelle Weise („Kürwille", Neigung zu einer Gesellschaft). Sprich: die Gesellschaft ist ihm zumindest auch Mittel zur Erreichung seines eigenen individuellen Ziels, als Einzelner oder als Organisation (vgl. Abb. 24.3).

Hier fassen die Autoren diese beiden Formen überwiegend in einem Begriff „Organisation" zusammen (vgl. Abb. 24.2). Dies deshalb, weil sich auch nach Tönnies die theoretisch reinen Begriffsbildungen von der praktischen, gelebten sozialen Welt strikt unterscheiden. Dort trifft man sie heute ausschließlich gemischt an.

Die soziale Ordnung einer Organisation soll nun für sich selbst handlungsorientierte und regulierende Bewertungsmaßstäbe bereitstellen, wie ein „Social Credit Rating" von Menschen. Hier auf der Basis von Aktivitäten, deren Träger sie sind.

Zur Entwicklung und Erhaltung gesellschaftlicher oder gemeinschaftlicher Moral (Q2) sind dazu auf der Mitgliederebene nach herrschender Meinung (vgl. Schacher und Fornefett 2010, S. 10; vgl. auch Lind 2000 und 2003) drei Bedingungen erforderlich:

1. Das Wissen über Normen und Werte muss erworben, verstanden und situativ interpretiert werden (Voraussetzung für Bewusstsein).
2. Der normative Geltungsbereich muss anerkannt werden (Voraussetzung für eine gewünschte Einstellung).
3. Die normativen Vorgaben müssen dauerhaft „korrekt" befolgt werden (Voraussetzung für eine entsprechende innere Haltung bildet das Training bzw. stete Wiederholung).

Diese drei durchaus anspruchsvollen Voraussetzungen für die Erlangung von Bewusstsein, Einstellung und schließlich Haltung, müssen nicht nur mit Inhalten, wie Normen und Werte, versehen werden. Vielmehr müssen sie von den jeweiligen Urteils- und Kompetenzniveaus der Mitglieder her ‚moderiert' werden.

Das heißt, Aktivitäten einer Organisation müssen bevor sie implementiert werden, stets erst einmal adäquat entworfen und geplant werden, beispielsweise in Form von Szenarien oder spielerisch (dazu näher Heinrich et al. 2018).

Denn jene Mitglieder in einem bis dahin überwiegend „objektiven Modus" (Q2) mit dezidiertem Regelwerk müssen in einen größeren soziokulturellen Kontext bzw. „identitätsstiftenden Modus" (Q4) erst einmal mit einbezogen werden. Allen Mitgliedern einer Organisation ist dazu inzidenter geboten, im jeweiligen Kontext den jeweils zur Verfügung stehenden Fähigkeiten von Mitgliedern empathisch zu begegnen, diese eher großzügig zu bewerten oder selbst bescheiden zu sein. Hierzu gehört zum Beispiel der Umgang mit Kritikern oder Widersachern der eigenen Moral.

Die Forderung nach einer kontextuellen Differenzierung resultiert immerhin daraus, dass mit hoher Wahrscheinlichkeit mancher Problematik nur im wissenschaftlichen Zusammenhang oder einem spezifisch fachlichen Umfeld angemessen Rechnung getragen werden kann; am Ende vielleicht lediglich noch in einem rein philosophischen Kontext. Andere Aufgaben fallen möglicherweise eher den jeweils Herrschenden oder Machtinhabern selbst zu, die diese delegierten. Oder sie fallen in Organisationen bzw. deren Einrichtungen an, denen es ausschließlich um die Sicherheit und Erhaltung ihres organisierten Systems geht (dazu näher Kritzner und Fornefett 2017, S. 36 ff.).

Die Art Außenseitern in einer Organisation ganz gleich in welchem Kontext zu begegnen, erfordert Rücksicht auf die Neigung oder Motivation des Aktivitätenträgers hinsichtlich der zwei bereits genannten Modi: einen „identitätsstiftenden" Modus (vgl. Q4) und einen „objektiven" Modus (vgl. Q2).

Ein identitätsstiftender Modus würde die Perspektive des Entscheiders durch Kontraste anreichern. Macht die Organisation dagegen den Anspruch geltend, eine weithin objektive Gesamtanalyse vor der Entscheidung über anstehende Handlungen vorzunehmen, so läuft sie Gefahr, die Entscheidung gefällt zu erhalten, ohne dass der eigene Kontext der Organisation in Beziehung zur Perspektive des Entscheiders gesetzt würde. Andersherum würde man dem Entscheider bei der Interpretation seiner Aufgabe Tür und Tor für eine Entscheidung im identitätsstiftenden Modus erlauben.

Die erste Frage nach einer möglichen Auflösung dieses Dilemmas führt stets zum Hinweis auf die geschlossene Organisation bzw. Gesellschaft oder eine stringent formulierte Aufgabe nebst aller notwendigen Vorgaben in Form eines Handbuchs oder eine feste Einbindung in einen Workflow oder auf ein ähnliches Instrument zur Schaffung von Prozesssicherheit.

Die Abb. 24.2 lässt die danach stets verbleibenden Unsicherheiten auch in geschlossenen Systemen mit freiem Willen der Systemleitung erahnen. Organisierte Systeme sind niemals deterministischer Natur (vgl. hierzu Abb. 24.4).

System-merkmale	Deterministische Systeme, Flows	Unorganisierte Systeme	Organisierte Systeme
Elemente / Variablen	wenige gerichtete	sehr viele unabhängige gleicher Art	mittlere Zahl, interdependent
Disziplin	Klassische Wissenschaften	Statistik	interdisziplinäre Wissenschaften zu komplexen Systemen
Prognose	grundsätzlich exakte Voraussagen möglich	statistische Wahrscheinlichkeit	Szenarien oder Muster-Voraussagen
Intervention	punktuell	stochastisch	kontextuell

Abb. 24.4 Vergleich deterministischer, unorganisierter und organisierter Systeme. (Quelle: eigene Darstellung)

Gemeinhin lässt sich damit an dieser Stelle bereits sagen, dass uns die fehlende Legitimität einer zusammenhanglosen Kontrastierung also spätestens, aber auch schon dann auffallen kann und wird, wenn sie die unbehagliche Relativierung bereits in der eigenen Zeit zum Gegenstand hat und mithin dem Selbstbild der Organisation normativ zeitlich (Q2) und nicht erst räumlich (Q4) zum Schaden gereicht.

Nur weil wir scheinbar stichhaltige Argumente im objektiven Modus für oder gegen eine Aktivität formulieren können, diese entsprechend als richtig oder falsch oder vielleicht auch als gut oder böse zu klassifizieren gewohnt sind, können hieraus noch keine Schlüsse gezogen werden, die mehr als das eigene geistige Umfeld im Modus der Identitätsstiftung anbetreffen.

Insofern ist ein standortloser Blick auf im Laufe der Zeit sich verändernder, organisierter Systeme illusionär. Eine Wertung bedarf stets eines ihr jeweils eigenen Bedeutungsgefüges.

24.5 Subjektorientierung und die Bewertung von Menschen

Mitglieder sind als „Subjekte" in dieser Sichtweise einerseits von einer Gesellschaft oder Gemeinschaft betroffen und stellen diese andererseits zugleich selbst her. (vgl. Tönnies 2019, S. 13 ff.; vgl. Bolte 1983, S. 15)

Derartige Denk- und Forschungsweisen sind Teil einer gelegentlich als „subjektorientierte Wende" bezeichneten partiellen Schwerpunktverlagerung in der Soziologie, die versucht, sich von betont „objektivistischen" oder „strukturellen" Konzepten wie Systemtheorie nach Luhmann, dem Marxismus oder Strukturalismus bzw. Strukturfunktionalismus abzusetzen. Gleichzeitig grenzt sie sich von Konzepten ab, die zwar ebenfalls individuell Handelnde ins Zentrum des Interesses stellen und in ähnlicher Weise eine Lösung des Dilemmas aus strukturellen und handlungsbezogenen Analysen anstreben, dabei aber von primär rational gesteuerten Handelnden ausgehen (Theorie der rationalen Entscheidung, „homo oeconomicus" etc.). Vergleichbar mit anderen Theorierichtungen wie die Kritische Theorie, Kritische Psychologie, Cultural Studies, Historische Anthropologie, Europäische Ethnologie oder Empirische Kulturwissenschaft (Volkskunde), geht die subjektorientierte theoretische Orientierung mit der Untersuchung konkreter alltäglicher Kontexte einher. Dies bedeutet am Ende eine Präferenz für qualitative Verfahren der empirischen Sozialforschung: nicht-standardisierte offene Interviews, biografische Interviews, Gruppendiskussionen oder Fokusgruppen, teilnehmende oder nicht teilnehmende Beobachtung, Fallstudien etc. (vgl. Langfeldt 2009).

Die subjektorientierte Soziologie fußt dabei auf der traditionsreichen, an den Idealen des Konstruktivismus und des Humanismus orientierten Kategorie des „Subjekts". Das heißt, auf dem Bedürfnis und der Fähigkeit von Menschen, Subjekt sein zu wollen und zu können. Wegen vielfältiger struktureller Zwänge bestehen jedoch nur begrenzte Möglichkeiten. „Subjektivität" kann sich auch als mehr oder weniger reflexiv gesteuerte oder als ‚hilflose' Widerständigkeit oder als nur wenig erfolgreiche Aneignung von Verhältnissen äußern. Ein plastisches Beispiel für eine solche Paradoxie bietet die Geschichte der „Brücke am Kwai". Ein Spielfilm von David Lean aus dem Jahr 1957, der auf dem gleichnamigen Roman von Pierre Boulle basiert.

Beispiel

Der Film handelt von einer Gruppe britischer Kriegsgefangener in einem japanischen Lager in Burma. Die Gefangenen sollen eine Eisenbahnbrücke errichten. Damit die Brücke termingerecht fertiggestellt wird, teilt der japanische Lagerkommandant auch die britischen Offiziere zur Arbeit ein. Der Kommandeur des gefangen genommenen Bataillons widersetzt sich diesem Befehl. Er beruft sich dabei auf die Genfer Konvention, die Japan allerdings nicht unterschrieben hat. Der Lagerkommandant reagiert mit drastischen Strafen, lenkt aber schließlich ein – die Offiziere werden von der körperlichen Arbeit befreit und erhalten Führungstätigkeiten. Der Kommandeur bemüht sich nun nicht mehr nur darum, dass seine Soldaten ihren Stolz und ihre Würde behalten und sich nicht wie einfache Sklavenarbeiter von den japanischen Bewachern erniedrigen lassen. Er will die Überlegenheit der britischen Soldaten beweisen, indem er eine technisch aufwendigere Brücke in kürzerer Zeit errichtet, obwohl er sich bewusst sein muss, damit dem Feind zu helfen. Die Aufgabe treibt die Soldaten zu Höchstleistungen, und die Brücke wird rechtzeitig fertiggestellt. Der Lagerkommandant muss nach der Fertigstellung die Überlegenheit der Gefangenen eingestehen. (näheres vgl. Krusche und Labenski 1987, S. 102 f.) ◄

Wichtig für den vorliegenden Beitrag ist es daher zu erkennen:

Im Falle einer Subjekt-orientierten Herangehensweise an Fragen einer Bewertung und Entwicklung von Menschen als Mitglieder einer Gesellschaft oder Gemeinschaft werden alle drei genannten Voraussetzungen (Bewusstsein, Einstellung, Haltung) durch das Niveau der „moralisch-ethischen Kompetenz" des Mitglieds selbst als auch das seiner jeweiligen Gruppe oder eben Organisation: Familie, Team, Gesellschaft, Gemeinschaft etc., wesentlich beeinflusst (dazu näher Kritzner und Fornefett 2014, S. 1–5).

Die Gesamtheit der vom Mitglied als gültig anerkannten und von ihm eingehaltenen Normen begründet dann die individuelle Moral und Ethik des Menschen. (vgl. Schacher und Fornefett 2010, S. 10; vgl. auch Lind 2000 und 2003) Für sie lassen sich vier Indikatoren zur Beurteilung finden:

1. Normen-/Wertewissen (auch über die eigene Moral hinaus).
2. Normen-/Wertebeurteilung (Gewissen oder äußerer Zwang).
3. Moralmotiviertes Verhalten (zur Beruhigung des eigenen oder fremden Gewissens).
4. Ethische Gefühle (zum Beispiel Alternativ-Schuld oder Zorn gegen sich selbst oder andere).

Anhand dieser vier Indikatoren ließe sich theoretisch ein bestimmtes Verhalten des Mitglieds als moralisch im Sinne seiner ethischen Relevanz beurteilen.

In der Praxis sind die Menschen jedoch sehr unterschiedlich. Ob beispielsweise eine Führungskraft mit jeweiligen Persönlichkeitseigenschaften erfolgreich ist, hängt wesentlich auch von der zu führenden Personengruppe oder dem historischen Kontext ab. So benötigte Alexander der Große ein anderes Persönlichkeits- und Führungsprofil für seine militärischen Erfolge in der Antike, als Gandhi für seine politischen Erfolge in der Moderne, von geografischen, kulturellen und anderen Bedingungen ganz absehen.

Die Persönlichkeitspsychologie zeigt, dass viele Persönlichkeitseigenschaften wesentlich erblich mitbestimmt sind. Bei vielen Persönlichkeitseigenschaften findet sich jedoch eine Umweltvarianz von bis zu 50 Prozent. Das bedeutet, dass die Hälfte der Merkmalsvariationen auf Umwelteinflüsse und nicht auf Erblichkeit zurückzuführen ist (vgl. Schacher und Fornefett 2010 m. w. Nw.).

Zu den wichtigsten Rahmenbedingungen gehören nach neueren Erkenntnissen aber auch das moralisch-ethische Niveau der Gruppe eines jeden Mitglieds oder Mitarbeiters wie jeder Führungskraft, ihres Teams, ihres Verantwortungsbereichs bzw. ihres Unternehmens. Um diese nicht nur bewerten, sondern auch fortentwickeln zu können, bedarf es nicht nur des Blicks auf den Einzelnen und auf seine Gruppe. Es bedarf auch des Blicks auf das größere Ganze. Nämlich auf deren Umfeld wiederum, das sich anders als sie, rascher und stärker verändert. Auf das die Mitarbeiter wirken und das auf sie zurückwirkt.

Das heute präferierte Ziel ist möglicherweise schon deshalb häufig nur die Akzeptanz vorgegebener Normen durch das Mitglied als ihn verpflichtende Normen. Dritte sollen sich schlicht an diese halten. Da bleibt wenig Raum für die ethische Entwicklung einer eigenen, einer offenen bzw. ständig fortzuentwickelnden Moral von innen heraus.

24.6 Was ist richtig oder falsch, gut oder böse?

Zur Erreichung des präferierten (äußeren) Ziels werden schließlich ganz allgemein drei Vermittlungsstrategien angewendet:

1. Argumentative Normenvermittlung (zum Schaffen eines gewünschten Bewusstseins).
2. Positive oder negative Verhaltensmodelle (zur Entwicklung der gewünschten Einstellung).
3. Belohnung oder Bestrafung (zum An-Trainieren der gewünschten Haltung).

Häufig wird in **Gesellschaften** der dritte Weg gewählt, der zwar kaum einen Beitrag zur Internalisierung liefert, aber kurzfristig normkonformes Verhalten produzieren kann. Die Mitglieder bleiben nachhaltig von extrinsischen Faktoren wie Kontrolle, Entzug von Belohnung, materiellen Anreizen oder Strafe abhängig. Verliert einer oder verlieren mehrere dieser Faktoren ihre verhaltenssteuernde Einflussmacht, kann normkonformes Verhalten nicht mehr sichergestellt werden. Abgesehen von der einhergehenden latenten Gefahr der

Dysfunktionalität verursachen Sanktionsandrohungen oder Belohnungsversprechen daher fortlaufend erhebliche Kontrollaufwendungen (vgl. Schacher und Fornefett 2010 m. w. Nw.).

Der erste und zweite Weg, vor allem aber der erste, wären demgegenüber in Kombination mit induktivem Kommunikationsverhalten, einer wertschätzenden, angstfreien Atmosphäre und klaren Verantwortlichkeiten geeignet, Normen und Werte zu einem Teil des moralischen Selbst des Mitglieds oder Mitarbeiters werden zu lassen (Entwicklung ethischer Fähigkeiten).

Das bedarf Zeit, die sich heute am ehesten noch **Gemeinschaften** nehmen. Sie erlauben sich selbst und ihren Mitgliedern Eigeninitiative und Mitbestimmung in anderer Form als Gesellschaften, deren Mitglieder gemeinsam ein in der Zukunft liegendes Ziel (Werte wie Wachstum bzw. Wohlstand) verfolgen. Letztere streben vornehmlich nach einem ‚Mehr‘ an materiellen oder immateriellen Dingen, während sich Gemeinschaften eher als Sachwalter ihrer materiellen oder immateriellen Dinge, Gegenstände oder Objekte verstehen. Sie bieten so den passenderen Rahmen für die Verwirklichung eines „Wesenswillens" ihrer Mitglieder.

Trotz aller demokratischer Feigenblätter wird jedenfalls weltweit noch immer vornehmlich die Be- und Verfolgung von ‚da oben‘ gewollter und bestimmter Normen bzw. Werte durch die ‚da unten‘ Mitglieder einer Organisation versucht. Nachhaltiger wäre dagegen aber auch für diktatorisch geführte Organisationen, die Internalisierung der Normen und Werte auf einem möglichst hohen ethischen Niveau des Menschen, ganz gleich ob als Mitglied oder als Mitarbeiter. Denn höhere moralische Urteilskompetenz korrespondiert mit höheren Stufen der kognitiven Entwicklung (vgl. hierzu im Hinblick auf die Unternehmensökonomie Oesterdiekhoff und Rindermann 2008 sowie Oesterdiekhoff 2006a, b m. w. Nw.).

Jean Piaget betrachtet bereits den Menschen als ein offenes System. Darunter versteht er einen Organismus, der sich wandelt, auf Einflüsse der Umwelt reagiert, sich anpasst und die Umwelt selbst beeinflusst. Der Mensch gliedert seine Welt. Das System bleibt offen. In diesem offenen System ist vieles möglich. Dennoch sind dem System bzw. Menschen Grenzen gesetzt, zum Beispiel die biologische Grenze. Zur Offenheit des Systems gehören auch Denkstrukturen und Gefühle, die für andere Menschen nicht ohne weiteres transparent sind (vgl. Piaget 1983, S. 34 ff.).

Nun streben Menschen als auch Organisationen nach einem ständigen Ausgleich. Sie versuchen ihr Gleichgewicht zu erlangen (Äquilibration). Dies geschieht durch Assimilation oder Akkommodation. Wenn dies misslingt, entsteht ein Ungleichgewicht (Disäquilibrium). Doch der Organismus strebt nach Erkenntnis bzw. hat ein Bedürfnis nach Erkenntnis dessen, was für ihn eine Bedeutung hat. Das Sein wird durch seinen Antrieb aktiv, das offene System entwickelt sich.

So entsteht nach Piaget Identität durch das ständige Streben nach Gleichgewicht und die Auflösung des Ungleichgewichts (vgl. Abb. 24.2 und 24.3).

Diese Gedanken sind auch für Unternehmen in der Form einer Gesellschaft relevant. Die individuelle Entwicklung des Menschen schreitet stetig voran über seine Intelligenz fördernde Sinnesrückmeldungen, die Förderung frühkindlichen Denkens, über den Erwerb von Dezentrierung, Reversibilität, Transitivität und anderes mehr im Denken hin zum formal-operationalen Denken durch Erwerb der Fähigkeit zum logischen Denken und der Fähigkeit, Operationen auf Operationen anzuwenden. Jede Stufe der kognitiven Entwicklung eines Menschen lässt sich dabei durch spezifisches Welt-und Selbstverständnis charakterisieren. So beinhaltet das formal-operationale Denken die Entstehung des reflexiven, abstrakten und theoretischen Denkens, des Denkens in kontrafaktischen Idealzuständen, Metareflexion, Dezentriertheit und die Entkoppelung des Denkens von der unmittelbaren Wahr-Nehmung.

Nur Mitglieder oder Mitarbeiter auf dieser Verarbeitungsstufe besitzen die Fähigkeit, Konzepte und Theorien umfänglich und in ihrer Vielfalt zu verstehen und fort zu entwickeln. Förderung der moralisch-ethischen Urteilskompetenz führt daher zugleich zur Förderung dieser geistigen (kognitiven) Leistungsfähigkeit des Mitarbeiters.

Weltweit haben Gesellschaften und Gemeinschaften, vergleichbar international tätigen Unternehmen, Menschen unterschiedlicher kultureller Prägung zu integrieren, um erfolgreich zu sein. Diese Integration wird zumeist inhaltlicher Art verstanden. So werden die Mitarbeiter im Hinblick auf Erwartbarkeiten und Konventionen geschult, zum Beispiel die richtige Art, eine Visitenkarte in Empfang zu nehmen (Japan). Bekannt ist wohl auch der unterschiedliche Umgang mit Pünktlichkeit in Lateinamerika, wie „la hora boliviana" (die bolivianische Zeit) (vgl. Zimbardo und Boyd 2009; vgl. Levine 1998).

Ein solches Wissen ist ein wichtiger Teilaspekt, um interkulturell handlungsfähig zu sein. Nur formal-operationales Denken ermöglicht ein Verständnis von Kausalität, Wahrscheinlichkeit, Ding und Subjekt-Objekt Trennung. Komplementär dazu ist es auch wichtig zu verstehen, auf welcher Entwicklungsstufe Mitglieder oder Mitarbeiter als auch Externe, Dritte, ihre Welt konstruieren. Je nach Entwicklungsstufe wird das grundlegende Weltverständnis festgelegt (vgl. Dux 1982, vgl. 1989). So benötigen Prediger, Anhänger im Sinne materieller oder immaterieller Leistungserbringer für das moralische System oder sonst erfolgsabhängige Mitglieder einer Organisation die vertrauten Kategorien richtig oder falsch, gut oder böse. Doch was heißt richtig oder falsch, gut oder böse? Jede Arithmetik, Logik, Methode, jedes Regelwerk oder eben jede Moral besitzen einen definierten Geltungsbereich, innerhalb dessen richtige oder falsche Aussagen, gutes oder böses Verhalten existieren, aber eben nicht darüber hinaus. Deshalb helfen Bewertungen im Rahmen nur eines Bezugssystems, wie eben nur einer Moral, in komplexen Welten für sich genommen kaum weiter.

Beispiel

Die Anekdote vom Bauer und seinem Sohn, die gemeinsam Pferde züchteten mag das Dilemma vor allem über die Zeit veranschaulichen: Eines Tages brach ihr wertvollster Hengst aus und lief davon. Die Nachbarn kamen, um ihr Bedauern auszudrücken. Der Bauer aber fragte sie: „Woher wisst ihr, dass dies ein Unglück ist?" Am darauffolgenden Tag kam der Hengst, begleitet von einigen Wildpferden zurück. Wieder kamen die aufmerksamen Nachbarn, um zu dem Glücksfall zu gratulieren, doch der Bauer fragte nur: „Woher wisst ihr, dass dies ein Glücksfall ist?" Am nächsten Tag wurde der Sohn beim Versuch, eines der Tiere zuzureiten, abgeworfen und er brach sich ein Bein. Wieder kamen die Nachbarn, um ihr Mitleid zu bekunden, doch der Bauer fragte wieder nur: „Woher wisst ihr, dass dies ein Unglück ist?" Kurz darauf kam es zu kriegerischen Auseinandersetzungen mit Rebellen, doch da der Sohn verletzt war, wurde er als Soldat nicht einberufen … ◄

Wir haben also, anders als bei der Frage der Entwicklung von Menschen oder ihren Organisationen, bei der Frage der Bewertung von Menschen, auch hinsichtlich ihrer Aktivitäten, immer das Problem der Vor- oder Rückwärtsbetrachtung in der Zeit neben der Beachtung des räumlichen Kontextes, in dem wir etwas bewerten.

24.7 Ist das, was wir tun, wirklich das, was wir sollen oder wollen? Welche Moral passt zu uns?

Trotz oder gerade wegen der immensen Informationsflut müssen Entscheidungen für die Zukunft unter wachsender Unsicherheit getroffen werden. Die Komplexität allen Geschehens nimmt zu. Andererseits verschärfen sich die Folgen von Fehlentscheidungen. Daher ist in den letzten Jahren das Risikomanagement zu einer Notwendigkeit geworden. Der Gesetzgeber hat sich in den vergangenen 30 Jahren mit immer neuen Gesetzen in das Geschehen eingeschaltet. Die Verantwortlichen in den Unternehmen haben geeignete Maßnahmen zu treffen, insbesondere ein Überwachungssystem einzurichten, damit den Fortbestand der Gesellschaft gefährdende Entwicklungen früh erkannt werden.

Hierfür müssen wir auf die Kompetenz von Führungskräften setzen. Doch weltweit scheint sich gerade das Prinzip der Dezentralisierung durchzusetzen. Mit der dahinersteckenden Logik der Delegation von Verantwortung geht eine Abflachung von Hierarchien einher. Je mehr Mitarbeiter zu ‚führen' sind, desto weniger Zeit hat der Manager sich dem Einzelfall zu widmen. Projektorganisation und Gruppenarbeit werden sich deshalb wohl weiter ausbreiten.

Fachkompetenz wird zunehmend von ganzen Teams und Gruppen verlangt.

Die Entpersonalisierung betrieblicher Entscheidungen hat aber zur Folge, dass nicht der Vorgesetzte bestimmte Veränderungen mehr erzwingt, sondern dass die Gruppe insgesamt erkennen muss, dass etwa die angestrebte Kundenzufriedenheit nur über ganz bestimmte Veränderungen erreicht werden kann. Wir brauchen möglicherweise auf allen Gebieten unseres Gemeinwesens wieder mehr Selbstständigkeit und den Willen, unsere persönlichen Angelegenheiten eigenverantwortlich zu regeln. Dazu benötigen wir Risikobereitschaft. Menschen, die über den Tellerrand ihrer eigenen kurzfristigen Interessen hinausblicken.

Um Werte zu realisieren bedarf es nicht in jedem Fall für jede Gruppe bindende Normen, die Verhaltensweisen vorschreiben oder untersagen. Vielmehr durchläuft der Einzelne und damit durchlaufen auch Gruppen ständige Entwicklungsphasen von außen wie innen. Wären also Führungskräfte mit Hilfe der Ergebnisse aus Analyse und Vergleich von „Ist" und „Soll" der moralischen Urteilsfähigkeit und ethischen Gesinnung ihrer Mitglieder oder Mitarbeiter in der Lage, den Systemcharakter ihrer Unternehmen zu verändern bzw. ihn neu zu bestimmen, – wenn sie dies wollten?

Es ist zwar anzuerkennen, dass in denjenigen Unternehmen, in denen aktive Personalentwicklung betrieben wurde, sich in den letzten Jahren das kognitive Wissen und Können verbessert hat. Auf der Strecke geblieben ist aber bei fast allen Unternehmen das affektive Wissen, Bewusstseinsbildung, Einstellung zur Arbeit oder Unternehmen etc. In dieser klassischen Sichtweise spielt die Beachtung der moralischen Urteilskompetenz der Mitarbeiter für Personalauswahl und Personalförderung noch keine Rolle. Obwohl möglicherweise schlüssig, wird nur die eine Seite der Medaille des wertorientierten Managements in den Fokus genommen und dadurch in unzulässiger Weise verkürzt.

Um das ganze Bild in Augenschein zu nehmen, bedarf es eines integrierten Gesamtansatzes (dazu näher Fornefett 2014).

Das chinesische Schriftzeichen für Risiko symbolisiert die Doppelbedeutung des Begriffs als Chance und Gefahr. Gemäß diesem Bild verringert eine Personalarbeit, die die moralische Urteilsfähigkeit und ethische Gesinnung weiterbildet und entwickelt, die Gefahren individuellen Fehlverhaltens (zum Beispiel Korruption, Diebstähle, Betriebsspionage) und erhöht zugleich die Chancen erwünschten Verhaltens der Mitarbeiter (zum Beispiel gelebte Firmenkultur, kooperative Austauschprozesse, Identifikation mit dem Gesamtunternehmen).

Zu diesem asiatischen Risikoverständnis von „sowohl als auch" im Gegensatz zu unserem binären abendländischen von „entweder oder", beinhaltet die vorgestellte Methode zur Entwicklung organisierter Systeme, Gesellschaften und Gemeinschaften, zugleich ein Messinstrument zur Verortung und Erfolgskontrolle. Da die reine Einwirkung auf Bewusstsein, Einstellung und Haltung eines Menschen am Ende durch Belohnung oder Bestrafung auf der Basis eines Scorings, Ratings, Punktespiegels etc. lediglich kurzfristige, keinesfalls nachhaltige Ergebnisse zu zeitigen vermag, sollten wir gemäß unserer zumindest in dieser Hinsicht weltweit ähnlichen Religions- und Weisheitslehren den Menschen stärker fördern, um zu einer insgesamt lernenden Organisation zu gelangen.

Waren in früheren Strukturen Kreativität und Originalität eher lästig, so sind sie in der lernenden Organisation unerlässlich (vgl. Wollert 2001). Die Zukunft benötigt danach ein selbstständiges Mitglied bzw. einen selbstständigen Mitarbeiter. Zugleich wird es abhängig beschäftigte Erwerbsarbeit auch in Zukunft geben müssen, wie auch Arbeits- und Geschäftsprozesse, die einem Plan folgen, der alle Aufgaben und Handlungen vorgibt, Menschen eingebunden in ein „Workflow-System". Von innen her formen Menschen sich selbst, ihr Team, ihren Verantwortungsbereich, am Ende ihr Unternehmen, bilden im Ganzen seine faktische Unternehmenskultur, die nicht identisch sein muss mit der kommunizierten. Deshalb besitzt jedes kleine Team, jeder Verantwortungsbereich, jedes Unternehmen, seine eigene Kultur, sucht und bildet seine eigene Identität.

Die Frage, nach welchen Werten, Normen und üblichen Verfahren sich erfolgreiches, Gesellschafts- oder Gemeinschaftsorientiertes Handeln ausrichten sollte, ist nicht neu. Solange eine Organisation Verhaltensweisen und Normen bereitstellt, die zu ihren Zielen hinführen, weisen sie einen funktionalen Charakter auf. Eine Organisation ist aber auch ein offenes geschichtliches Gebilde. Seine Erfolgsfaktoren sind nicht an jedem Ort oder zu jeder Zeit gleich. Sogar zentrale Ziele können einem adaptiven Wandel unterliegen, dies kann vormals funktionale Verhaltensweisen und Normen dysfunktional machen. Darauf muss angemessen reagiert werden (können).

Schwierig wird es darüber hinaus, wenn innerhalb einer Organisation konkurrierende Kulturtypen (Subkulturen) existieren und unterschiedliche Werthaltungen zu Zielkonflikten führen, die sich letztlich auch auf den wirtschaftlichen Erfolg auswirken. Eine zielkongruente erfolgversprechende Kultur gezielt zu schaffen, ist deshalb ein komplexer, nie endender Prozess, der sich nicht auf ein Social Credit Rating beschränken sollte. Denn Organisationen bzw. deren Mitglieder lernen schwer und vergessen nur langsam.

Literatur

Assenmacher, W. (1998): Konjunkturtheorie, 8., vollständig überarbeitete Auflage, R. Oldenbourg Verlag, München 1989, S. 89

Berkel, K., Herzog, R. (1997): Unternehmenskultur und Ethik. In Arbeitshefte Führungspsychologie Band 27. Heidelberg: Sauer.

Bolte, K.M. (1983): Subjektorientierte Soziologie – Plädoyer für eine Forschungsperspektive. In: K.M. Bolte/ E. Treutner (Hrsg.): Subjektorientierte Arbeits- und Berufssoziologie, Frankfurt/M.; New York: Campus

Deci, E.L. (1995): Why we do what we do: The dynamics of personal autonomy. New York: G.P. Putnam's Sons.

Deci, E.L., Koestner, R., Ryan, R.M., (1999): Examining the effects of extrinsic rewards on intrinsic motivation. Psychological Bulletin, 125, 627–668.

Der „mut" wurde von Prof. Dr. Lind in Konstanz entwickelt. Link: http://www.uni-konstanz.de/ag-moral/lind.htm (Letzter Abruf 9. Januar 2020)

Dux, G. (1982): Die Logik der Weltbilder. Sinnstrukturen im Wandel der Geschichte. Frankfurt/M.: stw

Dux, G. (1989): Die Zeit in der Geschichte. Ihre Entwicklungslogik vom Mythos zur Weltzeit. Frankfurt/M.: stw

Erben, R., Fornefett, A., Kessler, B. (2020): Aktivitäten-basierte Risikosteuerung, FIRM Jahrbuch 2020, S. 32ff.

Erben, R., Fornefett, A., Pauli, M. (2010): Integriertes Performance-und Liquiditätsrisikomanagement – Ein Ansatz für eine konsistente Steuerung von Portfolio- und Cashflow-Risiken, Risiko-Manager Ausgabe 16/2010, S. 20ff.

Fornefett, A. (2007): Raum und Zeit, Markt und Feld, Risiken und Chancen. Gedanken über strukturelle Ähnlichkeiten zwischen Physik und Nationalökonomie in „Zeitsprünge" Doppelheft1/2 2007, Klostermann Verlag Frankfurt a.M.

Fornefett, A. (2012): Subjektive und objektive Grenzen von Transparenz, in Everling, Oliver et al (Hrsg.): Transparenzrating, Springer Gabler Verlag, Wiesbaden 2012

Fornefett, A. (2014): Risikomanagement komplexer Systeme, RiskNet 2014, https://www.risknet. de/themen/risknews/risikomanagement-komplexer-systeme/

Heinrich, C., Kritzner, U., Fornefett, A. (2018): Wie Erwachsene verlorene Spielkompetenz zurückgewinnen können, in Magazin komplex, 1/2018, S 41ff.

Kant, I. (1787): Kritik der reinen Vernunft, 2. Buch, 2. Hauptst., 3 Abschn., § 4

Kant, I. (1900): Kritik der reinen Vernunft. Ausgabe der Preußischen Akademie der Wissenschaften. Berlin 1900ff., AAIV

Kertész, I. (2002): Galeerentagebuch. Rowohlt Taschenbuch Verlag, Reinbek, 2. Aufl. 2002, S. 13. – Ungar. Orig. Budapest 1992. Imre Kertész ist Literaturnobelpreisträger des Jahres 2002

Kritzner, U., Fornefett, A. (2014): „Bevor wir etwas Neues tun, müssen wir etwas Neues denken!", in Kredit & Rating Praxis, Heft 5/2014

Kritzner, U., Fornefett, A. (2017): Im Dialog vernetzen, in Magazin komplex, 1/2017, S.37ff.

Krusche, D., Labenski, J. (1987): Reclams Filmführer. 7. Auflage. Reclam, Stuttgart 1987, ISBN 3-15-010205-7

Langfeldt, B. (2009): Subjektorientierung in der Arbeits- und Industriesoziologie. Theorien, Methoden und Instrumente zur Erfassung von Arbeit und Subjektivität. Wiesbaden: VS.

Levine, R. (1998): Eine Landkarte der Zeit: Wie Kulturen mit der Zeit umgehen. München: Piper

Lind, G. (2000): Ist Moral lehrbar? Ergebnisse der modernen moralpsychologischen Forschung. Berlin: Logos

Lind, G. (2003): Moral ist lehrbar. Handbuch zur Theorie und Praxis moralischer und demokratischer Bildung. München: Oldenbourg

Newton, I. (1686): Mathematische Principien der Naturlehre. Mit Bemerkungen und Erläuterungen. Herausgegeben von Prof. Dr. J. Ph. Wolfers. R. Oppenheim, Berlin 1872, online.

Nietzsche, F. (1883/1999): Also sprach Zarathustra, vollständiger Text Bd. 1-4 bei Nietzsche Source (Colli/Montinari)

Oesterdiekhoff, G. (2006a): Archaische Kultur und moderne Zivilisation. Hamburg: LIT Verlag

Oesterdiekhoff, G. (2006b): Kulturelle Evolution des Geistes. Die historische Wechselwirkung von Psyche und Gesellschaft. Hamburg: LIT Verlag

Oesterdiekhoff, G., Rindermann, H. (2008) (Hrsg.): Kultur und Kognition. Die Beiträge von Psychometrie und Piaget-Psychologie zum Verständnis kultureller Unterschiede. Hamburg: LIT Verlag

Piaget, J. (1983): Das moralische Urteil beim Kinde. Stuttgart: Klett-Cotta

Schacher, U., Fornefett, A. (2010): Der Mitarbeiter: Chance oder Risiko? Die moralische Urteilsfähigkeit von Mitarbeitern entwickeln, Symposion Publishing Verlag 2010

Schopenhauer, A. (1859): Die Welt als Wille und Vorstellung. Bd.I, §1, S.33ff.

Tönnies, F. (2019): Gemeinschaft und Gesellschaft. 1880-1935., hrsg. v. Bettina Clausen und Dieter Haselbach, De Gruyter, Berlin/Boston 2019 (Ferdinand Tönnies Gesamtausgabe, Band 2)

Wagner, R. (2016): Wagnerwerkverzeichnis: WWV 111. Parsifal Text. Hrsg. Muslitz, H.: https:// wagnerlibretto.wordpress.com/2016/10/18/parsifal-libretto/#more-2982. Ende der Szene 2 von Aufzug 1

Wollert, A. (2001): Führen – Verantworten – Werte schaffen. Frankfurt/M.: FAZ

Wikipedia (2020): dort unter „elektrischer Widerstand" m.w.Nw.

Woll, A. (2008): Volkswirtschaftslehre, in Vahlens Handbücher der Wirtschafts- und Sozialwissenschaften 2008, S.343

Zimbardo, Ph., Boyd, J. (2009): Die neue Psychologie der Zeit. Heidelberg: Spektrum Akademischer Verlag 2009

Andreas Fornefett ist Senior Advisor der plenum AG, ein Beratungsunternehmen mit Sitz in Frankfurt am Main. Er ist zudem Aufsichtsratsvorsitzender der i.EPC AG, Kelkheim. Seine Schwerpunkte liegen heute in der Komplexitäts- und Risikomanagementberatung. Fornefett besitzt langjährige Erfahrung in der Entwicklung von Modellen und verfügt, nach Ausbildung zum Kriminalkommissar und Rechts- und Volkswirtschaftsstudium in Göttingen, inzwischen auch über ein breites naturwissenschaftliches Bildungsspektrum.

Gerd Rupprecht ist Vorstand der i.EPC AG, Kelkheim. Er ist zudem Gründungspartner der Rupprecht & Partner Consulting mit heutigem Sitz in Leipzig und verfügt über langjährige Erfahrung in der Initiierung und Moderation von Veränderungsprozessen als kreativem Teamprozess in verschiedenen DAX-Unternehmen und Start-Ups. Nach Studium der Volks- und Betriebswirtschaft an den Universitäten Tübingen und Augsburg war er zunächst in der Markt- und Meinungsforschung sowie Marketingberatung der Young & Rubicam Group tätig.

Uwe J. Schacher ist freiberuflicher Markt- und Sozialforscher. Die Schwerpunkte seines Forschungs- und Entwicklungsunternehmens Schacher & Partner U.S. Research (Markt- und Sozialforschung) in Höchstadt/Aisch aus Oberursel im Taunus liegen in der Konzeption und Umsetzung von Befragungen, auch mittels eigens entwickelter Testverfahren und Tools, sowie in den Bereichen Polykontexturale Logik, Strukturgenese und Sentimentanalyse. Schacher studierte in Nürnberg Soziale Arbeit und in Frankfurt Soziologie, Psychologie und Erziehungswissenschaft.

Nutzen und Grenzen eines multidimensionalen Sozialkreditmodells für Unternehmen

25

Bernhard Kessler und Kaifei Jin

> „Alles ist Zahl!"
> – Pythagoras von Samos
> „Vertrauen ist der Anfang von Allem."
> – Deutsche Bank

Zusammenfassung

In einer Welt der schnellen Transaktionen mit sehr vielen unterschiedlichen Interaktionspartnern verbessert die Zusammenlegung vieler unterschiedlicher Informationen zu einem Gesamtprofil die Einschätzung über dessen Verhalten und erleichtert damit den Handel. Die von Internetgiganten teilweise betriebene soziale Einschätzung zur Erstellung von Kaufinteressen im Internet unterscheidet sich von einem zentral betriebenen Sozialkreditsystem (wie es in China geplant ist) massiv. Der Erwartung an den Informationsgewinn und der technologischen Machbarkeit stehen diverse gesellschaftliche Kosten gegenüber, die zu verstehen und abzuwägen sind. In diesem Beitrag werden der Nutzen und die Kosten eines sozialen Unternehmensratings für Unternehmen verschiedener Größenordnung, Banken und Staaten untersucht. Dies hilft den Umgang mit einem Sozialkreditsystem für sich selbst, aber auch die Wirkung auf andere Beteiligte zu verstehen.

B. Kessler (✉)
Olching, Deutschland
E-Mail: kessler123@gmx.net

K. Jin
Oberursel, Deutschland
E-Mail: kaifei.jin.consulting@gmail.com

© Springer Fachmedien Wiesbaden GmbH, ein Teil von Springer Nature 2020
O. Everling (Hrsg.), *Social Credit Rating*,
https://doi.org/10.1007/978-3-658-29653-7_25

25.1 Bedarf für Sozialkreditrating (SKR)

Datenverfügbarkeit und Steuerungsbedarf

Mit der Globalisierung und der Digitalisierung erhöht sich nicht nur die Möglichkeit zu einer einmaligen Interaktion mit Unbekannten, sondern auch die Geschwindigkeit mit der wieder in der Anonymität verschwunden wird. Glaubwürdigkeit und Reputation kann allerdings nicht durch Lippenbekenntnisse bzw. eigenen Webseitentext oder gestellte Videos erzeugt werden, sondern benötigt eine Instanz, die in einem Bewertungssystem oder Zertifikat viele Aktivitäten zusammenführt oder Transaktionen prüft.

Was bei Amazon bereits in Sternen und erläuternden Kommentaren über die Erfahrungen mit einem Verkäufer ausgedrückt wurde, wurde von Alibaba (Sesame Credit) weiter ausgebaut und mit einer zentral vorgegebenen normativen Bewertung versehen (zum Beispiel Salat ist besser als Burger). Nun soll ein ähnliches erweitertes Scoring sowohl für Privatpersonen als auch für Unternehmen in unterschiedlicher Form u. a. in China eingeführt werden. Da Privatpersonen als einzelne Menschen einem besonderen Schutz unterstehen und die Eingriffe sowohl auf der Datengenerierung als auch Verwendung noch einmal anders zu sehen sind, konzentrieren wir uns in diesem Artikel nur auf das Unternehmensrating.

Xinyong wie das chinesische Sozialsystem heißt wörtlich übersetzt „Wende Vertrauen an!". Dieses Vertrauen, was die Interaktionen erleichtern soll, drückt sich in einer Zahl aus, die verschiedenste Informationen aus den eigenen Aktionen und dem Netzwerk aus der Vergangenheit aggregiert.

Dieser Artikel stellt dar, welche Chancen und Risiken es für die unterschiedlichen Nutzer und Beteiligten eines solchen Systems gibt. Dazu wird zunächst die grundsätzliche Funktionsweise beschrieben.

Anschließend werden in einem Abschnitt die Unterschiede gegenüber einem „klassischen" Kreditsystem dargestellt. Dann werden Funktionsweise und Daten beschrieben und diese danach anhand von Anwendungsfällen verdeutlicht. Damit wird dann das Kapitel vorbereitet, das die Chancen und Risiken eines solchen Systems jeweils aus Sicht der verschiedenen Stakeholder beschreibt.

Ein Fazit, das zentrale Anforderungen zusammenfasst und einen Ausblick auf den Mehrwert eines SKRs beschreibt, beschließt diesen Artikel.

25.2 Einführung eines multidimensionalen Modells für Unternehmen

Der Hauptzweck des klassischen Ratingmodells ist es, die Kreditwürdigkeit eines Kreditnehmers zu beurteilen. Das Ratingsystem ist dabei eine Ansammlung von Methoden, Prozessen, Datensammlungen, die dazu dienen, eine homogene Gruppe anhand bestimmter Charakteristika auf einer geordneten Skala einzuordnen (vgl. CRR (2017) Art. 142 (1), 1.). Zum Beispiel werden auch für Unternehmen mit wenigen Daten inzwischen schon quantitative Beurteilungsverfahren von Investoren angewendet (vgl. Everling 2014).

Der Kern des klassischen Ratingmodells ist ein Multifaktormodell inklusive Regression. Sie basieren größtenteils auf quantitativen Faktoren, wie zum Beispiel Gewinn und Verlust in der Vergangenheit, Unternehmenswert (zum Beispiel durch Diskontierung zukünftiger Cashflows), Eigenkapitalquote. Eine beschränkte Menge von qualitativen Faktoren kam jedoch auch in Betracht, zum Beispiel die Qualität der Unternehmensführung, die Geschäftsstrategie des Unternehmens und die Intensität der bzw. die Investition in Innovation. Sie werden in quantitative Faktoren skaliert und fließen in das Multifaktormodell ein, damit eine Regression möglich ist.

Das klassische Modell ist mit folgenden Nachteilen verbunden und daher in einigen Bereichen nicht zukunftsfähig:

1. Das Rating ist ein Kriterium zur Entscheidung über Kreditvergabe bzw. Bestimmung des Margins für einen zu erwarteten Verlust. Es bietet aber kein Gesamtbild eines Unternehmens. Selbst wenn man „nur" Kreditgeber ist, tritt man zunehmend auch gegenüber Dritten in eine Beziehung zu dem Unternehmen (Image Nachhaltigkeit etc.). In anderen Kontexten hilft das Kriterium der finanziellen Fähigkeit einen Kredit zu bedienen noch weniger für die Entscheidungen zur Aufnahme von (Geschäfts-) Beziehungen.
2. Für eine adäquate Beschreibung und Bewertung eines Geschäftspartners sind viele zum Teil qualitative Kriterien notwendig. Es ist weniger sinnvoll, eine große Menge von qualitativen Eingangsfaktoren in eine Skala zu transformieren anstatt gleich eine andere Art des Verfahrens (wie das Sozialkreditsystem) zu entwickeln.
3. Zwischen den im SKR zu betrachtenden Faktoren ist eine hohe bis sehr hohe Korrelation zu beobachten. Dies widerspricht den Basisannahmen für ein lineares Multifaktormodell, das eine niedrige (bis Null-)Korrelation zwischen den Faktoren voraussetzt. Eine Reduktion durch eine Hauptachsentransformation beraubt die Faktoren ihres Interpretationsgehalts.
4. Das klassische Ratingmodell ist aussagekräftig für Ausfallwahrscheinlichkeiten oder zur Beurteilung der finanziellen Lage eines Unternehmens. Ein SKR umfasst dagegen viel mehr Faktoren und hat das Hauptziel, ein Gesamtportrait eines Unternehmens darzustellen. Dabei ist ein Vergleich nur des Scores nicht ausreichend aussagekräftig. Das gleiche SKR für zwei Unternehmen kann aus ganz unterschiedlichen Gründen resultieren. Ein umweltfreundlicheres Unternehmen kann durch hohe Kosten für Umweltschutzmaßnahmen wahrscheinlicher ausfallen als ein anderes weniger umweltfreundliches Unternehmen, das allerdings stabilere Finanzzahlen liefert. Es bleibt also multidimensional, was im klassisch-kalibrierten Ratingsystem nicht möglich bzw. sinnvoll ist (Use-Test für 0-1-Entscheidung, Pricing etc.).
5. Im klassischen Ratingmodell werden KMU aufgrund mangelnder Kredit- und weiterer Informationen oft analog wie Privatpersonen, das heißt nach bestimmten Gruppenmerkmalen klassifiziert behandelt. Durch Betrachtung weiterer Faktoren im SKR können KMU individueller behandelt werden.

Im Folgenden stellen wir einen teilweise auf dem chinesischen SKR basierenden Ansatz vor, um zu zeigen, wie dieser die Nachteile eines klassischen multidimensionalen Ratingsystems reduziert.

25.2.1 Modellbeschreibung

Das multidimensionale Modell dient den Zielen, eine Gesamtbeurteilung eines Unternehmens anhand eines Gesamtscores, der sich aus Scores (Sozialratings)mehrerer Dimensionen zusammensetzt, zu beschaffen und Maßnahmen anhand dieser Scores abzuleiten. Die Visualisierung der Gesamtbeurteilung erfolgt in einem Radardiagramm. Das Modell stellt sich mit den in Abb. 25.1 beschriebenen Formeln dar.

Es ist zu empfehlen, so viele Dimensionen zu definieren, sodass eine Gesamtbeurteilung möglich ist, aber gleichzeitig die Beurteilungskosten massiv steigen. Zwischen Dimensionen muss die Korrelation nicht strikt Null sein. Ein Ereignis kann zum Abzug oder zur Erhöhung des Scores in mehr als einer Dimension führen.

Mögliche Dimensionen können zum Beispiel Kreditwürdigkeit, Compliance, Umwelt und Ethik sein (wird die Kreditwürdigkeit basierend auf finanziellem Rating als eine Dimension definiert, kann dies unmittelbar in einen Score umgewandelt werden). Ein potenzieller Anwender kann gezielt nach eigenem Zweck von dem Score einer oder mehrerer Dimensionen Gebrauch machen.

Positive und negative Ereignisse sind durch Self Assessment und Berichte der Unternehmen, regelmäßiges und stichprobenartiges Audit der zuständigen Behörden, Gerichts-

$$G = \sum_{i=1}^{I} D_i \qquad \text{mit } G = \text{Gesamtscore der } n \text{ Dimensionen und } D_i = \text{Score der Dimension } i$$

$$D_i = S - \sum_{j=1}^{J} N_j + \sum_{k=1}^{K} P_k - \sum_{l=1}^{L} \rho_{U,UF} \cdot N_l^{UF} - \sum_{m=1}^{M} \rho_{U,SC} \cdot N_m^{SC} + \sum_{o=1}^{O} \rho_{U,UF} \cdot P_o^{UF} + \sum_{q=1}^{Q} \rho_{U,SC} \cdot P_q^{SC} \qquad \text{mit}$$

S = gleiches Startguthaben jeder Dimension (Beispielsweise 100);
N = Scoreabzug eines negativen Ereignisses j;
P = Scoreerhöhung eines positiven Ereignisses k;
$\rho_{U,UF}$ = Korrelation der Scoreänderung zwischen einem Unternehmen und seiner Unternehmensführung;
$\rho_{U,SC}$ = Korrelation der Scoreänderung zwischen einem Unternehmen und weiteren Unternehmen innerhalb seiner Supply Chain;
N_{UF} = Scoreabzug eines negativen Ereignisses l der Unternehmensführung;
N_{SC} = Scoreabzug eines negativen Ereignisses m eines weiteren Unternehmens in der Supply Chain;
P_{UF} = Scoreerhöhung eines positiven Ereignisses o der Unternehmensführung;
P_{SC} = Scoreerhöhung eines negativen Ereignisses q eines weiteren Unternehmens in der Supply Chain;

$$M(G; D_i) = \begin{cases} Red\ List\ Maßnahmen & \text{wenn } G \geq G_{RL}\ \text{und}\ D_i \geq D_B \\ Bonusmaßnahmen & \text{wenn } D_i \geq D_B \\ Manusmaßnahmen & \text{wenn } D_i < D_M \\ Black\ List\ Maßnahmen & \text{wenn } G < G_{BL}\ \text{oder}\ D_i < D_{BL} \end{cases} \qquad \text{mit}$$

G_{RL} = Gesamtscore, die Red-List-Maßnahmen auslöst;
D_B = Score einer Dimension, die Bonusmaßnahmen einer Dimension auslöst;
D_M = Score einer Dimension, die Malusmaßnahmen einer Dimension auslöst;
G_{BL} = Gesamtscore, die Black-List-Maßnahmen auslöst;
D_{BL} = Score einer Dimension, die ebenfalls Black-List-Maßnahmen auslöst

Abb. 25.1 Formeln zur Bildung eines Social Scores. (Quelle: eigene Darstellung)

urteile sowie Überwachung verfügbarer Informationen im Internet und Social Media zu identifizieren.

Eine positive Korrelation ist in Betracht zu ziehen zwischen dem Sozialrating

- des Unternehmens und seiner Geschäftsführung sowie
- des Unternehmens und weiterer Unternehmen innerhalb seiner Supply Chain, wenn es einen beherrschenden Einfluss auf das andere Unternehmen hat oder die anderen Unternehmen (zum Beispiel Lieferanten und Abnehmer) wirtschaftlich von ihm abhängig sind (vgl. Artikel 4 Nr. 39 CRR (2017)).

Negative oder positive Ereignisse der Geschäftsführung oder eines Unternehmens der Supply Chain führen ebenfalls zum Scoreabzug oder zur Scoreerhöhung des beurteilenden Unternehmens. Nach dem Best Practice in China greift es insbesondere, wenn eine der Parteien auf der Blacklist landet.

Das Modell basiert auf historischen Daten. Auswirkungen der negativen und positiven Ereignisse haben zeitliche Gültigkeit. Die abgezogenen Scores sind wieder zurückzusetzen, wenn die Ursachen behoben werden und in gewisser Zeit, je nach Wichtigkeit der Ereignisse, nicht erneut auftauchen. Dies gilt auch für positive Ereignisse.

Ein detaillierter Maßnahmenkatalog für jede vordefinierte Scorestufe ist klar zu definieren. Sämtliche Vorschläge sind Abschn. 25.2.3 zu entnehmen. Werden die Schwellenwerte für eine Redlist überschritten oder die Schwellenwerte für eine Blacklist unterschritten, greifen die kollektiven Maßnahmen. Redlist-Maßnahmen sind u. a. bessere bzw. unmittelbare Zugänge zu Behörden und vergünstigte Konditionen bei öffentlichen Ausschreibungen. Blacklist-Maßnahmen sind zum Beispiel hohe Strafzahlung, Drohung zum Konkurs, weitere administrative Strafmaßnahmen, Flugverbot, Schnellzugfahrverbot der Geschäftsführer und eingeschränkte Reisemöglichkeit der Geschäftsführer ins Ausland (vgl. http://www.gov.cn/zhengce/content/2016-06/12/content_5081222.htm). Die Redlist und die Blacklist werden unverzüglich veröffentlicht und sind für alle Bürger zugängig (vgl. http://www.gov.cn/zhengce/content/2014-08/23/content_9038.htm).

25.2.2 Datenquellen

Mögliche Zulieferer der zugreifbaren Inputdaten können Finanzinstitute, Telekommunikationsunternehmen und weitere Unternehmen sowie Online-Plattformen sein, die gewisse Unternehmens-, Kredit- und Handelsinformationen sammeln und verarbeiten. Überdies stellen auch Gerichte, weitere Behörden (zum Beispiel Finanzamt, Zollamt und Umweltministerium) sowie Social-Media-Unternehmen Input bzgl. Verhalten der zu beurteilenden Unternehmen im realen Wirtschaftsleben und im Netzwerk direkt zur Verfügung.

Anhand des Beispiels in China lassen sich hauptsächlich drei Quellen identifizieren. Davon nimmt das neue chinesische Sozialkreditsystem, die National Credit Information Sharing Plattform, Daten entgegen. 75 Prozent der Informationen sind veröffentlicht:

1. Credit Reference Center of the People's Bank of China: Inhaber: Volksbank, Datenquelle: staatliche Geschäftsbanken, lokale Banken und Privatbanken, Hauptdatenin-
   halt: klassische Kreditdaten von Unternehmen und Privatpersonen
2. weitere private und staatliche Rating Plattformen oder in Mischform
3. Zentral- und Regionalregierung, Behörden sowie Gerichte
4. Baihang Credit, die Anteilseigner sind einerseits die NIFA als e. V. sowie weiteren
   Organisationen, die Tab. 25.1 zu entnehmen sind.

Es ist vorhersehbar, dass alle drei Quellen aus folgenden Gründen in naher Zukunft in
eine einheitliche konsolidierte Plattform zusammengeführt werden.

Herausforderungen bestehen dabei durch folgende Themen:

Tab. 25.1 Anteilseigner von Baihang Credit

Anteilseigner	Anteil	Quelldaten	Besonderheiten
National Internet Finance Association (NIFA) e. V.	36 %		People's Bank of China und China Banking Regulatory Commission Koordination und Aufsicht
Sesame Credit	8 %	Behörde der Industrie und Handel, Justiz sowie Zoll Betriebs- sowie Investitions-informationen eines Unternehmens im In- und Ausland Risiko zwischen nahstehenden Unternehmen	Kreditinformationen aus **Alibaba** Tochterunternehmen: Banken, Investmentfonds, Versicherung und Wertpapierhandelsunternehmen Ausfallinformation auf Alibaba-Plattformen wie Taobao etc.
Tencent Credit	8 %	KMU Kreditinformationen KMU Sozialmedieninforma-tionen	Kreditinformationen aus **Tencent** Tochterunternehmen: Banken, Investmentfonds und Versicherung
Qianhai Credit	8 %	Finanzinformationen: Banken, Versicherung, Investmentfond, Internet Finance	Kreditinformationen aus **Ping'an** Tochterunternehmen: Banken, Investmentfonds und Versicherung
Pengyuan Credit	8 %	Kreditinformationen von KMU und Privatpersonen	Klassische Ratingagenturen Anti-Fraud
Zhongchengxin Credit	8 %	Banken, Internet Finance, Supply Chain Finance, Konsumkredit, Forderungsmanagement	
Zhongzhicheng Credit	8 %	P2P Internet finance, Konsumkredit, Klein- und Mikrokredite	
Kaola Credit	8 %	Kreditinformationen von KMU und Privatpersonen	Internet finance
Huadao Credit	8 %	Retailkunden, leasing	

1. Es gibt mögliche Überschneidung der Daten.
2. Daten einzelner Kategorien sind lediglich in einer Quelle erhältlich (Validität).
3. Es gibt kein einheitliches Beurteilungsverfahren pro Quelle gibt, das heißt, die Scores sind auf der National-Credit-Information-Sharing-Plattform schwer konsolidierbar.

Besser funktionieren könnte bzgl. 3. zum Beispiel eine Anlieferung der Rohinformationen mit einheitlichem Datenformat.

25.2.3 Details des multidimensionalen Modells

Im Rahmen dieses Artikels schlagen wir sechs Dimensionen vor: Kreditwürdigkeit/wirtschaftliche Lage (vgl. klassisches Rating); Compliance (Geldwäsche, steuerrechtliche Angelegenheit etc.); Umwelt; weitere soziale Verantwortungen; Internetsicherheit und Ethik. Jeder Dimension werden fünf Stufen (eine Bonus-, eine „Gute", eine „Bestehen-" und zwei Malusstufen) zugeordnet. Die Stufenbreite A bis D ist einheitlich (vgl. Tab. 25.2).

Überschreitet ein Unternehmen die Untergrenze der Bonusstufe, wird es durch Bonusmaßnahmen belohnt. In den Stufen „Gut" und „Bestehen" greift in erster Linie keine Maßnahme, wobei Unternehmen auf der Stufe „Bestehen" intensiver und regelmäßig geprüft werden, weil sie je nach Scorestand in die Malusstufen rutschen können. In den Malusstufen greifen die Sanktionen.

Die folgenden Abschnitte basieren teilweise auf Best Practices des chinesischen Sozialkreditsystems.

25.2.3.1 Kreditwürdigkeit/wirtschaftliche Lage

Grundsätzlich weicht diese Dimension nicht von einem klassischen Ratingmodell ab. Analog wird das Unternehmen einer Ratingstufe von AAA bis D zugeordnet. Um genaues Rating zu ermitteln, kann auf das externe oder interne Rating zugegriffen werden. Um das Ergebnis einheitlich wie in anderen Dimensionen darstellen zu können, wird das Finanzrating in fünf Stufen skaliert (vgl. Dagong, das heißt analog zum S&P Rating; vgl. Tab. 25.3).

Eine Erweiterung im Vergleich zum klassischen Ratingmodell ist die Betrachtung der Daten aus weiteren Plattformen, die über Kreditinstitute hinausgehen, zum Beispiel Klein- und Mikrokredit oder Crowdfunding über P2B- oder P2P-Plattformen. Darüber hinaus stehen über Plattformen wie Taubao, Tencent oder Jingdong quantitative Daten wie Umsätze, Cashflows der KMU sowie qualitative Daten wie Qualität der Produkte und Businessverhalten im Internet zur Verfügung.

Tab. 25.2 Stufenergebnisse eines Social Scorings. (Quelle: eigene Darstellung)

#	D_i	Bezeichnung	Maßnahmen
I	$D_i > A$	Bonusstufe	Bonus oder/und Redlist
II	$B < D_i \leq A$	Stufe „Gut"	Keine
III	$C < D_i \leq B$	Stufe „Bestehen"	Intensive Prüfung
IV	$D < D_i \leq C$	Malusstufe I	Malus oder/und Blacklist
V	$D_i < D$	Malusstufe II	

Tab. 25.3 Skalierung des Finanzratings in die Stufen des Modells anhand
Dagong/S&P-Skala

#	Bezeichnung	Dagong Credit
I	Bonusstufe	AAA – AA-
II	Stufe „Gut"	A+ – BBB-
III	Stufe „Bestehen"	BB+ – B-
IV	Malusstufe I	CCC+ – C
V	Malusstufe II	D

Gleich wie im klassischen Ratingmodell führt ein gutes Finanzrating bzw. eine niedrigere Ausfallwahrscheinlich dazu, dass die Unternehmen besseren Zugang zum Finanzmarkt bzw. Banken bekommen und niedrigere Zinsen für Kredite zahlen müssen. Im Vergleich zur einseitigen Informationsquelle von Kreditinstituten ist die Beurteilung mit
umfangreichen Quellen wesentlich plausibler. Dagegen wird die Refinanzierung eines
Unternehmens mit schlechterem Finanzrating schwieriger. Dies betrifft insbesondere Unternehmen, die traditionell dank Kreditdaten der Banken besser geratet sind.

25.2.3.2 Compliance
Unter Compliance fallen grundsätzlich alle rechtlichen Angelegenheiten. Dies bezieht
sich sowohl auf Inländische als auch internationale Gesetze. Um eine Umsetzbarkeit zu
gewährleisten und gleichzeitig ein Gesamtbild bzgl. rechtlicher Lage eines Unternehmens
beschaffen zu können, besteht jedem Anwender des Modells die Möglichkeit, auf wesentliche relevante Gesetze einzuschränken. Die Maßnahmen sollen nicht die gerichtlichen
Strafen ersetzen oder dazu führen, dass die Unternehmen doppelt bestraft werden.

Über monetäre Maßnahmen hinaus können zuständige Behörden parallel durch administrative Maßnahmen ein Unternehmen belohnen oder bestrafen (vgl. Tab. 25.4).

25.2.3.3 Umwelt
Die Dimension Umwelt beurteilt das Unternehmen nach Einhaltung der Umweltschutzvorschriften sowie Nachhaltigkeit seiner Produktion (vgl. Tab. 25.5).

25.2.3.4 Soziale Verantwortung
Diese Dimension enthält alle Aspekte bzgl. sozialer Verantwortung eines Unternehmens
(vgl. Tab. 25.6).

25.2.3.5 Sicherheit
Ein wesentlicher Bestandteil dieser Dimension ist die IT-Sicherheit. Sie gilt nicht nur für
die Sicherheit der Daten innerhalb eines Unternehmens, sondern auch für seine Auswirkung auf die Gesellschaft (Stichwort Datenschutz). Überdies sind alle Aspekte bezüglich
Sicherheit von Mitarbeitern, Unternehmen und Gesellschaft unter dieser Dimension zu
berücksichtigen (vgl. Tab. 25.7).

25.2.3.6 Ethik und sonstige
Diese Dimension bezieht sich auf ethische und sonstige Aspekte, die nicht zu anderen
Dimensionen gehören (vgl. Tab. 25.8).

Tab. 25.4 Ergebnisse des Social Ratings in der Kategorie Compliance

Positive Ereignisse	Negative Ereignisse	Bonusmaßnahme	Malusmaßnahme
• Termingerechte Steuerzahlung • Termingerechte Einzahlung der Sozialversicherung für längere Zeit gute Sozialversicherungsnachweise • Termingerechte Zahlung von Zöllen • Gewährleistung mehr Wettbewerbe auf dem Markt • … *Rückerstattung abgezogener Scores* • Rechtzeitiges Nachkommen der steuerlichen Verpflichtung[a] • Rechtzeitiges Nachkommen der Sozialversicherungsverpflichtung • Rechtzeitiges Nachkommen der Zollverpflichtung • …	• Steuerhinterziehung • Verweigerung zur Zahlung vom Arbeitgeberanteil der Sozialversicherung • Betrugsfall bei Sozialversicherung[b] • Zollhinterziehung • Import oder Export von unerlaubten Gütern/Dienstleistungen • Organisierter Schmuggel[c] • Produktion ohne Erlaubnis • Erwerbung von Unternehmens- oder Produktlizenzen durch fiktive oder betrügerische Unterlagen[d,e] • Fehlende Zahlung von Gehältern (insb. bei Gastarbeitern) • verdächtige oder identifizierte Geldwäschevorfälle • Unerlaubtes Insider-Handeln • Wettbewerbsbeschränkung • …	• niedrigerer Steuersatz • Senkung des Sozialversicherungssatzes • Teilweise Übernahme des Arbeitgeberbetrags durch die Regierung • Schnelle Bearbeitung bei Ausfuhrerklärung und Entzollung • Senkung der Inspektionshäufigkeit • Bürgschaft des Zollamts • Bessere Chance beim öffentlichen Beschaffungswesen • Gewährleistung besseren Zugangs zum Finanzmarkt (IPO, Emissionen) • Verkürzung des Bearbeitungsprozesses beim Patentantrag[f] • Aufrechterhaltung „harmonischer" Arbeitsbeziehungen • …	• Beschlagnahme von illegalem Einkommen • Strafzahlung und Sanktionen • höherer Steuersatz • Höherer Sozialversicherungssatz • Einschränkung für Import und Export • Erhöhung der Inspektionshäufigkeit • Erhöhung der Lagergebühren • Abzug der Unternehmens- oder Produktlizenzen • Einschränkung bei Teilnahme an Ausschreibungsprozessen von Behörden • Eingeschränkter Zahlungsverkehr • Regelmäßige Überwachung des Zahlungsverkehrs • Einschränkung des Zugangs zum Finanzmarkt • Einschränkung des Zugangs zum Arbeitskräftemarkt • Tausch der betroffenen Manager • …

(Fortsetzung)

[a]https://www.creditchina.gov.cn/xinyongdongtai/buwei/201911/t20191113_175360.html

Tab. 25.4 (Fortsetzung)

[b]http://www.acfic.org.cn/ddgh/bwzc/201911/t20191118_145994.html

[c]http://mofcom.gov.cn/article/b/g/201804/20180402738074.shtml

[d]http://gov.cn/gongbao/content/2012/content_2144293.htm

[e]http://www.gov.cn/zhengce/content/2019-09/18/content_5430900.htm

[f]https://www.creditchina.gov.cn/xinyongdongtai/buwei/201905/t20190517_155874.html

Tab. 25.5 Ergebnisse des Social Ratings in der Kategorie Umwelt

Positive Ereignisse	Negative Ereignisse	Bonusmaßnahme	Malusmaßnahme
• Sparsamer Energieverbrauch i. V. zur Peer-Gruppe • Umweltfreundliche und Ressourcenschonende Produktion • Pro-aktive Zusammenarbeit während Inspektion und Untersuchung • Einsatz von Dienstwagen mit erneuerbarer Energie • …	• Wasser und Stromverschwendung • Wasser- und Luftverschmutzung • Fehlende CO_2-Zertifikate • Betrieb ohne entsprechende Lizenzen • Verweigerung während Inspektion und Untersuchung • Verweigerung der Durchführung von vorgeschriebenen Maßnahmen • …	• niedrigerer Strompreis • niedrigerer Wasserpreis • Steuererstattung • bessere Möglichkeit in öffentlichen Ausschreibungen • infrastrukturelle Unterstützung durch Staats- oder Regionalregierung • weitere administrative Bonusmaßnahmen • …	• monetäre Strafe und Sanktionen • Höherer Strompreis • Höherer Wasserpreis • Abzug der Produktionslizenzen • Tausch der Unternehmensführung • Drohung des Konkurses • …

Tab. 25.6 Ergebnisse des Social Ratings in der Kategorie Soziale Verantwortung

Positive Ereignisse	Negative Ereignisse	Bonusmaßnahme	Malusmaßnahme
• Innovation der Produkte oder Dienstleistung • Vollständige Informationsveröffentlichung für öffentliche Interessen • Rechtzeitige Offenlegung von korrekten Finanzinformationen[a] • aktive Teilnahme an sozialen Wohltätigkeiten • hohe Produktqualität • Pro-aktive Zusammenarbeit während Inspektion und Untersuchung • …	• mangelnde Produktqualität • Verfälschung der Finanzdaten[b] • Betrug • illegale Aktivitäten in gemeinnützigen Organisationen oder im Namen von Wohltätigkeitsorganisationen[c] • Werbung mit irreführenden Informationen[d] • Verweigerung der Veröffentlichung von Unternehmensinformationen • Verzögerung in Offenlegung • …	• Vereinfachung der Prozesse zur Unternehmensspende • infrastrukturelle Unterstützung bei Etablierung unternehmensinterner NGO's • Steuerbegünstigung • Bessere Werbungskanäle • …	• monetäre Strafe und Sanktionen • Abzug der Zertifikate der betroffenen Personen • …

[a]http://www.amac.org.cn/cms/contentcore/resource/download?ID=8748
[b]http://m.mof.gov.cn/tzgg/201910/P020191023289180169380.docx
[c]https://www.creditchina.gov.cn/xinyongdongtai/buwei/201908/t20190802_164141.html
[d]http://www.gov.cn/zhengce/content/2016-06/12/content_5081222.htm

Tab. 25.7 Ergebnisse des Social Ratings in der Kategorie Sicherheit

Positive Ereignisse	Negative Ereignisse	Bonusmaßnahme	Malusmaßnahme
• Vollständige Gewährleistung der Vertraulichkeit, Integrität und Verfügbarkeit von unternehmensinternen sowie kundenbezogenen Daten • Implementierung von hohen IT-Sicherheitsstandard • Regelmäßige Schulung der Mitarbeiter bzgl. IT-Sicherheitsthemen • Vollständige Einhaltung von Sicherheitsvorschriften für Gebäude (inklusiv Aufzüge) und Brandschutz • Whistle-Blowing zur Enthüllung von Sicherheitsvorfällen • Produktion von Lebensmittel mit hoher Qualität • …	• Verlust von Daten in Bezug auf die nationale Sicherheit • Verlust von vertraulichen Daten in gesellschaftlichen und öffentlichen Interessen • Nicht verhinderte Hackerangriffe • Erhebung, Verwendung, Veräußerung und Vernichtung personenbezogener Daten unter Verstoß gegen Vorschriften[a] • Gefährdung der öffentlichen Sicherheit • Sicherheitsverletzungen und Unfälle • Verstoß gegen Sicherheitsvorschriften für Gebäude (inklusiv Aufzüge)[b] oder Brandschutz[c] • Produktion von unsicheren Lebensmittel • Veröffentlichung illegaler oder verfälschter Informationen auf der Firmenwebsite • Manipulation der Klickrate oder Besuchvolumen bei Online Shops • Aufgedeckte Sicherheitslücken in Online-Zahlung, -Überweisung sowie Wertpapierhandeln • …	• Zertifizierung als Unternehmen mit hoher IT-Sicherheit • Zertifizierung als vertrauter Online Shop oder App • Weitere kollektive Bonusmaßnahmen wie: • Begünstigung während öffentlicher Ausschreibungen • besserer Zugang zum Finanzmarkt (z. B. IPO und Emissionen) für große Unternehmen • bessere Möglichkeit zum Antrag von Klein- und Mikrokrediten für KMU • … • …	• monetäre Strafe und Sanktionen • Zwang zur Nachbesserung • Veröffentlichung der Vorfälle im Internet • Mittel- bis langfristige Sperrung der Website oder/und der App[d] • Abzug der Lizenzen • Marktzutrittsbeschränkung oder -verbot • Sperrung des Gebäudes • Mittel- bis langfristige Schließung der Produktion von unsicheren Lebensmitteln • …

(Fortsetzung)

Tab. 25.7 (Fortsetzung)

[a]http://www.cac.gov.cn/rootfiles/2019/09/11/1569729939897372-1569729948315802.docx
[b]http://www.gov.cn/zhengce/content/2018-02/09/content_5265380.htm
[c]http://www.gov.cn/zhengce/content/2016-06/12/content_5081222.htm
[d]https://www.creditchina.gov.cn/xinyongdongtai/buwei/201911/t20191108_174769.html

Tab. 25.8 Ergebnisse des Social Ratings in der Kategorie Ethik

Positive Ereignisse	Negative Ereignisse	Bonusmaßnahme	Malusmaßnahme
• Gleichberechtigung für alle Geschlechter • Keine Diskriminierung jeglicher Art • Verhinderung von sexueller Belästigung am Arbeitsplatz (Prozesse und Maßnahmen) • Sichere Lebensmittel • Gute Lebensmittelqualität • Kein Plagiat und keine Fälschung • Keine Korruption • Beibehalten vollständiger Daten und Bewertungen bei Online-Portalen • Angemessene Arbeitszeit • Pro-aktive Zusammenarbeit während Inspektion und Untersuchung • …	• Geschlechtsdiskriminierung • Diskriminierung jeglicher Art • Sexuelle Belästigung am Arbeitsplatz • Unsichere Lebensmittel oder schlechte Lebensmittelqualität, insbesondere für Schulen und Kita[a] • Verstöße gegen das geistige Eigentum[b] • Spionage von Businessgeheimnissen • Korruption • Löschen negativer Bewertung bei Online-Portalen[c] • Veröffentlichung verfälschter oder unvollständiger Informationen auf der Firmenwebsite[d] • Lange Arbeitszeit • Teilnahme an illegalen Sportwetten • Verweigerung der Zusammenarbeit mit Aufsicht, Inspektion und Untersuchung • Verweigerung der Durchführung von vorgeschriebenen Maßnahmen • …	• Besserer Zugang zum Arbeitskräftemarkt • Bessere Möglichkeit beim Kreditantrag (inkl. Klein- und Mikrokredits) • Bessere Kondition bei Refinanzierung und Emission • Bessere Note in App-Bewertung oder Nominierung als Trust-Shop • Verkürzung des Bearbeitungsprozesses beim Patentantrag • Vorteile bei öffentlichen Ausschreibungsprozessen • …	• Bestellung weiblicher Vorstandsmitglieder • Geldstrafe und weitere finanzielle und nicht finanzielle Sanktionen • Entzug der Firmen- oder Produktionslizenz • Schlechtere Note in App-Bewertung • Einschränkung des Zugangs zum Finanzmarkt • Einschränkung des Zugangs zum Arbeitskräftemarkt • Einschränkung oder Verbot bei öffentlichen Ausschreibungsprozessen • Beschränkung der Spielertransfers

[a]https://www.creditchina.gov.cn/xinyongdongtai/buwei/201911/t20191113_175362.html
[b]https://www.creditchina.gov.cn/xinyongdongtai/buwei/201911/t20191112_175236.html
[c]https://www.mct.gov.cn/whzx/whyw/201910/t20191009_847147.htm
[d]http://www.mofcom.gov.cn/article/h/redht/201907/20190702878644.shtml

25.2.4 Darstellung anhand eines Beispiels

Das Gesamtbild eines Unternehmens anhand seines Sozialratings kann in einem Radardiagramm mit allen Dimensionen visuell komfortabel dargestellt werden. Allerdings deuten
die Dimensionen auf unterschiedliche Wichtigkeit bei unterschiedlichen Branchen hin und
die Stringenz in der Beurteilung ist pro Dimension für unterschiedliche Branchen anders.
Zum Beispiel spielt Kreditwürdigkeit bei einer Bank eine wesentlich größere Rolle als bei
einem Chemieunternehmen, wobei im Umkehrschluss die Umweltschutzvorschriften bei
dem Chemieunternehmen strenger eingehalten werden müssen als bei der Bank.

Es bestehen zwei Möglichkeiten, die unterschiedlichen Wichtigkeiten abzufangen:

1. Werden die Stufenhöhen/-breiten derselben Dimension für unterschiedliche Branchen
   unterschiedlich definiert oder
2. Ereignisse führen zu unterschiedlichen Höhe des Scoreabzugs oder -erhöhung.

Um die Vergleichbarkeit zwischen Unternehmen bzw. Branchen gewährleisten zu können, ist die Alternative 2 zu empfehlen, denn die Skala ist daher für unterschiedliche Branchen gleich und die dazu passenden Maßnahmen können einheitlich definiert und abgeleitet werden.

Das Sozialrating mit Scorebeispielen wird anhand einer Bank als Beispiel visuell dar-
gestellt. Das Startguthaben entspricht 100 Punkten pro Dimension. Die Visualisierung des
Sozialratings von Unternehmen in anderen Branchen können analog abgeleitet werden.
Lediglich kann die Gewichtung jedes Ereignisses anders sein. Das heißt, ausgehend vom
Scorewert 100 kann sich der Score positiv zu 110 (A) oder negativ bis zu 50 (D) in ver-
schiedenen Dimensionen entwickeln.

Die Bank B verweist ursprünglich auf ein normales Sozialrating mit Scores in den
Stufen „Gut" und „Bestehen". Somit lauten ihre einzelnen Scores wie in Abb. 25.2 beschrieben und grafisch verdeutlicht (weiter außen bedeutet besser).

Nach bekannt gewordenen negativen Ereignissen von Steuerhinterziehung und Lücken
in der Internetsicherheit landet die Bank in Dimensionen „wirtschaftliche Lage" und
„Compliance" in der Malusstufe I (vgl. Abb. 25.3). Sie steht unter intensiver Prüfung als

Bank B vor negativen Vorfällen

		Bewertung	A	B	C	D
W	Wirtschaftliche Lage	70	110	80	60	50
C	Compliance	70	110	80	60	50
U	Umweltfreundlichkeit	80	110	80	60	50
S	Soziale Verantwortung	80	110	80	60	50
I	Internetsicherheit	90	110	80	60	50
E	Ethik	80	110	80	60	50

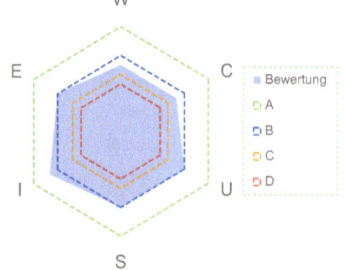

Abb. 25.2 Sozialrating einer Bank vor negativen Erlebnissen. (Quelle: eigene Darstellung)

Kandidat für die Blacklist. Da weder die Gesamtscore noch die Score in einer Dimension die Untergrenze unterschreitet, landet sie zunächst nicht direkt auf der Blacklist.

Durch stichprobenartige Prüfungen stellen sämtliche Behörden einen massiven Insider-Trade fest, woran mehrere Top-Manager teilnahmen. Außerdem fanden sie Korruptions- sowie Geldwäschevorfälle. Weitere Scores werden abgezogen von den Dimensionen „Wirtschaftliche Lage", „Compliance", „Soziale Verantwortung" und „Ethik", sodass der Gesamtscore die vordefinierte Untergrenze von 350 unterschreitet. Sowohl die Bank als auch die Top-Manager landen sofort auf der Blacklist. Um diese Tatsache in Veröffentlichung leicht erkennbar zu machen, brechen die Scores auf das niedrigste Niveau zusammen (vgl. Abb. 25.4). Sobald die Bank nach der Verbesserung oder Gültigkeit der negativen Ereignisse die Blacklist verlässt, werden die Scores auf die tatsächlichen Scores zurückgesetzt.

Tab. 25.9 fasst zentrale Charakteristika des chinesischen Sozialkreditsystems und dessen Anwendung zusammen:

Ein besonderer Vorteil aus Sicht der Ratingnehmer ist, dass mehr Daten verwendet werden und ein Upgrade des Scores möglich ist. Damit bieten sich für KMU und Neugründungen neue Möglichkeiten, aktiver einen positiven Score zu erreichen. Die vielen Freiheitsgrade/Wahlfreiheiten des Modells führen einerseits zu einer breiteren Datengrundlage, öffnen damit aber auch neue Spielräume für (Fehl-)Interpretationen. Die Modellwahl soll deshalb in Abschn. 25.3 noch einmal intensiver betrachtet werden, bevor anschließend eine Einwertung erfolgen kann.

Bank B nach negativen Vorfällen

		Bewertung	A	B	C	D
W	Wirtschaftliche Lage	40	110	80	60	50
C	Compliance	40	110	80	60	50
U	Umweltfreundlichkeit	80	110	80	60	50
S	Soziale Verantwortung	80	110	80	60	50
I	Internetsicherheit	60	110	80	60	50
E	Ethik	80	110	80	60	50

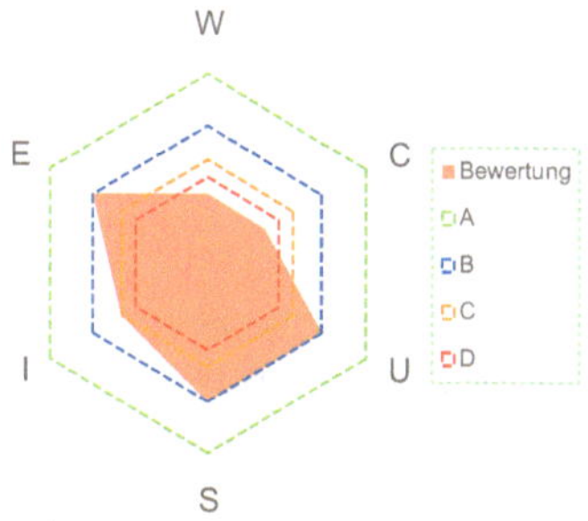

Abb. 25.3 Sozialrating einer Bank nach negativen Ereignissen. (Quelle: eigene Darstellung)

Bank B auf der Black List

		Bewertung	A	B	C	D
W	Wirtschaftliche Lage	10	110	80	60	50
C	Compliance	10	110	80	60	50
U	Umweltfreundlichkeit	10	110	80	60	50
S	Soziale Verantwortung	10	110	80	60	50
I	Internetsicherheit	10	110	80	60	50
E	Ethik	10	110	80	60	50

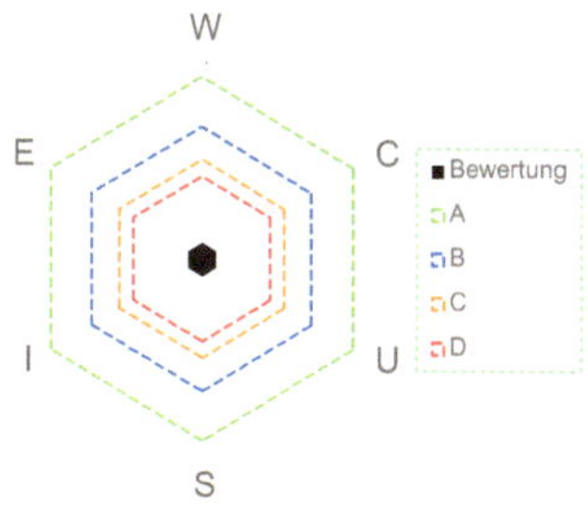

Abb. 25.4 Sozialrating einer Bank auf der Blacklist. (Quelle: eigene Darstellung)

Tab. 25.9 Übersicht der Besonderheiten eines Sozialkreditsystems

Eigenschaft	Klassisches Verfahren	Chinesisches SKR
Linearität	Ja	Ja
Dimensionalität der Auswertung	Nur eine Zielgröße (Kreditwürdigkeit)	Mehrere Subkategorien, aber auch Summierung zum Gesamtscore
Input Parameter	Zahlungshistorie, Kreditnehmercharakteristika	Sämtliche Interaktionen (wirtschaftlich, Umwelt, soziales Verhalten, …) Verfallsdatum von Daten
Korrelation zwischen Parameter	Niedrig bis Null	Unabhängig (Erhöhung der Flexibilität)
Einflussrichtung	Nur verschlechtern	Verschlechtern und verbessern
Einbeziehung qualitativer Merkmale	Eingeschränkt (Anpassung an einheitliche Metrik)	In größerem Umfang möglich durch verschiedene Submetriken
Indirekte Einflussgrößen (Netzwerkeffekt)	Nein – nur persönliche Parameter	Ja, Berücksichtigung von Effekten im Netzwerk
Behandlung KMU	Grundsätzlich als Mengengeschäft; Einzelbewertung grundsätzlich nicht möglich	Einzelbewertung möglich

25.3 Stärken und Schwächen des Modells

Das neue multidimensionale Modell bietet im Vergleich zum klassischen Modell mehrere Vorteile:

- Beschaffung eines Gesamtbildes des Sozialratings eines Unternehmens statt einer eingeschränkten Betrachtung von dessen Finanzlage
- Mehr Inputinformationen werden betrachtet, insbesondere für KMU
- Einfache Verwendbarkeit und Flexibilität: das Modell lässt sich von Zentral- und Regionalregierungen, Behörden, Banken sowie weitere Interessenten durch freie Auswahl sämtlicher oder aller Dimensionen zum bestimmten Zweck leicht verwenden
- Leichtere Skalierung der qualitativen Merkmale
- Unproblematische Abbildung der Korrelation zwischen Dimensionen/Faktoren
- Bessere Grundlage für ein ethisches Investment: mehr Finanzmittel werden in Unternehme mit besserem Sozialrating investiert, die nicht nur kreditwürdiger, sondern auch umweltfreundlicher, ethischer, sicherer und sozialverantwortlicher sind.

Allerdings verweist das Modell auch sämtliche Einschränkungen eines komplexeren Modells auf. Beispielsweise ist die Validierung anhand von eindeutigen Ereignissen ebenso schwieriger wie die Transparenz der Wirkungskette. Deshalb sollte es während der Verwendung weiterhin optimiert werden, zumal die Kritik vieler ausländischer Beobachter sich gerade auf „Fehler" in der Beurteilung stürzen werden.

25.3.1 Dimensionen

Ein Hauptproblem bei dieser Vorgehensweise ist es die richtige Auswahl an Dimensionen zu wählen, die zur Gesamtbeurteilung sinnvoll sind. Je mehr Dimensionen definiert werden, desto aussagekräftiger ist das Modell und desto expliziter und wirksamer sind die davon abgeleiteten Maßnahmen. Steigende Anzahl der Dimensionen sowie erhöhte Genauigkeit des Modells sind allerdings mit höheren Kosten durch Einsammlung und Verarbeitung des erhöhten Datenvolumens verbunden. Aber auch die Validierung und der Vergleich (beobachtbare Ereignisse, gemeinsames Verständnis von akzeptablen Regeln) werden schwieriger. Der Modellanwender muss für sich eine Balance zwischen Nutzen und Kosten finden.

Erhöhte Anzahl der Dimensionen bedeutet ebenfalls erhöhte positive Korrelation zwischen den gewählten Dimensionen. Die Auswirkung eines Ereignisses in allen relevanten Dimensionen korrekt darzustellen ist die zweite Hauptherausforderung für den Anwender.

Eine Entscheidung muss der Anwender ebenfalls im Voraus treffen, für wie lange eine bestehende Beurteilung gültig ist. Das Modell basiert auf historischen Daten. Wäre der Gültigkeitszeitraum zu lang, werden die Unternehmen benachteiligt, die sich rechtzeitig durch Eigeninitiative verbessern. Wäre der Gültigkeitszeitraum zu kurz, verlangt dies eine schnelle Aktualisierung der Daten, die mit hohen Aufwänden und einer hohen Herausforderung an die Technologie verbunden ist. Ebenfalls können keine stabilen Maßnahmen für den kurzen Zeitraum durchgeführt werden.

25.3.2 Bewertungskatalog

Die Hauptherausforderung für einen Bewertungskatalog stellt die Datenqualität dar. Derzeit liefern drei separate Datenquellen (vgl. Abschn. 25.2.2) Inputdaten an das chinesische Sozialkreditsystem. Solange diese nicht in einer einheitlichen und integrierten Plattform konsolidiert werden, besteht das Problem der Datenqualität sowie der Datenvollständigkeit. Ein Beispiel ist die Baihang Credit: Jede Teilnehmeragentur erhält eine gleichen Anteil von acht Prozent, wobei die angelieferte Datenmenge und der Umfang der Daten nicht einheitlich sind. Selbstverständlich kann die National Credit Information Sharing Plattform durch administrative Maßnahmen die Teilnehmer zwingen, so weit wie möglich alle Daten zur Verfügung zu stellen. Es ist allerdings nicht auszuschließen, dass die Teilnehmer sämtliche Daten lediglich für sich zu behalten.

Darüber hinaus besteht eine Manipulationsmöglichkeit für die Unternehmen, weil 75 Prozent der verarbeiteten Daten öffentlich verfügbar sind. Bewusst können sie deren Score durch bestimmte Aktivitäten verbessern, die sonst für keinen in der Gesellschaft Mehrwert bringen.

Im Umkehrschluss können die restlichen 25 Prozent nicht veröffentlichte Daten Unvorhersehbarkeit und Unsicherheit für alle beurteilten Unternehmen verursachen. Dies erweist sich auf der anderen Seite als Vorteil für die Behörden, weil die Unternehmen dadurch gezwungen werden, sich zum Erzielen der besten Scores in allen Dimensionen engagieren.

Ein weiterer Nachteil für die Unternehmen besteht darin, dass die Korrektur einer negativen Fehlbeurteilung nicht erzwingend gewährleistet wird. Selbst wenn die Korrektur stattfindet, nimmt der Prozess viel Zeit und Mühe zur Beschaffung der Evidenzen in Kauf. Während des Korrekturzeitraums finden sich die Unternehmen weiterhin in einer Beurteilungsstufe, die eventuell mit Malusmaßnahmen verbunden ist.

25.3.3 Maßnahmen

Für die Anwender (Regierung und Behörden), ist die Verantwortung der Aufsicht sowie der durchzuführenden Maßnahmen klar zu definieren. Anderenfalls wird der Verantwortliche im klassischen Sinne (zum Beispiel Zollamt) seine Aufgaben auf die National Credit Information Sharing Plattform übertragen lassen.

Auf der anderen Seite können die beurteilten Unternehmen durch ein negatives Ereignis doppelt oder mehrfach bestraft werden. Ein Beispiel ist der Verstoß gegen ein Umweltschutzgesetz. Während ein Unternehmen durch Gerichtsurteile bereits hohe Strafen zahlen muss, werden gleichzeitig Scores in Compliance, Umwelt und Kreditwürdigkeit abgezogen, was zu weiteren Malusmaßnahmen führt. Eine klare Definition der Grenze ist daher erforderlich.

25.4   Auswirkung und Nutzung für unterschiedliche Sektoren

Das Sozialkreditsystem nutzen verschiedene Akteure in unterschiedlicher Form und aus verschiedenen Interessen. Diese sollen hier im Folgenden dargestellt werden.

25.4.1 Staaten

Wenn der Staat ein neues Gut zur Verfügung stellt, nimmt er Einfluss auf die Handlungen der Akteure. Wie weitgehend die Folgen und Eingriffe des SKRs für in- und ausländische Akteure sind, hängt natürlich auch mit der konkreten Ausgestaltung der Sanktionen und den Möglichkeiten der Nachvollziehbarkeit der Regeln zusammen. In diesem Artikel sol-

Marktversagen

Wirtschafts- (Verteilen Knapper Güter) **politik** (Zentrale Festlegung)

Staatsversagen

Marktversagen:
1. Natürliche Monopole, bzw. durch Absprachen gebildete Kartelle
2. Externe Effekte und öffentliche Güter*
3. Unerwünschtheit des Marktergebnisses (Verteilungsgerechtigkeit) oder Unerwünschtheit des Handelns an sich (Organhandel, Drogen)

Staatsversagen:
1. Mangelnde Information und Flexibilität in den politischen Angeboten
2. Mangelnde Kompetenz, keine Verdrängung inkompetenter Anbieter
3. Bürokratisierungstendenz
4. Unerwünschte Vermeidungsreaktionen von Interventionen (Schwarzmärkte etc.)
5. Intertemporale Überschuldung durch asymmetrische Wahrnehmung der Lasten

*Bemerkung: Ein öffentliches Gut ist ein Gut, dass nicht rivalisierend ist (d.h. der Konsum von A beeinflusst nicht den Konsum von B) und nicht ausschließbar (d.h. A kann B nicht vom Konsum des Gutes ausschließen). Ein externer Effekt bedeutet, dass der Konsum von A sich nicht auf den von B auswirkt.

Abb. 25.5 Staats- und Marktversagen. (Quelle: eigene Darstellung in Anlehnung an Samuelson und Nordhaus 2016)

len grundsätzliche Parameter betrachtet werden, die für eine Ausgestaltung eines zentralen Systems von Bedeutung sind.

Der Vorteil eines unabhängigen, nicht käuflichen Regelsetzers ist es, den Wettbewerb um Regeln zu verringern. Wenn man sich viele (wohlfahrtsökonomische) „Renten" durch Lobbying oder sogar Korruption aneignen kann, dreht sich der Wettbewerb statt um Kunden um diese Regelsysteme. Aus der Wirtschaftstheorie sind positive und negative Effekte des Staatseingriffes hinreichend beschrieben (vgl. etwa Samuelson und Nordhaus 2016: Teil 2, Kapitel 9 sowie Teil 4, Kapitel 2).

Abb. 25.5 hilft auch im Fall eines SKRs zu beurteilen, inwieweit ein (berechtigtes und begründbares) Interesse an einem staatlichen Eingriff besteht.

Ein Sozialkreditsystem aggregiert dezentral vorliegende Informationen. Werden die Informationen frei verfügbar und ohne Kosten allen Interessenten zur Verfügung gestellt, stellt dies ein öffentliches Gut für diese dar (siehe oben).

Ratingagenturen wurden in der Finanzkrise von ihren Auftraggebern bezahlt. Der Wettbewerb hat dazu geführt, dass sie es sich meines Erachtens nicht so gut leisten konnten auf bestimmte Noten zu verzichten. Ein Bewertungssystem hat häufig einen Plattform- bzw. Netzwerkeffekt, das heißt, je mehr Informationen eingehen, umso weniger ist die Einzelinformation manipulierbar und umso besser wird der Aussagegehalt für alle.

Dies gilt jedoch nur, wenn Vergleichbarkeit (Aggregation) gegeben ist und die Transparenz so hoch ist, dass eine Manipulation und Ausnutzung des natürlichen Monopols nicht diesen positiven Effekt überwiegen. Während bei einer Einzelperson die Bewertung zu einem noch massiveren Machtgefälle (Staat/Bürger) führt, das sowohl normativ als auch aus Effektivitätsgründen suboptimal ist, besteht auch für Unternehmen die Gefahr einer zu starken Instrumentalisierung der Kräfte des Verfahrens für eine Lenkung der Wirtschaft. Eine Bewertung auf Basis lokal vorliegender Normen führt zwischen den Staaten zu einer stärkeren Abgrenzung (Bevorzugung lokaler Unternehmen). Innovationen im Verhalten und neue Marktentwicklungen (siehe oben: Staatsversagen) werden bei einem zentralen Verfahren sicher eher gebremst. Basierend auf der Ausgangslage muss man allerdings konstatieren, dass sowohl individuell als auch für Unternehmen die Akzeptanz von bestimmten Normen (und damit auch deren Wirksamkeit) sich kulturell stark unterscheidet.

In einer Situation, bei der persönliche Beziehungen und monetäre Zuwendungen vor allem auf dezentraler Ebene Positionen und politische Entscheidungen beeinflussen, kann ein datengetriebener Maßstab durchaus bereinigende Effekte haben. Allerdings wird die Grundthematik durch ein solches Werkzeug nicht beseitigt solange Input/Kontrollen für das System letztlich auch von diesen Personen mitgefüttert werden.

Defizite in der juristischen Durchsetzung von Neutralitätsgeboten auf alternativem Wege ohne entsprechende Institutionen mit Entscheidungsträgern durchzusetzen, ist ohnehin ein schwieriges Unterfangen.

Es steht zu befürchten, dass mit dem positiven Effekt der besseren Ausnutzung von Daten, dem Öl der Industrie 4.0, diverse Folgekosten wie zum Beispiel eingeschränkter Wettbewerb und Innovationen einhergehen. Die Form der Verwendung und auch die Offenheit/Transparenz des Systems (nicht nur von 75 Prozent der Faktoren) sind meines Erachtens eine entscheidende Komponente bei der Beurteilung der Effekte.

25.4.2 Banken

Kreditinstitute betreiben nicht nur Risikoaggregation und Risikotransformation, sondern sie bieten Kunden auch als Trusted Partner diverse Informationen zur Anlage und Investment an. Mit der möglichen Spezialisierung von Teilfunktionen ihrer Wertschöpfungskette gibt es neue Konkurrenten, Marktbegleiter und Partner.

Die veränderte Marktposition betrifft auch das Verhältnis zu ihren Kunden. Früher waren sie Bittsteller, die etwas beantragten, heute sind es Personen, die Orientierung suchen und einige Alternativen zu zentralen Informationen bereits kennen.

Ein zentrales, kostenloses Informationsportal verändert diese Position natürlich auch. Teile des Geschäftsmodells, der Ersteinschätzung auf Basis eigener Informationen, sind schon früher an Auskunfteien oder Poolanbieter übergegangen. Statt wie traditionell durch den Schutz der Intellectual Properties (Kundeninformationen) wird also auch hier der Schwerpunkt im Aufbau von Ökosystemen und der Vermittlerplattform liegen.

Die komplizierten Einigungsprozesse über das Pooling von Informationen wie beispielsweise bei einem zentralen Betrugsregister fallen durch eine zentrale Lösung weg und ermöglichen ggf. auch Lösungen für Erlösquoten oder andere Poolinglösungen. Der Wettbewerb in einem transparenteren Markt (sofern das System funktioniert und nicht politisch agiert) führt zu einer Konzentration auf andere Faktoren der Wertschöpfung.

Die Banken können durch das Sozialkreditsystem die Unternehmen besser beurteilen und daher die Ausfallwahrscheinlichkeit besser schätzen. Kreditnehmer, die unter dem klassischen Modell abgelehnt werden, können bessere Möglichkeit haben, Kreditzusagen zu bekommen, was die Qualität der Kreditportfolien erhöhen kann.

Wie bereits im vorherigen Abschnitt erwähnt, führt eine Monopollösung aber auch dazu, dass der Anreiz erlischt, sich selbst zu verbessern und dezentral nach besten Lösungen zu suchen. Dies gilt jedoch nicht nur für die ökonomischen (häufig kurzfristigeren Kriterien), sondern auch für Kriterien der nachhaltigen Unternehmensführung. Das dynamische Umfeld in dieser ganzheitlichen Unternehmenssicht für Banken zeigt Abb. 25.6.

Die Zentralisierung der Verantwortung, sich in diesen Themenfeldern eigene Gedanken zu machen, wird durch die Regeln und Erwartungen (inklusive Durchschau auf das Netzwerk) ersetzt.

Ob in einem Umfeld sich dynamisch ändernder Faktoren und Ziele eine zentrale Lösung diese Risiken erfassen, messen und steuern kann, sollte hinterfragt werden.

25.4.3 Industrieunternehmen

Aufgrund der Multidimensionalität des Sozialkreditsystems kann ein Industrieunternehmen bewusst durch Verbesserung der Leistung in einer bestimmten oder mehreren Dimensionen ein besseres Sozialrating erzielen und Vorteile durch Bonusmaßnahmen daraus ziehen. Dies fordert allerdings eine rechtzeitige Etablierung von Prozessen zur Analyse des Bewertungskatalogs, Durchführung von Self Assessment und Verbesserung des Sozialratings durch Eigeninitiative heraus.

Abb. 25.6 Anforderungen an eine nachhaltige Geschäftsstrategie. (Quelle: eigene Darstellung)

Ferner löst die Einführung des Sozialkreditsystems mindestens teilweise das Principle-Agent-Problem und motiviert Unternehmen, über Gewinnmaximierung hinaus langfristige Geschäftsstrategien und Ziele im Interesse weiterer Stakeholder in der Gesellschaft zu verfolgen. Insbesondere trägt die Betrachtung einer hohen Korrelation zwischen dem Sozialratings des Unternehmens und dem der Geschäftsführung hierzu bei. Strategien zur Erreichung rein kurzfristiger Gewinnziele der Geschäftsführung mit hohen langfristigen Kosten sind aufgrund Scoreabzug in weiteren Dimensionen suboptimal.

Dass das Sozialrating eines Unternehmens von dem anderer Unternehmen innerhalb seiner Supply Chain abhängig ist, worauf es beherrschenden Einfluss hat, wird dieses Unternehmen in eine Lage versetzen, Verantwortung von Strategien und Tätigkeiten der vom ihm abhängigen Unternehmen teilweise zu übernehmen. Kostensenkung und Ertragserhöhung sind keine alleinigen Ziele. Dies hat einen positiven externen Effekt und führt zu Verbesserung aller Unternehmen dieser Supply Chain. Das Ergebnis für die Gesamtgesellschaft ist positiv. Die Herausforderung hierbei ist, die genaue Korrelation zu messen, inwieweit das zu beurteilende Unternehmen seinen Einfluss ausüben kann.

25.4.4 KMU

Unter dem klassischen Modell werden die KMU aufgrund mangelnder Kreditdaten benachteiligt, weil sie oft als risikoreich eingestuft oder als Mengengeschäft behandelt werden. Durch Einführung des Sozialkreditsystems stehen einerseits mehr Daten aus innovativen Finanzierungsplattformen wie Klein- und Mikrokrediten, B2B-, P2P- und P2B-Krediten sowie Crowdfunding zur Verfügung. Andererseits können KMU durch deren relativ kleines Ausmaß rechtzeitig unternehmensweit Prozesse einführen, die sicherstellen, dass bessere Sozialratings erzielt werden können und daher bessere Möglichkeiten beim Kreditantrag entstehen.

Die Analyse der Kriterien im Sozialkreditsystem und die Einführung sowie Implementierung der Prozesse sind jedoch mit ziemlich hohen Kosten verbunden. KMU fehlen ebenfalls oft das notwendige Know-how und Ressourcen. Es ist zu empfehlen, rechtzeitig damit anzufangen und auf externe Ressourcen zuzugreifen. Dasselbe gilt insbesondere für ausländische Mittelstandsunternehmen.

25.5   Diskussion und Ausblick

Nicht nur vom Staat werden Verkehrssünder mit Punkten bestraft. Es gibt auch diverse private Initiativen, wie zum Beispiel das PAYD-Modell von Kfz-Versicherern oder das PAYL-Modell von Krankenkassen oder aber auch Kredittech (Social Media Daten für Kreditwürdigkeit), die soziale Faktoren verwenden.

Im Internet werden nicht nur Daten von Kaufentscheidungen ausgewertet (wie bei Bonussystemen), sondern es ist auch möglich, die Dauer des Aufenthalts und die Links sowie Anfragen zu einem umfassenden Persönlichkeitsprofil zusammenzuführen.

Beispielsweise können anhand des Interesses innerhalb eines Films (zum Beispiel bei wiederholtem Schauen von Szenen mit Gewalt oder sexuellen Handlungen) Präferenzen oder Fantasien abgeleitet werden (neben den bereits in Youtube unterbreiteten Vorschlägen auf der Basis von Auswertungen bzgl. bestimmter politischer Interessen).

Bisher verwenden das Social-Media-Betreiber nur, um gezieltere Werbung zu schalten, im Pricing (Diskriminierung auf Basis von Vorabinformationen über die Zahlungsbereitschaft) spielt es bisher nur vereinzelt eine Rolle (Hotelbuchungen, Flüge etc.). Die gestiegene Transparenz für die Käufer muss auch auswertbar bleiben. Wenn eine Seite alle Informationen über die Interessen und Ziele der anderen Partei besitzt, führt dies zu einer asymmetrischen Verhandlungssituation, die statisch (Preisdiskriminierung) effizient, dynamisch aber ineffizient ist.

Ein zentrales Sozialkreditsystem führt diverse Informationen zusammen und liefert im Gegensatz zu einem einfachen Ratingsystem nicht nur ordinale Werte, sondern immerhin eine mehrdimensionale Skala. Ratingsysteme oder Betrugserkennungssoftware können nicht komplett transparent gehalten werden, da sonst ein Engineering betrieben wird und das tatsächliche Verhalten dann doch anders ist. Beim Sozialkreditsystem wird demgegenüber möglichst das gesamte unternehmerische (sozial-)Verhalten abgebildet und Sanktionen erfolgen nicht nur durch einen Nichtvertrag (zu dem es immer noch Alternativen gibt), sondern wirken auf die gesamte Geschäftstätigkeit.

Es besteht also die Tendenz, dass ein Sozialkreditsystem sich zu einem Ersatz-Unternehmensstrafrecht entwickelt. In einem Strafrecht gibt es aber Verfahrensregeln, Verteidigungsmöglichkeiten, den Grundsatz, dass nur zum Zeitpunkt der Tat strafbare Handlungen auch bestraft werden können. Bei einem Scoringsystem können aber frühere Handlungen später umgewertet werden (Neukalibrierung) und sich damit negativ auswirken. Einen Wettbewerb der Unternehmen/Regel- bzw. Ökosysteme nur auf einen Systemwettbewerb (frei, dezentral, chaotisch vs. vorgegeben, zentral, organisiert) hinauslaufen zu lassen, ist meines Erachtens nicht optimal. Wenn eine weitgehende Optimierung des Verhaltens nach einem gemeinsamen Maßstab erfolgt, wirkt sich dies zwar positiv auf die Effizienz aus, wird aber mit einem Verlust an Effektivität erkauft (neue Ziele, Kreativität; vgl. Erben et al. 2020).

„Westlich orientierte" Unternehmen und kritische Denker tun gut daran, sich mit der technologischen Machbarkeit rechtzeitig auseinanderzusetzen. Außerdem sollten sie die Ausgangslage der verschiedenen Regelsysteme und Kulturen bedenken. „Östlich orientierte" kulturelle Wettbewerbsdenker sollten die Folgekosten eines zentral regulierten Systems mitdenken, selbst wenn die Vielzahl an Daten und Möglichkeiten von verbesserten Angeboten zunächst Chancen verheißen. Dieser Beitrag kann nur erste Anregungen zu einer solchen Debatte liefern, die im Interesse aller hoffentlich in den nächsten Jahren intensiv geführt wird.

Literatur

CRR (2017): Capital Requirements Regulation, Veröffentlichung der Europäischen Union.

Erben, R., Kessler, B., Fornefett, A. (2020): Aktivitäten-basierte Risikosteuerung für Konsistenz von Management und Führung, FIRM Frankfurt.

Everling, O. (2014): Clean Tech Rating, Frankfurt.

Samuelson, P. A., Nordhaus, W. (2016): Volkswirtschaftslehre: Das internationale Standardwerk der Makro- und Mikroökonomie, 5. Auflage, München.

http://www.gov.cn/zhengce/content/2016-06/12/content_5081222.htm

http://www.gov.cn/zhengce/content/2014-08/23/content_9038.htm

https://www.creditchina.gov.cn/xinyongdongtai/buwei/201911/t20191113_175360.html

http://www.acfic.org.cn/ddgh/bwzc/201911/t20191118_145994.html

http://www.mofcom.gov.cn/article/b/g/201804/20180402738074.shtml

http://www.gov.cn/gongbao/content/2012/content_2144293.htm

http://www.gov.cn/zhengce/content/2019-09/18/content_5430900.htm

https://www.creditchina.gov.cn/xinyongdongtai/buwei/201905/t20190517_155874.html

http://www.amac.org.cn/cms/contentcore/resource/download?ID=8748

http://m.mof.gov.cn/tzgg/201910/P020191023289180169380.docx

https://www.creditchina.gov.cn/xinyongdongtai/buwei/201908/t20190802_164141.html

http://www.gov.cn/zhengce/content/2016-06/12/content_5081222.htm

http://www.cac.gov.cn/rootfiles/2019/09/11/1569729939897372-1569729948315802.docx

http://www.gov.cn/zhengce/content/2018-02/09/content_5265380.htm

http://www.gov.cn/zhengce/content/2016-06/12/content_5081222.htm

https://www.creditchina.gov.cn/xinyongdongtai/buwei/201911/t20191108_174769.html

https://www.creditchina.gov.cn/xinyongdongtai/buwei/201911/t20191113_175362.html

https://www.creditchina.gov.cn/xinyongdongtai/buwei/201911/t20191112_175236.html

https://www.mct.gov.cn/whzx/whyw/201910/t20191009_847147.htm

http://www.mofcom.gov.cn/article/h/redht/201907/20190702878644.shtml

Bernhard Kessler ist Leiter der Competence Unit Risk & Compliance, sowie Managing Partner und Mitglied der Geschäftsleitung bei der plenum AG, einer mittelständischen Unternehmensberatung in der DACH Region. Zudem ist er im Fachbeirat des Frankfurter Instituts für Risikomanagement und Regulierung (FIRM) und in mehreren Arbeitskreisen der Risk Management Association (RMA) tätig. Er beschäftigt sich seit über 15 Jahren mit der Weiterentwicklung von Ratingverfahren, Risikomanagementmethoden – und deren Implementierung sowie der Weiterentwicklung von in- und externen Normen.

Kaifei Jin ist selbstständiger Unternehmensberater für europäische und chinesische Kreditinstitute. Seit über 14 Jahren ist er in den Bereichen Meldewesen und Risikomanagement tätig. Er lebt seit mehr als zwei Jahrzehnten in Deutschland.

Neurobiologische und psychologische Prozesse im Zusammenhang mit Sozialkreditsystemen

26

Heike R. Dahlmann

Zusammenfassung

Sozialkreditsysteme führen in Menschen zu neurobiologischen Reaktionen und daraus resultierenden Handlungsentscheidungen. Der vorliegende Artikel untersucht die Frage, wann und warum Sozialkreditsysteme ihre gewünschte Wirkung erzielen können und in welchen Fällen das weniger oder gar nicht der Fall ist. Soziologisch zeigt sich, dass Menschen schon immer Sozialkreditsysteme verschiedenster Art in ihre Gemeinschaften integriert haben. Neurobiologisch spielen vor allem der Hirnstamm, das Limbische System und das Stirnhirn eine Rolle in der Entstehung der Reaktionsmuster. Aus den ausgelösten psychologischen Prozessen können verschiedene, positive wie negative, Wirkungen auf das Individuum und die Gesellschaft abgeleitet werden. Je mehr emotionale Sicherheit und Vorteile ein Sozialkreditsystem dem Einzelnen oder der Gemeinschaft bringt, desto leichter wird es akzeptiert, je mehr Unsicherheit und innere Konflikte entstehen, desto höher ist die Ablehnung. Die geschilderten Mechanismen berücksichtigen mögliche (informations-)technische Ausgestaltungen und können auf jegliche Art von Sozialkreditsystemen bezogen werden.

26.1 Einleitung

Sozialkreditsysteme lösen in Menschen verschiedene neurobiologische Reaktionen und darauf beruhende Emotionen und Handlungsentscheidungen aus. Doch wann und warum haben sie die gewünschte Wirkung und in welchen Fällen nicht? Um dieses spannende Thema genauer zu untersuchen, zeigt der vorliegende Artikel auf, welche soziologischen

H. R. Dahlmann (✉)
IMTEI Strategie- und Managementberatung, Weinheim, Deutschland

© Springer Fachmedien Wiesbaden GmbH, ein Teil von Springer Nature 2020
O. Everling (Hrsg.), *Social Credit Rating*,
https://doi.org/10.1007/978-3-658-29653-7_26

Hintergründe den heutigen Sozialkreditsystemen zugrunde liegen und wo Menschen schon immer Sozialkreditsysteme verschiedenster Art in ihre Gemeinschaften integriert haben. Es folgt die Darstellung der neurobiologischen Grundlagen mit den anatomischen und physiologischen Zusammenhängen und den jeweils möglichen Reaktionsmustern auf Sozialkreditsysteme. Außerdem werden die weiteren psychologischen Prozesse und Verhaltensmechanismen in Relation gesetzt zu den Eigenschaften der Sozialkreditsysteme als externe Auslöser dieser Reaktionen. Daraus können verschiedene, positive wie negative, Wirkungen auf das Individuum und die Gesellschaft abgeleitet werden.

Die hier geschilderten grundsätzlichen Mechanismen, die Sozialkreditsysteme im Menschen bewirken, berücksichtigen mögliche (informations-)technische Ausgestaltungen der Systeme im Allgemeinen, es wird jedoch kein spezifisches, derzeit eingesetztes Sozialkreditsystem zugrunde gelegt.

26.2   Sozialkreditsysteme

In unseren modernen Gesellschaften finden wir eine Vielzahl von Sozialkreditsystemen, die staatlich initiiert, von Unternehmen ausgehend oder nur in kleinen Gruppen organisiert sein können. Allen gemeinsam ist das Ziel, in der beteiligten Gemeinschaft Individuen oder Organisationen eine bestimmte Reputation oder Vertrauenswürdigkeit zuzuordnen, auf der die Beziehung zu den anderen aufbaut. Mithilfe der Klassifizierung dieser Reputation werden Verhaltensweisen sowohl des Individuums oder der Organisation als auch der Gemeinschaft beeinflusst. Diese Beeinflussung durch Sozialkreditsysteme entsteht jeweils durch spezifische Reize im menschlichen Gehirn, auf die unter Einbeziehung weiterer Parameter eine individuelle Antwort erfolgt.

26.2.1   Terminologische und soziologische Grundlagen

Das Wort Kredit hat seinen Ursprung im lateinischen „credere", was „glauben" und „Vertrauen schenken" bedeutet. Heute verstehen wir unter Kredit vor allem im finanziellen Kontext das zeitlich befristete Überlassen von (Geld-)Werten mit vereinbarter Rückzahlung. Mit dem Begriff Sozial-Kredit wird der Fokus wieder mehr auf die ursprüngliche Frage nach der Vertrauenswürdigkeit eines Gegenübers im sozialen Gefüge gelegt.

Sozialkreditsysteme in diesem Sinne gibt es, genau genommen, seit es Menschen gibt und diese in Gemeinschaften leben. Von Anfang an wurde menschliches Verhalten unter- und zueinander eingeordnet, bewertet, sanktioniert oder gefördert. Dies war und ist wichtig, um Vertrauen untereinander aufzubauen und zu erhalten, eine der zentralen Dimensionen gesellschaftlichen Zusammenhalts. Um dieses hohe Gut zu schützen, wird gegebenenfalls auf Basis der gemeinschaftlichen Werte das Vertrauen auch entzogen. Jede Gesellschaft entwickelte stets ihre eigenen Systeme, je nach soziokulturellem Entwicklungsstand. Die Vertrauenswürdigkeit und die Erwartbarkeit erwünschten Verhaltens werden getestet, die Reputation ermittelt und so die weitere Beziehung gestaltet. Dieser Prozess wiederholt sich nicht nur in Gesellschaften, sondern auch in kleinen Gemeinschaften fortlaufend mit jeder Geburt eines Kindes und hält bis zum Tod des Individuums an. Es ist das uralte Prinzip von Lob und Tadel, von Belohnung und Bestrafung, und physiologisch letztlich von Reiz und Antwort.

26.2.2 Gestaltung des Gemeinwesens mittels Sozialkreditsystemen

Derartige Verhaltensweisen, Organisationen und Systeme finden sich grundsätzlich auch in der Tierwelt. Der Mensch unterscheidet sich jedoch von den Tieren dadurch, dass er neben instinktiven, meist angeborenen und/oder primär erlernten Verhaltensweisen seine Gedanken bewusst einsetzt, um (in diesem Fall mit Sozialkreditsystemen) eine von ihm ausgewählte Gegenwarts- und Zukunftsgestaltung vorzunehmen.

Dazu kommen mittlerweile die entwicklungsgeschichtlichen Fortschritte, die der Mensch im Gegensatz zum Tier erreicht hat, sowohl auf allgemeiner intellektueller Ebene als auch durch technische und inzwischen besonders informationstechnische Erfindungen. Der Mensch ist nunmehr in der Lage, sich von seinen persönlichen, unmittelbaren Möglichkeiten der Verhaltenskontrolle anderer völlig unabhängig zu machen und dies durch ein System durchführen zu lassen, das im einfachsten Fall von ihm kontrolliert und weitergenutzt wird und im komplexesten Fall als selbstlernende Künstliche Intelligenz wiederum umgekehrt auch vom Menschen völlig unabhängig agieren kann.

Derzeit arbeiten Anbieter von Sozialkreditsystemen in der Regel mit online betriebenen Rating- oder Scoringsystemen, die auf verschiedene Datenbanken zugreifen. Diese können entweder ausschließlich einen eingegrenzten, bestimmten Zweck verfolgen, um beispielsweise die Kreditwürdigkeit oder Verkehrsverstöße zu belegen, oder sie erweitern den Zweck auf umfassende Verhaltenskontrolle in diversen Bereichen.

26.2.3 Allgemeine psychologische Phänomene von Sozialkreditsystemen

Während der Mensch sich bei konkreten und abgegrenzten Systemen wenig im persönlichen Bereich beeinträchtigt sieht, kommt bei weitergehenden Kontrollsystemen schnell ein Gefühl der Überwachung auf. Die enormen technischen Möglichkeiten im digitalen Zeitalter lassen die Ausgestaltung eines Sozialkreditsystems nicht nur faszinierend, sondern oft auch bedrohlich wirken. Dies liegt besonders daran, dass die meisten Menschen das gesamte Ausmaß der dahinterstehenden Daten und ihrer Verarbeitung kaum noch erfassen können. Sie haben das Gefühl, nicht mehr wirklich zu wissen, was alles erhoben, gesammelt und wie es mit Algorithmen ausgewertet wird, und schon gar nicht, ob das Ergebnis ihren eigenen Vorstellungen entsprochen hätte.

Dies steht im Gegensatz zu typischen menschengemachten, analogen Vorgängen. Hier können die einzelnen Komponenten noch viel besser nachvollzogen werden. Ob analoge Vorgänge tatsächlich aber beispielsweise viel gerechter sind oder Missbrauch leichter verhindert werden kann oder dies nur ein „Gefühl" ist, muss an dieser Stelle offenbleiben. Denn auch bei ganz einfachen gesellschaftsinternen Sozialkreditsystemen gibt es nicht immer Transparenz. Menschen können sowohl als einzelne als auch im Gesellschaftskontext willkürliche Entscheidungen treffen. Immer wieder werden deshalb glücklicherweise diese Mechanismen von Menschen auch wieder hinterfragt und revidiert, insbesondere wenn die Entscheidungen zu offensichtlich extremen, dann als negativ empfundenen Ergebnissen führen oder geführt haben. Die Geschichte der Menschheit zeigt eindrucksvoll auf, wie Sozialkreditsysteme missbraucht werden und aus möglicherweise anfänglicher Faszination zuletzt massive Bedrohung fast aller werden kann. Die Gestaltung eines Sozialkreditsystems findet also stets auf einem schmalen Grat zwischen Akzeptanz und Ablehnung durch die betroffenen Menschen statt und vor allem emotionale Faktoren spielen bei dieser Entscheidung eine große Rolle.

26.2.4 Integration von Sozialkreditsystemen in die Lebenswelt

Interessant ist es, dass Menschen den Mechanismus von Sozialkreditsystemen nicht nur seit Urzeiten nutzen, sondern sie auch im Laufe der Geschichte ganz selbstverständlich in viele Lebensbereiche integriert haben. Überall findet soziale Kontrolle statt, und entsprechend abgestuft wird je nach bisher gezeigtem Verhalten für die Zukunft Vertrauen geschenkt oder entzogen. Es gibt Zeugnisse, analoge Beurteilungen und Online-Bewertungen, „Likes" und „Dislikes", und es gibt entsprechende Zuordnungsnachweise wie Kennzeichen, Ausweise und Ortungssysteme. Es scheint fast so, als ob menschliche Gemeinschaften ohne solche Verfahren gar nicht auskommen könnten und es im Rahmen der Möglichkeiten liegt, dass der Wunsch nach solchen, das soziale Gefüge regelnden Systemen, dem Menschen immanent ist und deshalb trotz gleichzeitiger Kritik immer wieder viele Befürworter findet.

Zur Klärung dieser These müssen die komplexen Vorgänge im menschlichen Gehirn etwas näher betrachtet werden und festgestellt, welche neurobiologischen und psychologischen Mechanismen menschlichen Verhaltensentscheidungen zugrunde liegen.

26.3 Neurobiologische Verhaltensgrundlagen

Die im Zusammenhang mit Sozialkreditsystemen wichtigsten Teile des menschlichen Gehirns sind vor allem drei Bereiche: der Hirnstamm, das Limbische System und das Stirnhirn. Hier entstehen die jeweiligen Anteile einer Reaktion auf die von der Außenwelt zur Verfügung gestellten und über unsere Sinne empfangenen Informationen. Durch komplexe Verschaltungen können diese drei Teile sowohl miteinander als auch mit allen anderen Regionen des Gehirns arbeiten. Im Ergebnis entsteht eine Handlungsempfehlung, die der Mensch in seinem Verhalten durch Sprache, Bewegung, Emotion usw. zu einer je nach Erfordernis mehr oder minder komplexen Reaktion umsetzt.

26.3.1   Instinkt – das Reptiliengehirn

Verhaltensbiologisch soll in die Entstehung einer Reaktion auf eine Gegebenheit zunächst der Begriff des Instinkts einbezogen werden. Der Duden schreibt hierzu, es handele sich um einen unbewusst gesteuerten, natürlichen Antrieb zu bestimmten Verhaltensweisen und eine ererbte Befähigung, in bestimmten Situationen ein bestimmtes (besonders lebens- und arterhaltendes) Verhalten zu zeigen. Auf einen bestimmten Reiz folgt eine bestimmte Reaktion. Instinktive Verhaltensweisen werden neurobiologisch im Hirnstamm abgerufen, dem phylogenetisch ältesten und durch seine Lage tief unten im Schädel am besten geschützten Gehirnteil. Dieser Teil ist für alle lebenswichtigen, ohne bewusste Steuerung ablaufenden Vorgänge verantwortlich und deshalb bei allen Wirbeltieren zwingend vorhanden. Bei niederen Wirbeltieren wie den Reptilien ist dies der überwiegende Teil des Gehirns, daher wird der Hirnstamm oft auch als Reptiliengehirn bezeichnet. Der Hirnstamm spielt bei allen Abläufen im Gehirn eine Rolle, entweder als Ausgangspunkt oder als Verbindungs- und Umschaltzentrale von Nerven und Nervenbahnen. Hirnstamm und Instinkt sind also grundsätzlich an menschlichen Verhaltensentscheidungen beteiligt, jedoch ist nicht unmittelbar festzulegen, in welchem Ausmaß. Aus dem Reptiliengehirn kommen schnelle, unmittelbare Reaktionen, bei denen Menschen nur bedingt bewusst entscheiden, eher spielen aus früheren Erfahrungen angelernte Sofortreaktionen die Hauptrolle. Manche Reaktionen sind so schnell, dass sie schon reflexhaft anmuten, allerdings sind die echten Reflexe im Rückenmark direkt verschaltet und nur eine zusätzliche Information gelangt in weitere Hirnregionen.

26.3.2   Sozialkreditsysteme und Instinkt

Instinktiv können Menschen ebenso wie Tiere auf Reize reagieren, die die gegenwärtige Situation betreffen, diese soll dadurch unmittelbar besser werden. Menschen vermeiden zum Beispiel Dinge, die ihnen gefährlich erscheinen und wechseln die Straßenseite beim Anblick einer merkwürdigen Person oder Situation. Andererseits kann der Instinkt in der gleichen Konstellation auch Neugier und Entdeckergeist aufrufen, wenn aus der Situation eine positive Erfahrung resultieren könnte. In diesem Sinne („Gefahr droht") gehört zu primär instinktiv ausgelösten Verhaltensweisen auch das reflexartige „Auf die Bremse treten" bei Anblick eines Verkehrsblitzers als Bestandteil eines der einfachsten sozialen Kontrollsysteme.

Daneben sind aber auch instinktive Verhaltensweisen bekannt, die in die Zukunft wirken. Allerdings werden beim Menschen und gerade im Zusammenhang mit Sozialkreditsystemen für längerfristig wirksame Handlungen sofort weitere Hirnregionen hinzuge-schaltet. Allein der Instinkt ist bei Menschen nicht in der Lage, ein wirklich langfristig arterhaltendes Verhalten abzurufen, da Menschen aus vielen verschiedenen Gründen nur noch sehr begrenzt Folgen und Tragweite des eigenen Handelns in der weiteren Zukunft unmittelbar abschätzen können. Zu diesen Gründen gehört mangelndes Wissen über die Zusammenhänge der Problematik, das Gefühl unzureichenden Einflusses oder fehlender Kontrolle auf die Situation beziehungsweise die Gesamtheit der weiteren Ereignisse ebenso wie mangelnde Vorstellungskraft der möglichen Resultate aus der persönlichen Aktion. Alle diese Faktoren sind durch die Komplexität des menschlichen Daseins be-dingt. So wird der Instinkt in vielerlei Hinsicht von Verhaltenseinflüssen aus höheren Hirnregionen wie Emotionen und Bewusstsein überlagert.

26.3.3 Soziale Interaktion – das Säugergehirn mit dem Limbischen System

Als evolutionär weit entwickeltes Säugetier besitzt der Mensch als nächste Stufe zusätz-lich zum Reptiliengehirn ein Säugergehirn. Auch dieser Teil des Gehirns ist entwicklungs-geschichtlich schon sehr alt und liegt gut geschützt in tieferen Regionen, unmittelbar ober-halb des Hirnstamms. In diesem findet sich unter anderem das sogenannte Limbische System. Diese Hirnregion ist anatomisch nicht exakt abgegrenzt, sondern bildet eine funk-tionelle Einheit für im weitesten Sinne soziale Interaktionen und Reaktionen.

Das Limbische System ist vor allem für Emotionen, Lernen und Antrieb zuständig. Angst, Wut und Trauer finden sich hier ebenso wie Lust, Liebe und Spiel, Glück und Un-glück. Positive Ereignisse rufen eine Dopaminausschüttung hervor, die in der weiteren Verarbeitung im Gehirn zur Ausschüttung opium-ähnlicher Stoffe, den sogenannten En-dorphinen, führt und damit zu Wohlbefinden und Glücksgefühlen. Nach diesem Zustand strebt der Mensch primär in allen seinen Handlungen. Auch das Gefühl, Vertrauen zu können und physische wie psychische Sicherheit zu erleben, wird positiv bewertet und es kommt zu einer Oxytocin- sowie Dopaminausschüttung. Umgekehrt möchte er negative Situationen vermeiden, die zu Furcht, einer Adrenalin- und Cortisolausschüttung und zu Stress mit allen seinen Folgen führen.

Das Limbische System ist intensiv mit weiteren Zentren verschaltet, aus denen es vor allem seine Informationen gewinnt, die es dann mehrfach einordnet, vergleicht, beurteilt und rückmeldet. Dies ist ein typischer Lernmechanismus, der gleichzeitig bestimmte Emotionen hervorruft und den Antrieb für weitere Handlungen entscheidet, der aber auch zeigt, wie komplex die Entstehung von Verhalten oberhalb von Instinkten und Reflexen durch das ständige Wechselspiel und die mehrdimensionale Verschaltung verschiedener Hirnregionen ist. Genau deshalb bestimmt das Limbische System in großem Ausmaß un-ser Verhalten und ist im Zusammenhang mit Sozialkreditsystemen immer aktiv.

26.3.4 Sozialkreditsysteme und Emotionen, Lernen und Antrieb

Über die Verarbeitung im Limbischen System werden alle Ereignisse, Gegebenheiten und damit auch Sozialkreditsysteme allgemein ebenso wie ihre detaillierten Bestandteile und Auswirkungen emotional empfunden. Durch den jeweiligen Anstieg von Hormonen und Botenstoffen im Gehirn kommt es zu einer positiven oder negativen, immer noch unbewussten, Bewertung. Eine Person fühlt sich zum Beispiel je nach Vorerfahrung möglicherweise sicherer (oder auch nicht) in der Beziehungsgestaltung zu ihrem Gegenüber durch die schlichte Existenz einer Videokamera oder eines Finanzratingscores.

Das Lernzentrum im Limbischen System ist unmittelbar mit der emotionalen Ebene gekoppelt. Lernen an sich im Sinne von Neues erfahren befriedigt ebenso das Lernzentrum wie die wiederholte Wahrnehmung und Verbesserung persönlicher Interessen und Vorlieben. Durch diese positive Stimulation wird Dopamin ausgeschüttet und es entstehen angenehme Gefühle. Der Mensch lernt außerdem, das betreffende Erlebnis als einen Erfolg für sich einzustufen. Negative Erfahrungen, sogar die Verweigerung von Lern- und Entwicklungsmöglichkeiten, wirken sich umgekehrt aus, also in Form von Stress und unangenehmen Empfindungen und dem Wunsch, solches zu vermeiden.

Ein Sozialkreditsystem macht sich diese Mechanismen zunutze, indem es neben der Kontrolle des Verhaltens meist relativ unmittelbar eine Belohnung oder eine Sanktion als Lernerfahrung einsetzt und damit erziehend auf die Beteiligten einwirkt. Es wird sowohl die Wiederholung erwünschten Verhaltens als auch die Vermeidung unerwünschten Verhaltens erlernt und weiter verstärkt. Sehr anschaulich findet sich diese Absicht im Fahreignungsbewertungssystem des Kraftfahrt-Bundesamtes in Deutschland, aber auch in den Testversionen der eine umfassende Verhaltenskontrolle beinhaltenden Sozialkreditsysteme in Teilen Chinas.

Mit Emotionen und Lernen unmittelbar verbunden ist der Antrieb. Durch die geschilderten Abläufe entsteht im Limbischen System die unbewusste Motivation zu einer Handlung oder der Vermeidung derselben. Positive Erfahrungen werden zusätzlich positiv rückgekoppelt und verstärkt, sodass der Mensch genau diese oder eine ähnliche Erfahrung erneut erleben möchte. Der Antrieb zu entsprechendem Verhalten steigt. Für negative Erfahrungen sinkt die Motivation zur Wiederholung und der direkte Antrieb dazu fehlt. Wird eine Vermeidungsstrategie gewählt, steigt wiederum der Antrieb für das als positiv empfundene Ergebnis der Vermeidung des Negativen. Ausnahmen und paradox scheinende Reaktionen kann es geben, wenn sich verschiedene Motivationsebenen für eine Handlung überschneiden, also beispielsweise neben einer Strafe für eine Handlung daraus auch eine als Belohnung empfundene erhöhte Aufmerksamkeit der Umwelt resultiert.

Sozialkreditsysteme erhöhen oder senken die Motivation und damit den Antrieb für eine Handlung in der Regel direkt durch ihre sofortigen oder zukünftig zu erwartenden Belohnungs- und Bestrafungsoptionen. Besonders Punktesysteme, weltweit genutzt vor allem in Geschäftsbeziehungen, beeinflussen intensiv Motivation und Antrieb.

26.3.5   Bewusste Informationsverarbeitung – das Stirnhirn im Großhirn

Für menschliches Verhalten spielt des Weiteren das Groß- oder Endhirn eine ebenso große Rolle wie die tiefer gelegenen Abschnitte. Das Großhirn ist das differenzierteste und am weitesten außen und oben gelegene Gehirnareal des Menschen, mit starker Morphe und Reifung in Kindheit und Jugend. Hier werden einerseits die Informationen aus der Umwelt wie Sehen, Hören, Fühlen usw. empfangen, verarbeitet und weitergeleitet, andererseits befinden sich hier die Ausgangspunkte von Motorik und Sprache. Zusätzlich finden sich im vordersten Teil des Großhirns, dem präfrontalen Cortex oder auch Stirnhirn, Funktionen des Gedächtnisses, der Handlungsplanung und des Bewusstseins.

Mit dem Stirnhirn werden der Verstand, das wertebasierte Sozialverhalten und eine ethische Entscheidungsfindung assoziiert. Das Stirnhirn verarbeitet durch seine Verschaltungen mit den anderen Hirnarealen die Impulse aus allen Regionen und kontrolliert sie, indem es, ebenso wie das Limbische System, eigene Impulse sendet, die Informationen immer wieder neu abgleicht und bewertet, und im ständigen Wechselspiel aktivierend oder hemmend tätig wird. Außerdem liegt hier die besonders wichtige Fähigkeit, Vergangenheitskontext und Zukunftserwartung in die Handlungsplanung zu integrieren, woraus die Antizipation, das heißt das Voraussehen der Konsequenzen einer Handlung, entsteht.

26.3.6   Sozialkreditsysteme und Antizipation

Die Fähigkeit zur Antizipation ist im Zusammenhang mit Sozialkreditsystemen besonders essenziell und eine der Grundlagen, warum sie überhaupt funktionieren können. Der Mensch benutzt Lernerfahrungen, um sie in den zeitlichen Kontext einzuordnen und um gegenwärtige Handlungen nicht nur für den Moment, sondern auch für erwartete spätere Ereignisse zu optimieren. Aus dem Stirnhirn kommt also die rationale Beurteilung einer Situation oder Sache, und in enger Zusammenarbeit mit den weiteren beteiligten Hirnarealen, vor allem dem Limbischen System, wird die Antizipation der Auswirkungen der jeweiligen Reaktion vorgenommen. Nicht nur für die Handelnden selbst, sondern auch für die Wirkung eines Sozialkreditsystems ist diese Antizipation in Bezug auf einen zukünftig zu erwartenden Vorteil, gleichbedeutend mit Glücksgefühl, oder Nachteil, gleichbedeutend mit Stress, der aus dieser gegenwärtigen Handlung resultiert, bei fast allen sozialen Interaktionen von Bedeutung.

26.3.7   Plastische Ausreifung des Gehirns durch soziale Interaktion

Das menschliche Gehirn bleibt ein Leben lang fähig zur Veränderung, sodass die Anpassung an neue Situationen oder Erlebnisse aus Sicht des Individuums immer wieder optimiert wird. Dies beschreibt die sogenannte Neuroplastizität. Besonders intensiv finden neuroplastische Entwicklungs- und damit auch Reifungsprozesse in den ersten Lebensjahren statt, und der wesentliche Auslösereiz dafür ist die soziale Interaktion. Gerade das Stirnhirn wird dadurch stark zu Wachstum angeregt, der Bereich, mit dem später die bewussten Entscheidungen geplant und getroffen werden. Diesen Prozess macht sich die Menschheit seit jeher in der Erziehung zunutze, es werden in der sozialen Interaktion erzieherische Impulse gesetzt, die im Idealfall zu einer entsprechenden neuronalen Abbildung und damit neben der physischen auch zur psychischen Ausbildung gewünschten Verhaltens führen. Da der Mensch lebenslang lernen kann, können ebenso lang auch erzieherische Anregungen in der sozialen Interaktion gegeben werden. Im persönlichen Miteinander von Mensch zu Mensch sind diese aufgrund der emotional möglichen Beziehungsebene zwar am stärksten wirksam, doch die Funktionsweise an sich kann auch stellvertretend durch ein technisches Sozialkreditsystem übernommen werden. Dies funktioniert umso besser, je größer der gesellschaftliche Rückhalt des Sozialkreditsystems ist und je positiver die damit verbundenen Emotionen sind.

26.4 Soziologische und psychologische Verhaltensmechanismen

Obwohl die neurobiologischen Abläufe im Gehirn recht eindeutig sind, ist es dennoch schwierig für Menschen, sich von vornherein, ohne langes Überlegen und langfristig, für alle nachhaltig und den gemeinsamen Werten entsprechend zu verhalten. In der Regel benötigt der Mensch immer wieder Kontrolle und Rückmeldung aus der Gesellschaft. Dies hat seine Ursache unter anderem in der vorhandenen Neuroplastizität, denn auf das Gehirn wirken ununterbrochen Eindrücke und Informationen ein, die es in permanentem Abgleich mit den bisherigen Erfahrungen weiterentwickeln oder verändern können. Die aktuellen Verhaltens- und Handlungsentscheidungen werden durch die persönlichen biologischen Voraussetzungen ebenso beeinflusst wie durch die jeweiligen äußeren Gegebenheiten. Dabei können die Ergebnisse für die Handlungsentscheidung sehr verschieden ausfallen, je nach Gewichtung der unterschiedlichen Faktoren, und durchaus auch im Widerstreit miteinander stehen. Um die daraus resultierenden soziologischen und psychologischen Phänomene geht es im Folgenden.

26.4.1 Handlungsentscheidungen – individuell und im gesellschaftlichen Abgleich

Aus den vielfältigen Mechanismen, die im Rahmen einer Entscheidungsfindung für eine Handlung ablaufen, wird schnell offensichtlich, dass es nicht nur eine festgelegte Standardhandlung geben kann, sondern schon allein für ein einziges Individuum eine Vielzahl an Handlungsmöglichkeiten entstehen wird. In der Gemeinschaft bringt dann jeder Mensch durch seine unterschiedliche genetische Ausstattung eine unterschiedliche primäre Anlage seines persönlichen Gehirns mit. Die anschließenden, lebenslangen Lernerfahrungen eines

jeden Menschen nach seiner Geburt sind ebenfalls grundsätzlich individuell und ihnen liegen zusätzlich, selbst bei gleichen Erlebnissen, verschiedene Verarbeitungs- und Speiche-
rungsprozesse im Gehirn zugrunde. Unterschiedliche Erziehungs- und Sozialisationsein-
flüsse in Kindheit und Jugend formen und prägen die Gehirnentwicklung bis in die einzelnen
Verschaltungen zwischen den Nervenzellen. Aber auch später bleibt das Gehirn plastisch
und bildet mit jeder Erfahrung oder Handlung neue Verbindungen. Ganz besonders intensiv
wirken dabei emotionale Erlebnisse, also solche, bei denen die tiefen Hirnstrukturen deutlich mitbeteiligt sind.

Die Entscheidung, wie jemand in einer bestimmten Situation reagiert, wird also häufig
zwar in Sekundenbruchteilen, aber immer wieder an sich neu gefällt. Je häufiger eine solche oder ähnliche Situation schon erlebt oder geübt wurde, desto schneller und unbewusster ist die Reaktion. Zu bewussteren Prozessen und Verzögerungen im Reaktionsablauf
kommt es, wenn sich beispielsweise ein als wichtig eingeschätzter Parameter im Gefüge
ändert. In beiden Fällen aber bleibt die Entscheidung aktuell und individuell.

Im Rahmen einer Gemeinschaft und deren vertrauensvollem Zusammenwirken sind
vor allem die schnellen, unbewussten Prozesse sehr wichtig, weil die Gemeinschaft sich
so auf fast automatisch ablaufende, relativ einheitliche Reaktionsmuster für das grundlegende soziale Miteinander weitgehend verlassen kann. Deshalb trainiert jede Gesellschaft
die von ihr erwünschten Verhaltensweisen im Rahmen von Erziehung, Wertevermittlung
und Vorbildern. Gleichzeitig kontrolliert sie (mittels Sozialkreditsystemen verschiedenster
Art) ihre Ergebnisse, denn einerseits wird ein junger Mensch, der noch wenige Erfahrungen abgespeichert hat, sich mit verschiedenen Handlungsoptionen ausprobieren und daraus lernen wollen, und andererseits sind eindeutige Lernerfahrungen wichtig, um im Gehirn klare Strukturen und schnelle Verbindungen herzustellen. Aber auch reifere Menschen
benötigen immer wieder einen Abgleich ihrer Werte und Einstellungen und der daraus
resultierenden Handlungen mit denen anderer, um sich selbst zu reflektieren und in der
Gemeinschaft zu positionieren.

26.4.2 Sozialkreditsysteme zur Konditionierung und zur Lösung eines Entscheidungsdilemmas

Ein kleines Kind muss während seines Heranwachsens außerdem die Fähigkeit zur kurz-
und längerfristigen Antizipation nachfolgender Resultate seiner eigenen Handlungen
überhaupt erst trainieren. Dabei sind unmittelbar eintretende Resultate verständlicher-
weise leichter zu erlernen als später eintretende, deren Zusammenhang mit der Ursprungs-
handlung zeitlich oder inhaltlich für ein Kind unter Umständen nur noch bedingt erkennbar ist. Diese Schwierigkeit, sowohl direkte und sofortige Konsequenzen einer Handlung
als auch spätere, möglicherweise nur noch indirekte Auswirkungen derselben überhaupt
zu erkennen, bleibt im Laufe des Lebens ebenso bestehen wie das sich eventuell anschließende Dilemma der Entscheidung zwischen Handlungsoptionen, die jeweils einen unterschiedlichen Vor- oder Nachteil für beide Möglichkeiten beinhalten. Als Individuum ist

der Mensch also oft im Zwiespalt zwischen einem unmittelbaren und einem späteren Ergebnis. Er tendiert dabei je nach persönlicher und situationsabhängiger Dominanz der verschiedenen Hirnstrukturen zu der einen oder anderen Reaktion, wobei tiefere und ältere Hirnstrukturen primär mehr Einfluss haben als jüngere. Grundsätzlich möchte der Mensch für sich stets das optimale Ergebnis erreichen und möglichst schnell und intensiv ein Glücksgefühl erzeugen, doch er kann nicht wirklich voraussehen, welche Handlungsoption dies tatsächlich erbringen wird, er kann sich nur an Wahrscheinlichkeiten und persönliche Einschätzungen aus seiner bisherigen Lernerfahrung halten.

Sozialkreditsysteme sollen auf die Lernerfahrung in einer bestimmten, allgemein akzeptierten Weise Einfluss nehmen, zumal gesellschaftlich erwünschte Verhaltensweisen oft eher mit langfristigen Vorteilen für alle oder zumindest die meisten Mitglieder und weniger mit kurzfristigen Vorteilen für einzelne oder eine kleine Gruppe verbunden sind. Dies bedeutet, es soll durch die konstante Rückmeldung und die soziale Kontrollfunktion ganz konkret ein Reaktionsverhalten konditioniert werden, das für die Gemeinschaft förderlich ist.

Dass kurz- und langfristige Folgen einer Handlung nicht immer gleichzeitig und unmittelbar positive Ergebnisse und damit Wohlbefinden im Menschen bewirken, ist einer der Hauptgründe für mangelnde Nachhaltigkeit in der Gestaltung unserer Lebenswelt. Der Mensch kann entweder den langfristigen Vorteil an sich nicht erkennen oder er schätzt ihn nicht als Vorteil für sich ein. Dies kann sogar allein aus der Tatsache resultieren, dass er auf den langfristigen Vorteil lange warten muss und er diesen deshalb als für ihn zu unsicher wertet. Tatsächlich verschieben Sozialkreditsysteme durch ihren Belohnungs- und Bestrafungsmechanismus diese Abwägungen beim Menschen. In der Handlungsentscheidung wird jetzt mindestens auf der rationalen Ebene der Abgleich mit gesellschaftlich erwünschtem Verhalten betont. Ein kurzfristiger persönlicher Vorteil könnte so plötzlich durch einen ebenso kurzfristigen Nachteil aufgehoben werden, sodass der langfristige Vorteil in den Vordergrund tritt oder überhaupt erst entsteht. Häufig erkennen Menschen erst an diesem Punkt, um wie viel größer der langfristige Vorteil auch für sie selbst eigentlich ist, weil sie durch den gesellschaftlichen Eingriff in ihre Gedankenwelt neue Zusammenhänge entdecken.

26.4.3 Soziale Kontrolle und Regeln – Findung, Gestaltung und Einhaltung

In Bezug auf die Lösung des Entscheidungsdilemmas von Handlungsoptionen hat also eine soziale Kontrolle auch Vorteile für den einzelnen. Dennoch empfindet der Mensch die Kontrolle oft als einschränkend, unter Umständen sogar als bedrohlich.

Um sich diesem Empfinden anzunähern, sollen die verschiedenen Gefühle näher analysiert werden, die Sozialkreditsysteme im Menschen hervorrufen. Dies erfolgt sowohl für Sozialkreditsysteme im Allgemeinen als auch im Hinblick auf die durch die digitalen Möglichkeiten außerordentlich ausgefeilten modernen Systeme im Speziellen.

Menschen sind soziale Wesen und leben am liebsten in Gemeinschaften. Prinzipiell
besteht unter den meisten Menschen dahingehend Einigkeit darüber, dass sich Gemeinschaften für ihr grundsätzliches Funktionieren Regeln geben müssen. Dies beruht auf
Jahrtausende alten Erfahrungen der Menschheit, die von Generation zu Generation weitergegeben werden und die sich auch im täglichen Zusammenleben immer wiederholen.
Regeln, die wir gut finden, geben Sicherheit, Vertrauen und ein Gemeinschafts- und Zusammengehörigkeitsgefühl. Sie erhöhen also unser Wohlbefinden und senken das Stress-
niveau. Allerdings gibt es schon bei dem Punkt der Findung der Regeln Dissens, ganz zu
schweigen davon, welche Regeln es denn sein sollen. Die Frage, was jeder gut findet und
was nicht, kann nie abschließend und eindeutig geklärt werden und bedarf auch immer
wieder der Überprüfung. Denn obwohl der Mensch Gemeinschaften präferiert, ist er doch
gleichzeitig immer ein Individualwesen. Jeder Einzelne hat durch seine unterschiedlichen
Lern- und Lebenserfahrungen und deren Intensität seine ganz persönliche und eigene Meinung. Dieser Mechanismus ist vom Individuum genauso auch auf kleine und größere Gemeinschaften übertragbar. Unterschiedliche Gesellschaften kommen so durchaus zu völlig
unterschiedlichen Ergebnissen bei der Regelfindung und auch in der zeitlichen Entwicklung ein und derselben Gesellschaft finden sich durch veränderte Umwelten andere Präferenzen. All dies hängt mit den oben genannten Mechanismen der Entscheidungsfindung
im Gehirn zusammen und dabei einerseits mit der Ausprägung der durch häufiges Benutzen am schnellsten verfügbaren Reaktionsmuster sowie andererseits der bewussten Einbeziehung neuer Parameter in die Überlegungen. Gerade Gewohnheiten, unbewusste gesellschaftliche Prägungen und bewusste Erziehung spielen deshalb eine große Rolle in der
gemeinschaftlichen Regelfindung. Nach dieser Regelfindung funktioniert eine Gemeinschaft durch die Einhaltung der Regeln durch alle Mitglieder der Gemeinschaft, auch
durch diejenigen, die von der jeweiligen Regel vielleicht nicht überzeugt sind. Dies wird
direkt oder indirekt von den einzelnen Mitgliedern eingefordert und überwacht. Wie diese
Überwachung nun stattfinden soll, unterliegt erneut den beschriebenen Mechanismen. Ein
Sozialkreditsystem entsteht also aus einer vielfältigen Regelfindung und -gestaltung im
Rahmen der Gesellschaft. Es wird deshalb immer die unterschiedlichsten Befürworter und
Kritiker haben und dabei die unterschiedlichsten Gefühle hervorbringen.

26.4.4 Positive Emotionen durch Regeln und Sozialkreditsysteme

Das Sicherheitsgefühl durch Regeln wird gefördert durch die Entstehung einer gewissen
Ordnung der Lebensumwelt des Menschen, von Dingen, auf die Verlass ist. Unabdingbar
ist dabei die Einhaltung und Durchsetzung des vereinbarten Rechts und der Gesetze, ansonsten geht der Vorteil der Regeln ebenso wie das Sicherheitsgefühl unmittelbar wieder
verloren. Auch der Wunsch nach Gerechtigkeit und der Gleichbehandlung aller wird befriedigt, dieses untermauert das Gemeinschafts- und Zusammengehörigkeitsgefühl und
trägt so indirekt wieder zur Sicherheit bei, denn eine starke Gemeinschaft kann sich gegen
innere wie äußere zerstörerische Faktoren besser wehren. Alle diese Punkte erfüllt ein

Sozialkreditsystem basierend auf gemeinschaftlich erarbeiteten Regeln und deren Überwachung. Auch Vertrauen untereinander wird durch diese Mechanismen gestärkt, der Begriff, der sich in der Namensgebung des Systems versteckt. Mit der Verhaltenskontrolle durch ein technisches System wird das regelkonforme Verhalten zusätzlich auch außerhalb unmittelbarer menschlicher Interaktion eingeübt und dieses durch das viel häufigere Abrufen im Gehirn so gebahnt, dass es zunehmend automatisiert ablaufen kann. Automatisierte Abläufe im menschlichen Nervensystem sind bezüglich einer definierten Reaktion nicht nur schneller, sondern auch erheblich zuverlässiger als bewusst gesteuerte.

26.4.5 Änderung der Entscheidungskriterien durch Sozialkreditsysteme

Die Nutzung eines Sozialkreditsystems berücksichtigt auch den Fakt, dass sich durch verschiedene äußere Umstände die Handlung eines Menschen in der gleichen Situation unmittelbar verändert. Auch dies beruht auf den oben geschilderten Abläufen zur Entscheidungsfindung, weil durch die weiteren Umstände die Abwägungsprozesse im Gehirn im Sinne einer Verschiebung der Vor- und Nachteile beeinflusst werden. Ein typisches Beispiel ist es, eine von der Gesellschaft als negativ klassifizierte Verhaltensweise, wie Müll einfach in die Umgebung zu entsorgen, doch durchzuführen, wenn es niemand bemerken kann. Jetzt überwiegt der Vorteil der Bequemlichkeit, also sich nicht mehr um den Müll kümmern zu müssen, das Problem unmittelbar gelöst zu haben, die Hände frei zu haben usw. den daraus resultierenden Nachteil, für alle die Umgebung verschmutzt zu haben und sich in einer unschönen Umgebung wiederzufinden. Wird aber die Handlung bemerkt und unter Umständen geahndet, überwiegen sofort der Nachteil der Scham vor der anderen Person und die mögliche Strafe wieder die Vorteile. Ähnlich ist es mit Handlungen, die entweder unmittelbar zwischen zwei Menschen oder zwischen Personen und anonymen Partnern wie Institutionen durchgeführt werden. Die Hemmschwelle für einen Vertrauensbruch ist viel höher unmittelbar zwischen zwei Menschen als dieselbe für eine Handlung zum eigenen Vor-, aber Nachteil für die andere Seite, wenn das Gegenüber keine Person ist oder die geschädigte Person nicht konkret identifiziert werden kann. Der Versuchung, den eigenen und direkten Vorteil zu wählen, wird hier widerstanden aufgrund des Nachteils der Scham und des zerstörten Vertrauensverhältnisses, in dieser Situation sogar mit der zusätzlichen Befürchtung, dass dies langfristig so bleiben könnte. Umgekehrt heißt dies jedoch auch, dass bei fehlender Kontrolle der Mensch durchaus Vorgaben übertritt, wissentlich und in Abwägung seiner persönlichen Vor- und Nachteile, sowohl für die unmittelbare als auch für die fernere Zukunft. Auf der Seite eines Belohnungseffektes für eine Handlung gelten die gleichen Grundsätze. Bekommt eine Person beispielsweise einen finanziellen oder sozialen Bonus für das korrekte Entsorgen des Mülls oder die Einhaltung von Vereinbarungen oder Vorgaben, wird sie schnell den erhöhten Aufwand im Moment auf sich nehmen, um langfristig den anderen Vorteil nutzen zu können.

26.4.6   Individualität und Sozialkreditsysteme

Neben dem Wunsch, einer Gemeinschaft anzugehören, hat der Mensch auch den Wunsch nach Individualität und der Berücksichtigung seiner individuellen Umstände, die ja seine ganz persönliche Meinung und Handlung bedingen. Ein Mensch fühlt sich selbstbestimmter und freier, wenn seine Individualität berücksichtigt wird. Ein Sozialkreditsystem schränkt die individuelle Handlungsfreiheit des Einzelnen per se ein und ist oft nicht in der Lage, individuelle Umstände in die Beurteilung miteinzubeziehen. Jedoch wird dies meist zunächst den Zielen der Gemeinschaft untergeordnet, die im besten Falle ja auch den Zielen des Einzelnen entsprechen sollten. Dadurch vermindert sich das aus dem Konflikt entstehende Unlustgefühl und das Wohlbefinden aus den positiven Aspekten kann in den Vordergrund rücken.

Dieses Abwägen vor Vor- und Nachteilen für sich selbst findet je nach Bedarf immer wieder statt und die Entscheidung kann sich je nach aktuellem Informationsstand und aufgetretenen Emotionen dazu zur einen oder anderen Seite hin verschieben. Die Stärke der Ausprägung der Gefühle ist zudem abhängig von der Sozialisation und den Erfahrungen des Individuums. Für den einen ist beispielsweise individuelle Freiheit und Selbstbestimmtheit ein hohes Gut, für den anderen steht das Zusammenleben in einer Gemeinschaft weit über den Bedürfnissen des Einzelnen. Gerade an diesem Punkt unterscheiden sich nicht nur Staaten und Gesellschaften, sondern sogar Parteien innerhalb von Gesellschaften sehr stark und damit auch ihre Akzeptanz von Sozialkreditsystemen.

26.4.7   Negative Emotionen und innere Konflikte
durch Sozialkreditsysteme

Hinter dem Freiheitsgedanken stehen noch weitere Gefühle, die im Zusammenhang vor allem mit umfassenden, modernen und auf künstlicher Intelligenz basierten Sozialkreditsystemen auftreten und deren Wahrnehmung als Nachteil sehr viel schwieriger durch die beschriebenen Vorteile auszugleichen ist. Dazu gehört das Unsicherheitsgefühl, tatsächlich die „richtigen" Handlungen zu tun, besonders wenn mit Fehlern unmittelbar eine Sanktion verbunden ist. In der Folge entstehen zusätzlich Ohnmachtsgefühle gegenüber einer ständigen Instanz, der der Mensch möglicherweise nicht ausweichen kann oder glaubt, nicht ausweichen zu können. Diese Gefühle werden deutlich stärker, wenn die eigene Einschätzung des Systems und seines Nutzens und die der dahinterstehenden Mehrheit oder ordnungspolitischen Macht nicht übereinstimmen. Eine weitere exponentielle Steigerung erfolgt bei Kombination der beiden vorgenannten Gefühle, besonders wenn die Definition der „richtigen" Handlung unklar bleibt und möglicherweise eine als massiv empfundene Sanktion droht. An diesem Punkt werden das gewünschte positive Ergebnis des Systems und die gefühlte Sicherheit in das Gegenteil verkehrt, in eine für den Einzelnen gesteigerte Unsicherheit. Solche Gefühle verursachen dem Menschen Stress, er versucht sie unbedingt zu vermeiden oder ihnen auszuweichen, wenn auch dies nicht möglich ist, werden Verdrängungsmechanismen in Gang gesetzt. Auch eine erfolgreiche Verdrängung aus dem Bewusstsein macht jedoch die Emotionen nicht unwirksam, auf Dauer belasten sie den Menschen verbunden mit allen weiteren möglichen psychischen und physischen Folgen.

Dies führt zu weiteren Fragen nach den Auswirkungen von Sozialkreditsystemen auf die Gesundheit des Menschen und der Überlegung, ob der Mensch einen geschützten Raum braucht, in dem er ohne jegliche Kontrolle sich ausschließlich nach seinen persönlichen Bedürfnissen verhalten kann, um gesund zu bleiben.

Aus vielerlei Erfahrung wissen wir, dass Mitglieder einer Gemeinschaft fast überall und ständig einer Kontrolle und Wertung durch die anderen unterliegen. Allerdings wird diese Kontrolle verschieden empfunden, je nachdem, als wie wohlwollend die dahinterstehende Instanz empfunden wird und wie frei sich der Mensch in seiner jeweiligen Umwelt fühlt. Dieses Freiheitsgefühl wird, analog der oben beschriebenen Zusammenhänge, von den verschiedensten Faktoren beeinflusst, vor allem aber, wie sehr sich derjenige mit den Regeln, Zielen und Absichten sowie der Gemeinschaft insgesamt identifiziert und wie einfach es ist, sich passend zu verhalten. Davon wiederum hängt das Gefühl der Beobachtung und der daraus resultierenden persönlichen Beeinträchtigung ab. Betrifft dies Situationen oder Handlungen, bei denen der Mensch keine Angst hat, Fehler zu machen, berührt ihn auch die Überwachung wenig. Betrifft dies aber Bereiche, in denen Unsicherheit bezüglich irgendeines der Parameter auftritt, kommt es zu Abwehrmechanismen, die primär auf Auflösung seines inneren Konflikts mit der Umwelt ausgerichtet sind. Kann der Konflikt auf Dauer nicht gelöst werden, wird der Mensch krank. Findet er eine für ihn selbst befriedigende Lösung, bleibt er gesund.

26.4.8 Emotionen aufgrund des technologischen Fortschritts der Sozialkreditsysteme

Ein weiterer zu beachtender Punkt in der emotionalen Reaktion auf Sozialkreditsysteme ist die technische Ausstattung derselben. Das darauf beruhende Vertrauen des Menschen ist abhängig von seinem tatsächlichen ebenso wie dem gefühlten Verständnis der Technik und der Kontrolle über das System und dessen korrekten Funktionieren.

Trotz aller unermesslichen Möglichkeiten des menschlichen Gehirns ist es in seiner Vorstellung von Dingen, die es nicht mit seinen fünf analogen Sinnen irgendwie erfassen kann und zu denen es ein ihm schon irgendwie bekanntes Ähnliches gibt, relativ begrenzt. Höhere mathematische, physikalische und informationstechnische Zusammenhänge müssen zusätzlich erlernt und trainiert werden. Neuronale Netze, künstliche Intelligenz und selbstlernende Systeme sind für Laien nur noch in den Grundzügen nachvollziehbar und entziehen sich immer mehr der menschlichen Vorstellungskraft. Der Mensch möchte aber in gewisser Weise das Gefühl haben, seine Umwelt im Großen und Ganzen verstehen zu können, dies bewirkt nach den beschriebenen Mechanismen eine Art gefühlter Kontrolle und Sicherheit. Diese Möglichkeiten werden ihm durch die technischen Fortschritte immer mehr entzogen, ganz besonders aber durch die informationstechnischen Entwicklungen. Dinge, die nicht begreifbar und einschätzbar sind, werden seit Urzeiten vom Menschen als bedrohlich eingestuft. Sozialkreditsysteme, die auf diesen Parametern beruhen, können vom Einzelnen, selbst von in diesen Bereichen fachgebildeten Menschen, kaum noch in allen Facetten erfasst werden und erzeugen so Kontrollverlustgefühle.

Jedoch nicht nur die auf Überforderung beruhende mangelnde Begreifbarkeit der einzelnen Abläufe hinter dem System, sondern auch ein Gefühl, einem nunmehr vom Menschen unabhängig agierenden maschinellen System ohne Empathie gegenüber zu stehen, lässt neben den Befürchtungen von Kontrollverlust auch Unsicherheits- und Ohnmachtsgefühle zum Tragen kommen und durch die empfundene Bedrohung außerordentlich ansteigen.

Die gleiche emotionale Reaktion kann, selbst bei grundsätzlicher Akzeptanz, auch aus unzulänglichen Funktionen resultieren. Der Mensch erlebt möglicherweise Falschverknüpfungen der erhobenen Daten oder das Auftreten von Fehlinformationen aus den Datenquellen. Die programmierten Konsequenzen sind völlig außerhalb seiner Kontrolle und damit auch außerhalb seiner Änderungsmöglichkeit. Je mehr dieser Fehler auftreten, desto größer wird die Skepsis bezüglich zukünftiger Ergebnisse und desto stärker steigt auch die Angst vor negativen Auswirkungen auf die eigene Person. Die Akzeptanz des Sozialkreditsystems sinkt dramatisch, weil es genau seinen zentralen Punkt verfehlt, nämlich Vertrauen aufzubauen.

26.4.9 Bewertung der gesellschaftlichen Legitimation von Sozialkreditsystemen

Neben der automatisierten Kontrolle und dem Belohnungs- oder Bestrafungsmechanismus aus dem Sozialkreditsystem an sich steht noch der weitere Maßnahmenapparat, der daraus abgeleitet werden kann. Hier verlässt die dahinterstehende Gemeinschaft möglicherweise den eigentlichen Pfad von ausschließlicher unmittelbarer und situationsbezogener Lob- und Tadelfunktion und versucht, das System zu weiterer Erziehung seiner Mitglieder zu nutzen. An dieser Stelle wird für jeden einzelnen die Frage wichtig, wer aufgrund welcher Legitimation wen kontrollieren oder gar erziehen darf und welche Maßnahmen für angemessen erachtet werden. Es muss im allgemeinen Diskurs geklärt werden, ob beispielsweise öffentliches Lob oder auch das Gegenteil öffentliches „Shaming" probate Mittel und tatsächlich wirkungsvolle Instrumente sind, gesellschaftlich erwünschtes Verhalten zu fördern und unerwünschtes zu reduzieren. An diesem Punkt unterscheiden sich erneut ganz erheblich die Bewertungen der Maßnahmen durch die Individuen und Gesellschaften aufgrund der unterschiedlichen individuellen Prägungen, Erfahrungen und jeweils bekannten Forschungsergebnisse. Gleichzeitig kann es schon allein durch die Tatsache, dass die Vielfalt von Möglichkeiten nicht mehr ausreichend berücksichtigt wird, dazu kommen, dass die Menschen sich dadurch eingeschränkt fühlen und dies zur Ablehnung des Sozialkreditsystems an sich führt.

26.4.10 Kreativitätsverlust und Entwicklungshemmung durch Sozialkreditsysteme

Je ausgeprägter und umfassender die Erziehung von bestimmten Verhaltensweisen durchgesetzt wird, desto intensiver kommt es zu einer Vereinheitlichung des Reaktionsverhaltens. Dies würde bedeuten, dass ein viele Bereiche des Lebens mit einbeziehendes Sozialkreditsystem die Kreativität des Einzelnen und konsekutiv die der Gesellschaft hemmt. Die Vielfalt von Reaktionsmöglichkeiten, die ein Mensch in einer bestimmten Situation entwickeln kann, wird durch die vorgegebene Beeinflussung der Reaktion im gewünschten Sinn, nach den vorgenannten Mechanismen reduziert. Wahrscheinlichkeiten von außergewöhnlichem Verhalten werden geringer. Dies ist in relativ engem Kontext, wie zum Beispiel einer konkreten Vertragssituation, unproblematisch, je umfassender und allgemeiner die Verhaltenskontrolle jedoch wird, desto größer werden die Auswirkungen eines gleichförmigen Verhaltens. Ungewöhnliche Verhaltensweisen oder Gedankengänge werden eher nicht erprobt, da sie weder einen Bonus bieten können noch die Person sicher sein kann, dass es nicht sogar zu einer Sanktion kommt. Der entsprechende Fall kann möglicherweise von einem technischen System nicht konkret oder nicht richtig erfasst werden oder ist nicht vorgesehen bzw. einprogrammiert. Allein diese Unsicherheit wird in der menschlichen Entscheidung überwiegend vermieden, weil kein Vorteil, sondern eher ein Nachteil erwartet wird. Da Kreativität für neue Ideen und Fortschritt unerlässlich ist, wäre durch deren Fehlen die Weiterentwicklung von Wirtschaft und Gesellschaft möglicherweise beeinträchtigt. Nur wenn völlig unbeeinflusster Raum für freies Denken, Ausprobieren und Entwickeln in ausreichendem Maße vorhanden ist oder zur Verfügung gestellt wird, kann sich Kreativität entfalten. Auch hier wird sich jede Gemeinschaft zur gleichzeitigen Risikobegrenzung entsprechende individuelle Regeln und Spielräume geben.

26.5 Fazit

Sozialkreditsysteme im Wortsinne werden seit Urzeiten von Menschen in der Gestaltung ihrer Gemeinschaften und für ein vertrauensvolles Zusammenleben genutzt. Der Mechanismus, auf dem diese Systeme beruhen, ist ein zutiefst menschlicher, nämlich die Anwendung von Lob und Tadel. Ursprünglich ausschließlich direkt von Mensch zu Mensch durchgeführt, werden heute daneben auch komplett digitale Systeme eingesetzt. Dennoch sind die Abläufe im menschlichen Gehirn primär recht einheitlich. Tief unten im Hirnstamm oder Reptiliengehirn kommt es zu instinktiven Reaktionen und im darüber gelegenen Säugergehirn mit dem Limbischen System werden die Grundzüge sozialer Interaktion mit Emotionen, Antrieb und Lernen beigesteuert. Das Großhirn als am weitesten außen gelegene Struktur und höchste Instanz sorgt schließlich nicht nur für die Aufnahme der Umweltinformationen, sondern auch für die Weiterleitung und Verarbeitung der Impulse aus den verschiedenen Hirnregionen und letztlich für die Handlungsausführung. Im

Stirnhirn erfolgt dabei neben der bewussten Einordnung und Bewertung der Situation auch die Antizipation der Handlungsfolgen. Im Ergebnis aller dieser Prozesse bevorzugt der Mensch Handlungen, die eigene positive Emotionen durch eine als persönlichen Vorteil empfundene Erfahrung hervorrufen. Genau an dieser Stelle setzen Sozialkreditsysteme jeglicher Art an. Sie bieten durch ihre Lob- oder Tadel-Funktion eine starke Lernerfahrung und beeinflussen gleichzeitig Emotionen und Antrieb des Menschen. Außerdem basieren sie auf der menschlichen Fähigkeit zur Antizipation, wenn die persönlichen Vor- oder Nachteile erst zu einem späteren Zeitpunkt wirksam werden.

Vor allem soziale Interaktion lässt das menschliche Gehirn zunächst plastisch ausreifen und später sich ständig durch neue Impulse verändern. So können Menschen ein Leben lang lernen und jedes Individuum besitzt durch seine persönlichen Erfahrungen in jedem Moment ein einzigartiges Gehirn. Allein schon aus diesem Grund ist eine jegliche Handlungsentscheidung stets ein individueller und aktueller Vorgang. Je nach Sozialisation des Individuums und der Gesellschaft bringen die Prozesse im menschlichen Gehirn unterschiedliche Ergebnisse hervor, in Gemeinschaften erfolgt durch die soziale Interaktion dabei ein fortlaufender interindividueller Abgleich. Für den Einzelnen kann auch durchaus ein Handlungsdilemma entstehen, vor allem zwischen kurz- und längerfristigen Ergebnissen. Sozialkreditsysteme können durch ihre konstante Rückmeldung auf bestimmte Handlungen ein konkretes Reaktionsverhalten konditionieren, ebenso das Entscheidungsdilemma zur gesellschaftlich erwünschten Seite verschieben.

Die Ambivalenz der Menschen gegenüber einer sozialen Kontrolle wird vor allem durch die Entstehung verschiedener Gefühle während der oben genannten Prozesse hervorgerufen. Schon die Regelfindung und -gestaltung sowie die Durchsetzung der Regeleinhaltung unterliegt vielfältigen Bewertungen. Sozialkreditsysteme bewirken positive Emotionen, wenn sie im Sinne der Regeln Sicherheit geben, Vertrauen fördern und das Gemeinschaftsgefühl stärken. Auch die Förderung von für die Gemeinschaft nachhaltigem Verhalten, ein für alle entstehender Vorteil, wird als positiv wahrgenommen. Dagegen wird eine Einschränkung der Individualität primär als negativ empfunden. Auch Unsicherheits- und Ohnmachtsgefühle können durch mangelnde Definition des „richtigen" Verhaltens bei gleichzeitiger Sanktionsandrohung und fehlender Ausweichmöglichkeit massiven Stress auslösen. Ohne Auflösung solcher inneren Konflikte mit der Umwelt wird der Mensch krank.

Der technologische Fortschritt der Sozialkreditsysteme lässt weitere emotionale Reaktionen zu. Kontrollverlustgefühle treten auf, wenn vor allem die informationstechnischen Entwicklungen die menschliche Vorstellungskraft weit übersteigen. Ängste vor einem maschinellen System ohne Empathie lassen die empfundene Bedrohung noch weiter wachsen. Auch Erfahrungen mit unzulänglichen Funktionen, möglicherweise außerhalb persönlicher Änderungsmöglichkeiten, führen wieder zu Unsicherheit- und Ohnmachtsgefühlen. Gerade die letztgenannten Punkte mindern die Akzeptanz von Sozialkreditsystemen dramatisch, weil der Vertrauensaufbau verloren geht.

Zu beachten ist auch die Frage nach der gesellschaftlichen Legitimation eines Sozialkreditsystems. Öffentliche Diskussion und Korrektur der Kriterien fördern ebenso wie die

Möglichkeit zu Wettbewerb und Vielfalt die allgemeine Akzeptanz. Auch die Weiterentwicklung von Wirtschaft und Gesellschaft ist abhängig von der Gestaltung und Ausprägung sozialer Kontrolle, da die Entfaltung von Kreativität auf freies Denken und Ausprobieren angewiesen ist.

Besonders im Hinblick auf die emotionalen menschlichen Reaktionen sollte der Einsatz eines Sozialkreditsystems stets wohlbedacht sein. Sicherheit und Vertrauen werden durch Übereinstimmung unter den Beteiligten über die darunterliegenden Regeln erzeugt. Gemeinschafts- und Zusammengehörigkeitsgefühle erhöhen das Wohlbefinden ebenso wie Individualität und Freiheitsgedanken, dies erfordert eine hohe Flexibilität auch in der Umgebung und der Handhabung des Systems. Die Berücksichtigung von Veränderungsoptionen trägt durch Vielfalt zur wirtschaftlichen und gesellschaftlichen Entwicklung bei und erhöht die individuelle Zustimmung zum System. Das grundsätzliche Verständnis und der Sinn können besonders durch eine Begrenzung des Umfangs des Sozialkreditsystems auf jeweils konkrete Bereiche erreicht werden. Belohnen ist durch den persönlichen Vorteil meist erfolgversprechender als Bestrafen und die Motivation zur Teilnahme am System steigt. Je höher die Akzeptanz unter den Beteiligten ist, desto besser fallen auch die Ergebnisse von Sozialkreditsystemen aus.

Die neurobiologischen und psychologischen Prozesse und Reaktionen wurden in diesem Artikel zum besseren Verständnis auf wesentliche Grundzüge der umfangreichen möglichen Abläufe im menschlichen Gehirn reduziert. Allein die genannten machen jedoch deutlich, dass in jeder Hinsicht divergierende Beurteilungen und Wirkungen von Sozialkreditsystemen entstehen können. Die unterschiedliche Gestaltung und Ausprägung derselben in verschiedenen Ländern und Kulturen zeigen dies heute schon eindrucksvoll.

Wie ausgeführt, ist die generelle Philosophie hinter jedem Sozialkreditsystem die Befriedigung des Bedürfnisses nach grundsätzlicher Ordnung und nach Sicherheit innerhalb einer Gesellschaft. Je nach Historie, Tradition und Kultur, aber auch dem aktuellen soziologischen und technischen Entwicklungsstand wird dies anders interpretiert, gestaltet und im Einsatz wahrgenommen. Die Unterschiede zwischen Ländern der EU und China sind genau aus diesen Gründen riesig, die USA beispielsweise befindet sich irgendwo dazwischen. Es kann dabei auch nicht endgültig geklärt werden, ob die letztlich erreichte Ordnung eher auf Angst (vor Bestrafung) oder auf Motivation (durch Belohnung) beruht, beides ist möglich und beides ist sogar parallel möglich. Davon abhängig ist die Unterstützung oder die Ablehnung des Systems und so wieder das Maß der damit erreichten emotionalen Sicherheit oder Unsicherheit. Weitere konkrete Forschung zu den einzelnen dargestellten Überlegungen wäre sehr wünschenswert.

Literatur

Bauer, J. (2019). *Wie wir werden, wer wir sind: Die Entstehung des menschlichen Selbst durch Resonanz*. München: Karl Blessing Verlag.

Chiao, J. Y., Hariri, A. R., Harada, T., Mano, Y., Sadato, N., Parrish, T. B., Iidaka, T. (2010). Theory and methods in cultural neuroscience. *Social Cognitive and Affective Neuroscience, 5*(2–3), 356–361. https://doi.org/10.1093/scan/nsq063

Engel, C., Singer, W. (2008). *Better Than Conscious?: DECISION MAKING, the HUMAN MIND, and IMPLICATIONS FOR INSTITUTIONS (Strüngmann Forum Reports)* (1. Aufl.). Cambridge, MA, USA: The MIT Press.

Instinkt. (2019). Abgerufen 8. Januar 2020, von https://www.duden.de/node/71494/revision/71530

Markus, H. R., Kitayama, S. (1991). Culture and the self: implications for cognition, emotion and motivation. *Psychological Review, 98*, 224–253.

McCabe, K. A. (2008). NEUROECONOMICS AND THE ECONOMIC SCIENCES. *Economics and Philosophy, 24*(3), 345–368. https://doi.org/10.1017/s0266267108002010

Pfeifer, W. et al. (2005). *Etymologisches Wörterbuch des Deutschen* (8.) (S. 730). München: Deutscher Taschenbuch Verlag.

Schäfers, A. T. U. (2019). Gehirn und Lernen – Gehirn. Abgerufen 8. Januar 2020, von https://www.gehirnlernen.de/gehirn/

Zak, P. J. (2004). Neuroeconomics. *Philosophical Transactions of the Royal Society of London. Series B: Biological Sciences, 359*(1451), 1737–1748. https://doi.org/10.1098/rstb.2004.1544

Dr. Heike R. Dahlmann beschäftigt sich mit der ganzheitlichen und nachhaltigen Gestaltung von (digitalen) Veränderungsprozessen in Unternehmen. Dabei denkt sie als promovierte Humanmedizinerin und innovative Strategieberaterin die Prozesse immer auch vom Menschen aus. Vor der Gründung ihrer IMTEI – Strategie- und Managementberatung war sie als Ärztin, Psychotherapeutin, Coach und Dozentin klinisch und wissenschaftlich tätig und leitete als Geschäftsführerin einen großen Sozialdienstleister. Sie ist Lehrbeauftragte an der Dualen Hochschule Baden-Württemberg.

Teil VII

Funktionen und Anwendungsbeispiele

Betriebswirtschaftliche Betrachtungen zum Social Credit Rating

27

Thomas Pache

Zusammenfassung

In diesem Beitrag wird das Thema Social Credit Rating aus konsequent betriebswirt-schaftlicher Unternehmensperspektive beleuchtet. Zukünftig zu erwartende Risiken, Kosten, aber auch die aus einem unbestechlichen Social Credit System resultierenden Chancen werden gegenübergestellt und Handlungsalternativen abgewogen, um Unter-nehmensentscheidern einen ersten Überblick zu geben.

Wenn wir uns in der westlichen Welt mit Social Credit Rating (SCR) beschäftigen, so geschieht dies häufig vor einem ethisch-moralisch-politischem Hintergrund. Nicht selten reduzieren wir unsere Betrachtungen des Chinesischen Social Credit Systems (SCS) ausschließlich auf natürliche Personen (Stichwort „DSGVO"), um dann, von einem scheinbar moralisch erhöhten Standpunkt aus, orwellistische Zukunftsszenarien zu beschwören.

Dieser Beitrag richtet seinen Fokus (ausgehend von der Fiktion eines vollständig umgesetzten SCS in der Volksrepublik China) jedoch nur auf den „unternehmens"-re-levanten Teil des zukünftigen SCS und teilt sich in acht Fragen und Antworten auf. Die unternehmerische Perspektive steht bei allen Betrachtungen, möglichen Risiken und Handlungsoptionen im Vordergrund.

27.1 Gibt es SCR-ähnliche Ansätze auch in Deutschland?

Ja, es existieren aber nur Insellösungen, die häufig nicht staatlich kontrolliert sind.

T. Pache (✉)
RiskPoint, Frankfurt am Main, Deutschland
E-Mail: thomas.pache@aon.de

© Springer Fachmedien Wiesbaden GmbH, ein Teil von Springer Nature 2020
O. Everling (Hrsg.), *Social Credit Rating*,
https://doi.org/10.1007/978-3-658-29653-7_27

Auch wenn wir dies eher unkoordiniert und siloweise tun, in Deutschland nutzen wir bereits eine Reihe von SCR-Parametern in unterschiedlichsten Geschäfts- und Verwaltungsprozessen zum Treffen teilweise gravierender Entscheidungen. Anders als bei dem hier betrachteten, umfassenden chinesischen Ansatz liegen in Deutschland wesentliche Teile der Datengenerierung und Verarbeitung in den Händen privater Unternehmen. Auch sind teilweise mögliche Ursache-/Wirkungszusammenhänge für die beurteilten (Unternehmen) teilweise nicht oder zumindest nicht ad ante sichtbar.

Am bekanntesten sind finanziell ausgerichtete „Bontätsratings", die von Unternehmen wie der Schufa (Bonität von Privatpersonen) oder Ratingagenturen wie S&P, Moodys oder Fitch (wirtschaftliche Stärke von Unternehmen) angeboten werden. Aber auch teilweise indirekt von Unternehmen wie Creditreform, Atradius, Hermes (Kreditwürdigkeit für Unternehmen) betriebene SCS beeinflussen wirtschaftliche Entscheidungen.

Im Bereich des sozialen Verhaltens von Privatpersonen werden polizeiliche Führungszeugnisse, das Verkehrszentralregister Flensburg, Vermieterbescheinigungen und Pre-Employment-Screenings, zum Beispiel in sozialen Netzwerken (auch wenn unter DSGVO-Gesichtspunkten zweifelhaft), genutzt.

Es gibt jedoch auch in Deutschland punktuell siloübergreifende Ansätze sowohl von privatwirtschaftlicher als auch von staatlicher Seite. Als Beispiel seien genannt:

- Das Geschäftsmodell der Firma Kreditech (www.kreditech.com/company). Kreditech vergibt im Falle eines positiv verlaufenden Internet-Screenings (Stichwort „Big Data") Kleinkredite an Privatpersonen.
- Das Gesetz zur Einrichtung und zum Betrieb eines Registers zum Schutz des Wettbewerbs um öffentliche Aufträge und Konzessionen (WRegG).
- Der seit dem 24.08.2017 geänderte § 44 Abs. 1 StGB: „Wird jemand wegen einer Straftat zu einer Freiheitsstrafe oder einer Geldstrafe verurteilt, so kann ihm das Gericht für die Dauer von einem Monat bis zu sechs Monaten verbieten, im Straßenverkehr Kraftfahrzeuge jeder oder einer bestimmten Art zu führen. Auch wenn die Straftat nicht bei oder im Zusammenhang mit dem Führen eines Kraftfahrzeugs oder unter Verletzung der Pflichten eines Kraftfahrzeugführers begangen wurde, kommt die Anordnung eines Fahrverbots namentlich in Betracht, wenn sie zur Einwirkung auf den Täter oder zur Verteidigung der Rechtsordnung erforderlich erscheint oder hierdurch die Verhängung einer Freiheitsstrafe oder deren Vollstreckung vermieden werden kann."

27.2 Was definiert den chinesischen SCR-Ansatz aus Unternehmenssicht?

Unbenommen der Diskussionen um die politische Dimension handelt es sich beim Chinesischen SCS um ein daten- und algorithmengetriebenes, holistisches System zur Kontrolle und Sanktionspriorisierung in Abhängigkeit von der Compliance von Unternehmen und natürlichen Personen

In dem Artikel „Eine lückenlose Compliance mit den geltenden Bestimmungen ist die beste Vorsichtsmaßnahme" vom 03.12.2019 in der Neuen Zürcher Zeitung nennt Markus Herrmann, Direktor bei Sinolytics, fünf Besonderheiten des Chinesischen SCS (vgl. Settelen 2019):

1. Erweiterung des bekannten Kriteriums „Kreditwürdigkeit" auf „holistische und moralisch betrachtete Rechtstreue"
2. Transparenz der Daten und Folgen für die Öffentlichkeit
3. Staatliches Bewertungsmonopol anstelle privater Unternehmen oder Stakeholder
4. Effizientere und konsequentere Ausnutzung von Daten verschiedener Quellen durch aktuelle Technologien (Big Data)
5. „Joint Punishment"-Ansatz, um auch juristische Personen nicht mehr nur in „Regulierungssilos" zu betrachten, sondern Kollektivsanktionen der gesamten Bürokratie zu legitimieren.

Verglichen mit der eher als zersplittert zu bezeichnenden SCR-Landschaft in Europa ist die Idee eines von staatlicher Seite ohne Gewinnerzielungsabsicht betriebenen, „überlagernden regulatorischen Metasystems zur Steigerung der Effektivität der Rechtsordnung und Rechtstreue unter Firmen" grundsätzlich begrüßenswert, vorausgesetzt, es gelingt, rechtsstaatliche Regeln, deren Umsetzung und die Generierung von Sanktionen in einem zentralen Computersystem sicher abzubilden.

27.3 Welche grundsätzlichen Chancen und Risiken aus Unternehmenssicht lassen sich bereits jetzt erkennen?

Eine wesentliche Chance stellt die zu erwartende höhere Rechtstreue im chinesischen B2B-Geschäftsverkehr dar. Ein wesentliches Risiko sind institutionell verhängte Wettbewerbsnachteile durch SCS-Complianceverletzungen.

Nach seiner vollständigen Inbetriebsetzung dürfte das SCS wesentlich dazu beitragen, dass sich die Vertragstreue chinesischer Unternehmen gegenüber ihren europäischen Partnern, aber auch untereinander deutlich verbessert. In einem transparenten und für alle Marktteilnehmer gleichermaßen geltenden SCS werden sich diejenigen Unternehmen zumindest temporäre Wettbewerbsvorteile erarbeiten können, denen es am schnellsten gelingt, ihr Unternehmens-Compliancesystem dazu zu bringen, die messbaren Parameter der im SCS abgebildeten Vorschriften bestmöglich zu optimieren. Legt man Frederik Herzbergs Zwei-Faktoren-Theorie zugrunde, verfügt das zukünftige SCS nicht nur über Hygienefaktoren, sondern auch über Motivatoren, um Unternehmen zu rechtskonformem Verhalten zu bewegen.

Die für das Rating zuständige Regierungsbehörde hat exemplarische Beispiele genannt: Häufigere Import-Export-Kontrollen können als Sanktion genutzt werden, um nicht vertrauenswürdige Unternehmen zu bestrafen. Basierend auf den Angaben der chinesischen Zollverwaltung lag die durchschnittliche Kontrollquote für ein Unternehmen, dem die Verwaltung nicht traut, im ersten Halbjahr 2019 bei 98,12 Prozent, während sie bei einem AEO-zertifizierten Unternehmen (das entspricht dem bestmöglichen Bewertungsergebnis unter der Zollauthentifizierungsbewertung) nur 0,5 Prozent beträgt. Ein rechtskonform handelndes Unternehmen wird sich also eine Menge administrativer Aufwendungen ersparen.

Die in der Natur eines derart umfassenden SCS liegende Komplexität wird unter anderem dadurch deutlich, dass die Verschlechterung eines Ratings aus „sachfremden" Bewertungsbereichen mittels der geplanten Algorithmen automatisch zu einer Herabsetzung des Zollauthentifizierungsratings führt. Damit ist das Zollauthentifizierungsrating der zentrale Baustein des SCS. Ihm kommt also gleichzeitig eine messende, eine fachliche und auch eine wesentliche Sanktionsfunktion zu (vgl. hierzu auch Abschn. 27.6).

Neben der Erlangung von Vorteilen, die häufig für natürliche Personen gelten, scheint das SCS-Anreizsystem für Unternehmen seine Schwerpunkte auf die Vermeidung von Nachteilen durch die Sanktionierung von Fehlverhalten zu setzen. Die Studie der Sinolytics GmbH und der European Union Chamber of Commerce in China nennt hier folgende beispielhafte Sanktionen:

Joint-Sanctions-Mechanismus
Wird ein Unternehmen aus Sicht des Zolls als nicht vertrauenswürdig eingestuft, so beeinflusst dies die Beurteilung anderer Regierungsbehörden bei der Entscheidung über andere Arten von Zertifikaten und Genehmigungen, wie zum Beispiel Umweltschutzgenehmigungen. Diese bei uns allenfalls als Ausnahme existierende sachverhaltsübergreifende (vgl. § 44 Abs. 1 StGB) bzw. allumfassende Sanktionierung von Fehlverhalten ist bekanntlich das Kernstück des SCS.

Unternehmensklassifizierung durch die staatliche Devisenverwaltung
Eine Herabstufung der Unternehmensklassifizierung führt auch hier zu strengeren Kontrollen der Devisentätigkeit. Derartige Kontrollen können durch unterschiedliche staatliche Institutionen durchgeführt werden: NDRC, SAFE, PBOC, China Securities Regulatory Commission (CSRC), Ministerium für Industrie und Informationstechnologie (MIIT), Finanzministerium (MOF), Handelsministerium (MOFCOM).

Ausschluss von öffentlichen Ausschreibungen
Unternehmen mit einem negativen Rating dürfen nur noch eingeschränkt an Ausschreibungen teilnehmen oder werden in schwerwiegenden Fällen komplett von der Teilnahme an Ausschreibungen für den öffentlichen Sektor ausgeschlossen.

Public Blaming and Shaming
Besonders bei Unternehmen, deren Geschäftserfolg maßgeblich von einem guten Ruf abhängt, dürfte eine öffentliche Schuldzuweisungen und „Schande" ein wirkungsvolles Mittel sein. Dieser virtuelle Pranger wird unter anderem durch die CreditChina-Website

(https://www.creditchina.gov.cn/) und das Nationale Unternehmens-Reputationssystem realisiert.

Auswirkungen auf die Unternehmensverantwortlichen und gesetzlichen Vertreter

Das chinesische SCS trennt nicht komplett zwischen natürlichen Personen und Unternehmen, sondern stellt eine Verbindung zwischen dem Ratingsystem für Unternehmen und dem für Privatpersonen her. So können Unternehmensverantwortliche sogar von Reisebeschränkungen betroffen sein, wenn ihr Unternehmen Zölle oder Bußgelder und Strafen nicht zahlt. Nach entsprechender gerichtlicher Anordnung kann dieser Personengruppe sogar verboten werden, Grundstücke zu kaufen oder Versicherungen mit hohen Prämien abzuschließen. Auch dürfen diese Personen keine neue Stelle als gesetzlicher Vertreter übernehmen, solange ihr Unternehmen „auf der schwarzen Liste" steht.

Glaubt man den Aussagen von Sinolytics GmbH und der European Union Chamber of Commerce in China, funktioniert diese Verbindung sogar bi-direktional. Je nachdem, wie konsequent die Verbindung zwischen dem Ratingsystem für Unternehmen und dem für Privatpersonen ausgebildet sein wird, könnten sogar ganze Unternehmen durch das Fehlverhalten ihrer Mitarbeitern diskreditiert werden, selbst wenn dieses Fehlverhalten nur privater und nicht beruflicher Natur war.

27.4   Wann wird es für die Unternehmen ernst?

Hier gibt es unterschiedliche Einschätzungen. Die chinesische Regierung hat die offizielle Einführung ihres SCS für spätestens Ende 2020 angekündigt.

Auch wenn in der öffentlichen Diskussion um das Chinesische Social Credit System meist nur das System für Personen Erwähnung findet, geht doch bereits aus den ersten Planungsdokumenten aus dem Jahr 2014 hervor, dass ein Unternehmens-SCS von Anfang an geplant war. Nicht nur die European Union Chamber of Commerce in China, sondern auch andere näher mit dem Entwicklungsstand vertraute Organisationen und Personen gehen davon aus, dass das SCS für Unternehmen Ende 2020 operativ geschaltet wird, während der Start des SCS für natürliche Personen sich vermutlich noch etwas verzögert.

Es wäre jedoch leichtsinnig daraus zu folgern, dass außerhalb der Testzonen keine „teil- oder unvernetzten" Segmente erprobt werden bzw. aktiv sind. Da derzeit noch deutliche Lücken in der Datenerfassung und -verteilung existieren, ist zu erwarten, dass die geplanten, über Algorithmen festgelegten Sanktionen oder Belohnungen für Unternehmen und deren Führungspersonal derzeit (Januar 2020) noch nicht oder noch nicht in vollem Umfang spürbar sein werden. Das bedeutet, dass es in Bezug auf die Ursachensetzung des Unternehmensratings zwar langsam ernst wird, SCS-Reaktionen/Auswirkungen jedoch noch nicht oder um einige Monate verzögert auftreten werden.

27.5   Wie verhält es sich mit der Kostenbelastung durch SCS-Konformitätsanstrengungen?

Erwartbar signifikant, aber alternativlos
Allein aufgrund der Vielzahl der beabsichtigten und bereits realisierten Bewertungs- und Sanktionsparameter des zukünftigen SCS wird es für Unternehmen kein leichtes Unterfangen sein, herauszufinden, welche der aus der derzeit ständig anwachsenden Zahl von Social-Credit-Richtlinien für das eigene Geschäftsmodell relevant sind. In seinem Interview mit der Neuen Zürcher Zeitung berichtet Markus Herrmann davon, „dass derzeit in hoher Kadenz neue Social-Credit-Richtlinien erlassen werden" (vgl. Settelen 2019). Neben der ständigen Beobachtung des in der Entstehung begriffenen SCS wird insbesondere die Adaption der Vorgaben im Unternehmens-Compliancesystem und die darauf aufsetzende Implementierung neuer Prozesse zu einem deutlichen Investitionsbedarf in das unternehmenseigene Compliancesystem und in SCS-Fachkräfte führen. Da eine gesetzeskonforme Unternehmensführung auch nicht an Dritte outgesourct werden kann (Compliance as a Service), wird es vermutlich „teuer". Laut Markus Herrmann, Direktor bei Sinolytics in Zürich gibt es für den Themenkomplex SCS sogar bereits eine neue „C-Suite"-Funktion, den Chief Social Credit Officer (CSCO) (vgl. Settelen 2019).

Allerdings lassen die zu erwartenden Belohnungs-/Bestrafungssystematik (Stichwort: „Joint Punishment"), die im ultimativen Fall zur Ausgrenzung von Unternehmen und damit zur Vernichtung ihrer Existenzgrundlage führen kann, die zu tätigenden SCS-Investitionen als das kleinere Übel erscheinen. Das gilt zumindest dann, wenn ein Unternehmen beabsichtigt, auch zukünftig seine Geschäftstätigkeit im Geltungs- und Wirkungsbereich des Chinesischen SCS erfolgreich fortzuführen.

Da nicht davon ausgegangen werden kann, dass es nach der „Scharfschaltung" des chinesischen SCS für Unternehmen eine Übergangzeit geben wird, empfiehlt es sich, alle notwendigen Maßnahmen baldmöglichst zu budgetieren und einzuleiten.

27.6   Welche technische Chancen und Risiken existieren?

Neben politischen Risiken gelten für das SCS dieselben technischen Chancen und Risiken wie für andere hochkomplexe IT-Systeme.

Es ist eine ganze Reihe eher operativer Risiken beim Betrieb des IT-gestützten SCS vorstellbar. Abstrakt betrachtet, können diese in Ursachen, Wirkungen und Schäden unterschieden werden.

Als Ursachen kommen technische Störungen oder Fehler der Software, aber auch mögliche strukturelle Unzulänglichkeiten in Betracht:

- Wie kann sichergestellt werden, dass es ein funktionierendes Business Continuity Management in den vom SCS abhängigen Behörden, Organisationen und Unternehmen gibt, für den Fall dass das System oder wesentliche Daten zumindest zeitweise nicht zur Verfügung stehen (zum Beispiel wegen eines Strom- oder Datenübertragungsnetzausfalls)?

- Gibt es Instanzen, an die sich ein betroffenes Unternehmen wenden kann, wenn es glaubt, von einer Verwechslung oder einem Berechnungsfehler der Entscheidungsalgorithmen benachteiligt worden zu sein?
- Da einige Fachbereiche, wie zum Beispiel die Steuerthematik, sowohl Sensor-, Verstärker- als auch Sanktionsfunktion haben, ist zumindest der Fall vorstellbar, dass die Herabstufung der steuerspezifischen Ratingwerte nicht nur zu einer steuerlichen Sanktion führt, sondern auch die einer Herabstufung anderer Ratingwertewerte initiiert, die dann wiederum ihrerseits eine negative Auswirkung auf das steuerspezifische Rating haben usw. Es müssen also bereits im SCS-Aufbau Maßnahmen ergriffen werden, um eine systemimmanente „Negativspirale" der Ratings betroffener Unternehmen zu vermeiden.

Hinsichtlich der Wirkung nutzt der Autor die Definition der für ein SCS unverzichtbaren Informationssicherheit. Diese wird ihrerseits durch drei Säulen garantiert: erstens die Verfügbarkeit, zweitens die Integrität und drittens die Vertraulichkeit von Daten und Systemen. Auch wenn der folgende Begriff nicht direkt dazu gehört, ist es insbesondere bei der Betrachtung eines SCS sinnvoll, auch den Begriff der Authentizität bei den Betrachtungen zu berücksichtigen. Damit ergeben sich folgende Informationssicherheitsverletzungen, denen sich verschiedenste Szenarien zuordnen lassen:

- Verlust der Vertraulichkeit: Sofern die SCS-Infrastruktur nicht nur Daten verarbeitet, die jedermann zugänglich sein sollen, müssen diese Daten geschützt werden. Gerade bei großen Datenmengen mit einer Vielzahl von Schnittstellen und Zugriffsberechtigten ist es jedoch an anderer Stelle zum Vertraulichkeitsverlust gekommen, wie zum Beispiel in Indien (vgl. Holland 2018; vgl. Sapkale 2019).
- Verlust der Integrität: Hier sind sowohl der Integritätsverlust des SCS selbst (zum Beispiel durch unberechtigte Zugriffe oder die Einflussnahme auf Prozesse der Datenverarbeitung) als auch der bestimmter Daten (zum Beispiel ein kriminelles Geschäftsmodell: Reputationsoptimierer; vgl. Koetsier 2018) denkbar.
- Verlust der Verfügbarkeit: Ein temporärer Verfügbarkeitsverlust des zentralen SCS ist aufgrund vielfältiger Ursachen, wie man sie von anderen IT-Systemen kennt, leicht vorstellbar. Hinzu kommt ein ggf. dauerhafter Verlust von Daten, den man zumindest mit geeigneten Back-up-Konzepten in Grenzen halten kann.
- Verlust der Authentizität: Hier sind neben den aus Science-Fiction-Filmen bekannten distopischen Vorstellungen von Identitätsübernahmen realer Personen (zum Beispiel durch Kaperung biometrischer Merkmale) oder auch aufgrund fehlerhafter Algorithmen, wie im Fall der bekannten chinesischen Geschäftsfrau Dong Mingzhu, die mit einem Werbefoto ihres Gesichts auf einem Bus verwechselt wurde (vgl. o. V. 2018), zu nennen.

Sofern dann in einer zukünftigen Ausbaustufe Algorithmen durch künstliche Intelligenz abgelöst und zum Beispiel Presseauswertungen etc. in die Gesamtbewertung einfließen, sind auch Fehlentwicklungen wie seinerzeit beim Microsoft-Chatbot (vgl. Steiner 2016) denkbar.

Für deutsche Unternehmen ergeben sich aber auch näherliegende Problemfelder: Wie verhält sich die auch für natürliche Personen geplante „Prangerfunktion" des SCS zu den geltenden Bestimmungen der Datenschutzgrundverordnung? Aus Unternehmenssicht wäre es hier wahrscheinlich sinnvoll, dass die EU eine ähnliche fiktionale Feststellung trifft, wie sie dies mit den USA (Stichwort: Privacy Shield) getan hat, da ansonsten ein unlösbarer Rechtsnormenkonflikt droht.

27.7   Wie sieht das SCS der Zukunft aus?

Technologiegetriebene Weiterentwicklungen (Stichwort „AI"), steigende Integration und mittelfristig Internationalisierung werden sowohl die Bedeutung als auch die Anfälligkeit des SCS steigern.

Um das SCS in dem angestrebten Rahmen und Umfang erfolgreich betreiben und weiterentwickeln zu können, müssen riesige Datenmengen in kurzen Zeiträumen sicher erfasst, gespeichert und verarbeitet werden können. Was zwangsläufig zu den Buzzwörtern Big Data und AI führt. In diesem Artikel verwendet der Autor den Begriff „Artificial Intelligence (AI)" zur Beschreibung sogenannter neuronaler Netze, oder des Deep Learnings und hält sich damit an in die in nächsten zehn Jahre erwartbaren AI-Entwicklungen anstelle der aus landläufigen Science-Fiction Betrachtungen.

Folgende Herausforderungen müssen gemeistert werden bzw. werden zumindest punktuell zu Problemen führen:

- Um einen Informationsfluss von und zu einer riesigen Anzahl von „Sensoren", Behörden und Infrastrukturbetrieben/Unternehmen gewährleisten zu können, muss ein hochperformantes und -verzweigtes Datennetzwerk errichtet und betrieben werden.
  - Sensoren und Erkennungs-/Zuordnungsalgorithmen (vgl. oben genannten „Gesichtserkennungsbug"-Beispiel)
  - Behörden (es müssen nicht nur Informationen ausgetauscht, sondern „konvertiert" werden); zudem müssen Arbeitsabläufe und Priorisierungen aus dem SCS heraus vorgegeben werden.
  - Infrastrukturbetriebe/Unternehmen (Datenlieferanten aber auch Teil der Exekutive; Stichwort „Reise-/Transportbeschränkungen" etc.)
- Um eine Echtzeitverarbeitung sämtlicher Daten realisieren zu können, muss ein zentral geregeltes (nicht gesteuertes) Computersystem entweder in Form zentraler Recheneinheiten oder aber, was wahrscheinlicher ist, in Form ergänzender autonomer, aufeinander abgestimmter Recheneinheiten errichtet und betrieben werden.

- Zur Sicherstellung der Integrität und Verfügbarkeit der zu erwartenden enormen SCS-Datenmengen muss ein manipulationssicheres Erhebungs-, Verarbeitungs-, Speicherungs- und Back-up-System errichtet und betrieben werden (Stichwort „Ticketing-Systeme").
- Sofern bestimmte SCS-Daten (zum Beispiel Auswertungs- oder Metadaten) Vertraulichkeit genießen sollen, müssen diese entsprechend „versteckt" bzw. verschlüsselt werden (wer genießt hier Vertrauensstellungen und wie werden diese auditiert?).
- Wenn selbstlernende Systeme zum Einsatz kommen sollen, was vor dem Hintergrund der zu bewältigenden Datenverarbeitungsaufgaben wahrscheinlich ist, müssen die neuronalen Netze akribisch trainiert werden, um unerwünschte Effekte wie zum Beispiel das Overfitting zu vermeiden.
  - Tücken der AI (Beispiele des „rassistischen" IBM Chat Bots)
  - Beispiel mit dem von AI zu bauenden Roboter

Es gilt als sicher, dass das SCS für Unternehmen in weniger als einem Jahr (zunächst in Form einer einem unvollständigen Puzzle gleichenden teilautomatisierten Ratinglandschaft) eingeführt werden wird. Die schiere Größe und Komplexität dieses Unterfangens wird dazu führen, dass das SCS mit der Zeit immer umfassender und integrierter werden wird, letztendlich wird es jedoch auch noch in mehr als zehn Jahren immer noch eine Baustelle sein, so wie Gaudis Kathedrale in Barcelona.

Sollte sich das SCS für Unternehmen tatsächlich als funktionierendes, transparentes und objektives Werkzeug zur Steuerung unternehmerischen und menschlichen Handelns erweisen, wird es einen immer größer und damit wichtiger werdenden Teil der Weltwirtschaft prägen.

In einem solchen Fall erwartet der Autor, dass es in anderen Staaten, Volkswirtschaften oder Handelsorganisationen, zum Beispiel der WTO (und natürlich im Rahmen der jeweiligen rechtlichen Rahmenbedingungen), Nachahmer finden wird.

27.8 Ist ein Risikotransfer als ein Baustein des Risikomanagements denkbar?

Mit Hilfe einer Versicherung könnten finanzielle SCS-Aufwendungen oder -Verluste allenfalls temporär und in begrenztem Umfang abgesichert werden.

Wie bereits zuvor ausgeführt, wird die permanente Nichtbefolgung vorgegebener Verhaltensmaßregeln durch ein Unternehmen zwangsläufig zu dessen Ausscheiden führen. Der hierdurch entstehende ultimative Schaden dürfte jedoch nicht versicherbar sein.

In einem bestimmten Rahmen dürften jedoch bekannte Unternehmensrisiken, wenn auch auf Basis einer neuen Ursache, versicherbar sein. Es sind dies: Reputationsverlust, politische Risiken, Betriebsunterbrechung und Contingent BI (Betriebsunterbrechung durch Ausfälle in der Lieferkette).

Denkbare Versicherungslösungen haben die folgenden beiden Prämissen gemein:

- Nicht versicherbar ist das wissentliche Abweichen von Gesetzen und Vorschriften bzw. die Pflichtverletzung durch die Repräsentanten der versicherten Unternehmen.
- In China wird die Versicherung finanzieller Folgen von SCS-Sanktionen (unabhängig von anderen versicherungstechnischen Restriktionen) höchstwahrscheinlich nicht zulässig sein. Daher können Versicherungslösungen wahrscheinlich nur für chinesische Tochterunternehmen ausländischer Eigentümer und dann auch nur über die Versicherung des „finanzielle Interesses der Muttergesellschaft" (Stichwort: FINC, Financial Interest Clause) geboten werden.

Reputationsverlust: Für die Absicherung der finanziellen Folgen eines Reputationsverlusts stellt sich zudem die Frage der Ursache-/Wirkungsallokation sowie der quantitativen Bemessung des Schadens, weil zum Beispiel ein andauernder geschäftlicher Nachteil keine abschließende Berechnung erlauben würde. Versicherer, die sich an solche Risiken herantrauen, werden sicherlich vorab Haftzeiten einführen.

Politische Risiken: Inwiefern ein transparentes von Gesetzen und Verordnungen getragenes Steuerungs- und Sanktionssystem als „politisches Risiko" gesehen werden kann, ist fraglich. Sollte, wie von einigen Kritikern befürchtet, das SCS von der chinesischen Regierung als Instrument politisch-wirtschaftlicher Manipulationen missbraucht werden, wäre eine solche Risikotransferlösung grundsätzlich denkbar. Dies könnte insbesondere bei transnationalen Produktionsprozessen von Interesse sein.

Betriebsunterbrechung: Sofern die Fortsetzung (von Teilen) der betrieblichen Tätigkeit eines Unternehmens als SCS-Sanktionsmaßnahme untersagt werden würde, könnten die daraus resultierenden finanziellen Verluste via FINC versicherbar sein.

Contingent BI: Diese Versicherungslösung deckt Betriebsunterbrechungen/-beeinträchtigungen für den Fall ab, dass zum Beispiel Zulieferer ausfallen und dadurch die Wertschöpfungskette des Versicherungsnehmers beeinträchtigt wird. Denkbar wären hier internationale Kooperationen oder transnationale Wertschöpfungsketten. Da die Komplexität aus Sicht der Versicherer sehr hoch ist, dürften der Aufwand zur Risikoanalyse und damit die zu erwartenden Prämien eher hoch sein.

Zusammenfassend kann festgehalten werden, dass SCS-spezifische Versicherungslösungen sicherlich nur von wenigen Versicherern angeboten und wahrscheinlich eher in Form von finanziell limitierten Erweiterungen bestehender Versicherungen auftreten werden.

27.9 Welche Handlungsalternativen ergeben sich nach Abwägung der bekannten Rahmenbedingungen und Restriktionen?

Alternative 1: „Verzicht auf Teilnahme am chinesischen Markt"
Abstinente Unternehmen müssen (zumindest kurzfristig) kein gesteigertes Augenmerk auf Gesetzeskonformität mit chinesischen Vorschriften legen. Die Opportunitätskosten durch

das Außerachtlassen eines solch großen und zukünftig noch wichtiger werdenden Marktes dürften mittel- bis langfristig jedoch gravierend sein.

Alternative 2: „Abwarten. Nichts wird so heiß gegessen, wie es gekocht wird."
Hierdurch spart ein Unternehmen kurzfristig die Aufwendungen zur Identifizierung und Erreichung gesetzeskonformen Verhaltens, wird aber höchstwahrscheinlich mittelfristig einen hohen wirtschaftlichen Preis, bis hin zur Aufgabe des Geschäftsbetriebs in China, zahlen müssen.

Alternative 3: „SCS-konforme Aufstellung des Unternehmens"
Diese Alternative scheint bei kurzfristiger Betrachtung die kostenintensivste zu sein. Sie ist jedoch wahrscheinlich die langfristig einzig erfolgversprechende, weil heutige Verhaltensmuster (inklusive Schmiergeldzahlungen etc.) in einem funktionierenden SCS kaum noch realisierbar sein dürften.

27.10 Wie lauten die Kernaussagen dieses Beitrags?

Die Kernaussagen dieses Beitrags lassen sich folgendermaßen darstellen:

- De facto gibt zwei SCS, das für natürliche Personen und das für Unternehmen.
- Beide SCS werden miteinander vernetzt sein, Wechselwirkungen sind beabsichtigt und erwünscht.
- Aus Sicht ausländischer Unternehmen wird die disziplinierende Funktion des SCS sowohl den chinesischen Markt als auch das Verhalten lokaler Geschäftspartner verlässlicher machen.
- Die chinesische Regierung plant die Einführung des SCS für Unternehmen bereits für Ende des Jahres 2020.
- Mit seinem SCS-Ansatz führt China ein neues Evolutionskriterium für Unternehmen ein, das die bekannten wirtschaftsdarwinistischen Mechanismen entscheidend verändert.
- Die dauerhafte Missachtung der Vorschriften des „Corporate Compliance Systems" durch ein Unternehmen wird mittelfristig unausweichlich zu dessen „Aussterben" im chinesischen Markt führen.
- Ob und inwieweit der Chinesische SCS-Ansatz politisch instrumentalisiert werden wird, wird sich erst in der Zukunft herausstellen und ist daher aus unternehmerischer Sicht (zunächst) irrelevant.
- Unternehmen, die in China erfolgreich sind oder werden wollen, kommen nicht umhin, sämtliche für sie relevanten Regeln zu identifizieren und danach zu handeln.
- Die Einführung einer SCS-Compliance ist für die im chinesischen Markt tätigen Unternehmen mit erheblichen finanziellen Belastungen verbunden, diese sind jedoch alternativlos.
- „Abwarten und Tee trinken" ist die falsche Strategie.

Eine gute Informationsgrundlage für Interessierte bietet die von *Sinolytics GmbH* und der *European Union Chamber of Commerce in China* veröffentlichten Studie „The Digital Hand – How China's Corporate Social Credit System Conditions Market Actors".

Abschließen möchte der Autor seinen Beitrag mit zwei Zitaten, die unsere Ausgangsposition als Europäer mit Blick auf das entstehende chinesische Social Credit Rating treffend beschreiben:

> „Von den Chinesen könnten wir einiges lernen. Man hat mir gesagt, sie hätten ein und dasselbe Schriftzeichen für die Krise und für die Chance." Richard von Weizsäcker
> „Man muss aufpassen, dieser Markt kann einen wirklich bestrafen." (o. V. 2020), Sergio Balbinot, Allianz Vorstandsmitglied über den chinesischen (Versicherungs-)Markt

Literatur

Holland, M. (2018): Indien: Wohl mehr als eine Milliarde Personendaten aus staatlicher Datenbank abgegriffen; unter: https://www.heise.de/newsticker/meldung/Indien-Wohl-mehr-als-eine-Milliarde-Personendaten-aus-staatlicher-Datenbank-abgegriffen-3934463.html, zuletzt abgerufen am 09.03.2020.

Koetsier, J. (2018): Hacking China's Social Credit: Cheaters Claim Millions In Cash, Instant Promotions, And Preferential Dating, unter: https://medium.com/@johnkoetsier/hacking-chinas-social-credit-cheaters-claim-millions-in-cash-instant-promotions-and-87ad89ed5c6f, zuletzt abgerufen am 09.03.2020.

o. V. (2018): Panne in China: Gesichtserkennung verwechselt Bus mit Fußgänger, unter: https://www.faz.net/aktuell/gesellschaft/menschen/gesichterkennung-in-china-verwechselt-bus-mit-fussgaenger-15905254.html, zuletzt abgerufen am 09.03.2020

o. V. (2020): Allianz steht bereit für Expansion in China – Gespräch mit Vorstandsmitglied Sergio Balbinot, unter: https://www.boersen-zeitung.de/index.php?li=1&artid=32291&titel=Allianz-steht-bereit-fuer-Expansion-in-China-%96-Gespraech-mit-Vorstandsmitglied-Sergio-Balbinot, zuletzt abgerufen am 09.03.2020.

Sapkale, Y. (2019): Aadhaar: Data of 78.2 million Persons Breached. Is UIDAI as Impenetrable as It Claimed to the SC?, unter: https://www.moneylife.in/article/aadhaar-data-of-782-million-persons-breached-is-uidai-as-impenetrable-as-it-claimed-to-the-sc/56887.html, zuletzt abgerufen, 09.03.2020.

Settelen, M. (2019): Eine lückenlose Compliance mit den geltenden Bestimmungen ist die beste Vorsichtsmassnahme; in: Neue Züricher Zeitung vom 03.12.2019.

Steiner, A. (2016): Künstliche Intelligenz – Zum Nazi und Sexisten in 24 Stunden, unter: https://www.faz.net/aktuell/wirtschaft/netzwirtschaft/microsofts-bot-tay-wird-durch-nutzer-zum-nazi-und-sexist-14144019.html, zuletzt abgerufen am 09.03.2020.

Thomas Pache verantwortet den Bereich Cyberversicherungen bei einem großen skandinavischen Assekuradeur. Er beschäftigt sich seit mehr als 25 Jahren mit Technologierisiken und deren Absicherungsmöglichkeiten. Pache ist Mitglied in den Arbeitsgruppen Cyber und IT-Haftpflichtversicherungen des Gesamtverbands der Deutschen Versicherungswirtschaft (GDV) sowie Dozent bei der Deutschen Versicherungsakademie (DVA) und Autor von Fachbüchern und Artikeln zu dieser Thematik.

China, das Corporate Social Credit System und die Implikationen für das Management deutscher Unternehmen

28

Thomas Solbach

Zusammenfassung

Der Aufsatz verschafft Unternehmern und Managern einen prägnanten Einblick in die äußerst vielschichtigen Implikationen für das Management deutscher Unternehmen, die aus der Einführung des umfassenden nationalen chinesischen Sozialkreditsystems für Unternehmen erwachsen. Neben einer Skizzierung der Kernelemente des neuen Ratingsystems wird das Thema auch ausführlich in die relevanten Aspekte der chinesischen Kultur eingeordnet.

28.1 Das Reich der Mitte, seine Gesellschaft und Kultur

28.1.1 Grundlagen des kulturellen Verständnisses

China begreift sich als diejenige Kulturnation der Erde mit der längsten erfolgreichen Geschichte. Die kulturellen Charakteristika im Reich der Mitte unterscheiden sich dabei fundamental von denjenigen in der westlichen Welt. Aufgrund der rasanten ökonomischen und technologischen Entwicklung der letzten Dekaden unterliegen die gesellschaftlichen Facetten ebenso signifikanten Wandlungen. Daher ist das Begreifen sowohl der historischen als auch der kontemporären chinesischen Kultur grundsätzlich als conditio sine qua non sowohl für den gelingenden Markteintritt als auch für den langfristigen Erfolg unternehmerischen Handelns in China anzusehen.

T. Solbach (✉)
FORTIS GmbH, Frankfurt am Main, Deutschland
E-Mail: solbach@fortis.gmbh

© Springer Fachmedien Wiesbaden GmbH, ein Teil von Springer Nature 2020
O. Everling (Hrsg.), *Social Credit Rating*,
https://doi.org/10.1007/978-3-658-29653-7_28

Im Folgenden werden diejenigen gesellschaftlichen und kulturellen Wesensmerkmale skizziert, die zum weiteren Verständnis im Zusammenhang mit dem Sozialkreditsystem und dessen Akzeptanz von Relevanz sind. Die Betrachtung ist dabei naturgemäß idealisiert. Die individuelle Lebenswirklichkeit der Menschen weicht von den Tendenzaussagen mitunter sehr deutlich ab, wobei Alter und Status (Einkommens- und Vermögenssituation) eine besondere Rolle spielen.

28.1.2 Gesicht als soziokulturelles Reputationskonzept

Der chinesische Kulturkreis verfügt über ein soziokulturelles Reputationskonzept für Individuen: 面子 (miànzi), das Gesicht. Dieser Begriff bezeichnet das Ansehen einer Person, ihren Ruf und die soziale Anerkennung und Wertschätzung, die ihr von anderen auf Basis ihrer realen oder zumindest scheinbaren Position, ihrer Leistung und ihres Verhaltens entgegengebracht wird. Je nach Konstellation können auch Position, Leistung und Verhalten von engen Bezugspersonen, insbesondere Familienangehörigen in die Beurteilung mit einfließen. Im Zeitablauf, quasi in jeder Situation und Interaktion mit anderen unterliegt das Ansehen einer sozialen Evaluation.

Chinesische Menschen legen traditionell großen Wert auf ihr in diesem Sinne verstandenes Gesicht. Ein Gesichtsverlust soll stets verhindert werden, wenn es irgendwie möglich ist. Eine der absoluten Prioritäten menschlichen Handelns liegt daher in der Wahrung des eigenen Gesichts, also der Aufrechterhaltung der Anerkennung, durch richtiges Verhalten und die Erfüllung der Anforderungen in den verschiedenen sozialen Rollen. Hierzu gehört auch die Wahrung des Gesichts anderer.

28.1.3 Erfolgsorientierung

Der generelle Wille, erfolgreich zu sein, ist in China auf allen Ebenen wesentlich ausgeprägter als in Deutschland. Es werden wirklich ambitionierte Ziele gesetzt, und die hierzu notwendigen Maßnahmen werden Schritt für Schritt, zumeist sehr schnell und vor allem konsequent umgesetzt. Und erreichter Erfolg wird mit Selbstverständlichkeit und ohne große Bescheidenheit nach außen gezeigt. Die chinesische Form der Erfolgsorientierung hängt letztlich auch mit der zuvor skizzierten Gesichtskultur zusammen und zeitigt in vielen Bereichen Auswirkungen.

So basiert beispielsweise oftmals ein beachtlicher Teil des Konsums auf der mit den gekauften Produkten verbundenen Signalwirkung und weniger auf einem realen Bedarf. China stellt nicht von ungefähr inzwischen die größte Luxuskonsumentengruppe der Welt.

Das individuelle Streben nach Erfolg im Allgemeinen und Wohlstand im Besonderen kommt auch in einer sehr starken Geschäftsorientierung zum Ausdruck. Praktisch in jeder Familie spielt Unternehmertum eine Rolle, oft eine zentrale, zumindest aber eine periphere. Das Verhalten vor allem auch im Geschäftsleben ist neben Umsetzungsstärke, Schnelligkeit und Ungeduld, markant auch durch Spontaneität, Flexibilität und Pragmatismus geprägt.

28.1.4 Qualitätspräferenz

Die Zahl der zahlungskräftigen Konsumenten in China steigt stetig. Solche Verbraucher möchten in allen Bereichen möglichst das beste Produkt kaufen, jedenfalls ein qualitativ hochwertiges. Hierbei ist die Wahrnehmung von Qualität entscheidend. Sie orientieren sich daher stark an aktuell noch meist westlichen Markenartikeln und möchten dabei absolut sicher gehen, dass sie Originalware beziehen. Die deutsche Qualitätskultur wird in China konsequenterweise sehr geschätzt.

28.1.5 Guanxi als soziokulturelles Netzwerkphänomen

Die zentralste Figur für die Blüte der chinesischen Kultur und damit zugleich eine der einflussreichsten Persönlichkeiten der Menschheitsgeschichte ist der Philosoph 孔夫子 (Kǒng Fūzǐ), Konfuzius. Ihm und seinen Schülern verdankt die Welt die Entwicklung des Konfuzianismus, „des ältesten und wirkmächtigsten Gedanken- und Wertesystems der Welt mit fundamentaler Prägekraft für die gesamte ostasiatische Region bis zum heutigen Tage" (Osten 2007, S. 19). Nach konfuzianischer Vorstellung ist der Mensch grundsätzlich beziehungsorientiert. Ferner kann der Aufbau einer starken Hierarchie interpersoneller Beziehungen zur sozialen und wirtschaftlichen Ordnung in der Gesellschaft beitragen (vgl. Luo et al. 2012, S. 142).

Vor diesem Hintergrund hat sich ein tiefverwurzeltes spezifisch chinesisches soziokulturelles Netzwerkphänomen herausgebildet: 关系 (guānxi), die Beziehung(en). Dieser Begriff bezeichnet sehr gute persönliche Beziehungen, die bewusst genutzt werden. Neben dem persönlichen Charakter und dem instrumentalisierten Einsatz von Guanxi lassen sich typischerweise noch weitere grundlegende Dimensionen dieses Phänomens beobachten: Vertrauen als Basis, Reziprozität des Verhältnisses mit impliziten gegenseitigen Verpflichtungen und Langlebigkeit der Beziehungen (vgl. Zhang und Keh 2010, S. 126; vgl. Luo et al. 2012, S. 142). Die Gesamtheit der individuellen Netzwerke aller Chinesen, also der Beziehungsgeflechte jedes Einzelnen, bildet gleichsam ein real existierendes soziales Netzwerk.

Die chinesische Gesellschaft ist bis heute deutlich mehr beziehungs- als regelorientiert. Die individuelle Einbindung in soziale Strukturen wie die Familie, die regionale Gemeinschaft sowie bildungsweg- und arbeitsplatzbezogene Gruppen ist traditionellerweise stark ausgebildet. Hinzu kommt, dass die umfeldbezogenen sowie auch die darüber hinaus absichtlich kultivierten Beziehungen zu anderen Menschen für alle Lebensaspekte (des Einzelnen) eminent wichtig sind. Dies gilt auch und besonders in der geschäftlichen Sphäre. Denn gerade wirtschaftliche bzw. administrative Entscheidungen basieren häufig auf Guanxi mit Geschäftspartnern bzw. Regierungsbehörden. Das Phänomen spielt also nicht nur auf der individuellen Ebene sondern auch auf der Ebene von Organisationen eine herausragende Rolle. Aus unternehmerischer Perspektive können insbesondere die Verbindungen der Führungskräfte als ein leistungsstarkes strategisches Werkzeug betrachtet werden, um Geschäftsabläufe zu erleichtern, Wettbewerbsvorteile zu erhalten und überragende Unternehmenserfolge zu erzielen (vgl. Luo et al. 2012, S. 139 f., 142).

Die Bedeutung von Guanxi ist daher bisher ungebrochen und hat mit der dynamischen ökonomischen Entwicklung eher noch zugenommen (vgl. Zhang und Keh 2010, S. 126). Menschen auf jeder sozialen Stufe und in jeder beruflichen Position versuchen Guanxi fortlaufend zu pflegen und weiter aufzubauen, um ihre jeweilige Situation ständig zu verbessern.

28.1.6   Technologieaffinität, insbesondere Digitalaffinität

Die kontemporären chinesischen Menschen mögen Modernität sehr. Sie sind jederzeit offen für Neues, probieren Innovationen gerne aus und nutzen intensiv und vertrauensvoll die ständig neuen technologischen Möglichkeiten, passend zu ihren Wesenszügen der Spontaneität, der Flexibilität und des Pragmatismus.

China kann sich daher heute wieder, jedenfalls was Technologieaffinität, insbesondere Digitalaffinität anbetrifft, als fortschrittlichste Zivilisation der Welt betrachten. Eine Sicht, die bis ins neunzehnte Jahrhundert hinein für den größten Teil der europäischen Geschichte bereits seinem durchaus berechtigten Selbstverständnis entsprach, wenn auch damals vor allem im Hinblick auf seine soziokulturelle Werteorientierung (vgl. Osten 2007, S. 18). Vor diesem Hintergrund liegt die enorme Bedeutung von Social Media und E-Commerce auf der Hand. Der chinesische Internethandel stellt inzwischen den weltweit größten elektronischen Markt dar.

Massive Investitionen in die digitale Forschung und Entwicklung sowie Infrastruktur haben den Grundstein gelegt für die beispiellose Metamorphose der chinesischen Gesellschaft in eine einzigartige Digitalgesellschaft, in der 24/7-Konnektivität, Mobilitätsprimat und All-in-one-Social-Media-Applikationen inklusive Zahlungssystemen bereits jetzt und FacePay schon fast gelebte Realität sind. Ein kosmopolitischer Beobachter könnte bei uns in Deutschland dagegen zunehmend den Eindruck gewinnen, dass wir in gewisser Weise in der alten Zeit stehengeblieben sind.

28.1.7   Autoritätsakzeptanz

China hat in seiner Geschichte sehr, sehr lange unter einer einzigen zentralen Autorität gelebt. Das Hierarchiebewusstsein war zu jeder Zeit deutlich ausgeprägt. Auch heute noch werden beispielsweise der Staatspräsident, die Regierung, die Chefs im Beruf, die Professoren, die Lehrer, sogar vorbildliche Mitschüler sowie Menschen der älteren Generation und dabei besonders die Eltern und Großeltern als Autoritäten akzeptiert, respektiert und anerkannt. Man sieht sich hierbei im Einklang mit der Natur und der hierarchischen Struktur ihrer komplexen biologischen Systeme.

28.1.8 Harmonie, Einheit und Stabilität als soziokulturelle Werte

Ein zentraler Ausfluss der Prägung der chinesischen Kultur durch Konfuzius und den Konfuzianismus ist die Vorstellung von einem Idealtyp des Menschen: 君子 (jūnzǐ), der Edle. Der edle Mensch ist würdevoll, verhält sich moralisch gut und strebt nach Harmonie, indem er sein sittliches Wesen mit der allumfassenden Ordnung in Einklang bringt. In engem Zusammenhang mit dem Wert der Harmonie stehen die Werte Einheit und Stabilität. So erklärt sich auch die fundamentale Bedeutung einer funktionierenden Familie und eines eigenen Hauses oder zumindest einer eigenen Wohnung. Es herrscht die feste Überzeugung vor, dass diese Form der Gemeinschaft und des Wohnens Glück und Sicherheit bringen.

Im Gegensatz zur weitverbreiteten Risikofreude bei geschäftlichen Entscheidungen, herrscht in wichtigen Aspekten des Privatlebens tendenziell eine teils drastische Risikoaversion vor, insbesondere was die Sicherheit von Leib und Leben angeht. Dies hat auch Implikationen auf das Konsumverhalten. Eine Folge ist, dass einmal zerstörtes Vertrauen in die Sicherheit bestimmter Produkte, vor allem aus dem Bereich Lebensmittel, nicht mehr wiederhergestellt werden kann.

28.1.9 Lernbereitschaft

Bei den Menschen in China existiert durchgängig eine enorme Lernbereitschaft. Für dieses Phänomen gibt es im Wesentlichen folgende Erklärungsansätze.

Die chinesische Gesellschaft ist quasi auf Bildung fokussiert. Dies läßt sich ebenfalls bereits auf Konfuzius zurückführen. So sah dieser in der Bildung u. a. eines der wesentlichen Mittel zur Erreichung des bereits erwähnten Ziels der Harmonie. Bis in die heutige Zeit hinein ist die chinesische Schul- und Hochschulausbildung ihrem hohen Stellenwert entsprechend inhaltlich breit gefächert und sehr anspruchsvoll.

Aufgrund der immensen Bevölkerungszahl des Landes befinden sich die Kinder und Jugendlichen schon früh in einer gesellschaftlichen Konkurrenzsituation, später konkret in einem harten Wettbewerb um die besten Studien- und Arbeitsplätze. Die jungen Menschen haben also einen mächtigen Anreiz, nachhaltig zu lernen und ständig lernbereit zu sein. Chinesische Eltern fordern und fördern ihre Kinder in diesem langen Prozess von Anfang an, wenn nötig und möglich auch mit massivem Einsatz finanzieller Mittel.

Schließlich spielt sogar die Natur eine äußerst wichtige Rolle in diesem Thema. Menschen beginnen schon sehr früh, ihre Muttersprache zu erlernen. Chinesisch ist eine Tonsprache mit logografischer Schrift. Die mit diesen Besonderheiten verbundenen Komplexitäten führen beim zeitgen Erlernen zu einem hohen Potenzial für die Entwicklung neuronaler Fähigkeiten. Dies wirkt sich für das gesamte Leben positiv auf die geistige Leistungsbereitschaft aus (vgl. Osten 2007, S. 20).

28.2 China und sein nationales Sozialkreditsystem für Unternehmen

28.2.1 Charakterisierung

Novissima Sinica, also das Neueste von China, titelte einst Gottfried Wilhelm Leibniz, der deutsche Universalgelehrte der frühen Aufklärung (Leibniz 1697). Sein in lateinischer Sprache verfasstes Werk über China und die Welt wirbt für einen geistigen, kulturellen, wissenschaftlichen und technologischen Austausch und ist daher von zeitloser Relevanz.

Das Neueste von China, das ist aktuell die beabsichtigte Entwicklung zu einem Reputationsstaat (vgl. Dai 2018). Die wichtigste Maßnahme zu diesem Zweck ist die Etablierung eines nationalen Sozialkreditsystems für Unternehmen bis zum Ende des Jahres 2020. Sechs Jahre zuvor hat die chinesische Regierung diesbezüglich ihren ambitionierten Plan veröffentlicht. Dessen konsequente Umsetzung schafft nun für alle in China tätigen Unternehmen erstmals einen umfassenden rechtlichen und regulatorischen Mechanismus.

Die Reputation jedes Unternehmens wird demnach durch den zielgerichteten Einsatz von Technologie mittels durchdachter Datenbankarchitektur und ausgeklügelter Algorithmen kontinuierlich und in Echtzeit bewertet und in einem Sozialkreditrating zusammenfassend ausgedrückt. Dieses dient dem Staat, seinen Behörden, wegen der zumindest partiellen öffentlichen Publizität aber auch allen Marktteilnehmern als Indiz, ob und inwieweit sich das Unternehmen in der Zukunft korrekt verhalten wird. In der Folge findet dann staatlicherseits eine direkte Belohnung von vertrauenswürdigen und Bestrafung von nicht vertrauenswürdigen Unternehmen statt. Es existieren also konkrete Anreize zu verantwortungsvoller Führung.

Basis für die Bewertung ist ein umfangreicher Datensatz, wobei die einzelnen Datenfelder durch einen unternehmensspezifischen Kriterienkatalog definiert werden. Die breite Fächerung erlaubt eine aussagekräftige Beurteilung. Die Daten werden aus verschiedensten Informationsquellen herangezogen. Es handelt sich zumeist um Vergangenheitsdaten, aber auch Gegenwartsdaten zum Verhalten des Unternehmens im Sinne der Gesetzes- und Regelkonformität werden einbezogen. Alle Daten des Sozialkreditsystems werden in einer zentralen integrierten Masterdatenbank gesammelt.

Mit der Initiierung dieses weltweit einzigartigen technologiegestützten Bewertungssystems stellt China einmal mehr seinen enormen Modernisierungswillen unter Beweis.

28.2.2 Ziele

Primäres Ziel der Einführung des nationalen Sozialkreditsystems mit seinem neuen Instrumentarium ist die bessere Gewährleistung einer landesweit wirksamen und schnellen Rechtsdurchsetzung sowie die Optimierung der Governance im Unternehmenssektor, aber auch im Staatswesen selbst. Letzteres soll durch den Einbezug nachgeordneter Behörden in das nationale System erreicht werden. Auch die Effizienz staatlichen Handelns soll dadurch gesteigert werden.

Unternehmen sollen generell zu einem Verhalten veranlasst werden, das als integer erachtet wird. Das Vertrauen zwischen Bürgern, Unternehmen und Regierung soll damit einhergehend zum Wohle des Gemeinwesens und insbesondere auch der Wirtschaft gefördert werden. Das Sozialkreditrating aller Unternehmen führt zu einer signifikant detaillierteren Informationsbasis. Dies kann die Einschätzung der Unternehmen durch die Konsumenten und untereinander verbessern und folglich die Entscheidungen der Marktteilnehmer erleichtern. Neben wirtschaftlichen können und sollen auch behördliche Entscheidungen in deutlich stärkerem Ausmaße als bisher evidenzbasiert erfolgen. Im Sinne des Verbraucherschutzes können unzuverlässige Anbieter in letzter Konsequenz auch aus dem Markt entfernt werden.

Die Umsetzung und ständige Weiterentwicklung des Sozialkreditsystems auch in technischer Hinsicht eröffnet China die Chance, sich einen Technologievorsprung im Bereich Big Data und KI zu erarbeiten. Schließlich ermöglicht das System dem Staat auch einen vollständigen Einblick in die Gesamtheit aller auf dem chinesischen Markt tätigen Unternehmen.

28.2.3 Kernelemente

28.2.3.1 Prozessablauf

Das Sozialkreditrating im Sinne des Verfahrens zur Reputationsbewertung eines bestimmten Unternehmens läuft jeweils in zwei Stufen ab: Im ersten Schritt bestimmt der Staat in Gestalt der für die verschiedenen einschlägigen Bereiche (wie zum Beispiel Steuern, Zoll, Kredite, Arbeitssicherheit, Umweltschutz, Produktqualität) zuständigen Behörden einen unternehmensspezifischen, beispielsweise von Branche und Produktportfolio abhängigen Kriterienkatalog für positives Verhalten mit mehr als zwei Dutzend Elementen, die wiederum aus je ungefähr zehn Unterelementen bestehen. Bezüglich dieser individuellen Liste der behördlichen Konformitätsanforderungen, die das Unternehmen einhalten sollte, um ein gutes Rating zu erhalten, gibt es die Möglichkeit der Interaktion mit dem Staat. Hierzu gibt es einen konkreten Ansprechpartner. Die Liste determiniert den für das Unternehmen relevanten, aus Datenfeldern bestehenden Datensatz.

Im zweiten Schritt sammeln die Behörden dann tatsächlich kontinuierlich und systematisch zumindest alle insofern festgelegten Informationen über das Unternehmen. Zusätzlich zu der jeweils behördeninternen Nutzung für eigene Bewertungen, fließen diese dann final in einer zentralen Masterdatenbank zusammen. Auf dieser Basis wird dann fortlaufend das Sozialkreditrating des Unternehmens bestimmt. In Abhängigkeit des diesbezüglichen Ergebnisses werden die staatlichen Stellen das Unternehmen dann in einem nachgelagerten Schritt belohnen oder bestrafen. Es ist zu erwarten, dass aufgrund der technischen Kapazitäten des Systems auf jedes festgestellte Fehlverhalten zeitnah reagiert wird.

28.2.3.2 Datenquellen

Die unternehmensbezogenen Informationen stammen entweder vom Unternehmen selbst (zum Beispiel in Form von Selbstauskünften) oder von den relevanten Behörden (zum Beispiel Finanzbehörden, Zoll, Ministerien, Provinzregierungen, Gerichten). Darüber hinaus können auch weitere unternehmensexterne Quellen der Datenbelieferung vorliegen.

Folgende Arten von Daten fließen ein: allgemeine Unternehmensinformationen, erteilte Genehmigungen, Finanzdaten, vorherige Ratings, behördliche Daten bzgl. der Gesetzes- und Regelkonformität, zum Beispiel ob eine Steuererklärung rechtzeitig abgegeben wurde, behördliche Protokolle die Ergebnisse von Inspektionen und Prüfungen betreffend, Analysedaten bzgl. der Produktqualität, aber auch Ratings anderer Unternehmen, die enge Geschäftspartner sind, sowie personenbezogene Informationen wie Qualifikationsstatus und persönliche Ratings des Top Managements.

Die integrierte Masterdatenbank fasst die Datensätze und behördenspezifischen Ratings aus bisher fragmentierten Datenbanken zentral zusammen.

28.2.3.3 Bewertung und deren Konsequenzen am Beispiel Steuern

Die Bewertungsmethodik soll zunächst am Beispiel des Bereichs Steuern simplifiziert dargestellt werden: Jedes Unternehmen startet ab erfolgreicher Registrierung in China quasi mit einem positiven Startwert, der einem neutralen B Rating entspricht. Die aktuelle Bewertung hängt dann jeweils vom Verhalten des Unternehmens ab und wird in ein fünfstufiges Rating (A bis D und >D) überführt. Dabei führt die Nichteinhaltung der Anforderungen zu einer Verschlechterung des Ratings. Beispielsweise folgende Tatsachen wirken sich tendenziell negativ aus: verspätete Abgabe einer Steuererklärung, verspätete Zahlung einer fälligen Steuer, unvollständige oder gar falsche Angaben gegenüber der Behörde sowie Steuerhinterziehung.

Vollumfänglich regelkonforme Unternehmen erhalten ein A Rating und werden mit Prämien belohnt. Normal regelkonforme Unternehmen verbleiben bei einem B Rating und werden neutral behandelt. Weniger regelkonformes Verhalten, welches sich in einem C Rating äußert, führt bereits zu Sanktionierungen in Form von mehr staatlicher Kontrolle, zum Beispiel durch häufigere Steuerprüfungen. Sinkt das Rating um eine weitere Stufe auf D, so werden zusätzlich Geldbußen verhängt. Wird ein Unternehmen noch tiefer, also mit einem Rating >D bewertet, so wird es schließlich auf eine Schwarze Liste gesetzt. Dies ist dann mit weitreichenden negativen Folgen verbunden.

Ein sogenanntes Blacklisting geschieht generell, also auch außerhalb des Bereichs Steuern, allerdings nur aufgrund gravierender Gesetzes- oder Regelverstöße. Beispiele für solche sind: Gefährdung der öffentlichen Sicherheit und Gesundheit, Betrug, Urkundenfälschung, Arbeitsunfälle verursacht durch die Verletzung von Arbeitsschutzvorschriften und Nichtzahlung von Löhnen und Gehältern.

28.2.3.4   Gemeinsame Belohnung oder Bestrafung

Das Prinzip der gemeinsamen Belohnung oder Bestrafung ist eines der wichtigsten Merkmale des Sozialkreditsystems. Vorbildliche Unternehmen, die Steuer-, Zoll-, Arbeitsschutz- und Umweltschutzvorschriften sowie sonstige Compliance Anforderungen, uneingeschränkt sehr gut erfüllen, gelangen auf Positivlisten, sogenannte Rote Listen. Dann sind beispielsweise folgende Belohnungen möglich: erleichterter Marktzugang, bevorzugte Berücksichtigung bei der öffentlichen Auftragsvergabe, attraktivere Kreditkonditionen, niedrigere Steuersätze, seltenere Inspektionen und Prüfungen, schnellere Bearbeitung in behördlichen Verfahren.

Das Sozialkreditsystem beinhaltet aber vor allem auch ein ausgeprägtes Sanktionsregime. Die Bestrafungsmechanismen sind dabei noch differenzierter ausgestaltet als die Belohnungsmechanismen. Die möglichen Sanktionen sind breit gefächert und im Zweifel auch sehr hart, aber ihrem Wesen nach immer noch bewusst niedrigschwelliger als solche des Strafrechts. Sie führen tendenziell zu einer Verringerung der Geschäftschancen und zu einer Erhöhung der Kosten sowie einer Vergrößerung der Prozesskomplexität.

Die Besonderheit des neuen Systems liegt in der Möglichkeit behördenübergreifender Sanktionierungsmaßnahmen. Das bedeutet, dass Unternehmen bei Fehlverhalten in einem Bereich von mehreren Behörden gemeinsam und in mehreren Bereichen simultan geahndet werden können, was die zu tragenden Konsequenzen signifikant verschärfen kann.

In Bezug auf besonders unzuverlässige Unternehmen, die auf einer Schwarzen Liste verzeichnet sind, sind folgende Sanktionen möglich: Ausschluss von der öffentlichen Auftragsvergabe, Ausschluss von Subventionen, schlechtere Kreditkonditionen oder Nichtberücksichtigung bei der Kreditvergabe, höhere Steuersätze, Importrestriktionen, häufigere und gezieltere Inspektionen und Prüfungen, schwierigere Erlangung sowie Entzug von Genehmigungen und Lizenzen, Einschränkungen bei Immobilientransaktionen, Reisebeschränkungen für gesetzliche Vertreter bis hin zu einem Ausreiseverbot. Letztes Mittel ist der Marktausschluss eines Unternehmens bzw. die forcierte Beendigung der Geschäftstätigkeit. Dieser Extremfall der Sanktionierung wird jedoch nur eintreten, wenn sich ein Unternehmen wiederholt weigert, gerichtliche oder behördliche Entscheidungen umzusetzen.

Die Verweildauer, während der ein Unternehmen auf einer Schwarzen Liste eingetragen ist, steigt mit der Schwere des zugrunde liegenden Fehlverhaltens. Sie kann in einem Intervall von drei Monaten bis zu einem Jahr oder von sechs Monaten bis zu drei Jahren liegen oder variabel höher und dabei vom Einzelfall abhängig sein.

28.2.3.5   Transparenz

Die in der zentralen Masterdatenbank aggregierten Bewertungen und zugrunde liegenden Daten stehen im Rahmen eines neu aufgesetzten ausgedehnten Informationsaustauschs allen Behörden zur Verfügung. Darüber hinaus stehen bestimmte Informationen sogar der gesamten Weltöffentlichkeit online zur Verfügung. Dies gilt insbesondere auch für Einträge auf einer Schwarzen Liste.

28.2.3.6 Gleichbehandlung in- und ausländischer Unternehmen

Chinesische und westliche Unternehmen werden im Sozialkreditsystem grundsätzlich gleich behandelt. Tendenziell steigt dadurch auch für deutsche Unternehmen die Chancengleichheit auf dem chinesischen Markt. Bei entsprechenden spezifischen Erfahrungen mit Governance, Risk & Compliance, insbesondere mit einem eigenen Compliance-Management-System kann sogar ein Wettbewerbsvorteil entstehen.

28.2.4 Beurteilung und Ausblick

Das neuartige Sozialkreditsystem für Unternehmen ist grundsätzlich sehr positiv zu beurteilen. Insbesondere die wechselseitig bessere Einschätzbarkeit ist ein mächtiger Vorteil. Denn damit steigt das Vertrauen, die alles entscheidende Größe für eine funktionierende Gesellschaft und eine florierende Wirtschaft. In der Folge werden Sicherheit und Geschwindigkeit insbesondere ökonomischer Transaktionen erhöht.

Die wichtigsten Voraussetzungen für den tatsächlichen Erfolg des Sozialkreditsystems im Sinne der genannten Ziele sind dessen aktive Nutzung und breite Akzeptanz. Vor diesem Hintergrund ist es äußerst vorteilhaft, dass dieses System geradezu perfekt mit den bereits skizzierten gesellschaftlichen und kulturellen Wesensmerkmalen Chinas kompatibel ist.

Zunächst erleichtert die inhärente Autoritätsakzeptanz und Harmonieorientierung eine durchgängig positive Resonanz auf das Sozialkreditsystem. Ferner knüpft der damit beschrittene fortschrittliche Weg an die mit dem Begriff des Gesichts bezeichnete traditionelle Reputationskonzeption an. Das Handeln der chinesischen Regierung ist dabei strategisch konsistent, programmatisch umfassend sowie ehrgeizig und lösungsorientiert in der Umsetzung. In kluger Weise wird an die erfolgsorientierten unternehmerischen Entscheidungsträger appelliert, zum einen in Bezug auf den Staat nicht nur Sanktionen zu vermeiden, sondern auch Prämien anzustreben, und zum anderen in Bezug auf die Öffentlichkeit auf die mit der Transparenz verbundene Signalwirkung und Konsequenz für die Reputation zu achten, zum Beispiel im Hinblick auf das Vertrauen der Geschäftspartner und die Präferenzen der Konsumenten.

Zwar kann eine Optimierung der Rechtsdurchsetzung formaler Gesetze und Verträge aus gesellschaftlicher Sicht zumindest partiell letztlich bereits durch die traditionelle informelle Netzwerksystematik des Guanxi erreicht werden. Der Staat möchte aber mit dem Sozialkreditsystem und dessen immanenter Stärkung der Evidenzbasierung von Entscheidungen bewusst einen Entwicklungsschritt weiter gehen. Tendenziell könnte die Bedeutung von Guanxi durch diese Maßnahme auf mittlere Sicht abnehmen.

Das Sozialkreditsystem ist nicht nur Norm und Führungsmechanismus. Es stellt auch ein geradezu gigantisches und komplexes informationstechnologisches Projekt dar, welches in dieser Form sicherlich nur in einer sehr intelligenten, lernbereiten und technologie-affinen Gesellschaft umzusetzen und ständig (zum Beispiel durch den Einbezug privater Bewertungssysteme) zu optimieren ist.

28.3   Implikationen für das Management deutscher Unternehmen

28.3.1 Strategisches Management

Unternehmern und Managern soll nun ein prägnanter Einblick in die äußerst vielschichtigen Implikationen für das Management deutscher Unternehmen verschafft werden, die aus der Einführung des neuen nationalen chinesischen Ratingsystems erwachsen. Dabei orientieren sich die Überlegungen im Wesentlichen an den relevanten betrieblichen Funktionen.

China entwickelt sich mehr und mehr zum ökonomischen Zentrum der Welt. Die Bedeutung des Landes sowohl als entscheidender Produktionsstandort für die Weltwirtschaft als auch als stetig wachsender Absatzmarkt, vor allem für die exportorientierte deutsche Wirtschaft, ist heute inzwischen für jedermann absolut evident. Auf den ersten Blick scheint das neue Sozialkreditsystem für Unternehmen nur ein weiterer, wenn auch bedeutender Bestandteil der rechtlichen Rahmenbedingungen für ökonomische Aktivitäten innerhalb Chinas zu sein. Daher unterschätzen selbst viele deutsche Unternehmen die immense Bedeutung der Veränderung für ihre Wettbewerbsfähigkeit. Denn der erwähnte Eindruck ist rudimentär. Vielmehr wird in den kommenden Jahren wohl kaum ein regulatorischer Mechanismus für den gelingenden Markteintritt und den langfristigen Erfolg unternehmerischen Handelns vor Ort in China eine zentralere Rolle spielen als das neue Ratingsystem. Aufgrund der immer stärkeren chinesisch-deutschen Interdependenzen ist es zudem vielfach auch für das Erreichen der wirtschaftlichen Ziele hierzulande erfolgskritisch.

Es gehört zu den elementaren Aufgaben der Unternehmensführung, auf Basis des Unternehmenszwecks und der Unternehmensziele und unter Berücksichtigung einer Analyse der Kernkompetenzen, der Wettbewerbssituation und der Rahmenbedingungen die Unternehmensstrategie zu definieren und somit die Ausrichtung der Organisation im Sinne der Zielerreichung zu bestimmen. Das Sozialkreditsystem ist ein übergreifender, nahezu alle Bereiche eines Unternehmens betreffender Mechanismus. Die Konsequenzen für die Wettbewerbsfähigkeit im Allgemeinen und die Reputation im Besonderen und damit für die Chancen zur Realisation der Unternehmensziele können aufgrund der bereits dargestellten Sanktions- und Transparenzwirkungen sehr massiv sein. Unter Umständen können bereits Fehlleistungen einer einzigen Organisationseinheit existenzielle Folgen zeitigen.

Vor diesem Hintergrund empfiehlt es sich für jedes Unternehmen mit zumindest indirektem Geschäftsbezug zu China dringend, dass das Top Management in eigener Verantwortung aktiv die Entscheidung trifft und umsetzt, sich auch selbst grundlegend mit diesem Thema zu beschäftigen, bei Beratungsbedarf auch professionelle Lösungsexpertise hinzuzuziehen, nötigenfalls Adjustierungen der Unternehmensstrategie vorzunehmen sowie für hinreichend viel Aufmerksamkeit auf das Sozialkreditsystem in der gesamten Unternehmensorganisation zu sorgen. Es reicht dabei nicht aus, nur die unmittelbar in der Verantwortlichkeit der Unternehmensführung stehenden Bereiche wie Compliance und Risikomanagement auf die neuen Herausforderungen zu fokussieren und mit entsprechenden Befugnissen auszustatten sowie die ordnungsgemäße und sachgerechte Implementierung aller notwendigen Maßnahmen durch diese final zu überwachen. Sondern jeder im Zusammenhang mit China einschlägige Bereich des Unternehmen sowohl dort als auch andernorts sollte durch die Unternehmensleitung entsprechend seiner Wichtigkeit in dieser Weise ausgerichtet werden, allermindestens jedoch durch sehr gute interne Kommunikation hinreichend sensibilisiert werden.

28.3.2 Regulatorisches Management, Compliance

Die Regeltreue eines Unternehmens in China, also die Einhaltung aller gesetzlichen Bestimmungen dort, gewinnt mit dem Sozialkreditsystem noch stärker an Bedeutung. Denn bereits graduelle Ratingunterschiede können signifikante direkte Implikationen für die spezifische Wettbewerbsfähigkeit haben. Dem Compliance kommt in diesem Zusammenhang daher eine entscheidende Rolle zu, um ein sehr gutes Sozialkreditrating zu erreichen und aufrechtzuerhalten.

Die für das Regulatorische Management in Form von Compliance in China zuständige Organisationseinheit sollte in Vorbereitung auf das neue System zunächst proaktiv die Gesamtheit der anzuwendenden Ratings sowie den vollständigen behördlich fixierten Kriterienkatalog der unternehmensspezifischen Konformitätsanforderungen eruieren, die konkret zur Berechnung der zu erfüllenden Bewertungen verwendet werden. Falls die getroffenen Festlegungen nicht im Sinne des Unternehmens sind, sollte die Einheit versuchen, diese in Verhandlungen mit dem zuständigen Behördenvertreter mittels konstruktiver Argumente zu optimieren. Es gilt, unbedingt ein bereits initial schlechtes Sozialkreditrating aufgrund unzutreffender Bewertungskriterien zu vermeiden.

Für das Unternehmen ist es dann natürlich wichtig, auch in der Zukunft jederzeit tatsächlich ein möglichst gutes Rating zu erzielen. Die zuständige Einheit sollte die jeweils relevanten Regeln daher weiterhin kontinuierlich identifizieren und exakt verstehen sowie in der aktuellsten Form kommunizieren und deren unternehmensinterne Einhaltung sicherstellen. Ferner sollten aus der jeweils neuesten Ausgestaltung des Sozialkreditsystems laufend die notwendigen Maßnahmen und Prozesse abgeleitet werden. Deren Implementierung sollte dann unter besonderer Berücksichtigung der analysierten hausinternen Schwachstellen in regelmäßigen Zeitabständen und ggf. noch stringenter als bisher überwacht werden. Im Sinne des Erfolgsmonitorings sollten schließlich auf allen relevanten Plattformen wiederholt Informationen über die jeweils vorhandenen Ratings und auch über eventuelle Einträge in Schwarzen Listen eingeholt werden. Im Falle negativer Scores oder Einträge sollte die Organisationseinheit unverzüglich aktiv werden. Falsche Bewertungen oder unberechtigte Maßnahmen sollten im Dialog mit den Behörden umgehend korrigiert werden. Berechtigte kritische Scores oder Einträge sollten zur unmittelbaren Einleitung von gegensteuernden Maßnahmen führen, um das Rating so schnell wie möglich wieder zu verbessern und negative Einträge wieder zu tilgen.

Grundsätzlich ist jedem Unternehmen zu empfehlen, ein unternehmensweites, vor dem Hintergrund des umfassenden Charakters des neuen Sozialkreditratings jedenfalls zumindest ein auf das Geschäft auf dem chinesischen Markt bezogenes Compliance-Management-System zu installieren. Ein bereits vorhandenes System sollte, selbst wenn es internationale Standards erfüllt, in seiner Wirksamkeit im Hinblick auf die neuen Verhältnisse optimiert werden. Die darin festgelegten Grundsätze und Maßnahmen sollten nötigenfalls verschärft werden.

Der Schwerpunkt der Anforderungen im Rahmen des Sozialkreditratings liegt bisher bei Fristen und sonstigen Formalien, deren Gewährleistung relativ leicht umgesetzt werden kann. Die für Compliance zuständige Einheit steht aber dennoch vor einer besonderen Herausforderung. Denn es werden auch Bewertungsergebnisse anderer Unternehmen, die enge Geschäftspartner wie zum Beispiel Lieferanten sind, sowie persönliche Ratings des Top Managements in die Bewertung mit einbezogen. Insofern sollten intelligente Anreizmechanismen entwickelt werden, um diese unternehmensexternen Aspekte zumindest einschätzen sowie optimalerweise steuern und kontrollieren zu können.

28.3.3 Risikomanagement

Aus offensichtlichen Gründen spielt die Aktivität des Risikomanagements im Zusammenhang mit dem Sozialkreditsystem in China in nahezu allen betrieblichen Bereichen eine äußerst wichtige Rolle, um die mit dessen regulatorischer Existenz verbundenen Risiken tatsächlich zu identifizieren, zu analysieren, zu bewerten, zu kommunizieren sowie im Zusammenspiel mit Compliance möglichst durch Maßnahmen zu beherrschen und durch Frühwarnsysteme mittels Risikoindikatoren zu überwachen. Die Risiken liegen konkret in der Möglichkeit des Eintritts ungünstiger Ratings oder Einträge, die zu negativen Auswirkungen in Form direkter Nachteile im Verhältnis zum Staat oder indirekter Nachteile gegenüber anderen Marktteilnehmern führen.

Das Ausmaß des Risikomanagements eines Unternehmens sollte sich naturgemäß am zu erwartenden, unter Umständen auch am im Extrem möglichen wirtschaftlichen Schaden aufgrund von Sanktionen orientieren, der sich materialisieren könnte, sofern bestimmte unsichere Umweltzustände eintreten. Bedeutende wirtschaftliche Folgen können erwachsen, wenn ein Unternehmen auf eine Schwarze Liste gesetzt wird. Die größten Auswirkungen ergeben sich aus dem Marktausschluss eines Unternehmens. Die fortlaufenden Prozesse des Risikomanagements sollten optimalerweise in andere Prozesse des Unternehmens integriert sein.

Die Risiken, die sich in den einzelnen betrieblichen Funktionen materialisieren können, deren Implikationen sowie im Übrigen auch die Chancen, die sich ergeben können,
werden in den folgenden Abschnitten überblicksartig skizziert.

28.3.4 Finanzmanagement

Das Sozialkreditsystem zeitigt im Bereich Finanzen für deutsche Mutterunternehmen und
deren in China aktive Tochtergesellschaften in der Gesamtschau tendenziell mehr Risiken
als Chancen. Hierbei sind vor allem zwei Fallkonstellationen betrachtenswert. Zunächst
ist der Abzug von finanziellen Mitteln aus China, zum Beispiel bei Gewinnausschüttungen
an die deutsche Muttergesellschaft grundsätzlich nur möglich, wenn das Tochterunternehmen auf der entsprechenden Roten Liste verzeichnet ist. Bei der Fremdkapitalbeschaffung
innerhalb Chinas führt ein solcher Eintrag zu attraktiveren Kreditkonditionen, während
der Eintrag auf einer Schwarzen Liste schlechtere Kreditkonditionen oder eine Nichtberücksichtigung bei der Kreditvergabe nach sich zieht.

Im angrenzenden Bereich Steuern sollte darauf geachtet werden, dass sich Tochterunternehmen in China absolut regeltreu verhalten, da selbst nur rein formale Fehler direkt zu
einer Verschlechterung des Sozialkreditratings und damit relativ unmittelbar zumindest zu
einer Profitabilitätsreduktion durch höhere Steuer- und Gebührensätze oder gar weitergehenden Konsequenzen führen.

Betroffen vom Sozialkreditsystem sind schließlich freilich nicht nur die betrieblichen
Funktionen klassischer Produktions- und Dienstleistungsunternehmen, sondern auch das
Kerngeschäft von Finanzdienstleistungsunternehmen und Immobiliengesellschaften. Es
ist daher für den Bereich Investmentmanagement zu erwähnen, dass negative Sozialkreditratings zu Einschränkungen führen können, im Financial Asset Management im Hinblick
auf die Zugangswege zu den chinesischen Kapitalmärkten für Portfolioinvestitionen und
im Real Estate Investment Management in Bezug auf Immobilientransaktionen in China.

28.3.5 Personalmanagement

Aufgrund des erwähnten Einflusses der persönlichen Ratings des Top Managements auf
das Sozialkreditrating des gesamten Unternehmens sollten intelligente Anreizmechanismen entwickelt werden, um diese aus Unternehmensperspektive externen Aspekte auch
ohne unzumutbare Eingriffe in die Privatsphäre der betreffenden Personen zumindest einschätzen sowie optimalerweise steuern und kontrollieren zu können. Ziel sollte die Vermeidung direkter Risiken für das Unternehmensrating sein.

Negative persönliche Ratings Einzelner bergen für das Unternehmen jedoch auch weitere, indirekte Risiken, so könnten diese Manager in ihren Aktivitätsmöglichkeiten eingeschränkt, zum Beispiel Reisebeschränkungen bis hin zu Ausreiseverboten unterworfen
werden. Ihnen könnte auch ihre Vertretungsmacht gegenüber staatlichen Behörden entzo-

gen werden. Bestimmten bisher unternehmensexternen Personen könnte auch die erfolgreiche Bestellung zum gesetzlichen Vertreter oder die Übernahme einer sonstigen Lei-
tungsfunktion verwehrt werden.

Dies wirkt sich auf die Personalselektion für derartige Positionen und die zugehörige
Vertragsgestaltung aus. Unternehmen sollten potenzielle leitende Angestellte im Vorfeld
noch intensiver überprüfen und beurteilen, wenn möglich sollten sie hierzu auch Einsicht
in die betreffenden Schwarzen Listen nehmen.

Anstellungsverträge sollten Bestimmungen für den Umgang mit negativen Auswirkungen persönlicher Ratings auf das Unternehmen enthalten. Deuten sich solche Entwicklungen an, sollte unter Umständen sogar die Möglichkeit erwogen werden, einen ähnlich
qualifizierten und autorisierten Manager bereits in Stellung zu bringen, der die Aufgaben
einer auf einer schwarzen Liste stehenden Person umgehend übernehmen könnte.

28.3.6 Beschaffungsmanagement

Für Produktionsunternehmen essenzielle Lieferketten sind heute aufgrund der fortge-
schrittenen Globalisierung eng verwoben. Auch die Produktion von Gütern in Deutschland hängt oft in sehr großem Ausmaß von der Zulieferung von Einzelteilen und Komponenten aus dem Ausland, insbesondere auch aus China ab.

Im Bereich Beschaffung sind daher Fragen aus dem Feld des Supply-Chain-Risikoma-
nagements von Bedeutung. Denn durch das neue Sozialkreditsystem kann sich ein
Lieferunterbrechungsrisiko materialisieren. Dies betrifft zumindest alle Teile, die direkt in
China hergestellt werden, und zwar auch dann, wenn der Produzent ein eigenes
Tochterunternehmen ist.

Der dem deutschen Unternehmen zuliefernde Hersteller in China könnte aufgrund eigener Unzulänglichkeiten oder auch durch negative Ratings seiner Geschäftspartner und
wichtiger Führungspersonen in der Konsequenz möglicherweise nicht mehr in Lage sein zu
liefern. Beispielsweise könnte die Situation eintreten, dass er nicht mehr produzieren, verkaufen oder exportieren darf. Es könnte auch sein, dass er nicht mehr produzieren kann,
weil er bestimmte Teile, die er für seine eigene Produktion benötigt, selbst nicht mehr nach
China importieren darf, oder weil er mangels staatlicher Aufträge oder unzulänglicher Kreditfinanzierung in wirtschaftliche Schwierigkeiten gelangt. Das Risiko wiegt umso schwerer, je wichtiger die Komponente ist und je eher die Beschaffung der Teile zeitlich auf die
Produktion abgestimmt ist. Das Risiko kann und sollte bei eigenen Tochtergesellschaften
selbstverständlich durch entsprechende Maßnahmen gemildert werden.

Generell ist zu empfehlen, dass die vertraglichen Vereinbarungen mit Lieferanten aus
China Bestimmungen enthalten, die sich mit den geschilderten Eventualitäten beschäftigen. Für Schlüsselkomponenten sollte unter Umständen sogar die Möglichkeit von Back-
uplieferanten erwogen werden. Diese Empfehlungen gelten im Übrigen analog auch für
Tochtergesellschaften deutscher Unternehmen in China und deren Verhältnis zu innerchi-
nesischen Lieferanten.

Es gibt auch eine wichtige positive Auswirkung des Sozialkreditsystems auf hiesige Unternehmen und ihre Töchter. Denn die neue Transparenz gibt ihnen ein Instrument für eine bessere Due Diligence an die Hand, zum Beispiel durch Ermöglichung der Einsicht in die Schwarzen Listen. Potenzielle Lieferanten können so im Vorfeld besser überprüft und eingeschätzt werden, bevor überhaupt eine Lieferbeziehung mit ihnen eingegangen wird. Schließlich kann ihr Sozialkreditrating während der Vertragsbeziehung jederzeit verfolgt werden.

28.3.7 Produktionsmanagement

Im Zusammenhang mit dem strategischen Plan Made in China 2025 strebt das Reich der Mitte nicht nur die Förderung der weiteren Entwicklung von technologischen Schlüsselindustrien an, sondern auch den generelleren Übergang zur Herstellung höherwertiger Produkte. Auch die im eigenen Land produzierten Güter sollen insofern immer mehr der Qualitätspräferenz der chinesischen Unternehmenskunden sowie der zahlungskräftigen inländischen Konsumenten entsprechen. Folgerichtig werden im neuen Sozialkreditsystem auch produktqualitätsbezogene Daten berücksichtigt. Hiervon sind auch die Tochtergesellschaften deutscher Unternehmen betroffen, die Investitions- und Konsumgüter in China fertigen.

Im Bereich Produktion sind daher Fragen aus dem Feld des Qualitätsmanagements von Bedeutung. Aufgrund der deutschen Qualitätskultur sind die eigenen Tochterunternehmen in China diesbezüglich üblicherweise sicherlich sowieso schon stark fokussiert. Durch das Sozialkreditsystem kommt nun aber noch ein zusätzlicher Anreiz insbesondere zu strikter Vermeidung grober Mängel in der Produktqualität hinzu.

28.3.8 Vertriebsmanagement und Marketing

Das Sozialkreditsystem birgt im Bereich Vertrieb und Marketing sowohl Chancen als auch Risiken für Unternehmen, die Ihre Waren und Dienstleistungen in China anbieten.

Unternehmen, deren produzierende Tochtergesellschaft vor Ort über ein gutes oder sogar sehr gutes Rating verfügt, könnten ermutigt sein, diese hohe Bewertung als Instrument des reputationsbasierten Marketings einzusetzen und beispielsweise in sozialen Medien zu teilen.

Im Falle negativer Ratings des Vertriebsunternehmens in China bestehen sowohl Reputations- als auch Disruptionsrisiken, und zwar grundsätzlich unabhängig davon, ob der Vertrieb über eine eigene Tochtergesellschaft oder über einen lokalen Partner erfolgt. Marken und Kundenbeziehungen könnten in beiden Fällen beschädigt werden, was ungleich schwerer wiegt als Vergleichbares im Westen, da Vertrauen in China deutlich schwieriger wiederzugewinnen ist. In einem solchen Fall sollte mit Maßnahmen des Reputationsmanagements versucht werden, den Ruf des Unternehmens bei den Kunden zumindest zu verbessern. Wird das eigene oder fremde Vertriebsunternehmen auf eine Schwarze Liste gesetzt, könnte es nicht mehr in der Lage sein, Produkte zu importieren oder Zahlungen an den Hersteller zu leisten.

Es ist daher auch im Vertrieb generell zu empfehlen, dass die vertraglichen Vereinbarungen mit Vertriebspartnern in China Bestimmungen enthalten, die sich mit den geschilderten Eventualitäten beschäftigen. Je nach Bedeutung des chinesischen Absatzmarktes bzw. einzelner Vertriebspartner sollte unter Umständen sogar die Möglichkeit der Diversifikation von Vertriebspartnern erwogen werden. Durch die neue Transparenz kann jedoch auch hier eine bessere Due Diligence potenzieller Partner im Vorfeld erfolgen. Ferner kann das Sozialkreditrating existierender Partner jederzeit nachvollzogen werden. Beides kann die skizzierten neuen Risiken reduzieren.

28.3.9 IT-Management

Operative Exzellenz in der IT erfordert eine konsequente Ausrichtung an den direkten und indirekten geschäftlichen und somit auch an den regulatorischen Bedürfnissen des Unternehmens. Im Hinblick auf das Sozialkreditsystem bedeutet dies zunächst einmal, dass das IT-Management jedes in China tätigen Unternehmens darauf zu achten hat, dass der eigene Bereich nicht selbst Ursache von Herabstufungen im Rating wird. Es ist also dafür zu sorgen, dass chinesische Tochtergesellschaften deutscher Unternehmen vor Ort auch in der IT alle einschlägigen Gesetze und Regeln strikt einhalten, zum Beispiel auch Datenschutzvorschriften.

Darüber hinaus kann IT durch eine enge Verzahnung mit Compliance und Risikomanagement eine wichtige Rolle bei der Unterstützung der breiten unternehmensinternen Anstrengungen zur Erreichung der Regelkonformität spielen. In diesem Zusammenhang macht es für das Unternehmen Sinn zu versuchen, die Sammlung der unternehmensbezogenen und sonstigen einfließenden Daten des staatlichen Systems zu duplizieren, um die darauf fußenden Berechnungsmechanismen durch eigene Datenanalysen so genau wie möglich nachzuvollziehen.

Das unternehmensweite IT-Management könnte und sollte den damit verbundenen Impuls auch zur übergreifenden Weiterentwicklung der Kompetenzen im Bereich Innovation und Digitalisierung nutzen, zum Beispiel in Bezug auf Big Data, Data Analytics, um damit zur Sicherstellung der technologischen Wettbewerbsfähigkeit des Gesamtunternehmens beizutragen.

28.3.10 M&A, Unternehmensgründungen

Das Sozialkreditsystem hat für wirtschaftliche Organisationen in Deutschland schließlich auch Auswirkungen in dem besonderen Bereich der Auslandsdirektinvestitionen. Hierbei sind im Wesentlichen drei Fallkonstellationen zu betrachten.

Bei der Beteiligung eines deutschen Unternehmens an einer originären Unternehmens-gründung in China, also im Falle der Gründung eines gänzlich neuen Unternehmens, sollte darauf geachtet werden, dass das neue Unternehmen von Anfang an alle Gesetze und Regeln in China vollumfänglich einhält, und dass bereits die initiale Bonitätseinstufung mindestens auf adäquatem Niveau liegt. Erfreulicherweise verschafft die Transparenz des Sozialkreditsystems dem deutschen Direktinvestor ein neues Instrument für eine bessere Due Diligence und nachfolgende Bewertungsüberwachung der potenziellen Beteiligungs-partner, aber auch möglicher Geschäftspartner und leitender Angestellter.

Bei Unternehmenstransaktionen in China spielt der neue regulatorische Rahmen in mehreren der typischen Prozessschritte eine Rolle. Der potenzielle deutsche Investor sollte bereits bei der Suche nach geeigneten Zielunternehmen (Target Search) in China darauf achten, dass nur solche Unternehmen in die engere Wahl kommen, die hinsichtlich des Sozialkreditsystems allermindestens unproblematisch sind. Spätestens bei der Evaluie-rung des ausgewählten Targets im Rahmen der Due Diligence sollte dann unbedingt veri-fiziert werden, dass das Zielunternehmen bezüglich Compliance und Risikomanagement hinreichende Prozesse implementiert hat, die weiterhin ein möglichst erfolgreiches Rating sicherstellen. Im Allgemeinen erstreckt sich die Due Diligence bei Unternehmenstrans-aktionen naturgemäß vor allem auf das Zielunternehmen. Aufgrund der besonderen Me-chanik des Sozialkreditsystems sollte diese aber auch auf enge Geschäftspartner sowie Mitglieder der Unternehmensleitung ausgedehnt werden. Die bereits erwähnten Transpa-renzvorteile durch das neue Ratingsystem können auch im Bereich M&A genutzt werden. Die insgesamt gewonnenen Erkenntnisse fließen dann ohne Zweifel auch in die Überle-gungen des Käufers zur Preisbildung und zur Vertragsausgestaltung hinsichtlich zu ver-langender Garantien und Haftungsfreistellungen ein.

Inzwischen finden auch zahlreiche Unternehmenstransaktionen aus China heraus statt. Dabei sind deutsche Unternehmen bekanntlich ein bevorzugtes Ziel chinesischer Auslands-direktinvestitionen. In solchen Fällen ist für die Verkäufer deutscher Zielunternehmen fol-gendes Risiko zu berücksichtigen: Die harten Sanktionierungsmöglichkeiten im Rahmen des Sozialkreditsystems können sich auch auf die juristische Fähigkeit zu und die faktische Möglichkeit von Geldtransaktionen auswirken. Ein potenzieller chinesischer Direktinves-tor könnte so für den Verkäufer unerwarteter Weise und ohne Absicht faktisch außer Stande sein, die Zahlung des Kaufpreises rechtzeitig und vollständig zu leisten. Insofern empfiehlt es sich für den Verkäufer, unbedingt großen Wert auf eine frühzeitige, im Umfang durchaus begrenzte Due Diligence in umgekehrter Richtung zu legen, um den möglichen Investor jedenfalls auch hinsichtlich seines Sozialkreditratings zu durchleuchten.

28.4 Ausblick

Das neue chinesische Sozialkreditsystem für Unternehmen ist auch für deutsche Hidden und Unhidden Champions von großer Zukunftsrelevanz. Es ist grundsätzlich sehr positiv zu beurteilen, denn es hat durch seine reputationsorientierte und anreizwirksame Konzeption, anspruchsvolle technologische Umsetzung und voraussichtlich breite gesellschaftliche Akzeptanz das Potenzial, die Rechtsdurchsetzung und Governance in China signifikant zu optimieren sowie wirtschaftliche Interaktionen durch Vertrauenssteigerung mittels Verbesserung der Informationsbasis der Akteure deutlich zu erleichtern.

Die Gesellschaft in Europa und insbesondere auch in Deutschland könnte das Sozialkreditsystem zum Anlass nehmen, neu über die Möglichkeiten der eigenen Entwicklung in verschiedensten Feldern, zum Beispiel in Bezug auf Lernbereitschaft und Technologieoffenheit nachzudenken, und diese dann stärker als bisher wahrzunehmen. Vielleicht realisiert sich ja auf diese Weise Leibniz Vision eines wechselseitigen und interdisziplinären Kultur- und Wissensaustausches mit China drei Jahrhunderte nach seinem Zukunftsentwurf.

Deutschland im Allgemeinen und die deutsche Wirtschaft im Besonderen genießen im Reich der Mitte eine einzigartig positive Reputation. Um die damit verbundenen Chancen noch erfolgreicher zu nutzen, ist es für deutsche Unternehmen absolut notwendig, sich auf der Basis gesteigerter interkultureller Kompetenzen umfassend auf die Implikationen des neuen Sozialkreditsystems vorzubereiten. Hierzu existiert entsprechende professionelle externe Lösungsexpertise.

Literatur

Dai, X. (2018): Toward a Reputation State: The Social Credit System Project of China, auf: https://ssrn.com/abstract=3193577, oder auf: https://doi.org/10.2139/ssrn.3193577

Leibniz, G. W. (1697): Novissima Sinica, historiam nostri temporis illustratura …, Stuttgart.

Luo, Y./Huang, Y./Wang, S. L. (2012): *Guanxi* and Organizational Performance: A Meta-Analysis, in: Management and Organization Review, Vol. 8, No. 1, S. 139–172, auf: https://doi.org/10.1111/j.1740-8784.2011.00273.x

Osten, M. (2007): Leibniz oder Chinesisch als Weltsprache?, in: Die Politische Meinung, Zeitschrift für Politik, Gesellschaft, Religion und Kultur, Nr. 446, Januar 2007, S. 18–20.

Zhang, J./Keh, H. T. (2010): Interorganizational Exchanges in China: Organizational Forms and Governance Mechanisms, in: Management and Organization Review, Vol. 6, No. 1, S. 123–147, auf: https://doi.org/10.1111/j.1740-8784.2009.00148.x

Thomas Solbach, Diplom-Kaufmann, Diplom-Volkswirt, ist Geschäftsführer der FORTIS GmbH (www.fortis.gmbh). Unter seiner Leitung schafft die Managementberatung für deutsche Champions Lösungen aus einer Hand für den gelingenden Markteintritt und langfristigen Erfolg unternehmerischen Handelns in China. Solbach verfügt über langjährige Erfahrungen in Führungspositionen im Investment Banking und in der Unternehmensberatung. Er ist mit einer Chinesin verheiratet.

Das chinesische Sozialkreditsystem für Unternehmen – Hintergründe und praktische Hinweise

Tim A. Fongern, Heng Wang und Lihong Yu

Zusammenfassung

Der vorliegende Beitrag erläutert die Hintergründe zur Einführung eines Sozialkreditsystems für Unternehmen in China. Die Autoren geben einen Überblick über die Auswirkungen des neuen Systems und liefern praktische Hinweise für Unternehmen, wie sie sich daran anpassen können.

29.1 Einführung

Für die chinesische Regierung ist das Sozialkreditsystem für Unternehmen ein wichtiger Bestandteil des marktwirtschaftlichen Systems sowie eines Systems zur sozialen Steuerung. Die schrittweise Einrichtung des Sozialkreditsystems basiert auf zahlreichen Gesetzen, Vorschriften, Industriestandards und vertraglichen Beziehungen. Auf der Grundlage von bonitätsrelevanten Informationen und durch die Unterstützung sogenannter Kreditinformationsanwendungen sowie Kreditdienstleistungssystemen sollen Bonität und Integrität der Gesellschaft insgesamt verbessert werden.

T. A. Fongern (✉) · H. Wang
Shen Heng Law Firm, Frankfurt am Main, Deutschland
E-Mail: t.fongern@shenghengls.eu; 13909886848@139.com

L. Yu
Sheng Heng Law Firm, Shenyang, China
E-Mail: international@shenghengls.com; lisayu@126.com

Dabei geht es der chinesischen Regierung vor allem um die Verbesserung in vier Bereichen:

- Staatliche Integrität,
- Soziale Integrität,
- Justizielle Glaubwürdigkeit und
- Bonität des Kreditmarktes.

Unter diesen Bereichen ist die Verbesserung staatlicher Integrität maßgebend und zugleich Schlüssel zum Verständnis des Sozialkreditsystems. Mit der Steigerung sozialer Integrität soll erreicht werden, dass die Bürger und Unternehmen aufrichtig miteinander umgehen. Denn nur eine auf Vertrauen basierende Gesellschaft könne soziale Harmonie und Stabilität sowie dauerhaften Frieden und Ordnung erreichen. Die Verbesserung der Glaubwürdigkeit der Justiz sei die Garantie, aber auch das Endergebnis von sozialer Fairness und Gerechtigkeit. Mit der Steigerung der Bonität des Kreditmarktes sollen Kosten gesenkt und das Marktumfeld insgesamt verbessert werden. Dies sei eine überlebensnotwendige Bedingung für eine nachhaltige Entwicklung des Kreditmarktes sowie dessen Marktteilnehmern.

Die Kreditwirtschaft ist ein wesentlicher Teil der Marktwirtschaft. Ohne Vertrauen in die Kreditwirtschaft kann die Marktwirtschaft nicht funktionieren. Seit seiner „Reform und Öffnung" hat China schrittweise marktorientierte Reformen des Wirtschaftssystems, eine rasche Entwicklung der Marktwirtschaft und eine erhebliche Verbesserung des Lebensstandards seiner Bürger herbeigeführt. Chinas Marktwirtschaftssystem hat derzeit jedoch noch Verbesserungspotenzial. Dies wird hauptsächlich auf das noch unvollkommene Sozialkreditsystem sowie ein schlechtes Vertrauensumfeld im Markt zurückgeführt. Das „Marktkreditsystem" befindet sich noch im Aufbau. Nicht selten kommt es zu Zahlungsrückständen, zur Umgehung von Forderungen oder zum Betrug bei Warentransaktionen. Das Sozialkreditsystem soll eingerichtet werden, um die Wirtschaft weiter zu entwickeln und eine gesunde Entwicklung der Marktwirtschaft zu fördern.

Angesichts der raschen Entwicklung von Wirtschaft und Gesellschaft nach der „Reform und Öffnung" ist die Einrichtung eines Sozialkreditsystems zu einem der wichtigsten Mittel geworden, um Transaktionssicherheit zu gewährleisten, eine geordnete wirtschaftliche Entwicklung zu fördern und soziale Harmonie und Stabilität zu gewährleisten. Dieser Artikel konzentriert sich auf die Rolle des Aufbaus des Sozialkreditsystems für die Marktwirtschaft und erörtert hauptsächlich die Auswirkungen auf Markteinheiten. Schließlich wird der Schluss gezogen, dass Markteinheiten Maßnahmen im Rahmen des Aufbaus von Sozialkreditsystemen ergreifen sollten, um sich in die gesunde und nachhaltige Entwicklung der Marktwirtschaft zu integrieren.

29.2 Rechtlicher Rahmen

Das Sozialkreditsystem basiert auf einer ganzen Reihe von Gesetzen, Verordnungen, Standards und Verträgen. Wichtige Elemente dieses Rechtsrahmens sind die Verbesserung der Kreditgesetzgebung, die schrittweise Einführung von Gesetzen, Vorschriften und Systemen im Zusammenhang mit dem Kreditwesen sowie die Schaffung einer Rechtsgrundlage und von Regeln in den Bereichen Erfassung, Erhebung, Anwendung, Sicherheit von Kreditinformationen sowie Schutz von Rechten und Interessen des Bürgers. Siehe dazu: „Regulations on the Management of Credit Information Industry", „Measures for the Administration of Tax Credits (Trial)", „Measures for the Administration of Credit Information in the Commercial Sector", „Measures for the Administration of Credit of Enterprise Statistics", „Measures for the Administration of Joint Punishment Targets for Commercial Credit", „Provisional Regulations on the Disclosure of Enterprise Information" (Order No. 654 of the State Council of the People's Republic of China), und „Several Opinions on the Use of Big Data to Strengthen Services and Supervision of Market Entities" (Guobanfa [2015] No. 51).

Darüber hinaus wurden „Mehrere Stellungnahmen zur Verwendung von Kreditunterlagen und Kreditberichten in Verwaltungsangelegenheiten" (Fakai Caijin [2013] Nr. 920) veröffentlicht und umgesetzt, um den Aufbau des Kreditsystems in der Marktwirtschaft schrittweise zu verbessern und die Erfassung und Veröffentlichung von Kreditinformationen im Markt zu standardisieren sowie Kredite zu realisieren.

Bei allen diesen Maßnahmen geht es vor allem um Offenheit und Weitergabe von Informationen, Erfassung, Erhebung, Weitergabe, Nutzung, Offenlegung und Aufbewahrung von Kreditinformationen von Marktteilnehmern, Förderung ehrlicher Funktionserfüllung von Marktteilnehmern im Einklang mit dem Gesetz, Stärkung des gemeinsamen Mechanismus zur Bestrafung von Fehlverhalten auf dem Markt und Beschleunigung der Einrichtung eines neuen kreditbasierten Überwachungsmechanismus sowie Stärkung der Kreditaufsicht im Marktsektor.

29.3 Beweggründe der chinesischen Regierung

In den genannten Rechtsquellen legt die chinesische Regierung Ihre Beweggründe zum Aufbau eines Sozialkreditsystems in der Wirtschaft dar. Vor allem gehe es um die Regulierung der marktwirtschaftlichen Ordnung, die Optimierung des Entwicklungsumfelds der Marktwirtschaft und die Senkung der Marktkosten.

29.3.1 Regulierung der marktwirtschaftlichen Ordnung

Die Stärkung der Strukturen im Kreditmarkt soll das Rechtsbewusstsein der Marktteilnehmer stärken und ihr Rechtsverständnis verbessern, die Marktregeln gemäß den Anforderungen der Gesetze und Vorschriften in marktwirtschaftlichen Tätigkeiten bewusst umzusetzen, den Grundsätzen des gegenseitigen Austauschs, der Gleichheit und des gegenseitigen Nutzens zu folgen und fair zu sein. Ein angemessener und geordneter Wettbewerb soll auch dazu dienen, den Markt mit authentischen Waren zu versorgen.

Regeln für die Geschäftsintegrität seien Ausdruck einer integren Geschäftskultur, welche die Gewohnheiten und Usancen des Geschäftsverhaltens umfasst, aber auch die Geschäftsethik und die Geschäftskreditsysteme. Durch die Stärkung des Marktkreditaufbaus könnten Marktteilnehmer ihre Reputation durch eigenes Handeln behaupten, ein gutes Image aufbauen, ihre Rentabilität und Wettbewerbsfähigkeit und ihr Entwicklungsumfeld bei der Teilnahme am Marktwettbewerb kontinuierlich verbessern.

29.3.2 Optimierung des Entwicklungsumfelds der Marktwirtschaft

Der Aufbau der Marktkreditwirtschaft sei zu stärken und die Marktteilnehmer zu lenken, um korrekte Werte zu etablieren, wie auch das Bewusstsein für soziale Verantwortung zu stärken. Mit Integrität zu handeln bedeute, Vorschriften zu beachten, die Marktregeln einer sozialistischen Marktwirtschaft einzuhalten und am Marktwettbewerb mit einer ehrlichen und vertrauenswürdigen Haltung teilzunehmen. Durch das gemeinsame Bemühen von Marktteilnehmern und der gesamten Gesellschaft würden die Werte der Ehrlichkeit und Vertrauenswürdigkeit gefördert und die Ordnung der Marktwirtschaft reguliert, um einer „Erosion von Marktteilnehmern" durch die negativen Auswirkungen der Marktwirtschaft entgegenzuwirken und eine gesunde Entwicklung der Marktwirtschaft zu unterstützen.

Marktwirtschaftliche Aktivitäten stellten externe Manifestationen des beruflichen Verhaltens in verschiedenen Branchen dar. Der Aufbau der Kreditwirtschaft spiele eine grundlegende Rolle bei der Aufrechterhaltung und Regulierung der Marktordnung. Böswillige Ausfälle und Betrug bei Marktaktivitäten würden hauptsächlich durch mangelnde Kreditwürdigkeit verursacht. Die Marktintegrität sei der Eckpfeiler der Existenz und der Entwicklung einer Marktwirtschaft und habe tief greifende Auswirkungen auf die Gestaltung der Branchenethik. Daher sei es notwendig, ein „Marktkreditsystem" einzurichten, Mitarbeiter in allen Branchen zu schulen und anzuleiten, um eine korrekte Weltanschauung, Lebenseinstellung und Werte zu etablieren, um Professionalität zu fördern, Industrienormen und -disziplinen strikt einzuhalten sowie um die Berufsethik kontinuierlich zu stärken.

Die Marktwirtschaft sei eine Kreditwirtschaft. Der Aufbau eines Sozialkreditsystems sei eine grundlegende Strategie zur Korrektur und Standardisierung der marktwirtschaftlichen Ordnung. Die Beschleunigung des Aufbaus eines Sozialkreditsystems trage dazu bei, Unehrlichkeit zu bekämpfen, finanzielle Risiken zu standardisieren und zu entschärfen, die finanzielle Stabilität und Entwicklung zu fördern, die soziale und wirtschaftliche

Ordnung aufrechtzuerhalten und die Rechte und Interessen der Bevölkerung mit Sozialmanagement und anderen öffentlich-rechtlichen Funktionen zu schützen. Der Aufbau der Marktkreditwirtschaft sei ein wichtiger Bestandteil beim Aufbau eines Sozialkreditsystems. Die Stärkung des Aufbaus der Marktkreditwirtschaft könne den Aufbau eines Sozialkreditsystems fördern.

29.3.3 Senkung der Marktkosten

Die rasche Entwicklung der Marktwirtschaft hänge von der Wissenschaftlichkeit der Entscheidungsfindung ab. Grundlage für die wissenschaftliche Entscheidungsfindung seien genaue und aktuelle Informationen. Ohne Marktvertrauen keine Glaubwürdigkeit. Unglaubwürdige Informationen könnten die Wissenschaftlichkeit der Entscheidungsfindung nicht garantieren, wenn über herkömmliche Informationskanäle erhaltene, fehlerhafte Informationen als Grundlage für die Entscheidungsfindung verwendet werden. Die materiellen und finanziellen Ressourcen für die Durchführung von Marktforschungen, um genaue und zutreffende Informationen zu erhalten, erhöhten die Entscheidungskosten. Durch die Stärkung des Marktkreditaufbaus und die Schaffung eines glaubwürdigen Marktumfelds könnten die marktbestimmenden Elemente in einen wirksamen Zustand überführt werden. Die Informationen, die über herkömmliche Informationskanäle bereitgestellt werden, könnten im Allgemeinen die Authentizität und Zuverlässigkeit der Informationen gewährleisten und somit zu wissenschaftlichen Entscheidungen verhelfen, was nicht nur die Sammlung von Informationen und Kosten verringere. Die Informationen könnten auch schnell und zeitnah übermittelt werden. Die wissenschaftliche Entscheidungsfindung von Regierungen und Marktteilnehmern werde garantiert und zudem würden die Entscheidungskosten gesenkt, was nicht nur der gesunden Entwicklung der Marktwirtschaft zuträglich sei, sondern auch soziale Ressourcen schone und die Interessen der Verbraucher schütze.

Durch die Stärkung des Marktkreditaufbaus könne die Glaubwürdigkeit der Marktteilnehmer gestärkt und ein beidseitiges Vertrauensverhältnis zwischen den an einer Transaktion beteiligten Parteien hergestellt werden. Integrität sei der Grundstein, um das Image des Marktes zu prägen und Glaubwürdigkeit zu erlangen. Marktteilnehmer verließen sich bei ihren Handelsaktivitäten auf Ehrlichkeit. Sie könnten nicht nur an Glaubwürdigkeit gewinnen, das Vertrauen ihrer Partner gewinnen und eine langfristige vertrauensvolle Beziehung aufbauen, sondern auch die Transaktionskosten senken, die Märkte erweitern, die Geschäftstätigkeit ausbauen, die Wirksamkeit der Marktaktivitäten verbessern und ihre Wettbewerbsfähigkeit steigern.

Durch die Stärkung des Marktkreditaufbaus könnten Zwischenglieder in der Wirtschaftstätigkeit abgebaut werden. In einem ehrlichen Marktumfeld müssten Marktteilnehmer trotz der Asymmetrie der Informationen im Wirtschaftsleben aufgrund des hohen Maßes an sozialer Integrität keine Informationen sammeln und identifizieren und müssten keine zusätzlichen Inspektionen der Produktqualität, Fälschungssicherheit, Verhandlungen und andere Aktivitäten durchführen. Mithin komme es zu einer Reduzierung der Zwischenverbindungen im Transaktionsprozess, zur Einsparung von Zeit und Transaktionskosten sowie zur Verbesserung der Qualität und Effizienz der wirtschaftlichen Abläufe.

Die Stärkung des Marktkreditaufbaus impliziere eine Senkung der Verwaltungskosten der Marktteilnehmer. Fehle die Integrität von Markteinheiten, wirke sich dies nicht nur auf die Herstellung eines guten Vertrauensverhältnisses aus, sondern auch auf das Image von außen. Die Stärkung des Aufbaus von Marktintegrität, der Aufbau eines Marktintegritätssystems, die Infiltration von Integritätskonzepten, -zielen und -anforderungen in die tatsächlichen Abläufe von Marktteilnehmern dienten dem Aufbau eines Verhältnisses von gegenseitigem Vertrauen und Verständnis. Es stärke den internen Zusammenhalt der Mitarbeiter und diene dem Aufbau eines guten Rufs nach außen. Mehr Integrität mindere nicht nur Managementschwierigkeiten und Managementkosten, sondern sei auch die Voraussetzung für die langfristige Entwicklung der Marktteilnehmer.

Die Stärkung des Aufbaus von Marktkredit erlaube die Senkung der Kosten für das staatliche Marktmanagement. In einer Marktwirtschaft müsste die Marktverwaltung gestärkt werden, um die Marktordnung aufrechtzuerhalten, die legitimen Geschäfte und den fairen Wettbewerb zu schützen, die Preise zu stabilisieren und die Rechte der Verbraucher zu schützen. Wenn die Integrität der Marktteilnehmer fehlt, werde der Markt ins Chaos geraten, und die Regierung eine Menge menschlicher, materieller und finanzieller Ressourcen investieren müssen, um Kriminalität zu verhindern und die Transaktionsordnung und Fairness aufrechtzuerhalten. Die Stärkung des Aufbaus sozialer Integrität werde die Schwierigkeit und die Kosten der Marktverwaltung verringern.

29.4   Auswirkungen des Sozialkreditsystems auf Marktteilnehmer

Durch die Einführung des Sozialkreditsystems sollen die Mechanismen einerseits zur Förderung und Motivation von Ehrlichkeit der Marktteilnehmer und andererseits zur Einschränkung und Bestrafung von Unehrlichkeit verbessert werden.

29.4.1 Förderung und Motivation von Ehrlichkeit

Um Ehrlichkeit zu fördern sollen während der Beantragung von Verwaltungslizenzen durch Marktteilnehmer verschiedene Regierungsabteilungen schrittweise drei aufeinanderfolgende Jahre komfortable Servicemaßnahmen wie „grüne Kanäle" und „Akzeptanz der Toleranz" für Amtskollegen mit typischer Integrität und ohne schlechte Kreditwürdigkeit umsetzen. Bei qualifizierten Verwaltungsmitarbeitern sollen hinsichtlich der zusätzlich zu den gesetzlich vorgeschriebenen Materialien vorgesehenen Unterlagen, wenn einige der Antragsunterlagen nicht vollständig sind und wenn ihre schriftlichen Zusagen innerhalb des angegebenen Zeitraums vorliegen, diese im Voraus akzeptiert werden, um die Verarbeitung zu beschleunigen.

Bei der Umsetzung verschiedener staatlicher Präferenzstrategien wie der Durchführung von Finanzkapitalprojekten und Präferenzstrategien für die Investitionsförderung sollen die Integrität der Marktteilnehmer und eine verstärkte Unterstützung Vorrang haben. Bei damit zusammenhängenden Transaktionen mit öffentlichen Mitteln wird von der Regierung empfohlen, Kredite und andere Maßnahmen an Marktteilnehmer gemäß Gesetz und Vertrag zu vergeben.

Marktdienstleistungsinstitute wie Finanzinstitute und kommerzielle Vertriebsinstitute sollen sich auf die Verwendung von Bonitätsinformationen zu Marktprojekten, Bonitätsnoten und Bonitätsbewertungsergebnissen beziehen und ehrlichen Marktprojekten den Vorzug geben, damit vertrauenswürdige Personen mehr Chancen und Vorteile auf dem Markt erhalten.

Ehrliche Marktteilnehmer sollen gute Kreditinformationen von Marktprojekten rechtzeitig auf Regierungswebsites und auf Websites von „Credit China" veröffentlichen und sich auf die Förderung ehrlicher Unternehmen bei Veranstaltungen wie Ausstellungen und bei Kreditinstituten konzentrieren. Sie sollen „Kredit" zu einem wichtigen Faktor bei der Allokation von Ressourcen auf dem Markt machen. Dazu gibt es einen Leitfaden für Kreditauskunfteien, um die Erfassung positiver Informationen von Marktteilnehmern zu verbessern und den Anteil der Anreizbewertungen für vertrauenswürdige Industriesektoren zu erhöhen, in denen sich Integritätsprobleme stärker konzentrieren.

29.4.2 Einschränkung und Bestrafung von Unehrlichkeit

Basierend auf dem Umgang und der Bewertung von nicht vertrauenswürdigen Verhaltensweisen gemäß den Gesetzen und Vorschriften werden die Regierungsstellen und sozialen Organisationen durch Informationsaustausch aufgefordert, gemeinsame Disziplinarmaßnahmen gegen schwerwiegende, nicht vertrauenswürdige Verhaltensweisen gemäß den Gesetzen und Vorschriften zu ergreifen.

Diese sollen sich auf die Bestrafung von Marktmissbräuchen konzentrieren, welche die Volkswirtschaft und den Lebensunterhalt der Menschen ernsthaft beeinträchtigen. Umfasst sind hauptsächlich:

- Handlungen, welche die Gesundheit und das Leben der Menschen ernsthaft gefährden, einschließlich schwerwiegender Unehrlichkeit in den Bereichen Lebensmittel und Medizin, ökologische Umwelt, technische Qualität, Produktionssicherheit, Brandschutz und obligatorische Produktzertifizierung, und die Marktgerechtigkeit ernsthaft untergraben,
- Wirtschaftliche Handlungen einschließlich Bestechung, Steuerhinterziehung und Betrug, böswilliger Schuldenhinterziehung, böswilliger Zahlungsrückstände für Waren oder Dienstleistungen, böswilliger Lohnrückstände, illegalem Fundraising, Vertragsbetrug, Pyramidensystemen, nicht lizenziertem Betrieb, Herstellung und Verkauf von gefälschten und minderwertigen Produkten und Vorsätzliche Verletzung von Rechten

des geistigen Eigentums, Ausleihe und Ausleihe von Qualifikationsangeboten, Abgabe von irreführenden Angeboten, falsche Werbung, Verletzung der legitimen Rechte und Interessen von Verbrauchern oder Anlegern in Wertpapieren und Termingeschäften, schwerwiegende Störung der Kommunikationsordnung im Cyberspace und Störung der sozialen Ordnung in der Versammlung,

- Handlungen, welche die Glaubwürdigkeit von Justiz- und Verwaltungsorganen erheblich beeinträchtigen, einschließlich schwerwiegender Unehrlichkeit, beispielsweise derjenigen Parteien, die nach einem Urteil oder einer Entscheidung des Justiz- oder Verwaltungsorgans den Vollzug der Vollstreckung verweigern oder der Vollstreckung entgehen und

- Handlungen, welche die Anforderung von zivilen Ressourcen ablehnen oder verzögern oder die Umwandlung der angeforderten zivilen Ressourcen behindern, nationale Verteidigungsinteressen gefährden oder gar nationale Verteidigungseinrichtungen zerstören.

Bei schwerwiegenden Unehrlichkeiten soll jede Abteilung diese als Hauptaufsichtsgegenstände aufführen und behördliche Auflagen und Disziplinarmaßnahmen im Einklang mit Gesetzen und Vorschriften treffen: Überprüfung von Genehmigungsprojekten für Verwaltungslizenzen, strikte Kontrolle der Ausstellung von Produktionslizenzen, Beschränkung der Genehmigung und Genehmigung neuer Projekte, Beschränkung der Ausgabe von börsennotierten Aktien und der Finanzierung oder der Ausgabe von Anleihen, Beschränkung der Kotierung und Finanzierung des nationalen Umlagerungssystems sowie Beschränkung der Einrichtung oder Beteiligung von Finanzinstituten und kleinen Darlehensfirmen, Finanzierungsgarantien, Risikokapitalfirmen, Internet-Finanzierungsplattformen und andere Institutionen, die Internet-Informationsdienste einschränken. Es sind streng die Beantragung von Finanzmitteln zu beschränken, die Teilnahme an damit zusammenhängenden Transaktionen mit öffentlichen Ressourcen und die Teilnahme am Franchising für Infrastruktur und öffentliche Versorgungsunternehmen. Unternehmen mit schwerwiegender Unehrlichkeit drohen Markt- und Branchenverbote, ebenso deren gesetzlichen Vertretern, Auftraggebern und registrierten Praktikern, die direkt für die Unehrlichkeit verantwortlich sind. Außerdem sind unverzüglich Ehrentitel zu widerrufen, die von den Unternehmen mit schwerwiegender Unehrlichkeit und deren gesetzlichen Vertretern, Direktoren, leitenden Angestellten, Handlungsbevollmächtigten, Aktionären und anderen Personen, die direkt für die Unehrlichkeit verantwortlich sind, erlangt wurden.

Bei schwerwiegenden Unehrlichkeiten sollen alle Regierungsstellen und Branchenverbände den einheitlichen Sozialkreditcode als Index verwenden und relevante Informationen rechtzeitig offenlegen, damit der Markt Unehrlichkeiten erkennen und Kreditrisiken vermeiden kann. Die relevanten Marktteilnehmer sind nachdrücklich aufzufordern, ihren gesetzlichen Verpflichtungen nachzukommen und Beschränkungen zu verhängen, so für die Veräußerung und den Erwerb von Immobilien, die Benutzung von Flugzeugen und Hochgeschwindigkeitszügen, die Buchung von Urlaubsreisen und Übernachtungen in Hotels je nach Sterneniveau und für andere verbrauchsintensive Verhaltensweisen. Dies gilt für schwerwiegend nicht vertrauenswürdige Unternehmen, die leistungsfähig sind, aber die Leistung verweigern oder geforderte Maßnahmen nicht umsetzen. Kreditbüros unterstützen dabei, Informationen über schwerwiegende Unehrlichkeit zu sammeln und in Kreditunterlagen und Kreditberichte einzubeziehen. Geschäftsbanken, Wertpapier- und Termingeschäftsinstitute, Versicherungsunternehmen und andere Finanzinstitute sind angewiesen, die Zinssätze für Kredite und Sachversicherungsprämien zu erhöhen oder die Bereitstellung von Krediten, Sponsoring, Underwriting, Versicherungen und anderen Dienstleistungen gemäß den Grundsätzen der Risikobegrenzung zu beschränken.

Die Festlegung und Verbesserung von Selbstdisziplinierungskonventionen und Berufsethikstandards für die Industrie soll der Förderung der Kreditbildung in der Industrie dienen. Der Leitfaden für Branchenverbände und Handelskammern zur Verbesserung des brancheninternen Mechanismus zur Erfassung und Weitergabe von Kreditinformationen und zur Aufzeichnung schwerwiegender Unehrlichkeit in den Kreditakten der Mitglieder soll Branchenverbände und Handelskammern ermutigen, mit qualifizierten Kreditinstituten und Drittanbietern zusammenzuarbeiten, um Bonitätsbewertungen von Mitgliedsunternehmen durchzuführen. Branchenverbänden und Handelskammern unterstützen bei der Umsetzung von Disziplinarmaßnahmen wie Abmahnung, Benachrichtigung über Unehrlichkeit, öffentliche Verurteilung, Ablehnung und Entlassung unehrlicher Mitglieder gemäß Branchenstandards, wie auch bei Vorschriften und Ernennungen.

Die Rolle verschiedener sozialer Organisationen soll voll ausgeschöpft werden. Die „sozialen Kräfte" werden dazu angeleitet, sich umfassend an gemeinsamen Disziplinarmaßnahmen gegen Unehrlichkeit zu beteiligen. Die Einrichtung und Verbesserung des Systems zur Meldung von Unehrlichkeit und die Aufforderung an die Öffentlichkeit, schwerwiegende Unehrlichkeit von Unternehmen zu melden, verlangen strikte Vertraulichkeit der Informationen von Berichterstattern. Relevante soziale Organisationen werden bei der Einreichung von Klagen von öffentlichem Interesse unterstützt, die sich gegen Verstöße gegen Gruppenrichtlinien richten, wie die Umwelt zu verschmutzen oder die legitimen Rechte und Interessen von Verbrauchern oder öffentlichen Investoren zu verletzen. Faire, unabhängige und qualifizierte soziale Institutionen sollen ermutigt werden, die öffentliche Meinung in Bezug auf Unehrlichkeit im Big-Data-Bereich zu überwachen und regionale sowie branchenweite Kreditanalyseberichte zu erstellen und zu veröffentlichen.

Schwerwiegende unredliche Verhaltensweisen im Markt sollen in den Kreditunterlagen von Unternehmen und Institutionen sowie in den persönlichen Kreditunterlagen ihrer gesetzlichen Vertreter, Auftraggeber und sonstigen, unmittelbar verantwortlichen Personen vermerkt werden. Bei der Durchführung gemeinsamer Disziplinarmaßnahmen gegen unehrliche Marktteilnehmer sollen die jeweiligen Verantwortlichen entsprechenden gemeinsamen Disziplinarmaßnahmen im Einklang mit den Gesetzen, Vorschriften und Richtlinien unterworfen werden. Durch die Einrichtung einer vollständigen Datenbank mit persönlichen Kreditunterlagen und eines gemeinsamen Disziplinarverfahrens sollen Disziplinarmaßnahmen gegen Unehrlichkeit umgesetzt werden.

29.5 Schlussfolgerungen und praktische Hinweise

Seit dem Aufbau des Sozialkreditsystems sind „Compliance" und „Geschäft" in China keine Entweder-oder-Entscheidung mehr, sondern müssen in Einklang gebracht werden, um überleben und sich weiterentwickeln zu können. In einigen Bereichen der Schlüsselaufsicht der Regierung sollten sich die Marktteilnehmer auf die Vorteile positiver Bonitätsergebnisse konzentrieren. Obwohl die vollständige Einhaltung aller Regeln und Vorschriften entsprechende Kosten verursacht, kann die Verwirklichung spezifischer Ratingziele den Unternehmen größere Vorteile bringen. Den in China aktiven Unternehmen empfehlen wir, interne Prozesse und Rahmenbedingungen an das Sozialkreditsystem anzupassen, mit den zuständigen Regierungsstellen offen zu kommunizieren und die eigenen Kreditbeziehungen rechtzeitig und regelmäßig zu überwachen.

29.5.1 Einrichtung interner Prozesse und Rahmenbedingungen

Marktteilnehmer können auf die Anforderungen des Sozialkreditsystems, insbesondere des Marktkreditsystems, reagieren, indem sie interne Prozesse und Rahmenbedingungen aufbauen, die den geltenden Gesetzen und Vorschriften entsprechen, und negative Ratings und entsprechende Strafen vermeiden. Marktteilnehmer müssen die spezifischen Anforderungen der Richtlinien des Sozialkreditsystems in Bezug auf die Branche und den Geschäftsumfang des Unternehmens verstehen, eine Vielzahl von Regierungsdokumenten und damit zusammenhängenden Gesetzen und Vorschriften selektieren und die relevanten Anforderungen an Ratings mit den aktuellen internen Prozessen und Rahmenbedingungen von Marktteilnehmern vergleichen. Weitere Verbesserungsbereiche sind zu identifizieren, um effektive Anpassungen zu planen und implementieren, wie auch sicherzustellen, dass das Unternehmen eine gute Bonität erzielt.

29.5.2 Kommunikation mit den Regierungsstellen

Marktteilnehmer sollten einen Berater oder einen Rechtsanwalt damit beauftragen, die erforderlichen Abklärungen vorzunehmen und mit den Regierungsstellen zu kommunizieren. In vielen Fällen (obwohl viele Richtlinien des Sozialkreditsystems klar sind) gibt es immer noch einige Fehler beim Verständnis und bei der Umsetzung durch die Marktakteure, weshalb Spezialisten mit den Regierungsstellen kommunizieren sollten. Zu diesem Zweck sollten Unternehmen mit den entsprechenden Regierungsstellen Kontakt aufnehmen und die Klärung der einschlägigen Bestimmungen fordern. Die Marktteilnehmer müssen auf die auf der offiziellen Website veröffentlichten statistischen Datentabellen achten, um zu vermeiden, dass sie es versäumen, Erklärungen abzugeben, oder sogar falsche und unwahre Daten fabrizieren, die in Statistiken aufgenommen werden sollen. Aufnahme in die Unehrlichkeitsliste und gemeinsame Bestrafung durch verschiedene Abteilungen wären die Konsequenz.

29.5.3 Überwachung von Kreditbeziehungen

Marktteilnehmer sollten weiterhin überwachen, ob ihre internen täglichen Abläufe dem Standard entsprechen, insbesondere, ob sie disziplinarische und strafbewehrte Verhaltensweisen aufweisen, und Änderungen der Anforderungen der amtlichen Stellen an Ratings berücksichtigen, die ihre eigenen Abläufe betreffen, um zeitnahe Anpassungen an internen Prozessen und Rahmenbedingungen vorzunehmen. Erkundigen Sie sich regelmäßig, was auf der schwarzen Liste oder der Überwachungsliste steht, und kommunizieren Sie rechtzeitig mit den zuständigen Regierungsstellen. Ergreifen Sie Maßnahmen wie die Behebung von Kreditstörungen, die Überprüfung von Verwaltungsangelegenheiten und Verwaltungsstreitigkeiten, um ihre legitimen Rechte und Interessen zu wahren.

Kreditbeziehungen sind im Geschäftsleben allgegenwärtig. Ohne Bonität ist eine gesunde und nachhaltige Entwicklung kaum möglich. Dies gilt auch für die Marktwirtschaft insgesamt. Daher, um das sozialistische Marktwirtschaftssystem zu verbessern, soll der Aufbau und die Verbesserung des Sozialkreditsystems beschleunigt werden, ein gutes Marktkreditumfeld geschafft und schließlich ein vertrauenswürdiger Anreiz- und Unehrlichkeitsbestrafungsmechanismus etabliert werden. Nur wenn die Marktteilnehmer sich ehrlich verhalten, können sie in der chinesischen Marktwirtschaft in Zukunft erfolgreich sein.

Literatur

范水兰:《企业征信法律制度及运行机制》,法律出版社2017年版。
李新庚:《社会信用体系运行机制研究》,中国社会出版社2017年版。
方乐华:《市场主体信用制度的法学思考:社会法、消费者权益保护法视角》,法律出版
    社2017年版。
刘肖原:《我国社会信用体系建设问题研究》,知识产权出版社2016年版。
纪森森:《信用经济:下一个10年红利风口》,电子工业出版社2018年版。
刘新海:《征信与大数据》,中信出版社2016年版。
张建华:《中国企业信用建设报告(2017–2018)》,中国法制出版社2017年版。
翟学伟:《中国社会信用:理论、实证与对策研究》,中国社会科学出版社2018年版。
连维良:发挥行业协会商会作用更加积极有效地推动社会信用体系建设,《中国信用》杂志,
    2017年第9期。
王军:积极健全纳税信用体系全力助推信用中国建设,《中国信用》杂志,2017年第3期。
连维良:加快推动社会信用体系建设迈上新台阶,《中国信用》杂志, 2017年第2期。
徐绍史:加快推进社会信用体系建设着力构建信用联合奖惩大格局,《中国信用》杂志,
    2017年第1期。
国务院,《社会信用体系建设规划纲要(2014–2020年)》, 2014年6月14日。

《 国务院办公厅关于运用大数据，加强对市场主体服务和监管的若干意见》，国办发
〔2015〕51号
《 关于在行政管理事项中使用信用记录和信用报告的若干意见》，发改财金(2013)920号

Tim A. Fongern ist Rechtsanwalt und Managing Partner Germany von Sheng Heng Law Firm. Fongern berät Unternehmen und Investoren in den Bereichen Wirtschaftsrecht und Streitbeilegung mit einem besonderen Fokus auf China-Deutschland-Geschäft. Bevor Fongern zu Sheng Heng Law Firm kam, arbeitete er sowohl für große internationale als auch namhafte mittelständische Wirtschaftskanzleien in Frankfurt am Main und Singapur. Er studierte in Mainz, Speyer und London.

Heng Wang ist chinesischer Rechtsanwalt (Lü shi), Gründungspartner und Direktor von Sheng Heng Law Firm sowie seit 2016 auch in Frankfurt am Main als niedergelassener chinesischer Rechtsanwalt zugelassen. Wang berät Unternehmen und Investoren in den Bereichen Finanzierung, Unternehmensbeteiligungen und Streitbeilegung. Wang studierte an der Dongbei Universität für Finanzen und Wirtschaft in China.

Lihong Yu ist chinesische Rechtsanwältin (Lü shi) und Associate bei Sheng Heng Law Firm. Sie berät Unternehmen und Investoren auf den Gebieten des internationalen Finanz- und Wirtschaftsrechts mit einem besonderen Schwerpunkt auf M&A-Transaktionen. Yu studierte an der Dalian Maritime University in China.

„Social Credit" bei der Auswahl von Mitgliedern des Aufsichtsrates einer Aktiengesellschaft

Dr. Yvette Bellavite-Hövermann

Zusammenfassung

Der Aufsichtsrat ist verpflichtet, der Hauptversammlung die Mitglieder des nächsten Aufsichtsrats vorzuschlagen, und ist damit de facto für die Auswahl der Aufsichtsratsmitglieder verantwortlich. Daraus können Schadenersatzansprüche in Form eines Auswahlverschuldens gegen ihn entstehen. Zudem können daraus auch Reputationsrisiken für die Unternehmen erwachsen, deren Tragweite sehr weitreichend sein können und die nicht unterschätzt werden sollten.

Digitalisierung, Vernetzung, wachsende Datenspeicher- und -verarbeitungskapazitäten verändern die industrielle Landschaft massiv und dringen in alle Lebensbereiche ein. Es gehört zur Professionalität eines Aufsichtsrats, sich dem nicht zu verschließen. Big Data-Analysen und -Technologien können im Auswahlprozess neuer Aufsichtsratsmitglieder zu einer besseren Auswahl führen sowie Haftungsrisiken für den Aufsichtsrates und letztlich auch für das Unternehmen reduzieren, auch wenn Algorithmen nur ein Element im Auswahlverfahren sein können und die letzte Entscheidung sowie die Haftung immer bei dem Gremium verbleiben muss.

Angesichts der Komplexität von Big Data-Technologie muss die Aufbereitung der Daten, auf die der Aufsichtsrat zugreift, von darauf spezialisierten Unternehmen geleistet werden, am besten in privatrechtlicher, nicht öffentlich-rechtlicher Form, jedoch unbedingt kontrolliert von einer staatlichen Aufsicht, zum Beispiel nach dem Vorbild der beaufsichtigen Ratingagenturen. Bewertung und Gewichtung der Informationen werden dabei eine ständige Herausforderung bleiben.

Eine solche Innovation bei der Auswahl von Aufsichtsratsmitgliedern wäre nicht vergleichbar mit dem Social Credit System der chinesischen Regierung, mit dem sie

Dr. Y. Bellavite-Hövermann (✉)
Frankfurt am Main, Deutschland

© Springer Fachmedien Wiesbaden GmbH, ein Teil von Springer Nature 2020
O. Everling (Hrsg.), *Social Credit Rating*,
https://doi.org/10.1007/978-3-658-29653-7_30

das ganze Land flächendeckend überwachen will. Sie wäre aber in jeden Fall auch weit von den engen Datenschutzbedenken entfernt, die in Deutschland vorherrschen. Das schwedische Beispiel zeigt, dass Transparenz von Personendaten durchaus mit Demokratie und Meinungsfreiheit vereinbar ist.

Social Credit ist kein Allheilmittel, bei der richtigen Ausgestaltung kann es jedoch ein wichtiges Tool sein bei der Auswahl von Aufsichtsratsmitgliedern und folglich dabei helfen, die Professionalisierung der Aufsichtsratstätigkeit weiter zu entwickeln.

Die Autorin dankt Dr. Katrin Burkhardt, Berlin, für die anregende Diskussion und die kritischen Anmerkungen sowie Felix Schatten, 01PC, Bad Soden, für die hilfreichen und interessanten Anregungen zu den Debatten betr. Social Media, Algorithmen und digitale Zukunft.

30.1   Einleitung

Das in China geplante und in Umsetzung befindliche Social Credit System (SCS) für Unternehmen und Privatpersonen ist vom Grundsatz „eine neuartige Anwendung von Big-Data-Technologie" (vgl. Geleitwort von Jörg Wuttke in diesem Buch), mit der laufend Informationen über das Verhalten von Unternehmen gesammelt und „durch undurchsichtige Algorithmen" (vgl. Geleitwort von Jörg Wuttke in diesem Buch) verarbeitet werden, um Konformität mit resp. Abweichungen von chinesischen Vorschriften festzustellen: „Basierend auf der Analyse erhalten Unternehmen in jeder von mehreren Dutzend Kategorien eine Bewertung, die sich dann zu einem Gesamtergebnis zusammenfügen." (vgl. Geleitwort von Jörg Wuttke in diesem Buch)

Wer glaubt, ein solches System hätte in westlichen Demokratien keinen Platz, sollte die heute schon starke Abhängigkeit von Ratings bedenken. Der verschiedentlich zu hörende Vergleich mit Ermittlungsinstrumenten bezüglich Kreditwürdigkeit bzw. Kapitaldienstfähigkeit, wie sie etwa die Schufa nutzt, greift allerdings zu kurz.

Von „Social Credit" reden inzwischen viele. Auch in Deutschland beschäftigt man sich nicht zuletzt in Unternehmen vermehrt damit. Ein solches System oder einzelne seiner Komponenten könnten, so die Hoffnung, u. a. den Aufsichtsrat einer Aktiengesellschaft bei der Nominierung künftiger Aufsichtsratsmitglieder unterstützen, und das wäre durchaus nötig, denn die Anforderungen an diese Vermittlungsinstanz zwischen den Eigentümern und dem Vorstand einer AG, ob börsennotiert oder nicht, wachsen und ihr Haftungsrisiko steigt.

Der vorliegende Beitrag versteht sich als Werkstattbericht. Er will erste Denkanstöße im Sinne eines Brainstormings geben und keinesfalls ein fertiges Konzept vorlegen. Es würde den Rahmen sprengen, sämtliche rechtlichen Aspekte bei der Wahl von Aufsichtsräten, der Geltendmachung von Schadenersatzansprüchen oder der derzeit gültigen Datenschutzbestimmungen auch nur darzustellen, geschweige denn zu diskutieren. Ebenfalls nicht intendiert ist eine Auseinandersetzung mit dem in Umsetzung befindlichen chinesischen Social Credit System (vgl. hierzu die verschiedenen Beiträge in diesem Buch) oder gar ein Plädoyer für dessen Übernahme. Vielmehr beschränkt sich der folgende Beitrag darauf, das SCS hinsichtlich seiner Tauglichkeit für die qualifizierte Besetzung von Aufsichtsräten abzuklopfen. Denn auf diesem Gebiet herrscht eine erschreckende Willkür, es gibt wenig bis keine Vorschriften für das Anforderungsprofil der Kandidaten und Kandidatinnen, die gesetzlichen Vorgaben sind sehr allgemein gehalten und die Möglichkeiten, selbst grobe Verstöße juristisch zu verfolgen, aufgrund dieser Sachlage fast aussichtslos. Es besteht mithin Handlungsbedarf, der einen grundlegenden Kulturwandel voraussetzt oder auch anstoßen wird.

Ein grundlegender Wandel geht mit einem Paradigmenwechsel einher, ohne den keine Revolution denkbar ist. Revolutionen werden nüchtern als institutioneller Systemwandel definiert – ein historischer Prozess, der Fortschritt und Entwicklung voraus- und zugleich in Gang setzt (vgl. Speich Chassé 2012). Vielleicht ist kein Bereich stärker dem Fluss ständiger Veränderung ausgesetzt als die Finanzindustrie, deren ganzes Wesen der Liquidität verpflichtet ist. Sie hat in den letzten 25 Jahren radikale Umbrüche erlebt, ganz unabhängig davon, wie man diese bewertet, ob man sie für richtig und gerecht hält oder nicht. Und nicht nur in diesen, schon die Einführung des deutschen Steuersystems 1919 durch den Reichsfinanzminister war eine echte Revolution. Erinnert sei an Basel I und II für die Eigenkapitalunterlegung von Kredit- und Handelsgeschäften der Banken, die Zulassung von Ratings und im Zusammenhang damit die Registrierungspflicht von Ratingagenturen, oder Aktiengesellschaften im allgemeinen betreffend etliche EU-Richtlinien und Gesetzesreformen in vielen wirtschaftlichen Bereichen (zuletzt die Aktionärsrechterichtlinie und das Umsetzungsgesetz ARUG II), den Deutschen Corporate Governance Kodex, die internationale Bilanzierung mit IAS/IFRS, deren regulatorische Anwendung für Risiko- und Kapitalsteuerung, Eigenkapitalbasis und Risikomanagement der Banken und, last but not least, das Bezahlsystem mit Internet.

Angesichts der Systemrelevanz des Finanzsektors und großen Konzernunternehmen ist er sicher gut beraten, auch bei den Auswahlkriterien für Aufsichtsratsmitglieder strengere und vor allem systematischere Richtlinien aufzustellen. Und hierbei könnte, so der von diesem Beitrag intendierte Denkanstoß, ein Social Credit System einen entscheidenden Schritt nach vorn bedeuten. Das Thema reiht sich in die Entwicklung klar in Richtung „Professionalisierung der Aufsichtratsarbeit" ein und dazu gehört auch eine Professionalisierung der Auswahl von Aufsichtsratsmitgliedern.

Im nächsten Kapitel werden zunächst die Eckdaten der rechtlichen Anforderungen hinsichtlich der Nominierung der Anteilseigner im Aufsichtsrat aufgeführt sowie die wesentlichen Auswahlkriterien für Aufsichtsratsmitglieder dargelegt. Abschn. 30.3 befasst sich dann mit Social Credit als Unterstützung im Auswahlprozess von Aufsichtsräten, und Abschn. 30.4 fasst schließlich die wesentlichen Gesichtspunkte zusammen.

30.2 Rechtliche Vorgaben für Nominierung und Wahl der Anteilseignervertreter

Der Aufsichtsrat (AR) ist gem. § 95 ff. AktG eines der drei Organe der AG; seine Rechte und Pflichten sind insb. in § 111 AktG geregelt. Der AR hat eine Wahlvorschlagspflicht betreffend die Anteilseignervertreter im AR.

Die Mitglieder des AR werden auf Vorschlag des AR von der Hauptversammlung gewählt (§§ 101 Abs. 1,119 Abs. 1 AktG).[1]

Der AR ist mithin selbst für seine Zusammensetzung und die in seinem Kreis vorhandenen fachlichen Kompetenzen verantwortlich. Zwar ist die Hauptversammlung grundsätzlich nicht an den Vorschlag des AR gebunden, folgt ihm aber in aller Regel (Ausnahme nach §§ 6 und 8 Montan-Mitbestimmungsgesetz). De facto findet die Auswahl der Anteilseignervertreter im AR durch den AR statt oder, wie es zum Teil immer noch Praxis ist, durch den Vorstand, der dem AR einen Vorschlag zur Weiterleitung an die Hauptversammlung unterbreitet (zur Wahlvorschlagspflicht/zum Wahlvorschlagsrecht, vgl. Redenius-Hövermann 2019, S. 146 m. w. N.). Da der AR nicht zuletzt die Arbeit des Vorstands kontrollieren soll, ist diese Praxis nicht mit guter Corporate Governance vereinbar.

Gute Corporate Governance verlangt, dass der AR einen Nominierungsausschuss bildet (D 12, alt 5.3.3. DCGK) (vgl. von Werder und Danilov 2018, S. 2003, wonach 100 Prozent der DAX 30-Unternehmen und 80 Prozent aller Unternehmen der Kodex-Empfehlung gefolgt sind), der ausschließlich mit Vertretern der Anteilseigner besetzt ist und dem AR geeignete Kandidaten für dessen Vorschläge an die Hauptversammlung zur Wahl von Aufsichtsratsmitgliedern benennt.

Die Satzung der AG kann persönliche Voraussetzungen der Aufsichtsratsmitglieder vorsehen (§ 100 Abs. 5 AktG). Diese dürfen allerdings nicht so festgelegt sein, dass das Wahlrecht der Hauptversammlung de facto auf bestimmte Personen eingeengt und damit unverhältnismäßig beschränkt wird (vgl. hierzu Redenius-Hövermann 2019, S. 151 m. w. N.).

Laut AktG soll die Vorschlagspflicht des AR bei der Wahl geeigneter Kandidaten für den AR (§ 124 Abs. 3 AktG) dafür sorgen, dass die vorgeschlagenen Kandidaten für die Aufsichtsratstätigkeit am besten geeignet sind (vgl. hierzu auch Bellavite-Hövermann et al. 2005, S. 17; vgl. Redenius-Hövermann 2010, S. 660, 662). Das bedeutet nach §§ 93, 116 AktG, dass er die Sorgfalt eines gewissenhaften und ordentlichen Geschäftsleiters walten lassen muss. Vorgeschrieben sind weiterhin Mindestkenntnisse und die Vermeidung dauerhafter Interessenskonflikte (§ 116 AktG).

AktG wie DCGK benennen verschiedene rechtliche oder statutarische Voraussetzungen, welche die Kandidaten erfüllen müssen. Allerdings kann davon im Rahmen des Comply-and-Explain gemäß § 161 AktG abgewichen werden. Das Aktiengesetz beschränkt sich auf einige formelle Anforderungen an Aufsichtsratsmitglieder; § 100 AktG schreibt Geschäftsfähigkeit sowie eine Höchstzahl an Mandaten als persönliche Voraussetzungen fest und verbietet Überkreuzverflechtungen (vgl. ausführlich Hüffer und Koch

---

[1] Ausnahme: Soweit sie nicht in den AR zu entsenden oder als Aufsichtsratsmitglieder der Arbeitnehmer nach dem Mitbestimmungsgesetz, dem Mitbestimmungsergänzungsgesetz, dem Drittelbeteiligungsgesetz oder dem Gesetz über die Mitbestimmung der Arbeitnehmer einer grenzüberschreitenden Verschmelzung zu wählen sind.

2018, § 100 Rn. 2 ff.). Bei Unternehmen mit besonderem öffentlichen Interesse bedarf es noch eines Mitglieds, das über Sachverstand in Rechnungslegung oder Abschlussprüfung verfügt (Hüffer und Koch 2018, § 100 Rn. 22).

Die Rahmenbedingungen für den Auswahlprozess der Aufsichtsratsmitglieder haben sich seit dem KonTraG von 1998 wesentlich verbessert (u. a. Höchstzahl der Aufsichtsratsämter, einmal im Kalendervierteljahr AR-Sitzung, AR erteilt Prüfungsauftrag).

Zu den allgemeinen Vorschriften kommen spezielle branchenspezifische Regelungen, überwiegend Nachweispflichten. So werden für die Beantragung der Bestellung von Aufsichtsratsmitglieder von BaFin-beaufsichtigten Unternehmen u. a. folgende Daten (sogenannte harte Faktoren) verlangt: Darstellung des Berufslebens mit Monatsangaben, Darstellung der Schul- und Berufsausbildung, Studium, Nachweise über die Teilnahme an Fortbildungen als Beleg für die erforderliche Sachkunde, Erklärung zu Straf- und Ordnungswidrigkeitenverfahren, gewerberechtliche Entscheidungen, vermögensrechtliche Verfahren, Angaben zur zeitlichen Verfügbarkeit, Führungszeugnis gemäß § 30 Abs. 5 BZRG, Europäisches Führungszeugnis sowie entsprechende Unterlagen aus dem Ausland, Auszug aus dem Gewerbezentralregister, Erklärung über Angehörigkeitsverhältnisse, Erklärung über Geschäftsbeziehungen, Angaben zu Mandaten/Mandantenbegrenzungen.

Merkblatt zu den Mitgliedern von Verwaltungs- und Aufsichtsorganen gemäß KWG und KAGB. In gleiche Richtung siehe Anforderungen der BaFin für Versicherungen: VAG-Merkblatt Mitglieder von Verwaltungs- oder Aufsichtsorganen vom 06.12.2018.

Gemäß der Hertie-Rechtsprechung aus dem Jahre 1982 müssen Mindestkenntnisse zum eigenständigen Verständnis und zur Beurteilung der beschriebenen Geschäftsvorgänge bei Amtsantritt vorhanden sein; zum Teil dürfen fehlende Mindestkenntnisse auch später erworben werden. Zudem darf auf internen oder externen Rat vertraut werden, allerdings muss durch den Aufsichtsrat immer noch eine eigene Plausibilitätskontrolle erfolgen (BGH vom 20.09.2011, II ZR 234/09 „ISION-Entscheidung"). Fachkenntnisse für die Beurteilung besonderer Probleme, etwa steuerliche, bilanzielle, vertriebliche, personalwirtschaftliche oder technische Spezialfragen sind nicht zwingend erforderlich, müssen aber ggf. durch hinzugezogene Sachverständige kompensiert werden. In einem solchen Fall sind die Mitglieder des AR verpflichtet, zuverlässige Erkundigungen einzuholen.

Der DCGK (Grundsatz 11) fordert über die „zur ordnungsgemäßen Wahrnehmung der Aufgaben erforderlichen Kenntnisse, Fähigkeiten und fachlichen Erfahrungen" hinaus die Einhaltung der gesetzlichen Geschlechterquote und dass der vorgeschlagene Kandidat den zu erwartenden Zeitaufwand aufbringen kann. Dem Kandidatenvorschlag soll ein Lebenslauf beigefügt werden, der über relevante Kenntnisse, Fähigkeiten und Erfahrungen Auskunft gibt. Ergänzt um eine Übersicht über die wesentlichen Tätigkeiten neben dem Aufsichtsratsmandat soll er für alle Aufsichtsratsmitglieder jährlich aktualisiert auf der Webseite des Unternehmens veröffentlicht werden.

Der AR muss Personen – unabhängig vom Geschlecht – zur Wahl in den Aufsichtsrat vorschlagen, die dem Interesse der Gesellschaft und dem Gebot einer möglichst effektiven Arbeit des Aufsichtsorgans entsprechen. Börsennotierte Gesellschaften, für die das Mitbestimmungsgesetz, das Montan-Mitbestimmungsgesetz oder das Mitbestimmungsergänzungsgesetz gilt, sind darüber hinaus verpflichtet, dass das unterrepräsentierte Geschlecht im Aufsichtsrat eine Quote von mindestens 30 Prozent erreichen muss (vgl. Redenius-Hövermann 2019, S. 148). Für die anderen vom Gleichstellungsgesetz erfassten Gesellschaften legt der Aufsichtsrat selbst Zielgrößen fest (vgl. Redenius-Hövermann 2019, S. 160).

Gem. Grundsatz C1 DCGK soll der Aufsichtsrat ein Kompetenzprofil für das Gesamtgremium erarbeiten und dabei Diversität und Unabhängigkeit, Branche, Geschäftsmodell, Größe und Kultur des Unternehmens berücksichtigen. Diese Empfehlung geht in die richtige Richtung, ändert aber noch nichts an der Feststellung, dass es derzeit über vage, sehr allgemein gehaltene Vorschriften hinaus keine rechtlichen und standardisierten Anforderungsprofile für zukünftige Aufsichtsräte gibt. Wie sich gleich zeigen wird, sieht die Sachlage in den Unternehmen selbst kaum besser aus.

30.3 Social Credit Rating zur Unterstützung des Aufsichtsrats im Auswahlprozess für den nächsten Aufsichtsrat

Da der AR für seinen Wahlvorschlag verantwortlich ist, sollte er für Verstöße entsprechend belangt werden können. Denkbar wären Auswahlverschulden, die etwa dann bejaht werden können, wenn ein AR einen Finanzexperten in den Prüfungsausschuss bestellt, der über keine speziellen Kenntnisse und Erfahrungen in Fragen der bankbetrieblichen Finanz- und Rechnungslegung verfügt, also nicht die nötige Qualifikation mitbringt. Bringt ein Aufsichtsratsmitglied selbst nicht die erforderlichen Mindestkenntnisse mit, kann ggf. ein Übernahmeverschulden vorliegen. Nicht zu leugnen ist allerdings, dass Auswahl- und Übernahmeverschulden eher theoretischer Natur sind, da der Anspruch in der Praxis meist an der Schadensbemessung scheitern wird.

Aktuell haben die meisten Unternehmen keine umfassenden, konkreten Anforderungsprofile für ihren AR. Vielmehr sind kaum überprüfbare Auswahlkriterien die Regel. Natürlich müssen unternehmensspezifische Gegebenheiten berücksichtigt werden, was eine standardisierte Festlegung erschwert. Unmöglich ist sie deswegen noch lange nicht.

Wie aber gestaltet sich der Auswahlprozess? Eine ältere Studie der Wirtschaftsprüfungsgesellschaft Deloitte hat dazu Aufsichtsratsvorsitzende börsennotierter Gesellschaften befragt. Immerhin zeigte sich dabei eine Entwicklung hin zu einer stärkeren Selbstbestimmung des AR. In mehr als der Hälfte (66,7 Prozent) der DAX-Gesellschaften formulierte der AR die Wahlvorschläge für den AR selbst und überließ die Auswahl nicht beispielsweise dem Vorstand. Lediglich 16,7 Prozent der Gesellschaften ließ sich durch externe professionelle Berater unterstützen. Während die Qualifikation und Fähigkeiten in allen Fällen beurteilt wurden, gaben nur 50 Prozent an, nach einem konkreten Anforderungsprofil entschieden zu haben (vgl. Seele 2007).

In Literatur und Praxis wird jedoch zunehmend über die Erstellung von Anforderungsprofilen diskutiert, also über schriftlich ausformulierte Qualifikationen, die für eine Funktion im AR unerlässlich sind, zum Beispiel Branchen- oder Führungserfahrung, Kommunikationsfähigkeit oder spezifische Fachkompetenzen. Ein positiver Nebeneffekt solcher Beschlüsse wird sein, dass sie die Außenwahrnehmung des Unternehmens bzw. sein Ansehen in der Öffentlichkeit verbessert (vgl. Hans Böckler-Stifung 2011). Vermutlich auch deshalb folgten 2018 bereits 76,4 Prozent der DAX 30-Unternehmen und 61,4 Prozent aller Unternehmen der DCGK-Empfehlung (nunmehr Grundsatz C1), die wie oben erwähnt erst 2017 einführt wurde (vgl. von Werder und Danilov 2018, S. 2003, die allerdings davon ausgehen, dass ausweichlich der Angaben der Unternehmen diese Empfehlung zumindest im DAX zu 100 Prozent befolgt werden wird).

Eines der Probleme bei der Ausformulierung oder Standardisierung von Kompetenzprofilen ist die Tatsache, dass es ganz wesentlich auf „weiche" Kriterien wie zum Beispiel Kommunikationsfähigkeit und andere soziale Talente ankommt. Und genau hier könnte ein Social Credit Rating den Aufsichtsrat im Auswahlverfahren unterstützen.

Durch Fortschritte in Wissenschaft und Technik lassen sich komplexe Systeme mit digitalen Methoden zunehmend effizienter analysieren. Längst werden Fähigkeiten und Potenziale von Bewerbern bei manchen Unternehmen durch Big Data mit eingeschätzt. Nicht erst seit gestern betrachten HR-Abteilungen neben den eigentlichen Bewerbungsunterlagen öffentlich zugängliche Informationen der Bewerber auf Facebook, Twitter oder anderen Plattformen und nutzen sie für die Entscheidungsfindung. Vor allem Softwarefirmen und so mancher internationale Großkonzern, etwa Goldman Sachs, greifen bei potenziellen neuen Mitarbeitern auf vorhandene Daten oder Informationen in den sozialen Netzwerken zu und feilen an ihrer Software für die Analyse der Bewerbereingänge.

Big Data-Technologien, gepaart mit algorithmischen Verfahren, versprechen neue Einblicke und prägen diese nachhaltig. Die dafür verarbeiteten Daten liegen nicht bei einzelnen Unternehmen, sie verteilen sich auf öffentlich Register, soziale Netzwerke, Suchmaschinen, private Websites und unzählige Foren. Dazu kommen von Bots und Crawlern erzeugte Analysedaten.

Auch in Deutschland stehen trotz des im Verhältnis zu anderen Staaten sehr weitgehenden Datenschutzes zahl- und aufschlussreiche personenbezogene Angaben öffentlich zur Verfügung, etwa aus Schuldnerverzeichnissen der zentralen Vollstreckungsgerichte oder über die Schufa oder bei der Lufthansa, soweit die Publikation personenbezogener Daten im Einklang mit der EU-Datenschutzgrundverordnung (DSGVO) und dem Bundesdatenschutzgesetz (BDSG) steht.

Erfassung und Auswertung immenser Datenmengen werden durch die stetig wachsenden und immer preiswerteren Speicher- und Rechenkapazitäten möglich. Eine Studie von KPMG schreibt Big Data-Analysen ein großes Potenzial zu, um Risiken zu reduzieren und u. a. im Personalprozess oder bei der Überwachung der Geschäftsentwicklung einen Mehrwert für Unternehmen zu schaffen (vgl. KPMG 2017).

Selbstredend steht bei alldem außer Frage, dass weder Auswahlverfahren noch andere Bereiche komplett mit Algorithmen abgedeckt werden können.[2]

Dafür sind die einzelnen Prozesse und Faktoren zu komplex, von der ethischen Problematik ganz abgesehen. Die letzte Entscheidung muss ein Gremium, nicht eine Software treffen.

Es geht lediglich darum, Entscheidungen vorzubereiten, die vorhandenen Daten aufzuspüren und so aufzubereiten, dass sie eine kleine Gruppe qualifizierter Personen in endlicher Zeit bewältigen kann. Maschinell erstellte Checklisten sind nützlich für eine schnelle Einschätzung der Kompetenzen von potenziellen Aufsichtsräten, nicht mehr, aber auch nicht weniger. Der AR kann seiner Verantwortung nicht mit dem alleinigen Einsatz von Algorithmen erfüllen, aber sie sind ein Instrument, das ihm die Arbeit erleichtert oder, wenn man die wachsende Unübersichtlichkeit ökonomischer Wechselwirkungen bedenkt, vielleicht in naher Zukunft überhaupt erst ermöglicht. Die Applikation von Algorithmen bringt rechtlichen wie ökonomischen Nutzen und gibt eine Art Hilfestellung, die im Hinblick auf gesteigerte Pflichten in der Krise und erhöhte Anforderungen an die Überwachungstätigkeit des AR äußerst willkommen sein dürfte.

Die Aufbereitung von Social Media Daten für die Berufung in einen AR (aber nicht nur dafür) könnten und sollten spezialisierte Unternehmen anbieten, denn die für die Erstellung von Verhaltensprofilen erforderlichen Extraktions-, Ladungs- und Transaktionsprozesse sind aufwendig und setzen Expertise voraus, wie sie nur in soliden Unternehmensstrukturen zu garantieren ist. Solche, nennen wir sie einmal so, Social Credit Agenturen müssten angesichts der Tragweite ihrer Aufgabe unbedingt strengen Regeln unterliegen, deren Einhaltung von einer staatlichen Aufsicht kontrolliert wird.

Vgl. BaFin: Ratingagenturen unter ESMA-Aufsicht; europaweit besteht nun für alle Ersteller von Kreditratings eine Registrierungspflicht. Vor der Registrierung müssen die Ratingagenturen ein umfangreiches Prüfungs- und Genehmigungsverfahren durchlaufen. Haben sie dieses Verfahren erfolgreich abgeschlossen, können sie mit ihrer Tätigkeit beginnen, die laufend von der ESMA beaufsichtigt wird.

Ihrer Einsetzung vorangehen müsste eine gesamtgesellschaftliche wie unternehmensrechtliche Diskussion, welche Informationen mit welcher Bewertung und Gewichtung in einem solchen System zu berücksichtigen wären. Hierüber Konsens zu erzielen, dürfte insbesondere in Deutschland auf gewaltige Hindernisse stoßen, selbst wenn das resultierende SCS nicht entfernt mit dem bestehenden chinesischen SCS vergleichbar wäre. Denn hierzulande sind derzeit wohl viele Daten frei verfügbar, aus denen sich Verhaltensprofile erstellen lassen, sie dürfen aber aufgrund der datenschutzrechtlichen Vorgaben nicht genutzt werden. Die Grenzen, die das allgemeine Persönlichkeitsrecht bzw. das Unternehmenspersönlichkeitsrecht steckt, sind eng und im Sinn der Meinungsfreiheit verfassungs-

---

[2]Allgemeine Definition: Ein Algorithmus ist eine formale Handlungsvorschrift zur Lösung von Instanzen einer bestimmten Problemklasse, Algorithmen sind Anwendungen, die im Netz besonders zum Sammeln und Auswerten von Daten genutzt werden; vgl. u. a. zur Theorie und praktischen Relevanz K. Mehlhorn, P. Sanders: Algorithms and Data Structures – The Basic Toolbox (Springer 2008), http://www.mpi-inf.mpg.de/~mehlhorn/Toolbox.html.

rechtlich geschützt. Doch das Beispiel Schweden zeigt, dass sich Demokratie und Meinungsfreiheit durchaus mit dem weitgehenden Abbau von Datenschutzvorkehrungen vereinbaren lassen. Dort erhält man mit einer Personennummer Auskunft über vielerlei persönliche Angaben, auch ohne berechtigtes Interesse, einfach so. Zum Beispiel ist die Steuererklärung jedes Bürger in Schweden öffentlich, Arbeitskollegen können das Gehalt des Chefs recherchieren, ein Käufer alle nötige Daten eines Gebrauchtwagens abrufen, die der Verkäufer gerade nicht zur Hand hat usw. (vgl. Reise 2013).

Den datenschutzrechtlichen Bedenken gegenüber steht einerseits die Verobjektivierung von Entscheidungen, die der Einsatz eines SCS böte, und andererseits die Aussicht, Unternehmensrisiken erheblich zu verringern, auch wenn SCS, das sei noch einmal betont, nur ein Element im Auswahlprozess sein kann und der AR sich nicht ausschließlich darauf stützen oder gar damit aus seiner Haftung schleichen darf – Haftungsfragen könnten durch SCS eher konkretisiert und dank verbesserter Einklagbarkeit verschärft werden. Es würde zu den künftig an Aufsichtsräte zu stellenden Anforderungen gehören, ein Social Credit Rating qualifiziert einschätzen zu können, um dem Sorgfaltsmaßstab in der Auswahl von Aufsichtsratsmitgliedern zu genügen.

Im Sinn eines effizienten Auswahlprozesses von Aufsichtsratsmitgliedern sollten sich Unternehmen und Gesetzgeber dringend mit der Implementierung von Algorithmen und Datenstrukturen befassen und neue Routinen bzw. Regelungen auf den Weg bringen. Die erforderlichen Informationen liegen im Netz bereit, ob freiwillig abgegeben oder zwangsläufig geliefert, weil man bestimmte Dienste in Anspruch nehmen will.[3] Ein solcher Gebrauch öffentlich zugänglicher Daten würde, das noch als letzte wichtige Anregung dieses Beitrags, Aufsichtsratsmitgliedern aufgrund der verbesserten Nachvollziehbarkeit ihrer Berufung mehr Vertrauen und Akzeptanz verschaffen. In jedem Fall ist anzunehmen, dass bereits gegenwärtig die bestehenden Technologien im Auswahlprozess von Aufsichtsratsmitgliedern unterstützen können.

---

[3] Hinweis: Eine Menge von Suchmaschinen wird auf Soziale Medien angewandt. Viele sind mit den geltenden Gesetzen, wie z. B. der DSGVO, unvereinbar und in zahlreichen Fällen auch unethisch (Schlagwort „Recht auf Vergessenwerden"). Solange sich ein Social Credit System aus diesen Datenquellen speist, werden die gesammelten Daten möglicherweise Lücken aufweisen. Es bietet sich an, die Daten im Rohformat, den Quellen mit Crawlern direkt, getaktet zu entnehmen und in unstrukturierten Datensätzen (Data-Lakes) zu hinterlegen. Nach Bedarf können diese strukturiert in Datenbanken (Datawarehouses) eingepflegt werden, um sie danach zum Beispiel durch Online Analytical Processing (OLAP) Systeme zu analysieren und visualisieren. Durch dieses Data-Mining, als Methode der Bewältigung der Big Data, in einem Zusammenspiel mit Computerlinguistik (die versucht die natürliche Sprache algorithmisch zu verarbeiten), ist ein Großteil der geschöpften Daten analysierbar. Weitere Methoden können sein, Sentiment Detection, Named-entity Recognition oder künstliche neuronale Netze. Ein Novum in der Speicherung von Daten verspricht die Blockchain Technologie. Zusammengefasst kann diese Technologie durch mathematische Formeln, Rohdaten oder referenzierte Daten unzertrennlich aufeinander aufbauen und diese auch dezentral verteilen, mit oder ohne öffentlichen Zugang. Referenzierte Daten können abänderbar und löschbar bleiben, DSGVO konform. Siehe u.a. Xu X, Weber I, Staples M, 2019, Architecture for Blockchain Applications. Springer International Publishing,Cam, S. 18.

30.4   Zusammenfassung und Ausblick

Der Aufsichtsrat der AG unterliegt einer Wahlvorschlagspflicht betreffend die Anteilseignervertreter. Das AktG sieht nur sehr allgemeine Vorschriften zur Bestimmung von Aufsichtsratsmitgliedern vor – entsprechend viel Wildwuchs gibt es. Bei BaFin-beaufsichtigten Unternehmen liegen deutlich konkretere Vorschriften vor. So unterliegt die Bestellung der neuen Aufsichtsratsmitglieder der Genehmigung durch die Aufsicht. Zwar passieren auch hier mitunter Fehler, aber durch die BaFin-Genehmigung wird in gewissen Grenzen sichergestellt, dass die Kandidaten über eine entsprechend Sachkenntnis, persönliche Eignung und ausreichend Zeit verfügen. Weiche Faktoren bleiben aber auch hier außen vor. Das Auswahlverfahren kann folglich optimiert werden. Eine Optimierungsmöglichkeit wären Big Data-Analysen und -Technologien, mit denen sich Risiken des Aufsichtsrates und letztlich auch des Unternehmens reduzieren ließen. Das Phänomen wachsender Datenmassen ist grundsätzlich nicht neu, doch die Anforderungen und Abfrageperformance bei gleichzeitig komplexerem Analysebedarf sind eine Innovation im Zusammenhang mit Big Data, die Aufsichtsräte unbedingt für den Auswahlprozess ihrer Nachfolger unbedingt nutzen sollten. Die dazu im Vorfeld erforderlichen aufwendigen Extraktions-, Ladungs- und Transaktionsprozesse werden bereits im Recruiting erprobt und sind in diesem Feld auf dem Vormarsch. Um sie für die Auswahl von Aufsichtsratsmitgliedern zu nutzen, müssten allerdings standardisierte Anforderungsprofile für zukünftige Aufsichtsratsmitglieder ausgearbeitet und mit regulatorischen Anforderungen abgestimmt werden, vergleichbar denen der BaFin für Kreditinstitute oder Versicherungsunternehmen. Das alles entlastet den Aufsichtsrat in seiner Auswahlverpflichtung selbstredend nicht, kann ihn lediglich bei den notwendigen Vorarbeiten unterstützen und verschafft ihm dank der Algorithmen Informationssicherheit. Die letzte Entscheidung bleibt bei dem Gremium bzw. der Hauptversammlung, die jedoch in aller Regel seiner Empfehlung folgt.

Die diesbezüglich seit einigen Jahren immer stärker formulierten Empfehlungen des DCGK gehen in die richtige Richtung. In vielen Bereichen liegen, auch in Deutschland, ausreichend öffentliche Daten für Big Data-Analysen vor, teils freiwillig, teils zwangsläufig abgegeben, weil man eine bestimmte Teilnahme in Anspruch nehmen will. Ihr breiter Einsatz wird in deutschen Unternehmen durchaus als Chance gesehen, jedoch noch nicht umfassend proaktiv in Prozessen angewendet. Die Auswahl der Aufsichtsratsmitglieder der Anteilseigner gehört eindeutig zu den vordringlichsten Anwendungsfeldern, und die Initiative dazu sollte vom AR selbst ausgehen.

Natürlich stellt der Einsatz von Algorithmen alle Beteiligten vor erhebliche Herausforderungen. Doch die steigende Aktualität der Frage nach Schadenersatzansprüchen bei der Verletzung von Rechten und Pflichten von Aufsichtsräten spricht unbedingt dafür, sich diesen Herausforderungen zu stellen. Es gehört zur Professionalität eines AR, sich mit Chancen und Risiken der zunehmenden Digitalisierung und Vernetzung auseinanderzusetzen.

Was die sehr weitgehende Transparenz und Nutzung von Personendaten angeht, ohne die das Instrument Big Data-Analyse ein stumpfes Schwert bliebe, braucht der Blick nicht ausschließlich gen China. Auch innerhalb der europäischen Union, gibt es Beispiele, so in Schweden, wo eine Offenlegung von Personendaten erfolgt, die wir in Deutschland so überhaupt nicht kennen und zunächst sehr gewöhnungsbedürftig erscheinen kann. Die Schweden selbst sehen es nicht mehr als bedrohlich an, dass jeder nach Registrierung per Personennummer ohne legitimierendes Interesse Auskunft über viele persönliche Daten erhält, etwa ihre Steuererklärung einsehen kann (vgl. Reise 2013). Der gläserne Mensch, wie in Schweden bereits erprobt, wäre ein Riesenschritt für Deutschland, der nicht nur historische und verfassungsrechtliche Hürden und Herausforderungen meistern müsste.

Social Credit ist kein Allheilmittel, bei der richtigen Ausgestaltung kann es jedoch ein wichtiges Tool sein bei der Auswahl von Aufsichtsratsmitgliedern und folglich dabei helfen die Professionalisierung der Aufsichtsrattätigkeit weiter zu entwickeln.

Literatur

BaFin Journal März (2020) „Generelle Billigung von Algorithmen durch die Aufsicht? Nein, es gibt Ausnahmen".

Bellavite-Hövermann, Y./Lindner, G./Lüthje, B. (Hrsg.) (2005): Leitfaden für den Aufsichtsrat, Stuttgart 2005.

Hans-Böckler-Stiftung (2011): Anlage Arbeitshilfe 10 Anforderungsprofile für Aufsichtsratsmitglieder, Berlin 2011, abrufbar unter https://www.mitbestimmung.de/assets/downloads/p_ah_ar_10_anl.pdf.

Hüffer, U./Koch, J. (2018): Kommentar zum AktG, 13. Aufl., München 2018.

KPMG (2017): Mit Daten Werte schaffen, Studie 2017, Report 2017, abrufbar unter https://home.kpmg/de/de/home/themen/2017/05/mit-daten-werte-schaffen%2D%2D-studie-2017.html.

Mehlhorn, K./Sanders, P. (2008): Algorithms and Data Structures – The Basic Toolbox, Wiesbaden 2008, http://www.mpi-inf.mpg.de/~mehlhorn/Toolbox.htmlabrufbar unter https://people.mpi-inf.mpg.de/~mehlhorn/ftp/Mehlhorn-Sanders-Toolbox.pdf.

Redenius-Hövermann, J. (2010): Zur Frage der Frauenquote im Aufsichtsrat, ZIP 2010, S. 660 ff.

Redenius-Hövermann, J. (2019): Verhalten um Unternehmensrecht, Tübingen 2019.

Reise, N. (2013): Mal kurz das Gehalt des Nachbarn checken, Spiegel 05.02.2013, abrufbar unter https://www.spiegel.de/karriere/gehaelter-in-schweden-maximale-transparenz-a-881340.html.

Seele, A. (2007): Rahmenbedingungen für das Verhalten von Aufsichtsratsmitgliedern deutscher börsennotierter Unternehmen – Eine ökonomische und verhaltenswissenschaftliche Analyse des Deutschen Corporate-Governance-Kodexes, 2007, abrufbar unter https://archiv.ub.uni-heidelberg.de/volltextserver/7878/.

Speich Chassé, D. (2012): Fortschritt und Entwicklung, Version: 1.0, in: Docupedia-Zeitgeschichte, 21.09.2012 http://docupedia.de/zg/chasse_fortschritt_v1_de_2012

von Werder, A./Danilov, K. (2018): Corporate Governance Report 2018: Kodexakzeptanz und Kodexanwendung, DB 2018, S. 1997 ff.

Dr. Yvette Bellavite-Hövermann ist Steuer- und Unternehmensberaterin. Zuvor war Dr. Bellavite-Hövermann über zwei Jahrzehnte in verschiedenen Führungspositionen bei Kreditinstituten insbesondere in den Bereichen Rechnungslegung/Risikocontrolling/Bankenaufsicht tätig. Auf diesen Gebieten hat sie zudem zahlreiche Beiträge veröffentlicht, darunter aber auch ein Leitfaden für den Aufsichtsrat. Dr. Bellavite-Hövermann hat Betriebswirtschaft und Volkswirtschaft studiert, im Versicherungswesen und Steuerrecht promoviert und eine Banklehre absolviert.

Einsatz von Emotional Data Intelligence für eine effektivere Handelsüberwachung

31

Heinz Ackermann, Jonas Krauß und Dr. Stefan Nann

Zusammenfassung

User-Generated Content (UGC) (Vickery und Wunsch-Vincent 2007) macht einen wesentlichen Teil der Kommunikation über soziale Medien aus. UGC, das den Austausch von Emotionen unterstützt, bezeichnen wir in diesem Zusammenhang als „emotionale Daten". Wir alle „produzieren" emotionale Daten, indem wir unsere Emotionen in Tweets, Forenbeiträgen, Blogs usw. zum Ausdruck bringen. Wir „konsumieren" auf der anderen Seite Emotionen, indem wir von geäußerten Gefühlen, Stimmungen oder Meinungen beeinflusst werden. Unsere Entscheidungen werden oft von diesen Daten oder Emotionen mit beeinflusst, was wiederum zu neuen Daten oder Emotionen führt. Die Entscheidungen können im Folgenden Verhaltensweisen oder Ergebnisse verändern. Wir bezeichnen die Analyse dieses Prozesses als „Emotional Data Intelligence". Dabei geht es um die Beantwortung der Frage, wie die unterschiedlichen Emotionen, die in öffentlichen digitalen Quellen permanent zum Ausdruck kommen, Entscheidungsprozesse beeinflussen können.

Emotionale Daten oder Stimmungsdaten aus digitalen sozialen Netzwerken können auf verschiedene Weise zur Marktbeobachtung genutzt werden. Stockpulse kooperiert mit führenden Börsen in Europa und den USA, um Erkenntnisse auf der Grundlage emotionaler Daten für die Handelsüberwachung zu liefern. Der Artikel soll einen kurzen Überblick über einige Anwendungsbeispiele in diesem Bereich sowie zur weiteren Monetarisierung von Stimmungsdaten geben.

H. Ackermann (✉) · J. Krauß · Dr. S. Nann
Stockpulse GmbH, Bonn, Deutschland
E-Mail: heinz.ackermann@stockpulse.de; jonas.krauss@stockpulse.de; stefan.nann@stockpulse.de

© Springer Fachmedien Wiesbaden GmbH, ein Teil von Springer Nature 2020
O. Everling (Hrsg.), *Social Credit Rating*,
https://doi.org/10.1007/978-3-658-29653-7_31

31.1   Einleitung

André Kostolany, einst ein überaus erfolgreicher und weltweit bekannter Börseninvestor, machte die berühmte Aussage, dass Fakten nur zehn Prozent der Reaktionen an der Börse ausmachen, alles andere sei offensichtlich Psychologie. Er glaubte fest daran, dass Emotionen Kurse stark beeinflussen können und wurde zu einem der renommiertesten und erfolgreichsten Investoren des 20. Jahrhunderts. Objektive Daten über seine Investitionen waren jedoch nicht der Schlüssel zu seinem Erfolg, seine Entscheidungsprozesse hingen zu einem großen Teil von Intuition und lange erarbeiteten Erfahrungen ab. 1999, als Kostolany starb, existierte das Internet bereits, allerdings steckten die heute allseits bekannten Sozialen Medien gerade erst in den Anfängen.

André Kostolany war nicht die einzige Person, die sich in früheren Zeiten bereits für die Psychologie der Märkte interessierte. Der mögliche Einfluss von Stimmungen und Gefühlen auf die Finanzmärkte wurde auch von John Maynard Keynes erkannt. Kostolanys' und Keynes' Intuitionen wurde mit dem Aufkommen des Internets und der neuen Technologien, die sich daraus entwickelten, greifbar und messbar. Eine dieser Technologien sind die Sozialen Medien (Social Media), die in der Regel aus Online-Räumen bestehen, in denen Menschen Informationen und Daten digital austauschen. Bei Stockpulse bezeichnen wir die Daten und Informationen dieses Austauschs als User-Generated Content (UGC, Nutzergenerierter Inhalt), was sich auf eine allgemeinere Darstellung von Social Media Inhalten bezieht.

Social Media und UGC sind Beispiele für Daten, die oft auch als wichtigste Ressource des 21. Jahrhunderts angesehen werden. Der britische Mathematiker Clive Humby prägte 2006 einen viel zitierten Satz: „Daten sind das neue Öl". Wir leben heute zweifelsohne in einer datengesteuerten Wirtschaft. Ständig werden Daten erstellt, ohne dass wir es merken. Vieles von dem, was wir täglich tun, findet heute im digitalen Bereich statt und hinterlässt eine immer größere digitale Spur, die gemessen und analysiert werden kann. Laut dem Global Web Index haben einzelne Internetnutzer im Durchschnitt 7,6 Social Media Accounts. We Are Social Ltd. und Hootsuite Inc. geben an, dass die Zahl der Social-Media-Nutzer zwischen September 2017 und Oktober 2018 um 320 Millionen gestiegen ist. Die Gesamtzahl der aktiven Social Media Nutzer beträgt 3,725 Milliarden (Stand 2018).

Die Nutzung mobiler Geräte, Online-Shopping, das Lesen von Online-Nachrichten und Blogs, die Nutzung von GPS-Diensten, die Verwendung von Sprachsteuerungssystemen und die Optimierung jedes Teils unseres Lebens mit Apps ist heute für fast jeden von uns Alltag. Ein größerer Datenkonsum bedeutet allerdings auch eine größere Bereitstellung eigener Daten. Menschen teilen gerne Daten in Social Media; egal, ob sie aus externen Quellen stammen oder in vielen Fällen auch persönlicher Natur sind. Häufig werden Menschen aufgrund der Daten, die sie konsumieren, von Emotionen beeinflusst (Guillory et al. 2011; Hancock et al. 2008; Kramer et al. 2014). Die Nutzung dieser emotionsgeladenen Daten erzeugt neue Emotionen für andere Menschen (Kramer 2012), wodurch ein Kreislauf entsteht, in dem Daten Emotionen erzeugen und neue Daten durch die gemeinsame Nutzung dieser Emotionen (zum Beispiel durch Social Media oder UGC) generiert werden.

31.2 Auswirkung von Stimmungsdaten auf den Finanzmarkt

Die Auswertung dieser großen Menge an (Stimmungs-)Daten erfordert einen enormen technischen und finanziellen Aufwand. Schließlich soll die Stimmung der Marktteilnehmer in Echtzeit und fortlaufend mit höchster Präzision erfasst werden. Nun besteht das Problem, dass niemand in die Köpfe aller Börsianer blicken kann, um deren Stimmungslage exakt zu erfassen. Aber mittlerweile gibt es eine intelligente Möglichkeit zu erfahren, wie die meisten Marktteilnehmer gerade „ticken" und was sie kurzfristig an den Finanzmärkten erwarten: Durch eine voll automatisierte Emotions- und Stimmungsanalyse. Täglich diskutieren unzählige Börsianer zum Teil heftig in Onlineforen, schreiben ihre Meinung in die Kommentarspalten großer Nachrichtenportale oder veröffentlichen kurze Tweets auf Twitter. Im Grunde ist das vergleichbar mit den angeregten Gesprächen und Debatten auf dem Börsenparkett in früheren Zeiten. Mit einem entscheidenden Unterschied: Die Diskussionen im Internet lassen sich öffentlich verfolgen, weltweit, rund um die Uhr und in Echtzeit. Zwar kann kein Mensch jeden Tag Abermillionen Forenbeiträge, Kommentare und Tweets in Echtzeit auswerten und Schlüsse daraus ziehen. Aber modernste Hochleistungsrechner und intelligente Software sind dazu in der Lage (vgl. Abb. 31.1).

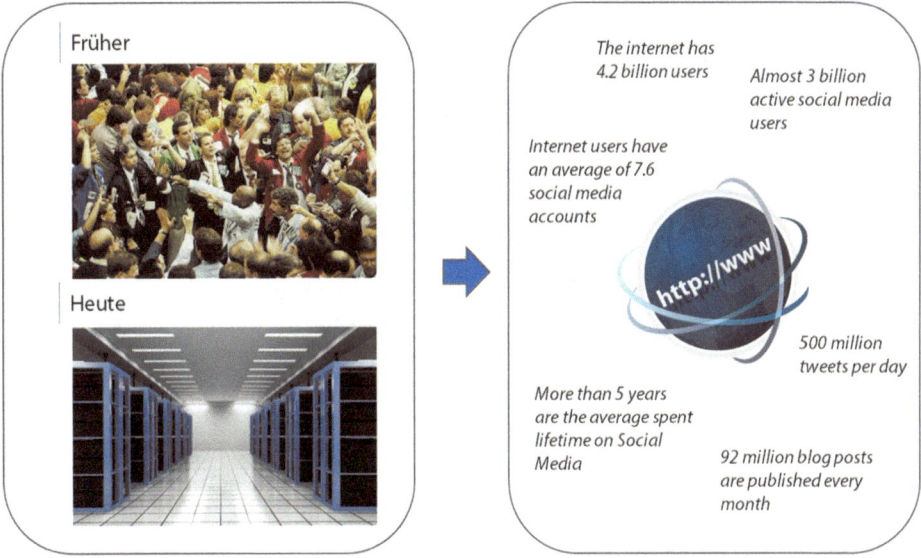

Abb. 31.1 Entwicklung der Kommunikation und Verlagerung der Diskussionen im Internet. (Quelle: eigene Darstellung)

Die Finanzmärkte sind in besonderem Maße von Daten abhängig. Informationen im Voraus oder selbst mit einem minimalen Zeitvorsprung vorliegen zu haben, kann den Return on Investment (ROI) erhöhen. So wird beispielsweise der Kurs einer Aktie oft als Ergebnis aller zu einem bestimmten Zeitpunkt verfügbaren Informationen betrachtet, wobei neue Informationen schnell nach Verfügbarkeit eingepreist werden (Markteffizienzhypothese (Fama 1970)). Das Kommunikationsverhalten in der Finanzindustrie hat sich in den vergangenen Jahren, wie in vielen anderen Branchen auch, aufgrund der technologischen Entwicklungen grundlegend verändert. Relevante Informationen und Daten werden größtenteils über das Internet und schließlich über die Sozialen Medien ausgetauscht. Da die Finanzmärkte durch schnell fließende Informationen gekennzeichnet sind, scheinen emotionale Daten aus diesen Quellen einen hohen potenziellen Prognosewert zu haben.

Es gibt viele Beispiele dafür, wie etwa Twitter bevorzugt zur Informationsweitergabe genutzt wird, um Emotionen hervorzurufen.

Beispiel

Elon Musk ist der Gründer und CEO von Tesla Inc. und oft auf Twitter aktiv. Am 7. August 2018 twitterte Musk, dass er „erwäge, Tesla für 420 US-Dollar je Aktie von der Börse zu nehmen. Finanzierung gesichert." Damals notierte die Tesla-Aktie bei rund 370 US-Dollar. Musks Tweet erhielt sofort viel Aufmerksamkeit. Nutzer, die ihm auf Twitter folgten, haben seinen ursprünglichen Tweet weiter getwittert und zahlreiche Nachrichtenagenturen veröffentlichten die Nachricht auf ihren Websites (vgl. Abb. 31.2). ◄

In diesem Beispiel sorgte ein einfacher Tweet in sehr kurzer Zeit für ein hohes Maß an Aufmerksamkeit. Nach schnellen ersten Reaktionen diskutierten viele Menschen (insbesondere Börsenteilnehmer) über die potenziellen Auswirkungen von Musks Nachricht auf

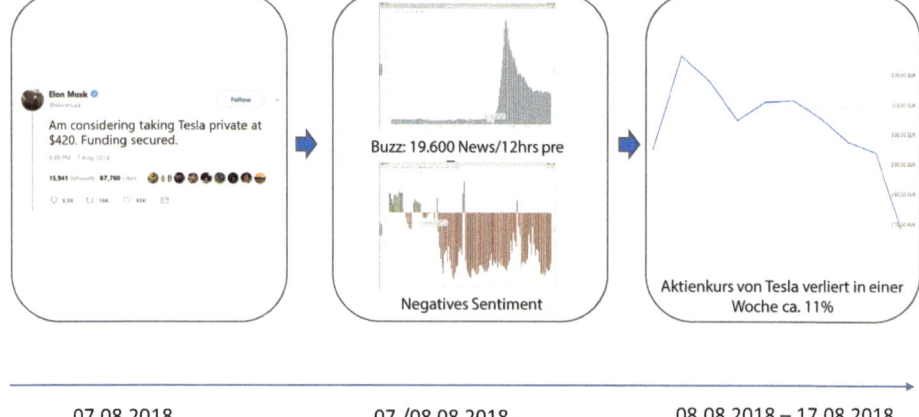

Abb. 31.2 Auswirkung eines Tweets auf die Entwicklung des Aktienkurses von Tesla. (Quelle: eigene Darstellung)

den Aktienkurs von Tesla. Die meisten Diskussionen waren spekulativ, da es an objektiven Informationen mangelte. Daher war der Inhalt der Diskussion meist subjektiv und emotional getrieben, die Reaktion auf Teslas Aktienkurs war allerdings sehr stark und folgte unmittelbar, mit einer steigenden Volatilität und Handelsvolumen. Die Börse NASDAQ sah sich gezwungen, den Handel mit Tesla-Aktien daraufhin für mehr als eine Stunde auszusetzen – eine sehr seltene und außergewöhnliche Vorgehensweise.

Nach einem kurzen und starken Anstieg des Aktienkurses direkt nach dem Tweet sank der Kurs in den folgenden Tagen, da die meisten Leute erwarteten, dass Elon Musk sein Unternehmen nicht einfach von der Börse nehmen konnte. Später berichtete die Financial Times über den wahren Grund für Musks Tweet. Musk bezog sich offenbar auf die Aussage eines saudi-arabischen Fonds, die Finanzierung zu unterstützen (Financial Times 2018). Unabhängig davon veranschaulicht das Beispiel jedoch die enormen Auswirkungen, die eine einzige Information, wenn auch tausendfach geteilt, in den Sozialen Medien auf die Börse haben kann. Das Beispiel zeigt auch, dass Emotionen Entscheidungen (in dem Fall den Verkauf einer Aktie) maßgeblich beeinflussen können.

31.3 Konzeptualisierung von Emotional Data Intelligence

„Emotional Data Intelligence" bezeichnet die Sammlung und Analyse großer Datenmengen, um Beziehungen zwischen verschiedenen Datenpunkten sinnvoll darzustellen und daraus eine optimale Entscheidungsfindung abzuleiten. Die Methoden und Werkzeuge von Data Intelligence basieren auf dem umfassenden Verständnis von Daten, das Aufdecken alternativer Lösungsansätze, das Erkennen von Problemen und das Identifizieren von Zukunftstrends zur Verbesserung von Entscheidungen.

Nutzergenerierte Inhalte (UGC) machen einen wesentlichen Teil der Kommunikation in den Sozialen Medien aus. Nach unserem Verständnis werden nutzergenerierte Inhalte als „emotionale Daten" bezeichnet, wenn ein Austausch an Emotionen stattfindet. Wenn die Methoden und Werkzeuge der Data Intelligence auf Daten angewendet werden, die Emotionen transportieren, bezeichnen wir diese Ergebnisse als „Emotional Data Intelligence".

Menschen produzieren emotionale Daten, das heißt, sie drücken ihre Emotionen über Tweets, Forenbeiträge, Blogs usw. aus, oder sie konsumieren sie, indem sie von Stimmungen, Gefühlen, Meinungen anderer beeinflusst werden. Forscher fanden heraus, dass Menschen oftmals von Emotionen beeinflusst werden, die auf den Daten basieren, welche sie konsumieren (Guillory et al. 2011; Hancock et al. 2008; Kramer et al. 2014; Kramer 2012). Das Teilen dieser Daten schafft neue Emotionen bis hin zu einem Kreislauf, in dem Daten Emotionen erzeugen und neue Daten durch das Teilen dieser Emotionen erzeugt werden. Entscheidungen hängen oftmals von diesen geteilten Emotionen ab, was wiederum zu neuen Daten führen kann, denn Entscheidungen können Auswirkungen auf das Verhalten der Nutzer haben. „Emotional Data Intelligence" sucht letztlich nach einer Antwort auf die Frage, wie all die verschiedenen Emotionen, die im Internet zum Ausdruck kommen, Entscheidungsprozesse beeinflussen (vgl. Abb. 31.3).

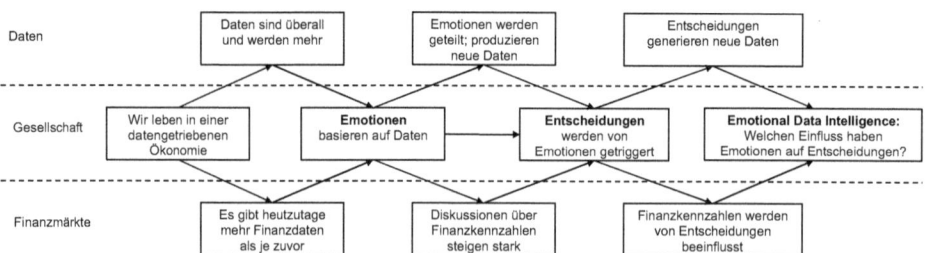

Abb. 31.3 Zusammenhang von Daten, Emotionen und Entscheidungen. (Quelle: eigene Darstellung)

31.4 Verwendung von Stimmungsdaten für die Handelsüberwachung

Stimmungsdaten aus digitalen sozialen Netzwerken können auf verschiedene Weise zur Marktbeobachtung eingesetzt werden. Stockpulse arbeitet mit führenden Börsen in Europa und den Vereinigten Staaten von Amerika zusammen, um Erkenntnisse auf der Grundlage emotionaler Stimmungsdaten für die Handelsüberwachung zu liefern. Die folgenden Beispiele zeigen einige Anwendungsfälle.

31.4.1 Aufdeckung von „Pump and Dump"-Schemata

Die Aufdeckung falscher, irreführender oder übertriebener Kommentare in den Sozialen Medien, die eine verdächtige Kursentwicklung nach sich ziehen, ist ein wichtiges Thema für die Handelsüberwachung. Social Media Daten oder allgemeiner alternative Daten und die Analyse großer Mengen dieser meist unstrukturierten Daten entwickeln auch in diesem Bereich eine zunehmende Bedeutung. Zur Klassifizierung dieser Art von Nachrichten kommen umfassende und ausgeklügelte Algorithmen zur Spam-Erkennung zum Einsatz. Sie erkennen beispielsweise, ob Benutzer oder Autoren in den Sozialen Medien falsche oder irreführende Informationen verbreiten.

Die Spam-Erkennung umfasst in der Regel mehrere Schritte. In einem ersten Schritt durchsucht ein Filteralgorithmus alle Nachrichten nach beleidigenden Wörtern oder Phrasen. Die meisten Nachrichten dieser Art lassen sich an ihrer unangebrachten und meist unflätigen Sprache erkennen. Weiterhin werden die Beziehungen zwischen Nutzern analysiert. Zum Beispiel können im Falle von Twitter die Analyse des Follower-Netzwerks der Nutzer oder die Überwachung von Interaktionen wie Likes, Erwähnungen oder Re-Tweets aufschlussreiche Informationen liefern. Für jeden Nutzer wird auf der Basis dieser Daten ein „Reputationsrang" oder eine „Autorenbewertung" berechnet. Der Einfluss einer Nachricht wird durch diese interne Autorenbewertung maßgeblich beeinflusst.

Darüber hinaus gibt es manuell kuratierte und verifizierte Social-Media-Nutzer (Finanzexperten oder renommierte Nachrichtenagenturen), die standardmäßig einen höheren

Reputationsrang besitzen. Elon Musk, Warren Buffet und die Nachrichtenagentur Bloomberg zum Beispiel werden dieser Kategorie zugeordnet, weil ihre Tweets und Beiträge typischerweise eine höheren Einfluss haben als die Beiträge von weniger bekannten Nutzern.

31.4.2 Twitter Influencer-Netzwerk und Alarmsystem

Einige Marktteilnehmer haben potenziell einen höheren Einfluss auf die Entwicklung der Aktienkurse als andere. Mit unserer Liste der kuratierten und verifizierten Social-Media-Nutzer (zum Beispiel Twitter-Konten von CEOs börsennotierter Unternehmen, einflussreichen Politikern, Journalisten, Analysten, Nachrichtenagenturen usw.) stellen wir bereits Kategorien von Nutzern zur Verfügung, die möglicherweise eine höheren Einfluss haben als normale Nutzer. Auf einzelnen Aktien-, Branchen- oder Sektorebenen bedarf es jedoch eines individuelleren Ansatzes, um die wirklich einflussreichen Social Media-Nutzer aufzudecken.

Betrachtet man das Tesla-Beispiel, so wissen wir, dass Elon Musk wahrscheinlich die wichtigste Person ist, der man folgen muss, wenn man über die relevantesten Nachrichten zum Unternehmen informiert bleiben will. Dieses Muster ist auch für andere Aktien, Branchen oder Sektoren von Bedeutung. Es wird einige Meinungsmacher für ein bestimmtes Gebiet geben, die unbedingt beachtet werden müssen, um so schnell wie möglich die wichtigsten Informationen zu erhalten. Auf die einzelne Aktie heruntergebrochen, wären dies die Twitter-Konten der Unternehmensleitung, bei denen die Suche beginnen sollte. In anderen Fällen können Analysten die Rolle der Experten für eine bestimmte Aktie oder Branche einnehmen.

Wir nennen diese Nutzer oder Autoren „Root-User". Sie sind die glaubwürdigsten Experten für eine Aktie oder ein Geschäftsfeld. Ein erster Schritt besteht darin, die relevanten Social-Media-Statistiken dieser Root-Nutzer permanent zu beobachten. Diese Statistiken können die Anzahl der gesendeten Tweets, Erwähnungen anderer Benutzer, Favoriten anderer Tweets oder Retweets beinhalten. In einem weiteren Schritt werden alle Benutzer, denen die Root-Benutzer folgen, ebenfalls analysiert und die gleichen Statistiken aufgezeichnet. Es ist auch möglich, dies für weitere Ebenen zu wiederholen. Auf diese Weise wird es einige hundert, möglicherweise sogar einige tausend Social Media-Accounts geben, die mit relativ hoher Wahrscheinlichkeit die größte Relevanz für die entsprechende Branche, Geschäftsfeld oder Aktie haben.

Wissen über diese „Influencer-Netzwerke" kann für Handelsüberwachungen von Börsenplätzen sehr wichtig sein, da dies die relevanten Quellen sind, um Nachrichten (auch Falschnachrichten) oder Gerüchten nach einer verdächtigen Preisbewegung schnell und ohne große Zeitverluste auf den Grund zu gehen.

Neben der Erkennung des Influencer-Netzwerks von Social Media-Nutzern für eine einzelne Aktie oder Branche könnte auch ein Frühwarnsystem für bestimmte Social Media-Accounts von hoher Relevanz sein. Sobald ein bestimmter Social-Media-Nutzer, der zuvor als relevante und glaubwürdige Quelle für diese Aktie identifiziert wurde, etwas über ein bestimmtes Unternehmen veröffentlicht, möchte die Handelsaufsicht Kenntnis darüber erlangen.

31.4.3 Überwachung von Schlüsselereignissen im Finanzmarkt

Wichtige Ereignisse an den Finanzmärkten können die Kursentwicklung zu einem großen Teil beeinflussen. Wir haben bei Stockpulse mehr als 150 marktrelevante Ereignisse definiert, die unsere Crawler in den Sozialen Medien beobachten. Jedes Ereignis ist durch charakterisierende Schlüsselwörter oder Phrasen gekennzeichnet. Jedes Schlüsselwort kann auch mit einem Faktor gewichtet werden, der die Bedeutung des Wortes für das Ereignis beschreibt. So kann beispielsweise das Ereignis „Merger & Acquisition" mit den folgenden Schlüsselwörtern beschrieben werden: „merger, merger & acquisition, merger approval, merger deal, reverse merger, m&a, transaction, approved". Schlüsselwörter können auch in mehreren Sprachen (Englisch, Deutsch und Chinesisch) angeboten werden. Letztendlich führt dies zu einer Verbindung zwischen Objekt (zum Beispiel börsennotierten Unternehmen) und Ereignis, wenn beide in derselben Nachricht erkannt werden.

Darüber hinaus betreibt Stockpulse umfangreiche Identifier-Mapping-Tabellen für Finanzinstrumente. Das Identifier-Mapping ist eine Art Zuordnung von relevanten Identifikatoren zu einem Instrument. Zum Beispiel würden Nachrichten, die über ein Analysten-Upgrade des Automobilherstellers Tesla diskutieren, mit dem Bloomberg-Ticker TSLA:US versehen, weil das Wort „Tesla" im Text erkannt wurde. Weiterhin würde das Ereignis „Anlysten-Upgrade" erkannt und es könnte eine Verbindung zwischen Ereignis und Aktie hergestellt werden. Verschiedene Prozesse aktualisieren regelmäßig alle Identifier-Mappings, zum Beispiel ISIN oder Tickersymboländerungen automatisch. Derzeit verwaltet das System mehr als eine Million Identifier-Mappings.

Für eine Handelsüberwachung ist es sehr wertvoll, Gerüchte oder Beiträge über Fusionen oder andere relevante Ereignisse in Echtzeit in den Sozialen Medien zu verfolgen und darüber sofortige Benachrichtigungen zu erhalten, wenn bestimmte Unternehmen oder Ereignisse plötzlich in den Fokus rücken. In Verbindung mit dem zuvor beschriebenen Frühwarnsystem können kritische Situationen sehr schnell erkannt werden und zu einer adäquaten Reaktion führen.

31.5 Weitere Anwendungsfälle aus der Praxis auf Basis von Stimmungsdaten

Zwei weitere Anwendungsfälle von Stockpulse sollen zudem verdeutlichen, wie aus der konsequenten, über Jahre hinweg kontinuierlichen Auswertung des UGC und einer darauf aufbauenden, regelbasierten Strategie ein echter Mehrwert auch für Anleger und Investoren generiert werden kann; und zwar (stark verkürzt dargelegt) im Grunde allein basierend auf von Marktteilnehmern ausgedrückten Emotionen.

Gemeinsam mit dem Frankfurter Index-Anbieter Solactive AG hat Stockpulse den Europe Big Data Sentiment Index (BDX Europe) entwickelt. Auf Grundlage der Auswertung von Social Media-Daten wird einmal pro Quartal ein Korb mit Aktien aus einem klar definierten Universum zusammengestellt und diese Aktien-Auswahl in der fortlaufenden Indexberechnung berücksichtigt. Im Falle des BDX Europe dient der Solactive Europe Total Market 675 Index als Aktien-Universum, aus dem die Aktien für den BDX quartalsweise herausgesucht werden.

Entscheidend dafür, ob eine Aktie für die kommenden drei Monate in den BDX Europe aufgenommen wird, ist die von Stockpulse gemessene Stimmung zu einem jeden Titel. Nur diejenigen Aktien, die in den vergangenen Monaten von den meisten Marktteilnehmern als besonders positiv bewertet und kommentiert wurden, sollen in den Index aufgenommen werden. Die Idee hinter dem Indexkonzept: Besonders "stimmungsstarke" Aktien entwickeln sich kurz- bis mittelfristig besser als der Gesamtmarkt.

Dass dieses Index-Konzept aufgeht, zeigt der Blick auf die bisherige Performance des BDX Europe: Alleine im Börsenjahr 2019 lag die Performance des Index bei +43 Prozent. Zum Vergleich: 2019 legten die beiden Leitindizes Dax und EuroStoxx 50 zwar ebenfalls stark zu, mit je +25 Prozent allerdings längst nicht so stark wie der BDX Europe. Auch gegenüber dem deutlich breiter aufgestellten Stoxx600 (+23 Prozent in 2019) hat der BDX Europe seine Stärke unter Beweis gestellt. Das gilt auch im langfristigen Vergleich, hier konnte das Index-Konzept die Benchmarks klar schlagen: Die kumulierte Performance des BDX Europe liegt bei +199 Prozent (21.09.2012 bis 31.12.2019). Im selben Zeitraum gelang dem Dax ein Plus von +79 Prozent und dem EuroStoxx 50 ein Zuwachs von +46 Prozent (vgl. Abb. 31.4).

Abb. 31.4 Europe Big Data Sentiment Index (BDX) im Vergleich zum Stoxx600 (Börsenjahr 2019)

Hinweis: Vor der erstmaligen Berechnung des Europe Big Data Sentiment Index am 14.03.2019 werden historische Performance-Daten auf Basis eines Backtests vorgehalten. Die Entwicklung des BDX ist unter der WKN SLA7W0 (ISIN DE000SLA7W00) auf Börsenportalen wie finanzen100.de oder onvista.de abrufbar. Der Index wird in Euro berechnet und börsentäglich von 9:00 bis 22:30 Uhr veröffentlicht. Der Index selbst ist für Anleger nicht direkt investierbar. Über ein Index-Zertifikat von UBS (WKN: UBS1BX – ISIN: DE000UBS1BX9) können Anleger jedoch unmittelbar an der Entwicklung des BDX partizipieren. Weitere Informationen zum BDX finden sich im Internet auf www. bdx-online.de.

Einen ähnlichen, regelbasierten Strategie-Ansatz verfolgt Stockpulse auch beim Anwendungsbeispiel „AktienSensor Deutschland". Seit Dezember 2018 gibt Finanzen100, mit 5,1 Millionen Unique User reichweitenstärkstes Finanzportal Deutschlands (AGOF Juli 2019) und eine Marke der FOCUS Online Group (Burda-Konzern), den „AktienSensor Deutschland" als Premium-Börsendienst heraus; der monatliche Abopreis beträgt (je nach Bezugsdauer) bis zu 79,99 Euro. Zuvor wurde dieser Börsendienst, im April 2014 gestartet, zeitweise beim FID Verlag publiziert. Stockpulse liefert im Auftrag von Finanzen100 jeden Mittwoch auf Sentiment-Analysen basierende Trading-Signale, die in einem Musterdepot umgesetzt werden und in das die Abonnenten des Premium-Dienstes Einblick erhalten. Das Konzept des „AktienSensor Deutschland" ist sehr einfach und für Privatanleger gut nachvollziehbar aufgebaut: Ist die Grundstimmung der meisten Marktteilnehmer positiv, wird auf steigende Aktienkurse gesetzt. Im umgekehrten Fall, bei einem überwiegend negativen, kurzfristigen Sentiment, wird auf fallende Notierungen gesetzt. In der Praxis geschieht dies durch die Aufnahme von maximal fünf Aktientiteln in das Musterdepot (bei Long-Signalen) oder fünf Derivaten (Short-ETFs und Short-Faktorzertifikate bei Short-Signalen). Seit Start am 9. April 2014 konnte das Musterdepot des AktienSensor eine kumulierte Performance von mehr als +188 Prozent erzielen (Stand: 31. Dezember 2019). Zum Vergleich: Für denselben Zeitraum lag die Entwicklung des Dax bei +39 Prozent und des EuroStoxx 50 bei +18 Prozent (vgl. Abb. 31.5).

Beide Anwendungsbeispiele zeigen eindrucksvoll, dass in UGC echte Mehrwerte und damit Monetarisierungsmöglichkeiten stecken können, sofern diese unstrukturiert und in gigantischer Menge vorliegenden Daten konsequent strukturiert und intelligent ausgewertet werden.

31.6 Entwicklung geht weiter: IR- und CEO-Monitoring auf Basis alternativer Daten

Stockpulse hat mit den in den vorherigen Abschnitten beschriebenen Anwendungsfällen für Anleger und Investoren sowie für Handelsüberwachungsstellen großer Finanzmarktplätze eine gute Marktposition erreicht hat und wird diese mit zusätzlichen Angeboten wie einem Germany Big Data Sentiment Index (BDX Germany) ausbauen. Zudem stecken

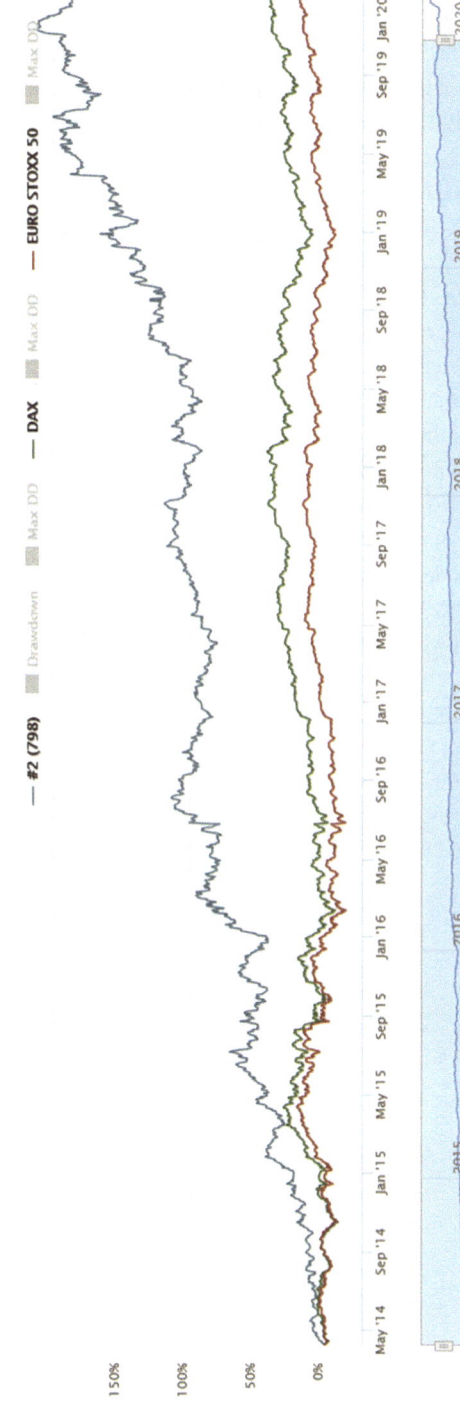

Abb. 31.5 Langfristige Entwicklung des "AktienSensor Deutschland" im Vergleich zum Dax und EuroStoxx 50 (09.04.2014 bis 31.12.2019)

weitere Produkte auf Basis alternativer Daten kurz vor der Marktreife bzw. werden schon in ersten Praxisanwendungen ausgiebig getestet.

So ist die Aufdeckung falscher, irreführender oder übertriebener Kommentare und Postings in den Sozialen Medien ist nicht nur für die Handelsüberwachungsstellen der Börsen absolut relevant, sondern lässt sich auch in weiteren Anwendungsfällen nutzen, auf die wir zum Ende des Beitrags noch kurz eingehen. Insbesondere Investor Relations-Abteilungen sind darauf angewiesen, so schnell wie möglich über potenziell kursrelevante oder imageschädigende Postings und Kommentare informiert zu werden, um im Zweifelsfall so schnell wie möglich reagieren zu können.

Ein erster Anhaltspunkt liefert dabei der Buzz-Wert für ein Unternehmen (vgl. Abb. 31.6). Steigt dieser Wert plötzlich deutlich an, so ist das ein Hinweis darauf, dass über ein Unternehmen, deren Produkte oder Dienstleistungen oder eben über die Aktie der Firma stärker als sonst üblich diskutiert wird. In einem zweiten Schritt ist dann eine schnelle Analyse des UGC über das Unternehmen notwendig. Hier zeigt sich dann sofort, ob die Diskussion in einem normalen Rahmen verläuft (etwa, weil ein neues Produkt besonders positiv bewertet wird oder zum Beispiel die Analystenmeinung zu einem Unternehmen im üblichen Rahmen unter Anlegern diskutiert wird) oder ob es sich um potenziell schädliche "Fake News" oder Gerüchte handelt. Bei dem Anwendungsfall (Echtzeit-Informationen für Investor Relations auf Basis von Stimmungsdaten) gehen wir bei Stockpulse ähnlich wie beim Anwendungsfall für Handelsüberwachungsstellen vor. Gleiches gilt für das von Stockpulse entwickelte Angebot eines Monitorings der CEO Reputation. Erste Praxistests zeigen auch hier eine hohe Relevanz für den täglichen Einsatz bei IR-Abteilungen und in Vorstandsetagen.

Bei den großen Konzernen gehört es mittlerweile fast schon zum guten Ruf, dass deren Geschäftsführer selbst auf Social Media-Kanälen präsent sind und das Unternehmen, dem sie vorstehen, vertreten und dabei auch vermarkten. Wie an den vorher besprochenen

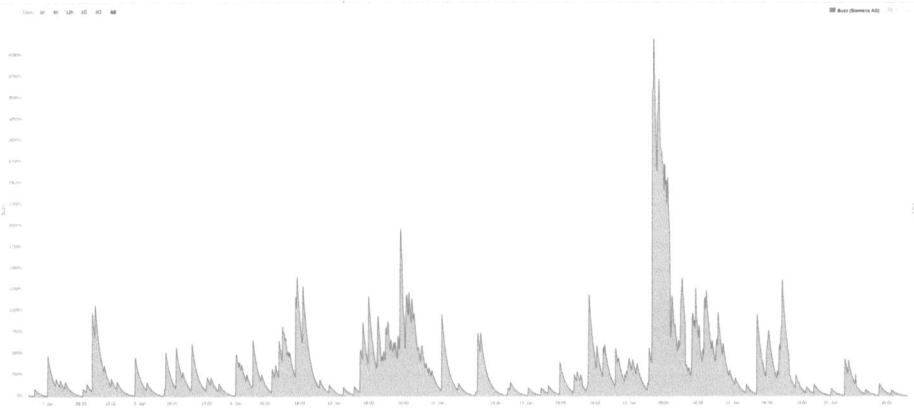

Abb. 31.6 Investor-Relations-Monitoring – Imageschädigende Diskussionen; Entwicklung des Buzz der Siemens AG während und nach dem Tweet von Joe Kaeser

Beispielen um Tesla-CEO Elon Musk klar wurde, ist dies jedoch eine zweischneidige Sache. Neben einer Steigerung des Bekanntheitsgrads und anderen positiven Aspekten, kann es durch unreflektierte Nutzung dieser Kanäle durch CEOs auch sehr negative Effekte für die Anteilseigner geben. Daher ist die Überwachung von Unternehmens-Accounts in Social Media-Kanälen ein weiterer wichtiger Baustein in der Kommunikationsstrategie börsennotierter Firmen. Mit der Stockpulse-Methode zur Bestimmung der CEO Reputation bieten wir auch für diesen Bereich eine Lösung an.

Da Stockpulse die Interaktionen zwischen Social Media-Nutzern in Echtzeit beobachtet, also beispielsweise Reaktionen auf Tweets von CEOs, ist es frühzeitig möglich, sich anbahnende kritische Kommunikationslawinen zu erkennen und im besten Fall ein Kommunikations-Desaster zu verhindern. So lassen sich die bereits vorgestellten Kennzahlen Buzz und Sentiment auch individuell für einzelne Autoren berechnen, sodass durch eine Überwachung des Buzz für einen bestimmten Account schnell klar wird, wenn Diskussionen sich verselbstständigen. So waren etwa im Januar 2020 in unseren Systemen klare Ausschläge im Buzz und Sentiment von Siemens AG und deren CEO Joe Kaeser zu erkennen, was auf die eher unglückliche Kommunikation von Kaeser mit der Fridays-for-Future-Repräsentantin Luisa Neubauer zurückzuführen war (vgl. Abb. 31.7).

Abb. 31.7 Tweet von Joe Kaeser, CEO von Siemens

31.7 Fazit

Daten sind heutzutage allgegenwärtig und die Datenmengen, die Tag für Tag produziert werden, wachsen stetig nicht zuletzt durch die rasante Verbreitung von Social Media und die Kommunikation von Nutzern auf Interaktionsplattformen im Internet. Mit den richtigen Tools und Methoden können diese öffentlich zugänglichen und nutzergenerierten Inhalte, meistens unstrukturierte Textdokumente oder Snippets, eingesammelt und für eine weitergehende Analyse strukturiert und aufbereitet werden. Wir bei Stockpulse bezeichnen diese Daten in diesem Zusammenhang als „emotionale Daten", das heißt, sie transportieren Emotionen über Tweets, Forenbeiträge, Blogs usw., weil Nutzer in ihren Beiträgen subjektive Meinungen veröffentlichen, die oft persönliche Emotionen beinhalten. Durch Methoden aus dem Natural Language Processing werten wir diese unstrukturierten Inhalte so aus, dass bestimmte Schlüsselworte und Phrasen erkannt werden, die Emotionen ausdrücken. Wissenschaftler haben in den vergangenen Jahrzehnten immer wieder belegt, dass Menschen von Emotionen beeinflusst werden können, die sie durch Social Media konsumieren. Emotionen beeinflussen weiterhin oft Entscheidungsprozesse, die wiederum Verhaltensweisen oder Ergebnisse verändern können. Wir bezeichnen diesen Prozess als „Emotional Data Intelligence", in dem letztlich nach einer Antwort auf die Frage gesucht werden soll, wie all die verschiedenen Emotionen, die in öffentlichen Online-Quellen zum Ausdruck kommen, Entscheidungsprozesse beeinflussen können.

Stockpulse entwickelt seit vielen Jahren Algorithmen und Methoden, die die Basis für Emotional Data Intelligence-Anwendungen sind und letztendlich in konkrete Produkte umgewandelt werden. Weltweit führende Börsenplätze nutzen unsere Daten zum Beispiel für die Handelsüberwachung, um die Transparenz des Marktgeschehens zu erhöhen. Auf Grundlage der Sentiment-Analysen wurde darüber hinaus Anfang 2019, gemeinsam mit dem Index-Anbieter Solactive AG, ein Aktien-Index aufgelegt, der „Solactive Europe Big Data Sentiment Index NTR", kurz: BDX Europe.

Diese und weitere Beispiele zeigen, wie wichtig neue digitale Kommunikations- und Interaktionsformen, wie zum Beispiel die Nutzung von Social Media, für reale Ereignisse und die Verbesserung von Entscheidungsprozessen sein können.

31.8 Über Stockpulse

Stockpulse ist ein deutsches Unternehmen aus Bonn, das sich auf die Bereitstellung von „Emotional Data Intelligence" (Stimmungsdaten) spezialisiert hat. Im Kern unterstützt und verbessert das Unternehmen die datengetriebene Entscheidungsfindung für Investoren, indem alternative Daten aus sozialen Medien mit Fokus auf den Finanzmarkt gesammelt und analysiert werden.

Während unverarbeitete und unstrukturierte Daten isolierte und begrenzte Informationen liefern, können über Emotional Data Intelligence Beziehungen hergestellt werden, sodass einzelne, zunächst nicht offensichtlich zusammenhängende Datenpunkte miteinander verbunden werden. Ein weiterer zentraler Aspekt ist die Messung der Emotionen im Rahmen dieser Verknüpfung. Sprachanalyse-Werkzeuge ermöglichen eine präzise Extraktion der verwendeten Emotionen in den Daten. Sogenannte Webcrawler scannen kontinuierlich tausende verschiedene Internetquellen nach relevanten Finanzthemen und die damit einhergehende Kommunikation rund um diese Themen. Täglich werden mehrere Millionen Tweets, Nachrichten, Forenbeiträge, News sowie Kommentare zu einzelnen News eingesammelt. Diese meist unstrukturierten Textdokumente werden mit Methoden des Natural Language Processing (NLP) und Methoden der Künstlichen Intelligenz (KI) verarbeitet. Ein Beispiel dieser Methoden, die bei StockPulse im Einsatz sind, ist Latent Dirichlet Allocation (LDA). LDA ist ein generatives, auf Wahrscheinlichkeiten basierendes Modell, mit dem abstrakte Themen aus Textdokumenten extrahiert werden können. Es untersucht versteckte semantische Strukturen innerhalb großer Textmengen und ermöglicht neue Analyseergebnisse für große Mengen an Daten. NLP wird auf Deutsch, Englisch und Chinesisch durchgeführt.

Zu den Kunden von Stockpulse zählen führende Hedgefonds, Vermögensverwalter, Banken, Börsen, Beteiligungsgesellschaften, Finanzdienstleister, Online-Broker, Nachrichtenportale, Forschungsunternehmen und Universitäten. Sie nutzen die Daten für unterschiedliche Zwecke, zum Beispiel suchen Hedgefonds nach Handelssignalen oder Handelsmodellen, die es ihnen ermöglichen, ihre bestehenden Portfolios oder Strategien zu optimieren. Die Börsen nutzen die Daten für die Handelsüberwachung oder die Marktbeobachtung. Beteiligungsgesellschaften wollen neue Investitionsmöglichkeiten auf der Grundlage alternativer Daten finden, die noch nicht für jedermann zugänglich sind.

Literatur

Fama, E.F. 1970. „Efficient capital markets: a review of theory and empirical work," Journal of Finance (25), pp. 383–417.

Financial Times 2018: Financial Times: „Saudi Arabia's sovereign fund builds $2bn Tesla stake", https://www.ft.com/content/42ca6c42-a79e-11e8-926a-7342fe5e173f, downloaded on March 11th, 2019

Guillory, J.; Spiegel, J.; Drislane, M.; Weiss, B.; Donner, W. and Hancock, J. (2011). „Upset now?: emotion contagion in distributed groups". Proceedings of the SIGCHI conference human factors computing system.

Hancock, J.T.; Gee, K.; Ciaccio, K. and Lin, J.M.H. (2008). „I'm sad you're sad: emotional contagion in CMC". Proceedings of the 2008 ACM conference computing supported cooperative work.

Kramer, A.D.; Guillory, J.E. and Hancock, J.T. (2014). „Experimental evidence of massive-scale emotional contagion through social networks". Proceedings of the National Academy of Science.

Kramer, A.D. (2012). „The spread of emotion via facebook". Proceedings of the SIGCHI conference human factors computer system, pp 767–770.

Vickery, G. and Wunsch-Vincent, S. (2007). „Participative Web and User-Created Content: Web 2.0". Wikis and Social Networking. http://browse.oecdbookshop.org/oecd/pdfs/free/9307031e.pdf, retrieved 12/31/2013.

Heinz Ackermann ist Betriebswirt und Absolvent der renommierten Kölner Journalistenschule für Wirtschaft und Politik. Als Privatinvestor ist er an mehreren Start-ups (Fintech, Industrie 4.0 und eLearning) beteiligt. Bei StockPulse unterstützt er u. a. in der Produktentwicklung und Vermarktung.

Jonas Krauß ist studierter Wirtschaftsinformatiker und hat einen Diplomabschluss von der Universität zu Köln. Er ist Geschäftsführer des Fintech-Unternehmens StockPulse in Bonn und dort verantwortlich für Systemarchitektur und Technologie. Während seiner akademischen Arbeit im Bereich der Sentiment-Analyse hat er außerdem einige wissenschaftliche Arbeiten auf internationalen Konferenzen veröffentlicht.

Stefan Nann hat Wirtschaftsinformatik an der Universität zu Köln und am MIT studiert. Seine Promotion im Bereich Betriebswirtschaftslehre hat er im Jahr 2020 abgeschlossen. Er ist außerdem Geschäftsführer des Bonner Fintech-Unternehmens Stockpulse und dort verantwortlich für den Vertrieb und Business Development.

Chinesische Verhältnisse auch in Deutschland? Gläserne Unternehmen und Digitale Unternehmensfinanzierung – Aktualität, Konsistenz und Integration von Informationen im Digitalen Scoring

32

Rainer Langen

Zusammenfassung

In den Markt der mittelständischen Unternehmensfinanzierung drängen mit großer Macht neue Finanzvermittler und -dienstleister, die alle den gerade stark wachsenden Markt der **Online-Kredite** erobern wollen. Neu gegründete **FinTechs** bieten besonders spezialisierte und kundenorientierte Finanzdienstleistungen an und machen damit den traditionellen Banken im Firmenkundengeschäft zunehmend Konkurrenz. **Prozesseffizienz** und **Digitale Vernetzung** sind die Schlagwörter der Zukunft. So unterschiedlich die Vertriebsfunktionen der einzelnen FinTech-Player auch sein mögen, so gleich und artverwandt sind jedoch bei allen die im Hintergrund ablaufenden **digitalen Prozesse.**

Geheimnisumwitterte **Algorithmen** steuern die Kreditentscheidung. Implementiert in komplexe IT-Programme liefern diese autonome Handlungsempfehlungen: Kreditvergabe, ja oder nein? Wenn ja, in welcher Höhe, mit welcher Kondition und welcher Laufzeit? Der **Kreditanalyst**, so wie man ihn von klassischen Banken kennt, ist letztlich „nur" noch für die **Prozessoptimierung** zuständig. Seine Aufgabe besteht zunehmend darin, automatisierte Risikosysteme zu installieren, diese zu verbessern, Sonder- und Spezialfälle mit dem Algorithmus zu trainieren und die Prozessabläufe mit den Kunden zu optimieren.

Automatisierte, lernfähige und objektivierte Risikomodelle (**Data Analytics**) sind das Herzstück jeder digitalen Kreditentscheidung. Über unterschiedlichste **Programmschnittstellen** werden alle relevanten Informationen aus unternehmensinternen oder externen Quellen herangezogen: aus der Buchhaltung, der Kontoführung (**Digital Account Check**), aus öffentlichen und privaten Registern oder Auskünften,

R. Langen (✉)
Rainer Langen & Partner Mittelstandsfinanzierung, Bad Kreuznach, Deutschland
E-Mail: langen@langenpartner.de

© Springer Fachmedien Wiesbaden GmbH, ein Teil von Springer Nature 2020
O. Everling (Hrsg.), *Social Credit Rating*,
https://doi.org/10.1007/978-3-658-29653-7_32

aus Marktdaten und Branchenvergleichen, aus Nachrichten via Internet, Youtube oder sonstigen Social-Media-Kanälen. Die Zusammenfassung dieser speziellen Informationen über Vernetzung, maschinenlesbare Daten und anschließender umfassender **Datenanalyse** ermöglicht ein Aggregationsniveau, das bisher manuell nicht möglich war. Ziel der **Vernetzung** ist es dann, über Cloud-basierte Lösungen und maschinenlesbare Datentechnik den automatisierten Austausch von Informationen zwischen mehreren, unabhängig voneinander agierenden Netzwerkpartnern zu ermöglichen.

Öffentliche Wahrnehmung, das Internet und die sozialen Medien bieten über Vernetzung und Digitalisierung einen schier unerschöpflichen Fundus an Fakten, Meinungen und Hintergrundinformationen, die in keinem Jahresabschluss eines Unternehmens stehen. Über verschiedenste Portale oder Medien lassen sich somit Informationen zur „**Sozialen Unternehmenskompetenz**" filtern. Damit ist die Grundlage für ein **Sozialkreditsystem** auch in Deutschland gelegt.

Von daher liegt in der Beschäftigung mit den Erfordernissen der d**igitalen Finanzkommunikation** eine wesentliche Voraussetzung für den Erhalt der unternehmerischen Konkurrenzfähigkeit. Es gilt, neue Wege zu gehen und schneller auf der Lernkurve zu sein als der Wettbewerb. Manches mag noch nach Zukunftsmusik klingen, beschränkt lediglich auf einige wenige standardisierte und unkomplizierte Sachverhalte. Aber lernfähige Algorithmen, automatisierte Prozessketten und ein vernetzter Datentransfer werden mehr und mehr völlig neue Finanzierungslösungen im Rahmen von **Kooperationspartnerschaften** ermöglichen – mit allen, die bereit sind, sich den neuen Realitäten der **Digitalisierung** zu stellen.

32.1 Der neue Markt der Online-Finanzierungen

In den Markt der mittelständischen Unternehmensfinanzierung drängen mit großer Macht neue Finanzvermittler und -dienstleister sowie bekannte Direktbanken, die alle den gerade stark wachsenden Markt der **Online-Kredite** erobern wollen. Auf inzwischen mehr als zwanzig deutschen Websites sowie verschiedenen Vergleichs- und Investorenportalen kann ein mittelständisches Unternehmen seine Finanzierungsanfrage online erstellen. Daraus erwachsen neue Chancen. Aber auch neue Risiken und Herausforderungen. Denn der Markt ist in starker Bewegung; jeder Teilnehmer sucht seine Position. Und dieser Markt folgt völlig neuen Regeln. Wesentlicher Vorteil der **Online-Finanzierung** ist die hohe Geschwindigkeit. Zu- oder Absagen erfolgen zumeist innerhalb von wenigen Tagen, manchmal sogar nur einigen Stunden. Propagiert wird vielfach der Firmenkredit auf Knopfdruck. Der mittelständische Schnellkredit ohne Sicherheiten, ohne Papierkram, ganz und gar digital. Alles ganz einfach und problemlos. Gut gemachtes Marketing? Ist das wirklich möglich? Und wenn ja, wie?

32.1.1 Digitale Evolution

Das Kredit-Marktvolumen für mittelständische Unternehmen mit einem Umsatz bis zu 50 Millionen Euro beträgt in Deutschland einschließlich des Segments der Selbstständigen und Freiberufler aktuell rund 320 Milliarden Euro und damit rund ein Viertel des gesamten Kreditmarkts. Im Jahr 2019 lag das Volumen der digitalen Finanzierer, gemessen an den insgesamt vergebenen Neukrediten bereits deutlich über einer Milliarde Euro. Zwar mag dies auf Anhieb ein noch sehr unbedeutender Marktanteil sein, das weitere Wachstum ist aber nicht mehr aufzuhalten. Oder ist dies nur digitales Wunschdenken?

Man erinnere sich: Der Online-Versandhändler **Amazon** hat mit mit seiner breiten Produktpalette in den letzten 20 Jahren die Strukturen im Versand- und Einzelhandel, insbesondere aber auch das Käuferverhalten drastisch verändert. Wer hätte sich Mitte der Neunzigerjahre vorstellen können, ein so komplexes Produkt wie eine Waschmaschine online zu bestellen? Und heute? Der Umsatz von Amazon ist seit Gründung von anfänglich einigen Millionen US-Dollar auf inzwischen weltweit mehr als 230 Milliarden US-Dollar gestiegen. Exponentiell, das heißt zunächst noch langsam, dann aber immer rasanter. Inzwischen boomt das Online-Geschäft auch in vielen anderen Märkten.Und ein Ende des Online-Wachstums ist noch lange nicht in Sicht.

Ein anderes Beispiel: Als Anfang der Neunzigerjahre erste **digitale Kameras** auf den Markt kamen, wurden diese weithin belächelt. Nichts für Profis. Nichts für anspruchsvolle Fotografien. Nichts für komplexe Lichtverhältnisse. „Wird sich nicht durchsetzen", war die einhellige Meinung der selbst ernannten Experten von gestern. Und heute? Wer fotografiert (außer einigen Nostalgikern) noch analog? Und ist nicht erst die digitale Fotografie in der Lage, Komplexitäten abzubilden, von denen der analoge Fotograf seinerzeit nur träumen konnte? Und wo sind die analogen Gralshüter und unternehmerischen Platzhirsche aus alten Zeiten geblieben? Zumeist vom Markt verschwunden. **Digitale Evolution**. Survival of the fittest. Ersetzt man das Wort Fotografie durch Finanzierung und Kamerahersteller durch Bank, mag jeder für sich entscheiden, wie die Markt-Evolution der Zukunft aussieht …, digital oder analog?

32.1.2 FinTechs kommen

Junge Unternehmen, sogenannte **FinTechs**, bieten besonders spezialisierte und kundenorientierte Finanzdienstleistungen (**Fin**ancial Services) mit Hilfe technologiebasierter Systeme (**Tech**nology) an – und machen damit den Banken unter anderem im Firmenkundengeschäft zunehmend Konkurrenz. Zunächst noch als additiver Mix ohne die Ambition, eine Hausbankenfunktion zu übernehmen. Zufrieden damit, die beste „Zweite Wahl" zu sein. Strukturell besetzt wird der **Online-Markt** derzeit durch Vergleichsportale (prägend ist hier die Zusammenarbeit mit einer Vielzahl von Kreditinstituten), Vertriebsplattformen (hier erfolgt die Kooperation nur mit einigen wenigen, ausgewählten Kreditinstituten),

Investorenplattformen (auf denen sich sowohl private als auch institutionelle Anleger registrieren lassen können) und bereits langjährig bekannte oder neu in den Markt eintretende Direktbanken.

Aber so unterschiedlich die Vertriebsfunktionen der einzelnen FinTech-Player als Intermediär zwischen den kreditnachfragendem Unternehmen und den jeweiligen Kapitalgebern im digitalen Kreditmarkt auch sein mögen, so gleich und artverwandt sind bei allen die im Hintergrund ablaufenden **digitalen Prozesse**. Am Anfang steht eine **KYC-Plattform** (Know-Your-Customer) zur Registrierung der Unternehmen und Aufnahme aller relevanter Basisdaten mit anschließendem Video-Ident-Verfahren des vertretungsberechtigten Geschäftsführers. Es folgen selektierende Scoring- und Risikoprozesse, Angebot und Vertragsabschluss sowie Auszahlung und Kreditverwaltung. Alles voll digital – ohne jeglichen persönlichen Kontakt in einer Bankfiliale.

32.1.3 Gläserne Unternehmen – Albtraum oder Realität?

Geheimnisumwitterte **Algorithmen** steuern die Kreditentscheidung. Implementiert in komplexe IT-Programme gelingt es mithilfe der Algorithmen Unternehmensinformationen, teils direkt zugänglich, teils aber auch im Markt weit und unabhängig voneinander verstreut, nach bestimmten Rastern in autonome Handlungsempfehlungen umzusetzen: Kreditvergabe, ja oder nein. Wenn ja, in welcher Höhe, mit welcher Kondition und welcher Laufzeit. Eindeutig. Unmissverständlich.

Im Backend der digitalen Finanzierungsabteilungen arbeiten künftig keine menschlichen Analysten mehr sondern nur noch emotionslose Algorithmen, die die Informations-Eingaben der potenziellen Kunden abarbeiten. Schreckgespenst eines jeden Unternehmers? Algorithmen, die alles wissen wollen, alles analysieren, alles vergleichen. Algorithmen, die vernetzt sind bis in die letzten Tiefen des Internets, denen jede Unstimmigkeit auffällt. Und wer weiß schon, wie ein Algorithmus Daten und Informationen im Einzelnen verarbeitet, welche Gewichtungen er vornimmt, wie er lernt und wie er seine Berechnungen und Vorgehensweisen verändert. Ein Algorithmus ist immer unpersönlich, leidenschaftslos, nach außen intransparent, und doch vielleicht objektiv, unbeeinflussbar, neutral?

Das gläserne Unternehmen – Albtraum oder Realität? Viele werden fragen: Steht uns jetzt auch in Deutschland ein chinesischen **Sozialkreditsystem** (Social Credit System) für Unternehmen ins Haus? Mit Beurteilungskriterien, die weit über die klassischen Wirtschaftsdaten hinausgehen. Welche Folgen hätte eine schlechte Bewertung? Gibt es dann vielleicht Listen von „nicht vertrauenswürdigen Unternehmen"? Wann würde hier ein Eintrag drohen? Mit welchen Konsequenzen? Mancher wird antworten: Wenn das der Preis der **Digitalen Finanzkommunikation** ist, dann verzichte ich lieber – und bleibe meiner bewährten Hausbank treu.

32.2   Was Unternehmen jetzt beachten müssen

Die Zeit drängt. In ein paar Jahren sieht die Finanzierungswelt anders aus; Veränderungen ergeben sich mit exponentieller Geschwindigkeit. Was heute noch undenkbar ist, wird morgen Standard sein. Und es gibt kein Zurück mehr. Die klassischen Banken werden, soweit sie nicht selbst schon im Markt über Kooperationen, Übernahmen oder Neugründungen aktiv sind, beim Thema „**Digitale Finanzierung**" nachziehen. Der eine früher, der andere später. Denn der **Margendruck, insbesondere im mittelständischen Kreditgeschäft** ist enorm. Hohe Personal- und Verwaltungskosten lassen die Kreditprozesskosten seit Jahren nach oben schnellen und damit die Deckungsbeiträge, insbesondere bei kleinteiligen Kreditvolumina, gegen Null, vielfach sogar ins Negative gehen. Die Entwicklung ist rasant. Wer jetzt stehen bleibt, den überholt die Geschichte … Deshalb: Je früher Unternehmer beginnen, sich mit der Thematik zu befassen, umso besser. Dazu gehört auch die Überprüfung und Umstellung von internen Prozessen, insbesondere in den Bereichen Finanz- und Rechnungswesen sowie Controlling. Aber Veränderungsprozesse benötigen Zeit; in den Köpfen der Menschen, in der Anpassung von IT-Systemen, in Veränderungen der Kommunikationsprozesse, im täglichen Miteinander …

32.2.1 Digitale Kreditprozesse: Maschine versus Mensch

Zunächst nochmals ein Blick hinter die Kulissen. Was ist das grundlegend „Neue" im **Digitalen Kreditvergabeprozess**? Der heute bei nahezu allen Kreditinstituten noch immens hohe manuelle Aufwand der Kreditbearbeitung wird, ja muss sich zwangsläufig aus Kosten- und Effizienzgründen deutlich verringern. Insbesondere die kostenintensive Beschäftigung eines Firmenkundenbetreuers über mehrere Wochen mit einer Kreditvorlage, die dann vielleicht doch nicht umgesetzt wird, gehört der Vergangenheit an. So manch ein Unternehmer kann ein Lied davon singen: Unterlagen, Unterlagen, und nochmals Unterlagen einreichen – aber am langen Ende dann doch keine Zusage. Bei Online-Finanzierungen sieht die digitale Zukunft anders aus: Bis zu **90 Prozent maschinelle Vorab-Ablehnungen** innerhalb von wenigen Stunden, allenfalls von ein paar Tagen werden die unmissverständliche Regel sein. Über das Erkennen von gelernten Mustern errechnet der Algorithmus auf Basis aller beigezogenen Informationen autonom eine Kapitaldienstfähigkeit nebst Ausfallwahrscheinlichkeit und beurteilt die relevanten qualitativen Faktoren. Passt dies alles nicht mehr in die Bandbreite der externen Investment-Prämissen, insbesondere auch des vorgegebenen „Risikoappetits" des Financiers, erfolgt die kompromisslose Ablehnung. Vorteil: Als Antragsteller weiß man sofort, ob man eine Finanzierungschance hat oder nicht. Nachteil: Eine schlechte Vorbereitung mit unzureichenden Unterlagen mündet auch dann in einer Absage, wenn die Kreditnehmerbonität eigentlich gut ist. Nachbesserungen oder Erklärungsversuche lässt ein streng digitalisiertes System nicht mehr zu.

Und bei den verbleibenden zehn Prozent der Kreditanfragen, die nicht dem sofortigen digitalen Auslese-Scoring zum Opfer gefallen sind, werden wohl immer noch vier von fünf ausschließlich digital und mit einem nur sehr geringen manuellen Aufwand bearbeitet. Schritt für Schritt wird somit der Faktor „Mensch" im Kreditvergabeprozess ersetzt. Der Kreditanalyst, so wie man ihn von klassischen Banken kennt, ist letztlich „nur" noch für die **Prozessoptimierung** zuständig. Seine Aufgabe besteht zunehmend darin, automatisierte Risikosysteme zu installieren, diese zu verbessern, Sonder- und Spezialfälle mit dem Algorithmus zu trainieren und die Prozessabläufe mit Kunden zu optimieren. Von Algorithmen ausgesprochene Handlungsempfehlungen müssen evaluiert werden, um gegebenenfalls die zugrunde liegenden Raster oder Muster anzupassen oder zu verändern. Letzteres bedeutet beispielsweise, getroffene Entscheidungen konsequent zu überprüfen und in ein internes Reporting-System zu überführen: War die Kreditvergabe „falsch", weil der Kunde später beispielsweise mit Zahlungsverzug negativ auffällt – oder war eine Kreditablehnung „falsch", weil der Kunde sofort einen Kredit beim Wettbewerb erhält? Lebenslanges Lernen gilt auch für Algorithmen!

32.2.1.1 API – das neue Zauberkürzel

Manche werden sagen: Natürlich sind maschinelle Rating- und Scoring-Systeme nichts Neues. Seit vielen Jahren schon greifen IT-basierte Bilanzanalysen oder betriebswirtschaftliche Analyse-Tools auf Jahresabschlüsse, unterjährige Erfolgsrechnungen oder Mehrjahresplanungen zurück. Natürlich nimmt jeder Kreditanalyst Einblick in Handelsregister, Büroauskünfte oder die neuesten Unternehmensnachrichten, veröffentlicht im Internet, der Fachpresse oder speziellen Informationsdiensten. Und natürlich fließen am Ende alle Komponenten einschließlich Kontoführungshistorie und sonstiger qualitativer Faktoren, wie zum Beispiel der persönliche Beurteilung des Firmenkundenbetreuers zum Thema „Management", über spezielle Gewichtungsfaktoren in eine abschließende Ratingnote ein.

Wo ist der Unterschied? Die Antwort lautet:

Erstens in der **Vernetzung der Informationen**. Beispielsweise können so Handelsregisterauszüge mit anderen Daten- und Nachrichtenbanken verknüpft werden, sodass ein viel umfassenderes Gesamtbild über Gesellschafterbeziehungen, Unternehmensverflechtungen und sonstige Aktivitäten des Managements entsteht.

Zweitens im Ausschluss jedweden persönlichen Eingriffs – sei es in der anfänglichen Eingabe von Daten in ein Scoring-System (mit hier und da immer wieder gegebenen Interpretations- und Auslegungsspielräumen), sei es in Ausübung eines menschlichen Veto-Rechts (Overruling) hinsichtlich maschineller (Teil-)Ergebnisse.

Und Drittens im Desinteresse der Maschine an allen denkbaren Cross-Selling-Effekten, wie Übernahme der Kontoführung, Devisengeschäft, Private Wealth Management oder Versicherungen – um nur einige Ertragsquellen klassischer Bankkonzerne zu nennen. Fin-Techs und ihre Algorithmen denken im Kreditgeschäft ausschließlich in Finanzkennzahlen, Bonitäten und vernetzten Informationen.

Automatisierte, lernfähige und objektivierte Risikomodelle (**Data Analytics**) sind das Herzstück jeder digitalen Kreditentscheidung. Über unterschiedlichste Programm-Schnittstellen, sogenannte **API** (**A**pplication **P**rogramming **I**nterfaces) ziehen sie wie eine Krake die relevanten Informationen aus allen möglichen unternehmensinternen oder externen Quellen: der Buchhaltung, der Kontoführung, aus öffentlichen und privaten Registern oder Auskünften, aus Marktdaten und Branchenvergleiche, aus Nachrichten via Internet, Youtube oder sonstigen Social-Media-Kanälen.

32.2.1.2   Prüfung der Datenstimmigkeit

Ziel von Vernetzung ist es, über Cloud-basierte Lösungen und die **XBRL-Technologie** (**EX**tensible **B**usiness **R**eporting **L**anguage), den automatisierten Austausch von Daten zwischen zwei oder mehreren Netzwerkpartnern zu unterstützen. Gelingt dies, weil die angeschlossenen Partner über diese Technologien verfügen, spielt das Layout der Informationen, zum Beispiel in einem Wirtschaftsprüfungsbericht oder einer Büroauskunft in den allermeisten Anwendungsfällen keine Rolle mehr. Somit können die Daten eines internen oder externen Berichts im Erscheinungsbild nach Belieben gestaltet sein, ohne dass dies auf die automatisierte Verarbeitung Einfluss hat. Dadurch ist zum einen die Darstellung sämtlicher Kreditparameter ohne jegliches menschliches Zutun möglich. Dies spart Zeit und Kosten. Zum anderen wird jedwede persönliche Interpretation oder Subjektivität hinsichtlich der relevanten Informationen ausgeschlossen. Und am Ende, nicht zu unterschätzen, bietet eine solche Vernetzung die Überprüfung aller Daten auf Unstimmigkeiten oder Unplausibilitäten im Echtzeit-Modus an. Alle auf Konsistenz geprüften Informationen fließen dann mittels IT-Integration zusammen und werden von einem „undurchsichtigen", für Außenstehende nicht fassbaren Algorithmus verarbeitet und einer ganzheitlichen Bewertung zugeführt. So entsteht ein einzigartiges, lernfähiges und datenbasiertes sowie streng vernetztes Bewertungssystem. Ein „lebender" Datenorganismus (**Open Data**), der auf Knopfdruck aktualisiert, ergänzt, Daten auf Konsistenz plausibilisiert und nach vorgegebenen Mustern überprüft. Das Hochladen von pdf-Dokumenten ist somit digitale Steinzeit; ebenso wie per Email oder Fax übersandte Daten – denn immer ist für eine zusammenfassende, ganzheitliche Analyse und Bewertung zusätzlicher manueller und erheblicher zeitlicher Aufwand erforderlich. Mit allen Problemen wie Fehleranfälligkeit oder Korrektur- und Abstimmungskosten. Deshalb: **Open Data** ist die Zukunft – maschinenlesbar und für jeden voll automatisierten Analyse-Verwendungszweck geeignet.

32.2.2   Welche Quellen und Informationen sind relevant?

32.2.2.1   Digitalisierung und Vernetzung

Digitaler Finanzbericht

Ein wesentliches Instrument, der **Digitale Finanzbericht** (DiFin; siehe auch www. digitaler-finanzbericht.de) wurde im März 2018 bei der Deutschen Bundesbank in Berlin vorgestellt. Mittels dieses Übertragungsstandards kann eine medienbruchfreie, elektronische Übermittlung von Jahresabschlüssen, Erfolgsrechnungen und anderen Finanzberichten zwischen Unternehmen, ihren Steuerberatern und Wirtschaftsprüfern sowie den „auf der anderen Seite" stehenden Financiers erfolgen. Dies bedeutet nichts anderes als die vermehrte Digitalisierung der den betriebswirtschaftlichen Ergebnissen zugrunde liegenden unternehmerischen Prozesse in Form von maschinenlesbaren Daten und Dokumenten. Und damit ergibt sich automatisch die Möglichkeit von zunehmenden Vernetzungen. Dies bedeutet: Neben der **Digitalisierung und Vernetzung** der Zusammenarbeit zwischen dem Unternehmen und seinen Finanz(ierungs)partnern gewinnt im Vorfeld bereits die Digitalisierung und Vernetzung der Zusammenarbeit zwischen dem Unternehmen und seinem Steuerberater oder Wirtschaftsprüfer zunehmend an Bedeutung. Ganz zu schweigen von der immensen Wichtigkeit einer zunehmenden Digitalisierung und Vernetzung der Zusammenarbeit zwischen dem Unternehmen und seinen Handelspartner, wie beispielsweise Kunden oder Lieferanten. Und hinter allen diesen „Neuerungen" steht ein Grundgedanke: Informationen, die bisher oftmals nur in analoger Form Partnern in einer bilateralen Geschäftsbeziehung zur Verfügung standen, nunmehr über eine multilaterale Vernetzung allen Partnern via Digitalisierung über entsprechende Plattformen, Clouds oder andere intermediäre Medien, und im Bedarfsfall just in time, elektronisch zur Verfügung zu stellen.

Digital Account Check

Ein weiteres Instrument zur Schaffung vollständiger Transparenz ist der **Digital Account Check** (DAC). Seit Anfang 2018 gilt auch in Deutschland die EU-Zahlungsdiensterichtlinie PSD2 (Payment Services Directive2). Danach müssen Kreditinstitute Schnittstellen für externe, sogenannte Kontoinformations-Dienstleister zur Verfügung stellen. Diese dürfen dann mit Erlaubnis des Kontoinhabers auf die jeweiligen Kontodaten zum Zwecke einer ausführlichen Auswertung zurückgreifen. Dies kann beispielsweise die Analyse von Kontosalden, Umsatzdaten oder auch Bar-Transaktionen betreffen. Insbesondere können hier auch zusammenfassende, institutsübergreifende Konsolidierungen erfolgen, soweit ein Verbraucher zum Beispiel über mehrere Konten bei verschiedenen Instituten verfügt. Gerade diese Aggregierung von Kontoinformationsdaten war bisher, wenn überhaupt, nur mit erheblichem manuellen Aufwand möglich.

Natürlich kann jeder Kontoinhaber seine Zustimmung zu solchen Auswertungen verwehren. Möchte er jedoch in der digitalen Welt einen Kredit oder ein Darlehen beantragen, so wird dies ohne seine Bereitschaft, externe Dienstleister in alle „Geheimnisse" seiner Konten blicken zu lassen, nicht mehr möglich sein. Und was für Verbraucher gilt, gilt für Unternehmen erst recht. Grundvoraussetzung einer **Digitalen Fianzkommunikation** ist somit die Zustimmung zur digitalisierten Übermittlung von (Konto-)Daten seitens der (Haus-)Bank an einen dritten Buchführungsdienstleister oder den Steuerberater via Service-Rechenzentrum oder API-Bankenschnittstellen. Damit einher geht der Kontozugriff in spezialisierten, zweckgebundenen Formaten. Zum Beispiel zur Aufgliederung von Überweisungssammelposten; oder zur Kategorisierung von Kontodaten, beispielsweise zur Aufgliederung nach Verwendungszwecken; oder zur Analyse von Zahlungsein- und ausgängen, zum Beispiel zur Vorhersage der Kapitaldienstfähigkeit; oder für Scoring-Systeme und Prognosemodelle, hier als Instrument zur Risikominimierung und letztlich zur Empfehlung an einen Kreditgeber zwecks daten- und informationstechnischer Unterstützung von dessen finaler Finanzierungsentscheidung. Vorteile für den Verbraucher oder mittelständischen Unternehmer? Ja, ein aufwändiger Versand von Kontoauszügen entfällt. Ja, garantiert wird eine beschleunigte Kreditentscheidung – aber alles hat seinen Preis und der heißt: **Vollständige Transparenz!**

32.2.2.2 Social Media

Ein im Bereich Unternehmensfinanzierung oftmals noch unterschätztes Feld ist die **Öffentliche Wahrnehmung, das Internet und die Sozialen Medien.** Sie bieten über Vernetzung und Digitalisierung einen schier unerschöpflichen Fundus an Fakten, Meinungen und Hintergrundinformationen, die in keinem Jahresabschluss stehen, die von keinem Steuerberater via Kurzfristiger Erfolgsrechnung übermittelt werden und die auch nicht auf der Homepage eines Unternehmens zu finden sind. Diese zumeist qualitativen Informations- und Bewertungselemente wurden in der Vergangenheit oftmals vom jeweiligen Firmenkundenbetreuer nach subjektiven Einschätzungen in eine Rating-Klassifizierung mit eingegeben – oder eben auch nicht! Nun arbeiten aber Algorithmen alles ab, was sie in den Tiefen des World Wide Webs finden: Reaktionen im Netz und sozialer Medien zu Preiserhöhungen oder Qualitätsmängel. Meinungen zu unerfüllten, zuvor aber firmenseits breit angekündigten Eigenschaften eines neuen Produktes. Informationen über rechtliche Verstöße, wie zum Beispiel Geldbußen wegen Compliance-Themen oder Abmahnungen wegen missbräuchlicher oder irreführender Werbung.

Über verschiedenste Portale oder Medien lassen sich Informationen zu einem Unternehmen und seiner „**Sozialen Kompetenz**" filtern. So finden sich über Zugriff auf unterschiedlichste Datenbanken schnell Bewertungen von (ehemaligen) Mitarbeitern, Youtube-Videos über Firmenevents oder Meinungen von zufriedenen oder unzufriedenen Kunden. Beispielhaft sei hier nur einmal auf das Thema Kinderarbeit in der Textilindustrie hingewiesen. Aber auch Informationen zum Verhalten oder zu Äußerungen von Top-Managern, Führungskräften oder gesetzlichen Unternehmensvertretern werden ausgewertet. Das Auftreten dieser Personen außerhalb des Unternehmens, beispielsweise bei politischen Aktivitäten oder durch die Veröffentlichung von (privaten) Fotos, und die damit verbundene öffentliche Wahrnehmung werden im Internet unwiderruflich dokumentiert und kommentiert. Öffentliches Auftreten und Verhalten im Netz sind somit nicht mehr nur Privatsache. Alles kann sich sowohl positiv als auch negativ auf das Ergebnis von Rating- und Scoring-Systemen auswirken. Denn über Digitalisierung, Verflechtungen und Schnittstellen wird es um ein Vielfaches einfacher, relevante Informationen abzugreifen, zu analysieren und in ein Bewertungssystem einfließen zu lassen.

So manch einer, der bereits erste Berührungspunkte zum Thema **Digitale Finanzierung** hatte, fragt sich vielleicht, wie es denn auf verschiedenen Kredit-Plattformen immer wieder möglich ist, nur mit Eingabe des Kreditwunschs und des Firmennamens eine erste schnelle Bewertung, etwa in der Form: „Ampel steht auf grün" oder in Sekundenschnelle eine Zinsindikation zu erhalten. Die Antwort klingt einfach: Plattformen sammeln zunehmend, auch im Realtime-Modus mittels entsprechende Schnittstellen, Informationen über ihre Zielkunden: alles, was zunächst einmal öffentlich zugänglich ist; über Auskunfteien, über das Web oder unterschiedlichste Datenbanken. Schnell kann somit ein erstes individuelles Risikoprofil des Zielkunden und eine Prüfung der Konformität mit der Risiko-Policy potenzieller Kapitalgeber oder Investoren und deren vorgegebenem „Risikoappetit" erstellt werden. Taucht der Zielkunde dann tatsächlich auf der Plattform auf, ist man vorbereitet ...

32.3   Wie kann man sich als Unternehmer vorbereiten?

Viele, gerade auch kleinere Unternehmen, wie Handwerksbetriebe oder Freiberufler und Selbstständige stehen nun vor großen Herausforderungen. Oftmals werden diese (noch) allzu gerne beiseitegeschoben. Und noch gibt es ja die klassische Hausbank, mit der man seit vielen Jahren erfolgreich zusammenarbeitet. So rettet einen vielleicht das Bekannte, weil hier ja traditionell alles beim Alten bleibt? Dies mag kurzfristig so sein, mittel- und langfristig sind die digitalen Entwicklungen mit allen ihren Konsequenzen aber nicht mehr aufzuhalten.

32.3.1 Fragen, die man stellen sollte

Stellt man sich den angesprochenen Themen, offen und neugierig, bereit für Veränderungen, so helfen am Anfang immer **Fragen**:

- Wo überall in unserer Medien-Welt sind welche Informationen über ein Unternehmen verbreitet oder gespeichert? Steht beispielsweise auch das (Firmen-)Video von der letzten Betriebsfeier im Netz?
- Auf welche Daten können auch Dritte zugreifen?
- Wie geht man mit dem Thema Informations-Wahrheit um (Fake News)?
- Wie kann man falsche oder nicht korrekte Daten und Informationen korrigieren?
- Welche rechtlichen Möglichkeiten gibt es, negative Einträge oder unpassende Berichterstattungen löschen zu lassen?
- Ab wann gibt es im Internet ein „Recht auf Vergessen"?
- Auf welchen Quellen beruhen Daten und Informationen. Sind diese Quellen vertrauenswürdig?
- Wie wird im Unternehmen die eigene Informationsweitergabe an Dritte überprüft und gesteuert?
- Sind alle Führungskräfte im Unternehmen über diese Themen und die „Neue Welt" via interner Kommunikation hinreichend informiert und instruiert – gerade auch, um für (persönliche) Handlungen zu sensibilisieren, die die Bewertung des gesamten Unternehmens beeinflussen könnten?

32.3.2 Herausforderungen, denen man sich stellen sollte

Viele Fragen – viele Antworten. Hieraus erwachsen für Unternehmer große **Herausforde-rungen**, insbesondere auch bei der Modernisierung, sprich Digitalisierung der eigenen Geschäftsprozesse:

- Wie muss sich ein Unternehmen intern mit seinen Prozessabläufen verändern, um auch in Zukunft kreditfähig zu sein? Sind hierfür neue Technologien erforderlich?
- Wie geht man mit digitaler Kommunikation um?
- Wie motiviert man Mitarbeiter für digitale Innovationen?
- Wie geht man mit dem Thema Datenschutz um (Vertrauen in Vertraulichkeit)? Wie mit dem Thema Cyber-Kriminalität (zum Beispiel Vertrauen in die unterschiedlichsten digitalen Video-Ident-Verfahren)?

Empfehlenswert ist bei diesen Themen eine verstärkte Zusammenarbeit mit externen (Finanzierungs-)Experten sowie dem eigenen Steuerberater. Denn durch Fehler oder Unzulänglichkeiten in der Prozesskette, im Kreditantrag oder den eingereichten Unterlagen wirken Unternehmen weniger vertrauenswürdig. Dies reduziert die Chancen auf eine positive Kreditentscheidung. Kaufmännische Sorgfalt ist also mehr denn je gefragt, denn Digitalisierung hebt im Kreditgeschäft schon gar nicht die „Gesetze der Schwerkraft" auf. Auch hier gilt es, informiert zu bleiben:

- Welche Kriterien gibt es, neben den landläufig bekannten Bonitätsrastern, die mit in die Bewertung eines Unternehmens einfließen?
- Wie erkennt man in digitalen Verträgen „Nebenbedingungen und Kleingedrucktes"? Erinnert sei hier beispielsweise an das Thema „Vorzeitige Kreditrückführung und Vorfälligkeitsentschädigung", das sich oftmals erst in den „hintersten" Vertragsformulierungen findet.
- Wen kann ich bei Unklarheiten fragen? Über welchen Kommunikationskanal? Spricht man mit dem Digitalen Finanzierer persönlich, per Mail, per Chat oder gar per Chatterbot, also einem textbasierten Dialogsystem, welches das Chatten mit einem maschinellen System erlaubt?
- Sind alle Daten des Unternehmens, die in externen Datenbanken gespeichert sind (beispielsweise denen von Auskunfteien), aktuell und konsistent?

32.4 Sechs Beispiele aus der Praxis

Beispiel

Die Auszahlung eines bereits genehmigten Kredits verzögerte sich um mehrere Tage, weil die Adresse des Geschäftsführers in seinem Personalausweis nicht mit der in einer

Auskunftsdatenbank übereinstimmte. Hintergrund: Der Geschäftsführer war erst wenige Tage zuvor umgezogen. Der maschinelle Abgleich der Daten führte zu einem „Veto", das heißt Auszahlungsstopp, weil die „Maschine" den Geschäftsführer nicht zweifelsfrei identifizieren konnte. ◄

Beispiel

Die Informationen zum beruflichen Werdegang einer Geschäftsführerin eines Maschinenbauunternehmens in einer Auskunftsdatenbank waren nicht aktuell. So hatte die Geschäftsführerin einige Jahre nach ihrer Ausbildung zur Kosmetikerin zwar noch erfolgreich ein betriebswirtschaftliches Studium abgeschlossen, die Information hierzu war aber datentechnisch nicht entsprechend verarbeitet. So kam es zur Sofortablehnung einer Kreditanfrage, weil innerhalb des vorgegebenen Rasters die ursprüngliche Berufsausbildung als nicht ausreichend zur Führung eines Wirtschaftsunternehmens mit mehr als fünfzig Mitarbeitern angesehen wurde.

Deshalb: Jeder Unternehmer sollte mindestens einmal im Jahr, auf jeden Fall aber vor einer Kreditanfrage, die über sein Unternehmen und seine Person bei den verschiedensten Auskunfteien gespeicherten Daten auf Richtigkeit und Aktualität überprüfen und erforderlichenfalls dann auch berichtigen lassen. Dies gilt insbesondere auch im Fall von Veränderungen in der Gesellschafterstruktur oder den Beteiligungsverhältnissen, der Geschäftsführung oder bei den Unternehmensverflechtungen. In jedem Fall hat der Unternehmer selbst immer eine Bringschuld sowie Aktualisierungspflicht. ◄

Beispiel

Die Bezeichnung „Pfändung" für ein Konto in der Summen- und Saldenliste einer Betriebswirtschaftlichen Auswertung (BWA) mit einem kreditorischen Saldo von über Zehntausend Euro führte zu einer Vorab-Kreditablehnung, weil dies ein nicht zu umgehendes Negativkriterium war. Tatsächlich stellte sich später aber heraus, dass es sich um eine von dritter Seite initiierte Lohnpfändung gegen einen nicht mehr im Unternehmen arbeitenden Mitarbeiter handelte. Die mit dem Begriff „Pfändung" einhergehende Kontodeutung hin zu laufenden Zwangsmaßnahmen gegen das Unternehmen lief somit de facto ins Leere. Hier wäre also eine andere, weniger irreführende Kontobezeichnung angebracht gewesen. ◄

Beispiel

Der Jahresabschluss eines Unternehmens in der Rechtsform einer GmbH enthielt auf der Aktivseite der Bilanz die Position: „Nicht durch Eigenkapital gedeckter Fehlbetrag in Höhe von 65.000 Euro." Den in den weiteren Erläuterungen einige Seiten später „versteckten" Hinweis, dass zum Bilanzstichtag für ein Gesellschafterdarlehen in Höhe von 100.000 Euro eine Rangrücktrittserklärung abgegeben und damit die Eigenkapital-basis positiv gestärkt wurde, konnte mangels Maschinenlesbarkeit nicht entsprechend verarbeitet werden. Somit kam es zu einer Kreditablehnung wegen Überschuldung der GmbH.

Deshalb: Digital eingereichte Unterlagen sollten bezüglicher textlicher Bezeichnun-gen oder zum Beispiel im Fließtext eingebundener quantitativer (Zahlen-)Informatio-nen immer auf deren durch ein maschinelles Analysetool gegebene Lesart hin überprüft werden, um Missverständnisse, Fehlinterpretation oder Nichtbeachtung von Anfang an zu vermeiden. ◄

Beispiel

Ein Restaurantbetrieb wies in den Kontoauszügen zahlreiche, teilweise auch nur klei-nere Bartransaktionen auf. Diese wurden im Rahmen eines Digital Account Checks aufgegriffen und in Summe zusammengefasst. Über einen Zeitraum von zwölf Mona-ten ergab sich ein Betrag von über 30.000 Euro. Der Kredit wurde abgelehnt wegen Anfangsverdacht auf Geldwäsche bzw. unversteuerter Einnahmen. ◄

Beispiel

Im Rahmen eines Digital Account Checks wurden über drei Monate hinweg die ange-gebenen Verwendungszwecke von Überweisungen analysiert. Dabei stellte sich heraus, dass es mehrfach Negativhinweise gab, die Formulierungen wie „Unsere verspätete Zahlung vom …", „Ausgleich Ratenrückstand …" oder „Erledigung Kontoüberzie-hung …" zu entnehmen waren. Der Kreditwunsch wurde daraufhin abgelehnt.

Deshalb: Zumindest für einen aktuellen Drei-Monatszeitraum müssen Kontoaus-züge zum Zeitpunkt der Kreditbeantragung frei von jeglichen Negativinformationen sein. Denn schriftliche Erläuterungen, warum etwas „unglücklicherweise" wie passiert ist, helfen wenig weiter. Der nach vorgegebenen Mustern entscheidende maschinelle Analyseprozess ist nun einmal völlig emotionslos, sprich ohne menschliches Verständ-nis für alle möglichen „Verstrickungen". ◄

32.5 Ausblick

Je besser die risikomäßige Bewertung eines Unternehmens ist, desto einfacher ist der Zu-gang zu Krediten, desto niedriger ist der Zinssatz, desto besser sind die sonstigen Kredit-konditionen wie Höchstbetrag, Laufzeit oder Gebühren. Mit allen hieraus erwachsenden unternehmerischen Möglichkeiten … und damit den Vorteilen für die eigene **Wettbe-werbsfähigkeit**. Die Beschäftigung mit den Erfordernissen der **Digitalen Finanzkom-munikation** ist somit direkte Voraussetzung für den Erhalt der eigenen Konkurrenzfähigkeit. Nun gilt es, neue Wege zu gehen, schneller auf der Lernkurve zu sein als der Wettbewerb und Finanzierungssituationen neu zu strukturieren.

Zurück zur Eingangsfrage: Gibt es chinesische Verhältnisse bald auch in Deutschland? Müssen auch wir uns mit einem umfassenden **Sozialkreditsystem** auseinandersetzen? Eines steht fest: Digitalisierung verbunden mit Vernetzung, maschinenlesbaren Daten und emotionslosen Algorithmen führt zunehmend in eine Welt von „gläsernen Unternehmen" – und dies auf allen Informationsebenen. Neue Dimensionen eröffnen sich, wenn der Zugriff auf weitere Informationen möglich ist, die nicht nur die klassischen Finanzzahlen betreffen. Seien dies zum Beispiel Informationen zur **Öffentlichen Wahrnehmung,** aus dem **Internet** oder den **Sozialen Medien.** Die Zusammenfassung dieser speziellen Informationen über Vernetzung, maschinenlesbare Daten und anschließender umfassender **Data-Analytics** ermöglicht ein Aggregationsniveau, das bisher manuell nicht möglich war. Gleichzeitig werden alle qualitativen Daten, die oftmals einer „persönlichen" Wertung unterlagen, durch algorithmische Muster in ihrer Gesamtheit einem „objektivierten", quantitativen Scoring zugeführt. In diesem spielt insbesondere auch die **Soziale Kompetenz** eines Unternehmens eine entscheidende Rolle. Somit wird das bekannte Bonitätsrating zunehmend von einem **Sozialen Scoring** ergänzt werden.

Für das Kreditgeschäft bedeutet dies: **Prozesseffizienz** und **Digitale Vernetzung** sind die Schlagwörter der Zukunft. Ein voll automatisierter und allein datenbasierter Antragsprozess bis hin zu algorithmischen Kreditentscheidungen wird neben klassischen Bonitätskennziffern auch auf aggregierte soziale Daten zurückgreifen. Damit ist die Grundlage für ein **Sozialkreditsystem** auch bei uns gelegt: Im ersten Schritt für die Vergabe von Krediten, aber im zweiten Schritt sind überall dort Weiterungen denkbar, wo „objektive" Entscheidungen auf breiter Datenbasis zu treffen sind, beispielsweise in der öffentlichen Vergabepolitik oder ganz allgemein in Kunden- und Lieferantenbeziehungen. Und über die unendlichen Möglichkeiten der Vernetzung mag es dann nur noch ein kleiner Schritt sein hin zu Listen von „nicht vertrauenswürdigen Unternehmen". Zugänglich für jeden, der diese käuflich erwerben will, so wie dies heute beispielsweise mit Adresslisten bereits der Fall ist.

Manches mag sich noch nach Zukunftsmusik anhören, beschränkt lediglich auf einige wenige standardisierte und unkomplizierte Sachverhalte. Werden aber alle diese Prozesse in der Nachbetrachtung konsequent einem maschinellen Lernen unterworfen, wird zunehmend auch das Handling von komplexeren Fällen möglich sein. Dabei wird der Automatisierungsprozess via Datentransfer mehr und mehr völlig neue Finanzierungslösungen im Rahmen von **Kooperationspartnerschaften** ermöglichen – seien dies solche mit andere FinTechs oder gemeinsam mit „aufgerüsteten" klassischen Banken, die bereit sind, sich den neuen Realitäten der **Digitalisierung** zu stellen.

Man mag gespannt sein auf die Zukunft.

Rainer Langen ist Diplom-Volkswirt und startete seine Karriere als Risikomanager und Verhandlungsführer im Firmenkundenkreditgeschäft einer deutschen Großbank. Inzwischen unterstützt er seit mehreren Jahren als Berater mittelständische Unternehmen in schwierigen Finanzierungsfragen. Er ist Autor der Fachbücher „Die Sprache der Banken" (2007) und „Finanzierungschancen trotz Bankenkrise" (2009) und hat schon früh auf das Erfordernis einer professionellen Kommunikation zwischen Unternehmen und Banken hingewiesen. In diesem Zusammenhang gründete Langen 2010 das Deutsche Institut für Kreditmediation (IKME) sowie als bundesweites Expertennetzwerk den Bundesverband der Kreditmediatoren e. V. (BdKM), dessen 1. Vorstand er ist. Langen ist zudem Mitherausgeber der in 2012 und 2013 erschienen Fachbücher „Finanzkommunikation – Chancen durch Kreditmediation" und „Basel III".

Social Credit Ratings in der Praxis – dargestellt am Beispiel des Wareneinkaufsfinanzierers

33

Hendrik Schütte und Maximilian Klein

Zusammenfassung

Deutsche Kreditinstitute befinden sich in einer Phase ausschlaggebender Veränderungen. Aufgrund der anhaltenden Niedrigzinsen ist das Geschäft der Fristentransformationen nur noch bedingt möglich. Zu einem bisher vernachlässigten Firmenkundensegment gehören die Geschäfts- und Gewerbekunden. Das Berliner Start-up VAI Trade nimmt sich dieses Segments an und setzt auf automatisierte Entscheidungswege. Die Zielgruppe der VAI Trade GmbH sind kleine Unternehmen des Mittelstands. Dies entspricht Unternehmen mit bis zu 50 Mitarbeitern und einem Umsatz zwischen 100.000 und 15 Millionen Euro. Einen besonderen Fokus legt die VAI Trade GmbH auf eCommerce-Händler. VAI Trade verwendet die Bonitätseinschätzungen der Auskunfteien als Vorselektierung der Finanzierungsanfragen. So können harte Negativmerkmale und eventuell bestehende Zahlungserfahrungen der Auskunfteien dabei helfen, das anfragende Unternehmen vorerst einzuschätzen. Da Einzelunternehmen keine Veröffentlichungspflicht ihrer Umsätze haben, liegen den Auskunfteien lediglich rudimentäre Informationen zu den Unternehmen vor.

Mit Blick auf die asiatischen Vorläufer der FinTech-Branche steht der Kurs fest. Themen wie die Nutzung unstrukturierter Daten werden weiter an Relevanz in deutschen Finanzunternehmen gewinnen. VAI Trade setzt aus diesem Grund auf die Akkumulation von Daten durch die Anbindung diverser Microservices der Kunden und die Speicherung und das Monitoring der Social Media Auftritte. Betrugsversuche werden hierdurch bereits abgewehrt und Ausfälle vermieden.

H. Schütte (✉) · M. Klein
VAI Trade, Berlin, Deutschland
E-Mail: hendrik.schuette@vaitrade.de; maximilian.klein@vaitrade.de

© Springer Fachmedien Wiesbaden GmbH, ein Teil von Springer Nature 2020
O. Everling (Hrsg.), *Social Credit Rating*,
https://doi.org/10.1007/978-3-658-29653-7_33

33.1   Abgrenzung des Einkaufsfinanzierers: VAI Trade GmbH

Deutsche Kreditinstitute befinden sich weiterhin in einer Phase ausschlaggebender Veränderung. Aufgrund der anhaltenden Niedrigzinsen ist das Geschäft der Fristentransformationen nur noch bedingt möglich. Aus diesem Grund wird der ertragreichen Ausnutzung von vorhandenen Potenzialen der Finanzwelt eine besondere Wichtigkeit zugeschrieben. Zu einem bisher vernachlässigten Firmenkundensegment gehören die Geschäfts- und Gewerbekunden. Das Berliner Start-up, VAI Trade, nimmt sich dieses Segments an und setzt auf automatisierte Entscheidungswege, um kleinen Mittelstandsunternehmen Warenfinanzierungen zu ermöglichen.

VAI Trade agiert bei der Finanzierung von Waren im Bereich des *Finetradings*. Bei dieser Art der Einkaufsfinanzierung kauft der Dienstleister die physischen Waren an und verkauft sie mit verlängertem Zahlungsziel an die Unternehmung. Durch den An- und Verkauf der Waren besteht bis endgültiger Zahlung des Unternehmens ein Eigentumsvorbehalt auf Seiten von VAI, welcher als Hauptsicherheit dient. Eine detaillierte Darstellung der Handelsbeziehung ist Abb. 33.1 zu entnehmen.

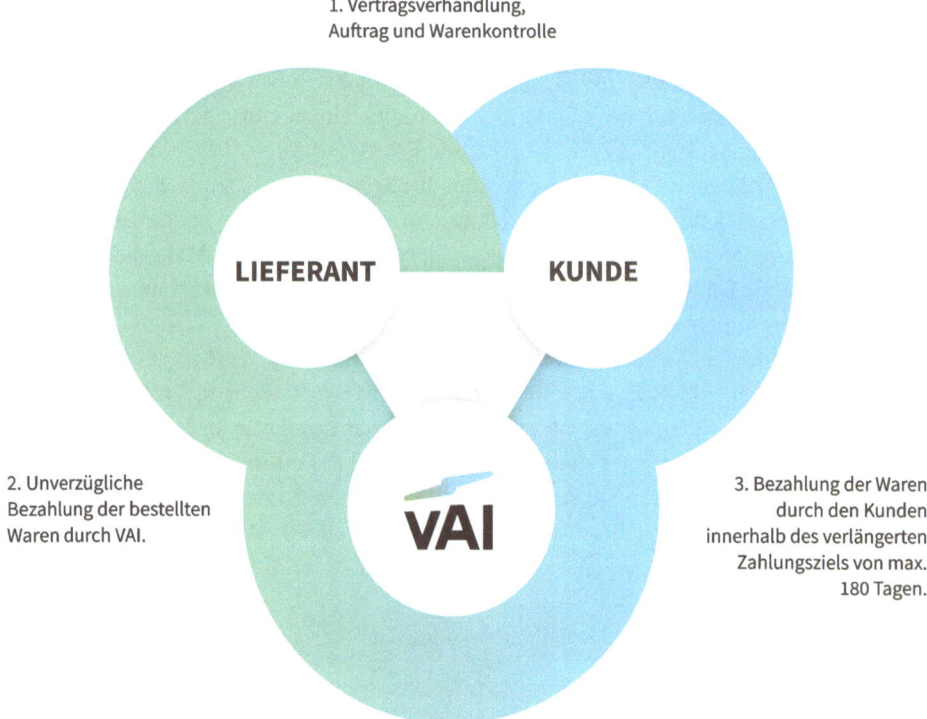

Abb. 33.1 Funktionsweise Finetrading. (Quelle: eigene Darstellung der VAI Trade GmbH, Berlin 2020)

Die Zielgruppe der VAI Trade GmbH sind kleine Unternehmen des Mittelstands. Dies entspricht Unternehmen mit bis zu 50 Mitarbeitern und einem Umsatz zwischen 100.000 und 15 Millionen Euro. Diese Definition ist deswegen so breit gefasst, weil die Vorfinanzierung von Waren eine immer wiederkehrende Herausforderung in dem genannten Unternehmen darstellt.

Einen besonderen Fokus legt die VAI Trade GmbH auf eCommerce-Händler. Das Marktsegment der „Onlinehändler" weist seit Jahren ein starkes Wachstum auf. Gerade in den DACH-Märkten ist der Onlinehandel-Bereich ein Wachstumsmotor der Wirtschaft. Hinzu kommt der Vorteil, dass dieser Zielgruppe bereits eine grundlegende Affinität zu digitalen Produkten unterstellt werden kann. Jedoch bleibt das Angebot und Produkt der VAI Trade GmbH offen für alle Branchen.

33.2 Die Notwendigkeit von Social Credit Scoring im Firmenkundengeschäft

Der direkte Vergleich zwischen dem traditionellen Firmenkundengeschäft eines Bankinstituts und dem Online-Geschäft der VAI Trade GmbH zeigt diverse Unterschiede. Grundlegende Differenzen ergeben sich bereits in der Klassifizierung der Vertriebsstrategie. Das klassische Firmenkundengeschäft legt die intensive Kundenberatung in den Hauptfokus. Somit ist das Firmenkundengeschäft sehr beratungsintensiv und ist als Individualgeschäft klassifizierbar. Die VAI Trade Strategie ist das Mengengeschäft.

Die Unternehmung setzt auf intuitive und reibungslose Prozesse, wodurch weniger menschliche Berührungspunkte mit dem Kunden entstehen. Durch den fehlenden Kontakt ergeben sich jedoch diverse Herausforderungen. Ein Beispiel ist die Diversität der Unternehmen. Die Geschäftszwecke der potenziellen Unternehmen unterscheiden sich in Art und Umfang. Zudem ergeben sich regionale Unterschiede, welche nicht wie bei dem traditionellen Filialkonzept einer Bank in einem persönlichen Gespräch leicht herausgefunden werden können. Folglich ergibt sich bei der Gegenüberstellung zum klassischen Finanzierungskonzept eine Verstärkung des ohnehin existierenden Informationsungleichgewichts zwischen Finanzierenden und Finanzinstitut.

Die Unterschiede zwischen Offline- und Online-Finanzgeschäft haben direkte Auswirkungen auf die Durchführung der Bonitätsanalyse der Unternehmen. So setzt sich eine klassische Bonitätsbewertung aus einer Zusammenfassung von Hard- und Softfacts zusammen (vgl. Eschbach et al. 2013, S. 50). Hierbei spricht man auch von materieller und persönlicher Kreditwürdigkeit. Die persönliche Kreditwürdigkeit gibt die Zuverlässigkeit und die Angemessenheit des Unternehmers wieder. Durch das persönliche Gespräch ergibt sich ein subjektiver Eindruck über das Auftreten des potenziellen Kreditnehmers. Verbunden mit dem Wissen über die Region und das Image der Unternehmung, bildet dieser Eindruck einen Bestandteil der *First Line of Defense*. Unter diesem Begriff wird ein Filter der Anfragen verstanden, durch welchen nur ein bestimmter Anteil von Finanzierungsanfragen tatsächlich weiter bearbeitet wird. In einem Beispiel bedeutet dies, dass ein

Unternehmer zunächst den Berater von seiner persönlichen Kreditwürdigkeit überzeugen muss. Sollte dies nicht gelingen, wird die Anfrage fallengelassen und ein potenzieller Ausfall vermieden. Im Onlinegeschäft fehlt dieser Schritt, wodurch digitale Finanzinstitute, wie VAI Trade, alternative Lösungsansätze finden müssen. In diesem Zusammenhang wird häufig das Social Credit Rating genannt. Ziel soll es sein, den persönlichen Eindruck des Unternehmers auf digitale Weise abzubilden und zu bewerten. Durch den Einsatz von diversen Datenquellen kann ein Profil des Unternehmers erstellt werden, welcher die persönliche und materielle Kreditwürdigkeit abdeckt. Im Folgenden werden Methoden und Funktionsweisen dargestellt, wie Informationslücken geschlossen werden können. Sie zeigen einen Einblick in die Prozesse und Vorgehensweisen des Start-ups. In der jüngsten Vergangenheit konnte VAI Trade durch die Standardisierung der Prozesse erhebliche Einsparungen in Zeit und Aufwand bewerkstelligen, sodass eine automatisierte Genehmigung in Echtzeit möglich ist.

33.3 Funktionsweise und Nutzung von Auskunfteien

Die wohl bekannteste Nutzung von vereinfachten Social Credit Ratings seitens der Kreditinstitute ist die Nutzung von Drittanbietern wie Schufa und Creditreform. Verglichen mit anderen westlichen Ländern, ergeben sich durch die hohen Datenschutzauflagen innerhalb Deutschlands diverse Einschränkungen bei der Nutzung von Auskunfteien. So sind lediglich juristische Personen zu einer Veröffentlichung von wirtschaftlichen Zahlen verpflichtet. Veröffentlicht werden vereinfachte Darstellungen des Jahresabschlusses zur freien Einsicht. Diese Daten werden im Anschluss von den Auskunfteianbietern ausgewertet. Da es sich bei den Zahlen lediglich um eine verkürzte Darstellung der Unternehmensdaten handelt, reichern die Auskunfteien jene durch weitere Datenquellen an.

Ein Teil des eigenen Datenbestandes der Auskunfteien wird durch die Anfragenden selbst geschaffen. Die Schufa erfasst zum Beispiel Geburtsdatum und Anschrift. Bei Mehrfachnennungen der gleich oder ähnlichen Daten und einer Validitätsprüfung werden die Informationen gespeichert und Änderungen erfasst. Ebenso arbeiten Telefonanbieter und Banken meist näher mit der Schufa zusammen, sodass Verträge genannt und eventuelle Zahlungsschwierigkeiten gemeldet werden. Zusätzlich arbeiten B2B-Auskunfteien, wie die Creditreform, mit dem Handels- und dem Insolvenzregister zusammen. So werden tagesaktuelle Negativmerkmale an die Auskunfteien gemeldet.

33.3.1 Nutzung von Auskunfteien im Anwendungsbeispiel VAI Trade GmbH

VAI Trade verwendet die Bonitätseinschätzungen der Auskunfteien als Vorselektierung der Finanzierungsanfragen. So können harte Negativmerkmale und eventuell bestehende Zahlungserfahrungen der Auskunfteien dabei helfen, das anfragende Unternehmen vorerst einzuschätzen. Die tatsächlichen Bonitätsnoten und Kreditlinien, welche die Auskunfteien veröffentlichen, haben für den operationalen Ablauf der VAI Trade GmbH eine geringe Auswirkung. So können keine festen Zusagen allein auf Basis der Auskunfteien getroffen werden. Dies ist zunächst dem Umstand geschuldet, dass ein Teil der Finanzierungsanfragen der VAI Trade GmbH von natürlichen Einzelunternehmungen kommen. Die Einzelunternehmen besitzen nicht die oben genannte Veröffentlichungspflicht. Somit liegen den Auskunfteien lediglich rudimentäre Informationen zu den Unternehmen vor. Dies wird zusätzlich durch die Größen der Unternehmen bestärkt. Durch die geringe Größe der individuellen Unternehmen wird es den Auskunfteien erschwert, einheitliche, aussagekräftige Daten standardisiert zu beschaffen. Eine flächendeckende Informationsbeschaffung ist derzeit noch nicht möglich und wäre mit hohem Aufwand und Kosten verbunden. Ergänzend handelt es sich bei den Antragstellern häufig um junge Unternehmen ohne große Unternehmenshistorie. Somit sind in nahezu allen Fällen die selbst eingeholten Daten des Finanzierers aktueller und aussagekräftiger als die der Auskunfteien. Dennoch werden die zusätzlichen Informationen der Auskunfteien genutzt, um die vorliegenden Informationen zu validieren und zu ergänzen.

33.3.2 Nutzung von Auskunfteien im Gesamtüberblick

Innerhalb von Deutschland sind die größten Auskunfteien die Schufa für Privatleute und die Creditreform für Unternehmen. Trotz intensiver Konkurrenz diverser, alternativer Anbieter besteht nach Einschätzung der Autoren weiterhin ein merklicher Qualitätsunterschied der beiden Marktführer gegenüber den Konkurrenten. Dieser Vorsprung kann bei der Creditreform den umfangreichen Analysearbeiten der Auskunftei zugeschrieben werden. So kommt es unter anderem dazu, dass die Creditreform sich darauf spezialisiert hat, Unternehmen in Deutschland selbstständig zu kontaktieren, um an neue, detaillierte Wirtschaftsdaten zu gelangen. Diese eingereichten Unterlagen werden dann seitens der Creditreform analog zu den öffentlichen Datenquellen analysiert. Das Einreichen zusätzlicher Unterlagen wird somit als eine freiwillige Leistung gesehen, welche die Bonitätseinschätzung der Unternehmung verbessern kann. So ist es auch kleineren Unternehmen durch das proaktive Handeln möglich, die Aussagekraft der Auskünfte zu verbessern.

Eine weitere Anwendung des Social Scoring Ratings bei Auskunfteien sind die direkten Zahlungserfahrungen. Ein Teil der deutschen Auskunfteien bietet neben den klassischen Bonitätsbewertungen zusätzliche Dienstleistungen an. So tritt beispielsweise die Coface gleichzeitig als Kreditversicherer und Inkassodienstleister auf. Die Creditreform bietet den deutschen Unternehmen ebenfalls einen Inkassoservice an und arbeitet zusätzlich auch als Factoring-Unternehmen. Ob und wie die aus den weiteren Geschäftszweigen erworbenen Informationen eine direkte Aussagekraft bezüglich der Bonitätsklassifizierungen haben, ist nicht genau beziffert.

Neben den direkten Zahlungserfahrungen arbeiten Auskunfteien zusätzlich auch mit indirekten Zahlungserfahrungen der Unternehmen. Dies bedeutet, dass die Möglichkeit für Unternehmen besteht, bei Anbietern wie Creditreform oder Creditsafe die Debitoren-zahlungen zu melden. So werden Informationen über die Zahlungsweise der Unternehmen ausgetauscht und über einen längeren Zeitraum betrachtet. Ein Unternehmen, welches partizipiert, erhält im Gegenzug Einblick in andere Meldungen von anderen Unternehmen. Hierdurch ergibt sich ein Gegenseitigkeitsprinzip, wodurch die Parteien einen Vorteil erlangen und innerhalb des Systems gehalten werden. Das Endergebnis ist eine Datenbasis für diverse Unternehmen, welche sich auf Dauer aktualisiert, wodurch ein besseres Risikomanagement möglich wird. Neben dem Debitorenregister Deutschland der Creditreform hat sich noch keine alternative Lösung flächendeckend in Deutschland durchgesetzt. Ein Grund hierfür könnten gesellschaftliche Problematiken sein.

33.3.3 Nutzung von Auskunfteien im Zukunftsausblick:

In der jüngsten Vergangenheit wurden datenschutzrechtliche Bedenken gegenüber dem Vorgehen der Schufa geäußert. Innerhalb des Privatkundenmarktes ist demzufolge zu vermuten, dass sich der Prozess des Social Scorings seitens der klassischen Auskunfteien zwar weiter verstärkt, sich jedoch auch mit Kritik und größeren kulturellen Problemen auseinandersetzen muss. Bei Betrachtung des Geschäftskundenmarktes treten im Zuge des Datenschutzes weniger Bedenken auf. In diesem Segment ist es daher leichter, einen Social-Scoring-Prozess in Deutschland zu etablieren. Das zuvor beschriebene Beispiel der Zahlungserfahrungen ist bereits ein Vorreiter. Das Prinzip setzt jedoch voraus, dass eine Auskunftei entweder selbst direkten Zugriff auf diverseste Zahlungen hat oder flächendeckende Meldungen von externen Unternehmen erhält. Dafür könnten die Daten von Zahlungsdienstleistern wie Paypal, Klarna und Amazon Pay als Informationsgrundlage dienen. Aus diesem Grund ist es nicht verwunderlich, dass ein Subunternehmen der Creditreform sich auf Zahlungsabwicklung von Unternehmen spezialisiert hat. Dies könnte bei einem größeren Marktanteil qualitative Bonitätsdaten akkumulieren.

Bei weiterer Außerachtlassung des Datenschutzes sind auch Organisationen wie Zhima Credit vorstellbar. Die chinesische Auskunftei ist Teil des Alibaba Konzerns. Dieser spezialisiert sich zunächst auf die Bonitätsbewertung von Privatleuten. Als Datenquellen dienen Zahlungserfahrungen von Versicherungen, Kreditanbietern und anderen Zahlungen über Alibaba oder Alipay. Zusätzlich wird vermutet, dass das Datingverhalten und die Bewegungsmuster der Nutzer ebenfalls eine Auswirkung auf die Bonitätsbewertung haben. Es ist somit nahezu gleichzusetzen mit dem bekannten Social Credit System des Landes, obwohl die Bewertung unabhängig und durch andere Datenquellen durchgeführt wird. Dennoch ist die Nähe zum chinesischen Staat deutlich. So besteht ein direkter Zugriff auf offizielle Dokumente der Regierung. Laut Aussagen der Unternehmung werden keine Informationen zurück an die Regierung übermittelt. Zudem sei angemerkt, dass die Teilnahme an Zhima Credit auf freiwilliger Basis erfolgt.

33.4 Nutzung unstrukturierter Daten

Der Begriff Big Data hat in den vergangenen Jahren viel Aufmerksamkeit auf sich gezogen und wird in Teilen bereits als allgemeines Buzzword genutzt (vgl. Vassakis et al. 2018, S. 4 f.). Allgemein gesprochen, versteht man unter Big Data eine Ansammlung diverser Datenpunkte. Der ausführende Sammler ist nicht entscheidend, sodass es sich sowohl um Großunternehmen als auch Regierungen handeln kann. Die gesammelten Daten stammen meist aus dem Internet, wobei in den häufigsten Fällen granularer klassifiziert werden kann. So können die Datensammlungen meist anhand der fünf Vs charakterisiert werden: Volume, Variety, Velocity, Veracity and Value. Ziel der Datenkumulation ist der Bau von Modellen, welche dazu genutzt werden könnten, in diversen Anwendungsbereichen Prognosen zu erstellen.

33.4.1 Nutzung von unstrukturierten Daten im Anwendungsbeispiel VAI Trade GmbH

Die Grundlage einer Big Data Scoring Engine ist eine ausreichende Ansammlung von Daten. Aufgrund des geringen Alters der VAI Trade GmbH und die bisher wenigen Ausfälle liegen derzeit keine hinreichenden Daten vor, um ein statistisches Modell für den allgemeinen Gebrauch zu entwickeln. Es werden aber bereits Vorbereitungen getroffen, um eine Analyse in der Zukunft produktionsreif zu nutzen. Aus diesem Grund werden bereits unterschiedliche Datenpunkte des Nutzers gesammelt. Unter anderem wird besonderer Fokus auf den digitalen Webauftritt des Kunden gelegt. Darunter fallen Informationen über die zugrunde liegende Technik der Website, öffentliche Bewertungen bei Portalen, Medienpräsenz, Social Media Aktivitäten und vieles mehr. Die gesammelten Informationen werden noch manuell ausgewertet und bewertet. Ziel ist es, die Bewertungen zukünftig automatisch zu treffen.

Es wird vermutet, dass bei isolierter Betrachtung digitaler Branchen auf längere Sicht eine Korrelation zwischen Auftritt und Bonität unterstellbar ist. Derzeit werden die vorhandenen Datenpunkte erfolgreich als Fraud-Indikator eingesetzt. Dieser Nutzen soll in der Zukunft verstärkt und ausgebaut werden.

33.4.2 Nutzung von unstrukturierten Daten im Gesamtüberblick

Innerhalb von Deutschland gibt es bereits Bestrebungen, anhand von unstrukturierten Daten Bonitätsentscheidungen zu treffen. Vorreiter wie das Hamburger Start-up Kreditech beschäftigen sich seit Jahren mit der Sammlung und der Auswertung von Social-Media- und Geolocation-Daten. In der jüngsten Vergangenheit war die Unternehmung mit viel Kritik konfrontiert. Laut Geschäftsbericht kann noch kein endgültiger Erfolg gefeiert werden. Zusätzlich kam es 2017 zu hohen Ausfällen aufgrund von nicht ausgereiften Algorithmen (vgl. Dohms und Schlenk 2019).

Im internationalen Umfeld der Auskunfteien finden sich bereits ausgereifte Prozesse. Ein Beispiel ist das Start-up mit Unicornstatus WeLab aus Hong Kong. Die durch Alibaba finanzierte Unternehmung spezialisiert sich auf die Auswertung von unstrukturierten Opt-in-Daten, welche innerhalb von Sekunden ausgewertet werden und zur Risikoentscheidung führen. Hauptfokus liegt somit auf der Bereinigung und Mustererkennung innerhalb der Daten. Die Muster werden dann zur Einschätzung des Fraud- und Kreditausfallrisikos genutzt.

33.4.3 Nutzung von unstrukturierten Daten im Zukunftsausblick

Inwieweit der Datenschutz in den Bereichen Datenakkumulation noch handeln muss, ist unklar. Dennoch steht der Kurs, welcher von den asiatischen Unternehmen gesetzt wurde, fest. Es liegt nahe, dass Big Data weiterhin an Relevanz in Finanzunternehmen gewinnen wird. So wird es vermutlich dazu kommen, dass zunehmend Fraud-Maßnahmen und im Anschluss Aussagen über die Bonität getroffen werden können. Dies folgt dem Schluss, dass jede Unternehmung und jede Person, welche im Web aktiv ist, Spuren und Daten hinterlässt. Somit sollte es innerhalb der nächsten Jahre denkbar sein, dass die Identität und die Legitimität von Kreditanfragen anhand von unstrukturierten Daten bewertet werden können.

Für eine Bonitätsbewertung anhand von gesammelten, frei zugänglichen Daten ist eine technische Weiterentwicklung erforderlich. Hier muss ein besonderer Fokus darauf gelegt werden, rassistische und systematische Neigungen von KI-gesteuerten Algorithmen zu vermeiden. Dieser Punkt wird besonders in Abschn. 33.6 diskutiert.

33.5 Implementation von Microservices

Das sogenannte *Account Linking* beschreibt den Prozess der Verbindung von Nutzerkonten zweier verschiedener Online-Dienstleister („Microservices") und dem anschließenden Datenaustausch. Das wohl prominenteste Beispiel dieses Prozesses ist die Registrierung bei einem Onlineanbieter durch die Nutzung des bereits vorhandenen Google- oder Facebook-Kontos. So ist es dem Endnutzer möglich, durch die Anmeldung mit seinem Google- oder Facebook-Konto ein Nutzerkonto bei einem Drittanbieter zu erstellen. Hierzu werden durch die explizite Anmeldung in das Hauptkonto Daten mithilfe der zur Verfügung stehenden API-Schnittstelle an den neuen Dienstleister gesendet, welcher im nächsten Zuge durch die zur Verfügung gestellten Daten ein Nutzerkonto erstellen kann. Dieses Vorgehen kann auch zum Austausch größerer Datenmengen oder zur Aggregation von Datenquellen genutzt werden.

Eine weitere Form des Account Linkings, welche innerhalb der jüngsten Vergangenheit verstärktes Aufsehen erregt hat und mitunter schlussendlich in einer EU Richtlinie „Payment Services Directive II" (kurz: PSD II) endete, ist der Austausch von Kontotransaktionsdaten. Vergleichbar zu „Google: Sign in" sind Banken durch die PSD-II-Richtlinie verpflichtet, auch Nichtbanken Zugang zu den Kontodaten zu ermöglichen. Dieser Prozess bietet primär FinTechs die Möglichkeit zur Verarbeitung der Bankdaten und/oder zur Validitätsprüfung der Kontoinhaber. Der Datentransfer wird mithilfe von Bafin-lizenzierten Servicedienstleistern durchgeführt. Die technische Umsetzung befindet sich derzeit in einer andauernden Weiterentwicklung. Der Hauptfokus liegt auf API-Schnittstellen der Banken. Da diese jedoch noch nicht alle Funktionsweisen abbilden können, werden zusätzlich weitere Methoden wie ScreenScraping des Online-Bankings unterstützend eingesetzt.

Der Einsatzbereich des Account Linkings beschränkt sich nicht nur auf die Auslesung der Bank-Transaktionsdaten. Es bestehen ebenso Möglichkeiten, durch die Nutzung von diversen APIs von Onlineservices an Daten zu gelangen. So können zum Beispiel sowohl die DATEV Buchhaltung, die Paypal-Umsätze als auch die Web Analytics durch ein Login übermittelt werden. Diese Daten müssen im Nachgang bereinigt und analysiert werden.

33.5.1 Nutzung von Microservices im Anwendungsbeispiel VAI Trade GmbH

Seit Mitte 2018 verarbeitet VAI Trade die Kontodaten seiner Kunden durch das Account Linking von Banktransaktionen. Durch die kurzfristige Natur der Wareneinkaufsfinanzierung bildet die Liquiditätsanalyse einen essenziellen Teil der Bonitätsbewertung der Unternehmen. Die Kontodaten werden im Zuge der Verarbeitung bereinigt, angereichert und analysiert, sodass eine sofortige Kreditentscheidung möglich ist. Ein besonderer Fokus bei der Bonitätsanalyse wird auf die negativen Anzeichen eines Kontos gelegt. So werden alle Transaktionen kategorisiert und gefiltert. Neben der Kategorisierung werden ebenso Verhaltensmuster erkannt und im Kontext der Kundenbranche gewertet. Einschneidende Ereignisse wie Darlehensauszahlungen und einmalige Verkaufserlöse werden ausgelesen und bereinigt. Die angereicherten Daten werden im Anschluss ausgewertet und resultieren in Finanzierungszusagen. Durch eine grafische Aufbereitung der selektierten Daten können auch höhere Finanzierungsvolumina innerhalb von wenigen Stunden gewährt werden. Die Analyse befindet sich in einer konstanten Entwicklung, sodass planmäßig auch höhere Einkaufslinien innerhalb von Sekunden entschieden werden können.

Neben der Verknüpfung der Kontodaten werden auch Verbindungen zu Shopify und Amazon Seller Central angeboten. Diese Daten geben einen zusätzlichen Einblick in das Ordermanagement und die Umsatzentwicklung der eCommerce-Unternehmen. So können anhand der Daten ebenfalls Einkaufslinien zur Verfügung gestellt werden, da die Bonität und die Einkaufsstärke des Unternehmens bonitäts seitig beleuchtet wird. Weitere Verknüpfungen von Dienstleistern wie Stripe, eBay o. ä. befinden sich in der Entwicklung.

Ein großer Vorteil des Account Linkings ist die Minimierung eines Betrugsrisikos. Durch die automatische Übermittlung der Daten ist eine Veränderung oder eine Imitation der Daten nahezu ausgeschlossen. Ein weiterer Vorteil bietet die meist konstante Verbindung zu den Diensten. Ist ein Bankkonto mit VAI verbunden, ist diese Verbindung nach Zustimmung des Unternehmers auch zukünftig gültig. So besteht für VAI die Möglichkeit, die Unternehmen konstant neu zu bewerten und die Einkaufslinien anzupassen. Dies führt nicht nur zu einem geringen Ausfallrisiko, sondern auch zu einem fairen Geschäftsverhältnis, welches ausgeweitet werden kann, sollte sich die Unternehmung positiv entwickeln.

33.5.2 Nutzung von Microservices im Zukunftsausblick

Durch die erst kürzlich umgesetzte PSD-II-Richtlinie befindet sich das Open-Banking-Konzept in Deutschland noch in einer frühen Entwicklungsphase. Dank der Richtlinie besteht Rechtssicherheit, sodass ein wesentlicher Grundstein der Ausweitung des Prozesses gesetzt ist.

Es liegt nahe, dass Verbindungen zu weiteren Diensten vergleichbar zu DATEV, Amazon, Shopify etc. ebenfalls zunehmen werden. Dies liegt an der steigenden Verbreitung von API-Schnittstellen. So ist es denkbar, dass zukünftig auch Warenwirtschaftssysteme auswertbar werden. Durch die dadurch entstehende Masse von diversen Datenpunkten ist es sinnvoll, dass sich Unternehmen bilden, welche diese Daten aggregieren und aufarbeiten. So kann dem Unternehmer ein ganzheitlicher Überblick über Marketing, Buchhaltung, Liquidität und Warenwirtschaft gegeben werden. Für externe Unternehmen würde ein solcher Dienstleister einen qualitativ hochwertigen Einblick in die wirtschaftliche Gesundheit der Unternehmung geben, welcher konstant betrachtet werden kann. Die Unternehmung wird hierdurch transparenter. Es bleibt abzuwarten, inwieweit sich die Finanzierungsbranche dahingehend entwickelt und wie die kulturellen Herausforderungen einer immer luzider werdenden Gesellschaft bewältigt werden.

33.6   Diskussion des Kundennutzen im Social Credit Scorings

Das Thema Daten wird weltweit, aber vor allem im europäischen Raum, mit sehr viel Skepsis und Kritik betrachtet und nahezu feindselig diskutiert. Der Umgang mit Daten in der Volksrepublik China wird als negative Diskussionsgrundlage fast aller Auseinandersetzungen mit diesem Thema herangezogen.

Die Diskussion über die Erhebung und Speicherung von Daten hat in der jüngsten Vergangenheit innerhalb der Europäischen Union dazu geführt, dass das Grundprinzip des Besitzes von Daten definiert wurde. Dies bedeutet, dass ein Endnutzer selbst Eigentümer seiner Daten ist und frei entscheiden darf, welcher Dienstleister Zugriff auf diese Daten hat. So ist es, wie bereits oben genannt, durch PSD II standardisiert möglich, die Bank-Transaktionsdaten zu teilen. Diese Definition dieser Grundlage ist bereits ein großer

Nutzen des Kunden, weil es weiterhin innerhalb des Einflussbereiches des Endverbrauchers liegt, ob und wie er seine Daten preisgibt. So besteht infolgedessen ein Einfluss auf Social-Scoring-Modelle.

Bezogen auf VAI Trade muss zunächst hervorgehoben werden, dass sich die folgenden Ausführungen ausschließlich auf Unternehmen beziehen. Zwar besteht bei kleinen Einzelunternehmern eine geringere Trennungsschärfe als bei größeren Konzernen, jedoch liegt es in der Verantwortung des Unternehmers, jene innerhalb eines vertretbaren Rahmens zu halten.

Da sich die Prozesse eines Social-Scoring-Ansatzes von den klassischen Bonitätsbewertungen unterscheiden, kommt es in der selektiven Einzelbetrachtung zu unterschiedlichen Ergebnissen. So können Nachteile weniger vorkommen. Diese Nachteile sind jeweils subjektiv zu sehen. Ziel einer Bonitätsbewertung ist nicht nur der Schutz vor Zahlungsausfällen, sondern auch der Schutz vor Überschuldung der Finanzierungsnehmer. Zusätzlich hätte eine fehlende Bonitätsbewertung gesamtwirtschaftliche Implikationen. Ohne ein angemessenes Modell müssten Finanzdienstleister Pauschalen zur Finanzierung von Ausfällen erheben. Dies würde dazu führen, dass Unternehmen mit guten Bonitäten mehr zahlen müssten und gesamtwirtschaftlich benachteiligt wären (vgl. Eschholz und Djabbarpour 2016, S. 78 f.). Hintergrund hier ist die kapitalistische Natur, dass Unternehmen mit guten Bonitäten expandieren sollen und schwächere Bonitäten sich zunächst restrukturieren und vorerst keine Mittel zur Verfügung bekommen.

Ein weiterer Kundennutzen von neuen Scoring-Modellen ist die Standardisierung und Objektivierung der Kreditentscheidungen. Die Entscheidungen können nur noch erschwert manipuliert werden, sodass der Einfluss weniger Personen schwindet und Diskriminierung reduziert werden kann. An dieser Stelle sei gesagt, dass standardisierende Modelle und AI-Algorithmen ebenfalls mit diskriminierenden Tendenzen arbeiten können. Dies kommt durch den sogenannten Selection Bias, welcher eine zugrunde liegende Voreingenommenheit der Trainingsgrundlage für das Modell beschreibt. So kann es unter anderem durch eine fehlerhafte Auswahl von Trainingsdaten, einer Manipulation des Entwicklers oder durch falsche Annahmen in den Kausalitäten zu rassistischen, frauenfeindlichen oder anderweitig diskriminierenden Tendenzen kommen.

Wie bereits beschrieben, stellt auch die soziale Akzeptanz einer immer digitaler werdenden Bonitätsanalyse ein großes Hindernis für eine breitflächige Implementation da. So ergeben sich beispielsweise für VAI Trade die Herausforderungen einer digitalen Antragsstrecke, weil ein Teil der Unternehmen weiterhin keine Verbindung mit dem Bankkonto erlauben will. So werden anstelle dessen PDF-Auszüge an die Unternehmung gereicht, wodurch zwar die identischen Informationen geteilt werden, der Unternehmer sich jedoch wohler mit einer vertrauten Methode des Datenaustausches fühlt. Eine solche Skepsis zu neuen Bewertungsmodellen untersuchte auch eine zum Thema Social Scoring durchgeführte Studie von PwC aus dem Jahre 2018. So haben lediglich 31 Prozent der befragten Endverbraucher der Studie jemals von Social Scoring gehört (vgl. Kleinschmidt und Hufenstuhl 2018, S. 6). Dennoch würden 71 Prozent der Verbraucher eine Fehlbewertung fürchten (vgl. Kleinschmidt und Hufenstuhl 2018, S. 8). Ebenfalls ergab die Studie, dass die Befragten die Schufa gegenüber den Social-Scoring-Algorithmen präferieren würden (vgl. Kleinschmidt und Hufenstuhl 2018, S. 9). Wie bereits in diversen Artikeln genannt, kann die Trennschärfe der durch die Schufa durchgeführten Methoden und der eines klassischen Social Scorings als schwindend gering bezeichnet werden. Der Autor vermutet hier, dass die Befragten zu dem bekannten System tendieren und das Neue kritisch hinterfragen. Zusätzlich wurde ein starker Fokus auf die Social-Media-Daten gelegt, wodurch die generelle Aussage zu einem Social Scoring in Frage gestellt werden kann. Dennoch kann die Skepsis zu einer Ausweitung der einbezogenen Daten zur Erstellung einer Bonitätsanalyse als Tendenz definiert werden.

Eine mögliche Lösung des Skepsis-Problems kann die Schaffung von Transparenz sein. Die meisten implementierten Modelle zur Bonitätsbewertung weisen eine Grundintransparenz auf. Dem Nutzer wird in den häufigsten Fällen nur das Endergebnis zur Verfügung gestellt. Zwar kann aus eigener Erfahrung des Autors gesagt werden, dass bereits bei einer klassischen Kapitaldienst- und Eigenkapitalberechnung der Banken Verständnisprobleme auf Kundenseite existieren. Dem wird bereits im Privatkundensegment durch die gesetzlichen Auflagen nach der Wohnimmobilienkreditrichtlinie (kurz: WKR) entgegengewirkt, stellt auf Unternehmensseite jedoch weiterhin ein Problem dar. Es ist damit zu rechnen, dass die Informationslücke weiter wachsen wird, je mehr Daten zur Bonitätsanalyse verarbeitet werden. Mit dieser Problematik ist auch VAI Trade konfrontiert. Aus diesem Grund wird ein transparenter Prozess angestrebt, ohne den Kunden einen zu großen Einblick in die Bonitätsanalyse zu geben, weil dies zu Betrugsversuchen führen könnte.

Ein Vorteil von Bonitätsbewertungen durch Social Scoring ist, dass Unternehmen und Verbraucher keine ausgiebige Kundenhistorie vorweisen müssen. Historisch wurde oft bei Kreditentscheidungen darauf beharrt, dass der potenzielle Kreditnehmer eine lange Historie von zurückgezahlten Krediten und eine ordentliche Kontoführung vorweisen kann. Somit waren Neukunden schlechter gestellt als Bestandskunden. Durch mehr Einsicht, mehr Informationen und ein besseres, automatisiertes Verständnis der Unternehmung, verliert die Kundenbeziehung an Werthaltigkeit. So kann beispielsweise eine Bank durch die Nutzung von PSD-II-Schnittstellen die gleiche Kontoauswertung vornehmen wie bei einem Bestandskunden und muss keine sechs Monate aktive Kundenbeziehung aufbauen, um ein vereinfachtes Schnell-Rating auf Kontobasis zu erstellen.

Anhand des wachsenden Marktes der FinTechs kann man einen weiteren Trend der neuen datenorientierten Finanzbranche ableiten. Es liegt nahe, dass durch mehr Daten und ein besseres Verständnis der Nutzer eine bessere Zielgruppensegmentierung/-definition möglich ist. Dies hat Auswirkungen auf die angebotenen Produkte und die gezieltere Ansprache.

Literatur

Dohms, H.-R./Schlenk, C. (Stand: 16.11.2019): Wie Kreditech fast 200 Mio. Euro verbrannt hat, unter: https://finanz-szene.de/eigene-artikel-von-finanz-szene-de/wie-kreditech-es-schaffte-fast-200-mio-euro-zu-verbrennen/, Finanz-Szene

Eschbach, R./Eschenbach, K./Langer, C. (2013): Rating und Finanzierung im Mittelstand, 1. Auflage, Springer Gabler, Wiesbaden

Eschholz, S./Djabbarpour, J. (2016): Scoring in der Finanzbranche", in Hoeren, T./Kolany-Raiser: Big Data zwischen Kausalität und Korrelation, 1. Auflage, LIT Verlag, Berlin

Kleinschmidt, P./Hufenstuhl, A. (2018): Ist Deutschland bereit für Social Scoring? Unter: https://www.pwc.de/de/finanzdienstleistungen/studie-ist-deutschland-bereit-fuer-social-scoring.pdf, zuletzt abgerufen am 24.03.2020.

Vassakis, K./Petrakis, E./Kopanakis, I. (2018): Mobile Big Data: A Roadmap from Models to Technologies, 1. Auflage, Springer International Publishing, Cham

Hendrik Schütte, Jahrgang 1995. Nach seinem dualen Studium bei einer mittelständischen Volksbank, übernahm er mit 21 Jahren die Ausweitung des Gewerbekundensektors. Nach einer Etablierung des neuen Geschäftsbereiches digitalisierte er 2017 als Teamleiter der Firmenkundenbank die Kreditantragsstrecke. Ende 2018 wechselte er als Head of Risk zu dem Berliner Start-Up VAI Trade GmbH. Seither ist er verantwortlich für die Konzeption und Entwicklung der internen Prozesse zur Bonitätsbewertung.

Maximilian Klein, Jahrgang 1984, ist gelernter Kaufmann Marketing-Kommunikation und studierter Journalist mit den Schwerpunkten Politik und Wirtschaft. Nachdem er für die deutsche Botschaft in Israel und als Autor und Journalist für den Deutschlandfunk Kultur und den WDR arbeitete, wurde er PR-Manager bei der VAI Trade GmbH.

Sicherheit messbar machen

34

Magnus Kneisel und Helmut Oppitz

Zusammenfassung

Die IT und deren Vernetzung nehmen zu. Damit steigt auch die Notwendigkeit und damit die einhergehende Komplexität, die Systeme sicher zu konzeptionieren und zu betreiben. Weitere Systemgruppen, wie die der operationalen Technologie (OT), beispielsweise Zeiterfassungssysteme, Industrieanlagen oder Steuersysteme, müssen Teil einer ganzheitlichen Betrachtung der IT-Landschaft sein.

Den sicheren Betrieb zu gewährleisten, fängt bei der Planung an und ist über den kompletten Lebenszyklus hinweg sicherzustellen. So werden mit CAD-Modellen der Infrastruktur Angriffssimulationen durchgeführt, um Schwachstellen bereits vor Implementierung zu erkennen, Schwachstellen vorab zu identifizieren und zu minimieren. Im laufenden Betrieb sorgt ein dauerhaftes Risikomanagement dafür, Gefahrenstellen zu priorisieren und das Bedrohungspotenzial richtig einzuschätzen.

Um das Risiko von Systemen korrekt und transparent zu beschreiben, wird ein objektiver Ansatz benötigt, der reproduzierbar und idealerweise auch automatisierbar ist, damit das Risiko im Zeitverlauf dokumentiert werden kann. Ansätze mittels quantitativer Stochastik, die die Widerstandsfähigkeit von Systemen in einem Messwert beschreiben, eignen sich dafür hervorragend. Wird dieser Messwert in das Risikomanagement eingeführt und regelmäßig bestimmt, entsteht eine belastbare Aussage über den Grad der IT-Sicherheit im Unternehmen.

M. Kneisel (✉) · H. Oppitz
securiThon GmbH, Sulzbach, Deutschland
E-Mail: magnus.kneisel@securithon.de; helmut.oppitz@securithon.de

© Springer Fachmedien Wiesbaden GmbH, ein Teil von Springer Nature 2020
O. Everling (Hrsg.), *Social Credit Rating*,
https://doi.org/10.1007/978-3-658-29653-7_34

34.1   Entwicklung der Informationstechnologie (IT)

Die Informationstechnologie (IT) ist ein Komplex, der sich seit den Anfängen der Schalt-
elemente in den 1940er-Jahren ständig weiterentwickelt. War das damalige Ziel noch der
Bau von Rechenanlagen, ist ein flexibel programmierbares IT-System heute Standard in
allen Branchen und Unternehmensbereichen.

Ein Unternehmen ohne IT-System ist heute längst nicht mehr vorstellbar. Die Systeme
passen sich dem jeweiligen Anwendungszweck durch Hardware- und Software-Anpassung
immer weiter an und werden dabei gleichzeitig immer leistungsfähiger. Die heutigen
IT-Systeme überragen die Leistung der Systeme von vor wenigen Jahren dabei um ein
Vielfaches.

Das führt dazu, dass eine Vielzahl der Unternehmensprozesse ohne digitale Schnittstelle nicht mehr umsetzbar ist. Deshalb stellt sich heute nicht mehr die Frage, ob überhaupt ein IT-System zum Einsatz kommt. Durch die Abhängigkeit der Prozesse von den
IT-Systemen rückt eine andere Frage in den Mittelpunkt; nämlich die nach den Schutzzielen eines IT-Systems. Vertraulichkeit, Integrität und Verfügbarkeit sind zentrale Anforderungen an jedes IT-System. Die Priorität der Schutzziele für das jeweilige System wird
dabei häufig durch den Verwendungszweck definiert.

Da sich Unternehmen und ihre Prozesse stetig weiterentwickeln, muss die elektronische
Datenverarbeitung und -speicherung auf fachlicher Seite mindestens Schritt halten. Sie erhält also ständig neue oder angepasste Anforderungen, welche es zu erfüllen gilt. Gleichzeitig erweitern sich die Möglichkeiten, die (nur) mit einem IT-System umsetzbar sind.

Die Schwierigkeit besteht irgendwann allein schon darin, den „Überblick zu behalten".
Allerdings ist es damit nicht getan. Durch die Abhängigkeit der IT-Systeme im Unternehmen gilt es über den Überblick hinaus, auch die Sicherheit (Widerstandsfähigkeit/Resili-
enz) der Systeme gegen jedwede Form des Missbrauchs oder der Korrumpierung sicherzustellen. Dabei wird die Sicherheit von IT-Systemen oftmals in eine Hardware-Sicherheit
und eine Informationssicherheit unterteilt. Die Hardware-Sicherheit beschreibt alle Maßnahmen rund um das Thema physikalischer Schutz, beispielsweise die Unterbringung in
Räumlichkeiten mit Zutrittskontrolle und Klimatisierung. Die Informationssicherheit beschäftigt sich mit allen Faktoren, die die Schutzziele des Systems (losgelöst von der Hardware) sicherstellen.

Ein sinnvoller erster Schritt für die Absicherung in Punkto Informationssicherheit ist
die Vereinfachung. Die Komplexität nimmt nicht nur bei den Systemen selbst, sondern
gerade bei der Vernetzung der Systeme zu. Durch verschiedene Anschaffungen pro Stand-
ort, verschiedenste und nicht interoperable Fachanwendungen und Systeme, wird das
Streben nach einem hohen Maß an Informationssicherheit nicht nur inhaltlich, sondern
auch organisatorisch zu einer gewaltigen Herausforderung. Um das Sicherheitsniveau
sinnvoll anzuheben, ist es deshalb ratsam, die Interoperabilität wiederherzustellen, dop-
pelte Systeme abzuschaffen und bei einer „geglätteten" Infrastruktur die Sicherheit direkt
mit einfließen zu lassen.

34.2 Schützenswerte Aspekte der IT

Um den Schutz von IT-Systemen im Unternehmen sicherzustellen, wurden in der Vergangenheit eigene Rollen geschaffen, die Sorge für den Schutz der Systeme tragen. Außerdem wurden Rollen und Verantwortlichkeiten getrennt. So ist es gängige Praxis, dass der IT-Betrieb, die Konzeptionierung von Systemen und die Aufrechterhaltung der Informationssicherheit von verschiedenen Verantwortlichkeiten sichergestellt werden, die alle eng zusammenarbeiten. Abhängig von der Größe der IT-Landschaft werden diese Verantwortlichkeiten (mindestens teilweise) von kompletten Teams übernommen.

Um die Informationssicherheit praktisch zu gewährleisten, ist die permanente Überwachung der Sicherheitsmechanismen und der aktuellen Bedrohungslage notwendig. Bei Bedarf, werden Maßnahmen umgesetzt, die auf eine geänderte Bedrohungslage reagieren. Dabei werden Informationen über die aktuelle Lage sowohl innerhalb als auch außerhalb des Unternehmens gesammelt. Daraus entsteht eine Flut an Informationen, die für sich allein ggf. ein zu lösendes Problem darstellt.

Sollen Sicherheitsinformationen effektiv den betroffenen Systemen zugeordnet werden, um notwendig gewordene Maßnahmen umzusetzen, setzt das eine Priorisierung innerhalb der IT-Landschaft voraus. Diese Priorisierung kann sich beispielsweise an den Unternehmensprozessen entlang entwickeln, sodass die IT-Systeme, welche für Kern- bzw. Wertschöpfungsprozesse benötigt werden, eine höhere Priorisierung erhalten, als Systeme, die beispielsweise lediglich für die interne Informationsverteilung genutzt werden.

Doch nicht nur entlang der Prozesskette lässt sich eine Priorisierung ableiten. Eine weitere Maßnahme, um die Infrastruktursicherheit zu gewährleisten, ist die Erstellung eines Informationssicherheitskonzepts. In diesem werden beispielsweise auch über die Systeme hinaus die Fragen nach den entstehenden Daten beantwortet. Abhängig von der Art und der Kritikalität der anfallenden Daten ergeben sich auch darüber eine Priorisierung bzw. neue Anforderungen an die Informationssicherheit.

Ein funktionierendes Informationssicherheitskonzept erfordert allerdings eine dauerhafte Dokumentation und Aktualisierung von IT-Systemen, also einen reibungslosen Lebenszyklus des IT-Systems von der Konzeptionierung, über den Betrieb bis hin zur Abschaffung des Systems. Dieser dauerhafte Prozess ist ressourcenintensiv, jedoch unerlässlich.

Ein gängiger Ansatz, um die Aufwände abzufedern, ist der Versuch, diese mit anderen Aufgaben zu bündeln. Hilfreich kann auch die Verknüpfung bzw. Nutzung anderer Systeminformationen sein. So liefert das Patch-Management beispielsweise wertvolle Informationen über die Systeme, die sich in ein Asset-Management überführen lassen.

In der Konzeptionierung des Sicherheitskonzepts kommt ein weiterer Aspekt der IT zum Tragen, nämlich das Systemrisiko. Wie hoch ist die Gefahr, dass die Schutzziele eines bestimmten Systems nicht eingehalten werden können? Wie hoch ist die Gefahr einer Korrumpierung des Systems durch einen externen Angreifer. Erklärtes Ziel der

Konzeptionierung ist es, die IT mit allen Daten zu schützen. Naheliegend ist es deshalb, das aktuelle Sicherheitsniveau zu bestimmen und aus dieser Information Rückschlüsse für das Konzept abzuleiten. Daten über die Gefährdung eines IT-Systems liefern dabei wertvolle Informationen zur langfristigen IT-Planung und sollten deshalb direkt im Sicherheitskonzept mit aufgegriffen werden.

34.3 Sicherheitsfragen in der IT

34.3.1 Allgemein

Es steht also fest: In jeder IT-Infrastruktur stellt sich die Frage nach der Sicherheit. Allein die Beantwortung der Frage nach den jeweiligen Schutzzielen in der IT ist allerdings nicht ausreichend und die Definition eines Informations-Sicherheitskonzepts nur ein erster sinnvoller Schritt, der allerdings nicht auf Dokumentenebene stehen bleiben darf. Es ist also notwendig, um das Sicherheitskonzept herum Prozesse zu etablieren, die die Informationssicherheit von Systemen langfristig sicherstellen.

Einer der wichtigsten Aspekte ist es, das Risiko kalkulierbar zu halten. Allgemein gilt Risiko als das Ergebnis von Schaden und Eintrittswahrscheinlichkeit. Der Schaden lässt sich hierbei häufig über einen finanziellen Schaden beschreiben. Bei der Beantwortung der Eintrittswahrscheinlichkeit tun sich jedoch viele schwer. Üblicherweise wird auf Klassifizierungswerte wie „niedrig", „mittel" und „hoch" zurückgegriffen. Nicht selten mit dem Ergebnis, dass bei der Frage nach der Eintrittswahrscheinlichkeit die goldene Mitte gewählt wird. Das ist nicht wirklich zielführend für eine anschließende Systempriorisierung.

34.3.2 Möglicher Bewertungsansatz für IT-Risiko

Um das IT-Risiko abzuleiten, gibt es verschiedenste Ansätze. Im Zuge einer ISMS-Implementierung (Informations-Sicherheits-Management-System) können unter anderem die IT-Assets und Schadensklassen für mögliche Risiken beschrieben werden. Ein anderer Ansatz ist es, das Risiko auf Basis des Patch-Managements abzuleiten, in dem geprüft wird, welche Patches grundsätzlich verfügbar, aber nicht in der Infrastruktur verteilt sind.

Ideal ist ein Ansatz dann, wenn er die notwendige Detailtiefe für die Betrachtung bietet und zugleich eine strukturierte Herangehensweise ermöglicht.

Gerade im Anfangsstadium der Risikobetrachtung ist es hilfreich, vom Groben ins Feine zu arbeiten, damit zu beginnen, Überblick zu gewinnen, Fakten zu schaffen und diese über die Dauer der Bewertung immer weiter zu konkretisieren. Das schafft Flexibilität in der Umsetzung und behindert wichtige Arbeitsabläufe nicht.

Eine effektive Methodik zur Umsetzung der obigen Punkte ist eine Resilienzanalyse, also eine Feststellung des Sicherheitsniveaus pro IT-System, auf Basis einer Infrastrukturmodellierung. Dabei werden die tatsächlichen IT-Systeme in ein Logikmodell überführt und dort auf ihre Sicherheit überprüft. Vorteil dieser Methodik ist, dass die tatsächlichen Systeme keinerlei Gefahren durch die Resilienzfeststellung im laufenden Betrieb ausgeliefert sind. Ein weiterer Aspekt ist, dass durch die logische Abstraktion das infrastrukturelle Konstrukt erneut hinterfragt wird. Dieser Ansatz erlaubt es, sowohl vom Groben ins Feine zu arbeiten, als auch einzelne Teilaspekte der Gesamtinfrastruktur zu untersuchen. Er ermöglicht eine Parallelisierung von mehreren Infrastrukturmodellen, die in einem späteren Schritt zusammengefügt werden. Die Detaillierung eines Modells ist immer dort sinnvoll, wo das erste/grobe Modell keine ausreichenden Antworten auf die Sicherheitsfrage liefert.

In der Modellierung wird die jeweilige Infrastrukturkomponente, auch Asset genannt, mit seinen logischen Funktionen beschrieben. Eine Funktion ist beispielsweise die IP-Adresse im Netzwerk, abgebildet mit dem Host-Symbol, ein anderes das Betriebssystem und wieder ein anderes die Netzwerkaktivitäten, die mit diesem System einhergehen; also Client- und Service-Verbindungen, die dazu genutzt werden, mit anderen Assets zu kommunizieren.

Abb. 34.1 zeigt einen Linux Web Server in einem lokalen Netzwerk, verbunden mit dem Internet. Diese Modellierung lässt eine Betrachtung pro Funktion zu. So wird der Apache Web Server beispielsweise noch einmal logisch in seine Funktionen unterteilt. Im Modell werden alle Aspekte des Systems beleuchtet, die für eine Resilienzbestimmung notwendig sind (vgl. Abb. 34.2).

Wirft man einen detaillierteren Blick auf den Web Server, so ist dieser wiederum mit mehreren logischen Informationen versehen. Der Dienst verfügt beispielsweise über ein Software-Paket. Darüber hinaus nutzt der Service für die Netzwerk-Kommunikation ein Authentifizierungsverfahren, dargestellt durch die AccessControl. Die Daten des Web Servers, hier mit Web content beschrieben, werden als Datastore erfasst. Über die Modellierung kann bei Bedarf deshalb ein sehr detailliertes Bild der Infrastruktur und ihrem Aufbau entstehen.

Im Modellierungsprozess werden nicht nur die einzelnen Funktionen der Infrastruktur den jeweiligen Modellierungsobjekten zugeordnet. Ziel der Modellierung ist die Bestimmung des Resilienzgrads des jeweiligen Assets, beispielsweise die des Linux Web Servers. Deshalb ist die Betrachtung von zwei Einflussgrößen auf die Resilienz zwingend notwendig, um den Resilienz-Grad zu bewerten. Nämlich zum einen, welche Angriffe bzw. welche Angriffsgruppen auf eine bestimmte Funktion der Infrastruktur möglich sind, zum anderen, welche Maßnahmen gegen eine solche Angriffsgruppe stehen. Am Beispiel einer Modellierung mit Hilfe der Software securiCAD©, des schwedischen Herstellers foreseeti AB, liefert die Software die logischen Angriffsgruppen für die Komponenten bereits mit. So müssen nur noch die Verteidigungsinformationen zum jeweiligen Asset erfasst werden.

In Abb. 34.3 sind die Angriffsgruppen (Attacks) und die Verteidigungsfunktionen (Defenses) des Host-Elements Linux Web Server zu sehen. Die Verteidigungsfunktionen auf der linken Seite fragen die relevanten Gegenmaßnahmen für die Angriffsgruppen auf der rechten Seite ab. So definiert das Attribut AntiMalware als Gegenmaßnahme beispielsweise die Güte der eingesetzten AntiMalware.

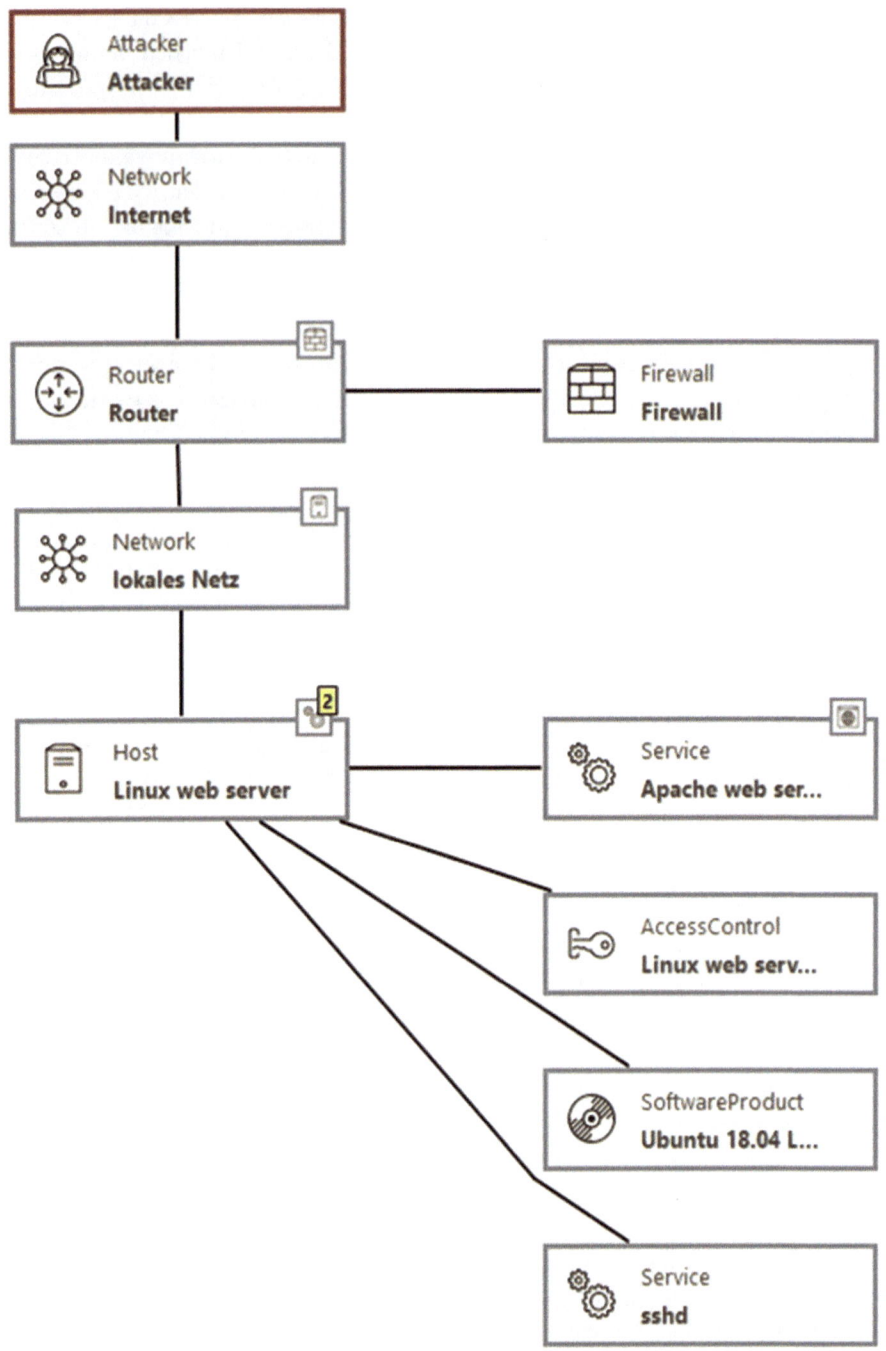

Abb. 34.1 Linux Web Server in einem lokalen Netzwerk, verbunden mit dem Internet. (Quelle: eigene Darstellung)

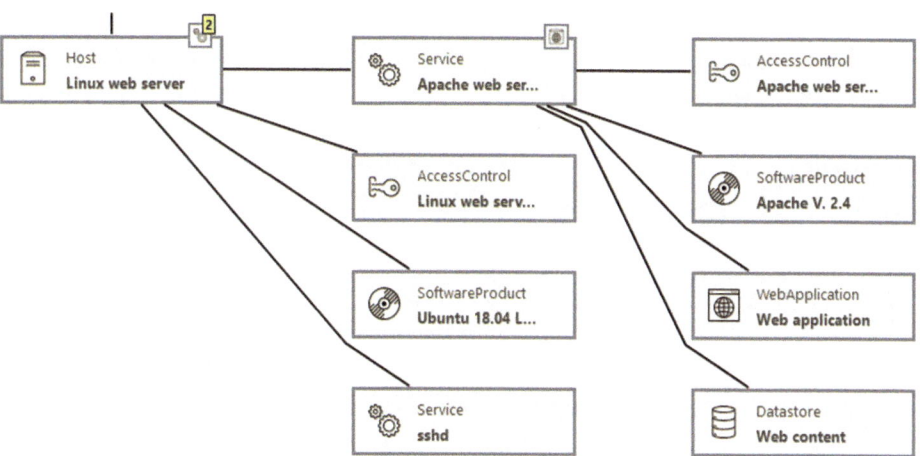

Abb. 34.2 Linux Server mit Apache Service Details. (Quelle: eigene Darstellung)

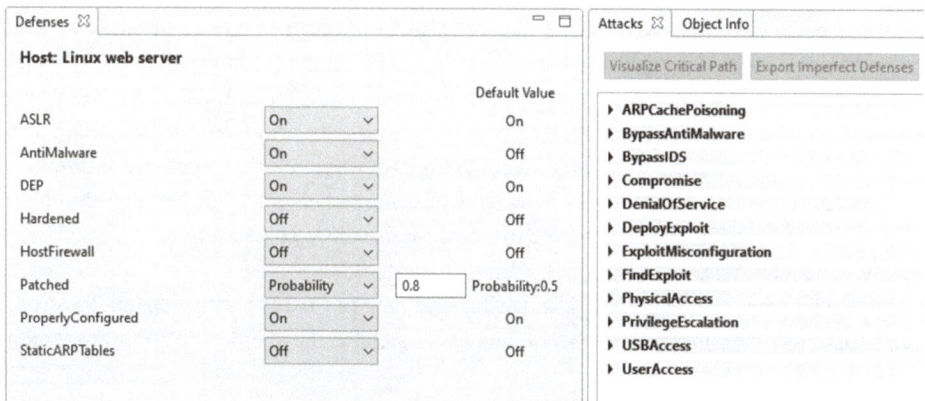

Abb. 34.3 Angriffsgruppen und Verteidigungsmaßnahmen des Webservers. (Quelle: eigene Darstellung)

Für die logische Modellierung ist es wichtig, dass Annahmen bei der Modellierung getroffen werden und diese mindestens für das komplette Modell identisch sind. Somit wird gewährleistet, dass der Parameter AntiMalware 100 Prozent (= „on") im gesamten Modell die gleiche Bedeutung hat. Das sorgt dafür, dass die Modellierung auch dann vergleichbar bleibt, wenn mehrere Personen parallel arbeiten.

Um die Infrastrukturresilienz zu bestimmen, wird eine Angriffssimulation im erstellten CAD-Modell der Infrastruktur durchgeführt. Diese Angriffssimulation schreitet vom definierten Angreifer (beispielsweise aus dem Internet kommend) hin zu den zu betrachtenden Systemen (hier des Web-Servers) alle möglichen Wege ab. Die möglichen Wege werden über die Netzwerkkommunikation sowie den Schutzgrad der jeweiligen Systemfunktion bestimmt. So fällt es dem Angreifer beispielsweise leichter, in eine Infrastruktur mit schlecht konfigurierter bzw. nicht aktueller Firewall einzudringen. Diese Parameter werden über die Simulation überprüft. Das Ergebnis soll eine möglichst ganzheitliche Sicht auf die technischen Gegebenheiten der Infrastruktur sein. Eine gute Simulation muss dabei mindestens Aufschluss über den Resilienzgrad des zu betrachtenden IT-Assets und der strukturellen Schwachstellen der Infrastruktur geben.

Durch die Simulation des Angreifers im logischen Abbild der Infrastruktur werden verschiedene Angriffswege bestimmt. Diese Wege lassen sich dazu nutzen, Maßnahmen für die Infrastruktur zu priorisieren, sodass zuerst Maßnahmen umgesetzt werden, die unmittelbar Einfluss auf einen identifizierten Angriffspfad haben (vgl. Abb. 34.4).

Die Simulation gibt darüber hinaus auch Aufschluss über das Resilienzniveau der jeweiligen Infrastrukturkomponente. Im Falle von securiCAD nutzt die Software einen selbst definierten Messwert namens Time to Compromise (TTC) welcher die Korrumpierungswahrscheinlichkeit (Max Success Rate) und die statistischen Aufwände eines Angreifers (Days) angibt, die benötigt werden, um das jeweilige IT-Asset zu korrumpieren (vgl. Abb. 34.5).

Auch wenn der Messwert hier ein statistischer Wert ist, also keinen Aufschluss darüber gibt, wie lange ein Angreifer in genau diesem Moment benötigen würde, um das Asset zu korrumpieren, so gibt er doch wertvolle Hinweise auf die Infrastrukturresilienz. Vor allem aber lässt der statistische Wert einen Vergleich zu: den Vergleich mit anderen IT-Assets innerhalb der Infrastruktur, um beispielsweise darüber eine priorisierte Abarbeitung von anstehenden Maßnahmen abzuleiten. Aber auch den Vergleich über mehrere Simulationen. Wird eine solche Simulation beispielsweise monatlich im Unternehmen durchgeführt, lässt sich am Resilienzwert Time to Compromise die Verbesserung bzw. Verschlechterung des Sicherheitsniveaus für den Web-Server ablesen.

Darüber lassen sich jedoch nicht nur Korrelationen mit bereits umgesetzten Änderungen in der Netzwerkinfrastruktur herstellen. Da es sich bei diesem Betrachtungsansatz um

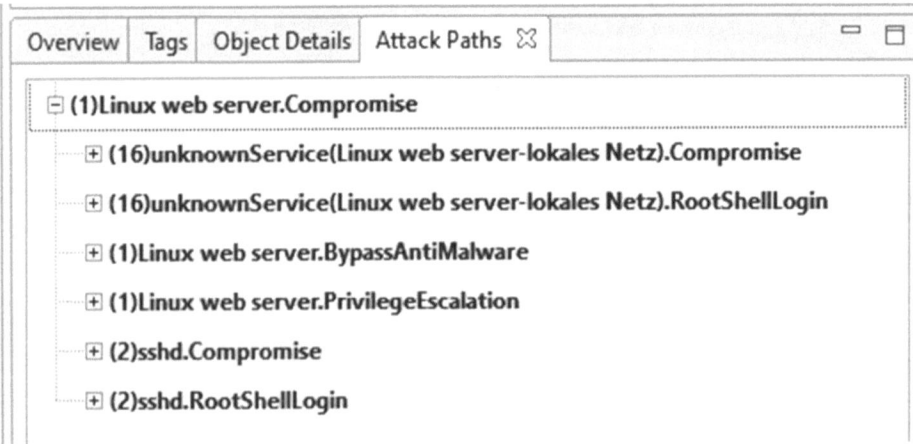

Abb. 34.4 Bestimmung der Angriffswege. (Quelle: eigene Darstellung)

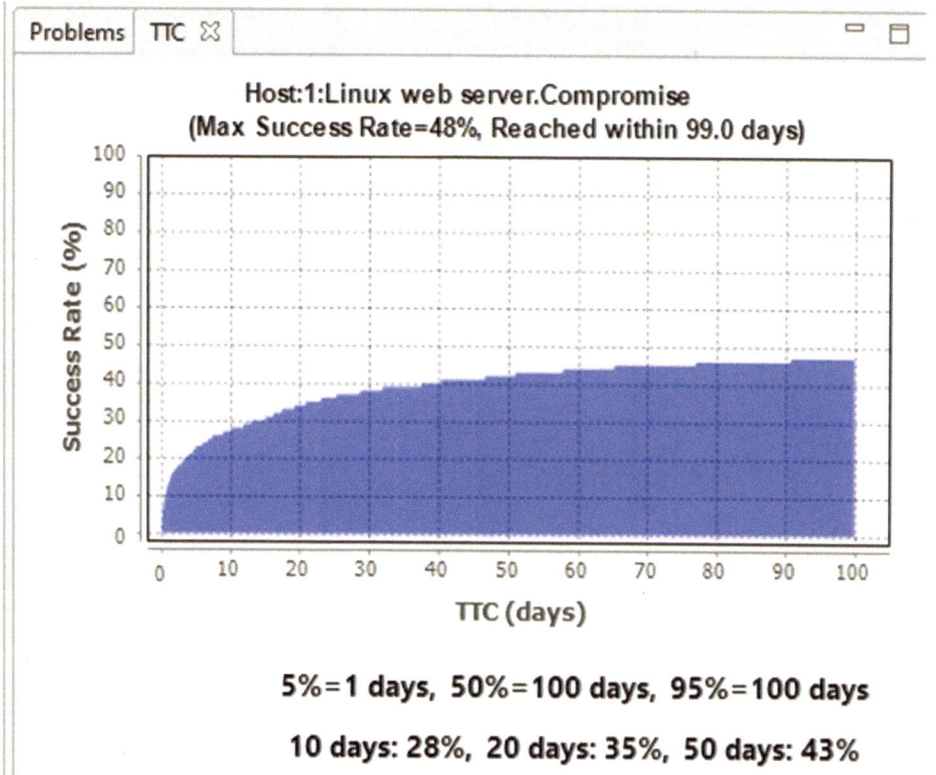

Abb. 34.5 Korrumpierungswahrscheinlichkeit in Abhängigkeit von den Aufwänden des Angreifers. (Quelle: eigene Darstellung)

ein Modell handelt, lassen sich Änderungen vor einer Umsetzung in der tatsächlichen Infrastruktur untersuchen und dank des gemeinsamen Nenners Time to Compromise sogar miteinander vergleichen. Komplexe Fragestellungen, wie beispielsweise die Verbesserung der Einführung einer neuen Firewall im Vergleich zur Netzwerksegmentierung oder Awareness-Kampagnen, können damit auf einen quantifizierbaren Wert abgebildet werden.

34.3.3 Abseits der Office-IT

Oftmals besteht die IT nicht mehr nur aus der sichtbaren Office-IT, sondern wurde über Cloud-Dienstleistungen erweitert oder es gibt eine Anbindung an Produktions- oder andere Techniknetzwerke.

Erste Hürde in Cloud-Infrastrukturen ist bereits die Transparenz des Aufbaus. Wo liegen meine Daten? Wie werden diese geschützt? Welche Systeme sind in die Löschung von Daten involviert?

Meist fällt schon der erste Schritt, auf den Cloud-Dienstleister zuzugehen und Auskünfte einzuholen, schwer. Der Aufbau der Cloud-Services auf Prozess- und Technik-
ebene wird in mehreren Abstimmungsrunden iterativ geklärt. Immer wieder kommen neue
Informationen dazu, mit dem Risiko, dass die Informationen nie auf einem gemeinsamen
Nenner basieren.

Dasselbe gilt für Unternehmensbereiche, die auf OT-Basis (Operational Technology)
betrieben werden, so etwa Industrieanlagen. Auch hier gibt es Dokumentationen zu den
Netzen, zu den Systemen und ggf. sogar für die Zusammenhänge.

Oftmals sind die Dokumentationen von klassischer Office-IT, Cloud-Infrastruktur und
OT-Systemen in einem so unterschiedlichen Detailniveau, dass eine übergreifende Darstellung und Aufbereitung unmöglich wird.

In solchen Szenarien lässt sich die Modellierung dazu nutzen, unterschiedliche Informationen und Detailtiefen miteinander zu verbinden. Eignen sich diese Modelle noch
nicht für eine aussagekräftige Angriffssimulation, können sie in jedem Fall dazu genutzt
werden, Wissen über die unterschiedlichen Infrastrukturkomponenten auf einen gemeinsamen Nenner zu bringen.

34.3.4 Bei besonders schützenswerten Gütern zusätzlich

Die Zahl der Angriffe auf kritische Infrastrukturen, also systemrelevante Infrastrukturen
der Länder, wie Strom, Wasser und Abwasser, steigt stetig. Im Jahr 2019 hat das BSI An-
griffen auf Energienetze eine „neue Qualität" attestiert.

Was unterscheidet eine kritische Infrastruktur (abgesehen von der Systemrelevanz) von
anderen Infrastrukturen? Der Systemmix! In den wenigstens Fällen bedeutet KRITIS eine
reine Büro-IT-Struktur. Stattdessen ist es häufig eine Verknüpfung von IT- und OT-
Systemen, ggf. erweitert um mechanische Komponenten. Bei diesen Komplexitäten ganzheitliche Lösungsansätze zu finden, bedarf deshalb einer umfassenderen Betrachtung.

Einer der wichtigsten Punkte in diesem Kontext ist, Zusammenhänge zu verstehen und
Abhängigkeiten zu identifizieren. Bedrohungen sind immer nur so relevant, wie Schwach-
stelle und Ausnutzungsmöglichkeit zusammentreffen. Um die Sicherheit langfristig zu
erhöhen, ist das Aufspüren der Schwachstellen deshalb nur ein Teil der Bestrebung. Nur
wer die Zusammenhänge der Systeme untereinander identifiziert, ist in der Lage, ein in-
einandergreifendes Sicherheitskonzept aufrecht zu erhalten.

Abhängig vom Alter der Maschinen verfügen diese bereits über eingebaute (neueste)
Sicherheitstechnik. Allerdings sind die Wartungs- und Austausch-/Abschreibungszyklen
in derartigen Infrastrukturen langfristiger zu sehen. Das bedeutet, dass neue Sicherheitstechnik häufig noch nicht zur Ausstattung der Maschine gehört. Die Wahl der Verteidigung
ist deshalb häufig eine passive, sodass die Verteidigung nicht die Maschine selbst übernimmt, sondern die darum gebaute Sicherheitsinfrastruktur. Um die Sicherheitsinfrastruktur so aufzubauen, dass sie die Arbeit der Maschine nicht behindert, die Sicherheit dennoch gewährleistet, ist die Identifikation der Kommunikation eine wichtige Aufgabe. Ist

die Maschinen- bzw. Systemkommunikation identifiziert, lassen sich passive Sicherheits-maßnahmen (zum Beispiel Organisation in Zellen bzw. abgetrennten Netzen) um die bestehende Architektur herum integrieren.

Ist ein sinnvolles Sicherheitskonzept erstellt, gilt es, dieses auf die Probe zu stellen. Eine Herausforderung in solchen Infrastrukturen ist, dass die gängigen aktiven Testmethoden wie Penetrationstests ggf. nicht oder nur unzureichend umgesetzt werden können, sofern nicht der richtige Ansprechpartner mit technischem Know-how in „beiden Welten" gefunden wird. Abhilfe kann deshalb auch hier eine Simulation schaffen, da alle Einflussgrößen der unterschiedlichen Infrastrukturen im Modell erfasst sind.

Um einen möglichst effizienten Weg hin zur Absicherung der Infrastruktur zu wählen, wird häufig die Entscheidung gegen ein großes Infrastrukturprojekt, zugunsten verschiedener kleinerer und besser kontrollierbarer Projekte gewählt. Identifizieren Sie System- und Prozessgruppen, die wenig Interaktion über die Gruppierung hinaus benötigen, und sichern Sie die Infrastruktur Stück für Stück weiter ab.

Die Checkliste in Kürze:

* Bedrohungen/Schwachstellen der Systeme identifizieren
* Zusammenhänge der Bedrohungen und der Systemmixe ableiten
* Wahrscheinlichkeiten für die jeweiligen Bedrohungen prüfen und den Fokus auf die gefährlichsten Stellen legen
* Sicherheitsarchitektur der IT- und OT-Systeme selbst überprüfen und geeignete Mittel zum passiven Schutz herbeiführen
* Zusammenhänge im Sicherheitskonzept berücksichtigen
* Überprüfung des Sicherheitskonzepts durch Simulation minimalinvasiv überprüfen
* Vor dem Start: sinnvolle Gruppierungen in Systemen und Prozessen für kleinere, besser kontrollierbarere Projekte nutzen.

34.4   Ableitung der Erfordernis „Güte der Sicherheit bestimmen + Festlegung eines Mindestsicherheitsniveaus"

Unabhängig von der gewählten Methodik zur Messung und Verbesserung der Informationssicherheit im Unternehmen gilt es, stets die unterschiedlichen Perspektiven zu berücksichtigen. Mit dem vorherigen Beispiel des Linux Web Servers lässt sich ein Ansprechpartner auf technischer Ebene sehr gut abholen. Die einzelnen Funktionen sind diskutierbar, die Auswirkungen von Änderungen lassen sich dadurch vom Ansprechpartner nachvollziehen.

Anders sieht dies meistens beim Management aus. Auf Managementebene ist eine technische Herangehensweise an das Thema oftmals nicht zielführend. Wichtiger ist es, das Thema Risiko als Fokusthema zu positionieren. Idealerweise stützt sich die Risikodiskussion um eine bzw. mehrere Kennzahlen, die eine interpretierungsfreie Diskussion zulassen.

Stellen Sie sich für einen Moment vor, es gäbe einen Zwischenfall in Ihrem Unternehmen, der im Home-Office eines Mitarbeiters stattgefunden hat. Dort wurde ein IT-System Ihres Unternehmens mit einer Schadsoftware befallen. Aktuell ist unklar, wie das passieren konnte. Es schwebt allerdings die Vermutung im Raum, dass die gewählte Sicherheitsinfrastruktur im Home-Office nicht bzw. nur ungenügend ist, was zu diesem Vorfall geführt hat. Da in unserer Annahme die Home-Office-Arbeitsplätze mit dem Hauptsitz logisch verbunden sind, auf den Systemen eine VPN-Verbindung Zugriff vom Home-Office auf die zentralen Server bereitstellt, besteht das Risiko der Verbreitung der Schadsoftware im Unternehmensnetzwerk.

Aus diesem Grund soll das Thema Home-Office-Risiko beleuchtet werden. Im Nachgang an den Schadensfall sind alle Stimmen von „das war ein Einzelfall und bedarf keiner Anpassung des Risikomanagements" bis hin zu „Home-Office ist als Risiko zu hoch und nicht tragbar" zu hören. Gerade in einer solchen Situation ist es hilfreich, auf eine Messmethodik zurückzugreifen, um der fachlichen Diskussion Anknüpfungspunkte und idealerweise Lösungsansätze mitzugeben.

Wird das Home-Office-Szenario auf Basis einer logischen Blaupause der Infrastruktur modelliert und die Resilienz der Systeme bestimmt, dann ist der erste Schritt für diese fachliche Diskussion die Definition des Home-Office-Arbeitsplatzes. Welche Anforderungen des Unternehmens müssen gegeben sein, damit das Arbeiten am heimischen Arbeitsplatz genehmigt wurde? Wird ausschließlich auf Firmen-Hardware zurückgegriffen oder besteht ein Infrastrukturmix aus privater und Unternehmens-Hardware, die sich im Zweifel schlechter kontrollieren und verwalten lässt?

Ist ein gemeinsamer Nenner beim Aufbau der Infrastruktur gefunden, so lässt sich der nächste Schritt angehen; nämlich die Bewertung der Infrastruktur. Abb. 34.6 zeigt einen einfachen Home-Office-Aufbau mit einem kleinen lokalen NAS-System als File-Server und einem Arbeitsplatzsystem im lokalen Netzwerk, verbunden über einen Router mit dem Internet.

Für die Resilienzbestimmung wird der potenzielle Angreifer mit dem Internet verbunden. Für die Bestimmung der Systemresilienz sind die beiden IT-Assets Workstation und File-Server relevant.

Für die Managementperspektive brechen wir die Infrastrukturresilienz nun auf Kennzahlen herunter, die sich aus der Angriffssimulation des Home-Office-Modells erzeugen lassen. Die eingeführten Kennzahlen lauten Angriffswahrscheinlichkeit (Probability) und Time to Compromise (in Tagen). Um einen visuellen Vergleich zu erhalten besteht zusätzlich die Möglichkeit, die beiden Werte miteinander zu multiplizieren und entlang einer farblichen Skala in Risikoklassen wie niedrig, mittel und hoch einzustufen.

Für eine Bewertung der Sicherheit ist es außerdem zielführend, Entscheidungsoptionen mitzuliefern, die als Diskussionsgrundlage für einen Lösungsansatz genutzt werden können.

Für das Home-Office wurde deshalb die Infrastruktur-Resilienz zum aktuellen Zeitpunkt bestimmt. Zusätzlich wurde nach infrastrukturellen Schwachstellen gesucht und

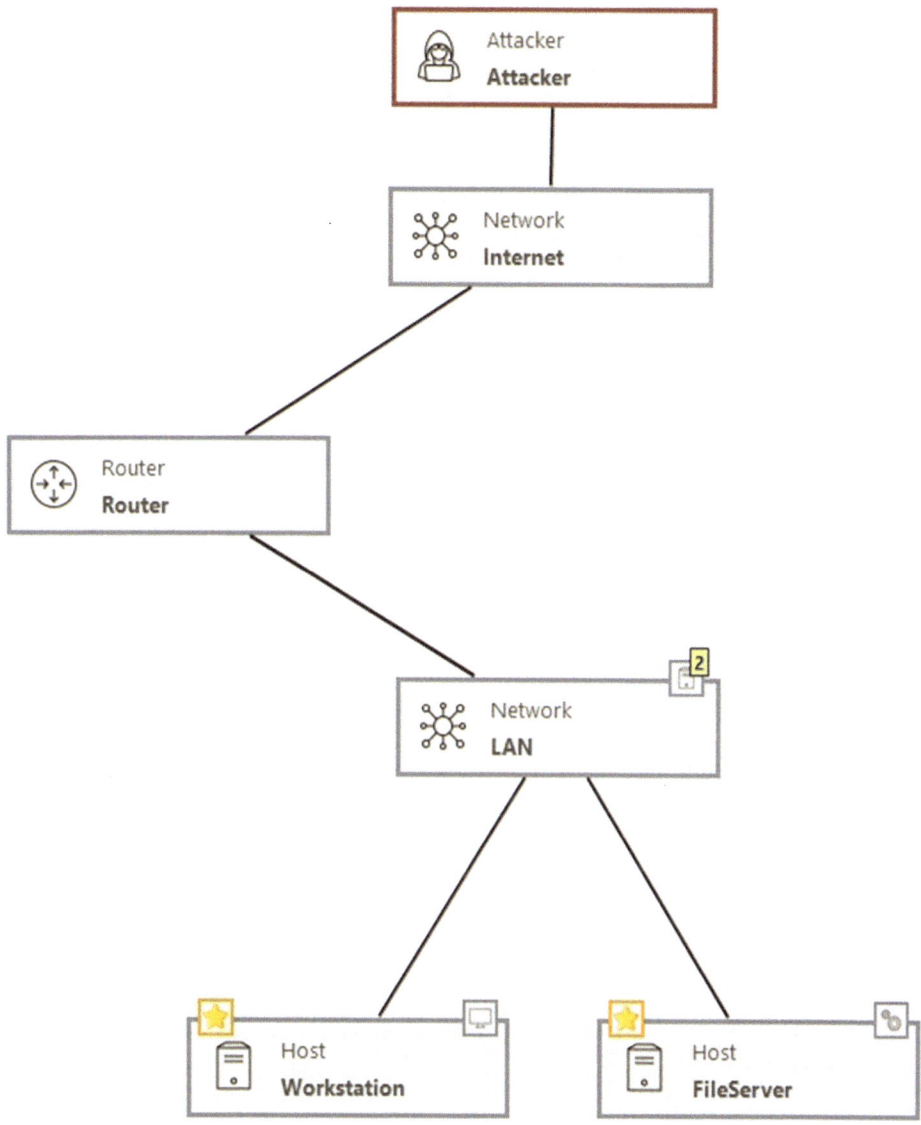

Abb. 34.6 Einfacher Home-Office-Aufbau. (Quelle: eigene Darstellung)

einige der Funde direkt mit Gegenmaßnahmen modelliert. Nachfolgend ist die Risiko-übersicht von drei Home-Office-Szenarien zu sehen.

Das linke Szenario (vgl. Abb. 34.7) zeigt die Ausgangslage. Das Home-Office ist schlecht geschützt. Beide Systeme sind innerhalb von wenigen Tagen durch einen poten-ziellen Angreifer erfolgreich angreifbar. Die Wahrscheinlichkeit eines erfolgreichen An-griffs liegt in beiden Fällen über 70 Prozent.

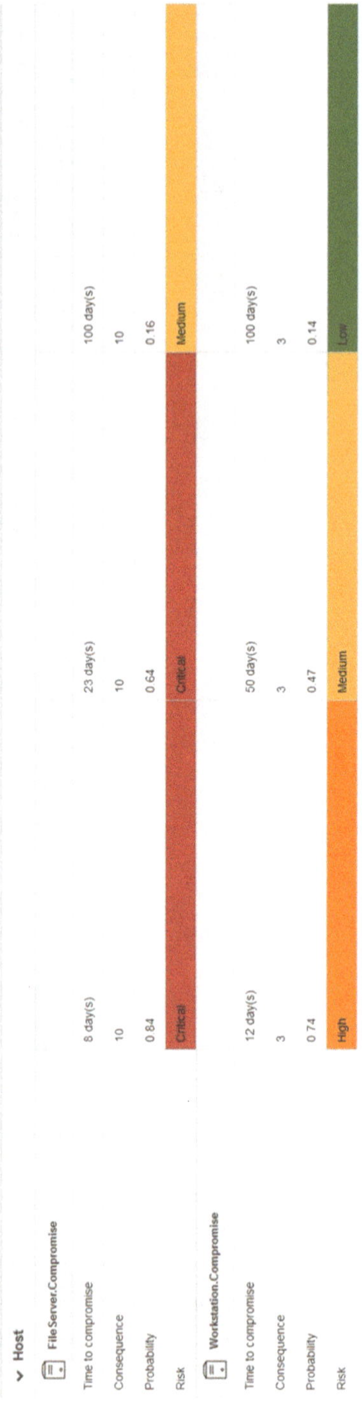

Abb. 34.7 Szenarioanalyse. (Quelle: eigene Darstellung)

Die Simulation der Ausgangslage hat einige grundsätzliche Schwachstellen in der Infrastruktur aufgedeckt. So fällt beispielsweise auf, dass die Home-Office-Systeme über keinen einheitlichen und zentral verwalteten Virenschutz verfügen. Einzelne NAS-Systeme hatten außerdem keine Passwort-Policy. Diese Schwachstellen wurden in einem Alternativszenario (hier in der Mitte von Abb. 34.7) verändert. Die Umsetzung der Maßnahme hat zur Folge, dass die Angriffswahrscheinlichkeit im Mittel um 20 Prozent verringert wird. Außerdem steigt der (statistische) Aufwand, den ein Angreifer benötigt, ebenfalls drastisch um im Mittel 25 Tage. Trotz dieser signifikanten Verbesserung ist das Home-Office auch in diesem Szenario noch mit einem kritischen Risiko belegt. Das liegt an der Vielzahl an offenen Schwachstellen.

Das dritte Szenario, rechts in Abb. 34.7, beschreibt das Home-Office mit einer eigenen Hardware-Firewall. Diese Maßnahme ist im Vergleich zur ersten Alternative deutlich budgetintensiver, liefert allerdings auch eine drastische Verringerung des Risikos.

Das grafische Beispiel bzgl. des Risikoverlaufs von verschiedenen Szenarien im Home-Office lässt sich als Diskussionsgrundlage nutzen, um beispielsweise die Anforderungen des Risikomanagements für das Home-Office abzuleiten. So lässt sich aus den verschiedenen Alternativen mit „noch zu hohem Risiko" und „sehr teurer Anschaffung zur Verringerung des Risikos" ein Mittelwert bilden, um das Restrisiko auf einen Schwellwert von 30 Prozent Restrisiko zu minimieren. Mit dieser Anforderung lassen sich alternative Szenarien aufbauen, die aufzeigen, welche infrastrukturellen Maßnahmen budgetverträglich umsetzbar sind.

Über einen quantifizierbaren Ansatz von Informationssicherheit, im Idealfall über allen Bereichen, also der Infrastruktur, den Prozessen und der Gebäudesicherheit, lässt sich so ein gemeinsamer Nenner zwischen Fachabteilung und Managementperspektive schaffen, welcher mittel- und langfristig fortgeschrieben werden kann und den Vergleich über weite Zeiträume zulässt. Ist dieser gemeinsame Nenner etabliert, lassen sich Anpassungen an der Infrastruktur ebenfalls auf diesen gemeinsamen Nenner bringen, sodass gerade Budget- und Strategieentscheidungen auch immer unter dem Aspekt der Informationssicherheit geführt werden.

Magnus Kneisel über 10 Jahre Erfahrung in der Informations-Sicherheit wirft der Autor Magnus Kneisel mit unzähligen Kryptografie- und Informations-Sicherheits-Projekten auf Kunden-Seite in die Waagschale. Von der Beratung bis zur Implementierung in komplexen und heterogenen Infrastrukturen zieht sich der rote Faden „sicherer werden". Und Verbesserungen in Punkto Sicherheit erreicht man umso einfacher, je besser man „messen" kann, wie sicher man ist, um Veränderungen benchmarken zu können. Threat Modeling im securiCAD-Ansatz bietet genau diese Möglichkeit, durch einen Messwert, in dem die Informations-Sicherheit quantitativ erfasst wird. Also Veränderungen messen, um die Informations-Sicherheit auf den richtigen Kurs zu bringen.

Helmut Oppitz mehr als 30 Jahre Erfahrung als Projektentwickler und Projektleiter in Technik-Projekten rund um die Themen Transformation, Digitalisierung, Datenschutz und Informations-Sicherheit schaffen die Basis, neue Herausforderungen mit Ruhe angehen zu können. Wer IT-Strukturen weiterentwickeln will kann das nicht tun, ohne die Informations-Sicherheit und den Datenschutz zu gewährleisten. Bei der stark ansteigenden Komplexität schafft nur Messbarkeit und objektive Bewertung von Sicherheits-Parametern eine perfekte Grundlage zu transparenten und jederzeit nachvollziehbaren Entscheidungen, was Diskussionen nicht nur abkürzt, sondern die Ergebnisse auch reproduzierbar macht. Diesen ganzheitlichen Ansatz lebt der Autor aus Überzeugung und ist immer wieder die treibende Kraft, um unter Beweis zu stellen, dass die Quantifizierung von Key Performance Indexes in Punkto Sicherheit genauso möglich ist, wie es in anderen Bereichen – beispielsweise in Finanz und Controlling – schon lange üblich ist.

Determinants of Consumer Credit Default in Romania: A Comparison of Machine Learning Algorithms

35

Ana Maria Sandica and Monica Dudian

Abstract

In this chapter, we investigate the separation power of several machine learning techniques and compared them with the benchmark logistic regression using real data from 17,520 private individuals of a Romanian commercial bank. In order to capture the financial crisis effect we equally divided the data in two samples prior and posterior the crisis and we compared 13 models in terms of misclassification Type I and II Errors. As the models aim to catch best the patterns in the *"default"* profile of a consumer credit borrower, we split the variables in socio-demographic factors (Social Rating) and financial factors (Financial rating) and conclude that *"default"* profile prior crisis is captured better by the linear models while the patterns of the financial crisis are captured better by the non-linear models. We found that accuracy ratio gives the better results on decision trees and ensembles based on decision trees such as adaptive boosting methods (Financial Rating) and Random Forest (Credit Rating, Social Rating) irrespective of the sample choice. The power of the model to classify the debtors using Social Rating, Financial Rating and the mix of these, the Credit Rating, depends on the trained data used. The Financial Rating's champion model's results are best on posterior crisis data, meaning that financial factors counted the most in detecting the patterns in *"default"* after the financial crisis. The order is not the same for Social Rating, where the best classification is obtained on prior crisis data meaning that classification considering the individual's creditworthiness is more difficult on posterior crisis *"default"* patterns.

A. M. Sandica
Bucharest, Sector 3, Rumänien

M. Dudian (✉)
The Bucharest University of Economic Studies, Bukarest, Rumänien
E-Mail: monica.dudian@economie.ase.ro

© Springer Fachmedien Wiesbaden GmbH, ein Teil von Springer Nature 2020
O. Everling (Hrsg.), *Social Credit Rating*,
https://doi.org/10.1007/978-3-658-29653-7_35

35.1 Introduction

The first credit rating grades were published in 1909 by John M. Moody and John Knowles Fitch founded in 1913, in New York, the Fitch Publishing Company. David Durand (1941) is the pioneer of credit scoring when he in 1941 applied discriminant analysis proposed by Fisher (1936) to classify prospective borrowers. In his paper published by the National Bureau of Economic Research examined about 7200 reports on good and bad installment loans made by 37 firms. After World War II broke out, many finances lacked the experts to perform the credit analysis as many skilled people in the field joined the war. Those companies then asked experienced experts to put down their knowledge in credit assessment in the form of guidance to help the relatively inexperienced make lending decisions. The statisticians that designed the scorecard in the early days hoped to model after the practice of insurance companies who scored applicants based on age and gender to calculate the premium. In the 1950s, attempts had been made to unify the automated credit decision making with statistical techniques to develop models that would help the making of credit decisions. But due to the deficiency of powerful computing tools, those models were substantially limited in sample size and model design. In that period, the first credit scoring consulting firm was founded by the mathematician Earl Isaac and engineer Bill Fair, named Fair Isaac. It was then when credit risk assessment became a key instrument in the decision making of the banking and financial institutions. The long term advantages of building a credible risk assessment are: reducing the cost of credit analysis, ensuring fast decision-making, guaranteeing credit collection and reducing possible risks (Huang et al. 2018).

The importance of credit risk assessment has become essential given the subprime mortgage crisis that began in 2007. Indeed, people realized that one of the main causes of that crisis was that loans were granted to people whose risk profile was too high. That is why, in order to restore trust in the finance system and to prevent this from happening again, banks and other credit companies have tried to develop new models to assess the credit risk of individuals more accurately (Charpignon et al. 2014). Therefore, a range of different statistical and artificial intelligence methods have been developed to support credit decisions. The variables included in the models of banks are more and more complex, but these models remain in the category of regulated systems distinct from the unregulated ones or "arbitrary, unaccountable, unregulated assessments" based on social media or non-transparent data (O'Neal 2016).

35.2 Literature Review

Altman (1968) introduced variables in a multivariate discriminant analysis and obtained a function depending on several financial ratios that has been considered a milestone in this field being a pioneer in creation of a scoring function. Ang, Chua and Bowling (1979) built a non-parametric credit scoring model based on a decision tree technique and reached the conclusion that linear credit scoring models are not the best solution, given that the relationship between some variables is not linear Hand and Henley (1997). The correlation between gender and credit scores in research shows potential drifts between males and females. For instance, females engage in more impulse buying according to Verplanken and Herabadi (2001) and typically earn less than males therefore affecting the ability to repay debt as presented by Ng, Eby, Sorensen and Feldman (2005). The conjunction between age and credit scores indicates that older individuals tend to be more risk averse and have higher economies than younger ones (Livingstone and Lunt (1992). Sarlija et al. (2009) estimated a logistic regression using application and transaction data of 50,000 customer accounts in Croatia over the period of 12 months. One of their conclusions is that females are more likely to have an account in *"default"* compared to male borrowers. In terms of borrower's age, the older it is, the lower the risk. Marjo (2010) modeled a unique dataset of 14,595 observations from one of Finland's largest consumer credit companies with more than 150,000 customers. Using logistic regression, one of his conclusions is that gender is a significant variable showing that female customers have much less difficulty in paying their debts and seem to *"default"* less often than men. Secondly, education was not identified as statically significant and in terms of age, younger clients tend to *"default"* more often. The first research documentation on marital status correlation to credit scores is back in 1993, when Tokunaga (1993) studied the behavioral patterns in consumer credit profiles and found that marital status has the potential to influence the financial problems. Ganopoulou, Giapoutzi, Kosmidou and Moysiadis (2013) using a sample set of applications from a large Greek financial institution estimated a probit model on two data sets, for 2007, respectively 2009. The main conclusion is that probability of *"default"* decreases for married people and older applicants while gender variable was not statistically significant for 2007 data sample. For after-crisis data, instead, the results indicated that only age was statically significant underling that the financial conditions became more difficult. They showed that the factors that influence the lending decision were related to the financial wealth of the respondent and not socio demographic. Kočenda and Vojtek (2009) using a logistic regression on loan database from a retail banking market from Czech Republic concluded that higher the education level leads to a lower probability of *"default"* and married clients are considered less risky from the credit perspective. The Gender variable was not considered in the model, as the information value for this variable was lower than the acceptable threshold.

Regarding the credit risk assessment, recent studies have shown that machine learning techniques (such as decision trees and support vector machines) show a better performance than the statistical model (such as logistic regression) and optimization method (Huang et al. 2018). According to the same authors, machine learning techniques do not require the assumption of variable distribution and can acquire knowledge directly from training data sets. Furthermore, artificial intelligence models perform better than the statistical ones especially when the credit risk assessment is nonlinear model classification (Huang et al. 2018). Addo et al. (2018) built binary classifiers based on machine and deep learning models on real data in predicting loan *"default"* probability. Their results indicate the fact that tree-based models are more stable than the models based on multilayer artificial networks. According to the authors, Random Forest and Gradient Boosting Model perform the best in terms of AuC and RMSE both on the validation and test set. Random Forest performs the best in terms of ROC curve and it is capable of distinguishing the information provided by the data and only retains the information that improves the fit of the model. Logistic Regression with Regularization selects more global and aggregate financial variables, while Random Forest and Gradient Boosting Model select detailed financial variables. Kruppa et al. (2013) used Random Forest, k-nearest neighbors, bagged k-nearest neighbors and Logistic Regression to estimate the consumer credit risk. Using test data on installment credits, they have shown that Random Forest outperformed a standard Logistic Regression model. Khandani et al. (2010) applied Generalized Classification and Regression Trees in order to build a robust forecast model for consumer credit *"default"* combining customer transactions and credit bureau data for a sample of a major commercial bank's customers.

In this chapter we aimed to analyze the importance of socio demographic factors in credit risk assessment using a real loan database from a retail bank from Romania. In order to detect the non-linearities in the risk profile of a borrower's creditworthiness we divided the data sample prior/posterior the financial crisis from 2008. In this respect we used 13 different models (4 methodologies linear and nonlinear) in order to assess the consumer credit risk using real data from 17,520 private individuals. Additionally, we run the models using data before and after the crisis taking into consideration qualitative and/or quantitative factors. The novelty of this research is given not only by the fact that we compare several credit risk assessments methods, but we also run the models on real data, emphasizing the importance of qualitative factors.

35.3 Methodology

Logistic Regression

If p_i is the probability that applicant i has defaulted, the purpose is to find w^* that best approximate:

$$p_i = w_0 + x_{1i} w_1 + x_{i2} w_2 + \ldots + x_{ip} w_p$$

The purpose is to find a function of p_i which could take values between 0 and 1 and one such function is the log of probability odds. The log likelihood function then is:

Decision Trees

Decision Trees (DT), or classification trees, are simple structures that can be used as classifiers (Abellán and Castellano 2017). According to the authors, DT can be used to predict the class value of an element by considering its attribute values when the elements are described by one or more attribute variables and by a single class variable. Abellán and Castellano (2017) noted that in this case each non-leaf node represents an attribute variable, the edges between that node and its child nodes represent the values of that attribute variable and each leaf node normally specifies an exact value of the class variable. DT are a non-parametric method, meaning that the system does not try to attempt to learn parameters upon which to score the applicants attributes; instead the system memorizes certain key characteristics about the data (Harris 2015). In order to predict the response, one should follow the decisions from the root node down to the leaf node. The method assumes to start with all input data and search for all possible binary splits on every predictor. Then select a split with best optimization criterion and repeat this step for the two child nodes. The number of splits allowed gives the complexity and size of the tree, hence in this research, we applied three types of DTs, namely Complex Tree, Medium Tree and Simple Tree.

The ensemble methods represent a combination among multiple techniques used in order to have a superior separation power than an individual model would have. There are two methods Bagging, also known as bootstrap method and Boosting. The first one optimizes on minimizing the variance and is used when the model overfits. The Bagged Trees use Breiman's (1996) 'random forest' algorithm and is based on a principle of creating many bootstrap replicas of the dataset and growing DT's on these copies. The prediction response is obtained by taking an averaged over predictions from individual DT's. Training data plays an important role as the trees are sensitive to specific data on which they are trained on. Random Forest is a type of Bagging classification that searches for the variable that minimizes the error from a random sample instead the entire all variables.

The boosted trees have as an ensemble method the algorithm AdaBoost with decision trees learners. The algorithm trains learners sequentially and for every learner there is the weighted classification error. For observations misclassified by learner the algorithm increases the weights and decreases the weights for observations correctly classified. The next learner is applied on data with the new weights. The learning rate plays an important role and one should select a lower number for the learning rate in order to converge to a good solution. Usually the learning rate below 1 is called "shrinkage" and in this paper we used a value of 0.1, which gives a slow rate of learning but converges to a better solution. The boosting principle for Random Under Sample Boosted trees is an extension of the adaptive boosting (AdaBoost) but instead of weighted classification error the algorithm uses weighted pseudo-loss. These can be used as a measure of accuracy in terms of classification from any learner.

In terms of advantages among these three ensemble methods, the prediction on speed is fast for adaptive boosting methods compared with medium for the random forest. The latter one also needs more memory usage and all three ensemble methods are sensitive to the number of learners and the adaptive boosting methods are also sensitive to the number of maximum splits.

Support Vector Machines

Another artificial intelligence method applied to evaluate the credit risk is the Support Vector Machine (SVM). It is a classification and regression tool that applies machine learning technique to maximize predictive accuracy while automatically avoiding over-fit to the data (Luo et al. 2017). According to the authors, it can learn both simple and highly complex classification models and employs using some mathematical procedures to avoid over-fitting. The non linear way to map the features from training data is done using kernel functions. The principle is to find the best hyperplane (the solution to the classification problem) that divides data points of the "*default*" class to "*non-default*" class, meaning the one with the largest margin between the two classes (Fig. 35.1).

The kernel function can be Linear, Gaussian, Quadratic or Cubic an it does not affect the prediction speed (fast) nor the memory usage (medium) but rather the model interpretability and flexibility. In this paper, we used six types of SVMs by changing the kernel function and the kernel scale (Fine Gaussian SVM, Medium Gaussian SVM and Coarse Gaussian SVM). The kernel scale controls the scale of the predictors on which it fluctuates significantly. Smaller the scale, higher the flexibility of the model, hence we use a classifier type with different kernel scale values obtained by using the number of predictors.

In Table 35.1 we present the specifics of the methodologies used to classify the private individuals in "*default*" and "*non-default*".

In order to assess the accuracy of the models' predictions we use confusion matrix as a reference for comparability. Confusion matrix reveals information about actual and predicted classifications done by a classification model. The matrix is a specific table layout with two rows and two columns and that reports the number of false positives, false negatives, true positives, and true negatives, with the following meanings (Table 35.2):

Sensitivity (True Positive Rate) = TP/TP + FN = power of the model

Specificity (True Negative Rate) = TN/FP + TN

The accuracy can be defined as the percentage of correctly classified instances (TP + TN)/(TP + TN + FP + FN).

Fig. 35.1 Support vector machine. (Source: author's computation)

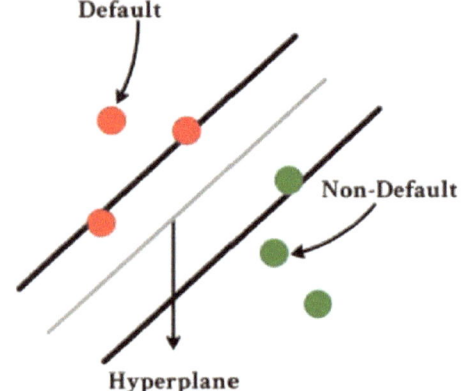

Table 35.1 The specifics of the methodologies used to classify individuals in *"default"* and *"non-default"*

Method	Specifics
Complex Tree	Maximum number of splits: 100 Split criterion: Gini's diversity index
Medium Tree	Maximum number of splits: 20 Split criterion: Gini's diversity index
Simple Tree	Maximum number of splits: 4 Split criterion: Gini's diversity index
Linear SVM	Kernel function: linear
Quadratic SVM	Kernel function: quadratic
Cubic SVM	Kernel function: cubic
Fine Gaussian SVM	Kernel function: Gaussian Kernel scale 1.2
Medium Gaussian SVM	Kernel function: Gaussian Kernel scale 4.7
Coarse Gaussian SVM	Kernel function: Gaussian Kernel scale 19
Boosted Trees	AdaBoost, maximum number of splits: 20, number of learners 30, learning rate 0.1
Bagged Trees (Random Forest)	Decision trees, number of learners: 30
RUS Boosted Trees	Random Forest, maximum number of splits: 20, number of learners 30, learning rate 0.1

Source: authors' computation

Table 35.2 Confusion matrix for binary classification

		Actual condition	
		Actual positive	Actual negative
Predicted condition	Actual positive	True positive (TP)	False positive (FP) (type I error)
	Actual negative	False negative (FN) (type II error)	True negative (TN)

Source: Baesens B. (2003). "Developing intelligent systems for credit scoring using machine learning techniques", no. 180 Basel Committee on

35.4   Data and Empirical Results

The database used contains 17,520 private individuals who were active in 2006 and defaulted in 2007Q3-2008Q2 (prior crisis) respectively 2008Q3-2009Q2 (posterior crisis).

Using the Basel (2004, 2005) definition, the observation period was for 12 months and all individuals that in the respective observation period had more than 90 days overdue were considered as *"default"* and the variable Trigger was marked as event. The proportion of *"default"* individuals in this portfolio is 50 percent, in order to have a balanced sample.

As the data sample is divided in two periods, before and after the 2008 crisis, a profile analysis of the *"default"* individuals shows (Table 35.3) how this profile has developed. The *"default"* profile before crisis shows the following characteristics: individuals with 29

Table 35.3 The profile analysis of the *"default"* individuals before and after the crisis

Criterion	Before crisis	After crisis
High school (%)	82.24	55.14
Marital status: single (%)	81.87	14.63
Marital status: married (%)	12.97	84.06
Seniority 6M-1Y (%)	24.54	20.82
Seniority 1–2 years (%)	35.00	34.13
Seniority 3–5 years (%)	14.68	23.88
Residence: With parents (%)	80.71	61.37
Residence: Own apartment (%)	10.68	23.77
Income bucket 1:<1500 (%)	65.43	45.32
Income bucket 2: 1500–3500 (%)	35.54	36.94
Average income	1498,00	2536,00
Average interest rate (%)	7.96	6.80
Repayment history: no due days (%)	6.26	10.87
Repayment history: in some cases, warnings required (%)	0.48	1.74
Average age	29	35
Average loan amount	21,543	46,442

Source: Authors' computation

average age, only with high school education, 81.87 percent single, having a seniority at the last job less than two years, living with parents and earning less than 1500 Lei (approximate 315 Euros).

The profile after the crisis shows similar core structure, but with significant changes such as: only 55.14 percent have high school comparing with 82.24 percent before crisis, only 14.63 percent are single while the remaining proportion of individuals are married, the seniority has increased after crisis, the average income is 2536 Lei (approximate 352 Euros) comparing with 1498 Lei (approximate 315 Euros), the average age is 35 and 23.77 percent have their own apartment comparing with 10.68 percent before crisis.

The repayment history indicates if the individual had same overdue amount in the past. The information is gathered from the Credit Bureau, and if the client has been delinquent in other banks the information is captured as such. No due days means he paid everything while no information means the client doesn't have other installments. After the crisis, the percentage of debtors that got some warnings in the past increased but much more increased the percentage of borrowers that paid in past and now have difficulties in paying.

Interesting to notice also the fact that the *"default"* individuals after crisis took loans with lower interest rate, which indicate that the financial burden is higher given also the higher debt to income values.

When analyzing the average loan amount we can see that almost doubled, which means that debtors that did not pay after crisis had in average higher loan amounts. As they were granted these high amount of loans, means their income was proportionally high, and apparently sensitive to financial shocks. This is correlated with seniority as well as the shift in the *"default"* profile show an increase in debtors with high seniority, translated to high income. As these clients had low probability of *"default"* at time of granting emphasizes the importance of an accurate classification and determine what influenced more the consumer to *"default"*, the financial burden, it's ability to pay given its social demographic profile or a combination of both?

When we compare the characteristics of the *"default"* individuals before and after the crisis we can observe that a change in the profile has changed. Before crisis the *"default"* profile is in line with the expectation: low income, medium seniority, pre-university studies, single and average age of 29. The reasons of not paying the installments are given by both soft and hard facts for this profile. When analyzing what happens with the profile after the crisis, we observe that the individual has with 50 percent more income in average, higher seniority, university studies, married and average age of 35. It is clear that the reasons of not paying the installments is given by the financial burden and uncertainty created by the financial crisis.

The *'default'* profile development indicates that individuals that defaulted after the crisis do not follow the same pattern as before the crisis (Fig. 35.2).

In order to test the difference in the *"default"* profile before and after the crisis, we used discriminant analysis on *"default"* data covering both data sample. In this way, we want to verify if the model is able to discriminate and create the two categories, prior and posterior the crisis. We used two fitting functions one that creates linear boundaries between these two classes (linear) and one that creates nonlinear boundaries (quadratic) to be able to capture non-linearity. As the previous analysis indicates that some nonlinearity in the

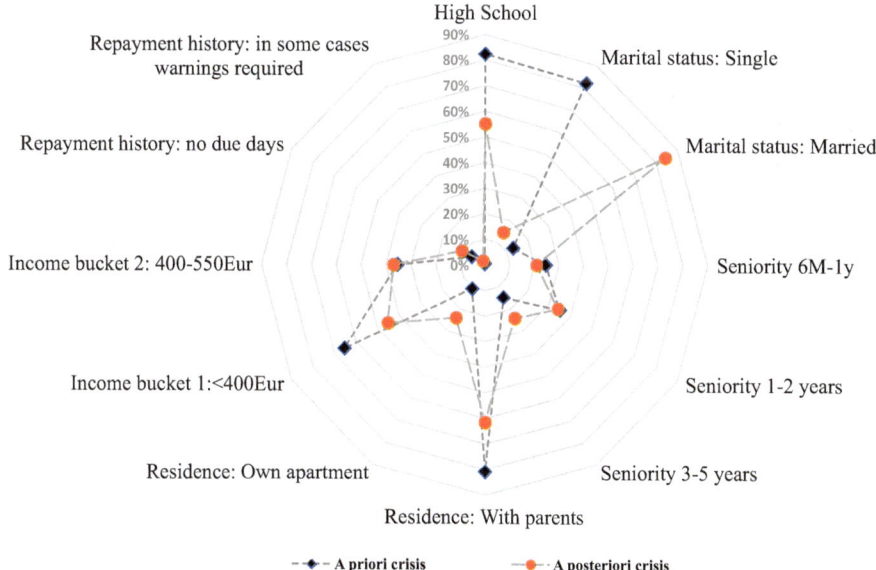

Fig. 35.2 The "default" profile of individuals before and after the crisis. (Source: authors' computation)

"*default*" profile after crisis exists, *we expect to have a higher number of individuals classified as after crisis when there are actually before crisis for the linear discriminant* (Table 35.4).

The separation power is 76.8 percent when using a linear component comparing with 75 percent when a quadratic one is used. In terms of accuracy, the model before crisis has correctly selected 87 percent of the "*default*" individuals, while only 67 percent were correctly classified as being defaults after the crisis. When using nonlinear function, the value of 69 percent indicates that more "*default*" individuals after crisis were correctly classified. *This shows that the "default" profile before crisis is captured better by the linear model while after the crisis is captured better by the non-linear model.* The nonlinearity also gives interesting results when checking the misclassified cases. 19 percent of the '*default*' individuals have been predicted as being after crisis when they actually were before crisis in comparison with 13 percent for the linear model.

In order to capture which of the factors influenced more we divided the variables in quantitative (related to the loan, income and payment history) and qualitative (education, marital status, age, residence, seniority). The first category is capturing a financial classification, rating of the individual, the second one gives as overview on his creditworthiness while combined gives as the full picture of the individual. The models were run using three combination of variables: one when all the variables have been included in the training data and this is what we name as Credit Rating, one only with the financial factors and this is the Financial Rating and the socio demographic variables are used for the Social Rating.

In order to determine how much of this profile structure could be captured by the model used to classify individuals in "*default*" and "*non-default*" we selected 13 methodologies linear and nonlinear (Complex Tree, Medium Tree, Simple Tree, Boosted Trees, Bagged Trees or Random Forest, RUS Boosted Trees, linear SVM, Quadratic SVM, Cubic SVM, Fine Gaussian SVM, Medium Gaussian SVM, Coarse Gaussian SVM and Logistic Regression). The aim is to train these classifier methodologies and observe if the learning process is sensitive to "*default*" profile pattern. One should take into account these nonlinearities and changes in the risk profile in order to be able to correctly assess individuals

Table 35.4 Discriminant analysis on "*default*" data

Model	Accuracy	Classified as before crisis-model predicts before crisis	Classified as before crisis-model predicts after crisis	Classified as after crisis-model predicts before crisis	Classified as after crisis-model predicts after crisis
Linear discriminant (%)	76.8	87.0	13.0	33.0	67.0
Quadratic discriminant (%)	75.0	81.0	19.0	31.0	69.0

Source: authors' computation

in time. As mentioned above the overall data sample is divided in two equal subsamples taking into account the crisis data point. As we noted some shifts in the portfolio on both quantitative factors and qualitative factors, the classification models have been applied including this division, too. Therefore, we have a twofold approach, one to capture the profile change before and after the crisis and one to see what influenced the most, the socio demographic or financial factors.

In order to have this subsamples and sub factors, given the size of the overall data sample, we used cross validation to cope with over-fitting. In this respect, the overall data sample is divided in five folds, for each one the model is trained using the out-of-fold data, and the model performance assessment is done using in-fold data.

Using all factors on combined data, amongst the models we obtained the highest value for accuracy for Random Forest on all three data samples (Table 35.5). If we compare the changes in the accuracy ratio prior and posterior crisis, Simple DT and Fine Gaussian SVM are the only models with an improvement. The poorest performance has been notified for SVM models irrespective of data choice.

Regarding the improvement of the socio demographic factors on separation power, Random Forest got the best results on all samples with or without the financial factors. Important here to note is the change in the accuracy ratio, when excluding these factors. The logistic regression before crisis has recorded an accuracy of 92.8 percent (Table 35.4) on the Social Rating in comparison with 81.8 percent accuracy on Credit Rating. In fact, the maximum accuracy obtained on Credit Rating models is lower than the minimum we got on Social Rating models. This is not the case when non-linearities are captured in the model, such as Random Forest, where the results on Social rating models registered an accuracy ratio lower comparing with Credit Rating models. When analyzing the results on after crisis data, we observe that all SVM models and logistic regression have higher accuracy ratio on Social Rating models than Financial Rating.

The results we obtained on Financial Rating models show that Boosted Trees outperformed on all data samples. The accuracy ratio is higher on full data sample (72 percent) than prior crisis (69.50 percent and lower than after crisis result (74.30 percent). This means that the posterior crisis *"default"* pattern is captured best by Boosted Trees and the model is able to learn this pattern given the 72 percent accuracy ratio on full data (Table 35.6).

Random Forest model's accuracy ratio on Social Rating (87.6 percent) on full data sample is lower than both prior and posterior crisis's results. This mean than the champion model on Social Rating classifies better the before *"default"* clients, then the after crisis and lat the mix of these patterns. This pattern is obsessed to all the models analyzed in this paper and explains that creditworthiness was easier to classify before crisis and more difficult when the profiles are not similar.

The power of the model to classify the debtors using Social rating, Financial Rating and the mix of these, the Credit Rating depends on the trained data used. The Financial Rating's champion model's results are best on posterior crisis data, meaning that financial factors counted the most in detecting the patterns in *"default"* after the financial crisis.

Table 35.5 Accuracy ratio results for credit rating

Models	Credit Rating	Credit Rating	Credit Rating	Conclusion
Data Sample	*Full*	*Prior Crisis*	*Posterior Crisis*	
Complex Tree	90.90 %	97.70 %	93.80 %	**Accuracy decreased**
Medium Tree	89.60 %	97.10 %	94.10 %	**Accuracy decreased**
Simple Tree	85.10 %	89.80 %	92.90 %	**Accuracy increased**
Logit	75.10 %	81.80 %	80.00 %	**Accuracy decreased**
Linear SVM	75.50 %	82.50 %	80.20 %	**Accuracy decreased**
Quadratic SVM	76.10 %	84.10 %	81.30 %	**Accuracy decreased**
Cubic SVM	76.60 %	83.80 %	81.20 %	**Accuracy decreased**
Fine Gaussian SVM	74.70 %	78.60 %	78.80 %	**Accuracy increased**
Medium Gaussian SVM	76.10 %	82.90 %	81.60 %	**Accuracy decreased**
Coarse Gaussian SVM	75.40 %	81.60 %	77.60 %	**Accuracy decreased**
Boosted Trees	89.60 %	97.80 %	94.50 %	**Accuracy decreased**
Random Forest	**91.60 %**	**98.00 %**	**94.70 %**	**Accuracy decreased**
RUS Boosted Trees	89.60 %	96.80 %	93.90 %	**Accuracy decreased**

Source: authors' computation

Using Social Rating the accuracy is higher on before crisis data showing that creditworthiness selection is captured better when no financial shocks are applied. The mix of factors, captured in Credit Rating, seem to follow the same pattern as Social Rating.

What is important to observe is not only the ranking in accuracy given by these models but also the change in the performance of the same model across different samples when using the same factors. The average accuracy ratio for Credit Rating on the sample prior to crisis is 88.65 percent and it decreases to 86.51 percent on the after crisis sample.

Table 35.6 Accuracy ratio results for financial and social rating

Models	Financial rating	Financial rating	Financial rating	Social rating	Social rating	Social rating
Data sample	*Full*	*Prior crisis*	*Posterior crisis*	*Full*	*Prior crisis*	*Posterior crisis*
Complex Tree (%)	68.40	67.90	70.60	86.80	93.80	90.20
Medium Tree (%)	68.70	68.90	71.20	86.20	93.70	90.90
Simple Tree (%)	68.70	68.40	71.30	85.10	89.70	89.50
Logit (%)	56.10	58.00	59.90	85.00	92.80	90.60
Linear SVM (%)	56.40	58.30	56.40	85.50	92.30	90.50
Quadratic SVM (%)	40.40	45.50	41.80	86.30	93.90	91.20
Cubic SVM (%)	50.10	40.90	45.80	86.30	93.00	90.80
Fine Gaussian SVM (%)	60.80	59.80	62.60	85.80	91.10	90.00
Medium Gaussian SVM (%)	59.40	59.20	61.80	86.10	93.40	90.80
Coarse Gaussian SVM (%)	57.20	58.60	58.80	85.00	93.00	89.10
Boosted Trees (%)	72.00	69.50	74.30	86.60	94.10	90.90
Random Forest (%)	67.70	67.00	71.30	87.60	94.20	91.30
RUS Boosted Trees (%)	68.60	69.00	71.10	86.20	93.60	90.80

Source: authors' computation

Notable is that all models recorded a decrease in accuracy ratio except Simple DT and Fine Gaussian SVM. Therefore, *our first conclusion is that the separation power decreased after the crisis.*

For Social Rating models the average of the accuracy ratio decreased with the same magnitude from 92.97 to 90.51 percent when comparing before and after the crisis and there is no record of an increase in accuracy ratio. On another hand, the average accuracy ratio for Financial Rating increased from 60.8 to 62.84 percent where almost all the models recorded an improvement except, linear and quadratic SVM model. For a linear model, we can observe that accuracy is increasing on quantitative factor models and decreasing on qualitative models, when assessing before and after crisis data samples. This is interesting from two points of view, firstly *the separation power increased on Social Rating models* therefore the models could assess better the *"default"* profile" after crisis and secondly, as *this relationship is not linear it has been best captured by the nonlinear models.*

For Credit Rating, the average power of detection of *'default'* individuals is 82.3 percent before the crisis comparing with 74.9 percent after the crisis; therefore, we can conclude that *capturing the true default profile was more difficult after the crisis.* This can be seen also in the *"default"* profile after the crisis when we observed that the average income for these individuals was higher and so is the age and the seniority. Therefore, if a model is trained to learn this profile on a training data prior crisis it would be problematic to catch the characteristics of an out of sample data.

To quantify the power of detecting the *"default"* profile after the crisis we calculated the sensitivity rate capturing the proportion of *"default"* individuals that have been correctly classified by the models used. In Fig. 35.3 we observe that for Social Rating models the hit rate increases on posterior crisis sample, while for Financial Rating for several models the probability of *"default"* detection decreased (Linear, Quadratic and Coarse Gaussian SVM). Interesting to mention is that for the logistic regression both Social and Financial Ratings improved in terms of power to predict Săndică (2017). Using the Social Rating the prediction of *"default"* cases increased on posterior crisis sample when using both linear and nonlinear models. For the Financial Rating the power of prediction decreased when using linear, quadratic and Coarse Gaussian SVM. *Therefore we can conclude that pattern the "default" profile is more accurate using Social Rating than Financial Rating when shifts in the profile are encountered.*

When we discuss about detecting the *"non-default"* patterns we consider as benchmark to compare among the models's results, specificity rate or selectivity. Higher the values for this ratio, lower the Error Type I is recorded. In Fig. 35.4 we show that for Social Rating, the Error Type I increased for all the models, except Fine Gaussian SVM. For the Financial Rating, the specificity increases posterior crisis for all SVM models except the one with Cubic kernel while all the decision trees and ensemble methods recorded an increase in Type I Error. Logistic regression recorded increases in this type of error for both Social and Financial Rating. The *"non default"* pattern after the financial crisis when using Social Rating slightly increased only for one SVM model, while the financial rating separation power improved for Random Forest and SVM models. The fact that the *"non default"* detection is better using financial rating shows that the "financial burden" the borrower had is captured by these models.

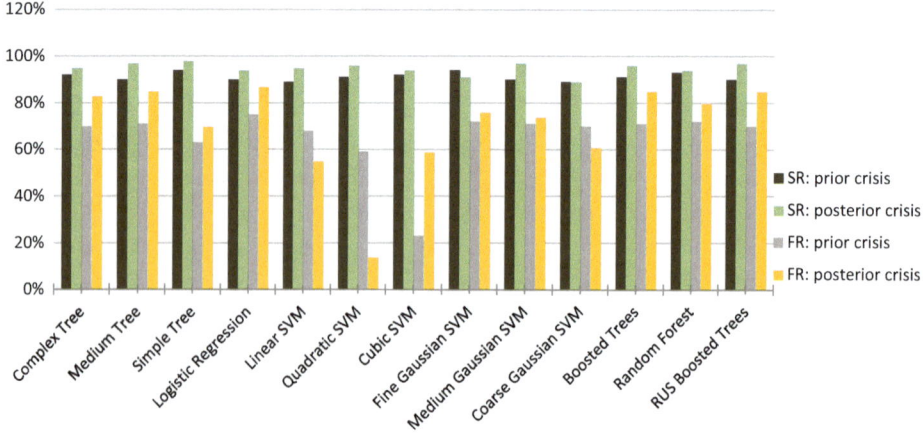

Fig. 35.3 Sensitivity (%) for social rating and financial rating: the "default" detection. (Source: authors' computation)

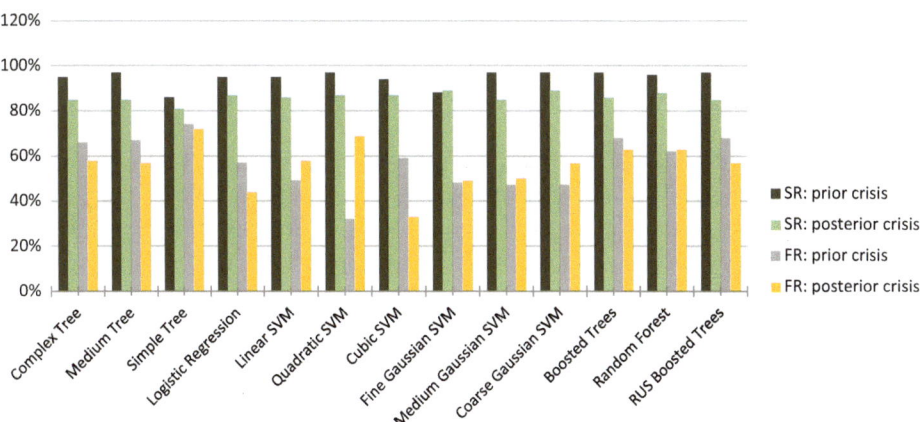

Fig. 35.4 Specificity (%) for social rating and financial rating: "non-default" detection. (Source: authors' computation)

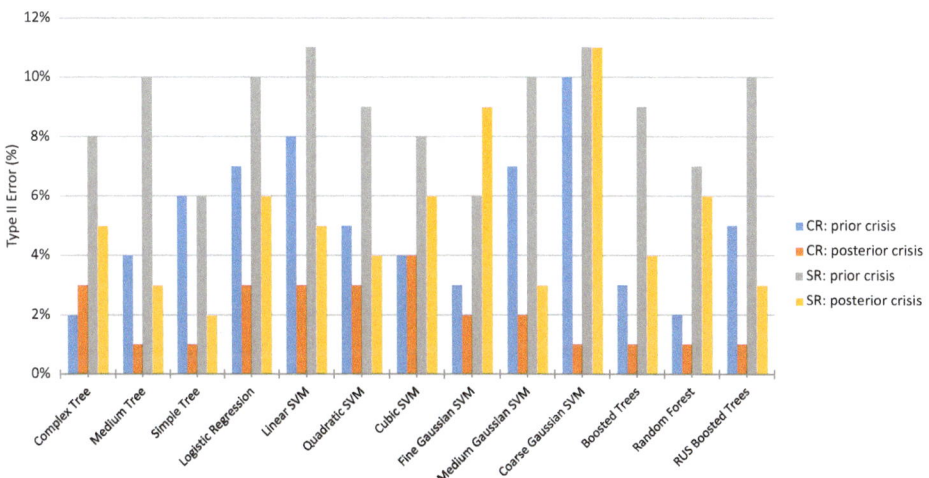

Fig. 35.5 Type II error (%) for social rating, financial rating and credit rating. (Source: authors' computation)

We compare the Type II Error for all these three categories of ratings as the Credit Rating incorporates also the financial information, giving us a glance of the importance of social rating component in the final credit rating. In Fig. 35.5 we plot the percentages on prior and posterior crisis data sample to have an indication of the split. As it is pointed out, for Credit Rating, the error increased for Complex Decision Tree or remain constant for Cubic SVM model. For Social Rating the results indicate that Fine Gaussian SVM had an increase in Type II Error while for Coarse Gaussian SVM a slightly change was obtained. The results for Logistic regression indicate that this type of error decreased on both Social and Credit Rating.

The results on posterior crisis demonstrate the ability of the models to learn the *"default"* pattern and correctly classified the clients. A maximum decrease in error on Credit Rating is recorded on a Coarse Gaussian SVM model while for Social Rating this is the worst of the models as false negative rate remained constant and high. When comparing the ratings among them, Decision Tree gives the same results while all the other models have recorded higher values for the Social Rating on prior crisis data. For the posterior crisis data there is no model for Social Rating with lower Type II Error than Credit Rating, indicate that these models could not accept the *"non default"* clients. Basically the *"non-default"* pattern was shifted, given the changes in their profile, and the Social Rating models were not able to capture this better than the rating with the financial factors.

35.5   Conclusions

On the basis of real data provided by a commercial bank, in this article, we compared the performance of several machine learning techniques and compared them with the logistic regression as a benchmark. We found that accuracy ratio gives the better results on decision trees and ensembles based on decision trees such as adaptive boosting methods and Random Forest. When comparing the accuracy ratio values before and after the crisis we observed that only the simple decision tree had a significant increase in the separation power (from 89.8 to 92 percent). In terms of benchmarking our results, in the graph on right, we highlight the values of accuracy ratio before and after the crisis and underline the intervals used in literature. As it can be pointed out all models based on decision trees method remain in the green area, all SVM models are in yellow area while logistic regression is also in yellow. Before and after crisis, Fine and Coarse Gaussian SVM are the only models that recorded values lower than the logistic regression. The average accuracy ratio including all factors on the sample prior to crisis is 88.65 percent and it decreases to 86.51 percent on the after crisis sample. Notable is that all models recorded a decrease in accuracy ratio except Simple DT and Fine Gaussian SVM. Therefore, *the separation power decreased after the crisis.*

In order to capture which of the factors influenced more we divided the variables in quantitative (related to the loan, income and payment history) and qualitative (education, marital status, age, residence, seniority). The first category is capturing a financial classification, rating of the individual, the second one gives as overview on his creditworthiness while combined gives as the full picture of the individual. The models were run using three combinations of variables: one when all the variables have been included in the training data and this is what we name as Credit Rating, one only with the financial factors and this is the Financial Rating and the socio demographic variables are used for the Social Rating.

The results we obtained on Financial Rating models show that Boosted Trees outperformed on all data samples. The accuracy ratio is higher on full data sample (72 percent) than prior crisis (69.50 percent and lower than after crisis result (74.30 percent). This means that the posterior crisis *"default"* pattern is captured best by Boosted Trees and the model is able to learn this pattern given the 72 percent accuracy ratio on full data. Random Forest model's accuracy ratio on Social Rating (87.6 percent) on full data sample is lower than both prior and posterior crisis's results. This mean than the champion model on Social Rating classifies better the before *"default"* clients, then the after crisis and lat the mix of these patterns. This pattern is obsessed to all the models analyzed in this paper and explains that creditworthiness was easier to classify before crisis and more difficult when the profiles are not similar. The power of the model to classify the debtors using Social rating, Financial Rating and the mix of these, the Credit Rating depends on the trained data used. The Financial Rating's champion model's results are best on posterior crisis data, meaning that financial factors counted the most in detecting the patterns in *"default"* after the financial crisis. Using Social Rating the accuracy is higher on before crisis data showing that creditworthiness selection is captured better when no financial shocks are applied. The mix of factors, captured in Credit Rating, seem to follow the same pattern as Social Rating.

References

Abellán J. and J. G. Castellano (2017). "A comparative study on base classifiers in ensemble methods for credit scoring", Expert System with Applications, no. 73, pages 1–10

Addo P. M., D. Guegan and B. Hassani (2018). "Credit Risk Analysis using Machine and Deep Learning Models", Risks, no. 6, 38

Altman, E.I. (1968). Financial Ratios, Discriminant Analysis and the Prediction of Corporate Bankruptcy. Journal of Finance,

Ang, J. S., Chua, J. H. And Bowling, C. H. (1979). The Profiles of Late-Paying Consumer Loan Borrowers: an Exploratory Study. Journal of Money, Credit & Banking (Ohio State University Press) 11, 222–226.

Baesens B. (2003). "Developing intelligent systems for credit scoring using machine learning techniques", no. 180

Basel Committee on Banking Supervision (2004). "International Convergence of Capital Measurement and Capital Standards", ISBN 92-9197-669-5, available online at https://www.bis.org/publ/bcbs107.pdf

Basel Committee on Banking Supervision (2005). "Studies on the Validation of Internal Rating Systems", working paper no. 14, ISBN 1561-8854, available online at https://www.bis.org/publ/bcbs_wp.14.pdf

Breiman, L. (1996). "Bagging Predictors." Machine Learning. Vol. 26, pp. 123–140.

Charpignon M.-L., E. Horel and F. Tixier (2014). "Prediction of consumer credit risk", CS229

Durand, D. (1941). Risk Elements in Consumer Instalment Financing. National Bureau of Economic Research, New York.

Fisher, R. A. (1936). The use of multiple measurements in taxonomic problems. Annals of Eugenics, 7(2): 179–188.

Ganopoulou, M., Giapoutzi, F., Kosmidou, K., & Moysiadis, T. (2013). Credit-scoring and bank lending policy in consumer loans. International Journal Financial Engineering and Risk Management, 1, 90–110.

Hand D. J. and W. E. Henley (1997). "Statistical classification methods in consumer credit scoring: a review", Journal of Royal Statistical Society: Series A. Statistics in Society, no. 160(3), pages 523–541.

Harris, T. (2015). "Credit scoring using the clustered support vector machine", Expert System with Applications, no. 42

Huang X., X. Liu and Y. Ren (2018). "Enterprise credit risk evaluation based on neural network algorithm", Cognitive System Research, no. 52, pages 317–324

Khandani A.E., A. Kim and A.W. Lo (2010). "Consumer credit-risk models via machine-learning algorithms", Journal of Banking and Finance, no. 34, pages 2767–2787

Kočenda, E., and Vojtek, M. (2009). Default Predictors and Credit Scoring Model. (CESifo Working Paper No. 2862)

Kruppa J., A. Schwarz, G. Arminger and A. Ziegler (2013). "Consumer credit risk: Individuals probability estimates using machine learning", Expert System with Applications, no. 40, pages 5125–5131

Livingstone, S. M., and Lunt, P. K. (1992). Predicting personal debt and debt repayment: Psychological, social and economic determinants. Journal of Economic Psychology, 13, 111–134

Luo, C., D. Wu and D. Wu (2017). "A deep learning approach for credit scoring using credit default swaps", Expert Applications of Artificial Intelligence, no. 65, pages 465–470

Marjo, H. (2010). The determinants of Default in consumer credit market. Aalto University School of Economics, available online at http://epub.lib.aalto.fi/en/ethesis/pdf/12299/hse_ethesis_12299.pdf

Ng, T., Eby, L.T., Sorensen, L. L., and Feldman, D. C. (2005). Predictors of objective and subjective career success: A metaanalysis. Personnel Psychology, 58, 367–408.

O'Neal, C. (2016). Weapons of Math Destruction: How Big Data Increases Inequality and Threatens Democracy, New York: Crown Publishers, 142–145.

Sarlija, N., Bensic, M., and Zekic-Susac, M. (2009). Comparison procedure of predicting the time to default in behavioural scoring. Expert Systems with Applications, 36, 8778–8788.

Săndică A.-M., (2017). "Determinants of Corporate Bankruptcy in Romania – A Comparison Approach"

Tokunaga, H. (1993). The use and abuse of consumer credit: Application of psychological theory and research. Journal of Economic Psychology, 14, 285–316.

Verplanken, B., and Herabadi, A. (2001). Individual differences in impulse buying tendency: Feeling and no thinking. European Journal of Personality, 15, S71–S83.

Dr. Ana Maria Sandica has been developing credit risk models for more than ten years. She started to study machine learning techniques during her master degree in Financial Econometrics (Dofin) and continued with completing a doctorate in the field of stochastic equilibrium models in Macroeconomy. Her thesis on postdoctoral research links the macroeconomic shock transmission mechanism in estimating the probability of bankruptcy for companies. She held a managerial role in model risk validation at a major German bank.

Prof. Dr. Monica Dudian is Professor of Economics at The Bucharest University of Economic Studies, where she received her PhD in Economics in 1999. She also held the position of Vice dean of the Faculty of Economics of The Bucharest University of Economic Studies, during 2001–2008. Her teaching is focused primarily on microeconomics and industrial organization. She manages research grants and performs research on country risk, credit risk, and industrial economics.

Teil VIII

Ausblick

Social Credit Rating – Und nun?

Bernd Thomsen

Zusammenfassung

Wir haben in dieser Publikation viel über Wesen und Inhalte des Social Credit Rating (SCR) gelesen, über Governance, Recht und Nachhaltigkeit, sowie Methoden, Modelle und Funktionen gelernt. Aber was heisst das jetzt – für Bürger wie für Unternehmen? Wie kann, wie wird SCR die Welt zukünftig verändern?

SCR ist im Rahmen der globalen digitalen Transformation und geopolitisch inmitten der die Welt verändernden Triade aus „Politischen Systemen", „Ökonomischen Systemen" und „Digitalen Systemen" zu begreifen. Als Digitales System ist es hierbei für SCR entscheidend, von welchen übergeordneten Werten diese Technologie in ihren Wechselbeziehungen mit Wirtschaft und Politik geleitet und mit welchem Ziel sie eingesetzt wird. In Demokratien können SCR-Systeme dazu beitragen, die Freiheit der Bürger zu bewahren. In Autokratien laufen sie Gefahr, zu einem Instrument lebensalltäglichen Geiseltums zu werden, das die individuelle Freiheit einschränkt. In der Wirtschaft kann SCR unternehmerische Übernahme von Verantwortung unterstützen, zum Beispiel im Kampf gegen den Klimawandel. In einem Staatskapitalismus ist die Befürchtung gerechtfertigt, dass SCR überwiegend dem totalitären Machtausbau dient. Es sollte also nicht darum gehen, SCR zu diskreditieren, sondern darum, die Absichten der für digitale Bewertungssysteme Verantwortlichen zu hinterfragen. Die aktuell zu beobachtende, weltweite Sinnkrise der liberalen Demokratien fällt mit einer Machtver-

B. Thomsen (✉)
THOMSEN GROUP, Hamburg, Deutschland
E-Mail: andrea.simon@thomsen.de

© Springer Fachmedien Wiesbaden GmbH, ein Teil von Springer Nature 2020
O. Everling (Hrsg.), *Social Credit Rating*,
https://doi.org/10.1007/978-3-658-29653-7_36

schiebung zugunsten repressiver Regime, die aktuelle Kapitalismusdepression mit einem Hype des Staatskapitalismus zusammen. Das Ausmaß beginnen viele Menschen gerade erst zu verstehen. SCR ist dabei nicht nur in seiner transformativen Wirkung, sondern auch für die westlichen Industrienationen ein ernst zu nehmender Faktor.

Am 11. Januar 2020 starb der erste Mensch an dem in China auffällig gewordenen Coronavirus, weltweit waren bei Redaktionsschluss dieses Beitrags 78.436 Infektionen und 2461 Todesopfer bekannt (o. V. 2020a). Die Zahlen stiegen weiter.

Um eine Epidemie innerhalb einer menschlichen Population oder sogar eine Pandemie möglichst schon im Keim zu ersticken, ist Tempo einer der wichtigsten Faktoren. Je schneller Krankheitsfälle bekannt werden, desto weniger Menschen können die ersten Infizierten anstecken.

Der 34-jährige Arzt Li Wenliang aus Wuhan, der am 7. Februar 2020 am Virus verstarb, hatte bereits am 30. Dezember 2019 in einer Online-Diskussionsgruppe vor dessen Ausbreitung gewarnt. Der Augenarzt, der ein Kind und eine schwangere Ehefrau zurücklässt, verfolgte dabei keine politischen Ambitionen. Dennoch stoppte ihn das System zunächst als „Unruhestifter" und zwang ihn, keine weiteren Informationen über das Coronavirus zu enthüllen. Die chinesische Zensur soll eingegriffen und seine Beiträge auf Socialmedia-Plattformen, in Webforen und Blogs gelöscht haben. Es vergingen wertvolle Wochen, bis die zuständige Regierung Wenliangs Warnung endlich zur Realität erklärte.

Was hat das Coronavirus mit Social Credit Rating zu tun? Wir können an diesem Beispiel ablesen, wie der SCR-Bewertung als digitales Element eine globale Relevanz im Rahmen auch der gerade stattfindenden weltweiten, digitalen Transformation zufällt, die wir bei der Gestaltung und dem Einsatz dieser neuen Technologien immer berücksichtigen müssen.

36.1   SCR in China – ein System zur totalen Kontrolle?

SCR dient in der Volksrepublik China aufgrund der weitgreifenden Nutzung digitaler Kontrollmöglichkeiten mutmaßlich vorrangig der Überwachung der Bürger und dem Machterhalt der Kommunistischen Partei. Das autokratische, politische System beruht auf Machtkonzentration und Repression. Das führt in der Bevölkerung zu Angst, die auch lokale Funktionäre packt, weil sie vermeiden wollen, Überbringer schlechter lokaler Nachrichten wie der Coronavirus-Erkrankungen in Wuhan zu sein. Sie übernahmen mutmaßlich nicht – wie demokratisch legitimierte Politiker – Verantwortung für das örtliche Geschehen. Weil sie Repressionen des Staates zu befürchten hatten, steuerten sie möglicherweise sehenden Auges in eine weltweite Gesundheitskrise.

Fünftausendfünfundneunzig. Das ist keine Zahl Neuinfizierter der Coronavirus Epidemie, sondern die Entfernung vom im Januar 0 Grad kalten, norddeutschen Flensburg bis in das dann ebenfalls 0 Grad kalte chinesische Kaschgar.

Die Nord-Süd-Ausdehnung Chinas beträgt von der nördlichsten Spitze bis zum Südzipfel mit rund 5500 Kilometer sogar noch etwas mehr. Das heisst, die Ausdehnung Chinas ist größer als die lange Strecke, die wir benötigen, um das Land, das 27-mal so groß ist wie Deutschland, zu erreichen. China besitzt mit 22.133 Kilometer Gesamtlänge die längste Landgrenze aller Staaten der Welt. Zurück nach Europa vom an der Grenze zu China gelegenen Mount Everest bis zum europäischen Mont Blanc sind es weniger als ein Drittel. Kurzum: China ist groß und weit weg.

Daher tun wir gut daran, uns nicht anzumaßen, näher dran zu sein, als wir es tatsächlich sind. Wenn beispielsweise der US-Sender CNN berichtet, China würde Gräber schänden, obwohl es sich vermeintlich um eine von Angehörigen gewünschte Umbettung von tradierten Feldgräbern, die regelmäßig von Tieren aufgewühlt wurden, in moderne Grabanalagen handelt. Oder wenn mangelhafte Übersetzungen aus dem Chinesischen einen falschen Eindruck erwecken, Chinesen interviewt werden, jene aber kaum Gelegenheit haben (u. a. weil die betreffenden Medien in China zensiert sind) zu entdecken, dass sie falsch übersetzt wurden, erhalten wir ein verzerrtes Bild der Lebenswirklichkeit in diesem fernen Land. Auch könnte das chinesische SCR eine konsequentere Umsetzung der SDG der Vereinten Nationen (Sustainable Development Goals, welche weltweit der Sicherung einer nachhaltigen Entwicklung auf ökonomischer, sozialer sowie ökologischer Ebene dienen sollen) liefern als in vielen westlichen Staaten. So respektvoll wie gegenüber den Bürgerinnen und Bürgern Chinas, so kritisch wie gegenüber medialer Berichterstattung und sozialen Medien - genau so unabhängig sollten wir auch weiterhin Fragen stellen (vgl. Thomsen 2019a). Zum Beispiel diese: Hat die machtpolitische Zielsetzung der chinesischen SCR-Politik zur weltweiten Verbreitung des Virus beigetragen?

Die Weltgesundheitsorganisation (WHO) lobte die Entschlossenheit, mit der die Regierung in Peking inzwischen versuche, das Virus in den Griff zu bekommen. In der Tat hat der Staat drastische Maßnahmen ergriffen und mehr als 50 Millionen Menschen praktisch unter Quarantäne gestellt. Eine ganze Provinz wurde isoliert. Ähnlich drakonische Maßnahmen gab es zuletzt bei der Spanischen Grippe, an der zwischen 1918 und 1920 weltweit 500 Millionen Menschen erkrankten und bis zu 50 Millionen starben. Dennoch stellt sich die Frage, inwieweit das Regime in China die bei Redaktionsschluss mögliche Pandemie tatsächlich zu verantworten hat?

36.1.1 Die Steuerungsmöglichkeiten von SCR

Das autokratische China ist Weltmarktführer beim Einsatz digitaler Kontroll-Software des 21. Jahrhunderts. SCR-Systeme prägen den Alltag der Chinesen. Mit Hilfe modernster Massenüberwachungstechnologien kann der Staat die Menschen beobachten und nahezu jede ihrer Bewegungen verfolgen. Die Sicherheitsbehörden im chinesischen Xinjiang speichern vom Kind bis zum Rentner systematisch Daten: von der Blutgruppe, über eindeutige Fotos und einen Scan des inneren Auges, bis hin zu Fingerabdrücken und DNA-Proben. Daher wäre es wohl ein Leichtes gewesen, die Bevölkerung frühzeitig über das Coronavirus zu informieren und über Verhaltensmaßnahmen aufzuklären. Das ist jedenfalls nicht passiert.

Zudem änderte laut Redaktion der Tagesschau, der ältesten, noch bestehenden offiziellen Nachrichtensendung des deutschen öffentlich-rechtlichen Rundfunks, die chinesische Gesundheitskommission im weiteren Verlauf der Epidemie nach eigenen Angaben mehrmals ihr Vorgehen: Menschen, bei denen die Lungenkrankheit zwar mit einem Test nachgewiesen wurde, die aber keine Symptome zeigen, wurden kurzerhand nicht mehr in die Statistik aufgenommen – ein Vorgehen, dass der Empfehlung der WHO klar widerspricht (vgl. o. V. 2020b).

Das Ziel der umfassenden Kontrollmaßnahmen über SCR besteht darin, die chinesische Gesellschaft umzuerziehen. Die machthabende Kommunistischen Partei belohnt wünschenswertes Verhalten mit „Punkten", nicht wünschenswertes Verhalten ahndet sie mit Entzug von Punkten. Die resultierenden Daten begründen die Vergabe der Punkte. Es geht um das „soziale Ansehen", auch interessiert sich der Staat zum Beispiel für politische Ansichten der Menschen. Ein gesunkenes Punktekonto ist gleichbedeutend mit alltäglichen Einschränkungen, die zu erheblichen gesellschaftlichen Nachteilen führen können.

36.1.2 Wie ist es, den ganzen Tag überwacht zu werden?

Keinen unbeobachteten Schritt tun zu können, bei jedem Wort wissen zu müssen, das alles gespeichert und im Zweifel gegen einen verwendet wird? In Shenzhen, einer Stadt im Südosten Chinas, kann man das bereits erleben. Bei einer Taxifahrt werden die Fahrgäste von mehreren Kameras begleitet: eine, die den Fahrer legitimiert, das Auto steuern zu dürfen; eine, die sein Fahrverhalten überwacht; eine, die den Insassen erlaubt, eine Fahrt überhaupt antreten zu dürfen – oder eben nicht, falls der SCR-Score, also der Punktestand, es nicht gestattet. Die Ermahnung einer automatischen Frauenstimme, etwas Verbotenes sofort zu unterlassen, ist eine der niedrigschwelligen Interventionen. Da nur der öffentliche Raum überwacht werden darf, erklärte die Regierung kurzerhand den Innenraum von diesen Fahrzeugen zum öffentlichen Raum (vgl. Hua 2018). Daher ist es verboten, die Kameras auszuschalten oder zu bedecken.

Verboten ist alles, was der Gemeinschaft nicht zuträglich sei, so die offizielle Version des Staates. Es ginge um Anstand und Achtung, Ethik und Moral, Ehrlichkeit und persönliche Integrität. Doch das autoritäre Regime demaskiert sich selbst: Über eine App, die an dunkelste Zeiten deutscher Geschichte in Hinblick auf Denunziation und selbst ernannte Blockwarte erinnert, können chinesische Bürger kriminelle Handlungen gegen finanzielle Belohnung melden. Staatsfeindliches Fehlverhalten sehen und den Behörden Bescheid geben, lautet das wirkliche Prinzip. Denn das meiste Geld wurde in diesem Fall, nach Angaben des deutschen Handelsblatts, das laut Vertrauensindex der Gesellschaft der führenden PR- und Kommunikationsagenturen in Deutschland (GPRA) als vertrauenswürdigstes Presseorgan in Deutschland gilt, für die Meldung illegaler politischer Aktivitäten gezahlt (www.gpra.de).

Als Fußgänger bei Rot die Fahrbahn überqueren? Auf einem öffentlichen Bildschirm am Straßenrand werden die Übeltäter angeprangert. Die Polizei kann ein Gesicht innerhalb von zwei Sekunden identifizieren und die Fotos auf ihre öffentliche Webseite stellen. Sie ist zudem in der Lage, im Handumdrehen zu orten, wo sich die Person in einer Millionenstadt aufhält. Und Apps chinesischer Anbieter, die eng mit dem Staat kooperieren, haben Zugriff auf staatliche Persönlichkeitsdaten, angeblich, damit der Login schneller ginge (vgl. Hua 2018).

Das Regime setzt die Daten zur Kontrolle und mutmaßlich zur Manipulation ihrer Bürger ein. Medien und alles, was im Internet erscheint, werden massiv zensiert. Landesweit gibt es zwei Millionen Zensoren, davon sogar bis zu 100.000 hauptamtliche. Ein Bericht des deutschen Außenministeriums beschreibt die Menschenrechtslage in China in vielen Facetten als katastrophal. Die Geschehnisse gehen weit über die seit Jahrzehnten unterschiedliche Definition von Menschenrechten hinaus. Ein vertrauliches Regierungspapier vom 22. Dezember 2019 belegt, dass die deutsche Regierung die Situation in China für bedrohlich hält. Seit bereits drei Jahren sei es in in Xijinang zu einer alarmierenden Zunahme von Kontrollen, Repressionsmaßnahmen und Diskriminierungen gekommen, die lückenlose digitale Überwachung und willkürliche Verhaftungen nach sich zögen. Anwälten würde der Kontakt zu ihren inhaftierten Mandanten untersagt, den Juristen drohe sogar Folter.

Insgesamt gäbe es Masseninternierungen von wohl über einer Million Menschen, Unterdrückung unerwünschter Ansichten sowie die Digitale Überwachung würde weiter zunehmen. Sippenhaft, repressive Limitierung von Religionsausübung, DNA-Identifikation und Speicherung, auch Indoktrinierungen ideologischer Art gehörten zum Alltag. Die Insassen von Internierungslagern müssten gemäß Augenzeugen derartige Umerziehungsmaßnahmen über sich ergehen lassen, die einer Gehirnwäsche gleichkommen, verbunden mit erzwungener Selbstkritik, Verleugnung von Religion und des Glaubens sowie eigener Herkunft. Aus den Lagern gäbe es Berichte über Misshandlungen, sexuelle Gewalt und sogar Todesfälle (vgl. Adelhardt et al. 2020).

Fakt ist: China beherrscht schon seit Jahren eine maßgeschneiderte Zensur. Informationen, die aus wirtschaftlicher Sicht Wertvolles enthalten, genießen ohne staatliche Behinderung höchste Transparenz. Politische Ansätze werden dagegen unterdrückt. Der moderne, digitale Überwachungsstaat unterdrückt Meinungsfreiheit, macht aber alles, um für den Staat ebenso wie für seine Bürgerinnen und Bürger steigenden Wohlstand reklamieren zu können.

Das Coronavirus kann am Ende nicht nur die Gesundheit, sondern auch die wachsende Prosperität der Bevölkerung gefährden, die die Volksrepublik als „Beruhigungspille" gegen ein mögliches Aufbegehren nutzt. 2049 will sie ihr 100-jähriges Bestehen feiern. Es könnte allerdings etwas dazwischenkommen, was der Kommunistischen Partei nicht gefallen dürfte. Mit zunehmendem Wohlstand werden sich die eineinhalb Milliarden Chinesen wie alle Menschen, die sich an etwas gewöhnt haben, fragen: „**Und nun?**" Dann könnte es passieren, dass das chinesische Volk sich nicht mehr in seinen freiheitlichen Rechten beschneiden lassen will und gegen die Unterdrückung aufbegehrt.

Das würde zwar die vor uns liegenden 29 Jahre totalitäre Staatskontrolle nicht vergessen machen, Hoffnung begründet es dennoch. Bis dahin wird die chinesische Regierung vermutlich weiterhin über eine Uminterpretation positiv konnotierter Gemeinschaft die freiheitlichen Rechte der Menschen wie auch ausländischer Unternehmen einschränken.

36.2 Digitale, ökonomische und politische Systeme – eine hilfreiche oder hemmende Ménage à trois

Die Demokratien der Welt, im Kern die Verteidiger menschenwürdiger Freiheiten, stehen unter enormem Druck. Einen demokratischen Einsatz von SCR erreichen sie nur, wenn sie die Ménage à trois, also die starke Dreiecksbeziehung zwischen Politik, Wirtschaft und Digitalisierung, die wie präzise Zahnräder ineinandergreifen, erkennen und gezielt für die Sicherung der freiheitlichen Werte einsetzen. Autokratien nutzen die Ménage à trois aus politischen, ökonomischen und digitalen Systemen bereits, um Ihre Stellung in der Welt zu stärken, im Fall China, um in den nächsten Jahren die globale Führungsrolle einnehmen zu können.

36.2.1 Digitale Systeme: digital versus digitalitär

Dass der chinesische Ministerpräsident Li Keqiang bereits 2016 auf der Big-Data Konferenz in Guiyang (Provinz Guizhou) ankündigte, bis 2020 auf chinesischen Servern rund 90 Prozent des gesamten Wissens der Welt dupliziert zu haben, ist nicht die wahre Gefahr. Der drohende Ersatz menschenwürdiger Freiheit durch menschenunwürdige Unterjochung ist es. Dies zu verbalisieren, ist für Menschen in Autokratien fast unmöglich. Deshalb sollten wir tunlichst auch Umfragen mißtrauen, die eine geschickt begründete, aber faktische Unterdrückung zur Sicherstellung einer totalitären Herrschaft gutheißen. Selbst in westlichen Demokratien leiden viele Marktforschungsmethodologien unter dem Faktor der gesellschaftlichen Erwünschtheit. Wer erinnert sich nicht an die überraschten Gesichtszüge des US-Präsidenten Donald Trump in der Wahlnacht des 8. auf den 9. November 2016. Er hatte zuvor den Umfragen geglaubt, die ihm keine Chance gaben.

Vor dem Jahr 2002, als weltweit erstmals mehr digitale als analoge Daten vorlagen, machten die unterschiedlichen Ausprägungen politischer Systeme an den Ländergrenzen halt. Die Ausprägungen digitaler Systeme sind dagegen schon heute so supranational wie die globalisierte Wirtschaftswelt. Deshalb nehmen künftig auch digitale Systeme eine Machtposition ein, deren gefährliches Entwicklungspotenzial die Wortkombination der

Begriffe „digital" und „autoritär" verdeutlicht: Die Möglichkeit eines Staates, seiner Ge-sellschaft den Aufbau eines allumfassenden Steuerungssystems zu verordnen, das der Staatsführung unumschränkte Herrschaftsmacht garantiert. Ein solches System könnte in China 2020 im ganzen Land zur Pflicht gemacht werden.

Mit SCR werden die Bürgerinnen und Bürger in ihrem täglichen Tun bewertet.

Die Daten dafür kommen größtenteils von Unternehmen, die an den Staat berichten. Zu den erfolgreichsten zählen heute Unternehmen, die eng mit dem Sicherheitsapparat kooperieren: allen voran Huawei, aber auch Cloudwalk, Hikvision, Megvii, oder auch Yitu. China nutzt die einflussreiche Marktstellung großer Technologiefirmen, um seine politischen und wirtschaftlichen Interessen zu verfolgen. In manchen Regionen ist die Technologie bereits in Regierungshand.

Die Datenanalyse und -prognostik von SCR wird in dem von China geplanten Umfang nur gelingen, wenn die Menschen in allen Bereichen ihres Lebens gescannt und analysiert werden. Und das kann zu metrischen Menschen führen und zu einer Gesellschaft, in der die Mathematik die Demokratie niederzwingt, in der Selbstbestimmung mithin keinen Platz hat.

Was gerade geschieht, wird sich nicht auf China und totalitäre Importländer begrenzen. Unter diesen ist China ohne Zweifel das vermögendste und innovativste, mithin auch das mit der größten Power im Bereich der Künstlichen Intelligenz.

Ist der oft milde lächelnde chinesische Präsident Xi Jinping der gefährlichste Gegner einer offenen Gesellschaft? Seine über SCR begünstigten Machtbestrebungen werden sich jedenfalls auf demokratische Gesellschaften auswirken, weil fortschrittliche Technologien immer auch das menschliche Bedürfnis nach Vereinfachung des Lebensalltags triggern. Gesichtserkennung aus China ließe sich hervorragend als eine Technologie der Data-driven Economy auch in andere, demokratische, Teile der Welt exportieren. Das bedeutet: Wir laufen Gefahr, dass uns vermeintliche digitale Annehmlichkeiten vergessen lassen, wo KI-gesteuerte Technologie unsere demokratischen Ideale unter Beschuss nimmt. Die chinesische Regierung nutzt eben diese unter anderem gegen Millionen Uiguren, einer muslimischen Minderheit.

Das digitale System ist innerhalb der Ménage à trois ein mächtiges Werkzeug, weil die Hand der Wirtschaft damit die Hand des Staates waschen kann: Unternehmen liefern der Regierung firmeninternes Wissen, unlimitiert sensible Informationen und Daten über Menschen oder Firmen für staatliche Datenbanken unvorstellbaren Umfanges.

Alibaba etwa zählt zu den oben genannten staatskapitalistischen Tech-Konzernen. Auch Tencent, die Firma, die allein mehr als zwei Milliarden Nutzer in Chat-Verläufen von WeChat, dem chinesischen WhatsApp, auswertet und Daten an die Kommunistische Partei liefert, die sie in den staatlichen Datenbanken zusammenführt. Damit gelingt der Regierung über das ursprünglich neutrale Werkzeug SCR ein Outsourcing nationaler Überwachung. Die Firmen übernehmen die Arbeit für die Partei. Auch Mitarbeiter und Zulieferer werden für „gutes", also auch parteikonformes Verhalten mit Ratingpunkten belohnt bzw. im gegenteiligen Fall bestraft. In der Folge müssen sich auffällige Firmen häufigeren staatlichen „Inspektionen" unterziehen. Sie haben bei Lizenzvergaben kaum noch eine Chance, bekommen Steuererhöhungen oder man kappt Subventionen sowie den Zugang zu öffentlichen Ausschreibungen. Gleichzeitig kann ein Angestellter einer in Misskredit gefallenen Firma für sich selbst weder Flüge oder Bahnfahrten mehr buchen, Versicherungen abschließen oder eine Wohnung erwerben. Und es wird wie bereits erwähnt, alles öffentlich gemacht. Firmen und Menschen, die bei dem in Ungnade gefallenen Unternehmen beschäftigt oder mit ihm geschäftlich verbunden sind, werden an den Pranger gestellt. Die öffentliche Hinrichtung, die am 17. Juni 1939 in Frankreich verboten wurde, ist digital basiert zurück. Schon durch die Zusammenarbeit mit Produzenten oder Serviceanbietern, die ein „absteigendes Unternehmen" beliefern, riskieren Firmen, ohne eigenes „Vergehen", mit einem Malus bestraft zu werden. Der an der Universität Trier lehrende Sinologe Sebastian Heilmann, von der Tagesschau als einer der besten China-Kenner Deutschlands bezeichnet, sagt: „Wer an der Spitze der Pyramide, der Datennahrungskette, steht, ist völlig klar: Das ist nicht das marktbeherrschende Digitalunternehmen, sondern das ist ganz eindeutig die Regierung und Partei in China." Die deutsche Außenhandelskammer hat nun ermittelt, dass sich 70 Prozent deutscher Firmen, die in China tätig sind, noch nicht einmal mit dem neuen System beschäftigt haben (o. V. 2020c). Sie haben damit weder eine Vorstellung davon, wie ihr jetziges Handeln unsere demokratische Zukunft beeinflussen könnte, noch haben sie einen Plan, um das zu verhindern.

36.2.2 Ökonomische Systeme: Staatskapitalismus versus soziale Marktwirtschaft

Ganz anders China. Die Volksrepublik verfolgt einen Plan, und zwar einen langfristigen. Und nutzt die oben beschriebene, systemische Triade geschickt dazu, auch und gerade, wenn es um die Wirtschaft geht. Planwirtschaft wird im Westen allzu gern belächelt, aber das Beispiel der Elektromobilität beweist, wie sie funktioniert. China investierte zunächst viel Geld in die Zukunftsbranche bis das Land eine global führende Rolle erreicht hatte, fährt nun aber die Subventionen herunter und überlässt die junge chinesische Automobilbranche dem Spiel der weltweiten Marktkräfte.

Nachdem China lange als globale Kammer billiger Massenproduktion galt, steht seit fünf Jahren „China 2025" auf der Agenda, um auch die industrielle Produktion zu digitalisieren. Den Anstoß gab offensichtlich eine Idee aus Deutschland unter dem Namen Industrie 4.0, die Bezeichnung für das deutsche Zukunftsprojekt, das 2011 in der Forschungsunion der Bundesregierung entstand. Der chinesische Staatskapitalismus könnte sich mit Hilfe digitaler Technologie nun auch den Weg zu einer produzierenden Supermacht ebnen.

Kapitalismus wird oft in einem Atemzug mit Demokratie genannt. Und in der Tat ist die Verwirklichung demokratischer Ordnungen bisher nur in kapitalistisch agierenden Ländern erfolgt. Die aktuelle Kapitalismusdepression ist von vielfältiger Kritik determiniert, auch Wirtschaftslenker selbst kritisieren: Konzerne fallen durch Steuervermeidung auf, globale Digitalfirmen sind nahezu oder tatsächlich monopolistisch, prekäre Arbeit nimmt zu. Ungleichheit, für die der Kapitalismus verantwortlich gemacht wird. Selbst höchste Privatheit unterliegt über lukrative Apps wie Tinder jetzt dem Prinzip des Kapitalismus. Ihm wohnt zwar Wachstum inne, doch gleichzeitig wirft er soziale und ökologische Probleme auf. In der Folge erfordert das mehr Sensibilität für Notwendigkeiten des Umweltschutzes sowie für die Gefahren von Rechtsextremismus. Wenn radikale Meinungsmacher gepusht vom Populismus zu politischem Einfluss gelangen, gehen parlamentarische Aktion und rechtsextreme Aggression Hand in Hand. Der Kapitalismus steht in der Kritik wie lange nicht. Das zeigt auch der globale Trust Barometer, der seit 20 Jahren das Vertrauen gegenüber Institutionen misst und für das Jahr 2020 weltweit über 34.000 Menschen in 28 Märkten befragte. Mehr als die Hälfte davon ist der Meinung, dass der Kapitalismus in seiner heutigen Form mehr schadet als nutzt, ein Weltphänomen also. Allerdings eines, das meist Demokratien zugeschrieben wird, und das, obwohl Kapitalismus auch in China existiert: als Staatskapitalismus mit vermeintlicher Menschenfreundlichkeit (durch steigenden Wohlstand) und faktischer Menschenfeindlichkeit, die gemäß Berichterstattungen Millionen Menschen in Lagern unterjocht und ihrer individuellen Menschenrechte beraubt.

Das wirtschaftliche System des chinesischen Staatskapitalismus ist durch eine enge Beziehung zwischen Staat und dem wirtschaftlichen System, also den Unternehmen, gekennzeichnet. Über politische Einflussnahme kann die Konkurrenz des Marktes, die bekanntlich das Geschäft belebt, protegiert oder geschwächt werden. Staatskapitalismus nutzt ein digitales Update der Planwirtschaft, das auch vom SCR geliefert wird und bereits in 18 autokratische Systeme der Welt exportiert wurde. (vgl. Strittmatter 2019).

Die soziale Marktwirtschaft, wie wir sie in Deutschland kennen, hat die Schwächen der, insbesondere im 18. und 19. Jahrhundert in Europa praktizierten, freien Marktwirtschaft, behoben. Mit dem Aufkeimen der Sozialen Frage im 19. Jahrhundert wurden dem System Grenzen aufgezeigt, da keine gerechte Einkommensverteilung erzielt werden konnte. Der Einbezug der Sozialen Frage in das wettbewerbliche Ordnungssystem ist Alfred Müller-Armack zu verdanken. Er verband das Prinzip der freien, durch den Wettbewerb dominierten Märkte mit dem Streben nach sozialer Sicherheit und sozialer Gerechtigkeit. In ihrem Zentrum steht die Würde des Menschen in Kombination mit Freiheit und Verantwortung. Die Entwicklung der Sozialen Marktwirtschaft geht einher mit dem Wiederaufbau in Deutschland nach dem Zweiten Weltkrieg. Müller-Armack erkannte die Notwendigkeit, eine neue Ordnung zu entwickeln. Dabei sollte die Marktwirtschaft den Rahmen bilden, aber „eben keine sich selbst überlassene, ungeregelte Marktwirtschaft, sondern eine bewusst [...] sozial gesteuerte" etablieren werden. Die Soziale Marktwirtschaft hat sich also zum Ziel gesetzt, im Sinne einer Sozialordnung, die Wohlfahrt und die Lebensqualität einer Gesellschaft zu erhöhen, oder wie Ludwig Erhard es populistisch formulierte: „Wohlstand für alle!" (vgl. Erhard 1975).

Das tönt auch aus Chinas Beruhigungsplan heraus, der aber mit den digitalen Möglichkeiten von SCR schon heute in eine abhängige Marktwirtschaft unter totaler Kontrolle führen kann. China nutzt Vorteile des Kapitalismus. Er schafft Wohlstand und beseitigt Armut.

Dennoch sollte man den Kapitalismus nicht idealisieren: Ja, er sorgt für Wohlstand, aber nicht automatisch für Gerechtigkeit oder eine gesunde Umwelt. Er fördert eine offene Gesellschaft, garantiert aber keine Chancengleichheit. Er ist ein Wirtschaftssystem, kein allgemeiner Beglückungsautomatismus. Er ist robust, aber nicht unverwundbar. Er wird zwangsläufig immer wieder auch disruptiv Unternehmen, Arbeitsplätze und ganze Branchen zerstören. Aber Kapitalismus zeichnet auch aus, was gleichzeitig seine Voraussetzungen sind: klare Eigentumsverhältnisse, Rechtssicherheit und Transparenz. Schutz des Eigentums bedeutet Schutz der Menschen. Wenn geschäftliche Konflikte mit staatlicher Gewalt gelöst werden, können enorme Vermögen ohne rechtliche Grundlage illegitim angehäuft werden. So ist der Kapitalismus beispielsweise in Russland von Beginn an daneben gegangen, was bis heute offenbar ist.

Kapitalismus und SCR haben also zwei Gemeinsamkeiten: Sie sind erstens niemals per se „gut" oder „böse". Und es kommt immer darauf an, was nicht nur die Wirtschaft, sondern auch die Politik daraus machen.

36.2.3 Politische Systeme: Demokratie versus Autokratie

Politisch betrachtet fällt SCR eine so große Bedeutung zu, weil sich mit dem Einsatz dieser Technologie Autokratien, deren Einordnung hier nicht eine festgeschriebene Verfassung (*de jure*), sondern die Lebenswirklichkeit (*de facto*) referenziert, einen menschenunwürdigen Vorteil verschaffen können.

Kein Wunder, dass der freiheitlich schlagende Puls höher schlägt, wenn eine digitale Diktatur die sicher geglaubte westliche Überzeugung zertrümmert, Demokratien befänden sich auf einem überlegenen Siegesmarsch. Auch das Pentagon warnt vor digitalen Diktaturen. Im 20. Jahrhundert bewiesen die Autokratien noch ihre Ineffizienz und vor allem Instabilität, da sie versuchten, alles von der Wirtschaft bis zur Gesellschaft zentral, vor allem aber analog zu steuern. Das hat sich durch die neueste Technologie geändert. Mit Big Data und KI bedient sich China der Algorithmen, die Daten fressen und keine Diät fürchten müssen. Denn die staatlich verordnete Datenvöllerei kennt in Autokratien kein Ende, während freie Demokratien sich Datenschutz verschreiben und sich auf diese Weise so schlank machen, dass sie ihr Gewicht in der Welt verlieren.

Das beschreibt das Ende eines Wettrennens, denn einer der beiden Läufer ist nicht mehr wiederzuerkennen. Er kam mit Kapitalismus um die Ecke, fusionierte staatspolitische Diktatur unveränderten Kontrollwahns (für den die Algorithmen wie geschaffen erscheinen) mit Wissen aus der Marktwirtschaft, fertig war der Staatskapitalismus. Dem digitalen Kontrollzwang kann sogar die Bevölkerung etwas abgewinnen. Im Vergleich zur Kaderwillkür erscheint SCR als unabhängiger und damit relativer Fortschritt. Die Erkenntnis „Pest statt Cholera" könnte nicht lang auf sich warten lassen.

Die Technologie ist längst da, auch SCR. Hier ist China bereits führend. Auch wenn das Reich der Mitte noch weit davon entfernt ist, stehen künftig 1,4 Milliarden Menschen als Datenquellen zur Verfügung – ein riesiges Kontrollpotenzial. Der neue politische Kampf hat begonnen. Die Freiheit favorisierenden Gesellschaften sollten also aufpassen, dass Demokratie nicht zum Auslaufmodell wird.

36.3   Wirtschaft als Katalysator für Wandel

Rechte und Pflichten sind die beiden Seiten einer Medaille namens Freiheit. Die Bürger demokratischer Staaten (Privatpersonen wie Unternehmenslenker) haben also nicht nur Rechte, sondern auch Pflichten, nämlich jene zur Erhaltung der demokratischen Grundrechte. Denn die persönliche Freiheit endet dort, wo sie die eines Anderen verletzt.

Vor diesem Hintergrund müssen wir uns fragen, ob die blinde Akzeptanz eines machtpolitisch gesteuerten SCR-Systems seitens der internationalen Wirtschaft, wie es im Geleitwort dieses Buches anklingt, wirklich alternativlos ist. Ist es wirklich zielführend im Hinblick auf die globale Entwicklung demokratischer Grundwerte, wenn sich Unternehmen der Forderung der Chinesen, sich ihrer SCR-Politik unterzuordnen, beugen? Nein! Das ist es nicht!

Firmen können sich der chinesischen Wirtschaftsmacht beugen. Sie können aber auch das Vertrauen der Gesellschaft, die sich auch aus Shareholdern, aktuellen und potenziellen Kunden sowie Mitarbeitern zusammensetzt, gewinnen, indem sie sich wehren. Es gehört zur unternehmerischen Vertrauensverpflichtung, sich überall auf der Welt für eine menschenwürdige Zukunft einzusetzen.

Das Vertrauen in die Übernahme von Verantwortung wird in Zukunft immer mehr zur Existenzsicherung von Firmen beitragen und beruht auf zwei Hauptfaktoren: Kompetenz und ethisches Verhalten. Der bereits genannte Trust Barometer zeigt: Aktuell wird weltweit keine Institution (ob in der Wirtschaft oder unter staatlicher Aufsicht, ob Nicht-Regierungsorganisation (NGOs) oder Medien) als kompetent und gleichzeitig ethisch einwandfrei angesehen. Es gilt nur die Wirtschaft als kompetent, während lediglich NGOs als ethisch eingeschätzt werden.

Diese Ergebnisse deuten darauf hin, dass die Menschen nur wenig Vertrauen in Unternehmen haben, einen Beitrag leisten zu wollen, um die dringendsten Herausforderungen unserer Zeit zu bewältigen. Die der Studie zugrunde gelegten Komponenten ethischen Verhaltens – Ehrlichkeit, Visionen, Zielstrebigkeit und Fairness – haben dabei die größte positive Auswirkung auf das Vertrauen. Das bedeutet, dass die kaufkräftige Gesellschaft auch von der Wirtschaft ethisches Verhalten und Übernahme von Verantwortung erwartet.

Unternehmenslenker haben nicht nur die Möglichkeit, nicht „Ja und Amen" zu den mit SCR einhergehenden Sanktionen zu sagen, sie haben die Pflicht, eine opportunistische Haltung zu unterlassen. Die Wirtschaft kann sogar ein Katalysator für den Wandel werden. Nehmen wir Chinas bereits angesprochene Beruhigungspille für die chinesische Bevölkerung: den Wohlstand. Dafür braucht das Land Produkte, auch die ausländischer Unternehmen. 2019 exportierte China zwar Waren im Wert von 2,5 Billionen US-Dollar, importierte aber zugleich für 2,1 Billionen US-Dollar. Chinesen lieben westliche Produkte, wenn diese plötzlich fehlen, würde das die KP unter Druck setzen.

Angenommen, ein westlicher Automobilhersteller würde nicht mehr in die Volksrepublik liefern, täte es seine Konkurrenz vielleicht dennoch oder wahrscheinlich sogar erst recht. Im Moment spielt der kapitalistische Wettbewerb China daher in die Hände. Hier steuert die Politik auch wirtschaftlich alles, im Westen dagegen gibt es viele Player. Es herrscht kein international bindender wirtschaftsethischer Standard. Spinnen wir das Auto-Beispiel weiter: Die Reputation des Herstellers für sein verantwortliches Handeln gegenüber Kunden, Mitarbeitern, allen Stakeholdern würde außerhalb Chinas deutlich steigen und er würde aufgrund des Vertrauensaufbaus mittelfristig in anderen Märkten die als nicht-integer wahrgenommenen Wettbewerber bzw. chinesische Anbieter verdrängen. Das ist heute zugegebenermaßen noch schwer vorstellbar, muss aber zusammen mit der sich verändernden gesellschaftlichen Anforderung der Menschen von morgen an unternehmerische Verantwortung gedacht werden.

Die Wirtschaft kann also nach ethischen Maßstäben Vertrauen gewinnen, wenn sie „NEIN!" sagt. Die chinesische Regierung wird die Unternehmen, die sich nicht beugen, aller Voraussicht nach ausschließen. Zunächst mag das erste Unternehmen, das eine Vorreiterrolle übernimmt, vielleicht allein dastehen. Doch andere, die bei der Konkurrenz steigende Zustimmungswerte seitens kritischer Konsumenten und daraus resultierende Umsatzzuwächse in anderen Ländern beobachten, könnten folgen.

Darüber hinaus können westliche Unternehmen auch im eigenen Land einen Beitrag leisten, um SCR im Sinne einer zukunftsgerichteten, freien Wirtschaft zu fördern. Beispiele, die Schule machen können: In Deutschland fördern Banken nachhaltiges Investment durch vergünstigte Kredite für ihre Kunden, die sich dem Lieferkettengesetz verpflichten, das Kinderarbeit verhindern, existenzsichernde Löhne garantieren und Umweltschäden vermeiden soll. Die größte Bank Europas, die HSBC, widmet sich verstärkt dem Thema Nachhaltigkeit. Sie trifft CO_2-basierte Risikoentscheidungen, interessiert sich also für den ökologischen Footprint ihrer Kunden. Auch das laut Fortune Global 500 umsatzstärkste Unternehmen und gleichzeitig mit über zwei Millionen Angestellten der größte privater Arbeitgeber der Welt, der US-Einzelhandelskonzern Walmart, zählt zu den Kunden des Londoner Bankhauses und unterwirft sich den nachhaltigen Ansprüchen, die durch Social Credit Rating in diesem Bereich gefördert werden.

Aber: Die Märkte allein werden es nicht schaffen, SCR als demokratisches und pluralistisches Instrument zu gestalten. Erst eine demokratisch verankerte und konzertierte Ménage à trois aus Digitalisierung, Wirtschaft und Politik sichert den Erhalt und die Entwicklung der freiheitlichen Werte.

36.4   Frostschutz für Freiheit

Wir reden über eine Technologie, die in den falschen Händen schon in ein paar Jahren unsere globale Freiheit beschneiden kann. Die Gefahr geht dabei allerdings nicht allein von Autokratien aus, sondern auch von westlichen Ländern, die Grenzen überschreiten, um nicht zurückzufallen. Getrieben vom Wettrennen mit dem vorauseilenden China lassen beispielsweise die USA (unter dem Twitter-Beifall Chinas auf dem Trump-Account) es zu, dass Bürger bei der Einreise das Kennwort ihres Computers offenbaren müssen. Auch den oft unbegrenzten Träumen des Silicon Valleys fehlt, angetrieben von chinesischen Erfolgen, bisweilen die ethische Fragestellung: „Tun wir, was wir tun könnten?" Das ist aber kurzsichtig. Um mittel- bis langfristig nicht zum Übernahmekandidaten eines eisigen Totalitarismus werden, benötigen wir daher einen „Frostschutz" für Demokratien. Die Ménage à trois weist dabei den Weg.

36.4.1 Was das politische System leisten kann

So wie der Wettstreit zwischen demokratischen, faschistischen und kommunistischen Gesellschaften das 20. Jahrhundert prägte, könnte der Kampf zwischen freiheitlich digitaler Demokratie und unfreier digitaler Diktatur das 21. Jahrhundert bestimmen. Demokratien werden ihn kaum für sich entscheiden können, wenn sie nicht supranational zusammenarbeiten, um ihre gemeinsamen freiheitlichen Grundwerte zu sichern. Nicht Chinas Stärke ist unser größtes Problem, sondern unsere eigene Schwäche. Die internationale Verständigung könnte innerhalb und außerhalb Europas deutlich besser sein. Europa zeigt sich oft unfähig, auf globale Herausforderungen, etwa beim Klimaschutz oder in Sicherheitsfragen, gemeinsame und überzeugende Antworten zu finden. Dazu gehört aber auch, dass wir uns mehr um Staaten an der europäischen Peripherie kümmern sollten. Zurzeit sind Chinas Angebote zum Beispiel für Ungarn und Griechenland so attraktiv, dass diese Länder einfach nicht nein sagen können. Die Demokratien der Welt müssen also zusammenhalten. Auch um den Kapitalismus mit politischen Mitteln zu neuen freiheitlichen Zielen zu verhelfen. Dies ist schwer, aber möglich.

In der Bundesrepublik ist die Harmonie von repräsentativer Demokratie und als Soziale Marktwirtschaft ausdefiniertem und damit optimiertem Kapitalismus gelungen, aber nationalstaatlich begrenzt. Sich damit zu begnügen, kann keine erfolgversprechende Strategie sein.

Es gibt gute Gründe, einen staatlichen Rüstungswettlauf nicht mit digitalen Waffen fortzusetzen. Für Autokratien ist Künstliche Intelligenz (KI) (vgl. Thomsen 2019b) die neue Geheimwaffe. Bereits zu Beginn des Jahres 2019 veröffentlichte das Handelsblatt eine Prognose, wonach der Anteil der KI am Bruttoinlandsprodukt 2030 in Nordeuropa keine zehn Prozent (9,9 Prozent), in den USA keine 15 Prozent (14,5 Prozent), aber in China über 25 Prozent (26,1 Prozent) betragen werde. Und wie steht es um die ehemaligen Weltmacht, der Wladimir Putin zu gern entsprechende Geltung zurückgeben würde? Die Eingabe „Künstliche Intelligenz Russland" in der Suchmaske des Wirtschaftsmagazins Handelsblatt ergab: 0 Treffer. Manche Beobachter sehen die Supermacht am Ende.

Vielleicht ist es, mit dem russischen Schicksal im Kopf und am Ende dieses Buches an der Zeit, eine unbequeme, wirklich beängstigende Frage zu stellen: Ähneln deutsche Bedenkenträgerschaft, die „German Technophobia", also die Angst vor dem Fortschritt, ebenso wie die Bürokratie-, Auflagen- und Regelflut zu sehr den verschlossenen Gittern, hinter denen vor 60 Jahren der russische Geheimdienst eine Staatsfeindin sperren ließ, deren Bedrohung so gewaltig schien, dass sie, so die Direktive der russischen Inlandsüberwachung, zwingend hinter Schloss und Riegel gebracht werden musste. Ihr Name: Fotokopiermaschine.

Fortan war die subversive Gefahr des Geheimnisverrats unter Kontrolle gebracht.

Ist die aktuelle, westliche Zukunftsangst ähnlich paranoid? Soviel ist sicher: Der Sowjet-Kommunismus ist Geschichte. Seine Bürokratie brach unter der eigenen Last zusammen. Auch mit Kapitalismus hat Russland eigene Erfahrungen gemacht. Jedoch auch jene machen müssen, dass ohne Rechtsstaatlichkeit Vieles nichts ist.

In China dagegen gibt es seit Jahrhunderten einen relativ gut funktionierenden Staat, nicht jedoch eine Tradition der Rechtsstaatlichkeit. In Indien ist es umgekehrt. Bis heute hat der Staatskapitalismus Chinas schon mehr Wohlstand geschaffen, die Behörden sind inzwischen bemüht, sicherere Eigentumsrechte zu entwickeln. In Indien fehlt es nicht an einem Rechtsverständnis, erst recht nicht an hoch motivierten jungen Menschen, aber Chaos – zum Beispiel in der Verwaltung – das entzieht dem Land Tempo. In allen drei Fällen, in Russland, China und Indien, ist aber nicht der Kapitalismus das Problem, sondern bei zweien eine repressive und bei dem anderen eine unzureichende Politik.

Eine fast beiläufige Bemerkung, die Cheng Chuanwei, ein Taxifahrer aus Shenzhen, der deutschen Korrespondentin Sha Hua anvertraute, verrät mehr als alles, was er zuvor im Sinne der Parteidoktrin fast druckreif formulierte: „Für Demokratie ist China – noch – nicht reif, erst müssen wir persönlichen Fortschritt erreichen und dann irgendwann werden wir sogar Demokratie wagen" (vgl. Hua 2018). „Sogar" – diese Partikel drückt bekanntlich eine Steigerung aus. Wenn das nicht die erträumte Freiheit und das prognostizierte Aufbegehren nach erfolgter Wohlstandserreichung beweist.

In einem der chinesischen Selbstdarstellungsvideos heißt es: „Der Schlüssel zu Chinas Erfolg liegt in seinem System der Demokratie." Noch aber klafft eine große Lücke zwischen Filmwelt und Realität. Ob aus der Corona- eine Legitimationskrise resultiert, die die Schließung der Lücke beschleunigt?

Die Diskussion zu SCR bewegt viele Menschen, die letztlich feststellen werden, dass das Bewertungssystem an sich nicht die Wurzel des Übels ist, sondern der totalitäre Diebstahl persönlicher Freiheit.

36.4.2 Was das digitale System leisten kann

Als Demokraten wollen wir keine Macht der Kontrolle. Warum fokussieren wir in dann in Europa nicht genau das, womit wir unsere Ziele und einen großen Wettbewerbsvorteil harmonisieren? Die technologische Antwort Europas auf totalitäre Kontrolle über SCR und die Antwort auf die Frage nach zukünftigen Erfolgschancen in globalen Märkten könnte heissen: Datensicherheit!

Digitale Infrastrukturen sind Voraussetzung für neue Wertschöpfungsketten und sichere Datenkanäle damit ein Geschäftsfeld der Zukunft. Es wird in absehbarer Zeit kaum noch

Produkte geben, die ohne digitale Lösungen ihre Wettbewerbsfähigkeit sichern können. Ob es um Kundennutzen oder Herstellungsprozesse geht: Daten sind Voraussetzung für Wirtschaftswachstum.

Wie kann Europa im Kampf zwischen China und den USA mithalten? Europa gilt im Vergleich zu den beiden Wettstreitern in der sogenannten Plattformökonomie als ausgesprochenes Leichtgewicht. Im Segment Business-to-Consumer (B2C), also bei Geschäftsbeziehungen zwischen Unternehmen und Konsumenten, könnte Europa bereits abgehängt sein, beim Business-to-Business (B2B), also bei Geschäftsbeziehungen zwischen Unternehmen, bestehen durchaus noch Chancen, eine Spitzenposition einzunehmen.

Doch B2C und B2B eint deren Notwendigkeit von Sicherheit. Zu den wichtigsten Hemmnissen der sogenannten Industrie 4.0 zählen für Unternehmen neben den Kosten (66 Prozent) die Anforderungen an den Datenschutz (65 Prozent) und die Datensicherheit (61 Prozent) (vgl. Riemensperger, F., Pfannes, P., Bongardt, S., 2020). Hier schlummern technologische Verschlüsselungspotenziale für eine neue Weltmarktführerschaft, die zudem perfekt zur Datensensibiltät der europäischen Demokratien passt. Während andere sich um digitale Systeme etwa für SCR kümmern, könnte Europa dafür sorgen, dass alle Anwender „Safetainty" genießen, also die unumstößliche Gewissheit, dass Ihre Daten (und letztlich damit auch freiheitliche Rechte und Pflichten) sicher sind.

Die Frage der Sicherheit erstreckt sich aber nicht nur auf zivile Felder, sondern auch auf militärische. In Zukunft finden Kriege zunehmend auch im virtuellen Raum mit Mitteln der Informationstechnologie statt. Ob Energie- und Wasserversorgung, wichtige Funktionen der Kommunikation oder des Finanzsystems – ein Cyberkrieg zerstört die Infrastrukturen des Gegners. Und das zum „Schnäppchenpreis".

Kaum eine Woche nach Amtsantritt der deutschen Verteidigungsministerin Annegret Kramp-Karrenbauer, offenbarte eine parlamentarische Anfrage im Deutschen Bundestag den Preis eines einzigen Panzers mit dem Namen „Puma". 350 davon waren in der Regierungszeit von Bundeskanzler Gerhard Schröder bestellt worden. Aktueller Stückpreis: über 17 Millionen Euro. Keine zehn Millionen Euro dagegen würde es kosten, beispielsweise in Österreich durch einen Cyberangriff auf den mitteleuropäischen Staat dessen zentrale Einrichtungen des (analogen) Militärs, die gesamte Infrastruktur von Krankenhäusern sowie die Strom- und Wasserversorger in Wien völlig lahmzulegen. Ein derartiger Angriff, so berichtete am 9. Februar 2019 die nach dem Vorbild der New York Times konzipierte, in Wien erscheinende österreichische Tageszeitung „Der Standard", werde irgendwann stattfinden! Damit rechne man beim österreichischen Heer (Al-Youssef und Sulzbacher 2019).

Sicherheit und Schutz gegen solche Attacken kann nur die bessere Technologie liefern. Hieß es früher: „Stell Dir vor, es ist Krieg und keiner geht hin", heißt es dann wohlmöglich „Stell Dir vor, es ist Krieg und keiner kommt durch". Ein Angriff, der dank europäischer Lösungen nicht durchs Netz geht, ist wie eine Vollsperrung der Datenautobahn für die Nachfolger des analogen „Puma". Europa muss nur JA sagen.

36.4.3 Was das ökonomische System leisten kann

Welche Learnings ergeben sich daraus: Wir können, anders als die Bevölkerung autokratischer Staaten „NEIN!" sagen und trotzdem gewinnen. Unternehmen demokratischer Staaten sind, wie oben ausgeführt, gefordert, ihre Integrität nicht an ein totalitäres System zu verkaufen. Und wir können von der Planwirtschaft lernen, dass gebündelte Zielsetzung und deren konsequente Umsetzung sehr wohl einen wichtigen Platz in unserer Wirtschaft einnehmen sollten, um nicht von China abgehängt zu werden. Staatliche Planungskonzepte, die sich auf globale wirtschaftliche Veränderungsprozesse fokussieren, scheinen in vielen Demokratien entweder gar nicht vorhanden zu sein oder im Gegenteil zu detailliert und aufgrund langwieriger Entscheidungsprozesse kurz nach Fertigstellung schon wieder veraltet. Für Staaten wie für Unternehmen gilt, dass das Referenzieren der zahlenbasierten Vergangenheit, besonders ausgeprägt bei Unternehmensberatungen, noch viel zu oft nicht den nächsten Schritt macht. Eine innovative Unternehmensplanung – ausgerichtet auf Dynamiken und valide Opportunitäten der Zukunft – kann messbar zum Gesamtgeschäftserfolg beitragen. Regierungsplanung ebenso, sie wird jedoch allzu gern, vielleicht unbewusst, als seit dem 18. Jahrhundert entstandenes Konzeptes planwirtschaftlicher Verfehlung diskreditiert, die Ressourcen zentral allokiert anstatt auf dezentrale Eigenverantwortung zu setzen. Letzteres ist tatsächlich sehr wichtig! Jedoch kommt es durchaus auf Planung an. Jene zur Vermeidung der Begrenztheit konventionellen Tuns. Stattdessen mit höchster Planungseffizienz und -effektivität (vgl. Thomsen 2019c).

Dann gilt es nur noch zu lernen, Erreichtes zu würdigen und immer wieder mit neuem Mut durchzustarten. Wie auf Startbahnen überall auf der Welt, auch in China, auf denen Airbus-Flugzeuge starten. Sie brachen die globale Boeing-Dominanz Amerikas in der zivilen Luftfahrt. Wer glaubt, die Todesmaschine Boeing 737 Max, deren zu große Triebwerke, die Software anders als gehofft nicht ausglich, sei der Grund dafür, denkt zu spät. Die Probleme begannen früher, nicht erst mit der überhasteten Reaktion auf den Markterfolg des Airbus 320 Neo. Probleme, die wegen der vorgenannten Triebwerk-Software-Konstruktion zu tödlichen Abstürzen in Indonesien und Äthiopien führten, bei denen insgesamt 346 Menschen in Boeing-Maschinen starben. Airbus fliegt Boeing davon, weil das amerikanische Management seit Jahren zu langsam, zu zögerlich bei Forschung und Entwicklung ist. Kein Wunder, dass Airbus nun in der zivilen Luftfahrt das modernere Fluggerät bietet. Wer fast dreimal so viel in Aktienrückkäufe steckt wie in Innovation, muss sich fragen lassen, ob Boeing tausende weinende Angehörige für lächelnde Aktionäre bewusst in Kauf nahm (vgl. Kort 2020).

36.4.4 Was wir alle leisten können

Wir müssen unsere Freiheit nutzen, um unsere Demokratie zu entwickeln, sodass sie den Anforderungen der Zeit standhält. Was es braucht, um die „German Technophobia" zu überwinden: Neue demokratische Leitplanken statt überbordenden Gesetzesterrorismus.

Jeder Wandel beginnt im Kopf. Geschwindigkeit ist ein gutes Beispiel dafür. Wir sollten anfangen, darüber nachzudenken, wie wir unter Wahrung unserer Werte den Wettbewerbsnachteil der schneckenartigen Langsamkeit politischer Entscheidungen und deren durch überbordende Regulierung gehemmte Realisierung endlich beseitigen. Wir können mutiger werden, kreativer und dadurch selbstbewusster. Wir müssen es nur wollen.

Selbst die 22,87 Millionen in Deutschland lebenden Eltern können unsere Freiheit auf ihre eigene Art bewahren, indem sie sich von traditionellen, nicht mehr zeitgemäßen Erziehungsmethoden wie Zwang, Strafe und Belohnung verabschieden und eine demokratische Haltung vorleben. Es geht um Ermutigen statt Bestrafen, um Handeln statt Reden und darum, Kindern so früh wie möglich die logischen Folgen ihres Verhaltens bewusst zu machen. So wird die Familie, die kleinste Demokratie der Welt, zur Keimzelle demokratischer Grundwerte. Und das ganz ohne Social Credit Rating.

Literatur

Adelhardt, C., Eckstein, P., Pinkert, R., Strozyk, J., Strunz, B. (2020): Alarmierende Verschlechterung, unter: https://www.tagesschau.de/investigativ/ndr-wdr/menschenrechtslage-china-101.html, zuletzt abgerufen am 11. März 2020.

Al-Youssef, M., Sulzbacher, M. (2019): Ziel: Brave Bürger – Überwachung und Kriege der Zukunft, unter: https://www.derstandard.at/story/2000097770298/ziel-brave-buerger-ueberwachung-und-kriege-der-zukunft, zuletzt abgerufen am 11. März 2020.

Erhard, L. (1975): Wohlstand für Alle. Düsseldorf 1975.

Hua, S. (2018): Die Überwachungsstadt – ein Besuch im chinesischen Shenzhen, unte: https://www.handelsblatt.com/unternehmen/it-medien/selbstversuch-die-ueberwachungsstadt-ein-besuch-im-chinesischen-shenzhen/22510404.html, zuletzt abgerufen am 11. März 2020.

Kort, K. (2020): Jetzt rächt sich, dass Boeing zu wenig in die Forschung investiert hat, unter: https://www.handelsblatt.com/meinung/kommentare/kommentar-jetzt-raecht-sich-dass-boeing-zu-wenig-in-die-forschung-investiert-hat/25460512.html?ticket=ST-6208638-yFdXxp3PsxFYowAC-GeVM-ap4, zuletzt abgerufen am 11. März 2020.

https://www.amazon.de/Soziale-Marktwirtschaft-nationale-internationale-Ordnung/dp/3879590842 Müller-Armack, Alfred, Symposion 1 – Soziale Marktwirtschaft als nationale und internationale Ordnung, Bonn 1978, S. 22–36, hier: S.23.

o. V. (2020a): Coronavirus: Wieder fast 100 Tote in China, unter: https://www.sueddeutsche.de/gesundheit/medizin-coronavirus-wieder-fast-100-tote-in-china-dpa.urn-newsml-dpa-com-20090101-200223-99-29252, zuletzt abgerufen am 11. März 2020.

o. V. (2020b): 16 Corona-Fälle in Deutschland, unter: https://www.tagesschau.de/inland/coronavirus-webasto-bundestag-101.html, zuletzt abgerufen am 11. März 2020.

o. V. (2020c): Totale Kontrolle – Was Chinas Sozialkreditsystem für deutsche Firmen bedeutet, unter: https://www.daserste.de/information/wirtschaft-boerse/plusminus/sendung/swr/china-sozialkreditsystem-100.html, zuletzt abgerufen am 11. März 2020.

Riemensperger, F., Pfannes, P., Bongardt, S., (2020): Top500-Studie Deutschland – Weltmarktführer von morgen, unter: https://www.accenture.com/_acnmedia/PDF-114/Accenture-Top500-Studie-Deutschland-Weltmarktfuhrer-von-morgen.pdf,

Strittmatter, K. (2019): China exportiert seine Überwachungstechnik – und den Totalitarismus. Europa muss endlich aufwachen, unter: https://chrismon.evangelisch.de/artikel/2019/46745/chinas-neuer-ueberwachungsstaat-findet-immer-mehr-anhaenger, zuletzt abgerufen am 11. März 2020.

Thomsen, B. (2019a): Sieben Dinge, die deutsche Firmen von China lernen können, unter: https:// www.handelsblatt.com/meinung/kolumnen/expertenrat/thomsen/expertenrat-prof-bernd-thomsen-sieben-dinge-die-deutsche-firmen-von-china-lernen-koennen/23895424.html, zuletzt abgerufen am 11. März 2020.

Thomsen, B. (2019b): Keine Angst vor der KI! Was Menschen Maschinen voraushaben, unter: https://www.handelsblatt.com/meinung/kolumnen/expertenrat/thomsen/expertenrat-prof-bernd-thomsen-keine-angst-vor-der-ki-was-menschen-maschinen-voraushaben/23856180.html, zuletzt abgerufen am 11. März 2020.

Thomsen, B. (2019c): Die Wirtschaft muss sich neu ausrichten, um nachhaltig erfolgreich zu sein, unter: https://www.handelsblatt.com/meinung/kolumnen/expertenrat/thomsen/expertenrat-prof-bernd-thomsen-die-wirtschaft-muss-sich-neu-ausrichten-um-nachhaltig-erfolgreich-zu-sein/24129300.html, zuletzt abgerufen am 11. März 2020.

Prof. Bernd Thomsen ist international renommierter Stratege, Innovations- und Zukunftsexperte. Als CEO leitet er die führende globale Management-Beratung mit Zukunftsexpertise, die seit 35 Jahren für alle wichtigen Industrien und als Regierungsberater tätig ist. Er lehrt in Asien und Europa. Thomsen misstraut dem Konsens in Zukunftsfragen, er sagt weltwirtschaftliche und gesellschaftliche Entwicklungen sowie neue Industrien voraus. Thomsen ist Kolumnist des deutschen Handelsblatts und Mitglied des Expertenrates. Er engagiert sich für Hochbegabte und führt eine Kinderstiftung, lebt in Miami und Hamburg. Thomsen liebt weltoffene Menschen und blauen Himmel.